RNA Technologies

Founding Editors

Jan Barciszewski
Volker A. Erdmann

Volume 14

RNA Technologies provides critical and comprehensive discussions of the most significant areas of RNA research, written by leading international authorities. Each volume in the series reviews both general topics, such as RNA interference, RNA biosynthesis and metabolism, microRNAs and diseases, transcriptome analysis and aptamers, as well as highly topical chapters on very recent findings in the field. Recurring themes of the series include the complexities associated with target recognition, RNA-protein interactions, analysis of genetic networks and the promise of small RNAs as cancer diagnostics and therapeutics in combating cancer.

RNA-based approaches have great potential to revolutionize molecular biology, cell biology, biomedical research and medicine. The nucleic acid-based molecules can be used to regulate the level of gene expression inside the target cells and their potential efficacy against several viruses, including SARS-CoV-2, and other chronic diseases represents a promising research area.

RNA Technologies publishes in both print and electronic format since 2010 and is intended for researchers from academia and industry, as well as graduates that look for a carefully selected collection of high quality review articles on their respective field of expertise.

The series is indexed in SCOPUS.

Jan Barciszewski
Editor

RNA Structure and Function

Springer

Editor
Jan Barciszewski
NanoBioMedical Centre of Adam
Mickiewicz University
Poznań, Poland

Institute of Bioorganic Chemistry
of the Polish Academy of Sciences
Poznań, Poland

ISSN 2197-9731 ISSN 2197-9758 (electronic)
RNA Technologies
ISBN 978-3-031-36389-4 ISBN 978-3-031-36390-0 (eBook)
https://doi.org/10.1007/978-3-031-36390-0

This Springer imprint is published by the registered company Springer Nature Switzerland AG
The registered company address is: Gewerbestrasse 11, 6330 Cham, Switzerland

Introduction

RNA Structure Meets Function

Current studies on RNA are flourishing, with the discovery of RNAs with new structures, a growing list of biological functions, and the development of RNA nanotechnologies. RNA is a central molecule in cell biology. Its activity is in various processes, from post-transcriptional maturation to regulation of gene expression and metabolite sensing.

RNA is a unique polymer. Like DNA, it can bind with great specificity to either DNA or another RNA through Watson-Crick base pairing. It can also bind specific proteins or small molecules. Contrary to DNA, RNA is an unstable molecule that is sensitive to degradation both through elevated temperatures and by RNases.

RNA-like proteins can catalyze chemical reactions. Some RNAs possess intrinsic enzymatic activity to catalyze different RNA modification reactions. These catalytic RNAs include certain self-splicing RNA transcripts, joining amino acids to make proteins, ribozymes, and RNase P, of which M1 RNA is an enzyme that maturates the 5' end of tRNA precursors.

RNA is also a unique informational molecule. In addition to carrying information in their linear sequences of nucleotides (primary structure), RNA molecules fold into intricate shapes. The pairing of local nucleotides forms secondary structures, such as hairpins and stem–loops, but interactions among remotely located sequences create tertiary structures. RNA structure influences the transcription, splicing, cellular localization, translation, and turnover of the RNA (Fig. 1).

Several classes of RNA molecules are involved in converting the information encoded in the cell's DNA into functional gene products. Messenger RNAs (mRNAs) are copies of individual protein-coding genes and serve as an amplified read-out of each gene's nucleic acid sequence. In eukaryotic cells, mRNAs are assembled from longer RNA transcripts by the spliceosome, which consists of spliceosomal RNAs and proteins. Spliceosomal RNAs help remove intervening sequences (introns) from pre-mRNA transcripts and splice together the mRNA segments (exons) to create a complex assortment of distinct protein-coding mRNAs from a single gene.

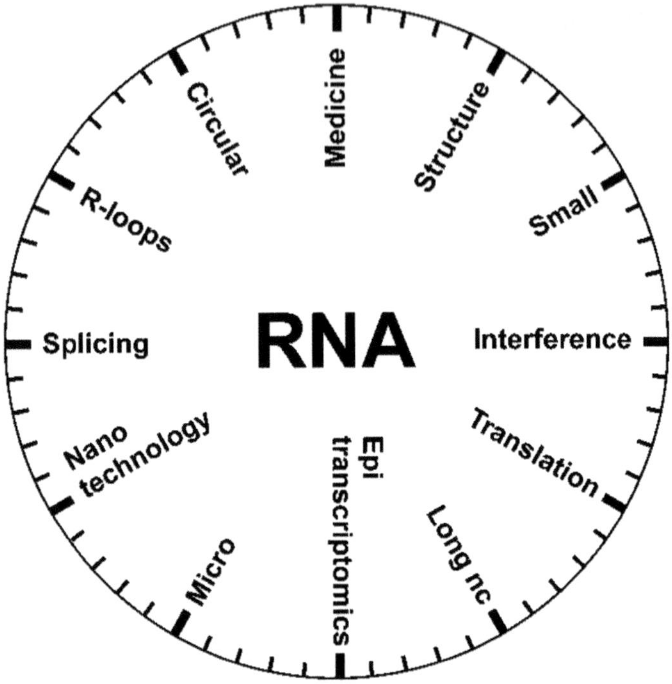

Fig. 1 RNA clock showing different activities of RNA molecules

Two categories of noncoding RNAs participate in assembling the proteins specified by mRNAs. Ribosomal RNA (rRNA) constitutes the core structural and enzymatic framework of the ribosome, the molecular machine that synthesizes proteins according to the instructions embedded in the sequence of an mRNA. Transfer RNAs (tRNAs) use their anticodon complementary base pairing to decode the three-letter genetic code in the mRNA. Each tRNA corresponds to a single amino acid that sequentially incorporates into a growing protein chain.

An enormous amount of data supports the view that RNA structure determines function. The best example of that is tRNA. It is the gold standard of molecular biology, multifunctional, and beautifully corresponds to its primary, secondary, and tertiary structure. More importantly, the crystal structure of different native tRNAs (76 nucleotides, 25 000 D) solved with a high resolution proved those observations.

The role of RNA structure is understood to have an impact on biological processes. Therefore, studies of RNA structure and function relationships are growing, aiming to test new hypotheses with original experiments and derive the correct conclusions. The functional diversity of RNA comes from its structural richness. RNA is intrinsically able to fold into complex three-dimensional structures, which post-transcriptional modifications can further modulate. Also, RNA can bind with other biomolecules and interact with ligands.

RNA molecules are biologically active (functional) only when folded into their native conformation, determined by a specific secondary and tertiary structure. The secondary structure of RNA plays an essential role in post-transcriptional regulatory processes, including splicing, localization, stabilization, and translation. Still, the tertiary structure is key to understanding the precise mechanisms behind RNA-protein recognition.

RNA is not always a faithful copy of DNA. High-throughput sequencing has revealed that the eukaryotic genome transcends beyond coding regions and uncovers many novel functional RNAs, known as long noncoding RNAs (lncRNAs). These RNA do not code for proteins and originate from sections of the genome once believed to be junk DNA. Evidence indicates that long noncoding RNAs play roles in many cellular processes, including guiding cell fate during embryonic development. However, there is ongoing research to reveal precisely how lncRNAs exert their influence. The basic mechanisms of action of these RNAs are only beginning to emerge. They are most likely exhibited due to specific RNA secondary and tertiary structures, as previously found for other RNAs, like rRNAs and tRNAs.

Once transcribed from the chromosomal DNA, most RNA molecules require structural or post-transcriptional modifications before they can function. For instance, ribosomal RNAs receive numerous chemical modifications for proper ribosome assembly and function. These modifications are introduced by enzymatic reaction in conjunction with specialized noncoding RNAs (called snoRNAs) that base pair with the rRNA and guide the modifying enzymes to precise locations on the rRNA. The chemical modification of RNA nucleotides affects several properties of RNA molecules, including nucleotide sequence, secondary structure, RNA–protein interaction, localization, and processing. Identifying and mapping RNA modification falls into a relatively new area of study called epitranscriptomics. With great hope, a better understanding of the interface between a single chemically changed nucleotide and a cellular function will pave the way toward developing novel diagnostic, prognostic, and therapeutic tools for managing various diseases.

Eukaryotic cells contain thousands of small RNAs associated with various RNA interference pathways. MicroRNAs, ca 22 nt long are produced from longer transcripts that have a hairpin structure, associate with a protein (Argonaute), and base pair specifically to mRNAs to inhibit their translation. A single miRNA can regulate the activity of hundreds of protein-coding genes. Therefore, miRNAs significantly impact the development and physiology of the cell.

Small interfering RNAs derive from any transcribed region of the genome, are associated with Argonaute proteins, and act directly upon the locus from which they are produced. Many endogenous siRNAs in eukaryotic cells specify the silencing of transposons and repeat sequences in the genome. Similarly, in animals, the Piwi-associated RNAs (piRNAs) promote genome integrity by silencing transposons and repeat sequences. Genome-wide RNA structure determination has relied heavily on computational predictions to create structural models for hypothesis testing. The RNA prediction algorithms have greatly advanced in their ability to propose more accurate secondary structures from both primary sequences and sequence covariation. These predicted structures are typically confirmed by secondary structure probing, which still serves as the gold standard of RNA structure determination.

In addition to secondary and tertiary structures, RNA can also form higher-order complexes due to canonical and noncanonical base pairing, which contribute to the structural versatility of the RNA molecule. This observation prompted many groups to design nanometer-sized tertiary RNA structures (nanoparticles) with the capacity to bind small compounds, other ligands, and chemotherapeutics and use them in chemotherapy. RNA nanoparticles efficiently target tumor tissues and can reduce the toxicity of various drugs. These nanostructures can be conjugated with many anti-cancer drugs and immunotherapy components.

RNA-based approaches can revolutionize molecular biology, cell biology, biomedical research, and medicine. The nucleic acid-based molecules can be used to regulate the level of gene expression inside the target cells and their potential efficacy against human pathogens.

Finally, one can quote Francis Crick's poem nicely characterizing RNA: "What are the properties of genetic RNA. Is he in heaven, is he in hell? That dammed, elusive Pimpernel."

Poznań, Poland Jan Barciszewski

Contents

Transfer RNA Fragments, from Structure to Function

Serafima Dubnov and Hermona Soreq

Contents

Abstract Transfer RNA molecules (tRNAs) are produced from numerous nuclear and mitochondrial genes and are primarily involved in bringing specific amino acid residues to polyribosomes and enabling correct elongation of polypeptide chains. Several different nucleases may degrade tRNA molecules into shorter oligonucleotide chains designated transfer RNA fragments (tRFs). It has recently been realized that these tRFs may resume diverse functions, including but not limited to enabling ribosomal activities, sperm cell differentiation, and interaction with messenger RNA transcripts (mRNA) carrying complementary sequence motifs. Such interaction can suppress the translation of those mRNAs and induce their degradation in a similar manner to that of microRNAs (miRs). Additionally, tRFs can interact with RNA-binding proteins and modulate translation processes and the functioning

S. Dubnov · H. Soreq (✉)
The Edmond and Lily Safra Center of Brain Science and the Life Sciences Institute, The Hebrew University of Jerusalem, 9190401 Jerusalem, Israel
e-mail: hermona.soreq@mail.huji.ac.il

J. Barciszewski (ed.), *RNA Structure and Function*, RNA Technologies 14,
https://doi.org/10.1007/978-3-031-36390-0_1

of transcription factors. This, in turn, can accelerate the division of some blood cell types, for example, in patients recovering from ischemic stroke. However, the full scope of tRFs structure–function relationships awaits further studies, addressing the specific conditions, cell types, tissues and organisms where tRFs are produced and function, as well as the dependence of their production on age, sex, health and disease. The rapidly accumulating knowledge about tRFs calls for approaching these issues and making the structure–function interrelationships of these intriguing molecules amenable for further exploration.

Keywords Brain disorders · Methods · RNA therapeutics · Short non-coding RNA · tRNA fragments

1 Introduction

Single cells, like sensible human beings, are carefully managing their budget, time and energy for achieving their desired goals while controlling their actions to ensure survival. Proliferation, growth, motility and other essential cellular processes require highly controlled protein synthesis, with some proteins being essential for achieving the desired goal, while others interfering with it. Hence, the machinery of protein production is the core of cellular homeostasis and its disruptions, such as aberrant protein accumulations or abnormally high activity of some proteins, endanger the survival of the whole organism, as is the case in neurodegenerative diseases or cancer. The process of protein synthesis is complex and time-consuming and requires a great investment of molecular energy (Kafri et al. 2016). Such investment in the production of a redundant protein might be a crucial mistake for a cell, disrupting its energy balance and engaging it in useless or even toxic molecular pathways. Furthermore, protein synthesis occurs at a time scale of minutes to hours (Brodsky 1975), limiting the cellular ability of rapid response to unexpected environmental stimuli. Hence, it is crucial for a cell to have additional, more efficient ways to control its homeostasis and guide the process of protein production.

Surprisingly, the major controller of protein synthesis appears to be RNA, which was initially thought to be only an intermediate molecule between DNA and protein in the process of gene expression, without exerting direct functions on its own (Crick 1970). Unlike that traditional perspective, multiple non-coding RNA families, such as microRNAs (miRs) and long non-coding RNAs (lncRNAs), emerged as playing essential roles in controlling cellular homeostasis and gene expression (Amaral et al. 2013). Transcription is a lot cheaper than translation in terms of molecular energy and time (Kafri et al. 2016), which gives RNA a great advantage over protein as a molecular regulator. Moreover, RNA turnover is usually faster compared to proteins (Belle et al. 2006; Wang et al. 2002), decreasing the risk of toxicity from overexpressing the regulator molecule. While avoiding translation spares a lot of energy, sparing transcription as well may be even more effective. This option can be enabled by the recently rediscovered family of non-coding RNA—short tRNA fragments (tRFs). TRFs are produced by specific cleavage of the pre-tRNA or mature tRNA molecules to smaller fragments. Hence, their synthesis only requires the breakdown of already existing transcripts, allowing a particularly rapid response, minimizing the requirement for the cellular energetic resources. Enrichment of mammalian short tRNA fragments was first identified in cancer and it was then considered a side effect of the high tRNA turnover rate (Borek et al. 1977).

The first evidence demonstrating that tRFs are not merely tRNA degradation products but rather functional fragments, capable of regulating transcription and translation, was published only a decade ago (Haussecker et al. 2010; Lee et al. 2009; Yamasaki et al. 2009). By now, tRFs are extensively studied in various biological contexts and their diverse roles in major biological pathways are rapidly being elucidated. However, many exciting questions, such as the evolutionary purpose of tRFs, the complex relationships of tRFs and miRs, and the cell-type specificity of different tRFs remain open and await further discoveries. Several excellent reviews have already been written about the tRFs (Anderson and Ivanov 2014; Fagan et al. 2021; Kim et al. 2020; Su et al. 2020), describing the current know-how about their biosynthesis, classification and functional roles. Here, we review the original classification of tRFs based on their biosynthesis and roles and assemble the existing evidence regarding the molecular functions exerted by each tRF type, seeking their structure–function relationships with a focus on the roles of tRFs in neurological disorders and brain-to-body signalling. Furthermore, we discuss methodological challenges in tRF research and describe several technical approaches developed specifically for studying these intriguing molecules.

2 tRF Biosynthesis

The structural classification of tRFs is primarily based on their mapping to the tRNAs of origin (Table 1). The synthesis of tRNAs starts with their transcription by RNA polymerase III, resulting in premature tRNA (pre-tRNA), including additional 5′ leader and 3′ trailer sequences to be subsequently removed by RNAse P and RNAse Z, followed by ligation of the cytidine–cytidine–adenosine (CCA) tail to the 3′ end by the enzyme tRNA nucleotidyltransferase (Maraia and Lamichhane 2011). These three modifications form the mature tRNA, ready to participate in amino acid transport. Strikingly, the process of tRNA synthesis already generates an independently functional short tRNA fragment, namely, tRF-1, representing the 3′ trailer sequence (Kumar et al. 2016). tRF-1 affects transcription, positively regulating cell proliferation and being evidently involved in multiple cancers (Lee et al. 2009). Notably, tRNA halves (or tiRNAs) were the earliest to be discovered and the most studied tRF type so far (Kumar et al. 2016). They are generated by the RNAse Angiogenin, cleaving the tRNA in the anticodon loop (Yamasaki et al. 2009). Other ribonucleases were also shown to be involved in the generation of tRNA halves, such as Rny1p in yeast (Thompson and Parker 2009) or RNase T2 in plants and humans (Megel et al. 2019). Notably, tRNA halves are the longest tRFs, ranging between 31 and

40 nucleotides. Multiple findings independently observe the involvement of tRNA halves in stress response, the regulation of apoptosis and cellular survival (Li and Hu 2012; Saikia et al. 2014).

While tRNA halves stand out as classified into known non-coding RNA families in terms of their structure, there are several tRF types that resemble microRNAs in length and molecular modifications. Of those, tRF-5s are fragments mapping to the 5′ half of the original tRNA, classified into three types (tRF-5a-c) based on their specific cleavage location (Table 1). Correspondingly, there are two types of tRF-3s, tRF-3a and tRF-3b, both mapping to the 3′ half of the tRNA of origin. Notably, both tRF-5s and tRF-3s include 5′ phosphate and 3′ hydroxyl groups, which, together with their size, makes them greatly similar to microRNAs and piRNAs, suggesting that they might have parallel functions (Couvillion et al. 2010). Furthermore, both 5′ and 3′ short tRFs are processed by Dicer, similarly to microRNAs (Babiarz et al. 2008), and by RNase T2, like tRNA halves (Megel et al. 2019).

Table 1 Classification and basic characteristics of the tRNA fragments (tRFs)

Type	Length (nucleotides)	Location in tRNA of origin	Restriction enzymes	Associated biological processes	References
tRF-1	16–27	3′ trailer sequence of pre-tRNA	RNAse Z	Cell proliferation	Kumar et al. (2016); Lee et al. (2009)
tRNA halves (tiRNAs)	30–40	3′ or 5′ half	Angiogenin Rny1p RNase T2	Stress response Apoptosis Epigenetic inheritance	Yamasaki et al. (2009); Thompson and Parker (2009); Megel et al. (2019); Li and Hu (2012); Saikia et al. (2014)
tRF-5a	14–16	5′ half	Dicer RNAse T2	Gene silencing	Kumar et al. (2014); Babiarz et al. (2008); Megel et al. (2019); Winek et al. (2020); Kim et al. (2017)
tRF-5b	22–24				
tRF-5c	28–30				
tRF-3a	18	3′ half		Gene silencing Translation enhancement	
tRF-3b	22				
i-tRF	~20	Anticodon loop	Unknown	Inhibition of tumour growth Neurodegeneration	Kim et al. (2020)

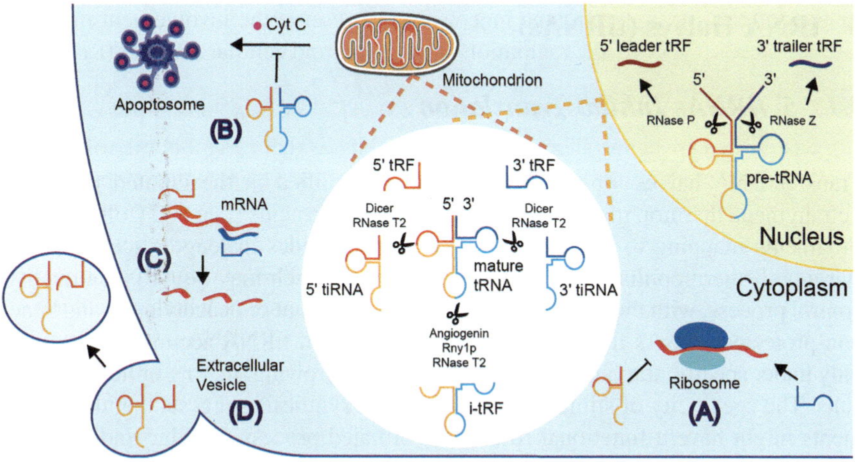

Fig. 1 Schematic overview of tRF biosynthesis and intracellular molecular pathways involving tRFs

Lastly, internal tRFs (i-tRF) include the whole anticodon loop (Kumar et al. 2016). i-tRFs are involved in breast cancer by sequestering the RNA-binding protein YBX1 (Goodarzi et al. 2015) and include an intron, which might be subjected to alternative splicing, producing a different fragment with potentially distinct functions. Further, mutated CLP1, which plays a major role in tRNA splicing, leads to the accumulation of intron-expressing i-tRFs, potentially linked to neurodegeneration (Schaffer et al. 2014). Apart from their classification by the location of cleavage, tRFs may be categorized by their genome of origin (nuclear or mitochondrial) or the amino acid associated with their parental tRNAs. However, in the present chapter, we consider the biosynthetic partition as the major one, aiming to summarize the existing evidence for the causal structure–function relationship of different tRF sub-families with a focus on their discovered biological roles and internal sub-classifications (Fig. 1).

5′ leader and 3′ trailer tRFs are cleaved from the premature tRNA (pre-tRNA) by Rnase P and Rnase Z correspondingly. Mature tRNAs give rise to 5′ and 3′ short tRFs, processed by Dicer and Rnase T2. Angiogenin, Rny1p and Rnase T2 generate longer tRNA halves, or tiRNAs. Lastly, i-tRF includes an anticodon loop and its processing pathway is yet to be established. (A) 5′ tiRNAs inhibit ribosomal translation, while a specific 3′ tRF derived from Leucine enhances ribosomal assembly. (B) Both 3′ and 5′ tiRNAs inhibit apoptosis by sequestering the cytochrome C released from mitochondria and thus preventing the apoptosome assembly. (C) 3′ and 5′ short tRFs, like microRNAs, arrest translation and induce mRNA degradation by binding to complementary sequence motifs in those mRNAs. (D) 5′ tRFs, both tiRNAs and shorter fragments, were identified in extracellular vesicles, showing their active transport between cells and tissues.

3 tRNA Halves (tiRNAs)

3.1 5′ tiRNAs Inhibit Translation

Transfer RNA halves, or tiRNAs, were first identified in the ciliated protozoan Tetrahymena thermophila deprived of essential amino acids (Lee and Collins 2005). Fragments mapping to tRNA halves, 30–35 nucleotides in length, accumulate in starving T. thermophila, peaking at 3 h starvation which may be part of a quality-control process, with the cleavage targeting the redundant or deacetylated, and hence non-protected, tRNAs for total degradation. However, tiRNA accumulation occurs only under specific starvation conditions but not in growing, mating or heat-shocked cells. The specificity of stimuli triggering tiRNA synthesis suggests that these fragments might have a functional role in the initiated process. Similar findings were later reported in bacteria (Haiser et al. 2008), fungi (Jöchl et al. 2008) and yeast (Thompson and Parker 2009). Primarily suggested involvement of tiRNAs in protein synthesis inhibition initiates with reducing the amount of tRNAs available for translation. However, high specificity of the cleavage reaction and of the parental tRNA molecules chosen for tiRNA generation, together with the fact that tiRNAs represent a very small fraction of the whole tRNA pool, suggests that tiRNAs may have other function(s), not directly related to tRNA degradation. Intriguing evidence supporting an independent role of tiRNAs as translation inhibitors emerged from human cell culture tests (Yamasaki et al. 2009). Transfecting cells with endogenous tiRNAs reduced protein synthesis in the treated cells, exposing a causal link between tiRNA activity and translational arrest. Importantly, 5′ halves alone led to this pronounced effect, while overexpressing the 3′ halves ensured protein levels similar to those of control cells.

In addition to proving the functional independence of 5′ tiRNAs, Yamasaki et al. (2009) pioneered a concrete molecular pathway of tiRNA generation in human cells. Surprisingly, the broadly studied ribonuclease Angiogenin, whose major function involves the promotion of angiogenesis, appears to be an essential nuclease for tiRNA production in mammalian cells and body fluids. Knockdown of angiogenin prevents both tiRNA accumulation and translation inhibition. tiRNA levels further depend on the Angiogenin inhibitor, RNH1, which sequesters the ribonuclease in the cytoplasm under physiological conditions and releases it in response to stress (Yamasaki et al. 2009). Further, a cascade of events leads to tiRNA-dependent translation inhibition (Emara et al. 2010; Ivanov et al. 2011). 5′ tiRNAs prevent translation by displacing the translation initiation eIF4G/A complex from mRNAs. This process depends on YB-1, a known cold shock domain protein (Ivanov et al. 2011). As a result, impaired translation leads to the assembly of stress granules, observed earlier as a downstream event of 5′ tiRNA overexpression (Emara et al. 2010).

5′ tiRNAs share several structural features that are essential for their capacity to inhibit translation and are missing in 3′ tiRNAs, indicating that such shared features may enable 5′ tiRNAs to interact with relevant proteins. Indeed, the 5′ terminated monophosphate modifications, absent in 3′ tiRNAs, emerged as crucial for the induction of stress granules (Emara et al. 2010). Furthermore, the capacity to inhibit translation is limited to tRFs originating from alanine and cysteine tRNAs, which are the only tRNAs featuring the terminal oligoguanine (5′-TOG) motif (Ivanov et al. 2011). Hybridizing the TOG motif to the 5′ methionine tiRNA is sufficient to potentiate its capacity for inhibiting translation and inducing stress granules. Finally, the secondary structure of the D-loop included in the 5′ tiRNA may also be essential for conferring the reported functions (Ivanov et al. 2011). Gradual truncation of the 3′ end of alanine tRNA-derived 5′ tiRNA damaged its capacity to inhibit translation, except for when it interfered with the D-loop (Ivanov et al. 2011). Furthermore, 5′ tiRNAs possess the ability to form intermolecular dimers and tetramers, likely contributing to their stability (Lyons et al. 2017). Importantly, the 5′ tiRNAs involved in translation inhibition are mostly nuclear genome-originated, and only a few mitochondrial tiRNAs were enriched following Angiogenin-mediated degradation (Haiser et al. 2008; Jöchl et al. 2008). Thus, similarly to other molecules controlling protein synthesis, such as transcription factors and microRNAs, translation-inhibiting tiRNAs also stem from the nucleus.

3.2 5′ tiRNAs Are Intercellular Communicating RNAs

While translation inhibition is a vastly important function, it affects one cell, rather than a whole population. Multiple findings suggest that 5′ tiRNAs are further involved in intercellular signalling, being transferred between different cells in extracellular vesicles (EVs), similarly to microRNAs (Chiou et al. 2018; Nolte-'t Hoen et al. 2012; Vojtech et al. 2014; Weng et al. 2022). Intriguingly, this transport in EVs seems to be specific to 5′ primed, but not 3′ primed tRFs (Chiou et al. 2018). 5′ tiRNAs were identified to be released in extracellular vesicles from the protozoan parasite Trypanosoma cruzi and consequently infect host HeLa cells, inducing a massive transcriptomic change (Garcia-Silva et al. 2014). Interestingly, only 5′ threonine and leucine-derived tiRNAs affected transcription in the host cells. Furthermore, 5′ tiRNAs packed in EVs serve as a means of intercellular communication within a multicellular organism. For example, immune response requires a carefully orchestrated cellular reaction relying on multiple intercellular communication pathways. Compatible with their predicted involvement in the immune response, 5′ tiRNAs are specifically enriched in hematopoietic tissues (Dhahbi et al. 2013). Moreover, leucine tRNA-derived 5′ tiRNAs selectively exported into exosomes during inflammatory reactions can repress T-cell activation (Chiou et al. 2018; Nolte-'t Hoen et al. 2012; Vojtech et al. 2014; Weng et al. 2022). Notably, T-cell-derived vesicles were more effectively enriched with tRFs than with microRNAs, possibly reflecting separate roles of these two small non-coding RNA families. Moreover, antisense oligonucleotides-mediated suppression of the most enriched 5′ tiRNAs enhanced T-cell activation, indicating that 5′ tiRNAs actively suppress inflammation (Chiou et al. 2018; Nolte-'t Hoen et al. 2012; Vojtech et al. 2014; Weng et al. 2022).

Intriguingly, a recent study by our group has found that both angiogenin and blood cell tRFs are massively induced by stroke, possibly taking part in regulating post-stroke immune deficiency (Winek et al. 2020). Supporting previous observations, microRNAs are downregulated following an ischemic event, with an especially sharp decline in microRNAs targeting the cholinergic transcripts, enabling preferential involvement of tRFs which replace microRNAs in immune response regulation and inflammation blockade (a 'changing of the guards' response). An alternative way to spread the tRFs within a cell population is to transfer the restriction enzyme responsible for their synthesis. For example, angiogenin is a secreted protein capable of invading mammalian cells, which is traditionally used to explain its potential to induce angiogenesis (Olson et al. 1998). Nevertheless, intercellular transfer of angiogenin might also transmit its ability to generate tiRNA. Specifically, tRNA restriction in the anticodon loop is a known protective mechanism in bacteria, that serves as a suicide attempt following viral infection for the sake of saving the rest of the population (Oberbauer and Schaefer 2018). Hence, bacteria spread plasmid-encoded ribonucleases to enable other cells to mimic this response (Oberbauer and Schaefer 2018).

3.3 5′ tiRNAs Modulate Epigenetic Inheritance

Apart from their intercellular communication role, the 5′ tiRNAs may also transfer information in an inter-generational manner. Two independent research groups find abundantly expressed small RNA fragments mapping to the 5′ half of their parental tRNAs in the sperm of mice treated with either low-protein or high-fat diet (Chen et al. 2016; Sharma et al. 2016). Sharma et al. (2016) hypothesize that these fragments are delivered to the mature sperm via extracellular vesicles, since they were not enriched in the testicles of the treated mice. Supporting this notion, transfecting healthy embryos with 5′ tiRNAs that were upregulated in the sperm caused severe metabolic disorders in the progeny of mice raised on a high-fat diet, independent of any genome modifications, such as DNA methylation (Chen et al. 2016; Sharma et al. 2016). Intriguingly, Sharma et al. (2016) suggest a concrete molecular pathway of tiRNA epigenetic modulation, whereby specific 5′ tiRNAs might interfere with the regulation of endogenous retroelements; and show that inhibiting the GCC anticodon of 5′-Glycine tiRNA anticodon downregulates around 70 mRNAs, all representing targets of the endogenous retroelement MERVL (Sharma et al. 2016). The authors conclude that MERVL activity modulation reflects a pathway whereby 5′ tiRNAs affect the metabolic changes in the progeny as a function of the paternal dietary conditions.

3.4 *Both 5′ and 3′ tiRNAs Inhibit Apoptosis*

Apoptosis is the natural cellular reaction to stress, when the cell has no resources for adaptation. It eventually leads to cellular death, and its aberrant activation is the basis of multiple diseases, such as neurodegenerative disorders, cancer, organ failure and others (Favaloro et al. 2012). Intriguingly, one of the mechanisms controlling apoptosis involves angiogenin-dependent tRNA breakdown to tiRNAs (Saikia et al. 2014). The cellular apoptosis is initiated by mitochondrial release of various signalling factors from the intermembrane space, including Cytochrome C. Once released into the cytoplasm, Cyt C binds to the apoptotic protease activating factor 1 protein (Apaf-1), forming a multimeric complex, which in turn activates the caspase-dependent apoptosis cascade (Zou et al. 1999). However, this mechanism can be inhibited by both 5′ primed and 3′ primed tiRNAs, or tRNA halves, which bind to Cyt C and thus prevent the formation of the apoptosome and the consequent cellular death (Saikia et al. 2014). Surprisingly, although mitochondria play a major role in initiating apoptosis, the vast majority of tiRNAs involved in apoptosis regulation are nuclear genome-originated (Saikia et al. 2014). Moreover, mitochondrial tiRNAs are not enriched under apoptosis, indicating that Cyt C complex formation occurs primarily in the cytoplasm.

4 Type I tRFs: miR-Like tRNA Fragments

While tiRNAs mostly function in a similar manner to long non-coding RNAs or even proteins, directly exerting their function by implementing the specific affinities of their three-dimensional structures, smaller tRNA fragments generated from mature tRNA in the cytoplasm, known as type I tRFs, act similarly to microRNAs, inhibiting the translation of mRNAs carrying complementary sequences (Kim et al. 2020). Indeed, the structural resemblance of short tRFs to microRNAs is striking. The two ncRNA families share the same length distribution (18–30 nucleotides), with some of them carrying the same chemical modifications, such as 5′ phosphorylation and 3′ hydroxylation (Kim et al. 2020). The similarity between short tRFs and microRNAs initially led to some ambiguity in their classification (Maute et al. 2013; Schopman et al. 2010). This part of our chapter is dedicated to molecular pathways of short tRF regulation with microRNA-like functions.

4.1 *Are MicroRNAs and tRFs Processed in the Same Way?*

The answer to the question above is currently inconclusive. Processing of microRNAs entails three major parts, namely cleavage by Drosha and DGCR8 within the nucleus, further processing by Dicer in the cytoplasm and assembly of RNA-Induced Silencing

Complex (RISC) with Argonaute (Ago) proteins (Shukla et al. 2011). Short tRNA fragments were shown to be independent of a Drosha/DGCR8 processing step (Babiarz et al. 2008); hence, they are likely to be generated by specific cleavage of mature tRNAs after their export to the cytoplasm. However, several studies show possible Dicer-mediated regulation of some tRFs (Babiarz et al. 2008; Cole et al. 2009; Haussecker et al. 2010; Maute et al. 2013). Interestingly, all the reported Dicer-dependent tRFs include a partial hairpin structure sequence originated from the parental tRNA. This may possibly explain the Dicer requirement, which cleaves out a similar hairpin loop from the pre-microRNA. Finally, tRFs tightly associate with Argonaut (Ago) proteins, forming RISC complexes and thus representing another type of RNA interference (Haussecker et al. 2010; Kumar et al. 2014; Maute et al. 2013). Multiple findings suggest other alternative ways of tRF generation, involving different cytoplasmic ribonucleases, other than Dicer, such as the RNAse T2 family of endonuclease (Megel et al. 2019) and angiogenin (Li et al. 2012). Thus, tRFs share part of the known processes leading to microRNAs production but are more flexible in their production pathway compared to microRNAs. Together with the absence of the transcription requirement, this makes tRFs a more sensitive and readily available molecular tool for acute stress responses.

4.2 Gene Silencing by Short tRFs

The association of tRFs with Ago proteins readily suggests translational inhibition of specific mRNAs with complementary sequences, indicating that tRFs participate in gene silencing. Indeed, antisense inhibition of selected tRFs increased the levels of targets with complementary sequences (Haussecker et al. 2010; Maute et al. 2013), whereas tRF overexpression inversely caused degradation of the complementary mRNAs (Kumar et al. 2014; Maute et al. 2013; Winek et al. 2020). Furthermore, many tRFs were initially considered to be microRNAs based on their functioning modes and were only later removed from the microRNA lists due to their sequence composition which identified them as fragments of tRNAs. Taken together, short tRFs function as RNA silencing actors, using the sequence complementarity, similar to microRNAs.

4.3 Leucine Derived 3′ Short tRF Enhances Translation

Compatible with the variety of their subtypes, short tRFs exhibit much more diverse capacities than microRNAs, since their role is not limited to gene silencing. Rather, a specific fragment derived from the 3′ end of Leucine tRNA enhances translation of the ribosomal proteins RPS28 and RPS15, promoting the maturation of the 18S ribosomal subunit (Kim et al. 2017). Excitingly, this mechanism of tRF action mimics the mechanism of gene silencing via complementary binding to the target mRNA. However, the pronounced effect of this process is completely reversed: rather than recruiting the RISC complex to inhibit translation, 3′-Leu-tRF interaction with its target mRNAs competes with their self-complementary motifs, thus unfolding their secondary structure and making the mRNA more accessible for the translation machinery (Kim et al. 2017). Inversely, 3′-Leu-tRF inhibition reduces RPS28 expression, which in turn impairs ribosomal biogenesis and induces apoptosis. Hence, some short tRNA fragments regulate the intracellular molecular pathways at a much more upstream level than the structurally similar microRNAs.

5 Type II tRFs

One of the prominent trends in the history of biological science is the discovery of high functional relevance of specific elements that were considered insignificant before. In this context, tRNA processing has been studied for decades in multiple experimental models, and the cleavage of 5′ leader and 3′ trailer sequences from the premature molecules has been a long-known fact (Deutscher 1984). However, only recently these sequences have been shown to act as independent small non-coding RNAs. A fragment representing the 3′ trailer sequence of the Serine tRNA precursor transcript, referred to as trf-1001, was found to promote cellular survival and cell cycle progression (Lee et al. 2009). Correspondingly, transcripts including the 5′ leader sequences of the Tyrosine tRNA precursor may rescue motor neurons from death induced by oxidative stress (Hanada et al. 2013). Taken together, these findings show that type II tRFs, previously considered as byproducts of tRNA processing, can play critical roles in cellular survival.

6 Internal tRFs (i-tRFs)

At the cancer front, internal tRNA fragments (i-tRFs) mapping to the anticodon loop of the precursor tRNA molecules (Kim et al. 2020) and derived from glutamate, glycine and asparagine tRNAs prevent the oncogenic action of the RNA-binding protein YBX1 by competitive binding to its targets, thus inhibiting tumour growth (Goodarzi et al. 2015; Kim et al. 2020). Compared to other tRF types, i-tRFs are expressed at higher levels in both human CSF and blood of healthy subjects (Paldor et al. 2022). Furthermore, human tRF profiles in both blood and cerebrospinal fluids vary with age and sex (Paldor et al. 2022).

7 Mitochondrial Genome-Originated tRFs

Apart from nuclear-originated tRFs, accumulating evidence shows biological relevance of eukaryotic mitochondrial-encoded tRNA genes. Those differ in sequence from the tRNAs encoded by the nuclear genome and feature a separate pathway of tRNA processing (Jöchl et al. 2008). Mutations in mitochondrial tRNA genes are responsible for various neuromuscular diseases (Lauber et al. 1991) interfering with protein synthesis. Given the anticodon specificity of the studied mutations, it is unlikely that the effect is caused by the lack of a specific tRNA. Rather, the real cause of the observed disease may lie in reducing the levels of the corresponding mitochondrial tRF fragments, but further research is required to test this hypothesis. Changes in mitochondrial tRF levels further occur following several treatments, such that mitochondrial tRF levels decline in cells cultured in iron-deficient medium, consistent with the observation that mitochondria use most of the cellular iron resources (Jöchl et al. 2008). However, it is not entirely clear if the reduction of mitochondrial tRFs is a direct result of the decrease in general mitochondrial activity, or if declined tRF levels lead to mitochondrial dysfunction. Lastly, the levels of mitochondrial tRFs change across various cancer types, with mitochondrial and nuclear tRFs targeting mRNAs from distinct biological pathways (Telonis et al. 2019).

Another intriguing possibility supporting the functional relevance of mitochondrial tRFs is their involvement in mitochondrial inheritance (Yaffe 1999). On one hand, tRFs were shown to affect epigenetic inheritance from the paternal side, being transferred to the embryo through sperm (Chen et al. 2016; Sharma et al. 2016). On the other hand, tRFs encoded by mitochondrial tRNA genes might be inherited from the mother, representing a novel epigenetic pathway. This phenomenon is not yet established and requires further study. In summary, although showing potential roles in regulating prominent disease-related pathways, mitochondrial tRNA fragments are currently understudied and further exploration is needed to establish their concrete functions and relation to nuclear tRFs.

8 Involvement of tRFs in Brain Disorders

Ever since tRNA fragments were first identified in the urine of oncological patients (Borek et al. 1977), their diagnostic potential and functional roles are extensively studied in cancer. The involvement of tRNA fragments in brain disorders, however, is still understudied, although multiple findings suggest that tRNA fragments are potential regulators in ageing-associated stress and related neurological disorders (Dhahbi et al. 2013). Loss-of-function mutations in the angiogenin gene were found in patients suffering from amyotrophic lateral sclerosis (ALS) several years before the discovery of angiogenin's role in tiRNA generation (Wu et al. 2007). The researchers mostly attributed this phenomenon to the role of angiogenesis in ALS progression, briefly mentioning an alternative possibility that angiogenin's loss of function may cause aberrant ribosomal RNA (rRNA) processing, eventually affecting disease progression. Consequent studies on mouse models of ALS showed that angiogenin delivery is neuroprotective and significantly increases the lifespan of the treated mice, promoting motor neuron survival (Kieran et al. 2008). Furthermore, exogenous 5′ alanine tiRNAs, generated by Angiogenin cleavage rescued the angiogenin-depleted motor neurons (Ivanov et al. 2014), but 5′ tiRNA-induced cell survival was interfered with by mutations in the C9ORF72 gene, the most common genetic cause of ALS, through the impaired construction of G-quadruplex structures (Ivanov et al. 2014).

Parkinson's Disease (PD) mouse models also exhibit reduced levels of angiogenin (Steidinger et al. 2011). Moreover, angiogenin treatment of a human neuronal cell line confers protection from toxin-induced death, probably via inhibiting apoptosis (Steidinger et al. 2011). A more recent study demonstrates modulated profiles of various tRFs in PD patients' CSF, proposing tRFs as novel potential diagnostic markers (Magee et al. 2019). Unlike other tissues, where 5′ tiRNAs appeared as prominent regulators, 3′ tiRNAs seem to be especially relevant in the brain. Neurons carrying mutations in the cleavage and polyadenylation factor 1 subunit (CLP1) were protected from stress-induced death by the 5′ phosphorylated tyrosine-derived 3′ tiRNA (Schaffer et al. 2014). CLP1 is involved in tRNA splicing, phosphorylating the 5′ end of the 3′ tiRNA following intron excision (Schaffer et al. 2014). The loss of function mutation in CLP1 exacerbated oxidative stress in patient-derived induced neurons. The toxicity was enhanced by overexpressing unphosphorylated 3′ tyrosine tiRNAs, and, in contrast, moderated by phosphorylated 3′ fragments (Schaffer et al. 2014). Notably, the corresponding 5′ tyrosine tiRNAs did not have any effect on the cells. On the other hand, 5′ primed short tRFs were reported as potential biomarkers of epilepsy. Specifically, three glycine, glutamate and alanine tRNA-derived 5′ short tRFs, all cleaved in close proximity to the 3′-end of the D-loop, are upregulated in the plasma of epileptic patients shortly before seizures, compared to healthy controls (Hogg et al. 2019).

Patients with psychogenic non-epileptic seizures failed to present such tRFs, making the identified tRFs powerful potential predictors of seizure onset. Also, ischaemic stroke patients presented massive increases in nucleated blood cells' tRFs, likely regulating post-stroke immune reactions via their microRNA-like capacity to block the cholinergic anti-inflammatory response (Winek et al. 2020, 2021). A range of post-stroke elevated tRFs were further shown to be induced by inflammatory response in both murine RAW 265.7 cells and human CD14+ monocytes, demonstrating their immune functionality. Moreover, overexpression of a mimic molecule of the short glycine tRNA-derived 3′ tRF modulated in post-stroke patients' blood reduced the levels of a known inflammatory agent, Zbp1 which carries a complementary sequence motif to the mimicked tRF (Winek et al. 2020). Supporting the hypothesis of high relevance of tRFs in the immune response regulation (Dhahbi 2015), this study demonstrates the involvement of tRFs in post-stroke immunity control. Taken together, all of these observations raise the possibility that tRFs represent a vastly important family of non-coding RNA, modulating crucial pathways of neuronal regulation and brain homeostasis.

9 Methodology Used to Study tRFs

Ever since the re-discovery of tRFs as biologically active entities, various experimental and computational methodologies have been adjusted or specifically developed for studying these molecules (Fig. 2). Several tRF features, such as their chemical modifications or the highly repetitive sequence of their parental tRNAs, make tRFs particularly challenging molecules for biological validations. Hence, tRF studies inspire the establishment of novel approaches, overcoming such obstacles in different ways, as briefly described below. The first study focused on tRFs employed polyacrylamide gel electrophoresis that exposed a specific band of 30–35 nucleotides in RNA extracted from starved T. thermophila (Lee and Collins 2005). To identify the exact sequences of the discovered fragments, these and subsequent authors sequenced the corresponding RNA clones (Haiser et al. 2008; Jöchl et al. 2008). A major limitation of this approach is that many tRFs are not amenable to cloning, possibly due to the chemical modifications or their secondary structure (Lee and Collins 2005). The next major approach quantifying the amount of specific tRFs is RNA Northern Blot (Chen et al. 2016; Haiser et al. 2008; Jöchl et al. 2008; Lee and Collins 2005; Sharma et al. 2016; Yamasaki et al. 2009) (Fig. 2), where electrophoresis separated small fragments from parental tRNAs by size. However, this procedure is complex and time-consuming, with its efficiency depending on many factors (Yang et al. 2022). Lastly, it might lack the sensitivity which is essential for detecting lowly expressed fragments.

Quantitative PCR (qPCR), unlike gel electrophoresis, can detect RNA with an extremely low expression baseline, as it involves massive amplification. Yet, qPCR by itself is insufficient for detecting the tRFs since their sequence is derived from

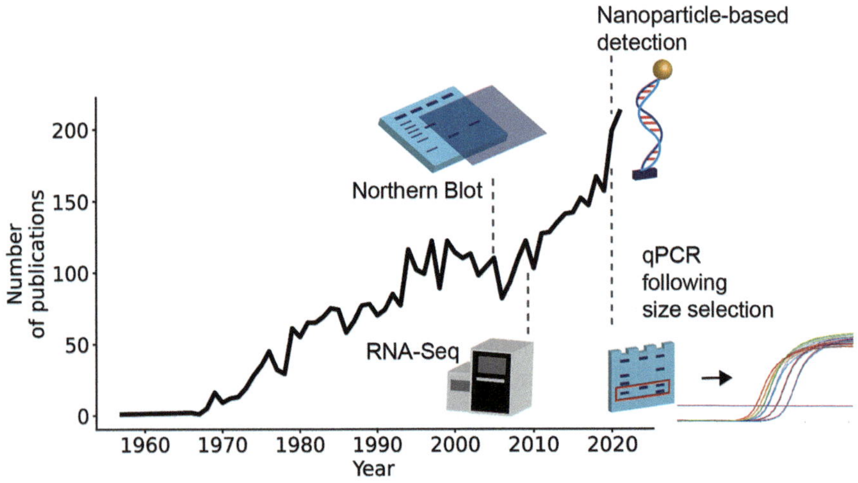

Fig. 2 PubMed timeline results showing the annual number of publications featuring "tRNA fragments"

their corresponding tRNAs of origin and because simple qPCR might also detect the paternal tRNA. To solve this difficulty, one may first perform size selection based on gel electrophoresis, then use qPCR detection of the relevant tRFs extracted from the gel among fragments smaller than 50 nucleotides in length (Winek et al. 2020) (Fig. 2). Unlike northern blot analysis, qPCR allows efficient quantification of multiple genes per experiment, providing an opportunity to detect lowly expressed fragments as well. However, a major limitation of qPCR-based tRF identification is its inability to detect and quantify chemical modifications, which represent an additional level of tRF regulation (Cozen et al. 2015).

Another efficient approach for selective tRF detection is based on amperometric quantification, using platinum nanoparticles (McArdle et al. 2020) (Fig. 2). Platinum nanoparticles are attached to probe nucleic acid sequences that are complementary to the target tRFs. Association by sequence complementarity allows for the specific marking of the tRFs of interest with the nanoparticles. Further injection of hydrogen peroxide leads to its reduction by nanoparticles, generating electrochemical current, which is proportional to the number of nanoparticles present. The number of nanoparticles hence indicates the concentration of the target tRFs. Unlike qPCR, amperometric quantification does not imply polymerase-based amplification, which makes it unbiased to chemical modifications interfering with polymerase activity. However, amperometry is more sensitive to the baseline level of expression of the target molecules, since it is less efficient in detecting lowly expressed transcripts (McArdle et al. 2020).

Lastly, next-generation sequencing revolutionized the field of tRF studies, allowing parallel quantification of thousands of fragments simultaneously and providing an unprecedented level of novel fragments detection and quantification (Cole et al. 2009; Haussecker et al. 2010; Lee et al. 2009) (Fig. 2). However, many tRFs are likely to be omitted by sequencing-based quantification due to the chemical modifications on their nucleosides, such as methylations. One way of overcoming this obstacle is eliminating such modifications and sequencing the pure RNA fragments (Zheng et al. 2015). Although efficient and relatively simple, this approach is problematic, since tRNA and tRF chemical modifications represent another layer of the induced molecular regulation, affecting the tRF generation, stability and function (Cozen et al. 2015). Alternatively, several methods have been developed to identify and measure RNA modifications (Cozen et al. 2015). Moreover, a novel computational pipeline has been introduced to deduce the chemical modifications from the standard RNA sequencing data (Yan et al. 2013).

Understanding the role of specific chemical modifications in tRF-induced biological regulation will take the RNA biology field to the next level, revealing the major principles of molecular organization.

Single-cell small RNA sequencing (scRNA-seq) is another anticipated technique for tRF research, that is yet to be developed. Single cell-based technology brought an unprecedented resolution to RNA biology, allowing to quantify RNA levels in single cells (Zheng et al. 2017). However, the most common scRNA-seq pipelines are designed for mRNA enrichment, amplifying the polyadenylated transcripts exclusively, thus omitting the quantification of small RNAs (Zheng et al.

2017). Existing protocols for single-cell small RNA detection are still limited by low-throughput, high-initial RNA concentration and the highly time-consuming procedure (Hücker et al. 2021). Hence, development of an efficient high-throughput single-cell small RNA-seq method will greatly contribute to tRF research, for revealing the cell type enrichment of specific fragments and possibly for exposing previously unknown cellular populations and developmental trajectories, characterized by small non-coding transcriptomes.

This graph, generated by PubMed, shows the number of publications per year featuring "tRNA fragments" (1957–2021). The trend shows that tRF research is becoming increasingly relevant with time. Vertical dash lines link various methods for tRF quantification to the approximate time points of the first studies implementing them. While the first papers mostly relied on the RNA northern blots (Lee and Collins 2005), nowadays, the majority of studies perform RNA-seq (Lee et al. 2009) with further validation using qPCR-based (Winek et al. 2020) or amperometric approaches (McArdle et al. 2020).

10 Conclusions

By summarizing the current knowledge about the recently rediscovered small non-coding RNA family of tRNA fragments (tRFs), we wished to share the currently available know-how regarding these intriguing molecules and their diverse activities. As written above, tRFs are potent regulators of multiple cellular processes and potential biomarkers for a range of diseases, including nervous system ailments such as stroke, epilepsy, ALS, PD and others (Anderson and Ivanov 2014; Kim et al. 2020; Su et al. 2020). Moreover, their involvement in promoting cellular survival and avoiding apoptosis introduces a prospect of using tRFs as efficient therapeutic agents (Li and Hu 2012; Saikia et al. 2014).

RNA therapeutics has recently become a very promising field, due to the international success of COVID-19 mRNA vaccines. In this context, tRFs present particularly advantageous molecules for clinical use due to their small size and low toxicity. In summary, tRF research is a new and exciting topic in RNA biology, and further scientific effort will undoubtedly reveal a great biological importance and clinical potential of this novel family of small non-coding RNAs.

Acknowledgements The authors acknowledge support for this study by the Israel Science Foundation grants no. 1016/18 and 3213/19 and by Keter Holdings (to H.S.), and PhD fellowship support to S.D. from the Kaete Klausner and Azrieli foundations.

References

Amaral PP, Dinger ME, Mattick JS (2013) Non-coding RNAs in homeostasis, disease, stress responses an evolutionary perspective. Brief Funct Genomics 12:254–278

Anderson P, Ivanov P (2014) tRNA fragments in human health, disease. FEBS Lett 588:4297–4304

Babiarz JE, Ruby JG, Wang Y et al (2008) Mouse ES cells express endogenous shRNAs, siRNAs, other microprocessor-independent, Dicer-dependent small RNAs. Genes Dev 22:2773–2785

Belle A, Tanay A, Bitincka L et al (2006) Quantification of protein half-lives in the budding yeast proteome. Proc Natl Acad Sci USA 103:13004–13009

Borek E, Baliga BS, Gehrke CW et al (1977) High turnover rate of transfer RNA in tumor tissue. Cancer Res 37:3362–3366

Brodsky WY (1975) Protein synthesis rhythm. J Theor Biol 167–200

Chen Q, Yan M, Cao Z et al (2016) Sperm tsRNAs contribute to intergenerational inheritance of an acquired metabolic disorder. Science 351:397–400

Chiou N-T, Kageyama R, Ansel KM (2018) Selective export into extracellular vesicles, function of tRNA fragments during T cell activation. Cell Rep 25:3356–3370

Cole C, Sobala A, Lu C et al (2009) Filtering of deep sequencing data reveals the existence of abundant Dicer-dependent small RNAs derived from tRNAs. RNA 15:2147–2160

Couvillion MT, Sachidanandam R, Collins K (2010) A growth-essential Tetrahymena Piwi protein carries tRNA fragment cargo. Genes Dev 24:2742–2747

Cozen AE, Quartley E, Holmes AD et al (2015) ARM-Seq AlkB-facilitated RNA methylation sequencing reveals a complex landscape of modified tRNA fragments. Nat Methods 12:879

Crick F (1970) Central dogma of molecular biology. Nature 227:561–563

Deutscher MP (1984) Processing of tRNA in prokaryotes, eukaryote. CRC Crit Rev Biochem 17:45–71

Dhahbi JM (2015) tRNA halves the next generation of immune signaling molecules. Front Immunol 6:74

Dhahbi JM, Spindler SR, Atamna H et al (2013) 5′ tRNA halves are present as abundant complexes in serum, concentrated in blood cells, modulated by aging, calorie restriction. BMC Genomics 14:298

Emara MM, Ivanov P, Hickman T et al (2010) Angiogenin-induced tRNA-derived stress-induced RNAs promote stress-induced stress granule assembly. J Biol Chem 285:10959–10968

Fagan SG, Helm M, Prehn JHM (2021) tRNA-derived fragments A new class of non-coding RNA with key roles in nervous system function, dysfunction. Prog Neurobiol 205:102–118

Favaloro B, Allocati N, Graziano V et al (2012) Role of apoptosis in disease. Aging 4:330–349

Garcia-Silva MR, Cabrera-Cabrera F, das Neves RFC et al (2014) Gene expression changes induced by Trypanosoma cruzi shed microvesicles in mammalian host cells relevance of tRNA-derived halves. Biomed Res Int 2014:305239

Goodarzi H, Liu X, Nguyen HCB et al (2015) Endogenous tRNA-derived fragments suppress breast cancer progression via YBX1 displacement. Cell 161:790–802

Haiser HJ, Karginov FV, Hannon GJ et al (2008) Developmentally regulated cleavage of tRNAs in the bacterium Streptomyces coelicolor. Nucleic Acids Res 36:732–741

Hanada T, Weitzer S, Mair B et al (2013) CLP1 links tRNA metabolism to progressive motor-neuron loss. Nature 495:474–480

Haussecker D, Huang Y, Lau A et al (2010) Human tRNA-derived small RNAs in the global regulation of RNA silencing. RNA 16:673–695

Hogg MC, Raoof R, El Naggar H et al (2019) Elevation of plasma tRNA fragments precedes seizures in human epilepsy. J Clin Invest 129:2946–2951

Hücker SM, Fehlmann T, Werno C et al (2021) Single-cell microRNA sequencing method comparison, application to cell lines, circulating lung tumor cells. Nat Commun 12:4316

Ivanov P, Emara MM, Villen J et al (2011) Angiogenin-induced tRNA fragments inhibit translation initiation. Mol Cell 43:613–623

Ivanov P, O'Day E, Emara MM et al (2014) G-quadruplex structures contribute to the neuroprotective effects of angiogenin-induced tRNA fragments. Proc Natl Acad Sci USA 111:18201–18206
Jöchl C, Rederstorff M, Hertel J et al (2008) Small ncRNA transcriptome analysis from Aspergillus fumigatus suggests a novel mechanism for regulation of protein synthesis. Nucleic Acids Res 36:2677–2689
Kafri M, Metzl-Raz E, Jona G et al (2016) The cost of protein production. Cell Rep 14:22–31
Kieran D, Sebastia J, Greenway MJ et al (2008) Control of motoneuron survival by angiogenin. J Neurosci 28:14056–14061
Kim HK, Fuchs G, Wang S et al (2017) A transfer-RNA-derived small RNA regulates ribosome biogenesis. Nature 552:57–62
Kim HK, Yeom J-H, Kay MA (2020) Transfer RNA-derived small RNAs another layer of gene regulation, novel targets for disease therapeutics. Mol Ther 28:2340–2357
Kumar P, Anaya J, Mudunuri SB et al (2014) Meta-analysis of tRNA derived RNA fragments reveals that they are evolutionarily conserved, associate with AGO proteins to recognize specific RNA targets. BMC Biol 12:78
Kumar P, Kuscu C, Dutta A (2016) Biogenesis, function of transfer RNA-related fragments (tRFs). Trends Biochem Sci 41:679–689
Lauber J, Marsac C, Kadenbach B et al (1991) Mutations in mitochondrial tRNA genes a frequent cause of neuromuscular diseases. Nucleic Acids Res 19:1393–1397
Lee SR, Collins K (2005) Starvation-induced cleavage of the tRNA anticodon loop in Tetrahymena thermophila. J Biol Chem 280:42744–42749
Lee YS, Shibata Y, Malhotra A et al (2009) A novel class of small RNAs tRNA-derived RNA fragments (tRFs). Genes Dev 23:2639–2649
Li S, Hu G-F (2012) Emerging role of angiogenin in stress response, cell survival under adverse conditions. J Cell Physiol 227:2822–2826
Li Z, Ender C, Meister G et al (2012) Extensive terminal, asymmetric processing of small RNAs from rRNAs, snoRNAs, snRNAs, tRNAs. Nucleic Acids Res 40:6787–6799
Lyons SM, Gudanis D, Coyne S et al (2017) Identification of functional tetramolecular RNA G-quadruplexes derived from transfer RNAs. Nat Commun 8:1127
Magee R, Londin E, Rigoutsos I (2019) TRNA-derived fragments as sex-dependent circulating candidate biomarkers for Parkinson's disease. Parkinsonism Relat Disord 65:203–209
Maraia RJ, Lamichhane TN (2011) 3′ processing of eukaryotic precursor tRNAs. Wiley Interdiscip Rev RNA 2:362–375
Maute RL, Schneider C, Sumazin P et al (2013) tRNA-derived microRNA modulates proliferation, the DNA damage response, is down-regulated in B cell lymphoma. Proc Natl Acad Sci USA 110:1404–1409
McArdle H, Hogg MC, Bauer S et al (2020) Quantification of tRNA fragments by electrochemical direct detection in small volume biofluid samples. Sci Rep 10:7516
Megel C, Hummel G, Lalande S et al (2019) Plant RNases T2, but not Dicer-like proteins, are major players of tRNA-derived fragments biogenesis. Nucleic Acids Res 47:941–952
Nolte-'t Hoen ENM, Buermans HPJ, Waasdorp M et al (2012) Deep sequencing of RNA from immune cell-derived vesicles uncovers the selective incorporation of small non-coding RNA biotypes with potential regulatory functions. Nucleic Acids Res 40:9272–9285
Oberbauer V, Schaefer MR (2018) tRNA-derived small RNAs biogenesis, modification, function, potential impact on human disease development. Genes 9:607
Olson KA, Verselis SJ, Fett JW (1998) Angiogenin is regulated in vivo as an acute phase protein. Biochem Biophys Res Commun 242:480–483
Paldor I, Madrer N, Vaknine Treidel S et al (2022) Cerebrospinal fluid and blood profiles of transfer RNA fragments show age, sex and Parkinson's disease-related changes. J Neurochem. https://doi.org/10.1111/jnc.15723
Saikia M, Jobava R, Parisien M et al (2014) Angiogenin-cleaved tRNA halves interact with cytochrome c, protecting cells from apoptosis during osmotic stress. Mol Cell Biol 34:2450–2463

Schaffer AE, Eggens VRC, Caglayan AO et al (2014) CLP1 founder mutation links tRNA splicing, maturation to cerebellar development, neurodegeneration. Cell 157:651–663
Schopman NCT, Heynen S, Haasnoot J et al (2010) A miRNA-tRNA mix-up tRNA origin of proposed miRNA. RNA Biol 7:573–576
Sharma U, Conine CC, Shea JM et al (2016) Biogenesis, function of tRNA fragments during sperm maturation, fertilization in mammals. Science 351:391–396
Shukla GC, Singh J, Barik S (2011) MicroRNAs processing, maturation, target recognition, regulatory functions. Mol Cell Pharmacol 3:83–89
Steidinger TU, Standaert DG, Yacoubian TA (2011) A neuroprotective role for angiogenin in models of Parkinson's disease. J Neurochem 116:334–341
Su Z, Wilson B, Kumar P et al (2020) Noncanonical roles of tRNAs tRNA fragments, beyond. Annu Rev Genet 54:47–69
Telonis AG, Loher P, Magee R et al (2019) tRNA fragments show intertwining with mRNAs of specific repeat content and have links to disparities. Cancer Res 79:3034–3049
Thompson DM, Parker R (2009) The RNase Rny1p cleaves tRNAs, promotes cell death during oxidative stress in Saccharomyces cerevisiae. J Cell Biol 185:43–50
Vojtech L, Woo S, Hughes S et al (2014) Exosomes in human semen carry a distinctive repertoire of small non-coding RNAs with potential regulatory functions. Nucleic Acids Res 42:7290–7304
Wang Y, Liu CL, Storey JD et al (2002) Precision, functional specificity in mRNA decay. Proc Natl Acad Sci USA 99:5860–5865
Weng Q, Wang Y, Xie Y et al (2022) Extracellular vesicles-associated tRNA-derived fragments (tRFs) biogenesis, biological functions, their role as potential biomarkers in human diseases. J Mol Med 100:679–695
Winek K, Lobentanzer S, Nadorp B et al (2020) Transfer RNA fragments replace microRNA regulators of the cholinergic poststroke immune blockade. Proc Natl Acad Sci USA 117:32606–32616
Winek K, Soreq H, Meisel A (2021) Regulators of cholinergic signaling in disorders of the central nervous system. J Neurochem 158:1425–1438
Wu D, Yu W, Kishikawa H et al (2007) Angiogenin loss-of-function mutations in amyotrophic lateral sclerosis. Ann Neurol 62:609–617
Yaffe MP (1999) The machinery of mitochondrial inheritance, behavior. Science 283:1493–1497
Yamasaki S, Ivanov P, Hu G-F et al (2009) Angiogenin cleaves tRNA, promotes stress-induced translational repression. J Cell Biol 185:35–42
Yan M, Wang Y, Hu Y et al (2013) A high-throughput quantitative approach reveals more small RNA modifications in mouse liver, their correlation with diabetes. Anal Chem 85:12173–12181
Yang T, Zhang M, Zhang N (2022) Modified Northern blot protocol for easy detection of mRNAs in total RNA using radiolabeled probes. BMC Genomics 23:66
Zheng G, Qin Y, Clark WC et al (2015) Efficient, quantitative high-throughput tRNA sequencing. Nat Methods 12:835–837
Zheng GXY, Terry JM, Belgrader P et al (2017) Massively parallel digital transcriptional profiling of single cells. Nat Commun 8:14049
Zou H, Li Y, Liu X et al (1999) An APAF-1·cytochrome c multimeric complex is a functional apoptosome that activates procaspase-9. J Biol Chem 274:11549–11556

Detection of RNA Structure and Interactions Using Nanopore Technology

Ashley Byrne and William Stephenson

Contents

Abstract RNA molecules are highly versatile and dynamic components within cells that orchestrate fundamental molecular biological processes, including information transfer and decoding, splicing, transcriptional regulation and translation. These vital processes are in large part determined by the complex structures and modular folding adopted by RNA. Nanopore technology has emerged as a powerful technique to interrogate native RNA directly at the single molecule level. In this chapter we will highlight and discuss applications of nanopore technology to investigate the conformation of RNA and its interaction partners. First, we will feature RNA translocation experiments which can extract key signatures of RNA size and global secondary or tertiary structure using nanopore electrical and temporal measurements. Second, we will focus on applications of nanopore technology to decipher RNA–ligand and RNA–protein interactions. Finally, we will summarize recent approaches using direct RNA nanopore sequencing in conjunction with chemical probing to obtain RNA

A. Byrne · W. Stephenson (✉)
Department of Microchemistry Proteomics Lipidomics and Next Generation Sequencing, Genentech, Inc., 1 DNA Way, South San Francisco, CA 94080, USA
e-mail: stephenson.william@gene.com

J. Barciszewski (ed.), *RNA Structure and Function*, RNA Technologies 14,
https://doi.org/10.1007/978-3-031-36390-0_2

structural profiles at high-throughput. Nanopore technology is well positioned to address the needs for single-molecule, high-throughput RNA detection and structural characterization.

Keywords RNA structure · Nanopore · Single molecule · RNA folding · RNA interactions · Chemical probing · Nanopore sequencing

Abbreviations

SiN	Silicon nitride
i.d.	Inner diameter
MspA	Mycobacterium smegmatis porin A
α-HL	α-Hemolysin
ZIKV	Zika Virus
MD	Molecular dynamics
NCp7	Nucleocapsid protein 7
UTRs	Untranslated regions
PM	Paromycin
HCV	Hepatitis C virus
IRES	Internal ribosome entry site
rRNA	Ribosomal RNA
mRNA	Messenger RNA
ONT	Oxford Nanopore Technologies
DMS	Dimethylsulfate
DMS-seq/MaP	Dimethylsulfate sequencing/mutational profiling
SHAPE-seq/MaP	Selective 2′-hydroxyl acylation analyzed by primer extension sequencing/mutational profiling
RT	Reverse transcription
SAFA	Semi-automated footprinting analysis
SVM	Support vector machine
AcIm	Acetylimidazole

1 Introduction

Nanopore technology is a promising approach to investigate the physical and chemical characteristics of biomolecules. Nanopores can be assembled from biologically derived sources (protein nanopores), fabricated de novo (synthetic or solid state nanopores) or constructed to contain both biological and solid-state components (hybrid nanopores) (Fig. 1). Despite the construction of a nanopore, the fundamental working principle of a nanopore sensor is largely the same across all nanopore types.

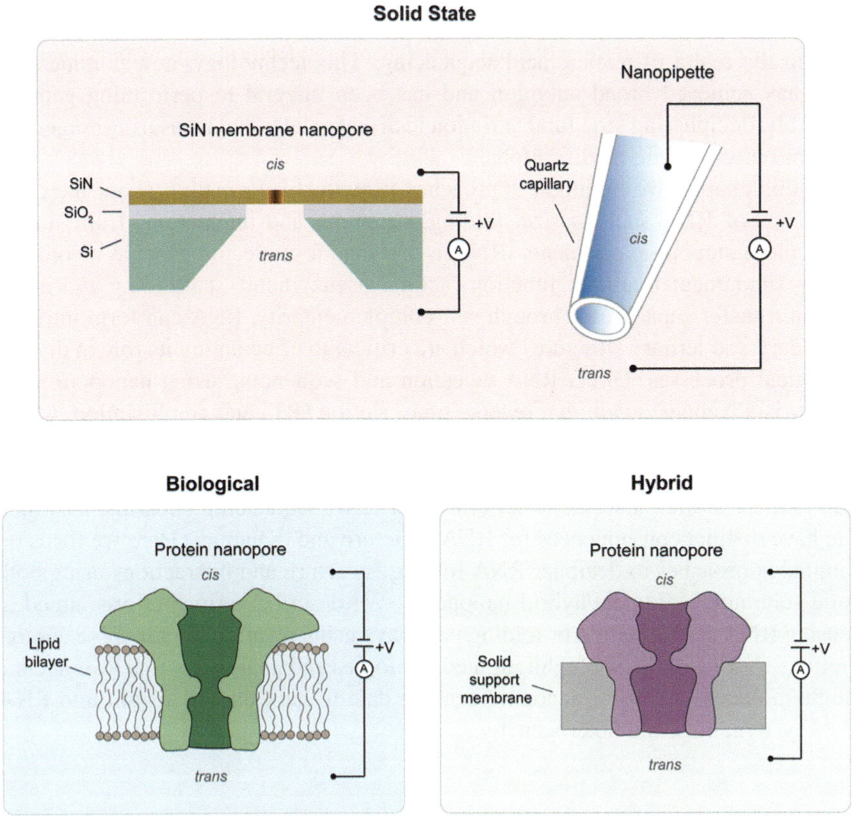

Fig. 1 Common nanopore technology variations. Solid state nanopore devices are typically made using micro- and nanofabrication techniques. Nanopipettes are generated by pulling heated quartz capillaries. Biological nanopores comprise a protein nanopore embedded in a lipid bilayer membrane. Hybrid nanopores consist of a protein nanopore situated within a solid support substrate

Nanopores measure fluctuations in ionic current across a nanoscale constriction. Electrical bias is typically applied across the nanopore which spans a membrane or support structure. Ion transport through the nanopore constriction is dictated by the applied voltage, solution properties and the nanopore dimensions. A baseline current (typically in the picoampere range) through the nanopore is established upon the application of a constant voltage (typically ~100's millivolts) across the *cis* and *trans* side of the nanopore. As a molecule translocates through the pore lumen it temporarily interrupts the baseline current flow through the nanopore constriction leading to complex ionic current blockage signatures reflecting the chemical and physical properties of the molecule. Nanopores, by virtue of their dimensions, are single-molecule sensors and thus are uniquely suited to interrogate molecular heterogeneity, a hallmark of their dynamic biological assemblies. Within the last decade, nanopore sensing has seen significant technological advancements particularly in the realm of nucleic acid sequencing. This technology, now commercialized, has achieved broad adoption and has been integral to performing genome assembly, deciphering structural variation in disease states and measuring epigenetic and epitranscriptomic modifications.

In this chapter we highlight approaches to extract information about the characteristics of RNA such as size, folding, structure, and interactions from single-molecule nanopore experiments. RNA is a dynamic molecule capable of orchestrating fundamental cellular functions acting in enzymatic, scaffolding and information transfer capacities. Through self-complementarity, RNA can form intricate secondary and tertiary structures which are critical to effectuating its role in diverse biological processes. Direct RNA detection and sequencing using nanopore technology is advantageous in that reverse transcription (RT) and amplification, which can be subject to distinct biases, are not required. This facilitates a rapid one-step detection in nanopore sensing applications and access to epitranscriptomic information (which is often lost due to RT) in direct RNA sequencing experiments which can have distinct consequences for RNA structure and dynamics. Here we focus on seminal approaches to decipher RNA folding, structure and interactions using both solid-state and biological/hybrid nanopores. We describe various efforts aimed at sensing RNA conformation or folding pathways using a variety of nanopore configurations. Additionally, we highlight methodologies and analyses which leverage the single molecule nature of nanopore sensing data to gain unique insight into RNA identity, dynamics and heterogeneity.

2 Extracting Global Features of RNA Molecules from Nanopore Translocation Experiments

2.1 RNA Detection and Sensing

Nanopore based platforms afford various means to measure and characterize nucleic acid translocation events. For example, ion blockage amplitudes can be used as a readout of nucleic acid characteristics such as size and topology or conformation (e.g., structure). Temporal analysis of individual translocation measurements can be used to infer unfolding energetics and kinetics. An early demonstration of nucleic acid characterization was shown to distinguish between a mixture of double-stranded (ds)DNA, dsRNA and a folded phenylalanine tRNA using silicon nitride (SiN) solid-state nanopore sensors (Wanunu et al. 2010). Ion blockage amplitudes varied depending on whether an A-form helix (RNA) or a B-form helix (DNA) was translocated through the nanopore. Further, the folded tRNA displayed larger ion blockage amplitudes and mean transport time, consistent with its more complex secondary structure relative to the dsDNA and dsRNA. The authors extended their approach to develop a platform for microRNA (miRNA) detection using p19 protein conjugated magnetic beads which permitted miRNA enrichment after hybridization with a capture probe. miRNA:probe duplexes were detected using the solid-state nanopore as a molecular counter. By analyzing the frequency of translocation events or counts compared against a calibration curve, the miRNA concentration was obtained.

An analogous approach was recently employed using the Mycobacterium smegmatis porin A (MspA) biological nanopore (Wang et al. 2021). The MspA nanopore has a large vestibule and a narrow constriction which allowed for a sensing mode called nanopore trapping/translocation which operates in conjunction with an asymmetric buffer configuration (*cis*: 1.5 M KCl, *trans*: 1 M $CaCl_2$) to slow the translocation time of nucleic acids thorough the pore (Wang et al. 2020). By investigating the unique current blockage amplitudes and mean dwell times (including oscillations within the nanopore lumen) of various RNA constructs, the authors were able to characterize distinct translocation signatures of RNAs ranging from a 22 nucleotide (nt) miRNA to the 120 nt 5S rRNA. Assisted by a random forest machine learning approach trained on current event features from experiments involving the individual RNAs, the authors were able to demonstrate classification accuracies in excess of 0.90 indicating the potential for reliable identification of small RNAs in simple mixtures.

2.2 *Investigation of RNA Folding*

Nanopore sensing has been greatly supplemented with the integration of optical tweezers which allow for the application and measurement of piconewton forces at nanometer length scales. Optical tweezers utilize focused laser light to trap and manipulate micron scale particles or beads. Multiple studies have demonstrated the utility of the nanopore-optical tweezers combination to investigate DNA dynamics (Keyser et al. 2006; Knust et al. 2017; Craig et al. 2019), but relatively few have employed this experimental configuration to examine RNA conformation. Van Den Hout and colleagues coated microparticles with RNA and using optical tweezers positioned them into close proximity with solid-state nanopore sensors to achieve piconewton force control of translocation events (Van Den Hout et al. 2010). Mechanical force based unfolding or force spectroscopy of RNA is a powerful method to probe the energetics and kinetics of secondary and tertiary structures (Tinoco et al. 2006; Bizarro et al. 2012; Stephenson et al. 2014), but the combination of nanopore manipulation with optical tweezers enables repeated measurements controlling both electrophoretic and mechanical forces which can be exploited to study nucleic acid-protein interactions including processing of DNA by molecular motors (Craig et al. 2019).

An alternative approach to controlled translocation of RNA to assess folding was demonstrated using a blocking scheme to prevent full translocation of the RNA through the nanopore (Zhang et al. 2017). Blocking was achieved by flanking the RNA between two DNA handles with the 3′ handle containing a biotin complexed with streptavidin protein which acts as a steric roadblock for complete translocation (Fig. 2). To demonstrate the advantage of this strategy, the authors profiled the folding of a pseudoknot from the gene 32 messenger RNA of bacteriophage T2 using an α-hemolysin (αHL) nanopore. The starting configuration of the nanopore/RNA complex (with the RNA unfolded on the *trans* side of the nanopore) ensured that the folding studies were initiated from the same single-stranded conformation. Folding proceeded on the *trans* side of the nanopore, but could be interrogated at different time points by switching the voltage bias across the nanopore to unfold the RNA. This allowed the authors to generate snapshots of RNA intermediates on the folding pathway of the T2 pseudoknot which were consistent with coarse grained molecular dynamics (MD) simulations.

RNA translocation through a nanopore can also be manipulated through direct changes to the molecular construct under study. For example, Niu and colleagues demonstrated the extreme anisotropy associated with unfolding the pseudoknot-containing exoribonuclease resistant xrRNA1 from Zika Virus (ZIKV) (Niu et al. 2020) (Fig. 3). In this approach, the leader sequence (which is used to initiate the

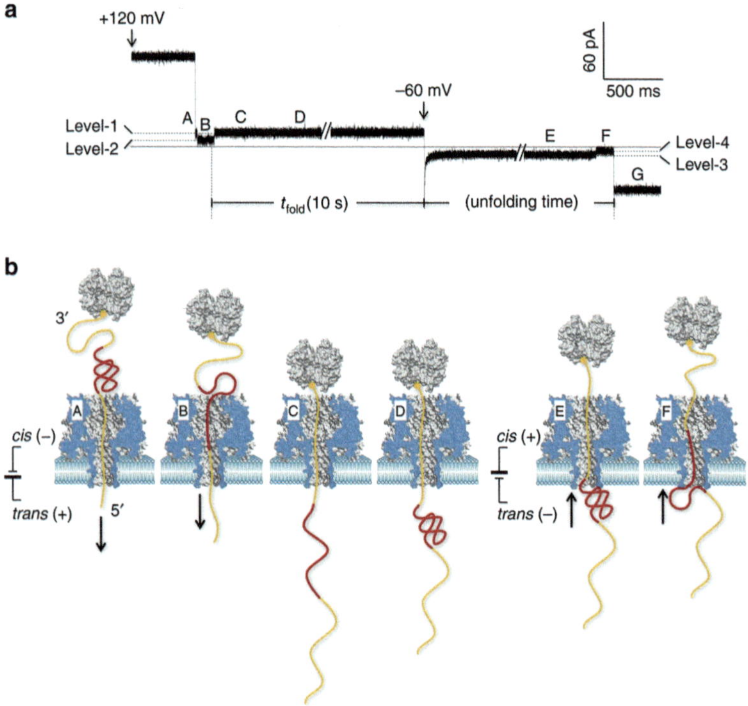

Fig. 2 **a** Nanopore current trace for T2 RNA pseudoknot formation. **b** Cartoon representation of trapping-folding-unfolding procedure utilizing biotin-streptavidin roadblock. Nature Communications (Zhang et al. 2017) Copyright 2017. (https://creativecommons.org/licenses/by/4.0/)

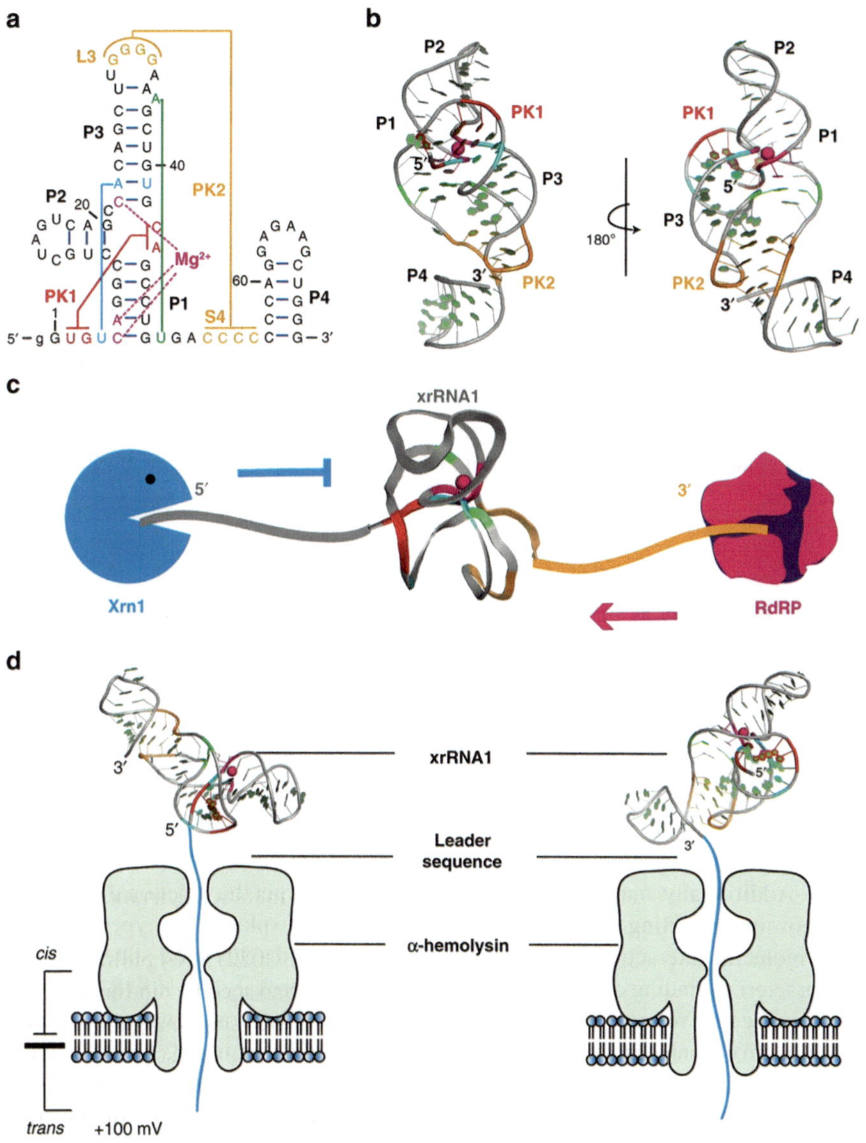

Fig. 3 Secondary (**a**) and tertiary (**b**) structures of the ZIKV xrRNA1. **c** Cartoon showing exonuclease resistance from Xrn1 ($5' \rightarrow 3'$) and traversal by RdRP ($3' \rightarrow 5'$). **d** Illustration showing the experimental configuration and directional unfolding of ZIKV xrRNA1 through the α-HL nanopore. From Nature Communications (Niu et al. 2020) Copyright 2020. (https://creativecommons.org/licenses/by/4.0/)

translocation of the RNA through the nanopore) was either attached at the 5′- or 3′-end of the RNA construct, thus biasing unfolding unidirectionally in the 5′ → 3′ direction (5′-leader) or the 3′ → 5′ (3′-leader) direction. The authors determined that xrRNA1 has extreme mechanical anisotropy as unfolding with the 5′-leader configuration resulted in a surprising mean dwell time in excess of 300 s whereas unfolding with the 3′-leader configuration led to a comparatively low mean dwell time of about 2 s. The dependence of the xrRNA1 structure on salt conditions was explored by generating a weakened construct amenable to translocation (xrRNA1-X) that retained the overall structure but displayed experimentally favorable dwell times. Mg^{2+}-dependent tertiary interactions and the overall ring-like structure of xrRNA1 were shown to be critical regulators of the overall stability of the RNA fold which is a common feature in the 3′-untranslated regions of mosquito-borne flaviviruses. These studies demonstrate that nanopore sensors have a key role to play in deciphering RNA conformations, empowering the dissection of folding intermediates and enabling directional unfolding providing unique insight into complex RNA structures.

3 Deciphering RNA Interactions with Nanopore Sensing

3.1 RNA–Small Molecule Interaction

RNA function is intimately coupled to its interaction with other biomolecules including DNA, protein and other RNA molecules. For example, riboswitches, often found within untranslated regions (UTRs), can bind with small molecule metabolites to regulate translation of genes associated with the metabolite (Lotz and Suess 2018). Additionally, many small molecule antibiotics target the functional centers of the ribosome inhibiting protein synthesis in bacteria. To explore these types of RNA–small molecule interactions, Wanunu and colleagues used a SiN solid-state nanopore to characterize binding of an aminoglycoside drug to a prokaryotic 16S rRNA A-site RNA molecule (Wanunu et al. 2011). By plotting the fractional change in current amplitude for translocation events the authors were able to quantify the fraction of bound and unbound molecules in the population as a function of paromycin (PM) concentration. This quantitative affinity curve was used to determine dissociation constant (K_d) values of the A-site RNA:PM complex at different ionic strengths by fitting the data to a simple 1:1 binding model. The K_d values were in good agreement with an orthogonal FRET-based assay, indicating the ability of sub-10 nm solid-state nanopores to characterize nucleic acid/drug binding.

RNA structure can also be heavily influenced by small-molecule binding. Shasha et al. explored the effects of drug compounds binding to the hepatitis C virus (HCV) internal ribosome entry site (IRES) IIa RNA domain (Shasha et al. 2014). Previous efforts using a FRET assay have shown that the HCV IRES IIa domain adopts a bent conformation, whereas addition of particular compounds increased the interhelical angle forcing the domain into a straight conformation. The authors examined translocation times through a 3 nm diameter pore in a SiN membrane observing shorter lifetimes in the IRES IIa drug-bound straight conformation versus the "apo" or drug-free bent conformation. Larger diameter nanopores led to fast pulses that were indistinguishable from drug bound complexes. Another group used an αHL-based biological nanopore system to explore the binding of adenine to the adenine-sensing riboswitch aptamer domain (ARS) (Lee et al. 2021). The ARS folds into a tuning fork structure upon adenine binding and regulates the production of adenosine deaminase (*add*) which is essential for bacterial metabolism, making it a prime antibacterial target. In this application the authors developed a nanopore-based screening approach of 40 natural compounds (split into 10 groups of 4) against ARS and selected groups which exhibited an increased frequency of translocation events associated with compound binding which were distinguished via elevated dwell times. Ultimately, the screen identified three novel ARS targeting compounds which were subsequently confirmed with nuclear magnetic resonance (NMR). These results suggest that nanopore based time-domain analysis in conjunction with appropriately sized nanopores can be a powerful approach to decipher RNA conformational changes resulting from drug or metabolite binding.

3.2 *RNA–Protein Interaction*

Beyond small molecule interactions, the interaction between RNA and protein has also been studied using nanopore based sensing. Niedzwiecki et al. demonstrated the detection and characterization of nucleocapsid protein 7 (NCp7) interaction with a stem-loop 3 (SL3) RNA aptamer derived from the packaging domain of the HIV-1 retroviral genome (Niedzwiecki et al. 2013). This approach utilized a solid-state SiN nanopore system with small (inner diameter, i.d. < 6 nm) and large (i.d. $= 7$–15 nm) pore diameters. In examining SL3:NCp7 complexes with the smaller i.d. nanopores, rather than seeing changes in dwell time or blockage current amplitude in comparison to the RNA only, the authors observed changes in the frequency of events. The inter-event times (τ_{on}) were used to construct a titration curve from which the K_d was derived and used to determine the binding affinity between various SL3 aptamer variants and NCp7. Alternatively, using the large i.d. nanopores the authors did observe changes in the dwell time and current amplitude of blockages. These observations suggested that in the experiment with small nanopores, translocation events were likely reflective of sampling unbound SL3 RNA, whereas the large nanopores likely permitted translocation of the SL3:NCp7 complex in addition to unbound SL3 RNA. These results show how nanopore sensing can be used to determine properties of biophysical interactions in a label-free manner and how these measurements are sensitive to the physical configuration of the nanopore used. Further, probing RNA–protein interactions may be adapted for sensitive quantification of disease biomarkers.

To date, nucleic acid-protein interactions studied with nanopores have relied on interpreting the energetics and kinetics of dissociation between the protein from the nucleic acid prior to translocation of the nucleic acid through the nanopore. An alternative strategy to investigate RNA–protein interactions might involve chemical crosslinking or fixation between RNA and protein followed by partial or near complete digestion of the protein. This could be facilitated with UV specific heterobifunctional linkers (Weidmann et al. 2021) and or UV crosslinking of RNA incorporated with metabolic labels such as 4-thiouridine (4sU) (Hafner et al. 2010). The presence of the resulting peptide adducts might be detected as complex current signatures in nanopore translocation experiments. In nanopore sequencing approaches (See Sect. 4), protein adducts on RNA may manifest as basecalling errors and/or exogenously detected modifications provided that the adducts are compatible with nanopore translocation. Mapping contact sites between the protein and RNA at the single molecule level would be transformative for elucidating ribonucleoprotein assembly and dynamics. The successful execution of this proposed approach however will ultimately rely on the efficiency and frequency of the chemical crosslinking reaction and throughput of the nanopore sensor or sequencer.

3.3 RNA–Ribosome Interaction

RNA–RNA interactions are crucial mediators of fundamental biological processes such as translation. During protein synthesis, engagement of ribosomal RNA (rRNA) with messenger RNA (mRNA) often occurs in a multiplicative fashion in which multiple ribosomes can translate a single mRNA molecule simultaneously. Readouts of rRNA loading on individual mRNAs can then be used to assess translational output. Raveendran et al. employed a nanopipette (a quartz pipette with a 60 nm diameter pore opening) to fingerprint 80S ribosomes and polysomes from a human neuronal cell line and Drosophila cells (Raveendran et al. 2020) (Fig. 4). A larger diameter pore was used in order to investigate the ribosomal complexes which are approximately 25–30 nm in diameter. The authors began by establishing a baseline dwell time and peak amplitude for monosome translocation events aided by sucrose gradient sedimentation profile separation. The authors showed discrimination of at least 4 polysomes compared with mRNA loaded with 1, 2 or 3 ribosomes from small sample volumes (3–5 μl). The authors note that further improvement might be achieved by additional tuning of the nanopore dimensions and performing chemical functionalization of the nanopore surfaces.

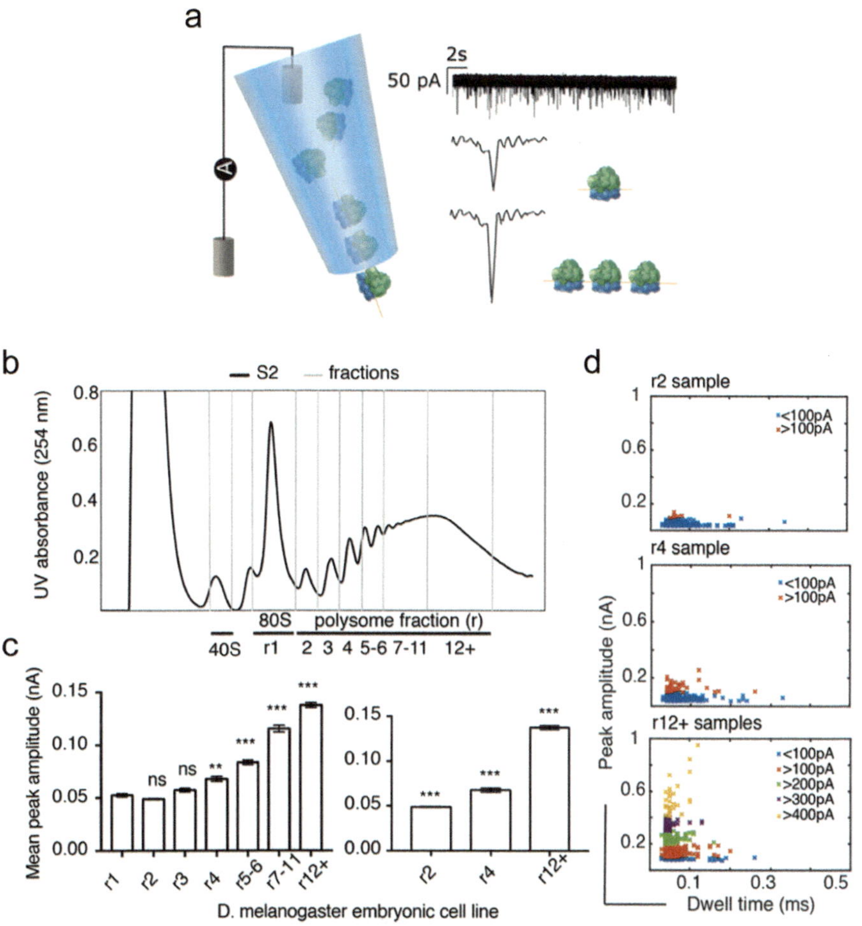

Fig. 4 **a** Illustration of nanopipette current measurement of 80S monosomes and polysomes. **b** Sucrose gradient fractionation profile showing ribosomal complexes separate based on sedimentation rate. **c** Mean peak current measurements of selected mono- and polysomal fractions. **d** Peak amplitude versus dwell time for samples from fractions r2, r4, and r12+. Adapted from ACS Sensors (Raveendran et al. 2020) Copyright 2020. (https://pubs.acs.org/page/policy/authorchoice_ccby_termsofuse.html)

3.4 *Engineering RNA-Interaction and Structure for Molecular Identification*

In contrast to measurements of endogenous RNA structure or interactions, RNA structure and interactions can be manipulated for detection and identification purposes in nanopore sensing experiments. Bošković and Keyser elegantly demonstrated this approach using a series of RNA–DNA nanostructures and/or protein labels termed 'structural colors' which were inspired by DNA origami technology

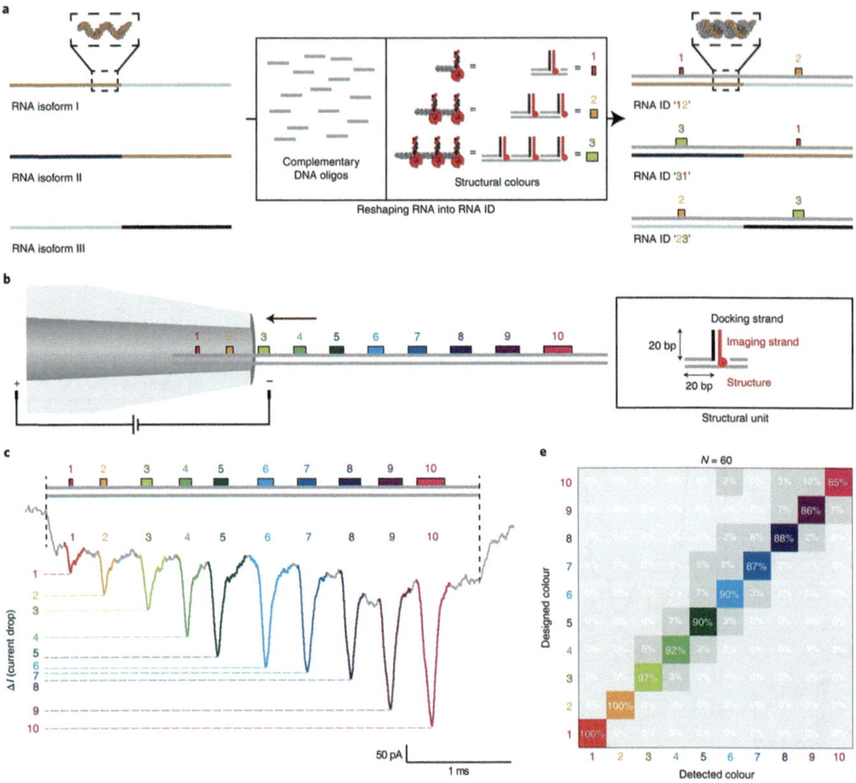

Fig. 5 ARTEMIS **a** Generation of RNA isoform identifiers (IDs). RNA IDs consist of an integer number of structural colors comprised of RNA/DNA nanostructures. **b** Nanopore 'microscope' showing translocation of an RNA complexed with 10 structural colors. **c** Ionic current trace of 10 structural colors on a single RNA molecule. **e** Error rates for assigning 10 structural colors, sample size (*N*) is 60 events. (Bošković and Keyser 2022) Reprinted/adapted by permission from Springer Nature Copyright 2022

(Bošković and Keyser 2022). The method, called Amplification-free RNA TargEt Multiplex Isoform Sensing or ARTEMIS, targets structural colors via sequence specific hybridization to long RNAs to form RNA identifiers or IDs. Structural colors consisted of a docking strand which partially anneals to the RNA target in addition to an imaging strand which contains a terminal monovalent streptavidin or DNA cuboid nanostructure. Structural colors (which consisted of an integer number of RNA–DNA nanostructures) were used along with standard complementary oligos to reform the RNA target structure to enable decoding with a solid-state glass capillary nanopore (Fig. 5). Ionic current signatures of translocation events correlated well with the molecular weight of the structural colors and were accurately assigned (first 4 structural colors: 97%, remaining 6: >85%). The authors demonstrated multiplexed RNA identification in a complex mixture in addition to classification of isoforms based on exon order, overall length and circularity. Finally, the authors used ARTEMIS to identify and quantify enolase mRNA and Xist lncRNA isoforms from total RNA using a combination of endogenous RNA structure and structural colors. ARTEMIS demonstrates promise to scale to ~10^{10} unique RNA IDs for transcriptome wide isoform detection and mapping of RNA motifs. While ARTEMIS sidesteps the need for RT and amplification, it does require RNA or genomic sequence information for the design of structural colors, thus additional development would be needed for detection of novel isoforms.

4 RNA Structure Probing with Nanopore Sequencing

4.1 Chemical Probing and Direct RNA Sequencing for RNA Structural Probing

Single-molecule nanopore sequencing of nucleic acids has seen vast improvements in accuracy and throughput over the last decade. Led by Oxford Nanopore Technologies (ONT), the commercial availability of nanopore instrumentation has significantly broadened access to long-read sequencing technology. Favored for its low-cost and small footprint, the flagship MinION device from ONT is a handheld nanopore sequencing instrument that can sequence DNA or RNA directly. Specifically, for direct RNA sequencing, a motor protein and series of DNA adapters are ligated to the RNA of interest, typically via a poly(dT) splint adapter (Fig. 6) (Garalde et al. 2018). The motor protein facilitates RNA translocation through the pore and acts as

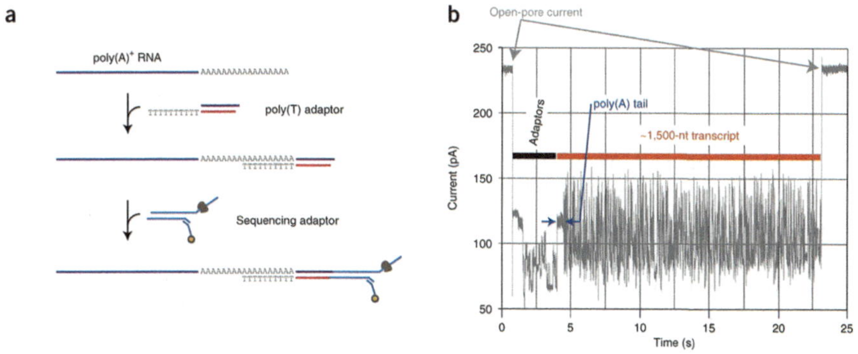

Fig. 6 Direct RNA sequencing library preparation from Oxford Nanopore Technologies. **a** A poly(dT) splint adapter is ligated to poly(A) RNA. Next, a sequencing splint adapter containing the motor protein is ligated before loading on the nanopore flow cell. Optionally, RT can be performed after the first adapter ligation and before the second sequencing adapter ligation. **b** Nanopore ionic current trace detailing the 3′- to 5′-translocation of a single 1.5 kb RNA molecule. The adapter sequences and poly(A) are visible at the start of the read. (Garalde et al. 2018) Reprinted/adapted by permission from Springer Nature Copyright 2018

a speed brake to slow translocation such that individual current events corresponding to roughly 5 ribonucleotides (i.e., a k-mer) situated at the nanopore constriction are observed. These current events called 'squiggles' can be processed according to a trained neural network k-mer model to produce a basecalled sequence. Importantly, the RNA molecule is sequenced directly (e.g., not through an intermediate cDNA) and theoretically the squiggle can report on any chemical modification present on the RNA. This feature has been successful in detecting native RNA modifications such as N6-methyladenine (m6A) and pseudouridine (Ψ) across diverse organisms (Workman et al. 2019; Parker et al. 2020).

Exogenous modifications introduced as nucleotide analogs or as a result of chemical probing can also be detected (Müller et al. 2019; Drexler et al. 2020; Stephenson et al. 2022). Chemical probing approaches such as DMS-seq/MaP (Dimethylsulfate sequencing/mutational profiling) and SHAPE-seq/MaP (Selective 2′-hydroxyl acylation analyzed by primer extension sequencing/mutational profiling) have been employed to dissect RNA structure at single nucleotide resolution using short-read sequencing (Merino et al. 2005; Ding et al. 2014). Chemical probes such as dimethylsulfate (DMS) or 2-methylnicotinic acid imidazolide (NAI) can be applied to cells or RNA to chemically modify the nucleobase (DMS) or the ribose sugar (SHAPE reagents) of unpaired or flexible sites within the RNA. Reverse transcription (RT) truncation or mutational profiling (MaP) events due to the presence of chemical adducts on the RNA can be identified and quantified via next-generation sequencing. This information is then used to constrain secondary structure predictions for RNA folding. MaP approaches, while powerful, are hindered by low adduct conversion rates using RT on chemically modified RNAs and additionally limited by short read lengths using Illumina sequencing platforms. Finally, MaP approaches are generally bulk averages of RNA structural ensembles, requiring much higher read depths and additional bioinformatic processing to attempt deconvolution of RNA structural states at the single-molecule level (Morandi et al. 2021; Olson et al. 2022).

To investigate RNA structural ensembles directly, researchers have used RNA chemical probing approaches in conjunction with single molecule direct RNA nanopore sequencing. Aw et al. performed chemical probing of various RNAs with the SHAPE reagent NAI-N3 (Aw et al. 2020) (Fig. 7). Sequencing of structure-directed chemically modified RNAs on the ONT platform revealed distinct current signatures compared to unmodified controls. Regions of reactivity were consistent with RT stops on similarly chemically modified RNAs assessed via semi-automated footprinting analysis (SAFA) gel readout. Modified bases in nanopore experiments were identified using a support vector machine (SVM) classification scheme called PORE-cupine. The authors first validated the method in vitro on the *Tetrahymena* RNA and further demonstrated with PORE-cupine that shared sequences in highly expressed RPS8 and RPL17 transcript isoforms could display structural differences highlighting the advantages of using the long-read approach for structural phasing. Notably the authors did identify limitations of the approach in that higher chemical modification rates led to higher errors during basecalling which ultimately led to poor mapping rates. Indeed, this is expected as the presence of multiple, large chemical adducts on the RNA molecule that will shift the current signal away from the canonical event level model values. Additionally, read throughput on chemically modified RNA sequencing runs were drastically compromised, signifying potential unfavorable interactions of bulky adducts with the motor protein.

To partially address motor protein processing during direct RNA sequencing of SHAPE modified RNAs, our group has employed the use of a SHAPE reagent with a more compact adduct. It was recently shown that acetylimidazole (AcIm) was an efficient SHAPE-seq reagent able to characterize the structure of the FMN riboswitch (Habibian et al. 2019). We have demonstrated, by benchmarking against other SHAPE reagents on extracted ribosomes, that AcIm is a veritable SHAPE-MaP reagent. We additionally showed that AcIm is useful for structural probing in direct RNA nanopore sequencing experiments exploring the multi-hairpin structure of an in vitro transcribed primary transcript of the 951 nt pri-miR 17~92 cluster

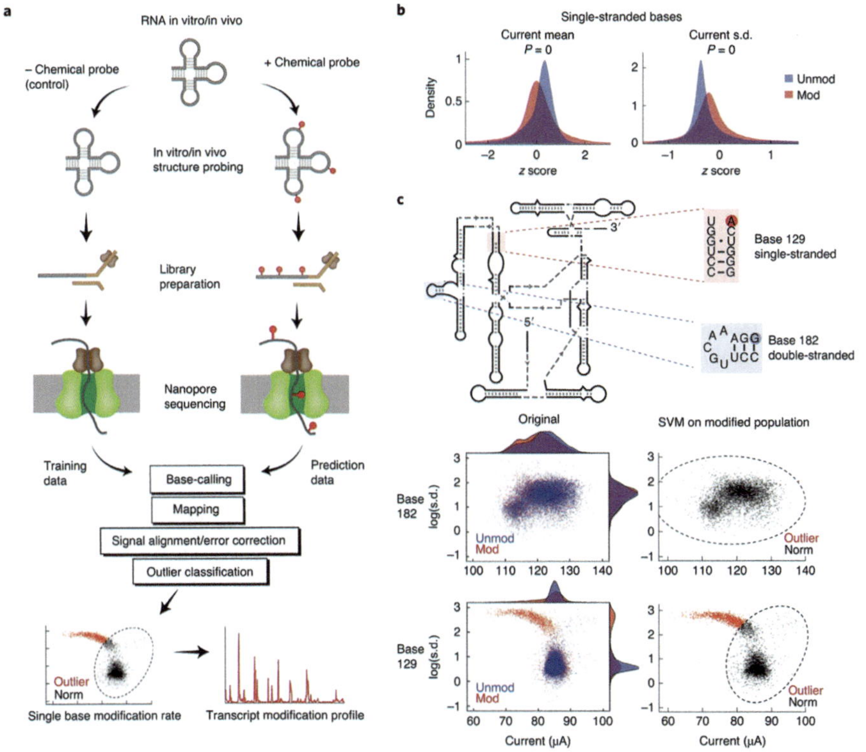

Fig. 7 PORE-cupine. **a** Schematic of chemical probing approach used with direct RNA nanopore sequencing followed by signal processing to determine RNA structure. **b** Normalized current and standard deviation distributions for single-stranded bases on *Tetrahymena* RNA modified with the NAI-N3 SHAPE probe. **c** Secondary structure of *Tetrahymena* RNA showing representative single- and double-stranded bases and their respective current and standard deviation of current plots before and after SVM classification. (Aw et al. 2020) Reprinted/adapted by permission from Springer Nature Copyright 2020

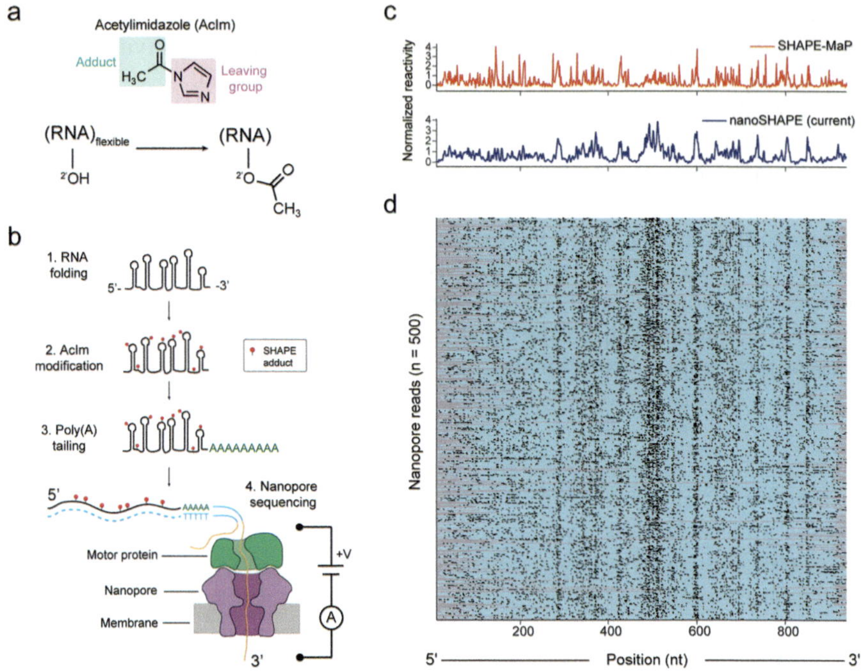

Fig. 8 NanoSHAPE. **a** Acetylimidazole reacts with the 2′-OH pf flexible sites in RNA. **b** NanoSHAPE comprises folding, modification with acetylimidazole, optional poly(A) tailing and direct RNA nanopore sequencing. **c** Normalized reactivity from short- (top) and long-read (bottom) structural probing of the pri-miR-17~92 RNA. **d** 500 single molecule nanopore reads of the pri-miR-17~92 RNA. Modified sites are shown in black, unmodified sites are shown in blue, grey regions denote poor signal to sequence mapping of reads. Adapted figure provided by the author (Stephenson et al. 2022)

(Fig. 8). The smaller adduct afforded by AcIm was more favorable for motor protein processing compared to the bulkier NAI SHAPE reagent (Stephenson et al. 2022).

Structure induced base modification and detection via direct RNA nanopore sequencing has also been demonstrated in a technique termed single molecule structure sequencing or SMS-seq (Bizuayehu et al. 2022). In contrast to the previously described approaches which utilized SHAPE reagents targeting the ribose of RNA, SMS-seq uses diethyl pyrocarbonate (DEPC) which modifies unconstrained adenine bases by imidazole ring opening. Using this approach, the authors profiled the thiamine pyrophosphate (TPP) riboswitch in the ligand-bound and unbound state in vitro. The authors developed an analytic framework which used a distance-normalized adjusted mutual information (dAMI) metric revealing which RNA bases in the riboswitch are contained within the same structural unit. A drawback of using DEPC for RNA structural probing is that it is only reactive towards adenine and is not suitable for in vivo assays. Despite the challenges and shortcomings of chemical probing on the direct RNA nanopore sequencing platform, these methods have demonstrated the ability to detect multiple structure-directed adducts at the single molecule level which will be crucial for interpreting molecular structural ensembles.

5 Conclusion

Nanopore technology has greatly advanced our understanding of RNA structure, folding, dynamics and interactions. As we have reviewed, nanopore characterization of RNA spans detection of simple RNA duplex formation to translocation of entire polysome complexes. In particular, nanopore sensing enables fine control over the solution conditions and nanopore dimensions which can be tuned towards the biological system under study. Addition of small molecules or proteins is straightforward in the nanopore-based label-free sensing approach. Nanopore methods are amenable to enzymatic, biochemical and or physical manipulation strategies which have revealed critical aspects of RNA secondary and tertiary structure and dynamics at the single molecule level. We further anticipate that solid-state and biological nanopore systems will eventually scale, enabling parallel measurement of RNA translocation events at much higher-throughput than currently achieved with single-channel devices.

Direct RNA sequencing has emerged as a strong candidate for high-throughput nanopore-based single molecule RNA analysis. Direct RNA nanopore sequencing is unrivaled in that the RNA molecule is measured directly in contrast to other sequencing technologies that rely on a sequencing-by-synthesis approach. Direct RNA sequencing permits the detection of endogenous and exogenous modifications, the latter of which, through chemical probing approaches, have been exploited to measure RNA secondary structure. Continued efforts are required to realize single molecule direct RNA structural probing using nanopore sequencing. Characterization of SHAPE reagents such as NAI, 2A3 (Marinus et al. 2021) and AcIm along with base-reactive reagents should be more thoroughly assessed, along with novel reagents, for compatibility with rapidly evolving direct RNA nanopore sequencing chemistries. Efforts to train neural-network basecalling algorithms to identify SHAPE adducts on RNA directly are currently hindered by the inability to generate or synthesize RNA molecules with adducts at defined positions (e.g., ground-truth sets). Thus, comparative raw signal analysis (unmodified vs. modified) is the primary strategy for detecting the presence of adducts as a result of nanopore based RNA structural probing. Continued efforts should be dedicated for further improving and optimizing nanopore raw signal analysis programs for RNA structural analysis.

Aided by developments in fabrication technology, innovative biochemical strategies and commercial availability, RNA analysis with nanopore technology is beginning to demonstrate significant promise. The nanopore platform is uniquely configurable and offers attractive characteristics in a variety of formats and experimental configurations. As single molecule nanopore sensing and sequencing technologies mature we can expect continued advancements towards deciphering RNA structure, folding and interactions which will have far reaching implications for understanding RNA function in healthy and disease states.

References

Aw JGA, Lim SW, Wang JX et al (2020) Determination of isoform-specific RNA structure with nanopore long reads. Nat Biotechnol 39:336–346

Bizarro CV, Alemany A, Ritort F (2012) Non-specific binding of Na+ and Mg2+ to RNA determined by force spectroscopy methods. Nucleic Acids Res 40:6922–6935

Bizuayehu TT, Labun K, Jakubec M et al (2022) Long-read single-molecule RNA structure sequencing using nanopore. Nucleic Acids Res gkac775 (Online ahead of print)

Bošković F, Keyser UF (2022) Nanopore microscope identifies RNA isoforms with structural colours. Nat Chem. https://doi.org/10.1038/s41557-022-01037-5 (Online ahead of print)

Craig JM, Laszlo AH, Nova IC et al (2019) Determining the effects of DNA sequence on Hel308 helicase translocation along single-stranded DNA using nanopore tweezers. Nucleic Acids Res 47:2506–2513

Ding Y, Tang Y, Kwok CK et al (2014) In vivo genome-wide profiling of RNA secondary structure reveals novel regulatory features. Nature 505:696–700

Drexler HL, Choquet K, Churchman LS et al (2020) Splicing kinetics and coordination revealed by direct nascent RNA sequencing through nanopores splicing kinetics and coordination revealed by direct nascent RNA sequencing through nanopores. Mol Cell 77:1–14

Garalde DR, Snell EA, Jachimowicz D et al (2018) Highly parallel direct RNA sequencing on an array of nanopores. Nat Methods 15:201–206

Habibian M, Velema WA, Kietrys AM et al (2019) Polyacetate and polycarbonate RNA: acylating reagents and properties. Org Lett 21:5413–5416

Hafner M, Landthaler M, Burger L et al (2010) Transcriptome-wide Identification of RNA-binding protein and microRNA target sites by PAR-CLIP. Cell 141:129–141

Keyser UF, Koeleman BN, Van Dorp S et al (2006) Direct force measurements on DNA in a solid-state nanopore. Nat Phys 2:473–477

Knust S, Kreft D, Hillmann R et al (2017) Measuring DNA translocation forces through MoS_2-nanopores with optical tweezers. Mater Today Proc 4:S168–S173

Lee DH, Oh S, Lim K et al (2021) Tertiary RNA folding-targeted drug screening strategy using a protein nanopore. Anal Chem 93:2811–2819

Lotz TS, Suess B (2018) Small-molecule-binding riboswitches. Microbiol Spectr 6:1–12

Marinus T, Fessler AB, Ogle CA et al (2021) A novel SHAPE reagent enables the analysis of RNA structure in living cells with unprecedented accuracy. Nucleic Acids Res 49:E34–E34

Merino EJ, Wilkinson KA, Coughlan JL et al (2005) RNA structure analysis at single nucleotide resolution by Selective 2′-Hydroxyl Acylation and Primer Extension (SHAPE). J Am Chem Soc 127:4223–4231

Morandi E, Manfredonia I, Simon LM et al (2021) Genome-scale deconvolution of RNA structure ensembles. Nat Methods 18:249–252

Müller CA, Boemo MA, Spingardi P et al (2019) Capturing the dynamics of genome replication on individual ultra-long nanopore sequence reads. Nat Methods 16:429–436

Niedzwiecki DJ, Iyer R, Borer PN et al (2013) Sampling a biomarker of the human immunodeficiency virus across a synthetic nanopore. ACS Nano 7:3341–3350

Niu X, Liu Q, Xu Z et al (2020) Molecular mechanisms underlying the extreme mechanical anisotropy of the flaviviral exoribonuclease-resistant RNAs (xrRNAs). Nat Commun 11:1–14

Olson SW, Turner AMW, Arney JW et al (2022) Discovery of a large-scale, cell-state-responsive allosteric switch in the 7SK RNA using DANCE-MaP. Mol Cell 82:1708-1723.e10

Parker MT, Knop K, Sherwood AV et al (2020) Nanopore direct RNA sequencing maps the complexity of Arabidopsis mRNA processing and m6A modification. Elife 9:e49658

Raveendran M, Leach AR, Hopes T et al (2020) Ribosome fingerprinting with a solid-state nanopore. ACS Sensors 5:3533–3539

Shasha C, Henley RY, Stoloff DH et al (2014) Nanopore-based conformational analysis of a viral RNA drug target. ACS Nano 8:6425–6430

Stephenson W, Asare-okai PN, Chen AA et al (2014) The essential role of stacking adenines in a two-base-pair RNA kissing complex. J Am Chem Soc 135:5602–5611

Stephenson W, Razaghi R, Busan S et al (2022) Direct detection of RNA modifications and structure using single-molecule nanopore sequencing. Cell Genomics 2:100097

Tinoco I Jr, Li PTX, Bustamante C (2006) Determination of thermodynamics and kinetics of RNA reactions by force. Q Rev Biophys 39:325–360

Van Den Hout M, Vilfan ID, Hage S, Dekker NH (2010) Direct force measurements on double-stranded RNA in solid-state nanopores. Nano Lett 10:701–707

Wang S, Wang S, Wang Y et al (2020) Retarded translocation of nucleic acids through α-hemolysin nanopore in the presence of a calcium flux. ACS Appl Mater Interfaces 12:26926–26935

Wang Y, Guan X, Zhang S et al (2021) Structural-profiling of low molecular weight RNAs by nanopore trapping/translocation using Mycobacterium smegmatis porin A. Nat Commun 12:3368

Wanunu M, Dadosh T, Ray V et al (2010) Rapid electronic detection of probe-specific microRNAs using thin nanopore sensors. Nat Nanotechnol 5:807–814

Wanunu M, Bhattacharya S, Xie Y et al (2011) Nanopore analysis of individual RNA/antibiotic complexes. ACS Nano 5:9345–9353

Weidmann CA, Mustoe AM, Jariwala PB et al (2021) Analysis of RNA–protein networks with RNP-MaP defines functional hubs on RNA. Nat Biotechnol 39:347–356

Workman RE, Tang AD, Tang PS et al (2019) Nanopore native RNA sequencing of a human poly(A) transcriptome. Nat Methods 16:1297–1305

Zhang X, Zhang D, Zhao C et al (2017) Nanopore electric snapshots of an RNA tertiary folding pathway. Nat Commun 8:1458

Mapping In Situ RNA–RNA Interactions with RIC-seq

Rong Ye, Zhaokui Cai, and Yuanchao Xue

Contents

Abstract Mammalian genomes encode large amounts of noncoding RNAs (ncRNAs), which tend to form intricate structures and interact with their target RNA molecules through complementary base pairing with the help of proteins. Mapping of intra- and inter-molecular RNA–RNA interactions (RRIs) is required to unravel the structure and targets of ncRNAs which are two essential aspects for understanding the molecular mechanisms of ncRNAs in various biological processes. At this frontiers, we recently invented RNA in situ conformation sequencing (RIC-seq) technology to profile protein-mediated RNA–RNA spatial interactions at single-nucleotide resolution in an unbiased manner. We have demonstrated that RIC-seq-identified RRIs are helpful for simultaneously deducing ncRNA structures and targets. Here, we summarize methods for probing RRIs and describe a step-by-step protocol for generating a successful RIC-seq library.

R. Ye · Z. Cai · Y. Xue (✉)
Key Laboratory of RNA Biology, Institute of Biophysics, Chinese Academy of Sciences, Beijing 100101, China
e-mail: ycxue@ibp.ac.cn

R. Ye · Y. Xue
University of Chinese Academy of Sciences, Beijing 100049, China

J. Barciszewski (ed.), *RNA Structure and Function*, RNA Technologies 14,
https://doi.org/10.1007/978-3-031-36390-0_3

Keywords RNA structure · RNA–RNA interaction · Noncoding RNA · Proximity ligation · High-throughput sequencing

Abbreviations

ncRNA	Noncoding RNA
rRNA	Ribosomal RNA
tRNA	Transfer RNA
piRNA	Piwi-interacting RNA
lncRNA	Long noncoding RNA
circRNA	Circular RNA
eRNA	Enhancer RNA
AMT	4′-Aminomethyltrioxsalen
RBP	RNA-binding protein
EMSA	Electrophoretic mobility shift assay
SPR	Surface plasmon resonance
FRET	Fluorescence resonance energy transfer
CLASH	Crosslinking, ligation, and sequencing of hybrids
RAP	RNA antisense purification
hiCLIP	RNA hybrid and individual-nucleotide resolution ultraviolet crosslinking and immunoprecipitation
RPL	RNA proximity ligation
MARIO	Mapping RNA interactome in vivo
PARIS	Psoralen analysis of RNA interactions and structures
SPLASH	Sequencing of psoralen crosslinked, ligated, and selected hybrids
LIGR-seq	Ligation of interacting RNA followed by high-throughput sequencing
COMRADES	Crosslinking of matched RNAs and deep sequencing
RIC-seq	RNA in situ conformation sequencing
RRI	RNA–RNA interaction
MNase	Micrococcal nuclease
pCp-biotin	Biotinylated cytidine (bis) phosphate
vRIC-seq	Virion RIC-seq
SHARC	Spatial 2′-hydroxyl acylation reversible crosslinking
ConA	Concanavalin A

1 Introduction

The human genome is pervasively transcribed to generate vast noncoding RNAs (ncRNAs) (Djebali et al. 2012; Hangauer et al. 2013). Many ncRNAs, such as miRNAs (microRNAs), piRNAs (Piwi-interacting RNAs), lncRNAs (long noncoding RNAs), circular RNAs (circRNAs), and eRNAs (enhancer RNAs), have been demonstrated to regulate gene expression at transcriptional or posttranscriptional levels (Chen and Xue 2016; Xue et al. 2020; Rinn and Chang 2012). Reconciling their crucial roles in gene regulation during various physiological processes, the aberrant expression of ncRNAs can lead to tumorigenesis as well as neurological, cardiovascular, and autoimmune diseases (Wapinski and Chang 2011; Esteller 2011). Due to lacking protein-coding potential, ncRNAs usually achieve their regulatory functions by forming sophisticated tertiary structures and interacting with other RNA molecules through base pairing with the help of diverse RNA-binding proteins (RBPs) (Hentze et al. 2018).

RNA–RNA interactions (RRIs) can be divided into intra-molecular and inter-molecular (Xue 2022). The intra-molecular RRIs happen within the same RNA molecules and can be used to infer RNA structures by combing them with various structure prediction algorithms (Cao et al. 2021; Ziv et al. 2018, 2020; Li et al. 2018; Sun et al. 2021; Siegfried et al. 2014; Watts et al. 2009), while inter-molecular RRIs happen between two proximally interacting RNA molecules and have been demonstrated to be valuable for deducing ncRNAs' targets in many studies (Cai et al. 2020; Nguyen et al. 2016; Lu et al. 2016; Sharma et al. 2016; Helwak et al. 2013; Engreitz et al. 2014). Both intra- and inter-molecular RRIs can be mediated by complementary base pairing and RBPs, which help place two RNA fragments into proximity. The most classic examples may be the miRNA-Ago2-target mRNA ternary complex and the piRNA-PIWI-target mRNA ternary complexes (Wang et al. 2022; Bartel 2004). The miRNAs and piRNAs in the complex can provide target recognition specificity through partial or complete base pairing. In contrast, RBPs can further wrap small RNAs and their targets to stabilize RRIs.

RRIs seem amazingly accurate in controlling many essential molecular processes, such as U1 small nuclear RNA (snRNA) can base pair with the 5′ splice site of precursor messenger RNAs (pre-mRNAs) to regulate splicing fidelity (Moore and Proudfoot 2009; Salzman et al. 2012). Similar to miRNAs and U1 snRNA, lncRNAs and eRNAs also use complementary base pairing to bind their cognate target RNAs. For example, TINCR, a lncRNA that regulates epidermal differentiation, contains several 25-nucleotide motifs that can reverse complementary to differentiation-related mRNAs. The potential RNA duplexes between TINCR and targets can further recruit STAU1 to stabilize those differentiation mRNAs (Kretz et al. 2013). In another example, a class of half-STAU1-binding site RNAs (1/2-sbsRNAs) contains an Alu element that can form imperfect base pairing with the reverse Alu element in target mRNAs in the 3′UTR regions to regulate their stability (Gong and Maquat 2011). In addition to RRIs-mediated RNA stability control, inter-molecular RRIs also function directly in transcription activation. For example, *CCAT1-5L*, a super-enhancer transcribed lncRNA, directly interacts with *MYC* promoter-associated RNAs with the help of hnRNPK to modulate long-range chromatin looping and *MYC* transcription (Cai et al. 2020).

Considering the crucial roles of ncRNAs in various biological processes, identifying their intra- and inter-molecular RRIs may enable investigators to infer their higher-order structures and targets globally, thus facilitating the dissection of the molecular mechanisms of ncRNAs in the development, differentiation, and disease. In this chapter, we will summarize the currently available methods for mapping RRIs. In addition, we will describe the RIC-seq technology in detail and discuss future perspectives of its applications.

2 Methods for Probing RNA–RNA Interactions

Many state-of-the-art methods have been developed to profile RRIs over the past decades, including biophysical, computational, and biochemical methods (Xu et al. 2022). These efforts have significantly expanded our understanding of RRIs in various physiological and pathological processes. RRIs are initially studied one by one in vitro but now can be globally profiled on a transcriptome-wide scale.

2.1 Low-Throughput Methods

Some simple but elegant biochemical or biophysical methods have been invented to validate RRIs (Table 1 and Fig. 1). For example, in electrophoretic mobility shift assay (EMSA), the interested RNA is first radiolabeled and incubated with the potential interacting RNA (Bak et al. 2015). Suppose these two RNA molecules can form a duplex in vitro. The investigators will observe a shifted band in non-denaturing polyacrylamide gel electrophoresis (PAGE) after autoradiograph (Fig. 1a). Furthermore, EMSA can evaluate the binding specificity and affinity of RRIs by introducing a second non-radiolabeled competitor oligonucleotide (Bak et al. 2015). Single-molecule Forster Resonance Energy Transfer (smFRET) method is another biophysical method to characterize RRIs (Zhao and Rueda 2009; Hardin et al. 2015). In smFRET, two RNA molecules under investigation are labeled with a donor fluorophore and an acceptor fluorophore, respectively (Fig. 1b). Because the emission spectra of the excited donor fluorophore overlap with the absorbance spectra of the acceptor fluorophore, energy transfer will occur through a dipole–dipole interaction if the two RNAs interact with each other in close proximity, which induces the fluorescence of the acceptor fluorophore (Stephenson et al. 2016).

In addition to in vitro methods, RRIs can be validated in vivo under various cellular conditions. Inspired by the yeast three-hybrid system, an RNA-hybrid system in *Saccharomyces cerevisiae* was developed for such purpose (Piganeau et al. 2006). In

Table 1 Summary of methods to profile RNA–RNA interactions

Method	Strategy	Principle	In vitro/ in vivo	Sample preparation	Ligation	Enrichment	Characteristic	Reference
EMSA	One to one	Biophysical, base-pairing	In vitro	Target RNA fragment preparation and labeling	N/A	N/A	Simple, rapid, and sensitive	Bak et al. (2015)
smFRET	One to one	Biophysical, base-pairing	In vitro	Fluorescence labeling	N/A	N/A	Detect RNA–RNA interaction at the single-molecule level	Zhao and Rueda (2009); Hardin et al. (2015)
Yeast RNA-hybrid system	One to one	Biophysical, base-pairing	In vivo	Reporter construction	N/A	N/A	Screen interaction partners of a target RNA	Piganeau et al. (2006)
PARS	N/A	Structure-specific enzymes digestion	In vitro	In vitro refolding, fragmentation	Adaptors ligation in solution	N/A	Profile the secondary structure of the mRNAs	Kertesz et al. (2010)
CLASH	All to all	RBP-mediated proximity ligation	In vivo	Target protein overexpression; UV crosslinking	In solution	Specific antibody pull-down	Identify RNA–RNA interaction mediated by specific RBP globally	Kudla et al. (2011)
RAP-RNA	One to all	Direct or RBP-mediated indirect RNA–RNA interaction	In vivo	Biotin-labeled probes preparation, crosslinking with formaldehyde or formaldehyde-DSG or AMT	N/A	Biotin-labeled probes pull-down	Identify interactome of specific RNA globally	Engreitz et al. (2014)

(continued)

Table 1 (continued)

Method	Strategy	Principle	In vitro/ in vivo	Sample preparation	Ligation	Enrichment	Characteristic	Reference
hiCLIP	All to all	RBP-mediated proximity ligation	In vivo	Target RBP overexpression, UV crosslinking	In solution	Specific antibody pull-down	Identify RNA–RNA duplexes mediated by STAU1	Sugimoto et al. (2015)
RPL	All to all	Proximity ligation	In vivo	N/A	In situ	N/A	Identify native RNA–RNA interaction in cells	Ramani et al. (2015)
MARIO	All to all	RBP-mediated proximity ligation	In vivo	Crosslinking with formaldehyde-Ethyl Glycol bis or UV	In solution	Biotin pull-down	Identify RNA–RNA interaction mediated by RBPs globally	Nguyen et al. (2016)
PARIS	All to all	Chemical-mediated proximity ligation	In vivo	UV crosslinking with AMT treatment	In solution	2D gel	Identify RNA–RNA duplexes globally	Lu et al. (2016)
SPLASH	All to all	Chemical-mediated proximity ligation	In vivo	UV crosslinking with biotin-labeled psoralen treatment	In solution	Biotin pull-down	Identify RNA–RNA duplexes globally	Aw et al. (2016)
LIGR-seq	All to all	Chemical-mediated proximity ligation	In vivo	UV crosslinking with AMT treatment	In solution	RNase R treatment	Identify RNA–RNA duplexes globally	Sharma et al. (2016)
COMRADES	All to all	Chemical-mediated proximity ligation	In vivo	UV crosslinking with psoralen-TEG azide treatment	In solution	Biotin-labeled probes pull-down	Map the structure of the virus RNA genome and identify viral RNA-host RNA interaction inside cells	Ziv et al. (2018, 2020)

(continued)

Table 1 (continued)

Method	Strategy	Principle	In vitro/ in vivo	Sample preparation	Ligation	Enrichment	Characteristic	Reference
RIC-seq	All to all	RBP-mediated proximity ligation	In vivo	Formaldehyde crosslinking	In situ	pCp-biotin selection	Capture RNA 3D conformation and identify RNA–RNA interaction in situ	Cai et al. (2020)
vRIC-seq	All to all	RBP-mediated proximity ligation	In vivo	Formaldehyde crosslinking	In situ	pCp-biotin selection	Capture RNA 3D conformation of the virus RNA genome	Cao et al. (2021)
SHARC	All to all	Chemical-mediated proximity ligation	In vivo	Crosslinking with bifunctional 2′-hydroxyl acylation reagents	In solution	2D gel	Reveal heterogeneous RNA structures and interactions	Van Damme et al. (2022)

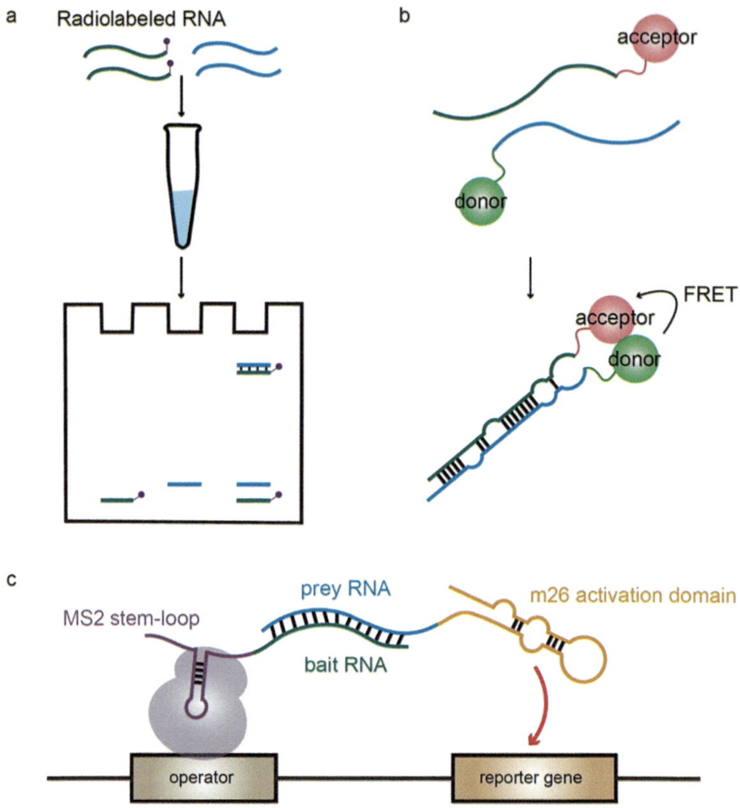

Fig. 1 Schematic diagram of low-throughput methods for detecting RRIs. **a** Electrophoretic mobility shift assay (EMSA). **b** Single-molecule fluorescence resonance energy transfer (smFRET). **c** Yeast RNA-hybrid system

this system, the activation of a reporter gene is controlled by an MS2 coat protein-fused operator (Fig. 1c). If bait RNA fused with the MS2 stem-loop could interact with the prey RNA fused with m26 (an RNA-based transcriptional activator), the m26 transcriptional activator will be tethered to the promoter region of the reporter gene, thus inducing reporter gene expression. These methods allow the investigation of specific RRIs in vivo. However, this RNA-hybrid method is limited to known RNAs and cannot determine the exact interacting mode.

2.2 *High-Throughput Methods*

To profile direct RRIs in a high-throughput way, many enzyme- and chemical-based methods have been developed in the past decade (Table 1). The first experimental approach for this purpose is the parallel analysis of RNA structure (PARS) (Kertesz et al. 2010), which uses RNase S1 and V1 to distinguish single- or double-stranded regions of in vitro refolded RNA and measure the conformation states of nucleotides through deep sequencing (Fig. 2a). To globally map RNA duplexes in living cells, three experimental methods, namely, PARIS (Psoralen analysis of RNA interactions and structures), SPLASH (Sequencing of psoralen crosslinked, ligated, and selected hybrids), and LIGR-seq (Ligation of interacting RNA followed by high-throughput sequencing) were developed in 2016 (Aw et al. 2016; Lu et al. 2016; Sharma et al. 2016).

For the PARIS method, cells are first treated with a psoralen derivative 4′-aminomethyltrioxsalen (AMT) and irradiated with UV 365 nm to crosslink RNA duplexes. After RNA fragmentation by S1 nuclease, the crosslinked RNA duplexes are purified using two-dimensional (2D) gel electrophoresis followed by proximity ligation with T4 RNA ligase (Lu et al. 2016). Finally, the resulting RNAs

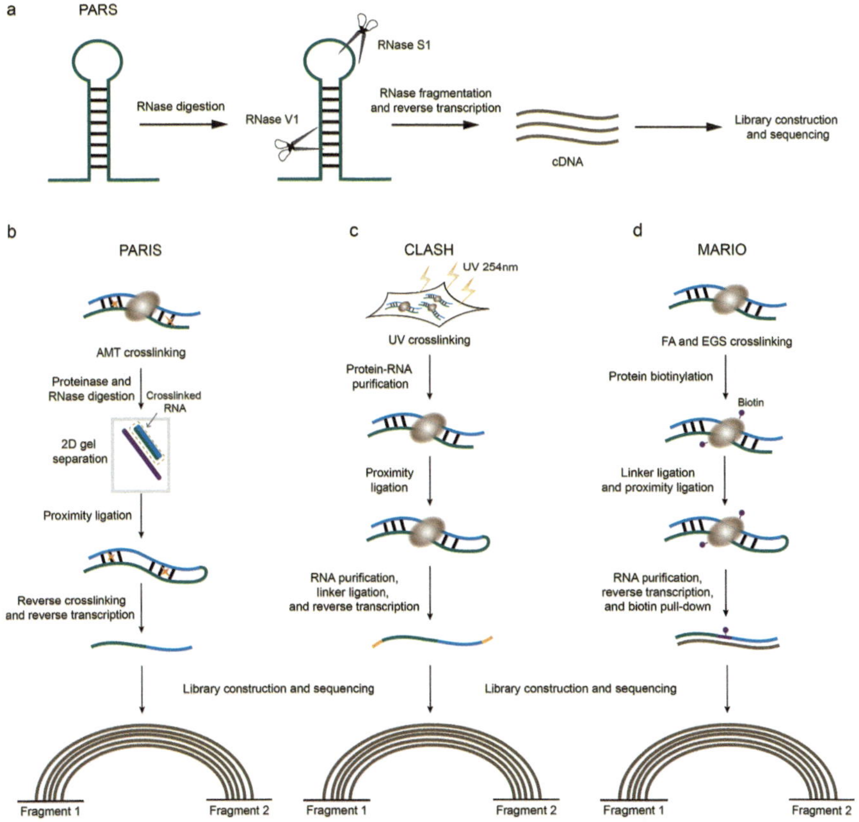

Fig. 2 Schematic diagram of high-throughput methods. **a** Parallel analysis of RNA structure (PARS). **b** Psoralen analysis of RNA interactions and structures (PARIS). **c** Crosslinking, ligation, and sequencing of hybrids (CLASH). **d** Mapping RNA interactome in vivo (MARIO)

are de-crosslinked by UV 254 nm irradiation and converted into libraries for deep sequencing (Fig. 2b). The other two methods, SPLASH and LIGR-seq, own similar workflows as PARIS and either adopt psoralen or AMT for crosslinking (Aw et al. 2016; Sharma et al. 2016). Unlike these methods, the spatial 2′-hydroxyl acylation reversible crosslinking (SHARC) method used bifunctional and reversible acylation reagents with flexible linkers to capture transcriptome-wide spatial RRIs for sequencing (Van Damme et al. 2022). All the above-mentioned methods can profile RRIs globally but with low detection capability for specific RNA molecule-organized RRIs. To address this limitation, the COMRADES method is first developed to capture specific RNA-associated RRIs from the whole RRI pools by using biotinylated reverse complementary oligos. The COMRADES have been demonstrated to be robust for uncovering the architecture of the Zika and SARS-CoV-2 virus RNA genome (Ziv et al. 2018, 2020).

Since intra- and inter-molecular RRIs are generally organized and stabilized by RBPs, which can hinder the accessibility of enzymes and chemicals, leading to the structural information deduced from the chemical-based methods mentioned above is incomplete. To determine RNA duplexes mediated by a defined RBP, CLASH, hiCLIP, and irCLASH were developed (Helwak and Tollervey 2016; Song et al. 2020; Sugimoto et al. 2015). CLASH is the first high-throughput method for studying RBP-mediated RRIs (Fig. 2c). The general workflow includes UV crosslinking of inter-molecular interacting RNAs to AGO1, ligation of two proximally interacting RNAs, and isolation of chimeric RNAs for deep sequencing. CLASH has been applied to detect different types of RRIs, such as small RNA interactions in bacteria (Waters et al. 2017), snoRNA–rRNA interactions in *S. cerevisiae* (Kudla et al. 2011), miRNA–mRNA interactions in humans (Helwak et al. 2013), and interactions between piRNAs and their targets in *C. elegans* (Shen et al. 2018). Later, hiCLIP and irCLASH methods introduced an additional adapter to improve the low efficiency of direct ligation of interacting RNAs and the chimeric reads assigning ambiguity in CLASH. However, all these methods are limited to a single "bait" protein.

In mammalian cells, at least 1,500 RBPs have been identified to be tightly associated with a vast repertoire of RNA molecules (Gerstberger et al. 2014). The panoramic view of the whole complement of RBPs in organizing cellular RRIs is still unclear. Seeking to address this limitation, an RNA proximity ligation (RPL) method was first developed (Ramani et al. 2015). However, RPL-detected RRIs appear limited to intramolecular RRIs due to lacking formaldehyde fixation. As a step forward, MARIO (Mapping RNA interactome in vivo) used formaldehyde and EGS (ethylene glycol bis (succinimidyl succinate)) to crosslink all the RBPs-mediated intra- and inter-molecular RRIs (Nguyen et al. 2016). After double crosslinking and RNA fragmentation (Fig. 2d), the proximally interacting RNA fragments are ligated in a dilute solution using a biotinylated RNA linker. The resulting linker-containing RNA chimeras are subsequently enriched with streptavidin magnetic beads and converted into libraries for deep sequencing. However, MARIO also suffered several limitations due to over-crosslinking, inefficient chimeric enrichment, and spurious ligation in a dilute solution.

3 Global Profiling In Situ RNA–RNA Interactions with RIC-seq

As lncRNAs tend to function around their transcription sites, simultaneously profiling their in situ intra- and inter-molecular RRIs becomes crucial for understanding the underlying functional mechanisms. For this purpose and trying to address the limitations of other high-throughput methods, we have developed RNA in situ conformation sequencing technology to map the whole complement of RBP-mediated RNA–RNA spatial interactions unbiasedly (Cai et al. 2020). As illustrated in Fig. 3, cells are first crosslinked with formaldehyde for 10 min at room temperature to fix RBPs-mediated RRIs. The fixed cells are subsequently permeabilized with a buffer containing several detergents, including Triton X-100, NP-40, and Tween-20. This step aims to maintain the cell "intact" and "punch holes" into cellular membranes to ensure all the subsequent enzyme treatment, pCp-biotin labeling, and proximity ligation happen in situ. Next, the single- and double-stranded RNAs protruding from the RBP complexes are randomly digested with micrococcal nuclease (MNase) to leave 3′ overhangs as phosphate groups and 5′ overhangs as hydroxyl groups. After quenching the MNase digestion with the buffer containing EGTA, the exposed 3′ overhangs are first dephosphorylated with alkaline phosphatase and further labeled with pCp-biotin. After removing the 3′ phosphate group from "Cp–biotin" by alkaline phosphatase, the 5′ overhangs are phosphorylated with T4 polynucleotide kinase (PNK). The resulting RNA fragments in proximity are ligated using T4 RNA ligase, while "C-biotin" is left to mark the junction position.

These steps mentioned above are performed in situ under native conditions. For the in vitro part of RIC-seq, cells are first digested with proteinase K for efficient RNA extraction. The total RNAs are subsequently fragmented into about 90 nt, and biotin-C-containing RNAs are captured with streptavidin magnetic beads. The enriched chimeric RNAs are converted into a DNA library for paired-end sequencing by sequentially performing the following steps: first-strand synthesis, second-strand synthesis, end-repairing, dA-tailing, adapter ligation, and PCR amplification. During the second strand synthesis, 20% of the dTTP is substituted by dUTP. The USER enzyme will digest the deoxyuridine-containing strand before PCR amplification to achieve strand specificity. Finally, the sequencing data can be processed and computationally analyzed as previously described (Cai et al. 2020).

3.1 Materials

3.1.1 Reagents

10 × PBS, pH 7.4, RNase-free
2-Propanol
3 M sodium acetate, pH 5.5

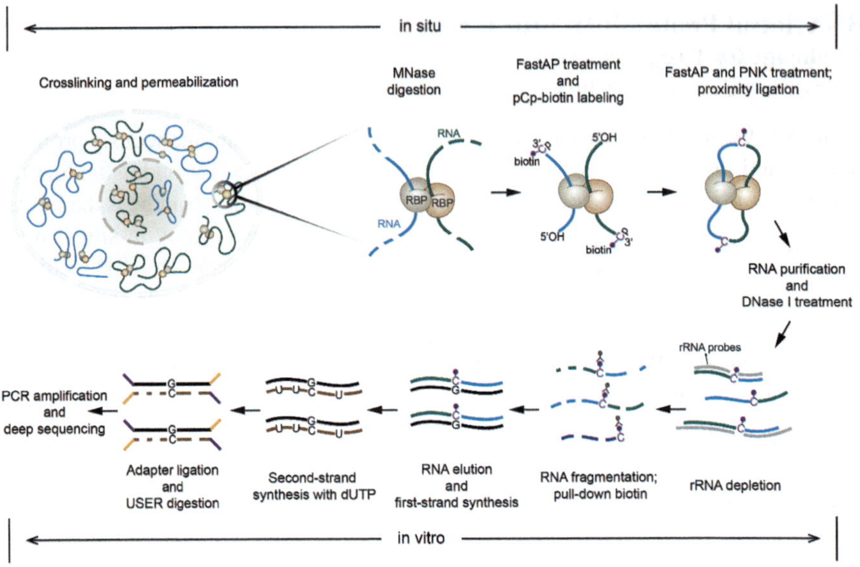

Fig. 3 Schematic diagram of RIC-seq technology. RIC-seq in situ part includes formaldehyde crosslinking, permeabilization, micrococcal nuclease (MNase) digestion, pCp-biotin labeling, and proximity ligation. The in vitro steps include rRNA depletion, RNA fragmentation, chimeric RNA enrichment, and strand-specific library construction. RBP: RNA-binding protein; pCp-biotin: biotinylated 3′,5′-bisphosphate cytidine

Acid phenol: chloroform, pH 4.5
Adenosine 5′-triphosphate (ATP, 10 mM)
Agencourt AMPure XP
Agilent high-sensitivity DNA kit
Agilent RNA 6000 nano kit
Biowest agarose
DNA polymerase I (10 U/μl)
dNTP set (dATP, dGTP, dCTP, dTTP, 100 mM, 4 × 0.25 ml)
dNTP solution mix (10 mM)
dNTP solution mix (25 mM)
dUTP (100 mM)
Dynabeads MyOne streptavidin C1
Ethylene glycol-bis(2-aminoethylether)-N,N,N′,N′-tetraacetic acid (EGTA)
Ethylenediaminetetraacetic acid disodium salt dehydrate ($Na_2EDTA \cdot 2H_2O$)
FastAP alkaline phosphatase (1 U/μl)
Formaldehyde solution (37%, wt/vol)
GelRed
IGEPAL CA-630 (also known as NP-40)
Invitrogen second-strand buffer
Klenow (3′ to 5′ exo-) (5 U/μl)

Klenow fragment (5 U/μl)
Micrococcal nuclease (300 U/μl)
Pierce™ RNA 3′ end biotinylation kit
Platinum Pfx DNA polymerase kit
Protease inhibitor cocktail (for mammalian cell and tissue extracts, DMSO solution)
Proteinase K (≥600 mAnson-U/ml)
Qiagen MinElute gel extraction kit
Qubit dsDNA HS assay kit
Qubit RNA HS assay kit
RNA clean and concentrator-5 kit
RNA from yeast
RNase H (5 U/μl)
RNaseZap RNase decontamination solution
RQ1 RNase-free DNase (1 U/μl)
Sodium dodecyl sulfate (SDS)
SUPERase•In™ RNase inhibitor (20 U/μl)
SuperScript II reverse transcriptase
T4 DNA ligase (Rapid) (600 U/μl)
T4 DNA polymerase (3 U/μl)
T4 polynucleotide kinase (10 U/μl)
T4 RNA ligase (10 U/μl)
TRIzol LS reagent
TURBO DNase (2 U/μl)
UltraPure DEPC-treated water
USER enzyme (1 U/μl)

3.1.2 Buffers (See Note 1)

1. 1 × PNK buffer: 50 mM Tris-Cl pH 7.4, 10 mM $MgCl_2$, and 0.2% NP-40 in DEPC-treated water. Store at 4 °C for up to 12 months.
2. 1 × PNK + EGTA buffer: 50 mM Tris-Cl pH 7.4, 20 mM EGTA, and 0.5% NP-40 in DEPC-treated water. Store at 4 °C for up to 12 months.
3. High-salt wash buffer: 5 × PBS and 0.5% NP-40 in DEPC-treated water. Store at 4 °C for up to 12 months.
4. Permeabilization buffer: 10 mM Tris-HCl pH 7.5, 10 mM NaCl, 0.5% NP-40, 0.3% Triton X-100, and 0.1% TWEEN 20 in DEPC-treated water. Store at 4 °C for up to 6 months. Add SUPERase•In™ RNase Inhibitor to a final concentration of 2 U/ml and dilute the protease inhibitor cocktail 1:100 to the buffer before use.
5. 1 × Micrococcal nuclease reaction buffer (1 × MN buffer): 50 mM Tris-HCl pH 8.0 and 5 mM $CaCl_2$ in DEPC-treated water. Store at 4 °C for up to 6 months.

6. Proteinase K buffer: 10 mM Tris-HCl pH 7.5, 10 mM EDTA, and 0.5% SDS in DEPC-treated water. Store at 4 °C for up to 6 months.
7. 5 × Hybridization buffer: 500 mM Tris-HCl pH 7.4, and 1 M NaCl in DEPC-treated water. Store at 4 °C for up to 12 months.
8. Solution A: 0.1 M NaOH and 0.05 M NaCl in DEPC-treated water. Store at 4 °C for up to 6 months.
9. Solution B: 0.1 M NaCl in DEPC-treated water. Store at 4 °C for up to 6 months.
10. 2 × TWB buffer: 10 mM Tris-HCl pH 7.5, 1 mM EDTA, 2 M NaCl, and 0.02% TWEEN 20 in DEPC-treated water. Store at 4 °C for up to 6 months.
11. PK buffer: 100 mM NaCl, 10 mM Tris-HCl pH 7.0, 1 mM EDTA, and 0.5% SDS in DEPC-treated water. Store at 4 °C for up to 6 months.
12. 1 × TBE: 890 mM Tris, 890 mM boric acid, and 20 mM EDTA in DEPC-treated water. Store at room temperature for up to 6 months.

3.1.3 Oligos and Primers

1. PEI adaptor A: /5Phos/GATCGGAAGAGCACACGTCT
2. PEI adapter B: ACACTCTTTCCCTACACGACGCTCTTCCGATCT
3. Illumina PE 1.0 primer: AATGATACGGCGACCACCGAGATCTACACTCTTT CCCTACACGACGCTCTTCCGATCT
4. Index primer: CAAGCAGAAGACGGCATACGAGATNNNNNNGT-GACTGGA GTTCAGACGTGTGCTCTTCCGATCT ("NNNNNN" denotes the 6 bp randomized barcode)

3.1.4 Equipment

0.22 μm filter unit
1.5 ml LoBind microcentrifuge tubes
5 and 10 ml Serological pipette
15 and 50 ml Centrifuge tubes
200 μl PCR tubes
Agarose gel electrophoresis
Agilent 2100 bioanalyzer
Cell scrapers
DynaMag™-2 Magnet
Gel tray
NanoDrop 2000c
pH meter
Qubit® 3.0 fluorometer
Razor blade
Refrigerated centrifuge and microcentrifuge
Safe Imager™ 2.0 blue-light transilluminator
Single-channel manual pipettes (0.1–2 μl, 2–20 μl, 20–200 μl, 100–1,000 μl)

Thermal cycler
ThermoMixer C
Tube revolver/rotator
Universal power supply
Vortex mixer

3.2 Methods

3.2.1 Adaptor Preparation

1. Mix 10 μl of PEI adaptor oligo A (100 μM) and 10 μl of PEI adaptor oligo B (100 μM) in a 200 μl PCR tube.
2. Incubate the oligo mixture at 70 °C for 10 min in a thermal cycler. Then, place the PCR tube on the benchtop to let the mixture cool to room temperature. The final concentration of the annealed adaptors should be 50 μM. Aliquot and store the mixture at −20 °C for up to several years.
3. Dilute 50 μM annealed adaptor to 2 μM for the RIC-seq experiment and store at −20 °C for up to a year.

3.2.2 Cell Crosslinking and Harvesting

1. Pour off the culture medium from the dish and wash the cells three times with 10–15 ml of ice-cold PBS (*see* **Notes 2 and 3**).
2. Add 10–15 ml of 1% (wt/vol) formaldehyde (freshly prepared with 1 × PBS) to the dish and crosslink the cells for 10 min at room temperature.
3. Add 2.5 M glycine to a final concentration of 0.125 M (1/20 volume) to terminate the crosslinking reaction and mix gently on a rocking platform for 10 min.
4. Wash the cells with 10–15 ml of ice-cold PBS three times.
5. Scrape the cells off the dish and transfer them to a 50 ml tube. Collect the cell pellets by centrifugation at 1,200 g for 10 min at 4 °C (*see* **Notes 4 and 5**).
6. Resuspend the cell pellets in 1 ml of ice-cold PBS and transfer the cells to a clean 1.5 ml Eppendorf tube.
7. Centrifuge at 600 g for 10 min at 4 °C to collect cell pellets.
8. Discard the supernatant and directly proceed with the next step or store the pellets at −80 °C.

3.2.3 Cell Permeabilization

1. For each sample, resuspend the cell pellets in 1 ml of cell permeabilization buffer and gently pipette up and down 15 times.
2. Incubate the cell suspension on ice for 15 min and invert the tube every 2 min.

3. Pellet the cells by centrifugation at 1,200 g for 5 min at 4 °C and discard the supernatant.
4. Wash the cell pellets three times with 600 μl of 1 × PNK buffer by rotating on a tube rotator at 4 °C for 5 min at 20 rpm. After every washing step, collect the cells by centrifugation at 1,200 g for 5 min at 4 °C. Note that in the following steps, wash the cell pellets in the same way as that described in this step by rotation and centrifugation.
5. Discard the supernatant thoroughly after the last centrifugation step.

3.2.4 RNA Fragmentation with MNase

1. First dilute MNase to 1:1000 with 1 × MN reaction buffer and then further dilute the solution to 1:10 with 1 × MN reaction buffer (*see* **Note 6**).
2. Add 200 μl of the final diluted MNase to the cell pellets and mix gently by pipetting up and down 15 times.
3. Incubate the reaction mixture at 37 °C in a ThermoMixer C for 10 min (mix for 15 s at 1000 rpm every 3 min).
4. Collect the cells by centrifugation at 1,200 g for 5 min at 4 °C.
5. Wash the cell pellets twice with 600 μl of 1 × PNK + EGTA buffer and twice with 600 μl of 1 × PNK buffer. Discard the supernatant thoroughly after the last wash.

3.2.5 pCp-Biotin Labeling (See Notes 7 and 8)

1. Prepare the following FastAP reaction mixture:

10 × FastAP buffer	10 μl
Fast alkaline phosphatase (1 U/μl)	10 μl
DEPC-treated water	80 μl
Total	100 μl

2. Add the reaction mixture to the cell pellets, mix gently by pipetting up and down 20 times, and incubate at 37 °C in a ThermoMixer C for 15 min (mixing for 15 s at 1000 rpm every 3 min).
3. Pellet the cells by centrifugation at 1,200 g for 5 min at 4 °C.
4. Wash the cell pellets twice with 600 μl of 1 × PNK + EGTA buffer, twice with 600 μl of high-salt wash buffer, and three times with 600 μl of 1 × PNK buffer (0.05% NP-40). Discard the supernatant after the last wash.
5. Prepare the pCp-biotin ligation mixture without 30% PEG 20,000 (*see* **Note 9**):

10 × RNA ligase reaction buffer	10 μl
RNase inhibitor (40 U/μl)	6 μl

Biotinylated cytidine (bis) phosphate (1 mM)	4 μl
T4 RNA ligase (10 U/μl)	10 μl
DEPC-treated water	20 μl
Total	50 μl

6. Add the mixture to the cell pellets and mix gently by pipetting up and down 20 times.
7. Add 50 μl of 30% PEG 20,000 to the cell suspension with a cut-off 200 μl pipette tip and mix gently by pipetting up and down 20 times.
8. Incubate the reaction mixture at 16 °C in a ThermoMixer C overnight (mixing for 15 s at 1000 rpm every 3 min).
9. On the following day, add 2 μl of T4 RNA ligase and 4 μl of 10 mM ATP to the Eppendorf tube and mix gently by pipetting up and down 20 times.
10. Incubate at 16 °C for an additional 3 h (mixing for 15 s at 1000 rpm every 3 min).
11. Wash the cell pellets with 600 μl of 1 × PNK buffer three times. Discard the supernatant completely after the last wash.

3.2.6 Proximity Ligation

1. Prepare the following FastAP mixture:

10 × FastAP buffer	10 μl
Fast alkaline phosphatase (1 U/μl)	10 μl
DEPC-treated water	80 μl
Total	100 μl

2. Add the reaction mixture to the cell pellets, mix gently by pipetting up and down 20 times, and incubate at 37 °C in a ThermoMixer C for 15 min (mixing for 15 s at 1000 rpm every 3 min).
3. Collect the cells by centrifugation at 1,200 g for 5 min at 4 °C.
4. Wash the cell pellets twice with 600 μl of 1 × PNK + EGTA buffer, twice with 600 μl of high-salt wash buffer, and three times with 600 μl of 1 × PNK buffer. Discard the supernatant thoroughly after the last wash.
5. Prepare the following PNK mixture and mix thoroughly:

10 × reaction buffer A	10 μl
10 mM ATP	15 μl
T4 PNK enzyme (10 U/μl)	10 μl

DEPC-treated water	65 μl
Total	100 μl

6. Add the reaction mixture to the cell pellets, mix gently by pipetting up and down 20 times, and incubate at 37 °C in a ThermoMixer C for 45 min (mixing for 15 s at 1000 rpm every 3 min).
7. Wash the cell pellets twice with 600 μl of 1 × PNK + EGTA buffer and twice with 600 μl of 1 × PNK buffer (0.05% NP-40). Discard the supernatant thoroughly after the last wash.
8. Prepare the following ligation mixture and mix thoroughly:

10 × RNA ligase reaction buffer	20 μl
RNase inhibitor (40 U/μl)	8 μl
T4 RNA ligase (10 U/μl)	10 μl
BSA (1 mg/ml)	20 μl
DEPC-treated water	142 μl
Total	200 μl

9. Add the mixture to the cell pellets and mix gently by pipetting up and down 20 times.
10. Incubate the reaction mixture at 16 °C in a ThermoMixer C overnight (mixing for 15 s at 1000 rpm every 3 min).
11. The next day, add 2 μl of T4 RNA ligase and 4 μl of 10 mM ATP to the reaction mixture and mix gently by pipetting up and down 20 times.
12. Incubate at 16 °C for an additional 3 h (mixing for 15 s at 1000 rpm every 3 min).
13. Wash the cell pellets with 1 × PNK buffer three times and discard the supernatant thoroughly after the last wash.

3.2.7 RNA Purification

1. Mix 200 μl of proteinase K buffer and 50 μl of proteinase K together, add the mixture to the cell pellets, and mix gently by pipetting up and down 20 times.
2. Incubate the reaction mixture in a ThermoMixer C at 37 °C for 60 min, 56 °C for 15 min, and then 22 °C for 1 min (mixing for 15 s at 1000 rpm every 3 min).
3. Add 750 μl of TRIzol LS to the mixture and homogenize by pipetting up and down 20 times. Incubate the mixture for 5 min at room temperature.
4. Add 220 μl of chloroform to the mixture and shake vigorously for 15 s. Incubate the sample at room temperature for 3 min.
5. Centrifuge the sample at 16,000 g for 15 min at 4 °C and transfer the aqueous phase to a new 1.5 ml Eppendorf tube.

6. Precipitate RNA by adding 1 μl of GlycoBlue and 500 μl of isopropanol and mix thoroughly by pipetting up and down 20 times. Place the sample at −20 °C overnight.
7. The next day, centrifuge the sample at 16,000 g for 15 min at 4 °C.
8. Discard the supernatant and wash the RNA pellet twice with 75% ethanol.
9. Centrifuge the sample at 16,000 g for 5 min at 4 °C.
10. Discard the supernatant thoroughly, air-dry the RNA pellet at room temperature for 2 min, and then dissolve the RNA in 20 μl of DEPC-treated water.
11. Determine the concentration of RNA with a NanoDrop 2000c. The expected yield of RNA is 50–80 μg. Take 20 μg of RNA to proceed with the following steps and store the remaining sample at −80 °C.

3.2.8 DNase I Treatment

1. Prepare the following DNase I reaction mixture in a 1.5 ml Eppendorf tube. Adjust the water volume based on the total RNA input.

Total RNA (20 μg)	*x* μl
10 × RQ1 DNase buffer	10 μl
RNase inhibitor (40 U/μl)	3 μl
RQ1 RNase-free DNase (1 U/μl)	5 μl
DEPC-treated water	82-*x* μl
Total	100 μl

2. Mix thoroughly and incubate in a ThermoMixer C at 37 °C for 20 min (mixing for 15 s at 300 rpm every 3 min).
3. Add 100 μl of DEPC-treated water to the sample.
4. Add 200 μl of acid phenol:chloroform to the mixture for RNA extraction, mix thoroughly by shaking, and centrifuge the sample at 16,000 g for 15 min at 4 °C.
5. Carefully transfer the aqueous phase to a clean 1.5 ml tube. Precipitate the RNA by adding 20 μl of 3 M NaOAc, 1 μl of GlycoBlue, and 500 μl of 100% ethanol. Mix well and place the Eppendorf tube at −20 °C overnight.
6. The next day, centrifuge the sample at 16,000 g for 15 min at 4 °C.
7. Discard the supernatant and wash the RNA pellet twice with 75% ethanol.
8. Centrifuge the sample at 16,000 g for 5 min at 4 °C.
9. Discard the supernatant completely, air-dry the RNA pellet for 2 min, and then dissolve the RNA in 6 μl of DEPC-treated water.
10. Take 0.25 μl of RNA and dilute it to 1:20 with DEPC-treated water to determine the length distribution of the RNA fragments with an Agilent 2100 Bioanalyzer.

3.2.9 Ribosomal RNA (rRNA) Depletion (See Note 10)

1. Prepare the following reaction mixture for rRNA depletion in a 200 μl PCR tube:

Total RNA	5.75 μl
Probe mix (2 μg/μl total)	10 μl
5 × hybridization buffer	4 μl
DEPC-treated water	0.25 μl
Total	20 μl

2. Mix thoroughly by pipetting up and down 20 times, and put the sample in a thermal cycler. Incubate at 95 °C for 2 min, then gradually cooldown to 22 °C at a rate of −0.1 °C/s, hold at 22 °C for 5 min, and immediately put on ice.
3. Prepare the following reaction mixture in a 200 μl PCR tube containing:

Probe/RNA mixture	20 μl
10 × RNase H buffer	3 μl
RNase H (5 U/μl)	5 μl
DEPC-treated water	2 μl
Total	30 μl

4. Mix thoroughly by pipetting up and down 20 times, incubate the sample at 37 °C for 30 min, and immediately put on ice.
5. The remaining rRNA probes in the sample should be removed by DNase digestion. Prepare the following DNase reaction mixture in a 200 μl PCR tube:

Sample from the last step	30 μl
10 × TURBO buffer	4 μl
TURBO DNase (2 U/μl)	5 μl
DEPC-treated water	1 μl
Total	40 μl

6. Mix thoroughly by pipetting up and down 20 times and incubate the sample at 37 °C for 30 min.
7. Purify the RNA with RNA Clean and Concentrator-5 kit following the manufacturer's instructions. Use 18.5 μl of DEPC-treated water to elute the RNA and collect rRNA-depleted RNA. The expected yield is about 17 μl. Save 1 μl for quantification with an Agilent 2100 Bioanalyzer. The expected concentration is 100–200 ng/μl (*see* **Note 11**).

3.2.10 RNA Fragmentation and pCp-Biotin Selection

1. Prepare the following reaction mixture in a 200 μl PCR tube:

RNA	16 μl
5 × first-strand buffer	4 μl
Total	20 μl

2. Mix thoroughly by pipetting up and down 15 times, incubating at 94 °C for 5 min in a thermal cycler, and putting on ice immediately.
3. For each sample, prepare 20 μl of C1 beads. Suspend the C1 beads in a vial by vortexing for at least 30 s and transfer the required amount of beads to a clean 1.5 ml Eppendorf tube (*see* **Note 12**).
4. Place the tube on the magnetic rack for 1 min and carefully discard the supernatant.
5. Add a volume of Solution A equal to the volume of the beads to the same tube and mix thoroughly by pipetting up and down 10 times.
6. Incubate the beads for 2 min at room temperature.
7. Place the tube on the magnetic rack for 1 min and discard the supernatant.
8. Repeat steps 5–7 once more.
9. Add an equal volume of Solution B to the tube and mix thoroughly by pipetting up and down 10 times.
10. Place the tube on the magnetic rack for 1 min and carefully discard the supernatant.
11. Prepare the 200 μl of mixture containing 50 μg of yeast RNA and 100 μl of 2 × TWB buffer. Add the mixture to the tube and mix gently by pipetting up and down 20 times.
12. Incubate on the tube rotator at 20 rpm for 60 min at room temperature.
13. Place the tube on the magnetic rack for 1 min and carefully discard the supernatant. Wash the beads with 600 μl of 1 × TWB buffer three times. Store the beads in 1 × TWB buffer until use.
14. Place the Eppendorf tube containing the C1 beads on the magnetic rack for 1 min and discard the supernatant.
15. Add 20 μl of fragmented RNA, 30 μl of DEPC-treated water, and 50 μl of 2 × TWB buffer to the C1 beads and mix gently by pipetting up and down 10 times.
16. Incubate on a tube rotator at 20 rpm for 30 min at room temperature.
17. Place the tube on the magnetic rack for 1 min and carefully discard the supernatant.
18. Wash beads with 600 μl of 1 × TWB buffer four times.
19. Place the tube on the magnetic rack for 1 min and carefully discard the supernatant.
20. Add 100 μl of PK buffer to the beads and mix gently by pipetting up and down 10 times.
21. Incubate at 95 °C in a ThermoMixer C at 1000 rpm for 10 min.

22. Place the tube on the magnetic rack for 1 min and then transfer the supernatant to a clean 1.5 ml Eppendorf tube.
23. Repeat steps 20–22 once more.
24. Briefly rinse the beads with 100 μl of PK buffer, place the tube on the magnetic rack for 1 min until the supernatant is clear, and transfer the supernatant to a clean 1.5 ml Eppendorf tube. The total volume of eluted RNA should be 300 μl.
25. Add 300 μl of acid phenol: chloroform to the RNA and mix by shaking.
26. Centrifuge at 16,000 g for 10 min at 4 °C and transfer the aqueous phase to a clean 1.5 ml Eppendorf tube.
27. Precipitate the RNA by adding 18 μl of 5 M NaCl, 1 μl of GlycoBlue, and 900 μl of 100% ethanol, mix by pipetting up and down 20 times, and place the tube at −20 °C overnight.
28. On the next day, centrifuge the sample at 16,000 g for 20 min at 4 °C.
29. Discard the supernatant and wash the RNA pellet with 75% ethanol twice.
30. Centrifuge the sample at 16,000 g for 5 min at 4 °C.
31. Discard the supernatant completely, air-dry the RNA pellet for 2 min at room temperature, dissolve the pellet in 10 μl of DEPC-treated water, and transfer the dissolved RNA to a clean 200 μl PCR tube.

3.2.11 First-Strand cDNA Synthesis

1. Add 0.5 μl of N6 primer (100 ng/μl) to the RNA and mix by pipetting up and down 10 times.
2. Incubate the 200 μl PCR tube at 65 °C in a thermal cycler for 5 min, and immediately put it on ice for at least 2 min.
3. Prepare the following master mix in a clean 200 μl PCR tube, multiplying by X based on your sample numbers:

5 × first-strand buffer	3 μl
dNTP mix (10 mM)	1 μl
100 mM DTT	0.5 μl
RNase inhibitor (40 U/μl)	0.5 μl
Superscript II (200 U/μl)	0.5 μl
Total	5.5 μl

4. Add the mixture to the sample from the last step and mix well by pipetting up and down 20 times. Place the PCR tube in a thermal cycler and incubate at 25 °C for 10 min, at 42 °C for 40 min, and at 70 °C for 15 min.

3.2.12 Second-Strand cDNA Synthesis

1. Prepare the following master mix in a 1.5 ml Eppendorf tube:

5 × second-strand buffer	10 μl
25 mM dNTP (dUTP/dTTP = 4:1, *see* **Note 13**)	0.8 μl
RNaseH (5 U/μl)	0.2 μl
DNA polymerase I (10 U/μl)	2.5 μl
DEPC-treated water	20.5 μl
Total	34 μl

2. Add the mixture to the sample from the last step, mix well by pipetting up and down 20 times, and incubate at 16 °C in Thermomixer C for 2 h (mixing for 15 s at 300 rpm every 3 min). During this step, equilibrate the AMPure beads at room temperature for 30 min.
3. Add 90 μl of AMPure beads to the sample, mix by pipetting up and down 10 times, and incubate at room temperature for 5 min.
4. Place the tube on the magnetic rack until the solution is clear.
5. Discard the supernatant and add 200 μl of freshly prepared 80% ethanol to the tube while in the magnetic rack. Rinse the beads briefly for 20 s and carefully discard the supernatant. Repeat once more with an additional 200 μl of 80% ethanol.
6. Elute the cDNA with 44 μl of Buffer EB from Qiagen by incubating at room temperature for 5 min.
7. Take 1 μl of the eluted cDNA to determine its concentration using a Qubit dsDNA HS Assay Kit according to the manufacturer's instructions. The expected concentration is 0.2–0.6 ng/μl.

3.2.13 End Repair

1. Prepare the following master mix in a 200 μl PCR tube:

10 × PNK buffer	5 μl
dNTPs (25 mM)	0.4 μl
T4 DNA polymerase (3 U/μl)	1.2 μl
Klenow fragment (5 U/μl)	0.2 μl
T4 PNK (10 U/μl)	1.2 μl
Total	8 μl

2. Add the mixture to the sample from the last step to bring the total volume to 50 μl, mix thoroughly by pipetting up and down 20 times, and incubate at 20 °C in a Thermomixer C for 30 min.
3. Purify the repaired cDNAs with 90 μl of AMPure beads as described in Sect. 3.2.12, and finally elute the cDNA with 20.5 μl of Buffer EB. Transfer 19.7 μl of the repaired cDNA to a clean 1.5 ml Eppendorf tube.

3.2.14 dA-Tailing

1. Prepare the following master mix in a 200 μl PCR tube:

10 × Blue buffer	2.3 μl
5 mM dATP	0.5 μl
Klenow (3′ to 5′ exo-) (5 U/μl)	0.5 μl
Total	3.3 μl

2. Add the mixture to the sample from the last step, mix thoroughly by pipetting up and down 20 times, and incubate at 37 °C in a Thermomixer C for 30 min.

3.2.15 Adaptor Ligation

1. Prepare the following master mix in a 200 μl PCR tube:

2 × rapid ligation buffer	1.4 μl
10 mM ATP	0.1 μl
Adapter (2 μM, *see* **Note 14**)	1 μl
T4 DNA ligase (Rapid, 600 U/μl)	1 μl
Total	3.5 μl

2. Add the mixture to the sample from the last step and mix well by pipetting up and down 20 times (*see* **Note 15**).
3. Incubate at 20 °C in Thermomixer C for 15 min.
4. Purify the product with 47.7 μl of AMPure beads as described in Sect. 3.2.12 twice, and finally elute the cDNA with 16 μl of Buffer EB.

3.2.16 PCR *Amplification (See Note 16)*

1. Prepare the following PCR mixture in a 200 μl PCR tube:

DNA	13.7 μl
Index primer (10 μM)	1 μl
Illumina PE 1.0 (10 μM)	1 μl
$MgSO_4$ (50 mM)	1 μl
10 × Pfx buffer	2.5 μl
dNTPs (25 mM)	0.4 μl
DEPC-treated water	2 μl
Platinum Pfx DNA polymerase (2.5 U/μl)	0.4 μl
USER enzyme (1 U/μl)	3 μl
Total	25 μl

2. Run the following program in a thermal cycler:
   37 °C for 15 min;
   94 °C for 2 min;
   11–13 cycles at
   94 °C for 15 s,
   62 °C for 30 s,
   72 °C for 30 s; and
   72 °C for 10 min.
3. Run 2% agarose gel in 1 × TBE buffer at 120 V for ~1 h.
4. Cut the smear band from 200 to 450 bp and purify the product with a Qiagen MinElute Gel Extraction Kit according to the manufacturer's instructions.
5. Determine the product concentration with a Qubit dsDNA HS Assay Kit according to the manufacturer's instructions. The expected yield is 20–100 ng.

3.3 Notes

1. Maintaining an RNase-free environment in the lab for handling RNAs is crucial. All glassware can be thoroughly washed and autoclaved before preparing RNA-related buffers. Users can also filter all buffers using a 0.22-μm filter to remove trace amounts of RNase. To avoid RNase contamination, spraying surface decontamination solutions such as RNAseZAP® onto the bench, pipettes, and other equipment before performing the RIC-seq experiment is also helpful. Wearing a mask and gloves is an efficient way to avoid RNase contamination.

2. For a standard RIC-seq experiment, we used approximately 1×10^7 cells (e.g., for the HeLa cell line) to yield a cell pellet of approximately 50–70 μl. Users should adjust the input cell numbers for different cell types or tissue samples according to the cell size and the RNA content. We strongly recommend that users prepare at least two biological replicates for each treatment condition. During the analysis, users can combine two highly correlated replicates to enlarge the detection power of low-expression RNAs.
3. Ensure that Mycoplasma does not contaminate the cells used for the RIC-seq experiment.
4. Scraping is only suitable for adherent cells. Suspension cells can be directly transferred to a 50 ml tube and washed by centrifugation at 1,200 g for 10 min at 4 °C. After pouring out the supernatant, the cell pellets can be resuspended in 10 ml of 1 × PBS containing 1% formaldehyde.
5. Users may note that many scraped cells float on the surface of the dish or PBS when harvesting adherent cells. This phenomenon is caused by cell surface tension and is the primary reason for cell loss at the cell harvesting step. Users could add 0.01% (vol/vol) NP-40 to the cell suspension to reduce the surface tension.
6. Diluting MNase at 1:10,000 is suitable for most human and mouse cell lines. We recommend titrating the dilution factor for other cell lines or tissues based on the RNA fragment distribution quantified by an Agilent 2100 Bioanalyzer after the proximity ligation step. Most of the RNA fragments should range from 200 to 2000 nt.
7. Users can set up a negative control without pCp-biotin labeling at this step to evaluate the specificity of the chimeric enrichment.
8. pCp-biotin should contain a 3′, 5′ phosphate on the ribose ring to accommodate ligation and a biotin linker on cytidine for downstream selection.
9. PEG 20,000 (30%) is very sticky during pipetting. Use a cut-off 200 μl tip to aliquot this reagent into the cell suspension. Homogenize the cell suspension containing 30% PEG 20,000 by gently pipetting and avoiding vortexing.
10. Ribosomal RNA (rRNA) should be removed using the RNase H-based rRNA depletion kit rather than the biotinylated hybridization probe-based kit. We recommend using the Ribo-off rRNA depletion kit or NEBNext® rRNA Depletion Kit (Human/Mouse/Rat). As the rRNA depletion kit is costly, we have followed a published protocol (Adiconis et al. 2013) and prepared homemade probes for rRNA removal. To this end, users need first to synthesize DNA oligos that are reverse complementary to rRNA and then mix them at an equal molar ratio.
11. Ribosomal RNA usually comprises approximately 90% of the total RNA mass. Users should quantify the depletion efficiency by comparing the concentration of total RNA before and after rRNA depletion. If an excessive yield of RNA is

obtained, users can either use a double amount of the rRNA probes and RNase H for rRNA depletion or repeat the rRNA depletion step once more.

12. As preparation of C1 beads will take at least 1.5 h, users can prepare it during the rRNA depletion step. Estimate how long rRNA depletion will take according to how many samples you will handle. Ensure that the samples will not stand on ice too long before selection by C1 beads.
13. To prepare the dUTP-containing dNTP mixture (dUTP/dTTP = 4:1), we typically dilute 100 mM dATP, dGTP, and dCTP to 25 mM and dilute dTTP and dUTP to 5 mM and 20 mM, respectively, in the same tube.
14. The yield of cDNA after the second-strand synthesis step determines the amount of adapter used. Users may calculate the concentration of cDNA based on the total mass and the mean size of fragments.
15. Ensure that T4 DNA ligase is the final reagent added to the reaction mixture. This small tip can prevent the self-ligation of the adaptors.
16. Testing the cycle number to prevent excessive PCR duplications in the final sequencing data is critical. We usually take 1 μl of the DNA template to titrate the optimal PCR cycle numbers by amplifying at least two different cycles (e.g., 10 and 14 cycles). After examining the strength of the smear bands on an agarose gel, users may choose an optimal cycle number that allows them to obtain enough product for sequencing.

4 Conclusion

In this chapter, we described a step-by-step protocol for constructing the RIC-seq library to facilitate its usage in mapping in situ RRIs. Considering intra- and intermolecular RRIs are the building blocks of ncRNA-mediated transcriptional and post-transcriptional regulations in almost all species of life, including viruses, bacteria, plants, and animals (Xu et al. 2022). Identifying in situ RRIs with RIC-seq can provide a panoramic view of the hidden layer of RNA regulations.

More than 90% of the human genome could be transcribed, which generates tens of thousands of ncRNAs. Although ncRNAs have been demonstrated to function directly in cell differentiation, cell proliferation, dosage compensation, and many other cellular processes (Xue et al. 2020), their functional mechanisms remain largely unknown because of their mysterious structures and interacting targets. The advancement of RIC-seq technology in unbiased profiling of all the RBPs-mediated RNA–RNA spatial interactions could deepen our understanding of the functionality of various types of ncRNAs, including lncRNAs, circRNAs, eRNAs, and so on.

Under normal conditions, RNAs often fold into higher-order structures and form complicated ribonucleoprotein complexes by associating with various RBPs to execute their biological functions. However, highly structured RNA molecules tend to undergo structural changes in response to environmental stimuli or stresses to keep cell and life homeostasis (Van Treeck et al. 2018; Kortmann and Narberhaus 2012). Mutations that disrupt RNA structures or prevent dynamic changes can cause dysregulation of gene expression and lead to diseases (Halvorsen et al. 2010; Haas et al. 2012). Profiling RRIs by RIC-seq in disease-related cell models or tissues from patients may help decipher the casual mutations that impair RNA structures and eventually provide candidate structure targets for developing small-molecule drugs.

5 Future Perspectives

RNA–RNA spatial interactions could provide new perspectives for understanding the regulatory mechanisms of ncRNAs in various biological processes. In addition to eukaryotic cells, we have devised a vRIC-seq (virion RIC-seq) technology by combining the principle of RIC-seq with the virions captured by Concanavalin A (ConA) beads. We have demonstrated that vRIC-seq is powerful for studying genomic RNA structures of SARS-CoV-2 and also seems applicable to other RNA viruses (Cao et al. 2021). Using vRIC-seq, we have successfully reconstructed the 3D architecture of the SARS-CoV-2 genome RNA inside the virion, which has a diameter of only ~80 nm. In the future, there is still plenty of room for improvement in RIC-seq and variant technologies to expand their capability in detecting RRIs.

First, the current RIC-seq method identifies RNA–RNA proximal interactions without knowing the exact involved RBP. Inspired by a protein-centric chromatin conformation capture method HiChIP (Mumbach et al. 2016), we recently developed a capture RIC-seq (CRIC-seq) method by simply integrating immunoprecipitation into the present RIC-seq protocol, enabling the global profiling of a single RBP-mediated RNA interactome (Ye et al. 2022). Using the capture RIC-seq strategy, researchers can obtain the RRIs mediated by a specific RBP and the profile of the RBP footprints on RNA molecules. Notably, using antibodies specific to subcellular compartment markers, the CRIC-seq method can be expanded to map the RRIs in the corresponding compartments in future studies. These efforts will bring novel insights into the RNA regulatory networks in a specific cellular compartment.

Second, the applicability of RIC-seq technology is limited for RNAs with relatively low abundance owing to its "all-to-all" nature. This limitation may restrict the utility of RIC-seq for studying ncRNAs since they are generally expressed at lower levels than messenger RNAs. If the RNA of interest expresses at a relatively lower level, the number of RIC-seq-detected RRIs may be insufficient for downstream analysis. Although deepening the sequencing depth could help, it is costly and requires more computing resources. Introducing pull-down steps using DNA probes reverse complementary to specific RNAs in RIC-seq can potentially overcome this limitation theoretically. This strategy will be beneficial for investigating lowly expressed RNAs such as circRNAs.

Third, a standard RIC-seq library requires 1×10^7 cells (e.g., for the HeLa cell line), thus restricting RIC-seq application to precious biological samples, such as embryos or tissues from patients. In addition, emerging studies at the single-cell level have revealed cellular heterogeneity in gene expression, which could raise differences in RNA conformation from cell to cell. Therefore, exploring the application of RIC-seq to low-cell input, even at the single-cell level, is another emergency.

With the improvements mentioned above, we hope RIC-seq and its variant technologies could become powerful tools for researchers interested in studying RNA structures and targets.

Acknowledgements We thank Drs. Changchang Cao and Di Wang for critical reading of this manuscript. This work was supported by the National Key R&D Program (2022YFA1303300), NSFC (32025008, 32130064, 91940306, and 81921003), and the K.C. Wong Education Foundation (GJTD-2020-06) to Y.X.

References

Adiconis X, Borges-Rivera D, Satija R et al (2013) Comparative analysis of RNA sequencing methods for degraded or low-input samples. Nat Methods 10:623–629

Aw JG, Shen Y, Wilm A et al (2016) In vivo mapping of eukaryotic RNA interactomes reveals principles of higher-order organization and regulation. Mol Cell 62(4):603–617

Bak G, Han K, Kim K-S et al (2015) Electrophoretic mobility shift assay of RNA–RNA complexes. In: Schmidt FJ (ed) RNA-RNA interactions: methods and protocols. Springer New York, New York, NY, pp 153–163

Bartel DP (2004) MicroRNAs: genomics, biogenesis, mechanism, and function. Cell 116:281–297

Cai Z, Cao C, Ji L et al (2020) RIC-seq for global in situ profiling of RNA-RNA spatial interactions. Nature 582:432–437

Cao C, Cai Z, Xiao X et al (2021) The architecture of the SARS-CoV-2 RNA genome inside virion. Nat Commun 12:3917

Chen J, Xue Y (2016) Emerging roles of non-coding RNAs in epigenetic regulation. Sci China Life Sci 59:227–235

Djebali S, Davis CA, Merkel A et al (2012) Landscape of transcription in human cells. Nature 489:101–108

Engreitz JM, Sirokman K, McDonel P et al (2014) RNA-RNA interactions enable specific targeting of noncoding RNAs to nascent Pre-mRNAs and chromatin sites. Cell 159:188–199

Esteller M (2011) Non-coding RNAs in human disease. Nat Rev Genet 12:861–874

Gerstberger S, Hafner M, Tuschl T (2014) A census of human RNA-binding proteins. Nat Rev Genet 15:829–845

Gong C, Maquat LE (2011) lncRNAs transactivate STAU1-mediated mRNA decay by duplexing with 3′ UTRs via Alu elements. Nature 470:284–288

Haas U, Sczakiel G, Laufer SD (2012) MicroRNA-mediated regulation of gene expression is affected by disease-associated SNPs within the 3′-UTR via altered RNA structure. RNA Biol 9:924–937

Halvorsen M, Martin JS, Broadaway S et al (2010) Disease-associated mutations that alter the RNA structural ensemble. PLoS Genet 6:e1001074

Hangauer MJ, Vaughn IW, McManus MT (2013) Pervasive transcription of the human genome produces thousands of previously unidentified long intergenic noncoding RNAs. PLoS Genet 9:e1003569

Hardin JW, Warnasooriya C, Kondo Y et al (2015) Assembly and dynamics of the U4/U6 di-snRNP by single-molecule FRET. Nucleic Acids Res 43:10963–10974

Helwak A, Tollervey D (2016) Identification of miRNA-target RNA interactions using CLASH. Methods Mol Biol 1358:229–251

Helwak A, Kudla G, Dudnakova T et al (2013) Mapping the human miRNA interactome by CLASH reveals frequent noncanonical binding. Cell 153:654–665

Hentze MW, Castello A, Schwarzl T et al (2018) A brave new world of RNA-binding proteins. Nat Rev Mol Cell Biol 19:327–341
Kertesz M, Wan Y, Mazor E et al (2010) Genome-wide measurement of RNA secondary structure in yeast. Nature 467:103–107
Kortmann J, Narberhaus F (2012) Bacterial RNA thermometers: molecular zippers and switches. Nat Rev Microbiol 10:255–265
Kretz M, Siprashvili Z, Chu C et al (2013) Control of somatic tissue differentiation by the long non-coding RNA TINCR. Nature 493:231–235
Kudla G, Granneman S, Hahn D et al (2011) Cross-linking, ligation, and sequencing of hybrids reveals RNA–RNA interactions in yeast. Proc Natl Acad Sci USA 108:10010
Li P, Wei Y, Mei M et al (2018) Integrative analysis of Zika virus genome RNA structure reveals critical determinants of viral infectivity. Cell Host Microbe 24:875-886 e5
Lu Z, Zhang Qiangfeng C, Lee B et al (2016) RNA duplex map in living cells reveals higher-order transcriptome structure. Cell 165:1267–1279
Moore MJ, Proudfoot NJ (2009) Pre-mRNA processing reaches back to transcription and ahead to translation. Cell 136:688–700
Mumbach MR, Rubin AJ, Flynn RA et al (2016) HiChIP: efficient and sensitive analysis of protein-directed genome architecture. Nat Methods 13:919–922
Nguyen TC, Cao X, Yu P et al (2016) Mapping RNA-RNA interactome and RNA structure in vivo by MARIO. Nat Commun 7:12023
Piganeau N, Schauer UE, Schroeder R (2006) A yeast RNA-hybrid system for the detection of RNA-RNA interactions in vivo. RNA 12:177–184
Ramani V, Qiu R, Shendure J (2015) High-throughput determination of RNA structure by proximity ligation. Nat Biotechnol 33:980–984
Rinn JL, Chang HY (2012) Genome regulation by long noncoding RNAs. Annu Rev Biochem 81:145–166
Salzman J, Gawad C, Wang PL et al (2012) Circular RNAs are the predominant transcript isoform from hundreds of human genes in diverse cell types. PLoS ONE 7:e30733
Sharma E, Sterne-Weiler T, O'Hanlon D et al (2016) Global mapping of human RNA-RNA interactions. Mol Cell 62:618–626
Shen EZ, Chen H, Ozturk AR et al (2018) Identification of piRNA binding sites reveals the argonaute regulatory landscape of the C. elegans germline. Cell 172:937–951 e18
Siegfried NA, Busan S, Rice GM et al (2014) RNA motif discovery by SHAPE and mutational profiling (SHAPE-MaP). Nat Methods 11:959–965
Song Y, Yang W, Fu Q et al (2020) irCLASH reveals RNA substrates recognized by human ADARs. Nat Struct Mol Biol 27:351–362
Stephenson JD, Kenyon JC, Symmons MF et al (2016) Characterizing 3D RNA structure by single molecule FRET. Methods 103:57–67
Sugimoto Y, Vigilante A, Darbo E et al (2015) hiCLIP reveals the in vivo atlas of mRNA secondary structures recognized by Staufen 1. Nature 519:491–594
Sun L, Li P, Ju X et al (2021) In vivo structural characterization of the SARS-CoV-2 RNA genome identifies host proteins vulnerable to repurposed drugs. Cell 184:1865-1883 e20
Van Damme R, Li K, Zhang M et al (2022) Chemical reversible crosslinking enables measurement of RNA 3D distances and alternative conformations in cells. Nat Commun 13:911
Van Treeck B, Protter DSW, Matheny T et al (2018) RNA self-assembly contributes to stress granule formation and defining the stress granule transcriptome. Proc Natl Acad Sci USA 115:2734–2739
Wang X, Ramat A, Simonelig M et al (2022) Emerging roles and functional mechanisms of PIWI-interacting RNAs. Nat Rev Mol Cell Biol. https://doi.org/10.1038/s41580-022-00528-0
Wapinski O, Chang HY (2011) Long noncoding RNAs and human disease. Trends Cell Biol 21:354–361
Waters SA, McAteer SP, Kudla G et al (2017) Small RNA interactome of pathogenic E. coli revealed through crosslinking of RNase E. The EMBO J 36:374–387

Watts JM, Dang KK, Gorelick RJ et al (2009) Architecture and secondary structure of an entire HIV-1 RNA genome. Nature 460:711–716
Xu B, Zhu Y, Cao C et al (2022) Recent advances in RNA structurome. Sci China Life Sci 65:1285–1324
Xue Y (2022) Architecture of RNA-RNA interactions. Curr Opin Genet Dev 72:138–144
Xue Y, Chen R, Qu L et al (2020) Noncoding RNA: from dark matter to bright star. Sci China Life Sci 63:463–468
Ye R, Hu N, Cao C et al (2022) CRIC-seq reveals positional rule of PTBP1-mediated long-range RNA looping in splicing regulation. bioRxiv, 2022.08.09.503273
Zhao R, Rueda D (2009) RNA folding dynamics by single-molecule fluorescence resonance energy transfer. Methods 49:112–117
Ziv O, Gabryelska MM, Lun ATL et al (2018) COMRADES determines in vivo RNA structures and interactions. Nat Methods 15:785–788
Ziv O, Price J, Shalamova L et al (2020) The short- and long-range RNA-RNA interactome of SARS-CoV-2. Mol Cell 80:1067-1077 e5

Unraveling RNA by Mechanical Unzipping

Paolo Rissone, Isabel Pastor, and Felix Ritort

Contents

Abstract We review the basic concepts and tools for mechanically unzipping RNA hairpins using force spectroscopy. By pulling apart the ends of an RNA molecule using optical tweezers, it is possible to measure the folding free energy at varying experimental conditions. Energy measurements permit us to characterize the thermodynamics of RNA hybridization (base pairing and stacking), the dynamics of the formation of native and kinetic (intermediates and misfolded) molecular states, and interactions with metallic ions. This paper introduces basic concepts and reviews recent developments related to RNA force thermodynamics, native and barrier RNA energy landscapes, and RNA folding dynamics. We emphasize the implications of mechanical unzipping experiments to understand non-coding RNAs and RNAs in extreme environments.

Keywords Single-molecule spectroscopy · RNA unzipping · Out-of-equilibrium thermodynamics · RNA free-energy landscape

P. Rissone · I. Pastor · F. Ritort (✉)
Facultat de Fisica, Universitat de Barcelona, Carrer de Martí i Franqués 1, Barcelona 08028, Spain
e-mail: fritort@gmail.com

F. Ritort
Institut de Nanociència i Nanotecnologia (IN2UB), Facultat de Fisica, Universitat de Barcelona, Carrer de Martí i Franqués 1, Barcelona 08028, Spain

J. Barciszewski (ed.), *RNA Structure and Function*, RNA Technologies 14,
https://doi.org/10.1007/978-3-031-36390-0_4

1 Introduction

RNA, the genome's dark matter, directly impacts biological diversity and life (Darnell 2011; Filipowicz 2022). RNAs can fold into multiple configurations stabilized by secondary and tertiary structures (Butcher and Pyle 2011; Herschlag et al. 2018), multivalent cations, and ligands (Pyle 2002; Woodson 2005; Draper et al. 2005; Bowman et al. 2012). The promiscuity of base pairing and stacking interactions makes RNA a unique biopolymer with many functions, from information carrier to regulatory and enzymatic activity. RNA exhibits a significant degree of heterogeneity at the sequence level and the conformational level (Treiber and Williamson 2001; Russell et al. 2002; Brion and Westhof 1997; Cruz and Westhof 2009). Upon folding, RNA can form native and non-native structures (such as misfolded and intermediates) (Woodson 2010), with critical roles at the level of genomic maintenance and the cellular function (Mattick and Makunin 2006; Aalto and Pasquinelli 2012), therapeutics (Esteller 2011; Matsui and Corey 2017) and diseases (Jain and Vale 2017; Blaszczyk et al. 2017; Zhao and Usdin 2021). Although the molecular forces operating in RNA are known, the role played by disorder at the structural and functional levels poses severe challenges to the life scientist who must cope with unprecedented complexity.

In recent years, a knowledge gap has appeared not only at the level of RNA transcriptomics but also at the level of non-coding RNAs (ncRNAs) and their remarkable variety of functions in concert with ligands and proteins (Stefani and Slack 2008; Aalto and Pasquinelli 2012; Mercer et al. 2009; Ma et al. 2013; Fatica and Bozzoni 2014; Statello et al. 2021).

Although the role of many RNAs remains unknown, new RNAs with new functionalities and structures are being discovered. Besides the much-studied tRNA, rRNA, microRNA, riboswitches, ribozymes, and artificially evolved RNAs, novel behaviors have been observed in response to environmental cues such as temperature (e.g., RNA cold denaturation (Mikulecky and Feig 2002, RNA thermometers (Loh et al. 2013), and in concerted action with proteins (catalytic complexes, chaperones, packaging, condensation, etc.).

Despite the enormous progress in next-generation sequencing and big data analysis, our current knowledge of RNA diversity is compromised by the limited accuracy, sensitivity, and specificity of available methods to detect different RNA conformations across RNA populations. Moreover, determining the folding pathways and the energetics of the various RNA structures is essential to understanding RNA function. Single-molecule techniques have represented a big step in addressing RNA complexity Ritort 2006; Seidel and Dekker 2007). Their great sensitivity and accuracy permit us to detect and measure the folding energies of rarely occurring conformations that escape detection by the standard bulk methods. Powerful techniques such as single-molecule FRET (Zhuang 2005; Alemán et al. 2008) and force spectroscopy (Ritchie and Woodside 2015; Bustamante et al. 2021) can monitor RNA conformational transitions in real-time More recently, solid-state nanopore microscopy for RNA target detection can analyze thousands of single RNAs without amplification offering exciting prospects (Henley et al. 2016; Bošković and Keyser 2022).

Compared to DNA, RNA exhibits more complex behavior. The replacement of deoxyribose for ribose and thymine for uracil makes RNA catalytic due to the reactive polarizable 2'-OH group of ribose. Ribose also induces large changes at the level of base stacking interactions between contiguous bases. In their double-stranded forms, nucleic acids (NA) form distinct right-handed double helices, B-form and A-form. Although DNA can adopt both A-form and B-form, RNA is mainly found in A-form with only a few rare exceptions (Shi et al. 2003). In DNA, bases are mainly parallel to the helical plane with an interphosphate distance of 3.4Å. In contrast, RNA bases are tilted by approximately 19 °C relative to the helical plane and the interphosphate distance is smaller (~2.8Å); these structural differences generate stacking between inter-strand bases and tighter water molecular bridges between phosphates and bases in RNA. Overall, base stacking tends to be stronger in RNA than in DNA. Base stacking is due to the Van der Waals attractive forces of the fluctuating dipole-dipole interactions between contiguous bases. Much weaker than the covalent nature of hydrogen bonding, the effect of the latter is minimized upon secondary structure formation due to the compensation effect of hydrogen bonding with water. Overall, base stacking and hydrogen bonding contribute equally to RNA helix stabilization, albeit the $1/r^6$ dependence of Van der Waals forces makes stacking strongly sensitive to the inter-base distance, r. Therefore, RNA structure strongly depends on RNA stacking between intra-strand and inter-strand bases, making RNA folding prediction a difficult problem.

Here we briefly review recent discoveries in the thermodynamics of RNA folding using force spectroscopy studies with laser optical tweezers (LOT). The paper is organized as follows. Section 2 describes three main experimental techniques to investigate RNA kinetics and thermodynamics at the single-molecule level. In Sect. 3 we focus on RNA force spectroscopy, the main experimental protocols, and a few selected examples. In Sect. 4 we discuss how RNA unzipping experiments make it possible to derive RNA thermodynamics at 0.1kcal/mol accuracy. Section 5 introduces the concept of the barrier energy landscape and the importance of stem-loops to stabilize a multiplicity of RNA kinetic structures. Finally, in Sect. 6 we digress about future perspectives in single-RNA manipulation.

2 The Power of Single RNA Manipulation

Single-molecule (SM) experiments permit us to study individual molecules' behavior and characterize the molecular properties' heterogeneity. On the contrary, in conventional bulk experiments, any observation is an average over a large population of molecules, making it impossible to observe ephemeral or transient events. During the last decades, the development of single-molecule techniques opened a new window to the rich phenomenology of RNA molecules. In particular, three useful single-molecule techniques are single-molecule fluorescence (Förster) resonance energy transfer (smFRET) (Alemán et al. 2008; Zhuang 2005; Ray et al. 2018; Chauvier et al. 2019; Ha et al. 1999; Zhao and Rueda 2009), solid-state nanopore microscopy

(Seidel and Dekker 2007; Cui et al. 2021; Bandarkar et al. 2020; Lee et al. 2021; Bošković and Keyser 2022) and force spectroscopy using optical tweezers (Woodside et al. 2006; Bustamante et al. 1991; Ritchie and Woodside 2015; Manosas and Ritort 2005; Wen et al. 2007; Zhuang 2005). First, we overview these techniques, focusing next on the optical tweezers experiments.

2.1 SmFRET Experiments

smFRET measurements measure conformational changes with a high temporal resolution (ms) (Joo et al. 2008; Ha 2001). This technique measures the distance between two fluorophores (donor and acceptor) at different positions along the biomolecule. The donor is located in the vicinity of the acceptor (distance less than 10 nm) and excited with an appropriate light wavelength. Part of the energy emitted by the donor is transferred to the acceptor through a non-radiative dipole-dipole interaction, causing the acceptor's emission. Donor-acceptor energy transfer is a quantum-mechanical effect due to the overlap between the donor emission spectrum and the acceptor absorption spectrum. The energy transfer efficiency, defined as the ratio $I_A/(I_D + \eta I_A)$ between light intensity emitted by the donor I_D and the acceptor I_A (η being a quantum yield correction factor), depends on the distance between the two fluorophores according to the equation: $E = 1/(1 + (R/R_0)^6)$, where R is the distance between both fluorophores and R_0 is the characteristic distance (Förster distance) where the efficiency is one-half (Lakowicz 2006).

The experiments can be carried out either on-surface-immobilized (Fig. 1a) or freely diffusing molecules. The former permits parallelized measurements during a long time (until photobleaching, typically up to tens of seconds) allowing the detection of slow conformational changes. smFRET requires biochemical modifications by attaching fluorophores to the molecule under study, which may interfere with the (unmodified) molecular behavior (Ha et al. 1999; Zhao and Rueda 2009). In addition, in the case of on-surface-immobilized smFRET experiments, the perturbations that the surface may exert on the molecule must be taken into account (Schmitz et al. 2015). There are several works where smFRET has been key to determining intermediate or misfolded states in RNA folding (Bartley et al. 2003; Bokinsky et al. 2003; Xie et al. 2004; Alemán et al. 2008; Bokinsky et al. 2003).

2.2 Nanopore Microscopy

In recent years, nanopore microscopy has shown to be a promising tool to address RNA complexity. Nanopores, either solid-state or biological ones, are extremely sensitive to the sequence and structure of biomolecules. This technique is based on the current established between two electrodes placed in different pools connected by a nanometric hole. The pools are filled with a salt solution and a charged biomolecule

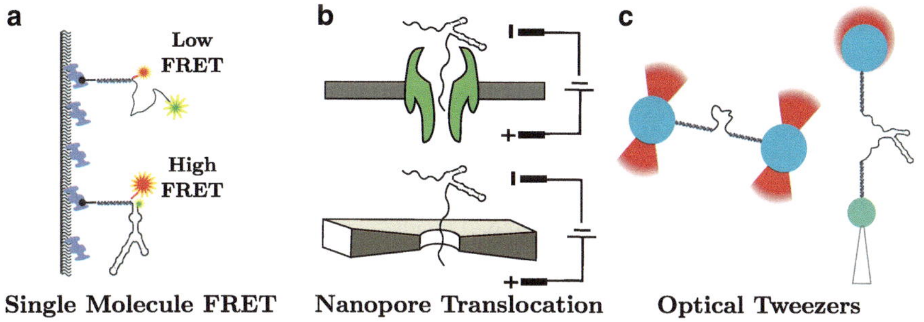

Fig. 1 Single-molecule techniques. **a** On-surface-immobilized smFRET experiment. **b** Nanopore microscopy uses a biological pore (top panel) and a solid-state pore (bottom panel). **c** Optical tweezers experiments in a double-trap geometry (left panel and in a single-trap geometry with a micropipette (right panel)

is placed in one of them. When a voltage difference is applied, the flow of ions through the nanopore results in an electric current. The flow of ions and the electric field between the electrodes makes the molecules flow through the nanopore. As molecular translation proceeds, the flow of ions is partially (or fully) obstructed, resulting in a reduction of the net current. These ion fluctuations depend on the biomolecule's properties. Usually, the biomolecule needs to rearrange to translocate, making nanopore translocation an important tool to screen different molecular conformations.

Traditionally, RNA translocation experiments use biological pores such as ion channel proteins (Fig. 1b, top). A widely used model is alpha-hemolysin (Butler et al. 2007; Sultan and Kanavarioti 2019; Meller et al. 2006) due to its large stability and small pore diameter that confers much sensitivity to detect individual nucleotides. Biological pores generally exhibit lower noise. However, solid-state nanopores (Fig. 1b, bottom) can operate at higher voltages and bandwidths, making it possible to achieve a better signal-to-noise ratio, key to detecting translocation events. Moreover, solid-state nanopores are pore-size and geometry-tunable and more robust than biological ones.

A key advantage of solid-state nanopore microscopy is that it does not require chemically modifying the biomolecule. However, there are some disadvantages such as the limited nanopores reproducibility and the high translocation speed.

2.3 Optical Tweezers

Among all single-molecule force spectroscopy techniques, optical tweezers have proved to be one of the most powerful for studying the complexity of nucleic acids, especially in the case of RNA. Laser optical tweezers (LOT) use a focused laser beam to optically trap a transparent microbead attached to one end of an RNA molecule. By attaching the other end to a surface, the RNA can be pulled by displacing the optically trapped bead by moving the laser. LOT can exert forces in the range of tenths to hundreds of piconewtons measuring energies with 0.1kcal/mol accuracy. LOT can monitor force and bead position, detecting conformational transitions with sub-millisecond temporal resolution.

There are different setups to carry out LOT experiments, such as dual-trap (Fig. 1c, left) and single-trap configurations (Fig. 1c, right). In the dual-trap configuration, the molecule is tethered between two trapped beads, while in the single-trap configuration, the molecule is tethered between two beads: one is optically trapped and the other is held by air suction at the tip of a micropipette. Usually, the dual-trap setup is more sensitive because the traps are formed from the same laser and have less drift. The single-trap micropipette configuration is used in many experiments, such as those presented in this review.

Independently on the setup, to avoid bead-bead interactions, the RNA molecule under study is inserted between two molecular linkers or spacers usually DNA-RNA hybrid handles (Wen et al. 2007; Manosas et al. 2007; White and Visscher 2011; Collin et al. 2005; Liphardt et al. 2001; Onoa et al. 2003; Martinez-Monge et al. 2022; Wu et al. 2014). As compared with other single-molecule techniques, the mechanical manipulation of RNAs allows monitoring the unfolding/folding of individual structural domains of large RNAs and their folding pathways. Moreover, RNAs can be pulled to detect short-lived intermediates and misfolded states, which can be modified by the ion concentration (Na^+, Mg^{+2}, etc.) or the temperature. In addition, RNA manipulation with optical tweezers permits measuring energies with 0.1 kcal/mol accuracy (Severino et al. 2019; Rissone et al. 2022; Martinez-Monge et al. 2022).

3 RNA Pulling in a Nutshell

Several pulling protocols can be implemented to study RNAs with optical tweezers. The most common are force-ramp, hopping, and force-jump experiments. In the following, we show examples of RNA pulling experiments with LOT.

In force-ramp protocols, the optical trap is repeatedly moved back and forth from the micropipette (Fig. 2a, top) with the force steadily increasing and decreasing, respectively. Upon increasing the force the RNA hairpin switches from its folded (native) state to the totally unfolded state and *vice-versa* upon releasing the force. A plot of the applied force versus the distance between the trap and the micropipette gives the force-distance curve (FDC) (Fig. 2a, bottom). In small RNA hairpins (typically a few tens of bases) that cooperatively fold and unfold, the rips in the FDC indicate unfolding or folding transitions between the native folded hairpin and the stretched ssRNA conformation. For longer RNA hairpins (a few hundred bases) the unfolding into ssRNA is a sequential process in which groups of base-pairs open. In this case, the FDC exhibits a sawtooth pattern depending on the molecular sequence (Fig. 5). Pulling experiments allows for the characterization of the unfolding/folding forces as well as the thermodynamics and kinetics of RNA molecules (see Sects. 4 and 5).

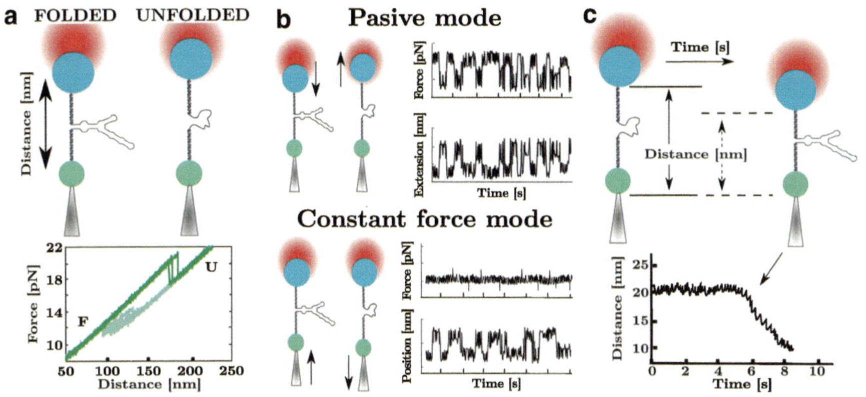

Fig. 2 Optical tweezers experiments. **a** Force-ramp protocol. The trap-pipette distance is changed at a constant velocity. The typical FDC obtained in force-ramp experiments is shown in the bottom panel. **b** Hopping protocols. Top: PM experiments. The optically trapped bead moves toward or away from the trap center as the molecule unfolds or folds, respectively. The right panels show force and extension variation with time (top and bottom, respectively). Bottom: CFM experiments. The force is kept constant by a feedback protocol compensating for extension changes of the molecule. The right panels show force and position variation with time (top and bottom, respectively). **c** Force jumps protocol. The force or the distance is quickly changed to a different value. The typical distance versus time signal is reported in the bottom panel

In hopping experiments, either the trap position or the force is kept constant, and the jumps of the molecule from the unfolded (folded) to folded (unfolded) state are monitored over time. There are two kinds of hopping assays: passive mode (PM) and constant force mode (CFM) experiments. In PM hopping experiments, the trap position is clamped (Fig. 2b, top), and the RNA hairpin hops between the unfolded and the folded state with a force jump at every transition. In PM both the force and the molecular extension change over time because the trapped bead relaxes to a new position at every RNA hop, and the force changes accordingly. The PM allows direct monitoring of the molecular transitions in a narrow range of forces close to the coexistence force where the RNA equally populates the native and unfolded states. PM is not suitable for long-time measurements in LOT in the single-trap configuration due to the uncontrolled movements of the pipette (drift effects). In contrast, in CFM hopping experiments, the force is kept constant with a feedback loop (Fig. 2b bottom), and the RNA extension is recorded as a function of the time. Here, the force is fixed at a value near the RNA coexistence force, allowing for measuring the transition between its folded/unfolded states and the lifetime of each state at the studied force.

The force-jump protocol is helpful in characterizing RNA irreversible processes. In this case (Fig. 2c), the force or the trap position is quickly changed to a new preset value and kept constant. The RNA lifetime is measured until a conformational transition is observed. Force-jump is an irreversible pulling protocol, requiring multiple experiments at different present force values to derive the unfolding and folding kinetics.

3.1 Brief History of RNA Pulling Experiments

Since the beginning of the 21st century, LOT experiments have been used in many works studying the structure and energetics of RNAs and their native, misfolded, and short-lived intermediate states, and the salt dependency of the RNA conformation. In Liphardt et al. (2001), it has been shown for the first time that by mechanically unzipping RNA hairpins, it is possible to derive its folding free energy (Fig. 3a). They studied three RNAs: a small hairpin (P5ab), a molecule with a three-helix junction (P5abcΔA), and a more complex molecule (P5abc), which in presence of Mg^{+2} ions folded into a stable tertiary structure. In particular, they found that P5abc folded into a stable tertiary structure through a short-lifetime intermediate.

Mechanical pulling experiments have also been carried out on longer RNAs: in Harlepp et al. (2003), it has been studied the 1540 nucleotides 16S ribosomal RNA

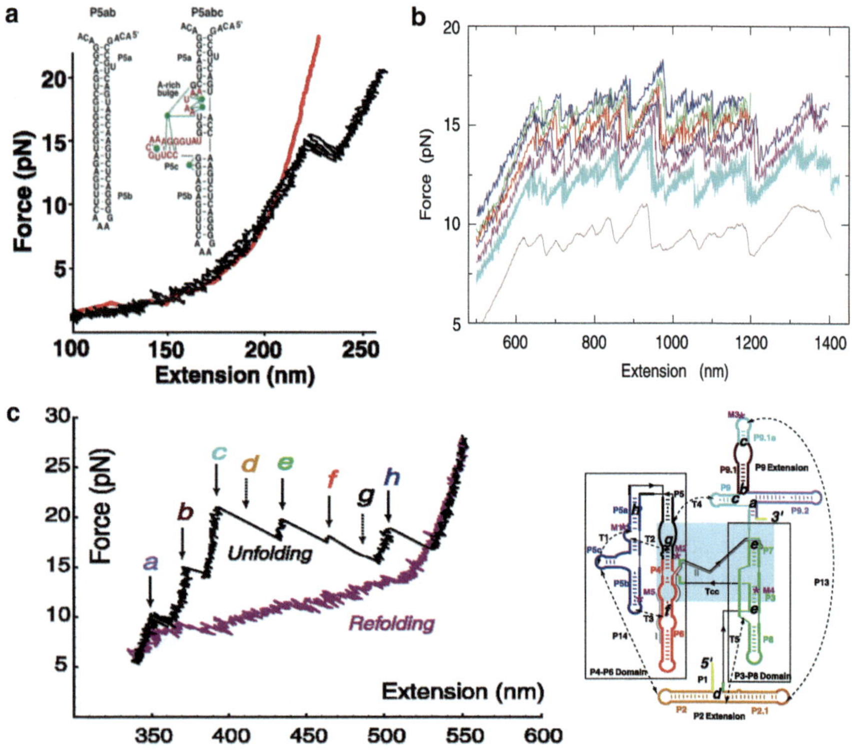

Fig. 3 Different results obtained in pulling experiments. **a** P5ab RNA force-extension curves in 10 mM Mg^2 (adapted from Liphardt et al. (2001)). Structure of P5ab and P5abc RNAs (inset). **b** 16S ribosomal FDCs, the colors represent successive pulling for the same molecule. Figure adapted from Harlepp et al. 2003. **c** Unfolding (black) and folding (pink) FDCs of the *T. thermophila* ribozyme (left panel). The letters indicate different kinetics barriers. The right panel shows the secondary structure. Figure adapted from Onoa et al. 2003

from *E. coli* (Fig. 3b). This large RNA exhibits a surprisingly well-structured and reproducible unfolding pathway under mechanical stretching. Similar results have been found in Onoa et al. (2003) by pulling the L-21 derivative of the *Tetrahymena thermophila* ribozyme (a 390 nucleotides RNA). This molecule featured a complex secondary structure with multiple unfolding intermediate states (Fig. 3c). To identify the different RNA structures, they measured the number of released base pairs along the FDCs of progressively large fragments of the RNA molecule. At the same time, they used mutants and anti-sense oligonucleotides to characterize these RNA structures further and measure their free energy of formation.

Another example of a biologically relevant RNA is the study of the operator rpsO of the RNA gene coding for the S15 subunit of 30S ribosomal protein from *E. coli* (Wu et al. 2014). This mRNA can fold into two spontaneously interchangeable structures: a double hairpin and a pseudoknot (Fig. 4a, left). Their work showed that the conversion from the double hairpin to the pseudoknot is stabilized by the interac-

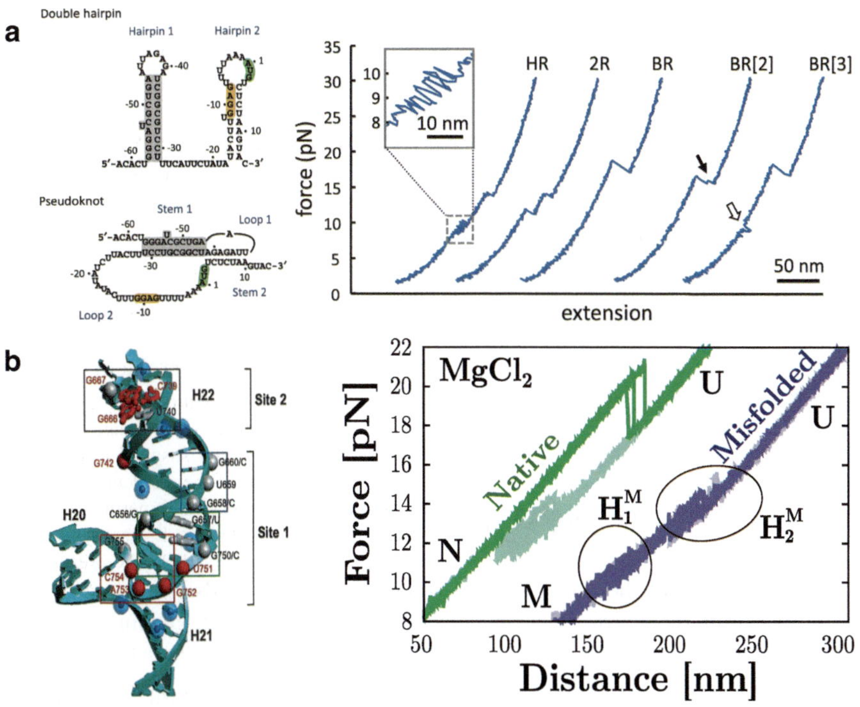

Fig. 4 Different results obtained in pulling experiments. **a** Left: the two structures proposed for rpsO RNA: double hairpin (top) and pseudoknot (bottom). Right: FDCs of the double hairpin (HR), pseudoknot (BR), and mixed transitions (2R). Figure adapted from Wu et al. (2014). **b** Left: The highly conserved crystal structure of the RNA three-way junction molecule from *T. thermophilus* Serganov et al. (2001). Right: FDCs for the native and misfolded structures in 10 mM Mg^{+2}. Figure adapted from Martinez-Monge et al. (2022)

tion between the two hairpins (Fig. 4a, right). Sequence mutations can modulate the interaction between the two hairpins.

Finally, in a more recent work (Martinez-Monge et al. 2022) RNA pulling experiments have been carried out to measure the non-specific and specific binding energies of magnesium to RNA. The RNA three-way junction (3WJ) containing the minimal binding site to protein S15 of the ribosomal RNA from *E. coli* was studied in monovalent and divalent salt conditions. This molecule can fold into a native structure that contains the 3WJ motif with specific Mg^{+2} binding sites or into a misfolded structure with a double hairpin structure where the 3WJ and the binding sites have been disrupted (Fig. 4b). By comparing the free energy of formation of these two structures with and without magnesium, it has been possible to determine the specific and non-specific energy contributions of Mg^{+2} binding to the RNA.

4 RNA Energetics at 0.1 kcal/mol Accuracy

The characterization of RNA thermodynamics is fundamental to understanding the promiscuity of behaviors observed in RNA, from the multiplicity of native structures (Gralla and Delisi 1974) to misfolding (Alemany et al. 2012). Only RNA exhibits such behavior despite DNA and RNA forming double-stranded helices. DNA unzipping is a fully reversible process (Fig. 5a) over a broad range of salt conditions and loading rates (Huguet et al. 2010; Bizarro et al. 2012). Instead, RNA unzipping presents strong irreversibility between the unfolding and refolding FDCs in the same experimental conditions (Fig. 5b). In this case, transient off-pathway misfolded structures appear during the unzipping-rezipping process, slowing down the hybridization reaction (Liphardt et al. 2001; Chen and Dill 2000; Woodson 2010). Characterizing these off-pathway structures competing with the native stem is a challenging problem (Rissone et al. 2022).

4.1 The RNA Free Energy of Formation

The hybridization of the double-stranded helix of nucleic acids (NA) is governed by the Watson-Crick pairing rules between nucleotides (adenine, guanine, cytosine, thymine or uracil) from opposite NA strands (Saenger 1984). These relations account for purine-pyrimidine bonding so that adenine can only bond to thymine (or uracil in the RNA case) and guanine to cytosine. As discussed before, the double helix is not only stabilized by base pairing but also by the stacking interactions between adjacent nucleotides. To account for these effects, the energetics of NA formation is usually described using the nearest-neighbor (NN) model (DeVoe and Tinoco 1962; Breslauer et al. 1986). According to this model, the base-pairing energy of two complementary bases depends on the base itself and the first neighbor located in the same strand along the $5' \rightarrow 3'$ direction. This gives 16 different possible

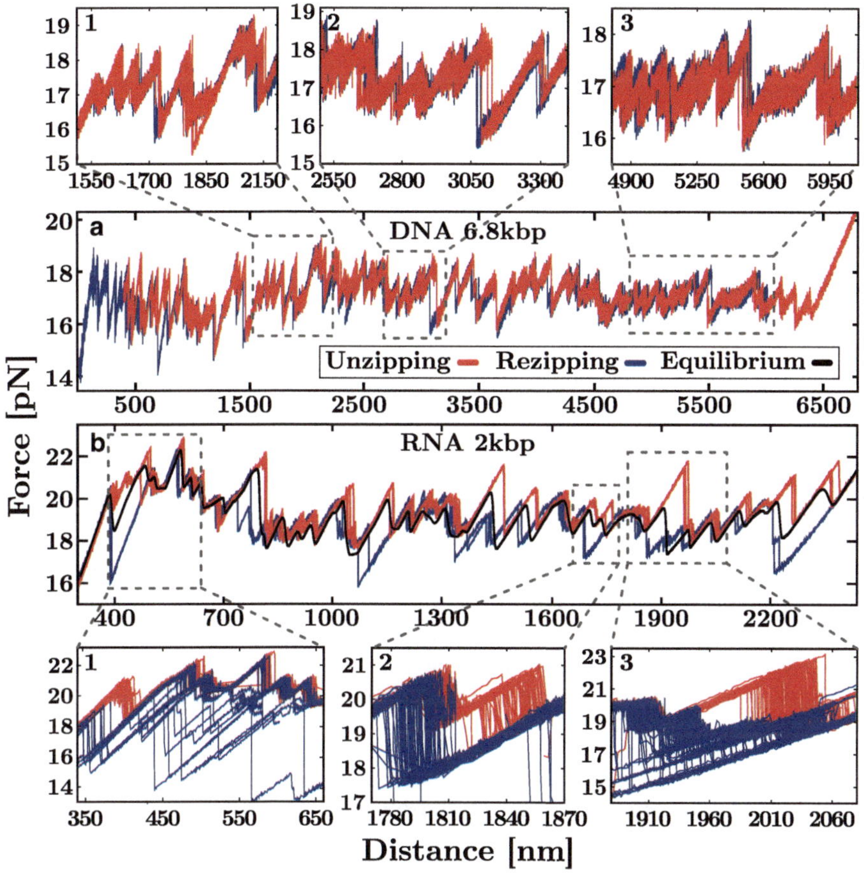

Fig. 5 Unzipping (red) and rezipping (blue) FDCs measured by pulling a 6.8 kbp DNA (**a**) and a 2 kbp RNA (**b**) hairpins in a 1M NaCl buffer at 25 °C. While DNA unzipping is a reversible process, RNA unzipping shows many off-pathway out-of-equilibrium states. The resulting hysteresis requires the computation of an equilibrium FDC (black) to measure the NNBPs' free-energies

combinations of nearest-neighbor base pairs (NNBPs): out of these, 6 are degenerate, and 2 can be expressed as a linear combination of the others (circular symmetry) (Huguet et al. 2017), leaving with only 8 independent parameters. The ten RNA NNBP are denoted as $XY/\overline{X}\,\overline{Y}$ (x-label in Fig. 7a) where X, Y = A, C, G, U, and $\overline{X}(\overline{Y})$ is the complementary base of X(Y) and $XY/\overline{X}\,\overline{Y}$ is the NNBP resulting from hybridizing dinucleotides $5' - XY - 3'$ and $5' - \overline{Y}\,\overline{X} - 3'$. The energies of $XY/\overline{X}\,\overline{Y}$ and $\overline{Y}\,\overline{X}/YX$ are equal due to complementary strand symmetry. For example, the RNA sequence $5'$ – CUUAGC – $3'$ forms a duplex with its complementary strand, $5'$ – GCUAAG – $3'$. According to the NN model the energy of hybridization of such a sequence equals $\Delta g_{CU/GA} + \Delta g_{UU/AA} + \Delta g_{UA/AU} + \Delta g_{AG/UC} + \Delta g_{GC/CG}$ with $\Delta g_{CU/GA} = \Delta g_{AG/UC}$ due to complementary strand symmetry. The derivation of the 10 (8 if circular symmetry is applied) NNBPs free-energies has been carried out by unzipping a 2.2 kbp and 6.8 kbp DNA and a 2 kbp RNA hairpins at different salt conditions (Huguet et al. 2010, 2017; Rissone et al. 2022).

In thermodynamics, the free-energy difference equals the mechanical work in a reversible process that requires the system to evolve along a sequence of equilibrium states. This is only applicable to the DNA case where the unzipping and rezipping FDCs do not show hysteresis (Fig. 5a). On the contrary, RNA unzipping is an out-of-equilibrium process in our experimental timescales: a large hysteresis is observed between unzipping and rezipping FDCs due to the formation of multiple long-lived (off-pathway) states (Fig. 5b). An equilibrium FDC (black line) had to be computed from the RNA unzipping/rezipping experimental data. This has been achieved by developing a statistical method based on the (extended) fluctuation relations (Bennett 1976; Jarzynski 1997; Shirts et al. 2003; Junier et al. 2009) that allowed for the computation of the equilibrium free energy during the isothermal unzipping process (see Supp. Info in Rissone et al. 2022). Finally, a Monte Carlo optimization algorithm has been tailored to relate the experimental data with the numerical FDC prediction, ultimately permitting measuring the NNBP energies in DNA and RNA (Huguet et al. 2010, 2017; Rissone and Ritort 2022). The experimentally derived values for RNA in sodium and magnesium are shown in Fig. 6a along with values reported in the literature (the Mfold set) (Freier et al. 1986; Mathews et al. 1999; Walter et al. 1994; Xia et al. 1998; Zuker 2003). As the latter are only available at 1M NaCl, the comparison required applying a correction to the measured NNBP energies.

4.2 Salt Dependency of the Hybridization Free Energy

The effect of a monovalent salt concentration, [Mon+] in molar units, on the hairpin free energy of formation is described by the relation

$$\Delta g_{0,i}[\text{Mon+}] = \Delta g_{0,i}[1\text{M}] - m \log[\text{Mon+}], \quad (1)$$

where $\Delta g_{0,i}[1\text{M}]$ is the free energy of formation of motif i at 1M of monovalent salt at zero force and $m = 0.10 \pm 0.01$ kcal/mol is the RNA NNBP-homogeneous monovalent salt correction (Bizarro et al. 2012). However, Eq. (1) only holds for monovalent ions and its extension to divalent ions requires to account for the effect of the divalent salt on the stabilization of the double helix.

RNAs are highly charged polyanions whose stability strongly depends on solvent ionic conditions. The ability of divalent ions, such as Mg^{2+}, to stabilize RNA structures at much lower concentrations than monovalent ions is known since the 70*s* (Cole et al. 1972). This effect is usually quantified by the so-called 100:1 rule of thumb which states that the concentration of divalent salt equals 100-fold that of monovalent salt. This phenomenological rule has been experimentally tested by unzipping the 2 kbp RNA hairpin at 500 mM NaCl and 10 mM $MgCl_2$ (Rissone et al. 2022). By plotting the RNA NNBP energy values in Na^+ versus those in Mg^{2+}

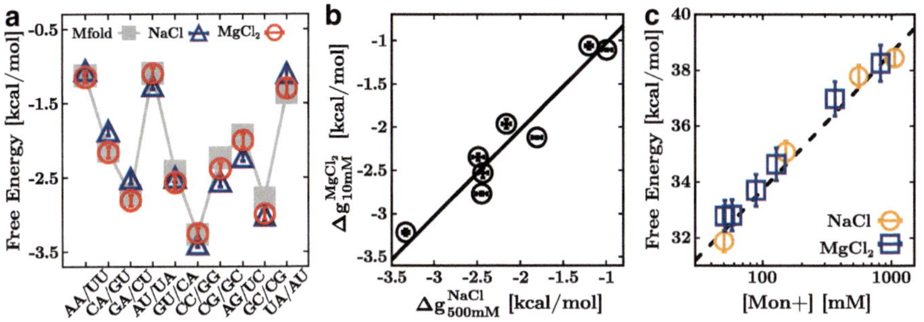

Fig. 6 RNA energetics. **a** NNBPs free-energies measured by unzipping a 2 kbp RNA hairpin at 500 mM NaCl (triangles) and 10 mM $MgCl_2$ (circles) (Rissone et al. 2022). Notice that the values are scaled to 1M of equivalent sodium concentration in order to compare with the literature (gray squares). **b** Experimental validation of the 100:1 rule of thumb. A fit to data (solid black line) gives 77($\pm$49):1 (Rissone et al. 2022). **c** Hairpin (total) free energy of formation measured in RNA unzipping experiments of a 20 bp hairpin (Bizarro et al. 2012) in sodium (blue squares) at 50 mM, 150 mM, 550 mM, 1050 mM and magnesium (orange circles) at 0.01 mM, 0.10 mM, 0.50 mM, 1 mM, 4 mM, 10 mM. A fit to data (dashed line) shows the logarithmic dependence of the salt correction. All results in magnesium are reported in sodium equivalents according to the 100:1 rule

(Fig. 6b), we demonstrate that the divalent salt concentration is equal to 77 ± 49 the one of monovalent salt, which is compatible with the phenomenological rule.

This result has also been validated by measuring the free energy of the formation of short RNA duplexes unzipped in a broad range of monovalent and divalent salt concentrations (Bizarro et al. 2012). By plotting these results versus the salt concentration, the data can be collapsed into a single master curve by multiplying the [Div 2+] by a factor ~80 (Fig. 6c). The agreement between the results proves that the RNA hybridization free energy of the hairpin, i.e., the energy of the hairpin native conformation, satisfies (within errors) the 100:1 rule of thumb. Finally, a fit to data (dashed gray line) proved the validity of the logarithmic salt correction to the formation energy in Eq. (1).

5 RNA Folding Kinetics

The typical unzipping patterns of DNA and RNA hairpins are very different, as shown in Fig. 5a and b. To characterize the strong irreversibility observed in RNA, it has been hypothesized that stem-loop structures form along the two unpaired RNA strands during the unzipping (rezipping) process. As the unzipping reaction progresses, forming such structures close to the junction (that separates the native stem from the unpaired ssRNA) slows down hairpin hybridization. Consequently, the system gets trapped into off-pathway metastable conformations generating the observed hysteresis (Fig. 7a, top-left).

This scenario can be modeled by introducing the barrier energy landscape (BEL) (Rissone et al. 2022; Rissone and Ritort 2022) that includes all combinations of a

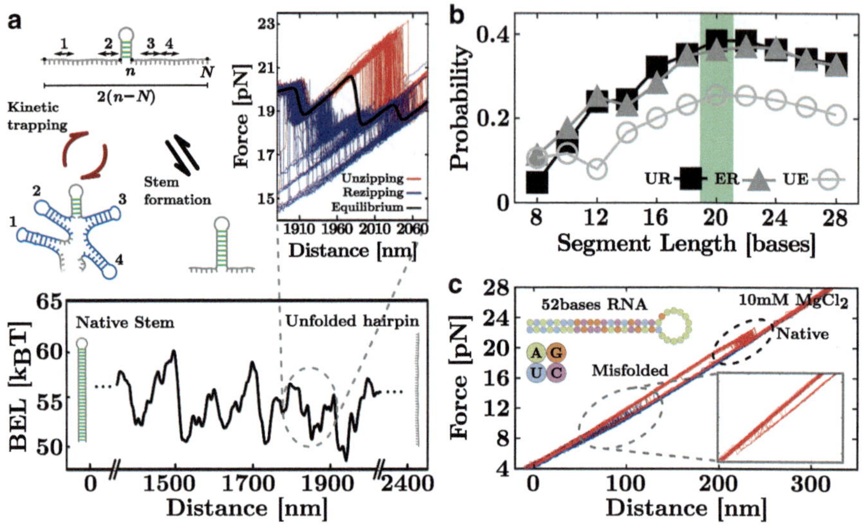

Fig. 7 RNA unzipping and rezipping kinetics. **a** Stem-loops BEL of a 2 kbp RNA hairpin. The stem-loops formation along the ssRNA drives the folding process by stabilizing off-pathway states that slow down the stem hybridization (top-left). The BEL correlates with the hysteresis measured along the FDC (bottom and top-right inset). Panel adapted from Rissone and Ritort (2022). **b** Correlation of the forming stem-loops with the UR, ER, and UE hysteresis (see text) as a function of stem-loops length, L. **c** Unzipping of a short (52 bases) RNA hairpin at 10 mM $MgCl_2$. The presence of divalent ions causes the molecule to misfold (inset)

number k of stem-loops ($k = 1, 2, \ldots K$, with K a maximum total number) that can form in the two unpaired RNA strands. Let n be the number of hybridized base pairs in a hairpin with a total number of N base pairs (Fig. 7a, top-left). Each unpaired strand contains N-n bases, and the two unpaired strands taken together contain 2(N-n) bases. Therefore, we can consider that stem-loops can form in a single unpaired strand of length 2(N-n). For a given number k of stem-loops of size L, we will consider them as randomly distributed along the 2(N-n) bases strand (Fig. A). For a given applied force f, the BEL is defined as

$$\Delta G_L(f) = -k_B T \log \sum_{k=0}^{K} \exp\left(-\frac{\Delta g_L(k, f)}{k_B T}\right), \tag{2}$$

where $\Delta g_L(k, f)$ is the total free-energy contribution of $k \geq 0$ stem-loops along the unpaired strand and $K = \lfloor 2(N-n)/L \rfloor$ is the maximum number of stem-loops (Rissone and Ritort 2022). Notice that for $k = 0$ no stem-loops are formed. In principle, stem-loops cannot overlap because only one structure can be formed with the same bases. For sake of simplicity, we will consider the approximated case where stem-loops can overlap.

The loop-BEL computed in Eq. (2) correlates with the amount of observed hysteresis (Fig. 7a, bottom and top-right inset), measured as the area between unzipping-rezipping (UR), unzipping-equilibrium (UE) and equilibrium-rezipping (ER) curves (Rissone et al. 2022). In particular, the study of this correlation as a function of the stem-loops size, L, shows that the stability of stem-loops increases with L reaching a maximum for $L \sim 20$ bases and gently decaying for larger sizes (Fig. 7b).

Stem-loops formation is a kinetic process that competes with native hybridization explaining the different behaviors observed in DNA and RNA unzipping. A main source of irreversibility in RNA is the higher stability of RNA stacking. In fact, the average NNBP RNA free-energy is $\langle \Delta g_0 \rangle^{\text{RNA}} \approx -2.2$ kcal/mol while for DNA $\langle \Delta g_0 \rangle^{\text{DNA}} \approx -1.7$ kcal/mol. This $\approx$0.5 kcal/mol difference could be sufficient for the RNA hairpin to slow down hybridization without the need for stem-loops forming along the unpaired strands making the loop-BEL unnecessary. However, there is mounting evidence that RNAs can form a multiplicity of structures as compared to DNA. In particular, in Rissone et al. (2022) it was found that a 52 bases native hairpin can also form alternative misfolded structures not predicted by secondary structure RNA numerical models. By competing with the native pathway of the 2.2kb RNA hairpin, the stem-loops slow down the hybridization reaction stabilizing the on-pathway intermediates in the FDC leading to the observed hysteresis.

Moreover, the observed irreversibility is enhanced in the presence of magnesium (Rissone et al. 2022). The metal ions induce higher flexibility to the RNA chain and higher stability to the RNA helices due to the coordination effect of the two positive charges and a reduction in the backbone's charge repulsion. An essential consequence of the reduced charge repulsion is that it allows more frequent close encounters between the different RNA segments, facilitating the formation of tertiary contacts (Tan and Chen 2009). Several studies pointed out that Mg^{2+} ions are more efficient than Na^+ in stabilizing RNA tertiary folds (Chu et al. 2007; Chen 2008; Draper 2008; Walter et al. 2008). The same phenomenon is not observed for DNA, where salt concentrations as high as 10 mM Mg^{2+} and 1 M Na^+ (equivalent concentrations as per the salt rule) induce similar foldings (Tan and Chen 2007). On the contrary, unzipping experiments at 10 mM Mg^{2+} on the previously mentioned 52 bases RNA hairpin showed that magnesium induces misfolding (Fig. 7c), whereas only the native conformation is present at the equivalent concentration of 1 M Na^+ (Rissone et al. 2022). In general, it has been observed that in mixed conditions of monovalent and divalent salt, low concentrations of the latter are sufficient to stabilize RNA tertiary structures in the presence of specific binding sites (Heilman-Miller et al. 2001).

6 Future Perspectives

Force spectroscopy is an exquisite tool to probe chemical interactions in biomolecules. The finely tuned balance between hydrogen bonding and stacking energies in nucleic acids makes them extremely sensitive to environmental changes, sequence mutations, and chemical modifications (Song and Yi 2017). Single-molecule fluorescence and nanopore microscopy are complementary tools that permit the detection and monitor conformational transitions at zero force. In contrast, force spectroscopy cannot probe molecular states at zero force because the end-to-end molecular extension cannot be detected. Therefore the combination of force with fluorescence (optical tweezers) (Whitley et al. 2017) and nanopores (optical trap nanopore) (Keyser et al. 2006; Trepagnier et al. 2007; Yuan et al. 2020) offers exciting prospects for the future.

Force spectroscopy permits the direct estimation of free-energy differences at room temperature by direct work measurements, $W = \Delta G$. A new direction of expansion is now possible using temperature-controlled optical tweezers (Mao et al. 2005; Mahamdeh and Schäffer 2009; De Lorenzo et al. 2015). Controlling temperature is often tricky, especially when the heating region is large compared to the typical micrometer-sized dimensions of the experimental trapping area. In this case, thermal expansion and convection effects lead to uncontrolled drift effects. In a recent setup (De Lorenzo et al. 2015), a heating laser can be switched on and off to controllably heat up the experimental trapping region without convection and drift effects. The instrument also operates at low temperatures by using an icebox kept at water-freezing temperatures (1–4 °C). This temperature-jump optical trap has been recently used to derive folding enthalpies, and entropies of DNA (De Lorenzo et al. 2015) and proteins (Rico-Pasto et al. 2022). The instrument permits measuring heat capacity changes, cold denaturation, and other previously inaccessible phenomena. Fascinating is the study of RNA at very low temperatures where thermal fluctuations are reduced, folding kinetics slowed down, and details of the molecular interactions intensified. By lowering the temperature, monitoring kinetics provides a natural microscope to amplify the finest energetic features driving RNA folding.

While DNA is often referred to as life's molecule (Frank-Kamenetskii 1993), RNA is the dark matter of the genome (Darnell 2011; Mattick and Amaral 2022), underlining how much we still do not know about this fascinating molecule. RNA presents such remarkable features that biophysical models must be refined to understand its many behaviors. We need physical models based on molecular energy landscapes to unravel the folding kinetics for RNA folding. Concepts borrowed from physics such as rugged free-energy landscapes (Kirkpatrick and Thirumalai 2015), and molecular replica symmetry breaking (Ritort 2022). Ideas from soft and condensed matter physics may come into play shortly to explain why RNA is so unique, in stark difference from the stable DNA counterpart. The critical question is understanding under which conditions DNA might behave as RNA. DNA lacks the reactive 2′–OH from ribose in RNA, impairing its catalytic activity. Yet, DNA might fold in some conditions as RNA does. Foreseeable experiments in the future are the study of folding kinetics of ssRNAs and differences with ssDNA (Viader-Godoy et al. 2021; Rissone and Ritort 2022). Other unexplored areas are awaiting discovery, such as RNAs with chemical modifications, RNAs at low temperatures, RNAs in crowded environments, etc. From custom-designed RNAs to biological RNAs (coding and non-coding), our knowledge of this fantastic molecule has never stopped growing and will be so for decades.

Acknowledgements P.R. was supported by the Angelo Della Riccia foundation. I. P. and F.R. were supported by Spanish Research Council Grant PID2019-111148GB-I00 and the Institució Catalana de Recerca i Estudis Avançats (F. R., Academia Prize 2018).

References

Aalto AP, Pasquinelli AE (2012) Small non-coding RNAs mount a silent revolution in gene expression. Curr Opin Cell Biol 24:333–340

Alemán EA, Lamichhane R, Rueda D (2008) Exploring RNA folding one molecule at a time. Curr Opin Chem Biol 12:647–654

Alemany A, Mossa A, Junier I et al (2012) Experimental free-energy measurements of kinetic molecular states using fluctuation theorems. Nat Phys 8:688

Bandarkar P, Yang H, Henley RY et al (2020) How nanopore translocation experiments can measure RNA unfolding. Biophys J 118:1612–1620

Bartley LE, Zhuang X, Das R et al (2003) Exploration of the transition state for tertiary structure formation between an RNA helix and a large structured RNA. J Mol Biol 328:1011–1026

Bennett CH (1976) Efficient estimation of free energy differences from Monte Carlo data. J Comput Phys 22:245–268

Bizarro CV, Alemany A, Ritort F (2012) Non-specific binding of Na+ and Mg2+ to RNA determined by force spectroscopy methods. Nucleic Acids Res 40:6922–6935

Blaszczyk L, Rypniewski W, Kiliszek A (2017) Structures of RNA repeats associated with neurological diseases. Wiley Interdiscip Rev RNA 8:e1412

Bokinsky G, Rueda D, Misra VK et al (2003) Single-molecule transition-state analysis of RNA folding. Proc Natl Acad Sci USA 100:9302–9307

Bošković F, Keyser U (2022) Nanopore microscope identifies RNA isoforms with structural colours. Nat Chem 1–7

Bowman JC, Lenz TK, Hud NV et al (2012) Cations in charge: magnesium ions in RNA folding and catalysis. Curr Opin Struct Biol 22:262

Breslauer KJ, Frank R, Blöcker H et al (1986) Predicting DNA duplex stability from the base sequence. Proc Natl Acad Sci USA 83:3746–3750

Brion P, Westhof E (1997) Hierarchy and dynamics of RNA folding. Annu Rev Biophys 26:113–137

Bustamante C, Marko J, Siggia E et al (1991) Entropic elasticity of λ-phage DNA. Proc Nati Acad Sci USA 88:10009

Bustamante CJ, Chemla YR, Liu S et al (2021) Optical tweezers in single-molecule biophysics. Nat Rev Methods Primers 1:1–29

Butcher SE, Pyle AM (2011) The molecular interactions that stabilize RNA tertiary structure: RNA motifs, patterns, and networks. Acc Chem Res 44:1302–1311

Butler TZ, Gundlach JH, Troll M (2007) Ionic current blockades from DNA and RNA molecules in the α-hemolysin nanopore. Biophys J 93:3229–3240

Chauvier A, Cabello-Villegas J, Walter NG (2019) Probing RNA structure and interaction dynamics at the single molecule level. Methods 162:3–11

Chen SJ (2008) RNA folding: conformational statistics, folding kinetics, and ion electrostatics. Annu Rev Biophys 37:197

Chen SJ, Dill KA (2000) RNA folding energy landscapes. Proc Natl Acad Sci USA 97:646–651

Chu VB, Bai Y, Lipfert J, Herschlag D et al (2007) Evaluation of ion binding to DNA duplexes using a size-modified Poisson-Boltzmann theory. Biophys J 93:3202–3209

Cole P, Yang S, Crothers D (1972) Conformational changes of transfer ribonucleic acid. Equilibrium phase diagrams. Biochem 11:4358–4368

Collin D, Ritort F, Jarzynski C et al (2005) Verification of the Crooks fluctuation theorem and recovery of RNA folding free energies. Nature 437:231–234

Cruz JA, Westhof E (2009) The dynamic landscapes of RNA architecture. Cell 136:604–609
Cui M, Ge Y, Zhuge X et al (2021) Recent advances in nanopore sensing. Chinese J Chem 39:2035–2043
Darnell JE (2011) RNA: life's indispensable molecule. Cold Spring Harbor Laboratory Press
De Lorenzo S, Ribezzi-Crivellari M, Arias-Gonzalez JR et al (2015) A temperature-jump optical trap for single-molecule manipulation. Biophys J 108:2854–2864
DeVoe H, Tinoco I Jr (1962) The stability of helical polynucleotides: base contributions. J Mol Biol 4:500–517
Draper DE (2008) Rna folding: thermodynamic and molecular descriptions of the roles of ions. Biophys J 95:5489–5495
Draper DE, Grilley D, Soto AM (2005) Ions and RNA folding. Annu Rev Biophys 34:221–243
Esteller M (2011) Non-coding RNAs in human disease. Nat Rev Genet 12:861–874
Fatica A, Bozzoni I (2014) Long non-coding RNAs: new players in cell differentiation and development. Nat Rev Genet 15:7–21
Filipowicz W (2022) RNA, the epicenter of genetic information: wirtten by John Mattick and Paulo Amaral. CRC Press, Boca Raton
Frank-Kamenetskii MD (1993) Unraveling DNA. VCH Publishers, Inc
Freier SM, Kierzek R, Jaeger JA et al (1986) Improved free-energy parameters for predictions of RNA duplex stability. Proc Natl Acad Sci USA 83:9373–9377
Gralla J, Delisi C (1974) Biological sciences: mRNA is expected to form stable secondary structures. Nature 248:330–332
Ha T (2001) Single-molecule fluorescence resonance energy transfer. Methods 25:78–86
Ha T, Zhuang X, Kim HD et al (1999) Ligand-induced conformational changes observed in single RNA molecules. Proc Natl Acad Sci USA 96:9077–9082
Harlepp S, Marchal T, Robert J, Leger J et al (2003) Probing complex RNA structures by mechanical force. Eur Phys J E 12:605–615
Heilman-Miller SL, Thirumalai D, Woodson SA (2001) Role of counterion condensation in folding of the Tetrahymena ribozyme. I. Equilibrium stabilization by cations. J Mol Biol 306:1157–1166
Henley RY, Carson S, Wanunu M (2016) Studies of RNA sequence and structure using nanopores. Prog Mol Biol Transl Sci 139:73–99
Herschlag D, Bonilla S, Bisaria N (2018) The story of RNA folding, as told in epochs. Cold Spring Harb Perspect Biol 10:a032433
Huguet JM, Bizarro CV, Forns N et al (2010) Single-molecule derivation of salt dependent base-pair free energies in DNA. Proc Natl Acad Sci USA 107:15431–15436
Huguet JM, Ribezzi-Crivellari M, Bizarro CV et al (2017) Derivation of nearest-neighbor DNA parameters in magnesium from single molecule experiments. Nucleic Acids Res 45:12921–12931
Jain A, Vale RD (2017) Rna phase transitions in repeat expansion disorders. Nature 546:243–247
Jarzynski C (1997) Nonequilibrium equality for free energy differences. Phys Rev Lett 78:2690
Joo C, Balci H, Ishitsuka Y et al (2008) Advances in single-molecule fluorescence methods for molecular biology. Annu Rev Biochem 77:51–76
Junier I, Mossa A, Manosas M, Ritort F (2009) Recovery of free energy branches in single molecule experiments. Phys Rev Lett 102:070602
Keyser U, Van der Does J, Dekker C, Dekker N (2006) Optical tweezers for force measurements on DNA in nanopores. Rev Sci Instrum 77:105105
Kirkpatrick T, Thirumalai D (2015) Colloquium: Random first order transition theory concepts in biology and physics. Rev Mod Phys 87:183
Lakowicz JR (2006) Principles of fluorescence spectroscopy. Springer
Lee DH, Oh S, Lim K et al (2021) Tertiary RNA folding-targeted drug screening strategy using a protein nanopore. Anal Chem 93:2811–2819
Liphardt J, Onoa B, Smith SB et al (2001) Reversible unfolding of single RNA molecules by mechanical force. Science 292:733–737
Loh E, Kugelberg E, Tracy A et al (2013) Temperature triggers immune evasion by Neisseria meningitidis. Nature 502:237–240

Ma L, Bajic VB, Zhang Z (2013) On the classification of long non-coding RNAs. RNA Biol 10:924–933

Mahamdeh M, Schäffer E (2009) Optical tweezers with millikelvin precision of temperature-controlled objectives and base-pair resolution. Opt Express 17:17190–17199

Manosas M, Ritort F (2005) Thermodynamic and kinetic aspects of RNA pulling experiments. Biophys J 88:3224–3242

Manosas M, Wen JD, Li PT et al (2007) Force unfolding kinetics of RNA using optical tweezers. II. Modeling experiments. Biophys J 92:3010–3021

Mao H, Arias-Gonzalez JR, Smith SB et al (2005) Temperature control methods in a laser tweezers system. Biophys J 89:1308–1316

Martinez-Monge A, Pastor I, Bustamante C et al (2022) Measurement of the specific and non-specific binding energies of mg2+ to RNA. Biophys J 121:3010–3022

Mathews DH, Sabina J, Zuker M et al (1999) Expanded sequence dependence of thermodynamic parameters improves prediction of RNA secondary structure. J Mol Biol 288:911–940

Matsui M, Corey DR (2017) Non-coding RNAs as drug targets. Nat Rev Drug Discov 16:167–179

Mattick J, Amaral P (2022) RNA, the epicenter of genetic information. CRC Press

Mattick J, Makunin I (2006) Non-coding. RNA. Hum Mol Genet 15:R17–R29

Meller A, Mathe J, Wanunu M, et al (2006) DNA and RNA unzipping using nanopore force spectroscopy. In: APS march meeting abstracts, pp B26–013

Mercer TR, Dinger ME, Mattick JS (2009) Long non-coding RNAs: insights into functions. Nat Rev Genet 10:155–159

Mikulecky PJ, Feig AL (2002) Cold denaturation of the hammerhead ribozyme. J Am Chem Soc 124:890–891

Onoa B, Dumont S, Liphardt J et al (2003) Identifying kinetic barriers to mechanical unfolding of the T. Thermophila Ribozyme. Science 299:1892–1895

Pyle A (2002) Metal ions in the structure and function of RNA. J Biol Inorg Chem 7:679–690

Ray S, Widom JR, Walter NG (2018) Life under the microscope: single-molecule fluorescence highlights the RNA world. Chem Rev 118:4120–4155

Rico-Pasto M, Zaltron A, Davis SJ et al (2022) Molten globule-like transition state of protein barnase measured with calorimetric force spectroscopy. Proc Natl Acad Sci USA 119:e2112382119

Rissone P, Ritort F (2022) Nucleic acid thermodynamics derived from mechanical unzipping experiments. Life 12:1089

Rissone P, Bizarro CV, Ritort F (2022) Stem–loop formation drives RNA folding in mechanical unzipping experiments. Proc Natl Acad Sci USA 119

Ritchie DB, Woodside MT (2015) Probing the structural dynamics of proteins and nucleic acids with optical tweezers. Curr Opin Struct Biol 34:43–51

Ritort F (2006) Single-molecule experiments in biological physics: methods and applications. J Phys Condens Matter 18:R531

Ritort F (2022) Molecular replica symmetry breaking. In: Spin glass theory and far beyond, World Scientific

Russell R, Millett IS, Tate MW et al (2002) Rapid compaction during RNA folding. Proc Natl Acad Sci USA 99:4266–4271

Saenger W (1984) Principles of nucleic acid structure. Springer

Schmitz AG, Zelger-Paulus S, Gasser G et al (2015) Strategy for internal labeling of large RNAs with minimal perturbation by using fluorescent PNA. ChemBioChem 16:1302–1306

Seidel R, Dekker C (2007) Single-molecule studies of nucleic acid motors. Curr Opin Struct Biol 17:80–86

Serganov A, Bénard L, Portier C et al (2001) Role of conserved nucleotides in building the 16 s rRNA binding site for ribosomal protein S15. J Mol Biol 305:785–803

Severino A, Monge AM, Rissone P et al (2019) Efficient methods for determining folding free energies in single-molecule pulling experiments. J Stat Mech: Theory Exp 2019:124001

Shi K, Pan B, Sundaralingam M (2003) Structure of a B-form DNA/RNA chimera (dC)(rG) d (ATCG) complexed with daunomycin at 1.5 å resolution. Acta Crystallogr D 59:1377–1383

Shirts MR, Bair E, Hooker G et al (2003) Equilibrium free energies from nonequilibrium measurements using maximum-likelihood methods. Phys Rev Lett 91:140601
Song J, Yi C (2017) Chemical modifications to RNA: a new layer of gene expression regulation. ACS Chem Biol 12:316–325
Statello L, Guo CJ, Chen LL et al (2021) Gene regulation by long non-coding RNAs and its biological functions. Nat Rev Mol 22:96–118
Stefani G, Slack FJ (2008) Small non-coding RNAs in animal development. Nat Rev Mol 9:219–230
Sultan M, Kanavarioti A (2019) Nanopore device-based fingerprinting of RNA oligos and micrornas enhanced with an osmium tag. Sci Rep 9:1–18
Tan ZJ, Chen SJ (2007) RNA helix stability in mixed Na+/Mg2+ solution. Biophys J 92:3615–3632
Tan ZJ, Chen SJ (2009) Predicting electrostatic forces in RNA folding. In: Methods in enzymology, vol 469. Elsevier, pp 465–487
Treiber DK, Williamson JR (2001) Beyond kinetic traps in RNA folding. Curr Opin Struct Biol 11:309–314
Trepagnier EH, Radenovic A, Sivak D et al (2007) Controlling DNA capture and propagation through artificial nanopores. Nano Lett 7:2824–2830
Viader-Godoy X, Pulido C, Ibarra B et al (2021) Cooperativity-dependent folding of single-stranded DNA. Phys Rev X 11:031037
Walter AE, Turner DH, Kim J et al (1994) Coaxial stacking of helixes enhances binding of oligoribonucleotides and improves predictions of RNA folding. Proc Natl Acad Sci USA 91:9218–9222
Walter N, Woodson SA, Batey RT (2008) Non-protein coding RNAs, vol 13. Springer Science & Business Media
Wen JD, Manosas M, Li PT et al (2007) Force unfolding kinetics of RNA using optical tweezers. I. Effects of experimental variables on measured results. Biophys J 92:2996–3009
White KH, Visscher K (2011) Optical trapping and unfolding of RNA. In: Single molecule analysis. Springer, pp 21–43
Whitley KD, Comstock MJ, Chemla YR (2017) High-resolution "fleezers": dual-trap optical tweezers combined with single-molecule fluorescence detection. In: Optical Tweezers. Springer, pp 183–256
Woodside MT, Anthony PC, Behnke-Parks WM et al (2006) Direct measurement of the full, sequence-dependent folding landscape of a nucleic acid. Science 314:1001–1004
Woodson SA (2005) Metal ions and RNA folding: a highly charged topic with a dynamic future. Curr Opin Chem Biol 9:104–109
Woodson SA (2010) Compact intermediates in RNA folding: annual reviews in biophysics. Annu Rev Biophys 39:61
Wu YJ, Wu CH, Yeh AYC et al (2014) Folding a stable RNA pseudoknot through rearrangement of two hairpin structures. Nucl Acids Res 42:4505–4515
Xia T, SantaLucia J Jr, Burkard ME et al (1998) Thermodynamic parameters for an expanded nearest-neighbor model for formation of RNA duplexes with watson-crick base pairs. Biochemistry 37:14719–14735
Xie Z, Srividya N, Sosnick TR et al (2004) Single-molecule studies highlight conformational heterogeneity in the early folding steps of a large ribozyme. Proc Natl Acad Sci USA 101:534–539
Yuan Z, Liu Y, Dai M et al (2020) Controlling DNA translocation through solid-state nanopores. Nanoscale Res Lett 15:1–9
Zhao R, Rueda D (2009) Rna folding dynamics by single-molecule fluorescence resonance energy transfer. Methods 49:112–117
Zhao X, Usdin K (2021) (DYS) function follows form: nucleic acid structure, repeat expansion, and disease pathology in FMR1 disorders. Int J Mol Sci 22:9167
Zhuang X (2005) Single-molecule RNA science. Annu Rev Biophys Biomol Struct 34:399–414
Zuker M (2003) Mfold web server for nucleic acid folding and hybridization prediction. Nucl Acids Res 31:3406–3415

Structured RNAs and Their Role in Biology and Therapeutics

Bogdan I. Fedeles and Vipender Singh

Contents

Abstract RNAs can form complex 3D structures to influence biology and disease. Small self-cleaving ribozymes and riboswitches are some well-characterized examples of structured RNAs, functional RNA sequences with 3D structures. In this chapter, we will discuss the structural features of hammerhead, hairpin, *glmS*, and twister small self-cleaving ribozymes that are relevant for their biological function, and specific biochemical studies that help elucidate the mechanisms of their self-cleavage reactions. We will also discuss the structural elements of the bacterial purine and thiamine pyrophosphate riboswitches responsible for recognizing specific ligands. Binding to their cognate ligands is an essential step in the regulation of gene expression by the riboswitches. Structured RNAs have also been targeted for developing drugs such as Ribocil and Risdiplam/Branaplam. These drugs are

B. I. Fedeles
Department of Chemistry, Department of Biological Engineering, and Center for Environmental Health Sciences, Massachusetts Institute of Technology, Cambridge, MA 02139, USA

V. Singh (✉)
Department of Biochemistry and Biophysics, Cardiovascular and Metabolic Diseases, Novartis Institute of Biomedical Research, MA 02139 Cambridge, USA
e-mail: vipender.singh@novartis.com

J. Barciszewski (ed.), *RNA Structure and Function*, RNA Technologies 14,
https://doi.org/10.1007/978-3-031-36390-0_5

notable examples of approved therapies for bacterial infections and spinal muscular atrophy, respectively, that target the RNA structures. In this chapter, we will discuss the targeting potential of riboswitches for developing antibacterial therapy and the mechanism of Ribocil recognition by the FMN riboswitch.

Keywords Ribozymes · Riboswitches · RNA structure · Pyrithiamine · Ribocil · RNA therapeutics

1 Introduction

RNA sequences can adopt complex 3D structures, despite the limited chemical repertoire of four nucleic acid bases, ribose sugar, and a phosphodiester backbone. Structured RNAs regulate a wide array of biological processes such as replication of RNA genomes, RNA splicing, protein synthesis, and regulation of gene expression in response to metabolic sensing. Small self-cleaving ribozymes and bacterial riboswitches are well-characterized examples of structured RNAs. Self-cleaving ribozymes, found in the sub-plant viruses, form structures that resemble proteins in their complexity, and can accelerate their cleavage and ligation reactions at rates that are comparable to protein enzymes. Bacterial riboswitches are another class of structured non-coding RNAs that can influence gene expression in response to ligand binding. For example, the thiamine pyrophosphate (TPP) riboswitch regulates the expression of genes involved in the thiamine metabolism in response to binding to the thiamine pyrophosphate ligand. The purine riboswitch regulates genes involved in the purine biosynthesis in response to binding to purines such as guanine and adenine. In this chapter, we will discuss structural biology and mechanisms of the autolytic reactions catalyzed by self-cleaving ribozymes and ligand recognition by the purine and the TPP riboswitches. Structured RNA has been targeted for developing therapeutics. For example, small-molecule drug candidates Risdiplam and Branaplam have been developed to bind the RNA-protein complex of the splicing machinery and induce alternative pre-mRNA splicing. Another promising direction has been the development of new antibiotics that target essential bacterial riboswitches. This chapter summarizes the challenges of this field and the importance of structural data that informs drug design. Ribocil, an approved antibacterial drug that targets the FMN riboswitch is discussed in more detail as a case study.

2 Structure and Catalytic Function of Small Self-Cleaving Ribozymes

Small autolytic ribozymes catalyze nucleolytic intramolecular self-cleavage reactions. The well-studied examples of small self-cleaving ribozymes are hammerhead, hairpin, *glmS*, Varkud Satellite (VS), and the twister ribozymes. These ribozymes have sizes typically in the range of 50–150 nucleotides and reside in genomes from sub-viral plant pathogens to eukaryotes. Small autolytic ribozymes catalyze cleavage of their own phosphodiester backbone. The self-cleavage reaction in most small ribozymes is initiated by a basic residue, typically the N1 of a catalytic guanosine, which activates a key 2′-hydroxyl nucleophile adjacent to the scissile phosphate (Fig. 1a). Twister ribozyme is an exception to this step; instead, it is proposed to utilize structural changes for the activation and positioning of the 2′-hydroxyl nucleophile in the in-line conformation to attack the phosphodiester backbone (Fig. 1d). The activation is followed by an internal transesterification reaction, in which the nucleophilic 2′-oxygen attacks the adjacent scissile 3′-phosphate to form two pieces of RNA, one containing the 2′, 3′-cyclic phosphate and the other the 5′-OH functional group (Fig. 1a). Except for the *glmS* ribozyme, small autolytic ribozymes do not require the participation of an external coenzyme to catalyze the self-cleavage reaction (Fig. 1b). Their reaction mechanism is similar to ribonucleoprotein RNase A, and it involves stabilization of bipyramidal oxyphosphorane transition state through different catalytic strategies, such as in-line atomic orientation, electrostatic neutralization, general base, and general acid catalysis (Singh et al. 2015). In the subsequent sections, we will discuss catalytic strategies employed by small self-cleaving ribozymes in the context of their active site structures.

2.1 Hammerhead Ribozyme

The Hammerhead ribozyme (HHR) was originally discovered in the satellite RNA of tobacco ringspot viruses. These viruses have a single stranded circular RNA genome (Prody et al. 1986). They replicate via a rolling-circular mechanism to generate linear concatemeric copies of the satellite RNA genome, which are subsequently cleaved into monomeric fragments. The cleavage points are highly specific at regular intervals and are embedded within the hammerhead RNA motifs. The autolytic reaction catalyzed by the HHR at the cleavage points resolves the monomeric fragments. Since its discovery in the above-mentioned viruses, HHR has been found in several other plant viruses or virus-like genomes involved in the rolling-circle replication (Symons 1997; Hutchins et al. 1986). Recent bioinformatics searches have revealed that the HHR sequences are present in the satellite DNA of newts and salamanders (Epstein and Gall 1987; Zhang and Epstein 1996), eukaryotic genomes of plants species like carnation (Daròs and Flores 1995), and *Arabidopsis thaliana* (Przybilski et al. 2005), and invertebrates like *S. mansoni* (Ferbeyre et al. 1998) and cave

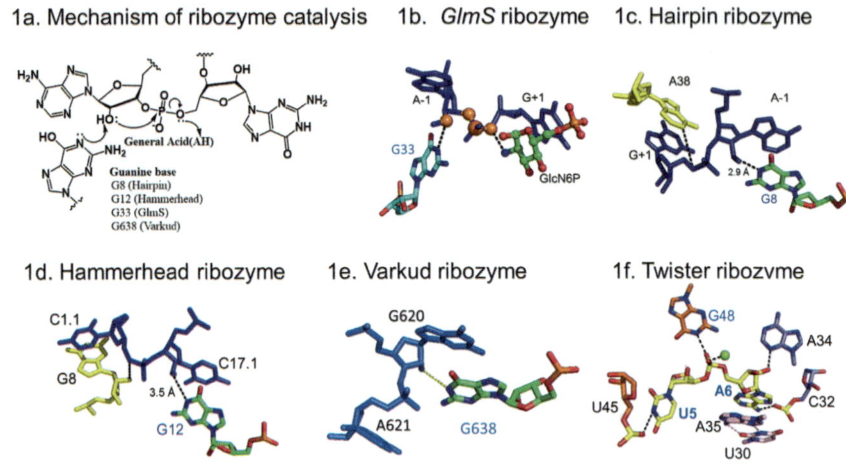

Fig. 1 Mechanism of the nucleolytic reaction catalyzed by small self-cleaving ribozymes (Singh et al. 2015). **a** Acid-base catalytic mechanism of small self-cleaving ribozymes and the positions of participating catalytic guanosines in the active sites of the respective ribozymes. **b** Catalytic guanosine (G33) in the active site of the *glmS* ribozyme is in close proximity to the 2′-hydroxyl of the A-1 nucleotide adjacent to the scissile phosphodiester bond. **c** The N1 of catalytic guanosine (G8) in the hairpin ribozyme is in close proximity to the 2′-hydroxyl of the A-1 nucleotide. **d** The N1 of catalytic guanosine (G12) in the hammerhead ribozyme is in close proximity to the 2′-hydroxyl of the A-1 nucleotide. **e** The N1 of the catalytic guanosine (G698) in the Varkud Satellite ribozyme is in close proximity to the 2′-hydroxyl of A-1 nucleotide. **e** was shared by Joe Piccirilli's laboratory at the University of Chicago (Suslov et al. 2015). **a–e** is adapted from reference (Singh et al. 2014) (CC BY 4.0), with permission from the RNA Society of Cold Spring Harbor Laboratory Press. **f** Displays the in-line alignment of the U5-A6 in the crystal structure of the *env22* twister ribozyme. **f** is adapted from (Ren et al. 2014) (CC BY 4.0)

crickets (Rojas et al. 2000). Deeper analysis of the genomic HHRs in plants showed that they are part of a new family of mobile genetic elements named retrozymes, non-autonomous retroelements with HHRs (Cervera et al. 2016). Variants of HHR are also found in the ancient family of Penelope-like elements (PLEs), a group of eukaryotic retrotransposons regarded as exceptional for encoding telomerase-like retrotranscriptases and spliceosomal introns (Cervera and De la Peña 2014).

Crystal structure of the full-length HHR from *S. mansoni* revealed that tertiary interaction in the outer regions of the RNA primes the ribozyme for catalysis (Figs. 1d and 2a) (Martick and Scott 2006). The full-length ribozyme has a γ-shaped fold, formed from stem II that aligns coaxially with stem III (Fig. 2a). The remaining portion is composed of the extended stem I and II which rejoins in an inverted lowercase γ, creating a helical bubble-like structure (Fig. 2a). The full-length structure also showed that the tertiary contacts between stem I and stem II organize the catalytic core (Fig. 2a) of the ribozyme. The organization of the catalytic core allows better alignment of the 2′-O nucleophile with the scissile phosphate to facilitate the self-cleavage reaction (Martick and Scott 2006).

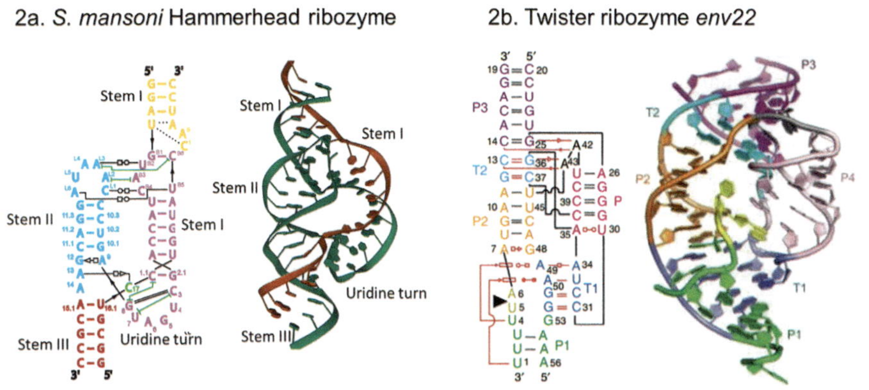

Fig. 2 Secondary and ternary structures of the hammerhead ribozyme (HHR) from *S. mansoni* and the *env22* twister ribozyme. **a** Left; Secondary structure of the hammerhead ribozyme from *S. mansoni*. **a** Right, Ternary structure of the HHR in the same orientation as the denoted in the secondary structure on the left. (**a**) was adapted from a reference (Martick and Scott 2006) and recreated using PyMol. **b** Left, Secondary structure of the Twister ribozyme *env22*. **b** Right, Ternary structure of the Twister *env22* ribozyme. The figure adapted from the reference (Ren et al. 2014)

The structural and the biochemical mechanism of HHR catalyzed self-cleavage reaction is well studied (Fig. 1d) (Chi et al. 2008; McKay 1996; Thomas and Perrin 2009). In the self-cleavage reaction, the N1 of the catalytic G12 residue acts as a general base to activate the 2′-OH of the residue at position 17 to generate the attacking nucleophile (McKay 1996; Thomas and Perrin 2008; Han and Burke 2005). Significant biochemical data support the role of G12 as a general base in the autolytic reaction. The G12 base shows a log-linear relationship of its activity with pH that does not plateau even at the highest pH, consistent with general base catalysis (Han and Burke 2005). Substitution of G12 with any other canonical nucleobase reduces the rate of the reaction by a factor of 10^2 to 10^5 (Chi et al. 2008; Ruffner et al. 1990). Adenine substitution results in a maximum reduction by a factor of 10^5 (Ruffner et al. 1990; Chi et al. 2008). The reduction in the rate of reaction, from G12A substitution, occurs despite better alignment between the N1 of A with the 2′-OH nucleophile (Singh et al. 2015). The distance between the N1 of G and the nucleophilic 2′-OH is 3.5 Å for G compared to 2.5 Å for A (Chi et al. 2008). A shorter distance for A, which contains the unprotonated N1, suggests that the deprotonated N1 is likely a preferred general base. However, the faster rate for G, relative to A, indicates that the electronic properties, namely, the pK_a of the N1 of G, drives the observed rate enhancement. The G12A substitution shifts the pK_a from ~9.5 to ~3.5 (Chi et al. 2008). In addition, substituting G12 with purine analogs, such as inosine, diaminopurine, or 2-aminopurine, that specifically change the pK_a of the N1 of G which alters the overall rate of the self-cleavage reaction. A maximum reduction of 10^3 was observed for 2-aminopurine (Han and Burke 2005; Singh et al. 2015). These data are consistent with the direct role of the N1 of G in the activation of the nucleophilic 2′-OH group. Additionally, the replacement of 2′-hydroxyl nucleophile with the electrophilic 2′-bromoacetamide group results in alkylation of the N1 position of G12 in a pH and Mg^{2+}-dependent manner suggesting that the N1 of G12 is the most basic functional group in the vicinity of the nucleophilic 2′-hydroxyl group (Thomas and Perrin 2008). The second half of the reaction involves protonation of leaving group oxygen by a general acid. Structural and biochemical data suggest that the 2′-OH of G8 is the general acid in the reaction (Han and Burke 2005). Altogether, these structural interactions in the catalytic core of the HHR ribozyme facilitate the self-cleavage reaction.

The unprotonated N1 of G12 is physiochemically better suited to participate in the reaction as a general base, which can be an anionic G form, generated through deprotonation at N1, or a minor tautomer of G, generated by the subsequent protonation of the anionic G at another site (Singh et al. 2015). It remains to be determined the extent to which the anionic or the minor tautomeric form of the catalytic guanosine participates in the catalytic step of the cleavage reaction (Singh et al. 2015). Taken together, the biochemical and structural data are consistent with a mechanism in which the N1 of the G12 residue directly participates in the self-cleavage reaction as a general base. The chemical mechanism of HHR catalyzed self-cleavage reaction is very similar to the mechanisms of other small self-cleaving ribozymes discussed below.

2.2 *Hairpin Ribozyme*

Hairpin ribozyme is another example of small self-cleaving ribozyme discovered in satellite RNAs of plant viruses such as Arabis mosaic virus (sArMV), tobacco ringspot virus (sTRSV), and chicory yellow mottle virus type 1 (sCYMV1) (Haseloff and Gerlach 1989; DeYoung et al. 1995). The single stranded viral genomes of these viruses replicate through a rolling-circle mechanism that creates a concatemer of linear genome sequences (Feldstein et al. 1989). The hairpin ribozyme utilizes its self-cleavage activity to resolve the product of rolling-circle replication and generate individual linear genomes. The subsequent ligation reaction produces the single stranded circular RNA genomes. The self-cleavage reaction is sequence specific and generates RNA products with 2′, 3′-cyclic phosphate and a 5′-OH terminus (Buzayan et al. 1986).

The mechanism of the hairpin ribozyme-catalyzed self-cleavage reaction is well studied. The cleavage reaction is initiated by the activation of a 2′-OH nucleophile by a catalytic guanosine. The G8 in the hairpin ribozyme is important for catalysis and acts as a general base (Fig. 1A and Fig. 1C) (Rupert and Ferré-D'Amaré 2001). The N1 of G8 is within the hydrogen binding distance from the 2′-OH (Rupert and Ferré-D'Amaré 2001). Substitution of G8 with an abasic site results in 1000-fold reduction in the rate of self-cleavage reaction, without altering the pH dependence of the reaction (Kuzmin et al. 2004). Its substitution with inosine or 2, 6-diaminopurine, which alters the pKa of its N1, changes the pH profile of the self-cleavage reaction (Pinard et al. 2001). The G8 residue participates in the reaction as a general base either in its canonical form or in the alternative form in which its N1 is not protonated. As mentioned above, the alternative ionized or minor tautomeric form of G, with unprotonated N1, can be achieved by realignment of hydrogen bonding or proton transfer, respectively (Singh et al. 2015; Kuzmin et al. 2004; Pinard et al. 2001).

Structural and biochemical studies have shown that the hairpin ribozyme utilizes a variety of catalytic strategies for its cleavage and ligation reactions (Cochrane and Strobel 2008a; Fedor 2000). These include precise substrate orientation, preferential transition state binding (Rupert and Ferré-D'Amaré 2001; Rupert et al. 2002), electrostatic catalysis, and general acid-base catalysis. The hairpin ribozyme can also function in *trans* in which a truncated 50 nucleotides, a minimal ribozyme, can catalyze the cleavage of the 14 nucleotides substrate (Hampel and Tritz 1989). Structurally, the hairpin ribozyme is formed from two helical stacks. The active site is in the cleft formed with contribution of nucleotides from both the helices (Fedor 2000). The nucleotide 3′- of the cleavage site, which is G + 1, is flipped out of its helix and is stacked between A26 and A30 in a pocket of the other helix (Cochrane and Strobel 2008a; Fedor 2000) (Fig. 1c). The G + 1 nucleotide is stabilized through the formation of a tertiary interaction with W-C base pair with C25 and additional hydrogen bonding to G36 and A38. These interactions are specific for G at the G + 1 position. Mutation of G + 1 with any of the other nucleotides results in a loss of catalytic activity, which can be partially restored by a compensatory mutation of C25 (Fedor 2000). The nucleotide 5′ of the cleavage site, A-1 stacks on top of G8. It forms a hydrogen bond with the exocyclic amine of A9. Structural studies have also shown that G8 allows the formation of an in-line conformation of the reactive groups for the self-cleavage reaction to occur. Its substitution perturbs this in-line alignment and alters the sugar pucker of A-1. G8 plays an essential role in organizing the active site for catalysis alongside A9, C25, A26, and A38 nucleotides (Cochrane and Strobel 2008a; Fedor 2000). The overall catalytic mechanism of the hairpin ribozyme is similar to the mechanisms of other self-cleaving ribozymes discussed in this chapter.

2.3 *Varkud Ribozyme*

The Varkud Satellite (VS) ribozyme is the largest among the small nucleolytic ribozymes. It was originally found in the mitochondrial satellite plasmid of *Neurospora* species. Its function is to process and resolve replication intermediates of the rolling-circle replication (Griffiths 1995). In mitochondria of the *Neurospora* species, transcription of 881 nucleotides long VS plasmid produces a multimeric RNA containing self-cleaving motifs or sites which are embedded in the ribozyme. The processing of intermediates at the self-cleavage motifs during rolling-circle replication, and their subsequent ligation generate circular monomeric RNAs from a multimeric transcript (Saville and Collins 1990).

The VS ribozyme catalyzes reversible site-specific cleavage and ligation reactions (Saville and Collins 1990). The ribozyme is folded into seven helical segments organized by a three-way helical junction (Suslov et al. 2015). The ribozyme core consists of helices 2–6 plus the substrate helix, the helix 1. Within each active site, the catalytic adenine and guanine nucleobases are positioned around the scissile phosphate to participate in acid-base catalysis (Fig. 1e). Crystal structure of the full-length VS ribozyme was solved at 3.1–3.3 Å resolution (Suslov et al. 2015). The structure revealed that the ribozyme is an intertwined dimer formed by an exchange of substrate helices. The exchange allows the substrate helix of one protomer to be docked into the catalytic helix of the second protomer (Suslov et al. 2015).

The VS ribozyme is *trans* acting and behaves much like an enzyme-substrate complex formed by protein enzymes. Biochemical data suggests that the site-specific adenosine (A756) and guanosine (G638) participate in acid-base catalysis (Fig. 1e). Site-directed mutagenesis of A756 with any other base results in a 300-fold or more reduction of its catalytic activity, whereas G638 to A substitution results in a 10^4 lower rate (Lafontaine et al. 2001; Sood and Collins 2002). Deletion of A756 results in 900-fold reduction in its activity, without affecting the ribozyme folding, suggesting that the A756 participates in the ribozyme catalysis (Lafontaine et al. 2001, 2002). A direct interaction between the A756 and the cleavage site was also established by the cross-linking studies using 4-thiouridine (Hiley et al. 2002). Nucleotide analog interference mapping (NAIM) studies also revealed that the ionization of A756 directly influences the VS ribozyme-catalyzed ligation reaction (Jones and Strobel 2003). Biochemical studies also suggest that Mg^{2+} ion is not needed in the autolytic reaction; instead, it plays a structural role in the VS catalysis (Maguire and Collins 2001; Lilley 2019; Murray et al. 1998). Strong ionic environments in the form of molar concentrations of monovalent cations Li^+ and NH_4^+ are shown to be sufficient for the ribozyme function (Lilley 2019; Murray et al. 1998). Furthermore, it was shown that addition of positively charged polyamines such as can significantly enhance the rate of VS catalyzed reaction (Olive and Collins 1998). Mutation complementation studies identified this spermine-enhanced cleavage as a *trans* reaction resulting from a symmetric association between two complete ribozymes (Olive and Collins 1998). Biochemical and structural studies have significantly enhanced our understanding of the VS ribozyme-catalyzed cleavage reaction.

2.4 *Twister Ribozyme*

Twister along with Twister-sister, Pistol, and Hatchet self-cleaving ribozymes were identified by bioinformatic searches of non-coding RNA databases using comparative genomic approaches (Roth et al. 2014; Weinberg et al. 2015). The sequences of twister ribozymes are widely distributed in bacteria, fungi, plants, and animals (Weinberg et al. 2015). These RNA sequences display self-cleavage activities both in vivo and in vitro that are comparable to the nucleolytic rates of other self-cleaving ribozymes (Liu et al. 2014; Ren et al. 2014). Although the biological function of Twister ribozymes is not understood, given their distribution in the non-coding regions of the genomes, they likely play a role in genetic regulation (Liu et al. 2014; Ren et al. 2014).

Comparison of nearly 2,700 twister ribozymes identified common secondary structure elements (Fig. 2b) (Ren et al. 2014). The core of the ribozyme is made up of a stem-loop interrupted by two internal loops, with a cleavage site located within loop L1 (Roth et al. 2014). Two ternary interactions, T1 and T2, fold into a double pseudoknot forming contacts that are critical for its catalytic activity. Loops 1 and 4 contain many conserved residues including the strongly conserved (>97%) ten residues (Roth et al. 2014). Two additional stem loops, P3 and P5, are observed in most but not all twister ribozymes. The P3 is present between T2 and P4, and P5 connects to the L2 internal loop. In general, the secondary structure of the twister ribozyme is highly conserved (Roth et al. 2014).

Crystal structures of the twister ribozyme from *Oryza sativa* and the *env22* twister ribozyme have been solved and they reveal common structural elements (Eiler et al. 2014; Liu et al. 2014; Ren et al. 2014). Both *O. sativa* and the *env22* twister ribozymes adopt a novel ternary fold generated through colinear stacking of helical stems. The novel ternary fold includes a pair of pseudoknots, two minor and two major grooves base triples, and two long-range non-canonical interactions (Fig. 2b) (Liu et al. 2014; Ren et al. 2014). Although the global architecture is similar, their active site does display differences that have implications for catalysis (Gebetsberger and Micura 2017). All base pairs in the conserved P1 segment in *O. sativa* form Watson–Crick base-pairing interactions (Liu et al. 2014). In contrast, in the *env22* ribozyme, this interaction is restricted to two central base pairs (Fig. 2b) (Ren et al. 2014), the other two nucleotides fold back and form base triples (Ren et al. 2014). The alignment of cleavage site nucleotides is also different between the *O. sativa* and *env22* twister ribozymes (Gebetsberger and Micura 2017; Ren et al. 2014). In the *env22* ribozyme, the cleavage site nucleotides, U5-A6, are in a near in-line conformation, poised for the catalytic activity (Ren et al. 2014). The modeled distance between the O2' of U5 and P-O5' in *env22* is 2.8 Å, and an in-line angle of 148°, consistent with the active conformation. In the *O. sativa* ribozyme, the alignment is different, the distance between the O2' of U5 and P-O5' is 2.9 Å, and an in-line angle is 81°, suggesting an inactive conformation. The third major difference is the presence/absence of Mg^{2+} ion (Gebetsberger and Micura 2017; Liu et al. 2014; Ren et al. 2014). The *env22* ribozyme has an inner-sphere Mg^{2+} that is directly coordinated to the *pro-S* non-bridging oxygen of the scissile phosphate, and an additional one coordinated to *pro-R* of the successive downstream phosphate groups. No Mg^{2+} was found in the *O. sativa* ribozyme. These observations suggest that the crystal structure of the *env22* ribozyme captures the active state of the ribozyme, whereas the *O. sativa* structure is in the inactive conformation (Ren et al. 2014; Eiler et al. 2014; Gebetsberger and Micura 2017; Liu et al. 2014).

The crystal structure of *env22* twister reflects the active state of the ribozyme (Ren et al. 2014). Therefore, the structural and biochemical studies of the *env22* twister ribozymes provide insights into its catalytic mechanism (Eiler et al. 2014; Gebetsberger and Micura 2017; Liu et al. 2014; Ren et al. 2014). The mechanism appears consistent with acid-base catalysis. The cleavage reaction rates exhibit a bell-shaped profile relative to pH, with a plateau around the pH of 6.5 (Liu et al. 2014). The reaction rate is dependent on the Mg^{2+} concentration (Roth et al. 2014). Despite the dependence, the biochemical data are not in favor of the direct participation of Mg^{2+} in the cleavage reaction (Roth et al. 2014). Instead, Mg^{2+} appears to contribute to the structural integrity of the cleavage site through its interaction with the non-bridging reaction (Ren et al. 2014; Roth et al. 2014).

The self-cleavage reaction in the *env22* ribozyme occurs at the U5-A6 site. The A6 is highly conserved and adopts a *syn* glycosidic alignment stabilized by stacking interactions, while not conserved U5, appears to adopt a flexible alignment to position these nucleotides in the in-line conformation for catalysis. The A6G mutation completely abolishes the catalytic activity whereas mutations at the U5 site are tolerated (Ren et al. 2014). Structural and mutational data also show that a conserved guanine, G48, participates in the *env22* twister ribozyme catalysis (Ren et al. 2014). The G48A mutation strongly reduces the cleavage activity of the ribozyme (Ren et al. 2014). The N1H of G48 forms a strong hydrogen bond with the non-bridging *pro-*R_P phosphate oxygen, suggesting its role in stabilizing a negative charge at the transition state (TS). The inner-sphere Mg^{2+} ion, which interacts with *pro-*R_s, could also stabilize the TS, but other biochemical data suggest that it largely plays a structural role in maintaining the integrity of the active site (Ren et al. 2014; Roth et al. 2014). These data collectively suggest that the twister ribozyme appears to follow S_N2-type mechanism for phosphodiester cleavage, which is facilitated by the in-line conformation of the participating nucleotide stabilized by the overall architecture of the ribozyme.

3 Riboswitches

Riboswitches are structured RNA elements that regulate gene expression in response to binding to a ligand. They are mostly found in the untranslated region (UTRs) of many bacterial mRNAs, where they regulate a variety of metabolic processes. Although found mostly in bacteria, they are present in all three domains of life including the eukaryotic genomes. Functionally, they act as molecular switches regulating gene expression via conformational changes in their three-dimensional structure. The conformation change is induced by the direct binding to a specific ligand. In general, riboswitches consist of two domains. The domain that binds to a ligand is called the aptamer domain, and the domain that regulates gene expression is called the expression platform. The aptamer domain selectively recognizes its small-molecule ligand. Binding of a ligand to the aptamer domain induces structural changes in the associated expression platform domain, which alters the expression of the downstream mRNA. The changes in gene expression typically include transcription termination or translation initiation, or more rarely, ribozyme-mediated mRNA degradation or the control of splicing. Riboswitches are a good target for the design of antibacterial drugs because they often regulate essential metabolic pathways in bacteria. While more than 40 riboswitches have been discovered, in this chapter we will only cover some of the best-characterized riboswitches such as the purine, thiamine pyrophosphate, and *glmS* riboswitches.

3.1 Purine Riboswitch

The purine riboswitch was found in the UTR sequences of mRNAs as a conserved regulatory feature associated with expression of genes involved in the purine metabolism (Christiansen et al. 1997). The aptamer domain of the purine riboswitch has emerged as a useful model system for studying various aspects of RNA structure and function (Porter et al. 2014). It has a multi-helix structure with side-by-side helical packing, allowing the formation of a three-way junction supported by a distal tertiary interaction through interloop interactions (Fig. 4a). These structural features are a highly recurrent theme in RNA biology (De La Peña et al. 2009; Porter et al. 2014).

The structures of four members of the purine riboswitch family have been solved (Porter et al. 2014). These riboswitches include *xpt-pbuX* guanine-responsive from *B. subtilis* (Batey et al. 2004), *add* adenine-responsive from *Vibrio vulnificus* (Serganov et al. 2004), *pbuE* adenine-responsive from *B. subtilis* (Delfosse et al. 2010) and 2'-deoxyguanosine-responsive from *M. florum* (Pikovskaya et al. 2011). Their aptamer domains share a common structural motif and ligand binding pocket (Fig. 4a). The binding pocket is located within the center of the three-way junction made up of J1/2, J2/3, and J3/1 (Fig. 4). The aptamer domain consists of three Watson-Crick paired regions called P1, P2, and P3, organized around the central three-way junction (Fig. 4). The two terminal loops L2 and L3 at the end of P2 and P3 stems, respectively, form a pseudoknot interaction between the GG invariant bases in L2 and CC in the L3 loop. The binding pocket located in the central three-way junction has conserved residues U47, U51, and C74 (Fig. 3a). These bases, along with the 2′-hydroxyl group of U22, create the network of hydrogen bonds that fully recognize the purine, adenine, or guanine ligands. The C74 is critical for purine ligand specificity (Fig. 3a) (Gilbert et al. 2006, 2009). The purine riboswitch discriminates between the adenine and guanine through a single base change in the pocket enabling Watson-Crick pairing to this nucleobase (Fig. 3a). The C74 base is present in the guanine and the U74 is present in the adenine riboswitch. Binding of the purine ligand induces structural changes to regulate expression of genes involved in the metabolism of purines.

The purine riboswitch also binds to a variety of non-natural purines such as hypoxanthine and xanthine, and synthetic analogs such as 2, 6-diaminopurine (Gilbert et al. 2009). Structural and biochemical studies of these ligands point to a chemical flexibility in the recognition of the alternative ligands. For example, it has been proposed that xanthine can bind to the purine riboswitch as a 2-enol minor tautomer (Gilbert et al. 2009). Crystal structures of the riboswitch with various ligands show that the pyrimidine at the 74-position (Y74) is critical for determining ligand specificity (Figs. 3a and 4a) (Batey et al. 2004; Serganov et al. 2004; Gilbert et al. 2006; Singh et al. 2015). The xanthine ligand forms a Watson–Crick base pair with C74 (Gilbert et al. 2009). Conserved residues at the U51, U47, and U22 positions contribute to ligand recognition (Fig. 3a). The carbonyl oxygens (O2) of C74 and U51 form hydrogen bonds with the 2-amino functional group of the native ligand guanine (Serganov et al. 2004). Hypoxanthine does not have the 2-amino group, thus binds

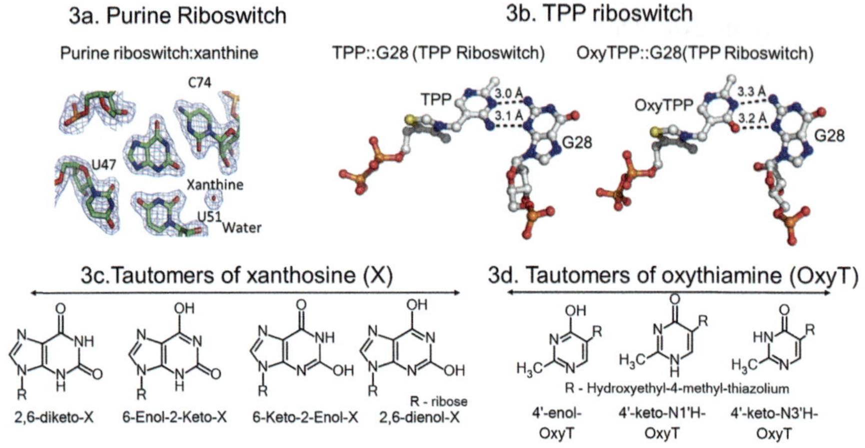

Fig. 3 Conformation of ligands in the binding pocket of the purine and the TPP riboswitches. **a** Top left, crystal structure of the purine riboswitch with xanthine in its binding pocket interacting with carbonyl oxygens of C74, U47, and U51, and a water molecule. 3c. all possible tautomeric forms of xanthosine. **b** Crystal structures of the TPP riboswitch showing interactions of its binding site G28 (guanosine at the 28th position) with the thiamine pyrophosphate (TPP) and oxythiamine pyrophosphate (OxyTPP) ligands. **d** Tautomeric forms of oxythiamine identified by NMR and vibrational spectroscopy (Singh et al. 2014; Fedeles et al. 2021). **a–d** is adapted from reference (Singh et al. 2015) (CC BY 4.0), with permission to use from the RNA Society of Cold Spring Harbor Laboratory Press

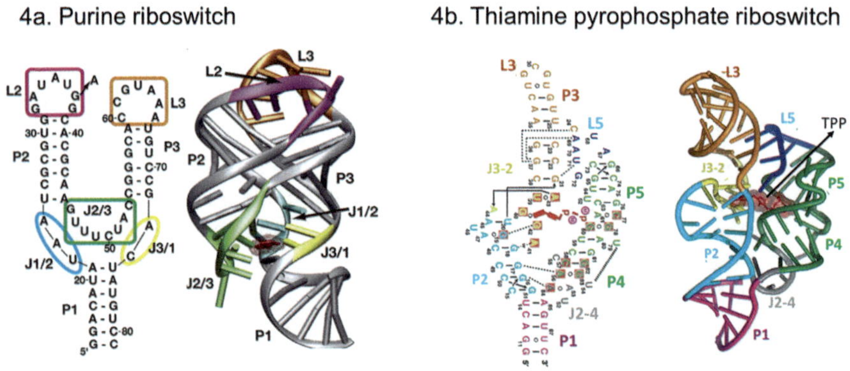

Fig. 4 Secondary and ternary structures of the purine and thiamine pyrophosphate (TPP) riboswitches. **a** Secondary and ternary structures of the aptamer domain of the *B. subtilis* xpt-pbuX guanine riboswitch. Three paired regions are labeled as P1, P2, and P3, terminal loops as L2 and L3, and junctions as J2/3. **a** is adapted from the reference (Stoddard et al. 2008), with permission to use from the RNA Society of the Cold Spring Harbor Laboratory Press. **b** Left, Secondary structure of the thiamine pyrophosphate riboswitch. 4B Right, Ternary structure of the TPP riboswitch bound to the thiamine pyrophosphate ligand. **b** is adapted from the reference (Serganov et al. 2006) (CC-BY-SA)

to the purine riboswitch with almost 200-fold less affinity compared with guanine (Batey et al. 2004). Besides the 2-amino functional group, the 6-keto, and N1 groups are also recognized by the riboswitch to form a Watson–Crick type of base-pairing (Batey et al. 2004; Gilbert et al. 2006; Serganov et al. 2004). These data suggest that the Watson-Crick interaction between the Y74 base and the ligand is critical for determining ligand specificity and the biological activity of the purine riboswitches.

3.2 GlmS Riboswitch/Ribozyme

The *glmS* ribozyme is the first known naturally occurring ribozyme that is also a riboswitch (Winkler et al. 2004). It binds to a small-molecule cofactor, glucosamine-6-phosphate (GlcN6P), to promote the self-cleavage reaction. It is found in the 5'-UTR region of glutamine-fructose-6-phosphate amidotransferase mRNA, encoded by the *glmS* gene (Cochrane et al. 2009; Winkler et al. 2004). It utilizes the self-cleavage reaction to regulate the expression of the *glmS* gene. A site-specific cleavage upon GlcN6P binding to the *glmS* ribozyme is correlated with the expression levels of the *glmS* gene. The GlcN6P ligand directly participates in the cleavage reaction as a general acid (Cochrane et al. 2009). Like other autolytic small self-cleaving ribozymes, *glmS* reaction mechanism involves an attack of the vicinal 2′-OH on the scissile phosphate to generate two fragments of RNA, one containing the 2′, 3′ cyclic phosphate and the other having a free 5′-OH group (Fig. 1a).

Multiple structural and biochemical studies of the *glmS* ribozymes suggest that its active site is poised for the self-cleavage reaction (Lim et al. 2006). The ribozyme undergoes a very little structural rearrangement during the self-cleavage reaction (Winkler et al. 2004; Cochrane et al. 2009; Hampel and Tinsley 2006; Klein and Ferré-D'Amaré 2006). The nucleotides flanking the scissile phosphate are in a splayed conformation that is stabilized mostly by hydrogen bonding and stacking interactions (Klein and Ferré-D'Amaré 2006). The A-1 nucleotide is oriented through hydrogen bonding with G33 and G57, where the G + 1 makes a tertiary hydrogen bond to G20 and stacks underneath A28 (Fig. 1B). These interactions allow A-1 and G + 1 to adopt an in-line conformation to facilitate the cleavage reaction in response to GlcN6P binding (Fig. 1B) (Klein and Ferré-D'Amaré 2006).

The active site of *glmS* is pre-organized for the self-cleavage reaction (Cochrane et al. 2009, 2007; Klein and Ferré-D'Amaré 2006). The structures of the pre- and post-cleaved stages of the self-cleavage reaction and bound to various effectors and competitive inhibitors show a very similar active site (Cochrane et al. 2009, 2007; Klein and Ferré-D'Amaré 2006). For example, it binds specifically to GlcN6P without undergoing any confirmation change (Cochrane et al. 2009; Klein and Ferré-D'Amaré 2006). A comparison of the pre-cleaved, containing the G33 mutation, and post-cleaved ribozymes show that their structures are very similar, the average RMSD value that is less than 2 Å (Cochrane et al. 2009). The structure of *glmS* with a competitive inhibitor is essentially identical to its structure with GlcN6A (Cochrane et al. 2009). Its structure with a non-natural competitive inhibitor like mannosamine-6-phosphate (MaN6P), with inverted C2-amine, is also identical to its structure with GlcN6P (Cochrane et al. 2009). While the stereochemistry of the C2-amine is inverted in MaN6P, the overall position of the sugar in the active site is similar to GlcN6A. These data suggest that the conformation of the *glmS* ribozyme is mostly unchanged during the self-cleavage reaction.

Structural and biochemical studies have also identified conserved guanosine, G33 in *glmS* from *B. anthracis* and G40 in *glmS* from *T. tengcongensis*, participating in the self-cleavage reaction as a general base to activate the 2′-hydroxyl nucleophile (Fig. 1B) (Cochrane et al. 2009, 2007; Klein and Ferré-D'Amaré 2006). G33 to A mutation leads to a significant reduction in its activity, by as much as 10^5, without much change in its structure (Cochrane et al. 2007; Klein et al. 2007). The N1 of G33 is within the hydrogen bond distance of the 2′-OH of A-1. Mutation of G33 with any other nucleobase results in a 10^3–10^5 fold reduction in the rate of cleavage in the presence of GlcN6P (Cochrane et al. 2009). Considering G33 to A mutation results in an almost identical active site, this indicates that the G33 role is to not just orient the 2′-OH but it directly participates in the self-cleavage reaction. Given the functional importance of G33 in the cleavage reaction and the structural proximity to the 2′-hydroxyl group, it was proposed that the N1 of G33 directly activates the 2′-OH nucleophile (Cochrane et al. 2007). However, the N1 of guanine has a $pK_a \sim 10$, making it a poor base to abstract a proton from the 2′-OH, with a pK_a of 13 (Cochrane et al. 2009; Singh et al. 2015). It's been suggested that deprotonation of the N1 of G33 through the formation of either the minor tautomer (6-enol) or an anion could generate an activated state of G33 that is more suitable as a general base (Cochrane et al. 2007; Cochrane and Strobel 2008a; Singh et al. 2015). However, the existence of tautomeric or ionic state of G33 has not been experimentally established (Singh et al. 2015).

The GlcN6P cofactor is essential for the self-cleavage activity of the *glmS* ribozyme (Winkler et al. 2004). In its absence, the ribozyme is ~10^4 less active (Winkler et al. 2004). The phosphate group of GlcN6P is critical for its binding. Whereas its amine group participates in the reaction as a general acid (Cochrane et al. 2009; Klein and Ferré-D'Amaré 2006). Its substitution with a closely related glucose-6-phosphate (Glc6P), in which the C2-amine is replaced by a hydroxyl group, also reduces the rate of reaction by ~10^4 (McCarthy et al. 2005; Winkler et al. 2004). The p*Ka* of the C2-amine of GlcN6P is 7.8, a value very similar to the observed pKa value of the self-cleavage reaction (McCarthy et al. 2005). In the crystal structure, the C2-amine of GlcN6P is hydrogen bonded to the 5-O leaving group (Cochrane et al. 2007; Klein et al. 2007). The structural location, the p*K*a, and the mutational studies suggest that the amine group of GlcN6P is responsible for the protonation of the leaving group (Cochrane et al. 2007; Klein et al. 2007).

Taken together structural and biochemical data support a mechanism in which the self-cleavage reaction is initiated by the base-catalyzed activation of the vicinal 2′-hydroxyl nucleophile by a catalytic guanine, G33 in *B. anthracis* (Fig. 1a). The activated 2′-oxygen attacks the scissile backbone phosphodiester bond to initiate the transesterification reaction. The amine group of GlcN6P through fine tuning of its p*K*a participates in the transesterification reaction as a general acid by donating a proton to the 5′-oxygen of the leaving RNA fragment. Bronsted coefficient 0.7 for the 5′-oxygen at the reaction site suggests a late transition state, in which the rate limiting step is the protonation of the 5′-oxygen by the amine group of GlcN6P (Viladoms and Fedor 2012).

3.3 Thiamine Pyrophosphate (TPP) Riboswitch

The TPP riboswitch negatively regulates the expression of genes involved in thiamine biosynthesis, phosphorylation, and transport (Sudarsan et al. 2003; Cheah et al. 2007; Winkler et al. 2002). The phosphorylated thiamin, TPP, is an essential cofactor for multiple enzymes in the Krebs cycle and the pentose-phosphate pathways, where it participates in making and breaking bonds between carbon and carbon, sulfur, and nitrogen (Frank et al. 2007; Cochrane and Strobel 2008b). The TPP riboswitch is one of the most abundant riboswitches. It is found in all three domains of life suggesting that it arose early in evolution (Cheah et al. 2007; Winkler et al. 2002). It is also found in as many as 49 human pathogens, making it a good target for developing antibiotics (Bocobza and Aharoni 2014; Panchal and Brenk 2021). It can be found upstream, downstream, or at splice-site junctions of these genes, indicating that it uses a diverse set of mechanisms for regulating gene expression (Barrick and Breaker 2007; Welz and Breaker 2007; Wachter 2010). These mechanisms can include mRNA decay, Rho-dependent transcription termination, alternative splicing, transcription termination through the formation of an intrinsic terminator, and translational inhibition through sequestration of ribosomal binding sites (RBS) (Winkler et al. 2002; Li and Breaker 2013; Barrick and Breaker 2007; Wachter 2010; Welz and Breaker 2007).

The biochemical and structural basis of ligand recognition is well studied for the TPP riboswitch (Fig. 4b) (Thore et al. 2006, 2008). The highly conserved secondary structure of the TPP-binding riboswitches consists of five helices, termed P1 to P5 (Fig. 4b) (Thore et al. 2008; Haller et al. 2013). The helix P1 is formed by the 5′ and 3′ end of the riboswitch RNA and is expected to be disrupted in the absence of TPP. Helices P1, P2, and P4 together with junctions J1/4 and J2/4 form the central three-way junction. Bulges J2/3 and J4/5 connect helices P2 with P3 and P4 with P5, respectively. Terminal loops L3 and L5 close helices P3 and P5.

Crystal structure of the aptamer domain of the TPP riboswitch has been solved with various ligands (Thore et al. 2008; Haller et al. 2013). These structures revealed that a conserved guanine, G28 in the TPP riboswitch from *Arabidopsis thaliana*, is critical for determining ligand specificity. The TPP ligand, commonly referred to as vitamin B1, is composed of three parts. These include the aminopyrimidine ring, a central thiazole ring, and a pyrophosphate group (Serganov et al. 2004). The *Arabidopsis thaliana* riboswitch binds to TPP in an extended conformation in which its aminopyrimidine ring is inserted into the pyrimidine sensor helix at the J2/3 bulge region by stacking between the bases G30 and A31, and it forms hydrogen bonding with the polar groups of G28 and G11 (Serganov et al. 2004, 2006; Thore et al. 2006). The guanine base of G28 adopts a "flipped out" conformation to facilitate hydrogen bonding with the aminopyrimidine ring of the TPP ligand (Thore et al. 2006; Serganov et al. 2006). The pyrophosphate group of TPP is coordinated with two Mg^{+2} ions and forms direct hydrogen bonds with G64, C65, and G66. The residues G66 and G48 coordinate to a putative bridging Mg^{2+} ion. The reactive part of the carbon in TPP is the C2 carbon of the thiazolium ring, which becomes a potent nucleophile through the formation of ylide in TPP-dependent enzymes. Interestingly, in the TPP riboswitch the central thiazole group is less constrained and only forms hydrophobic contacts with G72. The nucleotides that form close contact with TPP are highly conserved across various TPP riboswitch sequences.

A comparison of crystal structures of the TPP riboswitch from *Arabidopsis thaliana* with TPP and its oxidized form OxyTPP revealed that their hydrogen bonding interactions with G28 are largely identical (Thore et al. 2008). It was anticipated that the 4′-keto in OxyTPP, instead of 4′-amino as in TPP, would disrupt its hydrogen bonding interaction with G28. In the structure the amino group at the 4′-position of TPP acts as a hydrogen bond donor to the N3 atom of G28. However, OxyTPP and TPP form identical hydrogen bonding interactions with G28. To rationalize these data, it was proposed that the OxyTPP binds to the TPP riboswitch as a 4′-enol tautomer (Thore et al. 2008; Singh et al. 2014). Biochemical data showing tighter binding of OxyTPP at lower pH was consistent with this observation, enol tautomer would be more stabilized at lower pH (Thore et al. 2008). Furthermore, NMR and 2D-IR (2D infrared spectroscopy) also revealed that unlike the TPP ligand, OxyTPP exists in three different tautomeric forms including the 4′-enol tautomer that was proposed to bind to the TPP riboswitch (Singh et al. 2014). However, the 4′-enol tautomer of OxyTPP bound to the TPP riboswitch could be unambiguously established (Singh et al. 2015, 2014).

4 Targeting Riboswitches as an Antibacterial Strategy

The increasing threat from multi-drug resistant bacteria has been motivating efforts to develop new generations of antibiotics that, mechanistically, are different than the current antibacterial arsenal. Given their key role in regulating gene expression for various metabolic pathways, bacterial riboswitches are very attractive targets for new antibacterial agents (Blount and Breaker 2006; Serganov and Nudler 2013; Lünse et al. 2014; Giarimoglou et al. 2022). Detailed knowledge of the molecular mechanisms of riboswitches has been one of the key advances that enable the development of mechanism-based riboswitch inhibitors.

Riboswitches are widespread regulatory elements in bacterial mRNA, located mostly in the 5’UTR region. Their complex tertiary structures often harbor a binding site for their cognate ligand, which is often a key metabolic intermediate, such as an aminoacid (e.g., lysine), a vitamin cofactor (e.g., TPP, FMN), a nucleoside (e.g., guanine), etc. The binding of the ligand “locks” the riboswitch into a stable conformation, stabilizing one or more stem loops that directly impact the function of the mRNA. By their mode of action, the riboswitches can be categorized as on-switches and off-switches. The on-riboswitches naturally fold into a conformation that blocks transcription via a terminator loop, or translation via sequestration of the Shine-Dalgarno ribosome binding site. When the ligand binds, a new conformation develops which unfolds the terminator loop and allows full-length transcription, and/or exposes the ribosome binding site and allows translation to occur. These riboswitches are primarily used to respond rapidly to changes in the levels of select nutrients and/or stress factors. For example, one of the most recently discovered riboswitches is an on-switch that detects mM concentrations of sodium (White et al. 2022). By contrast, in the off-riboswitches, the inactive conformation is stabilized by ligand binding, which subsequently turns off gene expression. Off-switches often use as a ligand the end-product of a metabolic pathway, which upon binding, turns off the gene expression of an entire collection of genes involved in that pathway. This is the case for FMN, TPP, and lysine riboswitches.

While riboswitches are prevalent as regulatory elements of essential metabolic pathways in bacteria, many do not constitute viable antibacterial targets. Three types of considerations are important for selecting the most promising ones. First, the riboswitch needs to control an essential pathway, which, when blocked, inhibits the growth of the bacteria. Often, however, bacteria possess redundant enzymes, not all of which are regulated by the riboswitch. For example, targeting the lysine riboswitch in an effort to deplete cells of lysine is not bacteriostatic because many pathogens express isozymes in the lysine biosynthetic pathway that are not controlled by the lysine riboswitch. Moreover, lysine is an abundant metabolite that can be obtained directly from the host, making the inhibition of the lysine riboswitch an ineffective antibacterial strategy. The second consideration is the abundance of the riboswitch target. When the riboswitch is present on only a single gene transcript, a very high potency inhibitor is required, because partial inhibition potentially allows enough gene expression to maintain viability. By contrast, bacterial strains with the riboswitch at multiple loci are expected to be significantly more sensitive, because the inhibitory effect is compounded. Related to this, the third consideration concerns the development of resistance that renders riboswitch-targeting antibacterial ineffective. Most common mechanism of resistance appears to be the acquisition of mutations in the RNA sequence that changes the affinity of the riboswitch for the targeting ligand (Howe et al. 2015; Motika et al. 2020). Here, too, the abundance of the riboswitch is expected to play a key role. If several essential genes are regulated by the riboswitch, the acquisition of resistance mutations at all of them is an unlikely event. Not surprisingly, even for the same class of riboswitches, some sequences are more mutable than others, and therefore, some bacterial species develop resistant mutants faster than others. For example, when evaluating the inhibition of the FMN riboswitch with the synthetic ligand ribocil (see below), *E.coli* develop resistance at a frequency 1–2 orders of magnitude higher than other Gram-negative pathogens *Pseudomonas aeruginosa* and *Acinetobacter baumannii.*

Structural knowledge is also essential for the development of riboswitch inhibitors. Similar to the development of protein inhibitors based on the structure and properties of their ligand binding sites, riboswitches can be targeted with small molecules that are structural analogs of their cognate ligands (Warner et al. 2018; Hewitt et al. 2019). Moreover, just like some proteins are considered "druggable" compared to others, some riboswitch ligand binding pockets may offer more flexibility for medicinal chemistry than others (Hewitt et al. 2019). For example, the purine and lysine riboswitches form tight and almost completely closed pockets around their ligands, which limits the chemical and structural space that an inhibitor can utilize. By contrast, the FMN, TPP, and SAM riboswitches feature more open binding pockets that enclose only partially their ligands. This allows, in principle, the exploration of a wider range of potential molecules and scaffolds that diverge structurally from the cognate ligand. A great example for this phenomenon is Ribocil—a small-molecule FMN riboswitch inhibitor that does not immediately resemble the natural ligand of the riboswitch.

4.1 Ribocil

In bacteria, the concentrations of riboflavin (RF, vitamin B2) are maintained through regulation of the biosynthetic pathway and transport. This regulation relies primarily on the FMN riboswitches present at the 5' ends on the majority of relevant mRNA transcripts (Winkler et al. 2002). FMN is a negative regulator of the riboswitch: when abundant, it binds to the aptamer freeing a sequester loop, which causes early transcription termination and hides the ribosome binding site on the mRNA (Pedrolli et al. 2015). If FMN levels are low, then the aptamer adopts a different conformation that traps the sequester loop, allows the synthesis of the full-lengths transcripts, and exposes the ribosome binding site on the mRNA (Pedrolli et al. 2015). Genes involved in the FMN biosynthesis can now be expressed and translated.

The discovery of Ribocil came from a study investigating the targeting of the riboflavin biosynthetic pathway as a promising antibacterial strategy (Howe et al. 2015). It was observed that *E. coli* defective in the synthesis of RF are markedly less virulent, yield diminished bacterial titers, and cause limited morbidity in a mouse septicemia model. A screen of 57,000 small molecules with antibacterial activity identified one compound—later named Ribocil—whose antibiotic activity is inhibited by excess RF. The mechanism of this compound was suggested by the discovery of mutants resistant to Ribocil. Sequencing revealed that all mutants mapped to the FMN riboswitch located on the mRNA of key RF biosynthetic gene *ribB* (Howe et al. 2015, 2016). This observation suggested that Ribocil may be a mechanism-based inhibitor of the FMN riboswitch. The original compound was an enantiomeric mixture; only the S-enantiomer (known as Ribocil B) has the antibacterial activity, and indeed, it binds tightly the FMN riboswitch ($K_d = 13$ nM). Structure activity relationship studies further improved the binding of the molecule to the riboswitch, yielding Ribocil C with $K_d \sim 1.5$ nM (Fig. 5a, b).

Structural studies on the FMN riboswitch from the *Fusobacterium nucleatum impX* RF transporter, one of the best-characterized FMN riboswitches, revealed the stereochemistry of binding of Ribocil, and its key molecular interactions with the

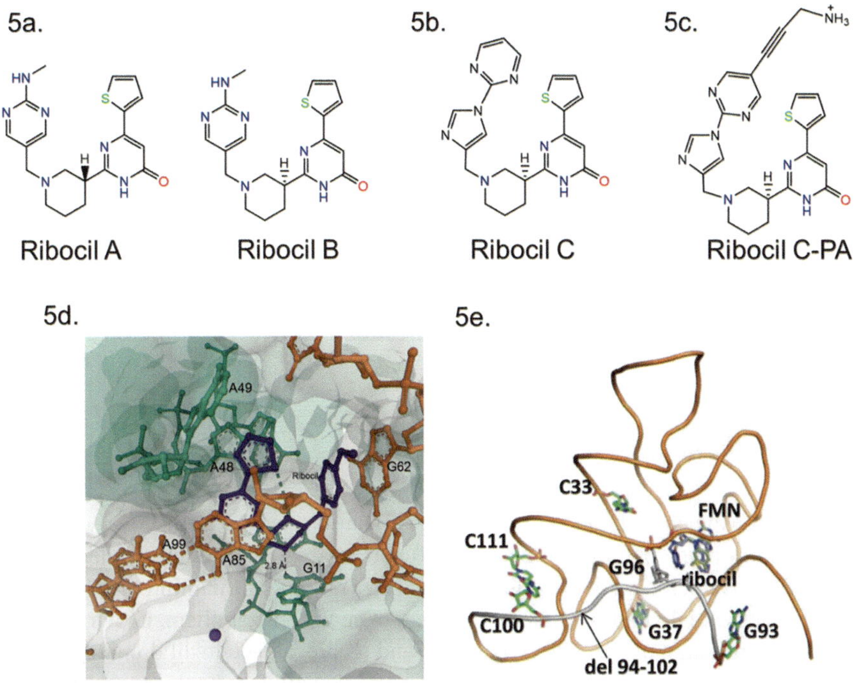

Fig. 5 Structure and mechanism of Ribocil—a mechanism-based inhibitor of the FMN riboswitch. **a** Structures of Ribocil A (R enantiomer) and Ribocil B (S-enantiomer of Ribocil). Only Ribocil B has inhibitory properties. **b** Structure of Ribocil C, an improved FMN riboswitch binder, based on the original molecular scaffold. **c** Structure of Ribocil-C-PA, a recently described (Motika et al, 2020) variant of Ribocil C, with superior antibacterial properties. **d** Co-crystal structure of Ribocil bound in the FMN riboswitch of *Fusobacterium nucleatum impX*. Figure drawn with PDB 3D View from PDB coordinates 5C45 and annotated based on reference (Howe et al. 2015). The Ribocil ligand is in dark magenta; the riboswitch chains are shown in green and orange. **e** Location of key residues that are mutated in FMN riboswitches resistant to Ribocil binding. Figure adapted from reference (Howe et al. 2016), with permission to use from Taylor and Francis Group (© 2016 Howe et al. Published with license by Taylor & Francis Group, LLC © 2016 Merck Sharp @Dohme)

RNA (Fig. 5d). Ribocil binds in the FMN binding site, but due to the structural differences between Ribocil and FMN, its structure is bent into a U-shaped conformation. Nevertheless, Ribocil does interact with the same key nucleotides as FMN, and a number of favorable H-bonding and pi-stacking interactions can be readily observed (Fig. 5d).

The identification of riboswitch mutants that confer resistance to Ribocil binding provided additional insight into its mechanism. Several mutants, namely, C111U, C100U, G93U, G37U, C33U, and D94-102 were characterized in terms of binding affinity to FMN and Ribocil (Fig. 5e). Generally, the mutants disrupted the binding of Ribocil much more than the FMN binding. The mutants in residues close to the active site disrupt the binding of the Ribocil and, to a lesser extent FMN, which can explain the Ribocil-resistant phenotype. More puzzling are mutants like C111U,

C100U and G93U, which are very far from the binding site. The binding of both FMN and Ribocil to these mutant riboswitches is unaffected. And yet, the mutants fail to inhibit translation and transcription in the presence of Ribocil. The best explanation here is that, in these mutants, the formation of the sequester loop is disrupted and the riboswitch is trapped into the ON state (Howe et al. 2015).

Further improvement on the Ribocil scaffold was demonstrated recently. By incorporating charged functional groups that enhance the ability of the molecule to accumulate inside bacteria, an even more potent version of Ribocil, named Ribocil C-PA was obtained (Fig. 5c) (Motika et al. 2020). The new molecule is even more potent against *E. coli* and other Gram-negative pathogens, as demonstrated in experiments on clinical isolates and mouse models of infection.

4.2 Pyrithiamine

The TPP riboswitch is another attractive target of antibacterial drug discovery because it is found in many pathogenic bacteria (Rupert et al. 2002). As mentioned above, the TPP riboswitch negatively regulates the biosynthesis of thiamin pyrophosphate (TPP), an essential vitamin cofactor involved in central metabolism. Given the high levels of expression of the TPP riboswitch, and its prevalence in all domains of life, small molecules that target this riboswitch could constitute potent, broad-spectrum antibiotics. Even before the structure and function of the TPP riboswitch were known, a compound named pyrithiamine (PT) was shown to have antibacterial and antifungal activity (Woolley and White 1943). Originally, PT was synthesized as a structural analog of thiamine (Robbins 1941) to aid in the study of thiamine binding and metabolism; these studies, however, eventually helped uncover the existence of the TPP riboswitch and its negative regulation by TPP. PT features a pyridine ring in the place of the central thiazole ring of thiamine (Fig. 5f). Just like thiamine, inside the cells, PT is phosphorylated to form pyrithiamine pyrophosphate (PTPP), which binds to the TPP riboswitch with a binding affinity that is similar to that of TPP; K_d's are 160 nM for PTPP and 50 nM for TPP for the TPP riboswitch from *B. subtilis* (Symons 1997, Thomas and Perrin 2008). Crystal structures of PTTP with the TPP riboswitch also revealed that its hydrogen binding interactions are identical to TPP and OxyTPP (Symons 1997, Thomas and Perrin 2008).

While PT provided the proof of concept that the TPP riboswitch can be targeted as an antibacterial strategy, a clinical candidate for a small-molecule targeting TPP riboswitch has yet to emerge. The biggest challenge here is that PT and other similar derivatives (amprolium, 3-deazathiamine) are also TPP antivitamins; they bind and inhibit TPP-dependent enzymes, not only in the target pathogens, but also in the host, thus causing toxicity (Tylicki et al. 2018). Nevertheless, multiple fragment-based and structure-based screens have identified promising hits for the TPP riboswitch (Thore et al. 2008; Warner et al. 2014). These hits show good binding affinity for the TPP riboswitch, which approaches that of PT, and correspondingly, antibacterial activity.

Therefore, the work toward a viable clinical candidate for a TPP riboswitch inhibitor is poised to succeed in the near future.

5 Conclusion

Structured RNAs play an important role in biology and disease. Structural and mechanistic understanding of RNA elements, such as riboswitches and ribozymes, helped us elucidate the rules RNA use to adopt complex three-dimensional structures. These RNAs resemble proteins in their complexity and are capable of catalytic rate enhancements and ligand specificity that is comparable to proteins. It was also recognized that many riboswitches are enriched in bacterial genomes, and therefore can be used for the therapeutic development of antibiotics. Ribocil is an approved antibiotic that binds to the FMN riboswitch to execute its antibacterial effect. Structured RNAs are also present in higher organisms and more recently have been targeted for a variety of indications in oncology and rare diseases.

Acknowledgements Vipender Singh is supported by the Novartis Institute of Biomedical Research, Cambridge MA 02139 USA. Bogdan I. Fedeles is supported by NIH grants R01-CA080024 and P30-ES002109.

References

Barrick JE, Breaker RR (2007) The distributions, mechanisms, and structures of metabolite-binding riboswitches. Genome Biol 8:1–19

Batey RT, Gilbert SD, Montange RK (2004) Structure of a natural guanine-responsive riboswitch complexed with the metabolite hypoxanthine. Nature 432:411–415

Blount KF, Breaker RR (2006) Riboswitches as antibacterial drug targets. Nat Biotechnol 24:1558–1564

Bocobza SE, Aharoni A (2014) Small molecules that interact with RNA: riboswitch-based gene control and its involvement in metabolic regulation in plants and algae. Plant J 79:693–703

Buzayan JM, Gerlach WL, Bruening G (1986) Non-enzymatic cleavage and ligation of RNAs complementary to a plant virus satellite RNA. Nature 323:349–353

Cervera A, De la Peña M (2014) Eukaryotic penelope-like retroelements encode hammerhead ribozyme motifs. Mol Biol Evol 31:2941–2947

Cervera A, Urbina D, de la Peña M (2016) Retrozymes are a unique family of non-autonomous retrotransposons with hammerhead ribozymes that propagate in plants through circular RNAs. Genome Biol 17:1–16

Cheah MT, Wachter A, Sudarsan N et al (2007) Control of alternative RNA splicing and gene expression by eukaryotic riboswitches. Nature 447:497–500

Chi Y-I, Martick M, Lares M et al (2008) Capturing hammerhead ribozyme structures in action by modulating general base catalysis. PLoS Biol 6:e234

Christiansen LC, Schou S, Nygaard P et al (1997) Xanthine metabolism in Bacillus subtilis: characterization of the xpt-pbuX operon and evidence for purine-and nitrogen-controlled expression of genes involved in xanthine salvage and catabolism. J Bacteriol Res 179:2540–2550

Cochrane JC, Lipchock SV, Smith KD et al (2009) Structural and chemical basis for glucosamine 6-phosphate binding and activation of the glmS ribozyme. Biochemistry 48:3239–3246
Cochrane JC, Lipchock SV, Strobel SA (2007) Structural investigation of the GlmS ribozyme bound to its catalytic cofactor. Chem Biol 14:97–105
Cochrane JC, Strobel SA (2008a) Catalytic strategies of self-cleaving ribozymes. Acc Chem Res 41:1027–1035
Cochrane JC, Strobel SA (2008b) Riboswitch effectors as protein enzyme cofactors. RNA 14:993–1002
Daròs JA, Flores R (1995) Identification of a retroviroid-like element from plants. Proc Natl Acad Sci USA 92:6856–6860
De La Peña M, Dufour D, Gallego J (2009) Three-way RNA junctions with remote tertiary contacts: a recurrent and highly versatile fold. RNA 15:1949–1964
Delfosse V, Bouchard P, Bonneau E et al (2010) Riboswitch structure: an internal residue mimicking the purine ligand. Nucleic Acids Res 38:2057–2068
DeYoung MB, Siwkowski AM, Lian Y et al (1995) Catalytic properties of hairpin ribozymes derived from chicory yellow mottle virus and Arabis mosaic virus satellite RNAs. Biochemistry 34:15785–15791
Eiler D, Wang J, Steitz TA (2014) Structural basis for the fast self-cleavage reaction catalyzed by the twister ribozyme. Proc Natl Acad Sci USA 111:13028–13033
Epstein LM, Gall JG (1987) Self-cleaving transcripts of satellite DNA from the newt. Cell 48:535–543
Fedeles BI, Li D, Singh V (2021) Structural insights into tautomeric dynamics in nucleic acids and in antiviral nucleoside analogs. Front Mol Biosci 8:823253
Fedor MJ (2000) Structure and function of the hairpin ribozyme. J Mol Biol 297:269–291
Feldstein PA, Buzayan JM, Bruening G (1989) Two sequences participating in the autolytic processing of satellite tobacco ringspot virus complementary RNA. Gene 82:53–61
Ferbeyre G, Smith JM, Cedergren R (1998) Schistosome satellite DNA encodes active hammerhead ribozymes. Mol Cell Biol 18:3880–3888
Frank R, Leeper F, Luisi B (2007) Structure, mechanism and catalytic duality of thiamine-dependent enzymes. Cell Mol Life Sci 64:892–905
Gebetsberger J, Micura R (2017) Unwinding the twister ribozyme: from structure to mechanism. Wiley Interdiscip Rev RNA 8
Giarimoglou N, Kouvela A, Maniatis A et al (2022) A riboswitch-driven era of new antibacterials. Antibiotics (Basel) 11
Gilbert SD, Mediatore SJ, Batey RT (2006) Modified pyrimidines specifically bind the purine riboswitch. J Am Chem Soc 128:14214–14215
Gilbert SD, Reyes FE, Edwards AL et al (2009) Adaptive ligand binding by the purine riboswitch in the recognition of guanine and adenine analogs. Structure 17:857–868
Griffiths A (1995) Natural plasmids of filamentous fungi. Microbiol Rev 59:673–685
Haller A, Altman RB, Soulière MF et al (2013) Folding and ligand recognition of the TPP riboswitch aptamer at single-molecule resolution. Proc Natl Acad Sci USA 110:4188–4193
Hampel A, Tritz R (1989) RNA catalytic properties of the minimum (-)sTRSV sequence. Biochemistry 28:4929–4933
Hampel KJ, Tinsley MM (2006) Evidence for preorganization of the glmS ribozyme ligand binding pocket. Biochemistry 45:7861–7871
Han J, Burke JM (2005) Model for general acid−base catalysis by the hammerhead ribozyme: pH−activity relationships of G8 and G12 variants at the putative active site. Biochemistry 44:7864–7870
Haseloff J, Gerlach WL (1989) Sequences required for self-catalysed cleavage of the satellite RNA of tobacco ringspot virus. Gene 82:43–52
Hewitt WM, Calabrese DR, Schneekloth JS Jr (2019) Evidence for ligandable sites in structured RNA throughout the Protein Data Bank. Bioorg Med Chem 27:2253–2260

Hiley SL, Sood VD, Fan J et al (2002) 4-thio-U cross-linking identifies the active site of the VS ribozyme. EMBO J 21:4691–4698
Howe JA, Wang H, Fischmann TO et al (2015) Selective small-molecule inhibition of an RNA structural element. Nature 526:672–677
Howe JA, Xiao L, Fischmann TO et al (2016) Atomic resolution mechanistic studies of ribocil: a highly selective unnatural ligand mimic of the E. coli FMN riboswitch. RNA Biol 13:946–954
Hutchins CJ, Rathjen PD, Forster AC et al (1986) Self-cleavage of plus and minus RNA transcripts of avocado sunblotch viroid. Nucleic Acids Res 14:3627–3640
Jones FD, Strobel SA (2003) Ionization of a critical adenosine residue in the Neurospora Varkud satellite ribozyme active site. Biochemistry 42:4265–4276
Klein DJ, Been MD, Ferré-D'Amaré AR (2007) Essential role of an active-site guanine in glmS ribozyme catalysis. J Am Chem Soc 129:14858–14859
Klein DJ, Ferré-D'Amaré AR (2006) Structural basis of glmS ribozyme activation by glucosamine-6-phosphate. Science 313:1752–1756
Kuzmin YI, Da Costa CP, Fedor MJ (2004) Role of an active site guanine in hairpin ribozyme catalysis probed by exogenous nucleobase rescue. J Mol Biol 340:233–251
Lafontaine DA, Wilson TJ, Norman DG et al (2001) The A730 loop is an important component of the active site of the VS ribozyme. J Mol Biol 312:663–674
Lafontaine DA, Wilson TJ, Zhao Z-Y et al (2002) Functional group requirements in the probable active site of the VS ribozyme. J Mol Biol 323:23–34
Li S, Breaker RR (2013) Eukaryotic TPP riboswitch regulation of alternative splicing involving long-distance base pairing. Nucleic Acids Res 41:3022–3031
Lilley DM (2019) Classification of the nucleolytic ribozymes based upon catalytic mechanism. F1000research 8
Lim J, Grove BC, Roth A et al (2006) Characteristics of ligand recognition by a glmS self-cleaving ribozyme. Angew Chem Int Ed Engl 45:6689–6693
Liu Y, Wilson TJ, McPhee SA et al (2014) Crystal structure and mechanistic investigation of the twister ribozyme. Nat Chem Biol 10:739–744
Lünse CE, Schüller A, Mayer G (2014) The promise of riboswitches as potential antibacterial drug targets. Int J Med Microbiol 304:79–92
Maguire JL, Collins RA (2001) Effects of cobalt hexammine on folding and self-cleavage of the Neurospora VS ribozyme. J Mol Biol 309:45–56
Martick M, Scott WG (2006) Tertiary contacts distant from the active site prime a ribozyme for catalysis. Cell 126:309–320
McCarthy TJ, Plog MA, Floy SA et al (2005) Ligand requirements for glmS ribozyme self-cleavage. Chem Biol 12:1221–1226
McKay DB (1996) Structure and function of the hammerhead ribozyme: an unfinished story. RNA 2:395
Motika SE, Ulrich RJ, Geddes EJ et al (2020) Gram-negative antibiotic active through inhibition of an essential riboswitch. J Am Chem Soc 142:10856–10862
Murray JB, Seyhan AA, Walter NG et al (1998) The hammerhead, hairpin and VS ribozymes are catalytically proficient in monovalent cations alone. Chem Biol 5:587–595
Olive JE, Collins RA (1998) Spermine switches a Neurospora VS ribozyme from slow Cis cleavage to fast trans cleavage. Biochemistry 37:6476–6484
Panchal V, Brenk R (2021) Riboswitches as drug targets for antibiotics. Antibiotics 10:45
Pedrolli D, Langer S, Hobl B et al (2015) The ribB FMN riboswitch from Escherichia coli operates at the transcriptional and translational level and regulates riboflavin biosynthesis. FEBS J 282:3230–3242
Pikovskaya O, Polonskaia A, Patel DJ et al (2011) Structural principles of nucleoside selectivity in a 2′-deoxyguanosine riboswitch. Nat Chem Biol 7:748–755
Pinard R, Hampel KJ, Heckman JE et al (2001) Functional involvement of G8 in the hairpin ribozyme cleavage mechanism. EMBO J 20:6434–6442

Porter EB, Marcano-Velázquez JG, Batey RT (2014) The purine riboswitch as a model system for exploring RNA biology and chemistry. Biochim Biophys Acta 1839:919–930
Prody GA, Bakos JT, Buzayan JM et al (1986) Autolytic processing of dimeric plant virus satellite RNA. Science 231:1577–1580
Przybilski R, Gräf S, Lescoute A et al (2005) Functional hammerhead ribozymes naturally encoded in the genome of Arabidopsis thaliana. Plant Cell 17:1877–1885
Ren A, Košutić M, Rajashankar KR et al (2014) In-line alignment and Mg^{2+} coordination at the cleavage site of the env22 twister ribozyme. Nat Commun 5:1–10
Robbins WJ (1941) The pyridine analog of thiamin and the growth of fungi. Proc Natl Acad Sci USA 27:419–422
Rojas AA, Vazquez-Tello A, Ferbeyre G et al (2000) Hammerhead-mediated processing of satellite pDo500 family transcripts from Dolichopoda cave crickets. Nucleic Acids Res 28:4037–4043
Roth A, Weinberg Z, Chen AG et al (2014) A widespread self-cleaving ribozyme class is revealed by bioinformatics. Nat Chem Biol 10:56–60
Ruffner DE, Stormo GD, Uhlenbeck OC (1990) Sequence requirements of the hammerhead RNA self-cleavage reaction. Biochemistry 29:10695–10702
Rupert PB, Ferré-D'Amaré AR (2001) Crystal structure of a hairpin ribozyme–inhibitor complex with implications for catalysis. Nature 410:780–786
Rupert PB, Massey AP, Sigurdsson ST et al (2002) Transition state stabilization by a catalytic RNA. Science 298:1421–1424
Saville BJ, Collins RA (1990) A site-specific self-cleavage reaction performed by a novel RNA in Neurospora mitochondria. Cell 61:685–696
Serganov A, Nudler E (2013) A decade of riboswitches. Cell 152:17–24
Serganov A, Polonskaia A, Phan AT et al (2006) Structural basis for gene regulation by a thiamine pyrophosphate-sensing riboswitch. Nature 441:1167–1171
Serganov A, Yuan YR, Pikovskaya O et al (2004) Structural basis for discriminative regulation of gene expression by adenine- and guanine-sensing mRNAs. Chem Biol 11:1729–1741
Singh V, Fedeles BI, Essigmann JM (2015) Role of tautomerism in RNA biochemistry. RNA 21:1–13
Singh V, Peng CS, Li D et al (2014) Direct observation of multiple tautomers of oxythiamine and their recognition by the thiamine pyrophosphate riboswitch. ACS Chem Biol 9:227–236
Sood VD, Collins RA (2002) Identification of the catalytic subdomain of the VS ribozyme and evidence for remarkable sequence tolerance in the active site loop. J Mol Biol 320:443–454
Stoddard CD, Gilbert SD, Batey RT (2008) Ligand-dependent folding of the three-way junction in the purine riboswitch. RNA 14:675–684
Sudarsan N, Barrick JE, Breaker RR (2003) Metabolite-binding RNA domains are present in the genes of eukaryotes. RNA 9:644–647
Suslov NB, DasGupta S, Huang H et al (2015) Crystal structure of the Varkud satellite ribozyme. Nat Chem Biol 11:840–846
Symons RH (1997) Plant pathogenic RNAs and RNA catalysis. Nucleic Acids Res 25:2683–2689
Thomas JM, Perrin DM (2008) Probing general base catalysis in the hammerhead ribozyme. J Am Chem Soc 130:15467–15475
Thomas JM, Perrin DM (2009) Probing general acid catalysis in the hammerhead ribozyme. J Am Chem Soc 131:1135–1143
Thore S, Frick C, Ban N (2008) Structural basis of thiamine pyrophosphate analogues binding to the eukaryotic riboswitch. J Am Chem Soc 130:8116–8117
Thore S, Leibundgut M, Ban N (2006) Structure of the eukaryotic thiamine pyrophosphate riboswitch with its regulatory ligand. Science 312:1208–1211
Tylicki A, Łotowski Z, Siemieniuk M et al (2018) Thiamine and selected thiamine antivitamins—biological activity and methods of synthesis. Biosci Rep 38
Viladoms J, Fedor MJ (2012) The glmS ribozyme cofactor is a general acid-base catalyst. J Am Chem Soc 134:19043–19049
Wachter A (2010) Riboswitch-mediated control of gene expression in eukaryotes. RNA Biol 7:67–76

Warner KD, Hajdin CE, Weeks KM (2018) Principles for targeting RNA with drug-like small molecules. Nat Rev Drug Discov 17:547–558

Warner KD, Homan P, Weeks KM et al (2014) Validating fragment-based drug discovery for biological RNAs: lead fragments bind and remodel the TPP riboswitch specifically. Chem Biol 21:591–595

Weinberg Z, Kim PB, Chen TH et al (2015) New classes of self-cleaving ribozymes revealed by comparative genomics analysis. Nat Chem Biol 11:606–610

Welz R, Breaker RR (2007) Ligand binding and gene control characteristics of tandem riboswitches in Bacillus anthracis. RNA 13:573–582

White N, Sadeeshkumar H, Sun A et al (2022) Na(+) riboswitches regulate genes for diverse physiological processes in bacteria. Nat Chem Biol 18:878–885

Winkler W, Nahvi A, Breaker RR (2002) Thiamine derivatives bind messenger RNAs directly to regulate bacterial gene expression. Nature 419:952–956

Winkler WC, Nahvi A, Roth A et al (2004) Control of gene expression by a natural metabolite-responsive ribozyme. Nature 428:281–286

Woolley DW, White AG (1943) Selective reversible inhibition of microbial growth with pyrithiamine. J Exp Med 78:489–497

Zhang Y, Epstein LM (1996) Cloning and characterization of extended hammerheads from a diverse set of caudate amphibians. Gene 172:183–190

5. Elute the cDNA with 5 µl of Buffer EB … temperature for 5 min.

7. Take 1 µl of the eluted cDNA … cDNA HS Assay Kit …

concentration is 0.2 …

3.2.13 … Repair …

RNA Versus Protein, How Structure Influences Targeting, a New Challenge for Drug Discovery

Alessandro Bonetti, Aurélie Lacroix, Emma Walsh, and Alice Ghidini

Contents

Abstract Unlike other large biomolecules, RNA carries two main tasks: an informational coding potential governed by its sequence and a catalytic/regulatory role, determined by its secondary and tertiary structure. During the last decade, a significant improvement in biophysical and biochemical techniques has enabled researchers to initiate exploratory studies on the relationship between RNA structure and its function. Among other technological improvements, the explosion of next generation sequencing (NGS) tools has allowed the transcriptome-wide investigation of RNA folding in cells. Deeper knowledge of 2D and 3D structures is extremely important for understanding the mechanisms of RNA function as well as for designing synthetic RNAs and the development of RNA-targeted drugs. RNA molecules can adopt specific 3D motifs that are now considered druggable and offer untapped potential to therapeutically modulate numerous cellular processes, including those linked to 'undruggable' protein targets. In parallel to the growing interest for the RNA targetome in the pharmaceutical sector, the in-silico modelling of RNA folds is

A. Bonetti · E. Walsh
Discovery Biology, Functional Genomics, AstraZeneca, Cambridge, United Kingdom

A. Lacroix · A. Ghidini (✉)
Sixfold Bioscience, Translation & Innovation Hub, London, UK
e-mail: alice@sixfold.bio

J. Barciszewski (ed.), *RNA Structure and Function*, RNA Technologies 14,
https://doi.org/10.1007/978-3-031-36390-0_6

developing complementary methods. Currently, RNA structure probing methods can only capture partial structure information. The ability to directly measure intact RNA structures could facilitate investigations of the functions and regulation mechanisms as well as druggability. In this chapter, we will summarize biophysical and biochemical strategies for determining RNA structures/motifs, including latest approaches that combine molecular biology strategies with NGS readouts.

Keywords RNA secondary structure · RNA tertiary structure · RNA biophysics · Next generation sequencing

1 Introduction to RNA Structures

DNA and RNA molecules possess very different functional roles. DNA molecules mainly contain the genetic code, whereas RNA molecules are involved in almost all life mechanisms: RNA translates the genetic code into proteins, can regulate gene expression or even catalyze biochemical reactions. It has been reported that about 85% of the human genome is transcribed into RNA, of which most has been estimated to be biologically functional, but only 2% is ultimately translated into proteins (Hendrix et al. 2005; Djebali et al. 2012). The structure of an RNA molecule mediates its activity, stability, and biological function. For example, the 3D structure of RNAzymes mediate their catalytic effects. It is therefore evident that studies on RNA structures can help understand biological processes.

From a chemical point of view, RNA differs from DNA in only two aspects: the presence of a 2′-hydroxyl group in the ribose and the lack of a methyl group in the uridine in most cases (thymine has been found in tRNA). These chemical changes impact the sugar puckering conformation (relative orientation of the base/phosphate/sugar), leading to changes in the physical and structural properties of the two polynucleotides. The 2′-hydroxyl group also plays a major role in nucleic acid interactions with water molecules and therefore conformation (Fingerhut 2021). Interestingly, RNA folds in a variety of tridimensional structures resembling protein folding. Figure 1 displays some examples of such RNA secondary structures. We will explore in the next section how RNA folds.

While the sequence defines the primary structure of RNA, the secondary structure of RNA is defined as any chain where the nucleotides are associated through base pairing. These interactions can be mediated through *canonical* Watson–Crick base pairing (A·U, G·C) with its complementary chain, in antiparallel orientation. Additional *non-canonical* base pairs have been reported, such as G·U base pairing (Wobble base pairing) or G·A pairs within internal RNA loops (Olson et al. 2019). More generally, each nucleotide within an RNA sequence contains charged, polar, aromatic groups, which can interact with almost every other nucleotide, or with their environment (Vicens and Kieft 2022).

The large variety of RNA's base-pairing interactions can be classified into 12 geometric base-pairing families, which can be divided based on Watson–Crick and

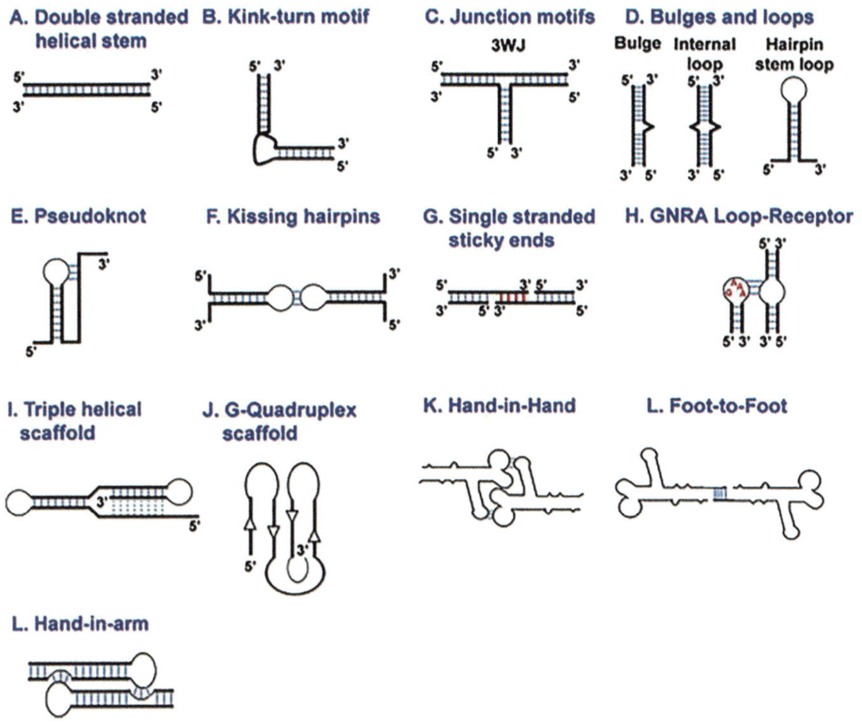

Fig. 1 RNA motifs. Some of the most common RNA motifs formed as a first step for assembly of larger order structures (Jasinski et al. 2017)

Hoogsteen base pairing, sugar, and glycosidic bonds' orientation (Fig. 2A) (Leontis and Westhof 2001). In all cases, the major driving force for base–base interactions is base stacking (Yakovchuk et al. 2006). The comparison with proteins is quite interesting: certain amino acids tend to interact preferentially with others aminoacids (e.g., hydrophilic residues on exterior, hydrophobic residues together) while RNA nucleotides do not exhibit the same selectivity (Stombaugh et al. 2009), and interactions are stabilized through H-bonding. Finally, it is worth noting that RNA can be post-transcriptionally modified. These modifications, such as base methylation, influence RNA's structure and function (Helm 2006; Fu et al. 2014; Incarnato and Oliviero 2017).

Just as proteins have helices and sheets, several distinct structural motifs have been described for RNA. Base–base interactions (paired or unpaired) mediate the formation of notable elements such as loops, mismatches, and bulges (Fig. 1), which then structurally define the thermodynamic stability of the secondary structure. Stem-loop structures, also called hairpins, can be external or internal and are the major building blocks for structured RNA. The stem-loop structure is composed of base–base interactions (Watson–Crick) ending with an unpaired loop. Non-canonical base pairs might play a more important role in the stem-loop structure, as more distantly

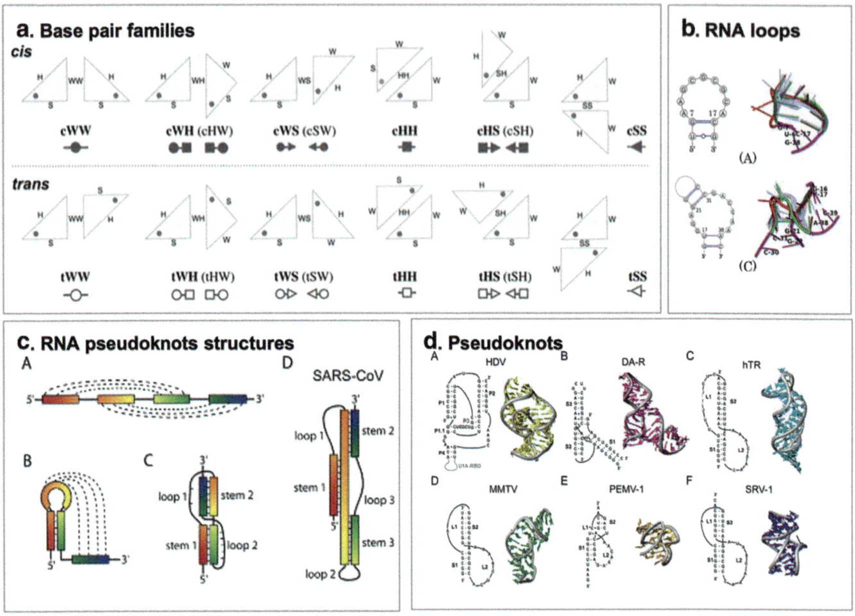

Fig. 2 RNA secondary structures. **a** 12 base-pair families have been described (Abu Almakarem et al. 2012). H stands for Hoogsten edges (squares), S for sugar edges (triangle), W for Watson–Crick edges (circles). Filled in symbols correspond to cis base pairs, while open ones are for trans base pairs. **b** Examples of sequences and structures (predicted in color, experimental in red) of RNA loops (Li et al. 2016). **c** Pseudoknot formation in the context of SARS-CoV. Dashed lines indicates base pairing. *A* Linear base-pairs elements. *B* Formation of the pseudoknots. *C* Pseudoknot fold. *D* SARS-CoV RNA pseudoknot is composed of three stems. D. RNA Pseudoknot structures and sequences in Nature. Stems and loops are numbered sequentially, unless otherwise noted. Structure coordinates were obtained from the Protein Data Bank (http://www.rcsb.org), and structural representations are produced using MOLMOL software (Staple and Butcher 2005)

located nucleotides may be brought in proximity and form less favorable base–base interactions. Stem-loop structures, also called hairpins, can be external or internal and are the major building blocks for structured RNA (Fig. 2b). Stem-loops are typically found in tRNA or in ribozymes. For example, the hammerhead ribozyme contains three stem-loop motifs (Scott et al. 2013). Pseudoknots, containing at least two stem-loop structures partially intercalated, are another important motif within RNA secondary structures (Fig. 2c). Several pseudoknots exist, but the H-type (hairpin type, involving bases in the loop of a hairpin, and often also referred as simple pseudoknots) is prevalent, yielding very complex yet stable structures. The H-type contains two stem regions connected by single-stranded loops, and often the stems are stacked which then give a quasi-continuous helix (Fig. 2d) (Brierley et al. 2007). Databases, such as Pseudobase++ (Taufer et al. 2009), have been created to group pseudoknots and understand the different types and structures existing (Legendre et al. 2018). Pseudoknots were first discovered in the turnip yellow mosaic virus

(Rietveld et al. 1982), and can be found in other viruses such as coronaviruses. Pseudoknots play a role in gene expression and can form the catalytic core of ribozyme or telomerases (Staple and Butcher 2005), highlighting the link between RNA structure and function.

Similarly to the protein data bank, nucleic acid structures have been grouped in various databases, with their number growing over years. Collection and publication of RNA structures (secondary and tertiary) within public databases, such as RNA CoSSMos 2.0 or RNA Frabase 2.0 (Popenda et al. 2010; Richardson et al. 2020), will also help scientists gaining more understanding of the characteristics of secondary motifs for the formation of tertiary structures, and collect data to improve computational predictions.

Several softwares have been developed to help predict RNA secondary structure, often calculating lower-free energy conformations. They are quite efficient and accurate for predicting highly constraining Watson–Crick base pairs (Bonnet et al. 2020; Vicens and Kieft 2022), but still lack resolution for less canonical structures, or structure changes in varying conditions (the effect of salts for example). As more and more RNA structures are being collected and published, along with the help of machine-learning tools, we expect RNA secondary structures prediction to greatly improve in the coming years (Zhao et al. 2021).

2 RNA Higher Order Structures

The next level of organization is defined as the tertiary structure. This is the association of secondary structure elements through various intramolecular interactions such as Van der Waals interactions or hydrogen bond formation.

For nucleic acids, the double helix structure is the predominant three-dimensional arrangement found in nature (Fig. 3a). Helical RNA (double-stranded) is mainly found in A-form and driven by canonical base pairing (while DNA adopts more predominantly a B-form helix). The backbone of dsRNA has six torsion angles plus one angle between the base and the sugar (compared to two for protein backbone structures), which allows RNA to adopt its helical form in almost any condition (Fig. 3b) (Murray et al. 2003). The configuration of each base contributes to the helical curve for a given structure. Other helical conformations have been described (B–, Z–, C–, L–…), including RNA triplexes (Ghosh and Bansal 2003). Tertiary RNA–RNA interactions can be categorized into relatively few structural motifs: pseudoknots (described previously), A-minor motifs, tetraloops, ribose zippers, kissing loops and coaxial stacking (Fiore and Nesbitt 2013). We shortly describe the relevance of all of them in the next section.

Within tertiary structures, new interactions between bases are of notice, such as the A-minor motif, which involves a canonical base pair and an adenine binding the minor groove (Fig. 3c). This motif was found in the large ribosomal subunit.

Tetraloops are four-base hairpin loop motifs, such as GNRA (N, any nucleotide and R, purine) (Fiore and Nesbitt 2013). Tetraloops are involved in tetraloop–receptor

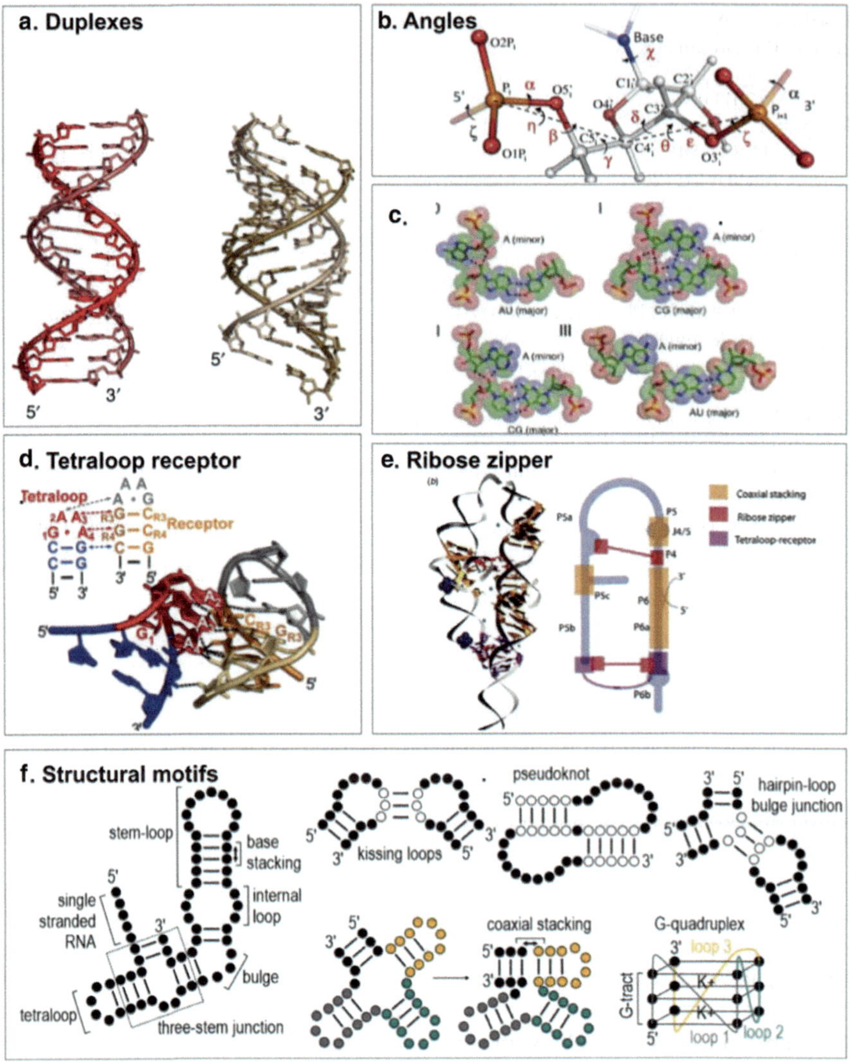

Fig. 3 Tertiary RNA structures. **a** Nucleic acids structures. B-form DNA (red) and A-form RNA (yellow) (adapted with permission from Reiter et al. (2008)). **b** Dihedral angles in nucleotides. The angle describes the rotation of the base relative to the RNA backbone, while the six other angles define the course of the backbone (adapted with permission from Frellsen et al. (2009)). **c** A-minor interaction subtypes from the large ribosomal subunit adapted with permission from Butcher and Pyle (2011)). **d**. GAAA tetraloop (red) interaction with tandem C:G base pairs (orange) in a helix minor groove, as within the hammerhead ribozyme (adapted with permission from Fiore and Nesbitt (2013)). **e** Motifs in the P4–P6 domain of the group I intron structure, including coaxial stacking, ribose zipper, and tetraloop-receptor interactions (adapted with permission from Hendrix et al. (2005)). **f** Common RNA structural motifs present in secondary and tertiary structures (adapted with permission from Taylor and Sobczak (2020))

interactions, where the loop motif meets a so-called receptor motif within an RNA duplex creating tertiary contact, stabilizing the entire structure (Fig. 3d). The ribose zipper is formed by consecutive hydrogen-bonding interactions between ribose 2′-hydroxyls within different positions of one strand or between RNA strands (Fig. 3e). Ribose zippers can link stem with loop chain segments (Tamura and Holbrook 2002). Kissing-loops are another major motif in RNA–RNA interactions (Carr and Marky 2019). Kissing loops are a tertiary structure defined by the interaction between the unhybridized nucleotides of two stem loops (Fig. 3F). The HIV-1 virus has two copies of genomic RNA that are brought together via kissing-loops interactions (Butcher and Pyle 2011; Mundigala et al. 2014).

Additionally, helix-helix interactions can happen. The interactions between elements of two separated helices can be categorized into two classes, one where the two contiguous helices stack on each other, or one where the two distant helices accommodate each other's grooves (Fig. 3d–f). The described structural flexibility allows RNA molecules to not only function as intermediates in genomic information transfer from DNA to proteins but also as a regulation and catalytic units. Helices can interact with one another through coaxial stacking in which terminal bases can stack resulting in an assembly of two macromolecules (Fig. 3f). Motifs such as kissing loops, three-way junctions (Tyagi and Mathews 2007) or quaternary structure formation (such as tRNA) are stabilized through helix-helix interactions (Fig. 3f). Considering coaxial stacking allows for the improvement of RNA structure predictions (Walter et al. 1994).

The conformation of any type of RNA structure is dependent on its environment including the concentration of ions in the media, the temperature, or the pH. It is also important to consider the presence of specific molecules or proteins in the environment, such as in the macromolecular crowded space of a cell. In particular, the presence and concentration of metal ions plays a major role in RNA folding (Draper 2004; Kolev et al. 2018). RNA molecules are highly negatively charged. Therefore, their folding into their tertiary structures involves close interaction of negative charges and a consequent significant charge–charge repulsion which is minimized by metal ions. Magnesium ions are critical to stabilize 3D structures (Kolev et al. 2018; Schauss et al. 2021). Such interactions with cations have remained hard to model and predict, though we are witnessing the emergence of more precise computational tools.

The energy involved in the RNA folding process includes several components (Draper 2004; Sun and Chen 2019). To understand how ions facilitate the formation of RNA tertiary structures, we need to quantify the different energetic components of this process, including the conformational entropy, the base-pairing/stacking interaction energy, the ion-mediated global electrostatic energy, and the free energy for specific tertiary contacts.

Finally, it is worth mentioning that RNA molecules can adopt different dynamic conformations, some conformations forming more rapidly than others, some being more stable, and therefore making up for the majority in the so-called RNA ensemble (Ganser et al. 2019; Vicens and Kieft 2022). These conformations can arise because of the presence of multiple energetic minima. RNA is not a static molecule and a lot of structures forming and unforming can occur overtime, mediated by transiting binding. Considering conformational dynamics when characterizing RNA secondary and tertiary structures is crucial in understanding biological processes better.

The elucidation of structural motifs has helped the development of computational tools to predict the formation of 3D structures. Different approaches have been used over the years. When a similar structure/sequence is available on a database, prediction *from the sequence* (de novo, force-field simulation such as simRNA), *from a template* (also called comparative modeling, and based on sequence alignment such as ModeRNA) or *using fragment libraries* (fragment assembly of different motifs founds, such as VfoldLA) have been developed (Rother et al. 2011; Li et al. 2020). When homology cannot be found, the 3D structure can be predicted according to the prediction of the 2D structure. For example, RNAfold (Vienna) can be used for 2D, followed by RNAComposer to assemble the structure (Lorenz et al. 2011). Other multidimensional models have been recently developed to improve computational predictions. As such, EvoClustRNA takes advantage of both de novo and comparative modeling for RNA 3D structure predictions (Magnus et al. 2019). In general, users tend to verify predictions using different tools available (comparing different structures). The accessibility of databases, the increasing number of data gathered around RNA structures, and the development of multi-dimensional computational tools should keep helping the development of robust softwares predicting structural conformations of RNA.

3 Biophysical Techniques to Measure and Characterize RNA Secondary Structures

Predicting and measuring tertiary RNA structures directly from the primary structure remains challenging, hence the interest to first characterize secondary structures. Physical, spectroscopy and chemical-based methods have been used, and we will describe herein the main tools currently available.

3.1 Physical Methods

Physical methods such as non-denaturing polyacrylamide gel electrophoresis (PAGE) are an easy and economical first-approach method to investigate secondary structures formation. Differences in molecule migration in the gel indicates formation of different structures (Fig. 4a) (Lilley 2004). Mutations within the structure and subsequent PAGE analysis can be used to gain structural insights (Jacques and Susskind 1991).

As one of the first spectroscopy methods established, thermal denaturation has been widely used to study duplex formation of RNA and DNA structures. These

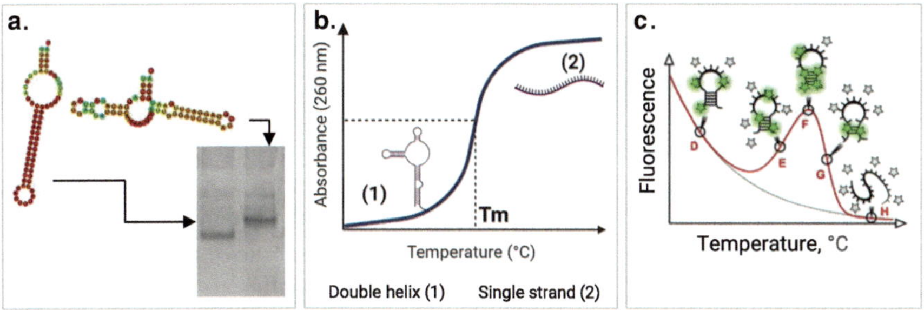

Fig. 4 Physical methods to study structures. **a** Native PAGE as a method to investigate secondary structures formation. Different RNA structures will run differently (despite having the same length) (reproduced with permission from (Baranovskaya et al. 2019). **b** Thermal denaturation. Typical melting curve for a double helix to random coil transition, which can be used to determine the melting temperature (T_m) (figure was created with Biorender.com). **c** Differential Scanning Fluorimetry. Use of a fluorescent reporter dye to track RNA structures melting and stability with high throughput (reproduced with permission from Silvers et al. (2015))

experiments are performed using the hyperchromicity associated with the folded-to-single stranded transition of nucleic acids. Practically, absorbance at 260 nm increases as temperature increases and RNA is melted from its helical form to single-stranded form (Fig. 4b) (Mergny and Lacroix 2003). This allows for the determination of the melting temperature T_m (i.e., the temperature at which 50% of the RNA remains structured) through simple UV–V is spectroscopy systems. Urea is an alternative to obtain denaturation parameters (Shelton et al. 1999). Other non-canonical structures, such a G-quadruplexes can be studied at other wavelength than the RNA maximum absorption peak (e.g 295 nm for G4s) (Mergny and Lacroix 2009). Thermal denaturation is commonly measured to confirm the formation of secondary structures, and to measure thermodynamics parameters such as the enthalpy/entropy contribution in duplex formation using Van't Hoff plots, or even to look at folding transitions (Mergny and Lacroix 2003). The hysteresis, i.e., the differences in melting temperature between folding and unfolding, can also provide information regarding thermodynamics and assembly pathways of duplexes (Harkness et al. 2018). Finally, and more recently, these experiments have been developed using fluorescent probes to perform Differential Scanning Fluorimetry (DSF) measurements (Fig. 4c) (Silvers et al. 2015). Indeed, intercalating dyes, such as SYBR Green or Ribogreen, can discriminate between single strands, unstructured regions and helical RNA, and fluorescence is observed only upon binding to 2D structures (Silvers et al. 2015).

Circular dichroism (CD) spectroscopy has also been widely adopted to monitor RNA secondary structures and has historically been one of the pioneering methods to discover new secondary structures (quadruplexes in particular). Indeed, just like UV–vis, CD is relatively inexpensive, uses low amounts of sample, and allows structural mapping of RNA. CD uses the differences between interaction of chiral molecules with circular polarized light (Berova et al. 2000). Nucleic acids bases are planar, but the asymmetry is given by the sugar of each nucleotide. Base interactions give rises to intense CD bands, making CD extremely sensitive to secondary structures (Fasman 2013). This method allows for the distinction between different conformations (each having a specific signature) or even look at folding transitions of discrete conformations (Kypr et al. 2009). Like UV–vis measurements, denaturation experiments or salt-dependency experiments, can be easily performed to unfold structures and confirm their formation (Fig. 5a). Recently, synchrotron radiation CD was used to extend the possibilities (le Brun et al. 2020). CD methods can also be applied to chemically modified oligonucleotides, for example, in the case of phosphornothioate strands (Clark et al. 1997). Interestingly, for proteins, online tools exist to predict structures from a CD spectrum (Sathyaseelan et al. 2020), while for nucleic acids the tool has been used primarily for qualitative structural assessment. Recently, Chaires et al., described the use of singular value decomposition method (SVD) to derive G-quadruplexes structures from their CD spectra (del Villar-Guerra et al. 2018). In another example, machine learning was used to develop webserver tools for RNA structural prediction from CD, called CD-NuSS (CD to nucleic acids secondary structure) with 85–87% accuracy (Sathyaseelan et al. 2021).

Besides CD and UV–Vis, new alternative spectroscopy methods have recently started to emerge. For example, RNA folding can be studied using optical tweezers (Bustamante et al. 2021) Typically, the RNA of interest is attached to two micron-sized beads, one being held in an optical trap (Fig. 5b). By monitoring the distance between the beads upon application of a mechanical force, one can study conformational change. The RNA molecule can also be forced to refold into certain conformations (Li et al. 2007). This single-molecule approach allows for the differentiation of various conformations and improvement of study of structural polymorphism. With tweezers, each molecule can be handled individually. Single-molecule methods based on fluorescence (such as smFRET) have also been used (Woźniak et al. 2008). By measuring FRET efficiency, one can retrace the geometry of the RNA structures.

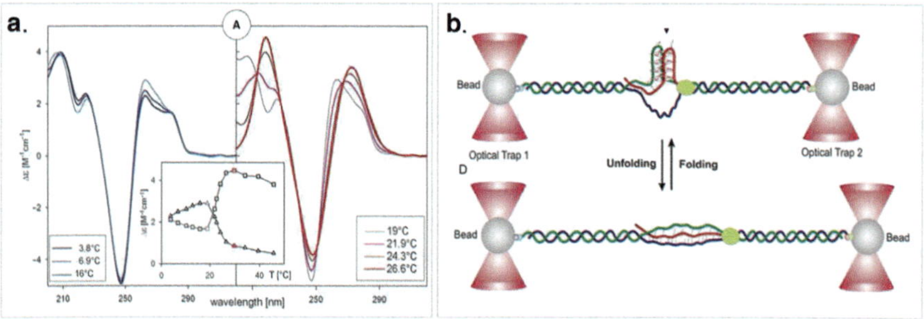

Fig. 5 Spectroscopy methods. **a** Circular dichroism. The blue panel reflects changes within double-helical structure and the red panel shows a helix–coil transition at different temperatures (adapted with permission from Kypr et al. (2009)). **b** Optical tweezers. Mechanical unfolding of structures formed in the G-core (reproduced with permission from Shrestha et al. (2014))

3.2 Enzymatic-Based and Chemical-Based Methods

Enzymatic-based method are an alterative to confirm or elucidate secondary structures of RNA, allowing scientists to scrutinize the effect of the surrounding environment (e.g., magnesium concentration, temperature) (Knapp 1989). The need to investigate secondary structures in physiologically relevant conditions strengthened the use of enzymatic reactions as complementary methods to spectroscopy methods. Furthermore, enzymatic-based techniques are relatively inexpensive and easily accessible. As an example, nucleases can be used to probe double-stranded regions, by analysing degradation products (Ziehler and Engelke 2000).

The most used agents for endonucleolytic cleavage are single-stranded specific RNases, such as RNase T1, RNase T2, RNase U2, RNase CL3, or RNase A, or the ds-specific RNase V1, each of them with their own specificity in terms of sequences/nucleotides (Fig. 6a) (Gilmer et al. 2021). RNase CL3 cleaves primarily 3′ to C-residues, while RNase V1 cleaves stacked residues with no nucleotide's specificity (Daou-Chabo and Condon 2009). Nuclease S1 is an endonuclease primarily cutting single-stranded oligonucleotides into mononucleotides. To illustrate this, we can cite Condon and Daou-Chabo who showed that RNase J1 could be used to confirm the structure of a tRNA (Fig. 6b) and subsequently probe an unknown mRNA structure (Daou-Chabo and Condon 2009). These methods have also been successfully applied for mapping protein–RNA interactions (Nilsen 2014). Although enzymatic-based methods are relevant tool in RNA structure exploration, have limitations in terms of their use in cells and in more relevant in vivo contexts.

Besides enzymatic footprinting, chemical degradation can also be used for techniques such as DMS (Dimethylsulphate) footprinting Tijerina et al. (2007). DMS causes methylation of adenosine, cytosine, and guanosine (Fig. 6c), which facilitates subsequent cleavage with chemical agents (Tijerina et al. 2007). The advantage of chemical footprinting is the compatibility with most buffer and in vivo systems, as well as the small size of the reagents compared to enzymes. The relative reactivity caused by folding states can be used to map structured sites. Chemical probe can target the base (for example DMS, acylation with SHAPE reagents such as N-methylisatoic anhydride, NAz nicotinoyl-azide), the sugar (e.g., radical formation) or the phosphate (e.g. ENU ethyl-nitrosourea) (Gilmer et al. 2021). Upon digestion by enzymes or chemical agents, fragments were originally studied using denaturing PAGE (in general involving 32P labeling). Methods such as capillary electrophoresis (CE) have then been used and accelerated the throughput and made the analysis more quantitative. As such, many structures have been validated using such methods: as an example, the entire HIV-1 RNA genome could be profiled with CE (Watts et al. 2009). Next generation deep sequencing and mutational profiling have recently been developed, and we will review those approaches in a dedicated section in this chapter (next-generation sequencing to identify RNA structures). Finally, it is worth noting that both chemical and enzymatic based methods can be used together to probe RNA folding (Maurin et al. 2015).

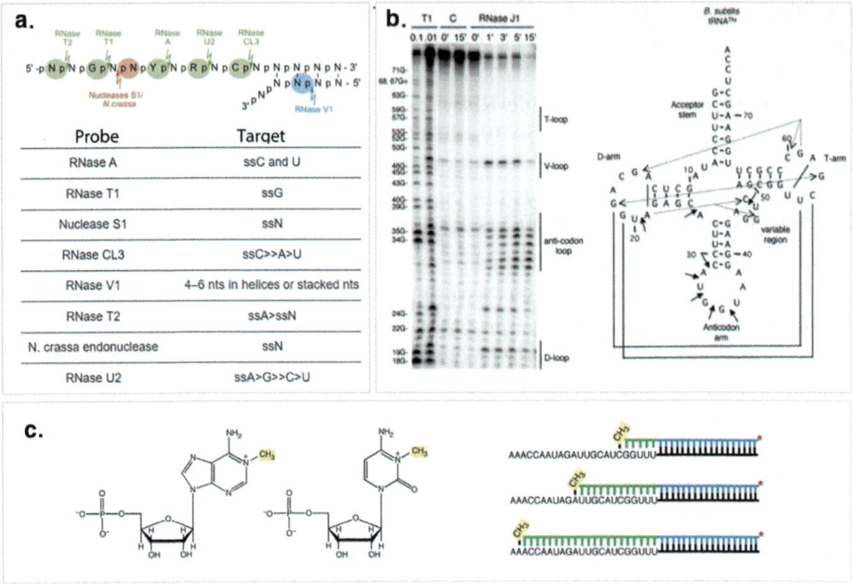

Probe	Target
RNase A	ssC and U
RNase T1	ssG
Nuclease S1	ssN
RNase CL3	ssC>>A>U
RNase V1	4–6 nts in helices or stacked nts
RNase T2	ssA>ssN
N. crassa endonuclease	ssN
RNase U2	ssA>G>>C>U

Fig. 6 Footprinting methods to probe structure. **a** Enzymatic-based methods to probe RNA structures (Gilmer et al. 2021). **b** Example of enzyme-based footprinting of RNA 3D structures: RNase J1 was used to probe the structure of *B. subtilis* tRNA (adapted with permission from Daou-Chabo and Condon (2009)). **c** Chemical footprinting methods illustrated with the use of DMS (adapted with permission from Tijerina et al. (2007))

4 Biophysical Techniques to Measure and Characterize RNA Tertiary and Higher Order Structures

4.1 X-ray Crystallography and Small-Angle X-ray Scattering (SAXS)

Two of the techniques which have been initially reported for the investigation of RNA 3D structures are X-ray diffraction and NMR.

X-ray crystallography has been by far the most used technique, initially developed for 3D DNA structures and hammerhead enzymes. The major drawback of the method is the low resolution, and it remains challenging to get an adequate level of structural information compared to similar techniques in proteins (Turnbull and Wu 2021). Accurate RNA 3D structure modelling at the typical diffraction limit of 2.5 Å resolution can be difficult since some features, such as the sugar atoms, are not easily positioned compared to more rigid groups like the phosphate backbone. The field has currently made great advancements, but the flexibility of RNA still represents the major hurdle. Currently RNA X-ray crystallography utilizes a variety of molecular engineering approaches such as the use of chaperone proteins (e.g. U1A and L7Ae) to bind to specific RNA motifs to stabilize the RNA crystal (Ferré-D'Amaré et al. 1998; Huang and Lilley 2013) or grafting of stable and intermolecular contact-prone RNA motifs (tetraloops and kissing loops) with their receptors or again introducing heavy or anomalously scattering atoms useful for de novo structure (Zhang and Ferré-D'Amaré 2014).

Flexibility of long RNAs is frequently the main reason for the failure of several structural study attempts. For this same reason, little is known about RNA native structure since high resolution characterization is, until now, mostly restricted to short RNA. Instead, long RNAs are unlikely to crystallize unless complexed with proteins or antibiotics, therefore an unperturbed structure cannot be elucidated through this technique (Gopal et al. 2012; Ma et al. 2022).

Small-angle X-ray scattering (SAXS) is a technique that allows for the study of macromolecule structure (Fig. 7a). The biggest advantages of SAXS are the sample preparation in solution, and the fast analysis, provided access to a synchrotron. SAXS data allows for the quantifications of interparticle interactions and most importantly to characterize ensembles of flexible structures such as RNA (Pollack 2011; Chen and Pollack 2016). It provides a diverse set of parameters that describe biomolecules, including global information about size and shape, intermolecular association, domain motion and flexibility.

Knowledge of sample conditions is essential for structural data generation since the resulting structural output is highly affected by salts, RNA concentrations and buffers. The large, negative charge of the RNA backbone can lead to strong repulsive forces between different molecules (Chen and Pollack 2016). These intermolecular interactions are easily detected by SAXS. The counterion atmosphere around RNA strongly affects interactions between RNAs, and the resulting SAXS profiles. Counterions also scatter and the ions those localized around the RNA contribute to the SAXS signal, primarily at the lowest scattering angles. Strong variations in scattering resulting from changes in electrostatic screening, such as when $MgCl_2$ are present. The effect of repulsive and attractive interactions is to modify the overall scattering profiles of the isolated molecules by introducing a structure factor.

The sample concentration dependencies of SAXS can hide monomer structure motifs therefore titration measurements are essential to characterize or eliminate intermolecular association. The conditions chosen for the experiment are dependent on the structural information of interest. High sample concentrations favor strong repulsion at low salts, whereas high salt concentrations increase the probability of interparticle attractions or oligomer formation. Frequently, SEC (size exclusion chromatography) gives complementary information to SAXS and is used to identify unplanned associations that can occur at the relatively high RNA concentrations used for SAXS.

A combinative approach including SAXS and NMR has successfully allowed the structural determination of the large RNAs of a trans-cleaving Neurospora Varkud Satellite ribozyme (Dagenais et al. 2021), through the structural determination of subensembles of the various subdomains (Fig. 7b). Currently most of the reported low-resolution 3D reconstructions of RNAs obtained by SAXS, have been a maximum of $\sim$200 nt long. For larger RNAs, it is technically challenging since the

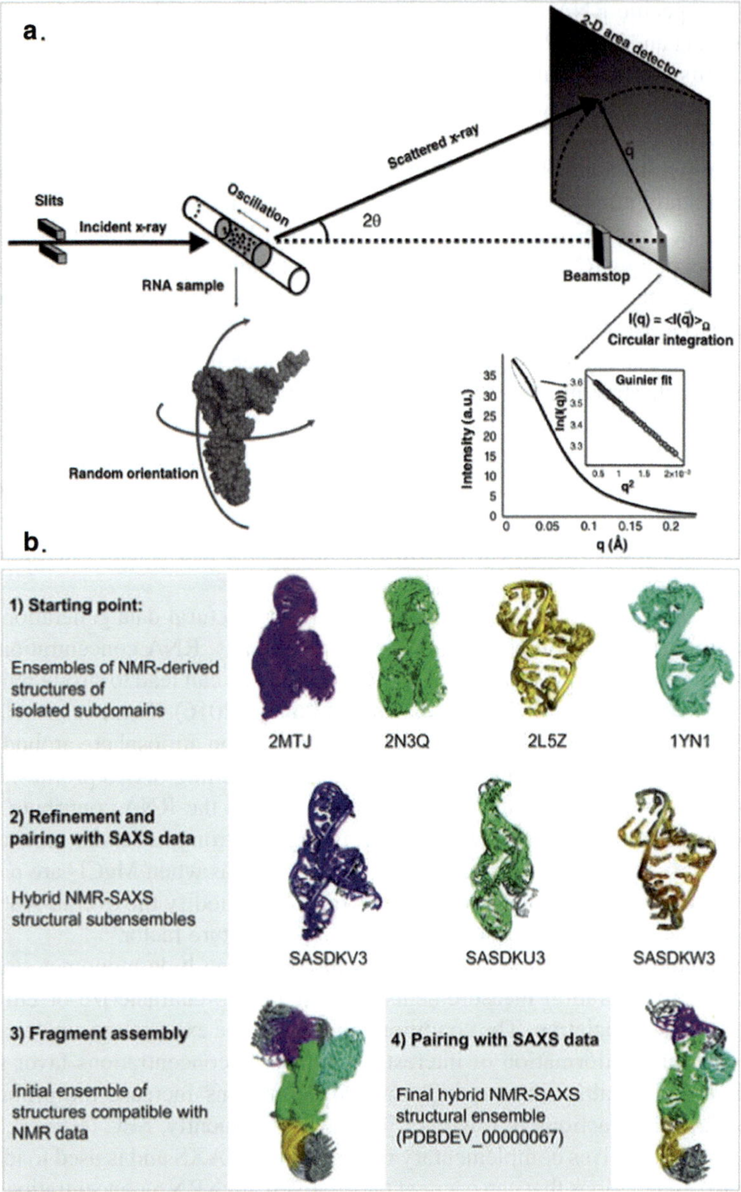

Fig. 7 SAXS experiment setup and fragment assembly strategy through NMR data. **a** Schematic of a SAXS experiment. The top panel represents a typical SAXS setup (reproduced with permission from Wiley and Sons and Copyright Clearance Center Chen and Pollack (2016)). An incident X-ray beam is collimated using sets of slits, then passes through a quartz capillary that contains a plug of the sample solution. X-Rays scattered by the sample are captured by a 2D area detector. The isotropic scattering pattern arises from the random orientation of the RNA in solution. The scattering profile of tRNA is plotted in the lower right

scattered intensity for overall molecule determination lies at very small angles, and close to the incident beam (Gopal et al. 2012).

4.2 Nuclear Magnetic Resonance (NMR)

The very same limitation, the RNA length, has also been haunting the NMR (Nuclear Magnetic Resonance) field. In the past 10 years only a relatively small number of large RNA structures have been determined through NMR (Barnwal et al. 2017).

Due to RNA limited chemical shift dispersion, NMR spectra of large and unlabelled RNAs are crowded and make ambiguous the resonance assignment process. One of the most common techniques to solve this issue is to incorporate isotopically enriched ^{13}C and ^{15}N rNTPs in the RNA allowing an NMR-based study of RNA structure, dynamics, and in some cases ligand binding (Zhang and Keane 2019).

Two major strategies that have helped technique advancement are segmental labeling and *divide and conquer* (Barnwal et al. 2017). The first allows for a simplification of the method minimizing external interference and is of particular interest when studying interaction of far apart regions of an RNA molecule. Segmental labeling helps reducing the number of resonances in NMR spectra, so that specific domains can be studied. The main challenge is the amount of segmental labeled RNA needed for the experiments. One of the simplest approaches is to transcribe segments of the full RNA with T7 RNA polymerase separately and subsequently ligate the segments (Nelissen et al. 2008; Duss et al. 2012). It is important to underline that this technique does not solve the NMR relaxation problem, but together with specific labeling such as 13C/15N/2H allows for heteronuclear triple resonance scalar correlated experiments reducing signal ambiguity. The second strategy, *divide and conquer* (Dagenais et al. 2021; Moudgal et al. 2022) is based on the determination of high-resolution structure of singular domains of a larger biomolecule. The technique has been widely used for proteins, allowing for reconstruction of large complexes. This technique is particularly suitable for RNAs that have no interactions between the stem-loops.

Another recent development has been reported for the use of magic angle spinning (MAS) solid-state NMR (ssNMR) in RNA structure resolution. MAS can be applied to non-soluble or non-crystalline biomolecules with a higher degree of complexity (Aguion and Marchanka 2021), making RNA an obvious candidate for the technique.

NMR chemical shifts have also been employed to study conformational changes in RNA ensemble by using previously published NMR residual dipolar coupling (RDC) data to guide the selection of conformers (Shi et al. 2020). The FARFAR-NMR approach described by Shi et al. can immediately be applied to previous RNA NMR structures, determined using conventional approaches, and generate dynamic ensembles information, where accuracies can be tested using the known chemical shift.

4.3 *Cryogenic Electron Microscopy (Cryo-EM)*

Due to the large number of intramolecular base-pairing, large RNA molecules have several structural combinations with the same energetic stability, hence ensembles of structures serve as a better representation, governed by statistical properties. Single-particle cryogenic electron microscopy (Cryo-EM) involves flash-freezing solutions of biomolecules followed by the electrons bombardment to produce microscope images at single molecule resolution. Single particles analysis cryo-EM has provided accessibility to challenging biological systems such as RNA, but unfortunately protein-free RNA structures accumulate in the Electron Microscopy Data Bank (EMDB) at a much slower rate compared to protein-nucleic acid complexes due to complex methodology development. To date, there are only three protein-free RNA cryo-EM structures determined at 4 Å or better resolution (Fig. 8b) (Zhang et al. 2019; Su et al. 2021).

A new strategy called ROCK (RNA oligomerization-enabled cryo-EM via installing kissing loops) has been recently shown by Liu et al. (Liu et al. 2022). Kissing-loop sequences are inserted into stem stretches (non-functional ones), resolving structural flexibility. Through ROCK, Tetrahymena group I intron sequence has been recently resolved at 2.98-Å, allowing for a de novo model of a complete RNA sequence.

5 Next Generation Sequencing Methods to Identify RNA Structures

RNAs have the intrinsic capacity to fold into secondary and tertiary structures to enable interactions with other molecules ranging from cations and complex metabolites to macromolecules such as nucleic acids and proteins. Although we have discovered the structure of transcripts like transfer RNAs and ribosomal RNAs some decades ago, systematic identification of the structure of RNAs has been challenging due to the lack of high-throughput experimental approaches necessary to determine their folding. However, novel methodologies have the potential to systematically assess secondary and tertiary structures of transcripts leveraging large datasets produced by next generation sequencing technology.

One of the most challenging aspects in designing drugs that target RNAs is the lack of a clear understanding about the relationship between primary sequence, structure, and function. A better understanding of the folding of primary sequences into functional motifs and their specificity will also help to identify molecules that interfere with either the processing of mature transcripts or the interaction with other macromolecules. Current in silico predictions for RNA folding are far from perfect and experimental validation of predicted motifs is still a necessary step. Furthermore, generation of large experimental datasets of RNA structural information with

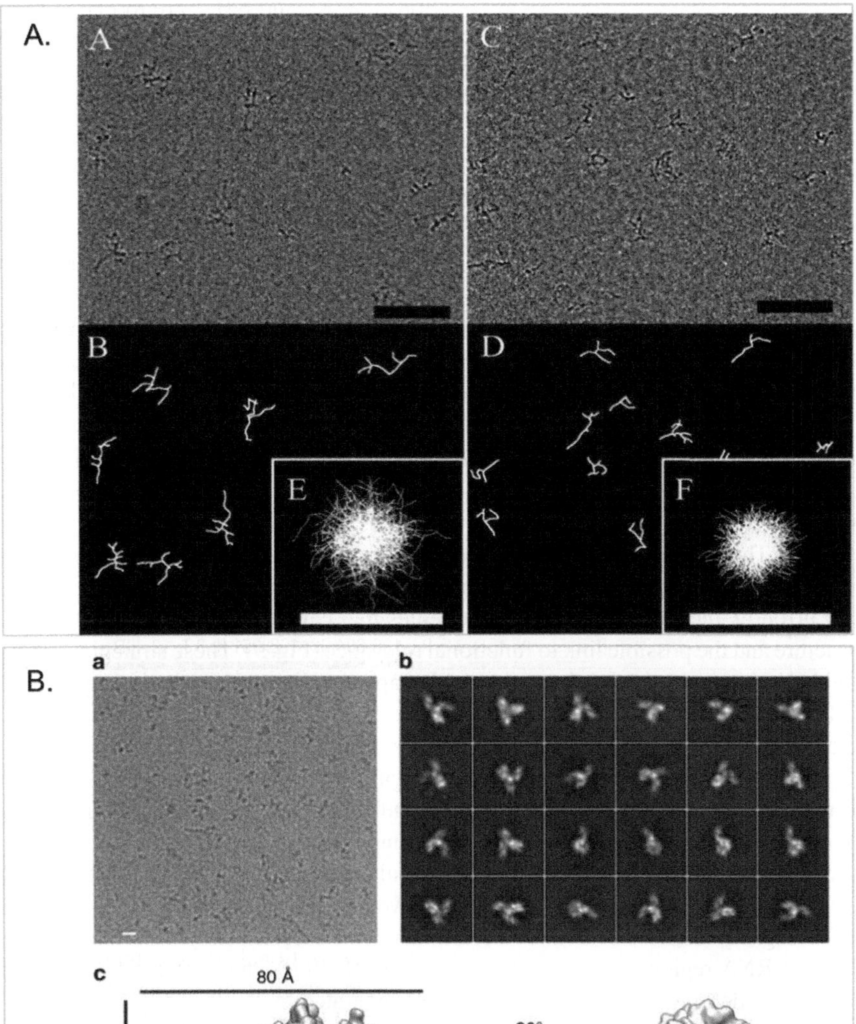
A.
A
C
B
D
E
F
B.
a
b
c
80 Å
70 Å
90°
Minor groove
Major groove

◀**Fig. 8** **A** Cryo-EM images of RNAs in films of freely suspended vitrified solution (reproduced with permission from (Gopal et al. 2012)). **A, C** Fourier band-pass filtered cryo-EM images of 1523-nt (**A**) and 975-nt. Individual RNA molecules with different configurations, suspended in random orientations, are seen as dark branched objects. **B, D** Traced skeletons of the molecules in panels A and C. For each RNA, the inset (**E, F**) depicts a hundred traced projections superimposed with their centres of mass in the registry. Scale bars, 60 nm. **B** Single-particle cryo-EM analysis of the apo SAM-IV riboswitch (reproduced with permission from Zhang et al. (2019)) a Representative motion-corrected cryo-EM micrograph. Scale bar represents 100 Å. **b** Reference-free 2D class averages. **c** Final 3D reconstruction in two different views

massive parallel sequencing is instrumental to developing more faithful model prediction. Ideally, machine learning algorithms have the potential to discover patterns in the data with minimal human intervention.

It is important to consider that RNAs do not fold into a single energy favorable secondary structure. Rather, the structure of a specific transcript can be described as an ensemble of several conformations, each with its own free energy. Conformations with low free energy are generally preferred as they have higher thermodynamic stability. Therefore, the RNA landscape in the ensemble is dominated only by a few conformations with adequate energetic stability.

Currently, there are two probing strategies that can be employed to study RNA structure and the possible link to functional relevance (Fig. 9). These strategies can be divided into enzymatic and chemical probing approaches, and both have the potential to be combined with high-throughput sequencing to gather a transcriptome-wide knowledge of RNA folding.

As discussed in the sections above, chemical, and enzyme-based approached can be used to gain insights on RNA structural conformations. Enzymatic footprinting is an in vitro approach employing specific ribonucleases (RNases) that cleave single- or double-stranded (ds) RNA nucleotides. Parallel analysis of RNA structure (PARS) uses enzymatic footprinting to map single- versus double-stranded regions of any RNA (Fig. 9a) (Kertesz et al. 2010). PARS uses two different enzymes, RNase V1 to digest dsRNA regions by cutting base-paired nucleotides and nuclease S1 to digest single-stranded regions of RNA. Fragments generated by the two different RNases are then used as inputs for the generation of high-throughput sequencing libraries to gather transcriptome-wide information of RNA folding. One limitation of the PARS approach is that it identifies RNA secondary structures in vitro that might significantly differ from how transcripts fold in vivo when interacting with other macromolecules and in the presence of different cations. Nevertheless, using this methodology in the budding yeast, Kertesz and colleagues retrieved structural information regarding thousands of messengers RNAs. Computational analyses revealed higher degrees of folding in coding regions of transcripts compared to untranslated regions, such as a three-nucleotide-cycle periodic structure signature found across coding regions (Kertesz et al. 2010).

Chemical probing utilizes chemicals that covalently modify residues of ribonucleotides present in specific configurations. One of the most common techniques used for chemical probing is Selective 2′-hydroxyl acylation by primer extension

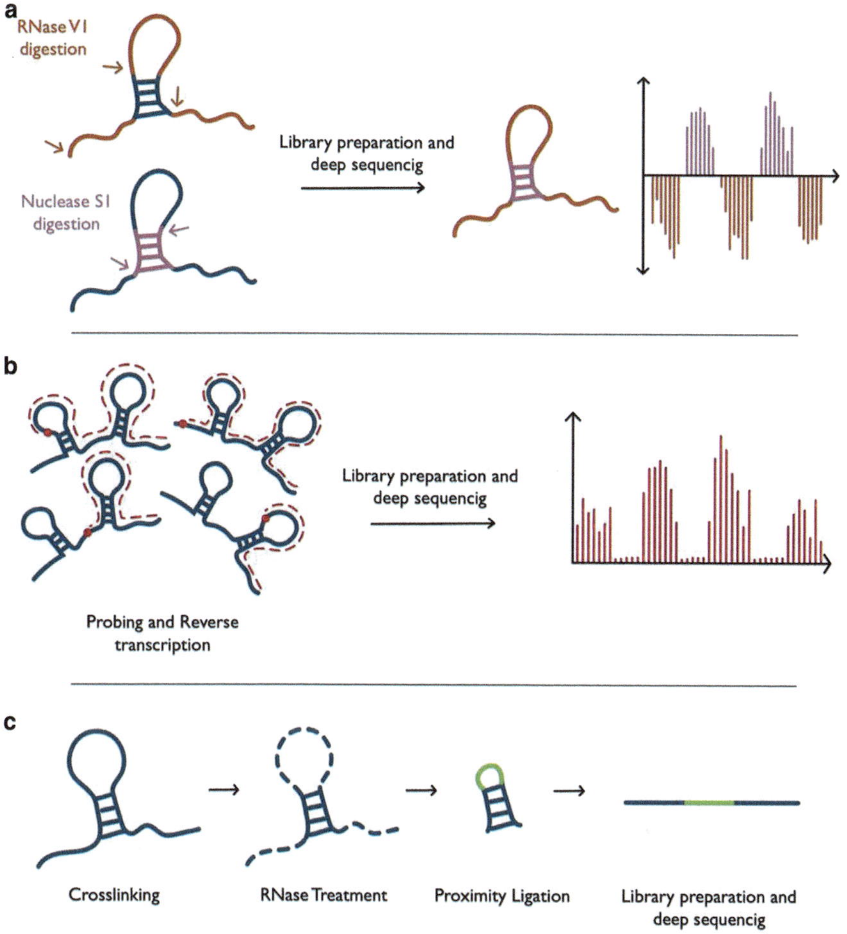

Fig. 9 Strategies to study RNA structure. **a** PARS uses enzymatic footprinting by cutting double stranded (RNase VI) and single stranded (Nuclease S1) regions of an RNA transcript and the fragments are sequenced to determine which regions are likely to be paired or unpaired. **b** SHAPE selectively acylates residues (red circle) in single stranded regions of an RNA molecule. When performing reverse transcription (here represented by the dotted red line) the acylation terminates the reaction and the resulting cDNA fragment is sequenced to determine which regions are single stranded. **c** PARIS first crosslinks paired RNA bases using psoralen-derivative 4′-aminomethyltrioxsalen (AMT) and photoactivation. The single-stranded RNA is then digested, and the fixed fragments are proximity ligated (here indicated by green line). After reversal of crosslink, fragments are sequenced to identify the base pairings of the duplexed RNA

(SHAPE) (Fig. 9b). SHAPE is a strategy that measures nucleotide reactivity by selectively acylating residues located in flexible and unpaired RNA regions (Wilkinson et al. 2006). Acylated residues cause the premature termination of reverse transcription reaction, resulting in truncated cDNA fragments that can be detected and analyzed with sequencing approaches. A significant improvement for the SHAPE technique has been the development of in vivo click-selective 2′-hydroxyl acylation and profiling experiment (icSHAPE) (Spitale et al. 2015). icSHAPE uses a chemical compound that is able to efficiently cross the cell membrane of living cells, then acylates residues located in unpaired regions and enables subsequent biotinylation of modified nucleotides for downstream enrichment. Captured transcripts are then used to generate libraries for high-throughput sequencing to interrogate RNA secondary structure at a transcriptome-wide level in living cells. Comparative analyses of the icSHAPE data obtained in mouse embryonic stem cells with purified RNA folded in vitro revealed different classes of structural elements controlling gene expression. Signatures encompassing translational start sites have conserved icSHAPE profiles both in vitro and in vivo conditions, suggesting that the folding of these elements is dictated by their primary RNA sequence. In contrast, transcript regions involved in binding to proteins or harbouring post-transcriptional modifications exhibit dynamic structural footprints (Spitale et al. 2015).

Approaches based on small-molecules-modification-induced probing generate rather limited information regarding RNA structure for each nucleotide base but do not directly identify specific base-pairing partners. Recently, several different methodologies that can capture intra- and intermolecular transcriptome-wide RNA–RNA interactions have emerged. Psoralen Analysis of RNA Interactions and Structures (PARIS) (Lu et al. 2016), psoralen crosslinked, ligated, and selected hybrids (SPLASH) (Aw et al. 2016) and LIGR-seq [ligation of interacting RNA followed by high-throughput sequencing (LIGR-seq) (Sharma et al. 2016), share a similar design that involves crosslinking in cells and proximity ligation. These methods employ a psoralen-derived crosslinking agent that traps RNA duplexes in living cells and enables the direct mapping of RNA regions engaging in intramolecular base-pairing (Fig. 9c). After crosslinking RNA is extracted and fragmented, it is followed by ligation of the two interacting strands in RNA duplexes. The resulting chimeric RNAs are sequenced, allowing the parallel identification of RNA regions engaging in intramolecular base-pairing. Apart from assessing double-stranded regions in multiple transcripts, these approaches can detect long-range duplexes and high order structures through the identification of intermolecular RNA–RNA interactions. Application of LIGR-seq technology to human cells revealed thousands of RNA-RNA interactions with the majority being mRNA-rRNA contacts that most likely originate from the process of translation (Aw et al. 2016). Similarly, LIGR-seq identified a complex intracellular RNA interactome, including the unexpected finding of a large number of contacts between mRNAs and small nucleolar RNAs (Sharma et al. 2016).

More recently, COMRADES (crosslinking of matched RNAs and deep sequencing) technology has introduced key improvements to study the dynamic structure of RNAs in cells (Ziv et al. 2018). Introduction of an affinity purification step using biotinylated oligonucleotides against the RNA of interest greatly enriches for interactions mediated by specific transcripts. Furthermore, a subsequent click reaction enables a second affinity purification of crosslinked regions, thus increasing the accuracy for structure probing. This dual enrichment enables deep structural assessment of RNAs of interest and therefore allows the identification of coexisting conformations (Ziv et al. 2018). COMRADES technology has been utilized to assess the structural complexity of viral RNA genomes during their intracellular cycle. Ziv and colleagues have identified the highly dynamic structure of Zika virus inside human cells (Ziv et al. 2018). The intraviral RNA–RNA interaction map revealed areas of compaction of the RNA genome and confirmed the presence of previously defined functional pseudoknots. Furthermore, downstream analyses identified the base–pair interaction of multiple host regulatory transcripts with the viral RNA genome (Ziv et al. 2018). Subsequently, the authors employed COMRADES workflow to study the folding dynamics of SARS-CoV-2 virus inside cells (Ziv et al. 2020). The identified RNA–RNA interactome highlighted long range contacts across the viral genome resulting in multiple coexisting conformations and the presence of site-specific interactions between SARS-CoV-2 genome and small nuclear RNAs encoded by the host (Ziv et al. 2020).

6 Conclusions

A striking lack of RNA structural knowledge compared to proteins, and the poor correlation between in vitro and in vivo structures, has channeled the focus of several research groups towards development of labeling chemical techniques associated with high-throughput sequencing technology (HTS). A particularly important topic and challenge for the study of RNA structures is to be able to obtain most of this information in in vivo conditions, until we can eventually reach the level obtained for protein predictions. With the in vivo adaptation of these technologies, it is currently possible to gain genome-scale structural information, which ultimately will lead to a deeper understanding of RNA intermolecular interactions with proteins and other RNAs (Solayman et al. 2022).

A deeper understanding and a more extensive data collection of RNA secondary and tertiary structures, with a specific attention to non-canonical interactions, is still in great demand.

Generation of high volumes of NGS datasets to enable AI-driven analyses for machine learning approaches to identify relevant structural conformations could slowly allow the field to gain information faster compared to the past years. Development of high throughput, high resolution and robust, standardized methods for experimental determination will help improving computational predictions.

References

Abu Almakarem AS, Petrov AI, Stombaugh J et al (2012) Comprehensive survey and geometric classification of base triples in RNA structures. Nucleic Acids Res 40:1407–1423. https://doi.org/10.1093/nar/gkr810

Aguion PI, Marchanka A (2021) Strategies for RNA Resonance Assignment by 13C/15N- and 1H-Detected Solid-State NMR Spectroscopy. Front Mol Biosci 8

Aw JGA, Shen Y, Wilm A et al (2016) In vivo mapping of eukaryotic RNA interactomes reveals principles of higher-order organization and regulation. Mol Cell 62:603–617. https://doi.org/10.1016/j.molcel.2016.04.028

Baranovskaya I, Sergeeva M, Fadeev A et al (2019) Changes in RNA secondary structure affect NS1 protein expression during early stage influenza virus infection. Virol J 16:162. https://doi.org/10.1186/s12985-019-1271-0

Barnwal RP, Yang F, Varani G (2017) Applications of NMR to structure determination of RNAs large and small. Arch Biochem Biophys 628:42–56. https://doi.org/10.1016/j.abb.2017.06.003

Berova N, Nakanishi K, Woody RW (2000) Circular dichroism: principles and applications. John Wiley & Sons

Bonnet É, Rzążewski P, Sikora F (2020) Designing RNA Secondary Structures Is Hard. J Comput Biol 27:302–316. https://doi.org/10.1089/cmb.2019.0420

Brierley I, Pennell S, Gilbert RJC (2007) Viral RNA pseudoknots: versatile motifs in gene expression and replication. Nat Rev Microbiol 5:598–610. https://doi.org/10.1038/nrmicro1704

Bustamante CJ, Chemla YR, Liu S, Wang MD (2021) Optical tweezers in single-molecule biophysics. Nat Rev Methods Primers 1:25. https://doi.org/10.1038/s43586-021-00021-6

Butcher SE, Pyle AM (2011) The molecular interactions that stabilize RNA tertiary structure: RNA motifs, patterns, and networks. Acc Chem Res 44:1302–1311. https://doi.org/10.1021/ar200098t

Carr CE, Marky LA (2019) Thermodynamic investigation of kissing-loop interactions. Biochimie 157:177–183. https://doi.org/10.1016/j.biochi.2018.11.012

Chen Y, Pollack L (2016) SAXS studies of RNA: structures, dynamics, and interactions with partners. Wires RNA 7:512–526. https://doi.org/10.1002/wrna.1349

Clark CL, Cecil PK, Singh D, Gray DM (1997) CD, absorption and thermodynamic analysis of repeating dinucleotide DNA, RNA and hybrid duplexes [d/r(AC)]12·[d/r(GT/U)]12 and the influence of phosphorothioate substitution. Nucleic Acids Res 25:4098–4105. https://doi.org/10.1093/nar/25.20.4098

Dagenais P, Desjardins G, Legault P (2021) An integrative NMR-SAXS approach for structural determination of large RNAs defines the substrate-free state of a trans-cleaving Neurospora Varkud Satellite ribozyme. Nucleic Acids Res 49:11959–11973. https://doi.org/10.1093/nar/gkab963

Daou-Chabo R, Condon C (2009) RNase J1 endonuclease activity as a probe of RNA secondary structure. RNA 15:1417–1425. https://doi.org/10.1261/rna.1574309

del Villar-Guerra R, Trent JO, Chaires JB (2018) G-quadruplex secondary structure obtained from circular dichroism spectroscopy. Angew Chem Int Ed 57:7171–7175. https://doi.org/10.1002/anie.201709184

Djebali S, Davis CA, Merkel A et al (2012) Landscape of transcription in human cells. Nature 489:101–108. https://doi.org/10.1038/nature11233

Draper DE (2004) A guide to ions and RNA structure. RNA 10:335–343. https://doi.org/10.1261/rna.5205404

Duss O, Lukavsky PJ, Allain FH-T (2012) Isotope labeling and segmental labeling of larger RNAs for NMR structural studies BT—Isotope labeling in Biomolecular NMR. In: Atreya HS (ed). Springer Netherlands, Dordrecht, pp. 121–144

Fasman GD (2013) Circular dichroism and the conformational analysis of biomolecules. Springer, US

Ferré-D'Amaré AR, Zhou K, Doudna JA (1998) A general module for RNA crystallization11Edited by D. E. Draper. J Mol Biol 279:621–631. https://doi.org/10.1006/jmbi.1998.1789
Fingerhut BP (2021) The mutual interactions of RNA, counterions and water-quantifying the electrostatics at the phosphate–water interface. 12880 I Chem Commun 57:12880. https://doi.org/10.1039/d1cc05367a
Fiore JL, Nesbitt DJ (2013) An RNA folding motif: GNRA tetraloop–receptor interactions. Q Rev Biophys 46:223–264. https://doi.org/10.1017/S0033583513000048
Frellsen J, Moltke I, Thiim M et al (2009) A probabilistic model of RNA conformational space. PLoS Comput Biol 5:e1000406
Fu Y, Dominissini D, Rechavi G, He C (2014) Gene expression regulation mediated through reversible m6A RNA methylation. Nat Rev Genet 15:293–306. https://doi.org/10.1038/nrg3724
Ganser LR, Kelly ML, Herschlag D, Al-Hashimi HM (2019) The roles of structural dynamics in the cellular functions of RNAs. Nat Rev Mol Cell Biol 20:474–489. https://doi.org/10.1038/s41580-019-0136-0
Ghosh A, Bansal M (2003) A glossary of DNA structures from A to Z. Acta Crystallogr Sect D 59:620–626. https://doi.org/10.1107/S0907444903003251
Gilmer O, Quignon E, Jousset A-C, et al (2021) Chemical and enzymatic probing of viral RNAs: from infancy to maturity and beyond. Viruses 13
Gopal A, Zhou ZH, Knobler CM, Gelbart WM (2012) Visualizing large RNA molecules in solution. RNA 18:284–299. https://doi.org/10.1261/rna.027557.111
Harkness VRW, Avakyan N, Sleiman HF, Mittermaier AK (2018) Mapping the energy landscapes of supramolecular assembly by thermal hysteresis. Nat Commun 9:3152. https://doi.org/10.1038/s41467-018-05502-z
Helm M (2006) Post-transcriptional nucleotide modification and alternative folding of RNA. Nucleic Acids Res 34:721–733. https://doi.org/10.1093/nar/gkj471
Hendrix DK, Brenner SE, Holbrook SR (2005) RNA structural motifs: building blocks of a modular biomolecule. Q Rev Biophys 38:221–243. https://doi.org/10.1017/S0033583506004215
Huang L, Lilley DMJ (2013) The molecular recognition of kink-turn structure by the L7Ae class of proteins. RNA 19:1703–1710. https://doi.org/10.1261/rna.041517.113
Incarnato D, Oliviero S (2017) The RNA epistructurome: uncovering RNA function by studying structure and post-transcriptional modifications. Trends Biotechnol 35:318–333. https://doi.org/10.1016/j.tibtech.2016.11.002
Jacques J-P, Susskind MM (1991) Use of electrophoretic mobility to determine the secondary structure of a small antisense RNA. Nucleic Acids Res 19:2971–2977. https://doi.org/10.1093/nar/19.11.2971
Jasinski D, Haque F, Binzel DW, Guo P (2017) Advancement of the emerging field of RNA nanotechnology. ACS Nano 11:1142–1164. https://doi.org/10.1021/acsnano.6b05737
Kertesz M, Wan Y, Mazor E et al (2010) Genome-wide measurement of RNA secondary structure in yeast. Nature 467:103–107. https://doi.org/10.1038/nature09322
Knapp GBT-M in E (1989) [16] Enzymatic approaches to probing of RNA secondary and tertiary structure. In: RNA Processing Part A: General Methods. Academic Press, pp 192–212
Kolev SK, Petkov PS, Rangelov MA et al (2018) Interaction of Na+, K+, Mg2+ and Ca2+ counter cations with RNA. Metallomics 10:659–678. https://doi.org/10.1039/c8mt00043c
Kypr J, Kejnovská I, Renčiuk D, Vorlíčková M (2009) Circular dichroism and conformational polymorphism of DNA. Nucleic Acids Res 37:1713–1725. https://doi.org/10.1093/nar/gkp026
Le Brun E, Arluison V, Wien F (2020) application of synchrotron radiation circular dichroism for RNA structural analysis BT—RNA Spectroscopy: Methods and Protocols. In: Arluison V, Wien F (eds). Springer US, New York, NY, pp 135–148
Legendre A, Angel E, Tahi F (2018) Bi-objective integer programming for RNA secondary structure prediction with pseudoknots. BMC Bioinform 19:13. https://doi.org/10.1186/s12859-018-2007-7
Leontis NB, Westhof E (2001) NOMENCLATURE PROPOSAL Geometric nomenclature and classification of RNA base pairs

Li B, Cao Y, Westhof E, Miao Z (2020) Advances in RNA 3D Structure Modeling Using Experimental Data. Front Genet 11

Li J, Zhang J, Wang J et al (2016) Structure prediction of RNA loops with a probabilistic approach. PLoS Comput Biol 12:e1005032. https://doi.org/10.1371/journal.pcbi.1005032

Li PTX, Bustamante C, Tinoco I (2007) Real-time control of the energy landscape by force directs the folding of RNA molecules. Proc Natl Acad Sci 104:7039–7044. https://doi.org/10.1073/pnas.0702137104

Lilley DMJ (2004) Analysis of Global Conformational Transitions in Ribozymes BT—Ribozymes and siRNA Protocols. In: Sioud M (ed). Humana Press, Totowa, NJ, pp 77–108

Liu D, Thélot FA, Piccirilli JA et al (2022) Sub-3-Å cryo-EM structure of RNA enabled by engineered homomeric self-assembly. Nat Methods 19:576–585. https://doi.org/10.1038/s41592-022-01455-w

Lorenz R, Bernhart SH, Höner zu Siederdissen C, et al (2011) ViennaRNA Package 2.0. Algorithms for Molecular Biology 6:26. https://doi.org/10.1186/1748-7188-6-26

Lu Z, Zhang QC, Lee B et al (2016) RNA duplex map in living cells reveals higher-order transcriptome structure. Cell 165:1267–1279. https://doi.org/10.1016/j.cell.2016.04.028

Ma H, Jia X, Zhang K, Su Z (2022) Cryo-EM advances in RNA structure determination. Signal Transduct Target Ther 7:58. https://doi.org/10.1038/s41392-022-00916-0

Magnus M, Kappel K, Das R, Bujnicki JM (2019) RNA 3D structure prediction guided by independent folding of homologous sequences. BMC Bioinformatics 20:512. https://doi.org/10.1186/s12859-019-3120-y

Maurin T, Melko M, Abekhoukh S et al (2015) The FMRP/GRK4 mRNA interaction uncovers a new mode of binding of the Fragile X mental retardation protein in cerebellum. Nucleic Acids Res 43:8540–8550. https://doi.org/10.1093/nar/gkv801

Mergny J-L, Lacroix L (2003) Analysis of thermal melting curves. Oligonucleotides 13:515–537. https://doi.org/10.1089/154545703322860825

Mergny J-L, Lacroix L (2009) UV Melting of G-Quadruplexes. Curr Protoc Nucleic Acid Chem 37:17.1.1–17.1.15. https://doi.org/10.1002/0471142700.nc1701s37

Moudgal N, Arhin G, Frank AT (2022) Using unassigned NMR chemical shifts to model RNA secondary structure. J Phys Chem A 126:2739–2745. https://doi.org/10.1021/acs.jpca.2c00456

Mundigala H, Michaux JB, Feig AL et al (2014) HIV-1 DIS stem loop forms an obligatory bent kissing intermediate in the dimerization pathway. Nucleic Acids Res 42:7281–7289. https://doi.org/10.1093/nar/gku332

Murray LJW, Arendall WB, Richardson DC, Richardson JS (2003) RNA backbone is rotameric. Proc Natl Acad Sci 100:13904–13909. https://doi.org/10.1073/pnas.1835769100

Nelissen FHT, van Gammeren AJ, Tessari M et al (2008) Multiple segmental and selective isotope labeling of large RNA for NMR structural studies. Nucleic Acids Res 36:e89–e89. https://doi.org/10.1093/nar/gkn397

Nilsen TW (2014) RNase Footprinting to Map Sites of RNA–Protein Interactions. Cold Spring Harbor Protocols 2014:pdb.prot080788. https://doi.org/10.1101/pdb.prot080788

Olson WK, Li S, Kaukonen T et al (2019) Effects of noncanonical base pairing on RNA folding: structural context and spatial arrangements of G·A Pairs. Biochemistry 58:2474–2487. https://doi.org/10.1021/acs.biochem.9b00122

Pollack L (2011) SAXS studies of ion-nucleic acid interactions. Annu Rev Biophys 40:225–242. https://doi.org/10.1146/annurev-biophys-042910-155349

Popenda M, Szachniuk M, Blazewicz M, et al (2010) RNA FRABASE 2.0: an advanced web-accessible database with the capacity to search the three-dimensional fragments within RNA structures. BMC Bioinformatics 11:231. https://doi.org/10.1186/1471-2105-11-231

Reiter NJ, Maher LJ III, Butcher SE (2008) DNA mimicry by a high-affinity anti-NF-κB RNA aptamer. Nucleic Acids Res 36:1227–1236. https://doi.org/10.1093/nar/gkm1141

Richardson KE, Kirkpatrick CC, Znosko BM (2020) RNA CoSSMos 2.0: an improved searchable database of secondary structure motifs in RNA three-dimensional structures. Database 2020:baz153. https://doi.org/10.1093/database/baz153

Rietveld K, van Poelgeest R, Pleij CWA et al (1982) The tRNA-Uke structure at the 3′ terminus of turnip yellow mosaic virus RNA. Differences and similarities with canonical tRNA. Nucleic Acids Res 10:1929–1946. https://doi.org/10.1093/nar/10.6.1929

Rother M, Rother K, Puton T, Bujnicki JM (2011) ModeRNA: a tool for comparative modeling of RNA 3D structure. Nucleic Acids Res 39:4007–4022. https://doi.org/10.1093/nar/gkq1320

Sathyaseelan C, Vijayakumar V, Rathinavelan T (2021) CD-NuSS: A Web Server for the Automated Secondary Structural Characterization of the Nucleic Acids from Circular Dichroism Spectra Using Extreme Gradient Boosting Decision-Tree, Neural Network and Kohonen Algorithms. J Mol Biol 433:166629. https://doi.org/10.1016/j.jmb.2020.08.014

Sathyaseelan C, Vinothini V, Rathinavelan T (2020) Secondary structural characterization of the nucleic acids from circular dichroism spectra using extreme gradient boosting decision-tree algorithm. bioRxiv 2020.03.16.993352. https://doi.org/10.1101/2020.03.16.993352

Schauss J, Kundu A, Fingerhut BP, Elsaesser T (2021) Magnesium Contact Ions Stabilize the Tertiary Structure of Transfer RNA: Electrostatics Mapped by Two-Dimensional Infrared Spectra and Theoretical Simulations. J Phys Chem B 125:740–747. https://doi.org/10.1021/acs.jpcb.0c08966

Scott WG, Horan LH, Martick M (2013) Chapter One - The Hammerhead Ribozyme: Structure, Catalysis, and Gene Regulation. In: Soukup GABT-P in MB and TS (ed) Catalytic RNA. Academic Press, pp 1–23

Sharma E, Sterne-Weiler T, O'Hanlon D, Blencowe BJ (2016) Global Mapping of Human RNA-RNA Interactions. Mol Cell 62:618–626. https://doi.org/10.1016/j.molcel.2016.04.030

Shelton VM, Sosnick TR, Pan T (1999) Applicability of Urea in the Thermodynamic Analysis of Secondary and Tertiary RNA Folding. Biochemistry 38:16831–16839. https://doi.org/10.1021/bi991699s

Shi H, Rangadurai A, Abou Assi H et al (2020) Rapid and accurate determination of atomistic RNA dynamic ensemble models using NMR and structure prediction. Nat Commun 11:5531. https://doi.org/10.1038/s41467-020-19371-y

Shrestha P, Xiao S, Dhakal S et al (2014) Nascent RNA transcripts facilitate the formation of G-quadruplexes. Nucleic Acids Res 42:7236–7246. https://doi.org/10.1093/nar/gku416

Silvers R, Keller H, Schwalbe H, Hengesbach M (2015) Differential Scanning Fluorimetry for Monitoring RNA Stability. ChemBioChem 16:1109–1114. https://doi.org/10.1002/cbic.201500046

Solayman M, Litfin T, Singh J, et al (2022) Probing RNA structures and functions by solvent accessibility: an overview from experimental and computational perspectives. Brief Bioinform 23:bbac112. https://doi.org/10.1093/bib/bbac112

Spitale RC, Flynn RA, Zhang QC et al (2015) Structural imprints in vivo decode RNA regulatory mechanisms. Nature 519:486–490. https://doi.org/10.1038/nature14263

Staple DW, Butcher SE (2005) Pseudoknots: RNA Structures with Diverse Functions. PLoS Biol 3:e213

Stombaugh J, Zirbel CL, Westhof E, Leontis NB (2009) Frequency and isostericity of RNA base pairs. Nucleic Acids Res 37:2294–2312. https://doi.org/10.1093/nar/gkp011

Su Z, Zhang K, Kappel K et al (2021) Cryo-EM structures of full-length Tetrahymena ribozyme at 3.1 Å resolution. Nature 596:603–607. https://doi.org/10.1038/s41586-021-03803-w

Sun L-Z, Chen S-J (2019) Predicting RNA-Metal Ion Binding with Ion Dehydration Effects. Biophys J 116:184–195. https://doi.org/10.1016/j.bpj.2018.12.006

Tamura M, Holbrook SR (2002) Sequence and Structural Conservation in RNA Ribose Zippers. J Mol Biol 320:455–474. https://doi.org/10.1016/S0022-2836(02)00515-6

Taufer M, Licon A, Araiza R et al (2009) PseudoBase++: an extension of PseudoBase for easy searching, formatting and visualization of pseudoknots. Nucleic Acids Res 37:D127–D135. https://doi.org/10.1093/nar/gkn806

Taylor K, Sobczak K (2020) Intrinsic Regulatory Role of RNA Structural Arrangement in Alternative Splicing Control. Int J Mol Sci 21

Tijerina P, Mohr S, Russell R (2007) DMS footprinting of structured RNAs and RNA–protein complexes. Nat Protoc 2:2608–2623. https://doi.org/10.1038/nprot.2007.380

Turnbull AP, Wu X (2021) Studying RNA–Protein Complexes Using X-Ray CrystallographyX-ray crystallography BT - Protein-Ligand Interactions: Methods and Applications. In: Daviter T, Johnson CM, McLaughlin SH, Williams MA (eds). Springer US, New York, NY, pp 423–446

Tyagi R, Mathews DH (2007) Predicting helical coaxial stacking in RNA multibranch loops. RNA 13:939–951. https://doi.org/10.1261/rna.305307

Vicens Q, Kieft JS (2022) Thoughts on how to think (and talk) about RNA structure. Proc Natl Acad Sci 119:e2112677119. https://doi.org/10.1073/pnas.2112677119

Walter AE, Turner DH, Kim J et al (1994) Coaxial stacking of helixes enhances binding of oligoribonucleotides and improves predictions of RNA folding. Proc Natl Acad Sci 91:9218–9222. https://doi.org/10.1073/pnas.91.20.9218

Watts JM, Dang KK, Gorelick RJ et al (2009) Architecture and secondary structure of an entire HIV-1 RNA genome. Nature 460:711–716. https://doi.org/10.1038/nature08237

Wilkinson KA, Merino EJ, Weeks KM (2006) Selective 2′-hydroxyl acylation analyzed by primer extension (SHAPE): quantitative RNA structure analysis at single nucleotide resolution. Nat Protoc 1:1610–1616. https://doi.org/10.1038/nprot.2006.249

Woźniak AK, Schröder GF, Grubmüller H et al (2008) Single-molecule FRET measures bends and kinks in DNA. Proc Natl Acad Sci 105:18337–18342. https://doi.org/10.1073/pnas.0800977105

Yakovchuk P, Protozanova E, Frank-Kamenetskii MD (2006) Base-stacking and base-pairing contributions into thermal stability of the DNA double helix. Nucleic Acids Res 34:564–574. https://doi.org/10.1093/nar/gkj454

Zhang H, Keane SC (2019) Advances that facilitate the study of large RNA structure and dynamics by nuclear magnetic resonance spectroscopy. WIREs RNA 10:e1541. https://doi.org/10.1002/wrna.1541

Zhang J, Ferré-D'Amaré AR (2014) New molecular engineering approaches for crystallographic studies of large RNAs. Curr Opin Struct Biol 26:9–15. https://doi.org/10.1016/j.sbi.2014.02.001

Zhang K, Li S, Kappel K, et al (2019) Cryo-EM structure of a 40 kDa SAM-IV riboswitch RNA at 3.7 Å resolution. Nat Commun 10:5511. https://doi.org/10.1038/s41467-019-13494-7

Zhao Q, Zhao Z, Fan X et al (2021) Review of machine learning methods for RNA secondary structure prediction. PLoS Comput Biol 17:e1009291

Ziehler WA, Engelke DR (2000) Probing RNA Structure with Chemical Reagents and Enzymes. Curr Protoc Nucleic Acid Chem 00:6.1.1–6.1.21. https://doi.org/10.1002/0471142700.nc0601s00

Ziv O, Gabryelska MM, Lun ATL et al (2018) COMRADES determines in vivo RNA structures and interactions. Nat Methods 15:785–788. https://doi.org/10.1038/s41592-018-0121-0

Ziv O, Price J, Shalamova L et al (2020) The short- and long-range RNA–RNA Interactome of SARS-CoV-2. Mol Cell 80:1067-1077.e5. https://doi.org/10.1016/j.molcel.2020.11.004

Probing the RNA Structure-Dependent RNA Regulations and Functions

Chang Liu, Xinying Wu, Tiffaney Hsia, Guoping Li, and Junjie Xiao

Contents

Abstract RNAs form complex structures in vivo to perform diverse functions. These RNA structures dynamically change in response to internal cellular regulatory factors. Latest technological innovations have allowed us to detect RNA structures on a large scale with recent analyses of RNA structure revealing how dynamic RNA structures mediate RNA regulatory functions. In this chapter, we review the latest technological advances in RNA structure probing and analysis. We also outline the technical challenges of RNA structure probing methods and discuss future directions.

Keywords RNA structure · RNA sequence probing · SHAPE · DMS-seq · Structure-seq

C. Liu · X. Wu · J. Xiao (✉)
Cardiac Regeneration and Ageing Lab, Institute of Cardiovascular Sciences, Shanghai Engineering Research Center of Organ Repair, School of Life Science, Shanghai University, Shanghai 200444, China
e-mail: junjiexiao@shu.edu.cn

T. Hsia
Department of Neurosurgery, Massachusetts General Hospital and Harvard Medical School, Boston, MA 02114, USA

G. Li
Cardiovascular Division of the Massachusetts General Hospital and Harvard Medical School, Boston, MA 02114, USA

J. Barciszewski (ed.), *RNA Structure and Function*, RNA Technologies 14,
https://doi.org/10.1007/978-3-031-36390-0_7

Abbreviations

ADAR1	Adenosine deaminase acting on RNA1
CDS	Coding sequence
Dance-MAP	Deconvolution and annotation of ribonucleic conformational ensembles
DMS	Dimethyl sulfate
FragSeq	Fragmentation sequencing
hnRNPs	Heterogeneous nuclear ribonucleoproteins
MaP	Mutational profiling
NAI	2-Methylnicotinic acid imidazolide
PARS	Parallel analysis of RNA structures
RBP	RNA binding protein
RNP	Ribonucleoprotein
SHAPE	Selective 2′-hydroxyl acylation analyzed by primer extension
UTR	Untranslated regions

1 Introduction

In cells, both coding RNAs (messenger RNAs, mRNAs) and non-coding RNAs (small RNAs and long non-coding RNAs, lncRNAs) are fundamental to biological function. Due to intramolecular base pairing, RNA molecules form complex structures, such as bulges, loops, hairpins, and junctions (Bevilacqua et al. 2016), which have been found to associate with their unique biological functions. Moreover, intermolecular base pairing occurs as well, binding RNAs such as microRNAs and target mRNAs together, a mechanism that is essential for the subsequent regulation of target mRNA expression. Growing evidence reveals the essential role of RNA structure in downstream RNA expression and function, especially for long RNAs like mRNAs and lncRNAs. The regulation of RNA structures facilitates their interactions with other molecules such as RNAs and proteins, maintaining cellular homeostasis. Yet, despite this critical correlation, there is much to be discovered regarding the relationship between RNA structure and function as technology continues to advance to improve the detection and quantification of these features.

Due to the progress in RNA structure probing technologies, structure-dependent RNA functions have been heavily explored in recent years (Underwood et al. 2010; Kertesz et al. 2010; Lucks et al. 2011; Ding et al. 2014; Rouskin et al. 2014; Siegfried et al. 2014; Spitale et al. 2015). Although the initial RNA structure probing techniques were restricted to in vitro detections (Wan et al. 2011), recent technological advances enable transcriptome-wide RNA structure probing in vivo (Ding et al. 2014). In this chapter, we review the brief history and advances of RNA structure probing technologies. We also focus on the links between RNA structures and their biological functions. Finally, we discuss the parameters that affect the RNA structures.

2 RNA Structure-Probing Methods

Transfer RNA (tRNA) was the first biological RNA with a delineated primary, secondary and tertiary structure. The sequence of $tRNA^{Ala}$ was discovered in 1965 (Holley et al. 1965) with other tRNA sequences determined soon after. Upon comparison across sequences, only one common fold was found, familiarly known as the cloverleaf structure. With the advancement of computing, comparative analyses of RNA structures have also been explored with respect to phylogeny, providing a powerful avenue for defining RNA structural evolution (Gesell and Schuster 2014). Due to the stable structure and multiple modifications on tRNA, which inhibit reverse transcription (RT), further development of the methods improved quantification of tRNA. In 2015, the development of tRNA sequencing technology, termed as AlkB-facilitated RNA methylation sequencing (ARM-seq) and DM-tRNA-seq (Cozen et al. 2015; Zheng et al. 2015), made it possible to quantify tRNA in a high-throughput manner. To overcome the challenges presented by tRNA modifications, base methylations were removed via demethylase treatment. While these methods increased tRNA readout in sequencing libraries, information on the modified sites of interest were removed as well. To solve this problem, modification-induced misincorporation tRNA sequencing (mim-tRNAseq) was developed (Behrens et al. 2021). This method enables RT read through of modified sites and retains the modification information on tRNAs.

RNA structure has also been probed by methods such as fiber X-ray diffraction, optical rotary dispersion, and nuclear magnetic resonance (Bevilacqua et al. 2016). Shortly after tRNA sequences were defined, the structure of tRNA was determined using crystal X-ray diffraction, greatly enriching the understanding of the tRNA landscape (Kim et al. 1974). However, these methods were not suitable for high throughput study of RNA, driving the development of technologies that can directly identify the structure of RNA on a large scale. Currently, two types of reagents can be used to directly identify RNA structure: RNases and chemical probes.

Certain RNases recognize and cut RNA at certain structural regions and therefore can be used for RNA sequence detection. For example, RNase V1 recognizes and cuts RNA in double-stranded regions, while RNases A, T1, S1 and P1 recognize and cut RNA in single-stranded regions. RNase cleavage-based methods, such as parallel analysis of RNA structures (PARS) (Kertesz et al. 2010) and fragmentation sequencing (FragSeq) (Underwood et al. 2010), have pioneered large scale RNA structure probing. However, above mentioned technologies are not amenable for RNA structure probing in vivo as these RNases are not membrane permeable. Another strategy uses small molecules to identify and modify structured RNA regions to replace RNases. Since small molecules are capable of passing through the cell membrane, these methods have been adopted to probe RNA structure in vivo and have successfully identified up to tertiary and quaternary RNA structures (Zafferani and Hargrove 2021).

Two types of chemical probes (base-specific and ribose-specific) have been widely used for RNA structure detection in vivo. These chemical probes act by blocking reverse transcription, resulting in truncated cDNA, or causing cDNA mismatches, which can be subsequently mapped by RNA-seq reads. Base-specific probes target base-pairing interactions, exemplified by adenosine and cytidine reactions with dimethyl sulfate (DMS), (Ding et al. 2014). Base-specific probing followed by sequencing, such as DMS-seq (Rouskin et al. 2014) and Structure-seq (Ding et al. 2014), mark adenine and cytosine residues with DMS and facilitate residue-specific detection. In contrast, ribose-specific probes, as used in selective 2′-hydroxyl acylation analyzed by primer extension (SHAPE) technique, mark the four common nucleotides (Spitale et al. 2013), improving throughput and potential for downstream applications. Given its specificity towards the sugar-phosphate backbone, SHAPE followed by RNA sequencing (SHAPE-seq) thereby interrogates conformational changes across all RNAs, allowing for precise mapping of RNA structures (Lucks et al. 2011).

To enhance the structure-probing accuracy, optimization strategies have been pursued in different probing methods. For example, in vivo click SHAPE (icSHAPE) enriches for modified RNAs to decrease unwanted RT stop signals through selective biotin-streptavidin RNA segment isolation and has refined SHAPE reagents (Fig. 1) to include 2-methylnicotinic acid imidazolide with the addition of an azide to the nicotinicacid ring (NAI-N3) acylation (Spitale et al. 2015). In further development, icSHAPE combined with mutation profiling (icSHAPE-MaP) increases quantification by identification of RT mutations (Luo et al. 2021). Despite its utility, icSHAPE and other mutation-recognizing RNA structure detection strategies remain limited by sequencing depth to identify precise mutation sites. Recently, a new technique has been developed for RNA structure analysis based on Nanopore sequencing, naming the consequent RNA structure analysis "pore-cupine" (Aw et al. 2021). This method labels unpaired bases with chemical probes to generate specific electrical signals as they pass through the biological nanopore. Following comparison with a control group, the positions of these specific electrical signals can be appreciated, thereby identifying the unstructured regions, and delineating the structure of RNA.

While there have been many advances in in vivo RNA structure probing, these technologies are hindered by limitations, preventing the complete study of the RNA structure. Thus far, chemical probing has been constrained to single stranded templates, necessitating further work to modify double-stranded RNA regions for evaluation. Furthermore, signal quantification has not exceeded binary assessment, preventing absolute interpretation of a lack of probing signal. Current studies are bounded by speculation, attributing lost signal to a variety of causes including double-stranded RNA structure or protein-binding (Bevilacqua et al. 2016). As such, further development is needed to expand the capacity of in vivo RNA structure probing.

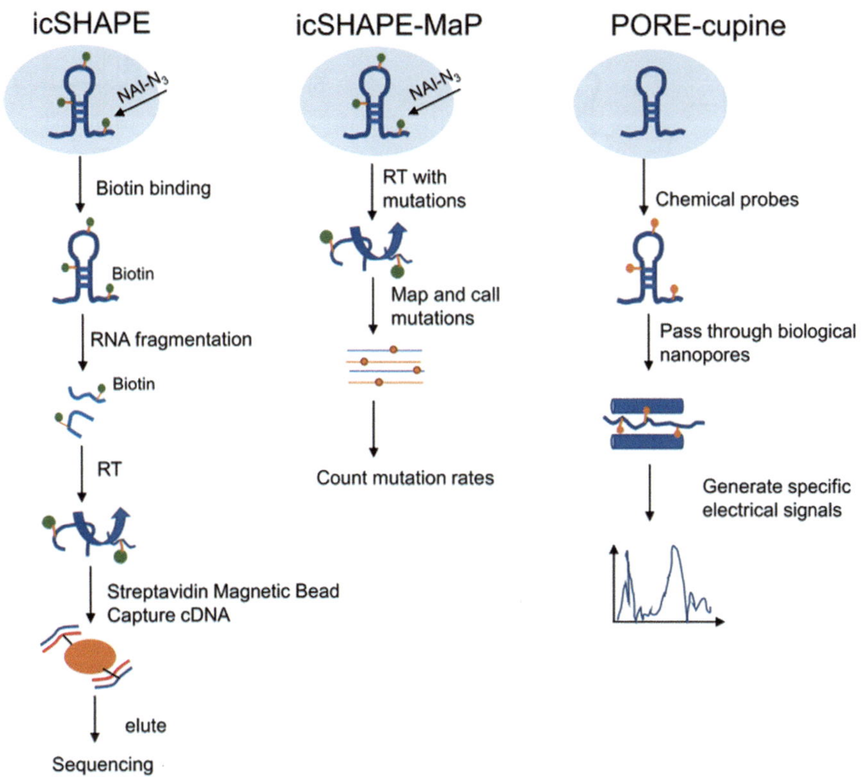

Fig. 1 Examples of in vivo RNA structure probing methods. icSHAPE: in vivo click SHAPE, icSHAPE-MaP: icSHAPE combined with mutation profiling, NAI: 2-methylnicotinic acid imidazolide, NAI-N3: NAI with the addition of an azide to the nicotinicacid ring (at position 2), PORE-cupine: RNA structure analysis using nanopore sequencing

3 Relationship Between RNA Structures and RNA Function

As revealed by RNA structure studies, RNA structure modulates its interactions with other RNAs or protein partners, subsequently affecting RNA processes such as RNA degradation, mRNA splicing, RNA polyadenylation, and translation (Fig. 2). Thus, studying its structure and interactions with other RNAs and proteins within the cell becomes essential to understanding the molecular mechanism of RNA function. mRNA structures have been shown to affect translation efficiency and, as a result, gene expression. For example, RNA G-quadruplex structures in the 5′ untranslated regions (5′ UTR) modulate translation by inhibiting cap-dependent translation and promoting cap-independent translation (Arora et al. 2008; Lammich et al. 2011; Morris et al. 2010). Interestingly, the RNA G-quadruplex is found to inhibit cap-dependent translation only when it is located within 50–100 nt of the cap (Kumari

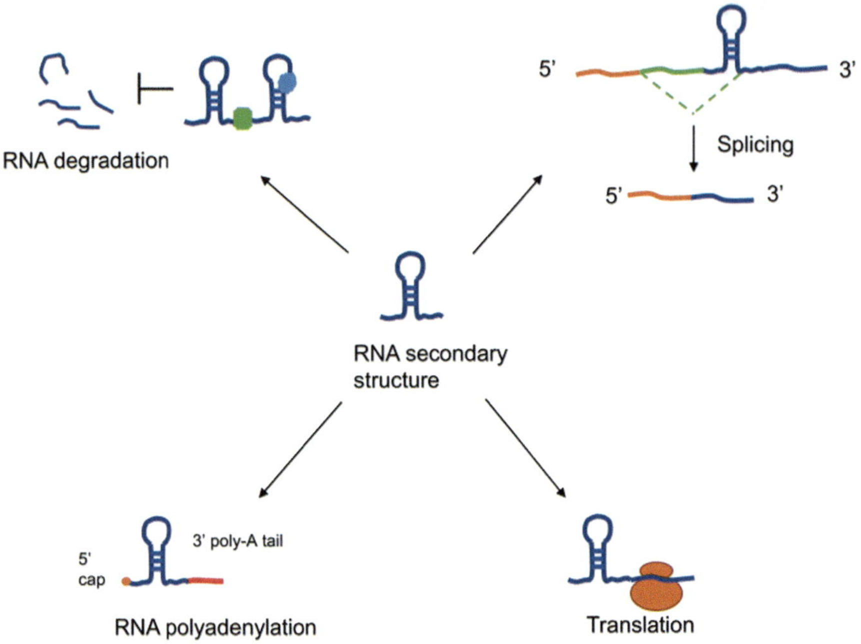

Fig. 2 Relationship between RNA structures and their functions

et al. 2008), suggesting the importance of its position within the 5′UTR for translation regulation. In contrast, the RNA G-quadruplex is necessary for internal ribosome entry site (IRES) activity in cap-independent translation initiation (Morris et al. 2010). As such, mutations disrupting this structure can abolish IRES mediated translation initiation.

The folding free energies of the 5′ UTR have also been shown to influence post-transcriptional regulation and translation (Ringner and Krogh 2005). These studies, conducted in yeast models, showed that 5′UTRs have higher free energy of structural folding as compared to other sequences in yeast with these regions demonstrating a bias for weak folding or low folding complexity. This evidence further substantiates that mRNAs with less structure in the 5′UTR have higher translation efficiency. RNA translation is also affected by other regions on the mRNA, as regulated by the sequence and resulting structures. A recent study on zebrafish embryos further supports the correlation between the mRNA structure of 3′UTR and translation efficiency (Shi et al. 2020). In detail, mRNA 3′UTRs contain regions with variable structures, and these structures facilitate RNA binding proteins (RBPs) binding which stabilized the mRNA. The ribosomal binding site (RBS) also demonstrates this phenomenon, with changes in structural stability regulating translation efficiency (Mustoe et al. 2018). To begin with, the free energy of the RBS structure as it unfolds is negatively correlated with translation efficiency. During translation, the RBS structure is further affected by the loading of mRNA on the ribosome 30S subunit.

RNA–protein interactions have also gained clarity as technology has advanced, enabling mapping of the RNA binding proteome (Wheeler et al. 2018). RBPs are critical for mRNA and lncRNA genetic expression. Not only do their interactions with RNA form a complex network that regulates cellular homeostasis, RBPs have also been found to regulate miRNA and mRNA binding, affecting RNA-RNA interactions and downstream activity such as disease development (Gebauer et al. 2021; Ciafre and Galardi 2013). Recent studies of RNA structure with regards to the RNA-binding proteome confirmed RBP binding to specific regions in the 5′ and 3′ UTRs and the presence of RBP binding motifs on coding sequences (CDS) (Casas-Vila et al. 2020; Mukherjee et al. 2019). These studies have elucidated the RNA structure-regulated RBP interaction and the role of RNA structure on its biological function, showing that RBPs interact preferentially with double-stranded RNAs (Sanchez de Groot et al. 2019).

Alternative splicing occurs on approximately all mammalian mRNAs. The spliceosome assembles on splice sites to catalyze intron removal on pre-mRNAs. The presence of RNA folding, and therefore structures, at these sites have been shown to affect splicing. Through this avenue, RNA structure can influence splicing through regulation of spliceosome assembly and as a binding motif for splicing factors (Buratti and Baralle 2004; McManus and Graveley 2011). Many proteins participate in the regulation of RNA splicing. These regulators include serine/arginine (SR) factors (promoting) and heterogeneous nuclear ribonucleoproteins (hnRNPs) (inhibiting). The assembly of the spliceosome relies on the binding of protein factors on single-stranded regions of pre-mRNAs. Therefore, RNA folding inhibits spliceosome assembly by hiding the binding sites of splicing enhancers. In contrast, RNA structure, which hides the binding sites of splicing repressors, can also enhance splicing. Studies have revealed that the N6-methyladenosine (m^6A) modification weakens the hairpin secondary structure of RNA to promote heterogeneous nuclear ribonucleoprotein C (HNRNPC) binding and splicing regulation (Liu et al. 2015). Recent evidence suggests that 5-methylcytosine (m^5C) is a prevalent modification on HIV-1 RNAs and that the methyltransferase NSUN2 is the major writer of m^5C (Courtney et al. 2019). Inactivation of NSUN2 represses m^5C modification on viral RNAs, leading to inhibition of viral replication. Removal of m^5C disrupts the alternative splicing, suggesting the pivotal role of m^5C in viral RNA processing. Another study has identified pseudouridines on splice sites and RBP binding sites of human mRNA (Martinez et al. 2022). This study showed that pseudouridine synthases regulate pre-mRNA alternative splicing, indicating the role of pseudouridylation in pre-mRNA processing.

The structures of RNA domains influence the RNA function as well. The 5′ cap and the 3′ poly(A) tail play pivotal roles in maintaining stability of mRNA and regulating translation (Geisberg et al. 2014; Moqtaderi et al. 2018). Eukaryotic mRNAs are modified by a 5′ cap structure composed of a N7-methylated guanosine (m^7G). The cap structure protects mRNA from 5′ to 3′ exonuclease cleavage. Moreover, the cap structure also enables the binding of the ribosome with mRNAs and the recruitment of translation initiation factors. The 3′ poly(A) tail can bind with poly(A)-binding protein C (PABPC), which protects mRNA from degradation by exonuclease. The 3′ poly(A) tail structure also facilitates translation via cooperation with the 5′ cap (Gallie 1991). It is interesting to note that RNAs with higher structured sequences in the 3′ UTR tend to decay faster. For example, the RBP ELAV like RNA binding protein 1a (Elavl1a) is found to stabilize maternal mRNAs via binding to the 3′ UTR, while RNA structures in this region inhibits Elavl1a binding and leads to the decay of the mRNA (Shi et al. 2020).

4 Cellular Internal Environment Elements That Affect RNA Structures

Cellular environment elements, such as RBPs, translation machinery, and RNA modifications, can significantly influence the RNA structure and function (Fig. 3). For example, the RBP Staufen homolog I (STAU1) binds to structured mRNA regions containing multiple short helices in the 3′UTR and CDS (Ricci et al. 2014). Upregulation of STAU1 in the CDS increases ribosome densities, suggesting the regulatory role of STAU1 on translation via binding to structured sites in CDS. In contrast to STAU1, the inhibitory splicing factor HNRNPC preferentially targets unstructured uridine tracts. HNRNPC binding sites highly overlap with structural variation sites, suggesting that HNRNPC binding causes RNA structural change (Sun et al. 2019). There is also evidence of gene regulation via RBP-induced RNA structural modifications. For example, RBPs affect the binding of miRNA through alteration of the template RNA structure (Wang et al. 2021). In detail, structural alterations on lin-28 homolog A (LIN28A), as caused by pumilio RNA binding family member 2 (PUM2), enables miR-30 binding and subsequently inhibits the expression of LIN28A.

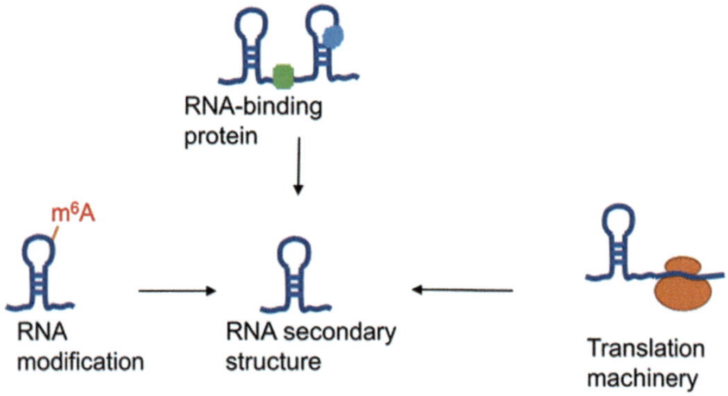

Fig. 3 Cellular environment elements that affect RNA structures

Within the cellular internal environment, the translation machinery guides RNA structure as well (Fig. 3). The ribosome, a ribonucleoprotein (RNP), possesses helicase activity to initiate translation and serves as an important regulator of mRNA structure (Beaudoin et al. 2018; Mustoe et al. 2018). Previously, the relationship between translation efficiency and RNA structure was observed in the developmental process of zebrafish embryos (Beaudoin et al. 2018). Translation efficiency and CDS accessibility is decreased after pateamine A treatment (inhibition of translation initiation), suggesting that translation initiation and ribosomal binding are essential to remodel mRNA for increased binding accessibility.

Post-transcriptional mRNA modifications, particularly methylation, is also shown to influence RNA structure (Bevilacqua et al. 2016). One of the most prevalent modifications on mRNAs is the m^6A modification, with over 12,000 m^6A sites currently identified (Dominissini et al. 2012). Investigation of m^6A modification has identified its regulation of structural switches and its effect on RNA–protein interactions (Liu et al. 2017) through exposure of protein binding sites (Lewis et al. 2017). It has been found that m^6A affects RNA structure via destabilization of RNA duplexes (Roost et al. 2015). Indeed, the regions containing the m^6A-modified sites tend to be single-stranded, yielding low structural thermodynamic stability (Spitale et al. 2015). It is important to note that covalent modification of mRNA also impacts structure-influenced RNA function, a relationship similarly found in mRNA composed of the four common nucleotides. Another post-transcriptional modifier, adenosine deaminase acting on RNA1 (ADAR1), also influences the fate of RNA structures. ADAR1 is an enzyme that is involved in adenosine-to-inosine editing (A-to-I editing) in double-stranded RNA. Recent evidence suggests that A-to-I editing catalyzed by ADAR1 stabilizes a large amount of double-stranded RNA (Solomon et al. 2017). As such, loss of ADAR1 is found to decrease the ratio of double-stranded RNA to single-stranded RNA, further supporting the regulatory role of ADAR1 on global RNA secondary structure through A-to-I editing.

5 Challenges and Future Directions

In recent decades, RNA structure-probing technologies have achieved substantial progress and have transformed the landscape of understanding. Information generated with these technologies have uncovered complex processes driven by RNA structural dynamics in vitro and in vivo. Nonetheless, challenges still exist to be solved by further innovations. Despite extensive biological study, the RNA structural landscape is not fully characterized, limited by current experimental technology. With the refinement of current methods and development of new tools for RNA structure probing, complex structural properties, up to quaternary structures, can be elucidated (Solayman et al. 2022). However, commonly practiced RNA structure validations are still based on functional rescue experiments, in which nucleotide mutations are constructed to disrupt RNA structures (Zubradt et al. 2017). Such methods can only provide an indirect avenue to obtain information on the RNA structure–function relationship and require introduction of synthesized RNA. Other validation approaches currently employ biophysical imaging modalities to monitor RNA structural change. Generating 3D structural information through two dimensional quantification, these methods include atomic force microscopy (Schon 2016), x-ray crystallography (Shi 2014), cryogenic electron microscopy (Zhang et al. 2019), single molecule fluorescence resonance energy transfer (Sung and Nesbitt 2020; Stephenson et al. 2016), and nuclear magnetic resonance spectroscopy (solid and liquid state) (Orlovsky et al. 2020; Yang and Wang 2018). While they allow for visualization and manipulation of RNA molecules in vivo without marking (Uroda et al. 2020), biophysical imaging is constrained to low throughput quantification. Therefore, development of high-throughput RNA structure visualization approaches under physiological conditions will largely accelerate the progress of RNA structurome study.

One of the major challenges in mapping RNA structure and function includes improving the accuracy, sensitivity, and capacity of high throughput study, performed via sequencing. Current sequencing studies remain limited to template reactivity and the complexity of RNA structural constructs as well as sequencing depth, requiring refinement of analysis to achieve true-positive signal detection. While some advances have been made by integrating the results of standard quantitative approaches (Mitchell et al. 2019), the study of the RNA structurome has begun to shift towards the collaboration of wet lab and computational modeling (Yu et al. 2022) and will benefit immensely from further development. Building on traditional computational methods (Han and Kim 1993; Tahi et al. 2002; Engelen and Tahi 2010; Bellaousov and Mathews 2010; Legendre et al. 2018) to utilize the rigor of machine learning (ML) for predictive modeling of RNA structure (Andronescu et al. 2014; Xia et al. 1998; Turner and Mathews 2010; Zakov et al. 2011) and incorporation of thermodynamic study into new ML models (Andronescu et al. 2007, 2010; Tang et al. 2008; Akiyama et al. 2018), the combination of lab-based study and computation have vastly improved the predictive modeling of RNA structures. Further innovation to include newly discovered structural phenomena within the modeling parameters will greatly expand and enrich our understanding of RNA structure-dependent function. RNA structural analysis has already provided vital insight into the fundamental biological processes that govern cellular function, homeostasis, and disease development. Through development of our analytical capacity, further elucidation of RNA structure and its complex interactions with the inter-cellular environment can contribute to a broad spectrum of applications, ranging from basic science to translational investigation such as candidate drug screening.

Authorship Contributions Junjie Xiao and Guoping Li organized the structure of this chapter. Xinying Wu searched the literature. Chang Liu and Tiffaney Hsia drafted and critically revised this chapter.

Funding This work was supported by the grants from National Natural Science Foundation of China (82020108002 and 82225005 to JJ Xiao, 82000287 to CL), the grant from Science and

Technology Commission of Shanghai Municipality (21XD1421300 and 20DZ2255400 to JJ Xiao), the "Dawn" Program of Shanghai Education Commission (19SG34 to JJ Xiao).

Declaration of Interests All the authors of this chapter declare no competing interests.

References

Akiyama M, Sato K, Sakakibara Y (2018) A max-margin training of RNA secondary structure prediction integrated with the thermodynamic model. J Bioinform Comput Biol 16:1840025
Andronescu M, Condon A, Hoos HH et al (2007) Efficient parameter estimation for RNA secondary structure prediction. Bioinformatics 23:i19-28
Andronescu M, Condon A, Hoos HH et al (2010) Computational approaches for RNA energy parameter estimation. RNA 16:2304–2318
Andronescu M, Condon A, Turner DH et al (2014) The determination of RNA folding nearest neighbor parameters. Methods Mol Biol 1097:45–70
Arora A, Dutkiewicz M, Scaria V et al (2008) Inhibition of translation in living eukaryotic cells by an RNA G-quadruplex motif. RNA 14:1290–1296
Aw JGA, Lim SW, Wang JX et al (2021) Determination of isoform-specific RNA structure with nanopore long reads. Nat Biotechnol 39:336–346
Beaudoin JD, Novoa EM, Vejnar CE et al (2018) Analyses of mRNA structure dynamics identify embryonic gene regulatory programs. Nat Struct Mol Biol 25:677–686
Behrens A, Rodschinka G, Nedialkova DD (2021) High-resolution quantitative profiling of tRNA abundance and modification status in eukaryotes by mim-tRNAseq. Mol Cell 81(1802–1815):e1807
Bellaousov S, Mathews DH (2010) ProbKnot: fast prediction of RNA secondary structure including pseudoknots. RNA 16:1870–1880
Bevilacqua PC, Ritchey LE, Su Z et al (2016) Genome-wide analysis of RNA secondary structure. Annu Rev Genet 50:235–266
Buratti E, Baralle FE (2004) Influence of RNA secondary structure on the pre-mRNA splicing process. Mol Cell Biol 24:10505–10514
Casas-Vila N, Sayols S, Perez-Martinez L, et al. (2020) The RNA fold interactome of evolutionary conserved RNA structures in S. cerevisiae. Nat Commun 11:2789
Ciafre SA, Galardi S (2013) microRNAs and RNA-binding proteins: a complex network of interactions and reciprocal regulations in cancer. RNA Biol 10:935–942
Courtney DG, Tsai K, Bogerd HP et al (2019) Epitranscriptomic addition of m(5)C to HIV-1 transcripts regulates viral gene expression. Cell Host Microbe 26(217–227):e216
Cozen AE, Quartley E, Holmes AD et al (2015) ARM-seq: AlkB-facilitated RNA methylation sequencing reveals a complex landscape of modified tRNA fragments. Nat Methods 12:879–884
Ding Y, Tang Y, Kwok CK et al (2014) In vivo genome-wide profiling of RNA secondary structure reveals novel regulatory features. Nature 505:696–700
Dominissini D, Moshitch-Moshkovitz S, Schwartz S et al (2012) Topology of the human and mouse m6A RNA methylomes revealed by m6A-seq. Nature 485:201–206
Engelen S, Tahi F (2010) Tfold: efficient in silico prediction of non-coding RNA secondary structures. Nucleic Acids Res 38:2453–2466
Gallie DR (1991) The cap and poly(A) tail function synergistically to regulate mRNA translational efficiency. Genes Dev 5:2108–2116
Gebauer F, Schwarzl T, Valcarcel J et al (2021) RNA-binding proteins in human genetic disease. Nat Rev Genet 22:185–198

Geisberg JV, Moqtaderi Z, Fan X et al (2014) Global analysis of mRNA isoform half-lives reveals stabilizing and destabilizing elements in yeast. Cell 156:812–824

Gesell T, Schuster P (2014) Phylogeny and evolution of RNA structure. Methods Mol Biol 1097:319–378

Han K, Kim HJ (1993) Prediction of common folding structures of homologous RNAs. Nucleic Acids Res 21:1251–1257

Holley RW, Apgar J, Everett GA et al (1965) Structure of a ribonucleic acid. Science 147:1462–1465

Kertesz M, Wan Y, Mazor E et al (2010) Genome-wide measurement of RNA secondary structure in yeast. Nature 467:103–107

Kim SH, Sussman JL, Suddath FL et al (1974) The general structure of transfer RNA molecules. Proc Natl Acad Sci USA 71:4970–4974

Kumari S, Bugaut A, Balasubramanian S (2008) Position and stability are determining factors for translation repression by an RNA G-quadruplex-forming sequence within the 5' UTR of the NRAS proto-oncogene. Biochemistry 47:12664–12669

Lammich S, Kamp F, Wagner J et al (2011) Translational repression of the disintegrin and metalloprotease ADAM10 by a stable G-quadruplex secondary structure in its 5'-untranslated region. J Biol Chem 286:45063–45072

Legendre A, Angel E, Tahi F (2018) Bi-objective integer programming for RNA secondary structure prediction with pseudoknots. BMC Bioinformatics 19:13

Lewis CJ, Pan T, Kalsotra A (2017) RNA modifications and structures cooperate to guide RNA-protein interactions. Nat Rev Mol Cell Biol 18:202–210

Liu N, Dai Q, Zheng G et al (2015) N(6)-methyladenosine-dependent RNA structural switches regulate RNA-protein interactions. Nature 518:560–564

Liu N, Zhou KI, Parisien M et al (2017) N6-methyladenosine alters RNA structure to regulate binding of a low-complexity protein. Nucleic Acids Res 45:6051–6063

Lucks JB, Mortimer SA, Trapnell C et al (2011) Multiplexed RNA structure characterization with selective 2'-hydroxyl acylation analyzed by primer extension sequencing (SHAPE-Seq). Proc Natl Acad Sci USA 108:11063–11068

Luo QJ, Zhang J, Li P et al (2021) RNA structure probing reveals the structural basis of Dicer binding and cleavage. Nat Commun 12:3397

Martinez NM, Su A, Burns MC et al (2022) Pseudouridine synthases modify human pre-mRNA co-transcriptionally and affect pre-mRNA processing. Mol Cell 82(645–659):e649

McManus CJ, Graveley BR (2011) RNA structure and the mechanisms of alternative splicing. Curr Opin Genet Dev 21:373–379

Mitchell D 3rd, Assmann SM, Bevilacqua PC (2019) Probing RNA structure in vivo. Curr Opin Struct Biol 59:151–158

Moqtaderi Z, Geisberg JV, Struhl K (2018) Extensive structural differences of closely related 3' mRNA isoforms: links to Pab1 binding and mRNA stability. Mol Cell 72(849–861):e846

Morris MJ, Negishi Y, Pazsint C et al (2010) An RNA G-quadruplex is essential for cap-independent translation initiation in human VEGF IRES. J Am Chem Soc 132:17831–17839

Mukherjee N, Wessels HH, Lebedeva S et al (2019) Deciphering human ribonucleoprotein regulatory networks. Nucleic Acids Res 47:570–581

Mustoe AM, Busan S, Rice GM et al (2018) Pervasive regulatory functions of mRNA structure revealed by high-resolution SHAPE probing. Cell 173(181–195):e118

Orlovsky NI, Al-Hashimi HM, Oas TG (2020) Exposing hidden high-affinity RNA conformational states. J Am Chem Soc 142:907–921

Ricci EP, Kucukural A, Cenik C et al (2014) Staufen1 senses overall transcript secondary structure to regulate translation. Nat Struct Mol Biol 21:26–35

Ringner M, Krogh M (2005) Folding free energies of 5'-UTRs impact post-transcriptional regulation on a genomic scale in yeast. PLoS Comput Biol 1:e72

Roost C, Lynch SR, Batista PJ et al (2015) Structure and thermodynamics of N6-methyladenosine in RNA: a spring-loaded base modification. J Am Chem Soc 137:2107–2115

Rouskin S, Zubradt M, Washietl S et al (2014) Genome-wide probing of RNA structure reveals active unfolding of mRNA structures in vivo. Nature 505:701–705

Sanchez de Groot N, Armaos A, Grana-Montes R et al (2019) RNA structure drives interaction with proteins. Nat Commun 10:3246

Schon P (2016) Imaging and force probing RNA by atomic force microscopy. Methods 103:25–33

Shi Y (2014) A glimpse of structural biology through X-ray crystallography. Cell 159:995–1014

Shi B, Zhang J, Heng J et al (2020) RNA structural dynamics regulate early embryogenesis through controlling transcriptome fate and function. Genome Biol 21:120

Siegfried NA, Busan S, Rice GM et al (2014) RNA motif discovery by SHAPE and mutational profiling (SHAPE-MaP). Nat Methods 11:959–965

Solayman M, Litfin T, Singh J et al (2022) Probing RNA structures and functions by solvent accessibility: an overview from experimental and computational perspectives. Brief Bioinform 23

Solomon O, Di Segni A, Cesarkas K et al (2017) RNA editing by ADAR1 leads to context-dependent transcriptome-wide changes in RNA secondary structure. Nat Commun 8:1440

Spitale RC, Crisalli P, Flynn RA et al (2013) RNA SHAPE analysis in living cells. Nat Chem Biol 9:18–20

Spitale RC, Flynn RA, Zhang QC et al (2015) Structural imprints in vivo decode RNA regulatory mechanisms. Nature 519:486–490

Stephenson JD, Kenyon JC, Symmons MF et al (2016) Characterizing 3D RNA structure by single molecule FRET. Methods 103:57–67

Sun L, Fazal FM, Li P et al (2019) RNA structure maps across mammalian cellular compartments. Nat Struct Mol Biol 26:322–330

Sung HL, Nesbitt DJ (2020) DNA hairpin hybridization under extreme pressures: a single-molecule FRET study. J Phys Chem B 124:110–120

Tahi F, Gouy M, Regnier M (2002) Automatic RNA secondary structure prediction with a comparative approach. Comput Chem 26:521–530

Tang X, Thomas S, Tapia L et al (2008) Simulating RNA folding kinetics on approximated energy landscapes. J Mol Biol 381:1055–1067

Turner DH, Mathews DH (2010) NNDB: the nearest neighbor parameter database for predicting stability of nucleic acid secondary structure. Nucleic Acids Res 38:D280-282

Underwood JG, Uzilov AV, Katzman S et al (2010) FragSeq: transcriptome-wide RNA structure probing using high-throughput sequencing. Nat Methods 7:995–1001

Uroda T, Chillon I, Annibale P et al (2020) Visualizing the functional 3D shape and topography of long noncoding RNAs by single-particle atomic force microscopy and in-solution hydrodynamic techniques. Nat Protoc 15:2107–2139

Wan Y, Kertesz M, Spitale RC et al (2011) Understanding the transcriptome through RNA structure. Nat Rev Genet 12:641–655

Wang J, Zhang T, Yu Z et al (2021) Genome-wide RNA structure changes during human neurogenesis modulate gene regulatory networks. Mol Cell 81(4942–4953):e4948

Wheeler EC, Van Nostrand EL, Yeo GW (2018) Advances and challenges in the detection of transcriptome-wide protein–RNA interactions. Wiley Interdiscip Rev RNA 9

Xia T, SantaLucia J Jr, Burkard ME et al (1998) Thermodynamic parameters for an expanded nearest-neighbor model for formation of RNA duplexes with Watson-Crick base pairs. Biochemistry 37:14719–14735

Yang Y, Wang S (2018) RNA characterization by solid-state NMR spectroscopy. Chemistry 24:8698–8707

Yu B, Li P, Zhang QC et al (2022) Differential analysis of RNA structure probing experiments at nucleotide resolution: uncovering regulatory functions of RNA structure. Nat Commun 13:4227

Zafferani M, Hargrove AE (2021) Small molecule targeting of biologically relevant RNA tertiary and quaternary structures. Cell Chem Biol 28:594–609

Zakov S, Goldberg Y, Elhadad M et al (2011) Rich parameterization improves RNA structure prediction. J Comput Biol 18:1525–1542

Zhang K, Li S, Kappel K et al (2019) Cryo-EM structure of a 40 kDa SAM-IV riboswitch RNA at 3.7 A resolution. Nat Commun 10:5511

Zheng G, Qin Y, Clark WC et al (2015) Efficient and quantitative high-throughput tRNA sequencing. Nat Methods 12:835–837

Zubradt M, Gupta P, Persad S et al (2017) DMS-MaPseq for genome-wide or targeted RNA structure probing in vivo. Nat Methods 14:75–82

Probing Techniques of Secondary and Tertiary RNA Structure and a Case Study for RNA G-Quadruplexes

Johanna Mattay

Contents

Abstract Like proteins, an RNA is only functional when it is folded into its native conformation and adopts a specific secondary and tertiary structure. Hence, the analysis of RNA structure is essential to understand the cellular roles of distinct RNA molecules. Technical approaches used to study RNA structure comprise bioinformatics tools, structural probing, and biophysical methods to integrate sequence and 3D structure information. In this review, I focus on structural probing techniques of RNA secondary structure. I discuss basic enzymatic and chemical probing techniques, and present novel approaches in combination with high-throughput sequencing. A focus is laid on SHAPE techniques and its various developments and applications. Finally, at the example of RNA G-quadruplexes, it is highlighted how an array of probing techniques can be combined to study a specific RNA structural motif in vitro and in vivo.

J. Mattay (✉)
Institute of Biochemistry, University of Münster, Corrensstr. 36, 48149 Münster, Germany
e-mail: johanna.mattay@gmx.de

J. Barciszewski (ed.), *RNA Structure and Function*, RNA Technologies 14,
https://doi.org/10.1007/978-3-031-36390-0_8

Keywords RNA secondary structure · Chemical probing · DMS · SHAPE-seq · SHAPE-MaP · Enzymatic probing · G-quadruplex

Abbreviations

mRNA	Messenger RNA
rRNA	Ribosomal RNA
tRNA	Transfer RNA
ncRNA	Noncoding RNA
snRNA	Small nuclear RNA
snoRNA	Small nucleolar RNA
miRNA	MicroRNA
lncRNA	Long noncoding RNA
nt	Nucleotide
G4	G-quadruplex
ATP	Adenosine triphosphate
RBP	RNA-binding protein
bp	Base-pairs
ssRNA	Single-stranded RNA
dsRNA	Double-stranded RNA
DMS	Dimethyl sulfate
CMCT	1-Cyclohexyl-(2-morpholinoethyl) carbodiimide metho-*p*-toluene sulfonate
kethoxal	2-Keto-3-ethoxy-butyraldehyde
NMIA	*N*-Methylisatoic anhydride
1M7	1-Methyl-7-nitroisatoic anhydride
1M6	1-Methyl-6-nitroisatoic anhydride
NAI	2-Methylnicotinic acid imidazolide
BzCN	Benzoyl chloride
•OH	Hydroxyl radical
s^2U	2-Thio-uridine
2-deaza-A	2-Deaza-adenosine
UTR	Untranslated region
CDS	Coding sequence
RT	Reverse transcription
cDNA	Complementary DNA
EDTA	Ethylenediaminetetraacetic acid
DBCO	Dibenzocyclooctyne
NGS	Next generation sequencing
m^6A	N^6-Methyladenosine
hnRNP	Heterologous nuclear ribonucleoprotein
carboxyPDS	Carboxypyridostatin
PDS	Pyridostatin

PCR	Polymerase chain reaction
SAXS	Single-angle X-ray scattering
AFM	Atomic force microscopy
NMR	Nuclear magnetic resonance spectroscopy
cryo-EM	Cryogenic electron microscopy
ASO	Antisense oligonucleotide
LNA	Locked nucleic acid
ALS	Amyotrophic lateral sclerosis
3D	3-Dimensional
FRET	Förster resonance energy transfer
smFRET	Single-molecule FRET

1 Introduction

RNA as a central molecule in biology covers functions from posttrancriptional processing over regulation of gene expression to metabolite sensing. Apart from mRNAs, most RNAs are not translated into protein (noncoding RNAs, ncRNAs), including the abundant rRNAs and tRNAs for ribosomal function. In addition, small nuclear and nucleolar RNAs (snRNAs, snoRNAs) mediate RNA processing steps, microRNAs (miRNAs) control RNA turnover, and long noncoding RNAs (lncRNAs) are regulators of RNA function and biogenesis. To perform their diverse functions, RNAs must fold into their native structures in a cellular environment. Hydrogen bonds from base-pairing and π-stacking of the aromatic ring bases define the RNA secondary structure elements. Long-range interactions to the sugar-phosphate backbone and between distant bases are crucial for tertiary structure. Determining RNA structure is challenging: an RNA of a given nucleotide sequence can adopt multiple low-energy states, with the preferred conformation being dependent on protein binding, ionic environment, nucleobase modifications, and other cellular conditions. Thus, the analysis of RNA structure and mechanisms of RNA folding are crucial to understand the fascinating cellular functions and regulation of RNAs.

2 Secondary and Tertiary RNA Structure

2.1 Hierarchical Folding of RNA

Merely four nucleotides and a highly charged negative phosphate backbone make it challenging for RNA to fold into energetically favourable conformations. RNA folding in 3-dimensional space follows a hierarchical order (Brion and Westhof 1997; Westhof et al. 1996). First, short independently stable helices form rapidly by Watson/Crick base-pairing. Second, these secondary structure elements undergo tertiary interactions and higher order structures.

Secondary structure elements are uniform: typically, an RNA is composed of a set of short A-form helices of max. 10 bp length, in which the majority of nucleotides are comprised (Russell 2008). The stability of each base-pair is dictated only by Watson/Crick hydrogen-bonding and stacking with the directly adjacent bases.

The primary interaction level of an RNA helix is coaxial stacking: two adjacent helices separated by a phosphodiester bridge stack end-to-end on each other to a colinear arrangement (Butcher and Pyle 2011; Walter et al. 1994). In tRNA, for example, the D-stem coaxially stacks with the anticodon stem, and the T-stem chooses the acceptor arm as the stacking partner, forming the cloverleaf structure (Fig. 1a) (Quigley and Rich 1976). The choice of stacking partners is determined by sequence since the two helix end-standing base-pairs stack via their aromatic bases. Stacking partners can be altered by mutations and non-canonical base-pairs (Sutton and Pollack 2015; Walter and Turner 1994; Yesselman et al. 2019).

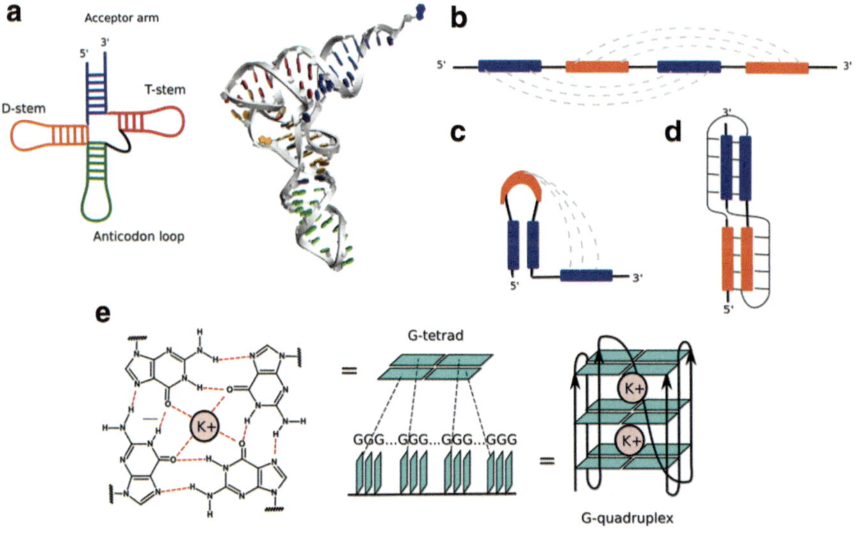

Fig. 1 Common secondary and tertiary structures of RNA. **a** Secondary structure of tRNA. The D-stem/T-stem and the acceptor arm/anticodon loop coaxially stack to form the 3D cloverleaf structure. Shown is the tertiary structure of yeast $tRNA^{Phe}$. PDB: 1EHZ (Shi and Moore 2000). **b** Long-range tertiary interactions between separate helices. **c** Pseudoknot structure by interaction of a hairpin loop with a ss region. **d** Coaxial stacking of helices, here in the form of two pseudoknot helices. Adapted from (Butcher and Pyle 2011). **e** G-quadruplex. Four guanines from a G-rich strand assemble through Hoogsteen base-pairing. Here, three of these planar G-tetrads stack upon each other to form a G-quadruplex. Shown here is a unimolecular G4 with parallel strand direction. G4s can also form intermolecularly from separate RNA strands or DNA/RNA hybrids. G4s can have antiparallel strand direction, comprise 2–5 stacked G-tetrads, and repeats of G4s can stack with each other to higher order structures

Tertiary interactions form between rigid secondary structures and flexible single-stranded regions (Fig. 1b). Long-range interactions are governed by non-canonical base-pairs such as Hoogsteen, C:A or A:G, electrostatic interactions to the sugar-phosphate backbone, and π-stacking of bases. 29 non-Watson/Crick interactions have been described. In a few cases, tertiary elements can form through Watson/Crick base-pairing and are thermodynamically as stable as secondary structures.

The free energy released to form a short RNA helix can reach 10 kcal/mol, and a GC-rich 10mer duplex can reach a dissociation half-life of 100 years (Turner 1989). Thus, RNA has the problem to become kinetically trapped in stable, but misfolded secondary structure intermediates. Their free energy can vary by only 0.5 kcal/mol from the native structure, as exemplified for $tRNA^{Phe}$ (Jaeger et al. 1989). Due to the multiple loose, transient tertiary interactions, native structures are often not thermodynamically favoured over competing tertiary structures. The RNA folding problem is more serious for long RNAs, which can result in slow folding times up to the minute scale (Weeks 1997).

In vivo, there are two regulatory mechanisms thought to prevent RNAs from misfolding and kinetic traps (Incarnato and Oliviero 2017; Shcherbakova et al. 2008). First, RNA polymerase kinetics, i.e., directionality, velocity, and pausing, guide the order and speed of folding events during transcription (Heilman-Miller and Woodson 2003; Lai et al. 2013; Schroeder et al. 2002). Second, many RNA-binding proteins may act as chaperones to stabilize folding intermediates. They can bind either in a passive way, e.g., hnRNPs like the U1 protein and ribosomal proteins, or actively through ATP hydrolysis, as seen for DEAD-box helicases (Herschlag 1995; Russell 2008; Weeks 1997).

2.2 Examples of Tertiary Structure Motifs

Coaxial stacking is the basis of several tertiary motifs, e.g., kissing loops and pseudoknots. Kissing loops form when the loops of two helices base-pair with each other. The L-shape of tRNA results from a kissing loop between the D-stem and the T-stem (Fig. 1a) (Quigley and Rich 1976). In pseudoknots, a loop region of an RNA helix forms Watson/Crick interactions with a single-stranded region outside of this helix (Fig. 1c) (Russell 2008). The A-minor motif is a triple helical structure in which an A interacts via Hoogsteen base-pairing with both nucleotides of a GC base-pair (Butcher and Pyle 2011). It is a building block for tetraloop interactions and kink turns (Keating et al. 2008; Klein et al. 2001). Ribose zippers glue together other motifs by 2′ OH hydrogen-bonding between backbone RNA strands (Tamura and Holbrook 2002).

G-quadruplexes (G4s) are stable tertiary structures that assemble from stretches of guanine repeats (Fig. 1e). Four Gs in a four-stranded arrangement assemble to a tetrad through Hoogsteen base-pairing. Two or more of these planar G-tetrads then stack upon each other to a G-quadruplex, which is stabilized in the centre by a K^+ ion. RNA G4s (G4s) are found in the UTRs of mRNA, in 5' introns of pre-mRNA, in ncRNAs such as the telomer-associated lncRNA TERRA, and in expansion segments of rRNA (Collie et al. 2010; Kharel et al. 2020; Mestre-Fos et al. 2019, 2020). G4s are most commonly known to act as transcriptional roadblocks in R-loops. However, they cover diverse functions such as the modulation of translation and splicing and the involvement in liquid–liquid phase separation.

3 Techniques to Study RNA Secondary Structure

3.1 RNA Structure Prediction in Silico

The nearest-neighbour model finds those base-pairings in an RNA sequence that undergo minimal free energy change (ΔG^0) upon folding (Mathews 2004; Xia et al. 1998). The thermodynamically most stable structure is determined based on hydrogen bonding energies of the base-pair and stacking with the adjacent bases. A second method of structure prediction relies on phylogenetic alignment of orthologous sequences and analysis of covariation sites (Russell 2013). A further development, the maximum expected accuracy, relies only on highly probable (>99%) single– and double-stranded regions and these high-confidence base-pairs are used to assemble the most accurate structure (Lu et al. 2009).

3.2 Enzymatic Probing Techniques

Early mapping techniques exploited endoribonucleases for sequence-specific cleavage of RNAs. The enzymes cut at a specific nucleotide (e.g., RNase T1, RNase A) or are nonselective (RNase I, nuclease S1) (Fig. 1d). While most enzymes prefer single-stranded RNA (ssRNA), RNase V1 targets double-stranded RNA (dsRNA) (Fig. 1d) (Ziehler and Engelke 2001). A drawback is the low resolution: some sites cannot be accessed by the sterically demanding enzymes. Therefore, a combination e.g. of RNase T1 (G), RNase A (C, U), and RNase V1 (dsRNA) gives a detailed secondary structure footprint. Due to the nature of enzyme catalysis, enzymatic probing is not suitable for quantifying the extent of cleavage and thus cannot quantify probed sites. Enzymatic probing coupled to high-throughput sequencing relies on the same principles as those for chemical probing (see 0) (Table 1).

Table 1 Probing reagents and associated high-throughput sequencing techniques for chemical and enzymatic probing of RNAs

Method	Reagent	Application	Specificities	Reference
Chemical probing reagents				
Chemical probing	DMS	In vitro	Unpaired A (N^1), C (N^3), G (N^7)	Waduge et al. (2019)
Chemical probing	CMCT	In vitro/in vivo	Unpaired G (N^1), U (N^3)	Harris et al. (1995)
Chemical probing	kethoxal	In vitro/In vivo	Unpaired G (N^1, $2{-}NH_2$)	Harris et al. (1995)
SHAPE	NMIA	In vitro	Ribose 2′ OH (flexible nt)	Merino et al. (2005)
SHAPE	1M7	In vitro/In vivo	Ribose 2′ OH (flexible nt)	Mortimer and Weeks (2007)
SHAPE	NAI	In vivo	Ribose 2′ OH (flexible nt)	Spitale et al. (2013)
SHAPE	BzCN	In vitro	Ribose 2′ OH (flexible nt)	Mortimer and Weeks (2008)
Chemical probing	• OH	In vivo	Solvent-exposed nt	Costa and Monachello (2014)
Sequencing-based chemical probing techniques				
SHAPE-Seq	1M7	In vitro/In vivo	IVT RNA with barcode seq	Lucks et al. (2011)
ChemModSeq	1M7, NAI, DMS	Ex vivo	Probability of RT drop-off rate for each nt Assembly of complexes	Hector et al. (2014)
SHAPE-MaP	1M7	Ex vivo	Introduction of noncomplementary nucleotides by RT at 2′ O-adduct sites De novo RNA motif discovery	Siegfried et al. (2014)
MAP-Seq	DMS, CMCT, 1M7	In vitro	IVT RNA with barcode seq	Seetin et al. (2014)
RING-MaP	DMS	In vitro	IVT RNA Mutation frequencies in single transcript to calculate correlation coefficients Through-space interactions Multiple conformations for single RNA	Homan et al. (2014)
Structure-Seq	DMS	In vivo	RNA structure ensembles related to protein function 3nt periodicity in highly translated CDS	Ding et al. (2014)

(continued)

Table 1 (continued)

Method	Reagent	Application	Specificities	Reference
DMS-Seq	DMS	In vivo	Higher number of mRNAs unfolded in vivo than in vitro	Rouskin et al. (2014)
CIRS-Seq	DMS, CMCT	Ex vivo/ deproteinized	mRNA 5′/3′ UTRs and lncRNAs highly structured RBP binding site prediction	Incarnato et al. (2014)
icSHAPE	NAI-N_3	In vivo	Biotin-clickable handle for affinity purification m^6A site prediction	Spitale et al. (2015)
HRF-Seq	• OH	Ex vivo	Determines ribose accessible surface area	Kielpinski and Vinther (2014)
MOHCA-Seq	• OH	In vitro	IVT RNA Mutate-and-map structural modeling	Cheng et al. (2015)
Sequencing-based enzymatic probing techniques				
PARS	RNase V1, nuclease S1	In vitro, ex vivo	Prediction of ss or ds conformation of base Structured CDS in mRNAs Structured AUG sites correlate with low translation	Kertesz et al. (2010)
PARTE	RNase V1	In vitro	RNA folding stability based on melting temp of dsRNA regions ncRNAs have distinct folding energy	Wan et al. (2012)
Frag-Seq	Nuclease P1	In vitro	Novel structures in ncRNAs Cutting score for each site	Underwood et al. (2010)

3.3 Chemical Probing Techniques

3.3.1 Base-Specific Chemical Probing

Chemical probing can assess any RNA region (Incarnato and Oliviero 2017). Nucleotides not engaged in base-pairing or tertiary interactions react with small electrophilic probes and are probed proportional to their accessibility (Chillón and Marcia 2020). It allows quantitative analysis because the number of modification products is directly proportional to the reactivity of the nucleotide. Dimethyl sulfate (DMS) methylates adenosine in N^1 and cytidine in N^3 position as well as guanosine in N^7 (Fig. 2a) (Wells et al. 2000). As the Watson/Crick interface is altered by methylation in A and C, but not in G, DMS probing identifies unpaired A and C in primer extension assays. Complementary, 1-cyclohexyl-(2-morpholinoethyl) carbodiimide metho-*p*-toluene sulfonate (CMCT) acylates guanosine in N^1 and uridine in N^3 position (Fig. 2b). A combination of DMS and CMCT is often used to assess the flexibility of all nucleotides in an RNA.

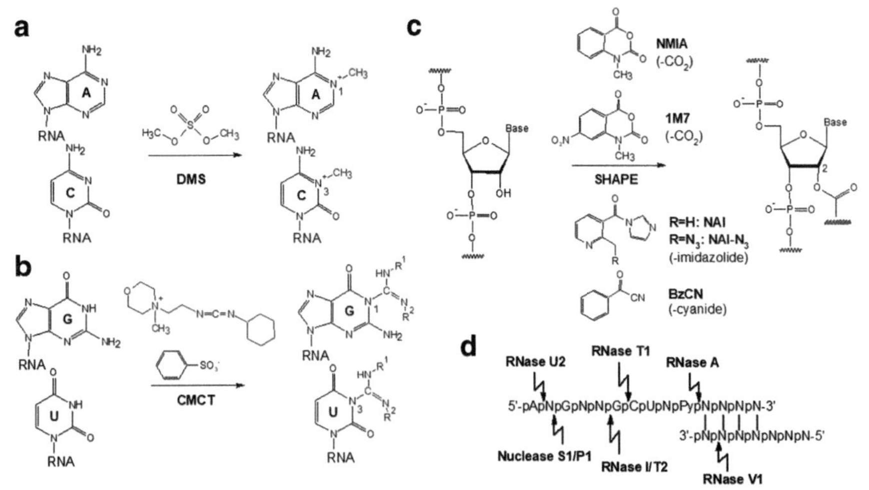

Fig. 2 Structural probing reagents. **a-c** Chemical probing. All reagents probe for non-base-paired nucleotides in ssRNA. **a** DMS methylates adenine and cytosine bases. **b** CMCT reacts with guanine and uracil bases. **c** SHAPE reagents for acylation of flexible 2′ OH groups of the ribose backbone: NMIA, 1M7, NAI, BzCN. **d** Enzymatic probing. RNases and nucleases have different cleavage selectivity either at 3′ or 5′ of the phosphodiester bond, respectively. With the exception of RNase V1 specific for dsRNA, all enzymes cut ssRNA

Chemical probes are used on in vitro-transcribed (IVT) RNA or on selected cell-extracted and purified targets. The cell-permeable DMS allows treatment in vivo in numerous organisms including bacteria, yeast, and mammalian cells (Wells et al. 2000). A structural map of the mouse lncRNA Xist was obtained by treating cells with DMS (Fang et al. 2015a). The in vivo use of CMCT and 2-keto-3-ethoxy-butyraldehyde (kethoxal) usually requires prior cell permeabilization (Harris et al. 1995). Kethoxal probes for G by ring formation between the N^1 and the 2-amino group. Of note, novel glyoxal and kethoxal derivatives can enter cells without permeabilization (see 0).

3.3.2 High-Throughput Readout of Chemical Probing

During reverse transcription (RT) of the probed RNA, the introduced modification blocks DNA polymerase from read-through and extension of the cDNA strand, and the enzyme drops off one nucleotide before the reacted site. For enzymatic probing, the cleaved RNA fragments directly result in cDNAs truncated at the site of cleavage due to polymerase run-off. To rule out background termination of RT that can occur on untreated RNA due to secondary structure or natural base modifications (Ziehler and Engelke 2001), controls omitting the probe are compulsory.

In its traditional form, RT was performed with 5′-^{32}P-labelled primers, cDNA fragments were separated on sequencing gels and compared to ddNTP-sequencing standards, and band intensities were quantified by autoradiography (Das et al. 2005). The use of capillary electrophoresis immensely accelerated automation. Fluorescent peaks of the probed substrates are quantified by priming the cDNAs with a fluorescent marker (Mitra et al. 2008; Vasa et al. 2008). Today, next-generation sequencing (NGS) technology allows genome-wide analysis of any RNA as opposed to a handful of in vitro targets for traditional readouts.

DMS Probing Coupled to Sequencing

After the pioneering work by Lucks et al. for NGS-based chemical probing (SHAPE-Seq, see 0) (Lucks et al. 2011), the MAP-Seq approach (multiplexed accessibility probing) implemented high-throughput sequencing for DMS probing (Seetin et al. 2014). Alternatively, RING-MaP (mutational profiling) relies on the incorporation of noncomplementary nucleotides during RT. The RT conditions and polymerase are chosen in a way that DNA polymerase reads through the DMS-adducts instead of stopping (see SHAPE-MaP 0). Mutations inserted at the sites of adduct formation are recorded simultaneously on one transcript and are used to analyse interdependencies of DMS-reactive sites and to calculate correlation coefficients (Homan et al. 2014). Thus, transient nucleotide interactions through space can be determined as well as RNA interaction groups (RINGs), which make up the multiple conformations a single RNA can adopt in solution (Table 1).

While the above techniques are limited to IVT RNA or a few purified single targets, Structure-Seq was the first application for genome-wide DMS probing in vivo (Ding et al. 2014). Applied to total RNA, it is used to identify RNA structural ensembles that can be associated with general protein functions. Structural characteristics of mRNAs were determined such as a 3 nt-periodicity in codons of highly translated mRNAs, and alternative polyadenylation sites based on high or low structured regions were discerned. The method DMS-Seq probes native RNA structure directly in DMS-treated yeast cells (Rouskin et al. 2014). It revealed a lower number of structured mRNAs in dividing cells. DMS-Seq data from ATP-depleted cells implied that mRNA structuring is restricted by ATP-dependent helicase unwinding steps. To keep mRNAs predominantly unfolded in vivo might be advantageous for the cell to provide a uniform structure to mRNAs for ribosome accession and translation. Nevertheless, hundreds of structured domains were also found in mRNAs.

A method that combines DMS and CMCT treatment of isolated RNA deproteinized with Proteinase K (ex vivo) has proven useful to infer an unexpected high structuring for lncRNAs. Equally, the 5′ and 3′ UTRs of mRNAs were found to be highly structured, while low structuring and thus good accessibility was found at ribosome binding sites and stop codons. CIRS-Seq (chemical inference of RNA structure) could also verify the 3 nt-periodicity of mRNA codons and the binding site of the RNA-binding protein Lin28a (Incarnato et al. 2014). Structure-seq2 provides an improved library preparation protocol to improve overall sequencing read coverage and quality using hairpin adapters for decreased ligation bias. Biotinylated dCTPs during RT are used to replace gel purification steps and remove unwanted ligation by-products (Ritchey et al. 2017).

Glyoxal derivatives have been developed for in vivo probing of RNA targets as alternative to DMS. The molecules successfully entered rice, *Bacillus subtilis*, and *Escherichia coli* cells to modify solvent-exposed G, C, and A residues (Mitchell et al. 2018). Keth-seq employs an azide-modified kethoxal, N_3-kethoxal, to probe G bases on a transcriptome-wide scale (Weng et al. 2020). It entered mouse embryonic stem cells in 1 min and successfully probed their RNA secondary structures.

3.3.3 Non-Base-Specific Chemical Probing

Principles of SHAPE

The reactivity of the ribose backbone can be exploited if it is not engaged in secondary or tertiary interactions, e.g., duplexes, Hoogsteen base-pairing or RNA triple helices. The 2′ OH group is reacted with an electrophilic probe, *N*-methylisatoic anhydride (NMIA), to form a SHAPE adduct (Fig. 2c). SHAPE, or selective 2′-hydroxyl acylation analysed by primer extension, probes any nucleotide of an RNA. During primer extension, the 2′ O-acylation induces the polymerase to fall off one nucleotide before the modified one (Fig. 3a). SHAPE readout is quantitative: the reactivity score (usually 0–1) of each site is directly proportional to local flexibility, i.e., the more SHAPE adducts are formed, the less constrained and more flexible is the nucleotide. Importantly, the reactivity map only corresponds to nucleotide flexibility but not to solvent accessibility (Gherghe et al. 2008; Merino et al. 2005).

Mechanistically, SHAPE reactivity of the 2′ OH group is increased by rare ribose C3′ or C2′ endo conformations and by electronegative and proximal substituents that serve as base catalysts for 2′ OH deprotonation and stabilise the tetrahedral transition state (McGinnis et al. 2012). Hence, reactivity is also strongly influenced by RNA modifications. For instance, the substituents in 2-thio-uridine (s^2U) and 2-deaza-adenosine (2-deaza-A) decrease nucleophilicity because of electronegativity effects and increased distance to the 2′ OH.

SHAPE probing data including folding constraints can be implemented into algorithms to predict RNA secondary structure, e.g., in RNAstructure (Mathews et al. 2004; Rice et al. 2014). Even pseudoknots whose stability factors are poorly understood due to transient tertiary and protein interactions can be predicted (Hajdin et al. 2013).

SHAPE Reagents for in Vitro and in Vivo Use

The most common SHAPE reagent, 1-methyl-7-nitroisatoic anhydride (1M7) (Fig. 2c), is a more electrophilic derivative of NMIA with shorter reaction times (70 s versus 20 min for NMIA) (Mortimer and Weeks 2007). 1M7 can be combined with derivatives of slightly different SHAPE reactivity. For instance, the human lncRNA

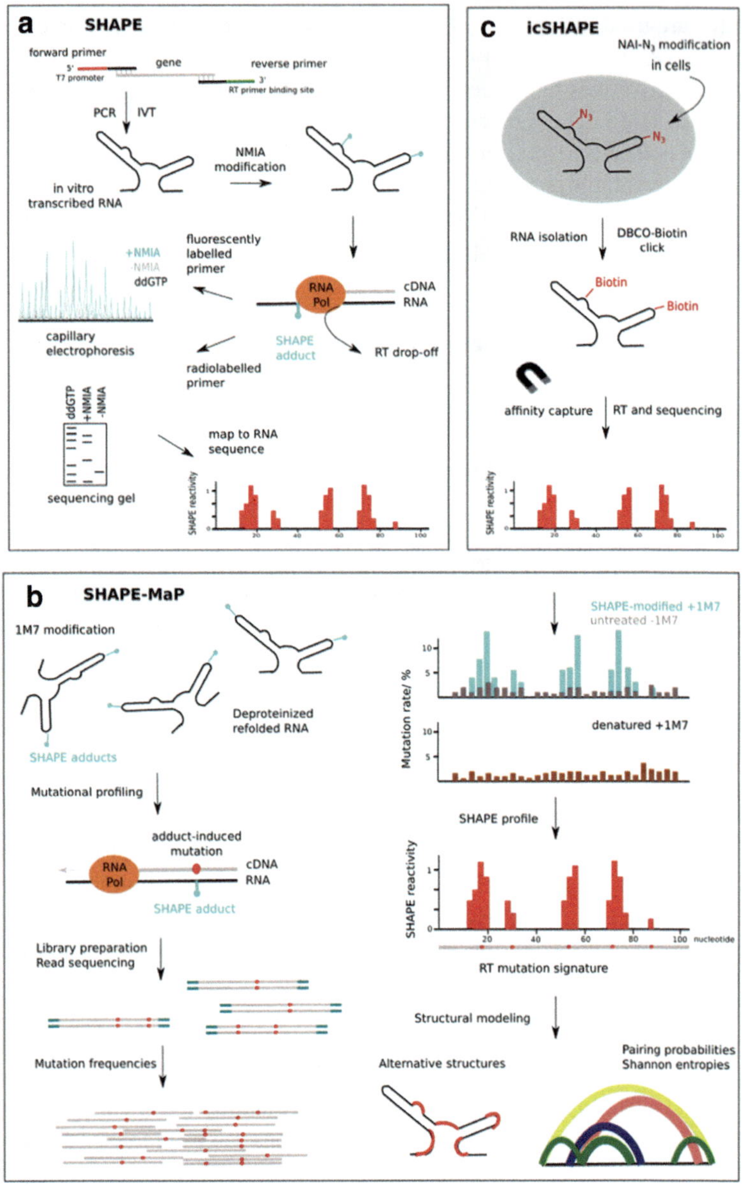

MEG3 was probed with NMIA, 1M7 and its regioisomer 1-methyl-6-nitroisatoic anhydride (1M6), and its secondary structure map was confirmed by DMS probing (Uroda et al. 2019). Combinatorial incorporation of SHAPE data from NMIA and 1M6—which detect noncanonical (NMIA) and tertiary (1M6) interactions based on differential kinetics and stacking interactions, respectively—can accurately predict secondary structures of RNAs that are difficult to model (Rice et al. 2014). The

◂**Fig. 3** Structural mapping techniques coupled to massively parallel sequencing analysis. **a** SHAPE: The target RNA is in vitro transcribed from a PCR template. 2' OH acylation with NMIA causes reverse transcriptase to stop at the modified site. The radiolabelled cDNA fragments are quantified and sequenced by gel electrophoresis, or, if a fluorophore-labelled RT primer is used, by capillary electrophoresis. Finally, SHAPE reactivities are plotted back onto the RNA sequence. **b** SHAPE-MaP: 1M7-modified RNA induces mutations that are inserted by a DNA polymerase at sites of 2′ O-adducts during RT. Mutation frequencies from sequencing reads are converted to SHAPE reactivities, plotted on an RNA secondary structure map or used for prediction of tertiary structure elements. Adapted from (Siegfried et al. 2014). **c** icSHAPE: RNA is modified in vivo with cell-permeable NAI-N_3. Isolated RNA is then treated in vitro with DBCO-Biotin in a copper-free azide-alkyne click reaction. Biotinylated transcripts are enriched on streptavidin beads for RT and NGS analysis

shotgun (3S) approach has been used for mouse RepA, a repeat element of the lncRNA Xist responsible for X chromosome-silencing in females. Herein, the generated fragments are probed individually by 1M7 SHAPE and their reactivity profiles are compared to that of the full-length transcript (Liu et al. 2017b; Novikova et al. 2013).

1M7 is commonly used for in vivo application in bacteria and eukaryotes, e.g. for 16S rRNA in *E. coli* and MEG3 in human fibroblast cells (McGinnis and Weeks 2014; Tyrrell et al. 2013; Uroda et al. 2019). However, the first SHAPE reagent that was developed exclusively for in vivo use was 2-methylnicotinic acid imidazolide (NAI) (Fig. 2c). NAI circumvents the pitfalls of NMIA of low solubility, cross-reactivity, and short half-life (Spitale et al. 2013). Comparison of in vitro and in vivo SHAPE profiles of 5S rRNA revealed functionally important nucleotides that differ in reactivity due to tertiary or protein interactions in cellular ribosomes.

Another SHAPE reagent, benzoyl cyanide (BzCN), was developed to probe RNA folding dynamics on a timescale of 1–2 s (Fig. 2c). As an example, RNase P forms several tertiary motifs including a tetraloop-receptor motif and a T-loop from A-minor interactions. When RNase P was probed with BzCN in 5 s-intervals during in vitro folding, the kinetics of the folding intermediates and a hierarchical folding pathway could be derived (Mortimer and Weeks 2008, 2009).

In Vivo SHAPE Probing Coupled to Sequencing

Shape-Seq. SHAPE-Seq combines 1M7 probing of an IVT RNA with deep sequencing of the aborted cDNA fragments (Lucks et al. 2011). During RT, a 4 nt-barcode unique for each RNA species is introduced. For a mixture of mutant transcripts, here of RNase P, subtle conformational variations can be analysed. The method is limited as each RNA species of interest has to be generated from a 3′-extended PCR template to contain an RT primer binding site and the barcode template sequence. ChemModSeq combines NGS of random hexamer-primed cDNAs and a novel algorithm for calculating RT drop-off rates and their probabilities to be caused by SHAPE adducts for each nucleotide position (Hector et al. 2014). It is suited to study RNA conformational dynamics during assembly of complexes and could elucidate structural intermediates of yeast 40S and 80S ribosome biogenesis. Thus, it overcomes obstacles typically encountered in cryo-EM of heterogeneous and instable particle purification.

SHAPE-MaP. The widely used method SHAPE-MaP is based on the incorporation of noncomplementary nucleotides at the sites of 2′ O-modification (Siegfried et al. 2014). In this mutational mapping (MaP), the rate of SHAPE adduct formation is directly converted to mutation frequencies by read counting (Fig. 3b). To obtain a SHAPE-MaP profile with relative SHAPE reactivity for each position, data from the untreated (–1M7) sample is subtracted from data from the treated (+1M7) sample after normalisation to a 1M7-treated denatured RNA control. If the RNA was probed e.g. in presence and absence of a ligand, the conformational changes during ligand coordination can be profiled by calculating the SHAPE difference of + ligand versus–ligand conditions. In addition, calculation of pairing probabilities and Shannon entropies can refine alternatively structured domains or regions with multiple conformations in equilibrium, and even discover RNA motifs de novo. Based on high Shannon entropies, three pseudoknots were predicted in the HIV-1 genome in regions hitherto unknown to contain defined RNA motifs (Siegfried et al. 2014).

Alternative SHAPE protocols. The in vivo click SHAPE method (icSHAPE) uses an azide-containing NAI reagent ($NAI{-}N_3$) (Fig. 3c) to click a biotin moiety to the modified nucleotides. The biotin handle is used for affinity capture and enrichment, followed by RT and NGS (Fig. 3c) (Spitale et al. 2015). icSHAPE sequencing data can be used to predict N^6-methyladenosine (m^6A) modification sites more accurately than based on the DRACH sequence motif only. In icSHAPE data from cells expressing m^6A methyltransferase METTL3, the methylated m^6A sites show higher SHAPE reactivity compared to cells depleted for METTL3. This is because m^6A disrupts base-pairing in duplex helices and leads to more unstructured regions.

Recently, SmartSHAPE was developed to probe low abundant RNA specimen from primary or immune cells to decrease the input amount of RNA from 1 μg to 1 ng (10^5 cells) (Piao et al. 2022). As improvements to the original icSHAPE protocol, RNaseI digestion of artifact truncated RNAs improved true positive RT stop signals and on-bead library preparation further increased RNA yield. By profiling the RNA structure landscape of two intestinal macrophage cell lines in mice, it was demonstrated that RNA structural changes directly regulate immune responses.

SHAPE coupled to direct RNA sequencing. Nanopore sequencing has advanced the detection of natural RNA modifications including ribose 2′ O-methyl (Nm) and pseudouridine (ψ) by measuring differences in current signal and dwell time between modified RNA and unmodified control of the same sequence. Methods combining SHAPE and long-read direct RNA sequencing have demonstrated the applicability of Nanopore sequencing to detect chemical modifications introduced exogenously. This was demonstrated for modification by the SHAPE reagent 1-acetylimidazole (AcIm) which forms small 2′ O-acetyl adducts (NanoSHAPE) (Stephenson et al. 2022), and in $NAI\text{-}N_3$-probed human RNA to phase combinations of structures between isoforms (Aw et al. 2021). Novel model-free algorithms further allow the identification of similar and conserved RNA structures in different organisms by direct comparison of their SHAPE reactivity profiles (Morandi et al. 2022). This can be helpful in the context of finding druggable and unique RNA targets.

Hydroxyl Radical Probing

RNA is treated with an amount of hydroxyl radicals (•OH) that is equivalent to provoke one cleavage event per molecule on average. The extent of backbone cleavage is then proportional to the solvent-accessible surface of each nucleotide (Mitra et al. 2008; Vasa et al. 2008). •OH radicals are generated in situ with Fenton reagents such as H_2O_2 and Fe(II)-EDTA or by synchrotron X-ray beams (Götte et al. 1996; Sclavi et al. 1997). In combination with NGS, HRF-Seq (hydroxyl radical footprinting) and MOHCA-Seq (multiplexed •OH cleavage analysis with paired-end sequencing) allow high-throughput analysis of RNA on a genome-wide scale (Table 1) (Cheng et al. 2015; Kielpinski and Vinther 2014). HRF-Seq of tumour suppressor MEG3 in combination with SHAPE revealed two pseudoknot regions that interact to form a kissing loop motif. This conformational change results in activation of the p53 pathway and cell cycle arrest (Uroda et al. 2019).

4 Probing Techniques to Study the RNA G-Quadruplex Motif

Probing reagents can also be designed to recognize a specific structural motif. Small molecule ligands, antisense oligonucleotides, and antibodies can be applied to modify, isolate, or visualize the structural motif and in certain cases to stabilize or disrupt the secondary structure. In the following, the RNA G-quadruplex structure serves as a model to present how different approaches and probing techniques can be combined to comprehensively study a distinct motif and its biology in cells.

4.1 In Silico Prediction of G4s

Prediction of DNA G4 structures from G-rich consensus sequences ($G_{3+}N_{1-7}G_{3+}N_{1-7}G_{3+}N_{1-7}G_{3+}$) has been performed computationally (Puig Lombardi and Londoño-Vallejo 2020). The presence of 700,000 DNA G4s that were found in the human genome by G4 probing coupled to high-throughput sequencing (Chambers et al. 2015) as compared to only 375,000 predicted loci (Huppert and Balasubramanian 2005) has yet again demonstrated the limitations of in silico prediction. The high false-negative rate is mainly due to non-G sequence variations in the consensus sequence and to regulatory factors in vivo that govern the equilibrium between folded and unfolded states.

4.2 Visualisation of RNA G4s by Immunolabelling

While G4s are well-known secondary structures in DNA, the first evidence for G4s formation in RNA was given by visualization with a G4-specific antibody, BG4 (Biffi et al. 2014). In fixed human cells, incubation with a FLAG-tagged BG4 revealed fluorescent BG4 foci in the cytoplasm, which were indicative of RNA G4 structures. By an increase in cytoplasmic foci, but not in nuclear signal, it was also shown that the RNA G4-specific probe carboxypyridostatin (carboxyPDS, see 0) could exclusively detect cytoplasmic RNA G4s when it was applied to living cells prior to fixation.

4.3 Chemical Probing of RNA G4s Coupled to Sequencing

G4-seq. Shortly after, G4-seq was the first method to map RNA G4s on a transcriptome-wide scale. The method makes use of reverse transcriptase stalling induced by fully folded G4 structures (Fig. 4a). To identify RT read drops, isolated RNA is treated under G4-favourable conditions (K^+ or stabilising ligand, e.g., pyridostatin (PDS), BRACO-19) to allow for G4 folding. This sample is compared to a normalization control obtained under G4-unfavourable conditions (Li^+) (Kwok et al. 2016; Yang et al. 2018). 3300 to 11,000 G4 sites were detected in human mRNAs under physiological K^+ conditions or with PDS, respectively. The majority were found in the 5′ and 3′ UTRs and were enriched in polyadenylation signals. This is consistent with a role in transcriptional and translational regulation and mRNA processing. Since the technique requires total or polyA-enriched RNA, the binding of proteins or other endogenous ligands is not taken into account. This opens the debate as to whether the identified G4 sites are actually formed physiologically.

G4-DMS-seq. To overcome this problem, G4-DMS-seq adds a DMS treatment step prior to the G4-seq protocol. Since the N^7 positions of the guanines are hidden when constrained in a G-tetrad, stable G4 structures are protected from DMS methylation, and will produce RT stops (Fig. 4c). A reduction in RT stops upon DMS treatment in vivo (+DMS) and a similarity to the in vitro/–K^+ conditions led the authors to conclude that G4s are mainly present in their unfolded state in mammalian cells (Guo and Bartel 2016).

G4-SHAPE. For the same purpose of capturing G4s in their physiologically folded state, the authors developed G4-SHAPE. Herein, NAI was shown to preferentially react with the exposed 2′ OH group of a loop-adjacent G in a stable G4 structure (Fig. 4d). From low SHAPE reactivity profiles, the authors concluded that G4s adopt a globally unfolded state in vivo (Guo and Bartel 2016). However, multiple fluorescent imaging studies have now verified the dynamic folding and unfolding of RNA G4 structures in cells (see 0). Of note, SHAPE probing of G4 candidates and RNAfold analysis have proven that alternative stable secondary structures compete with G4 folding. In the same work it was suggested that G4 formation also affects long-range tertiary folding (Kwok et al. 2016).

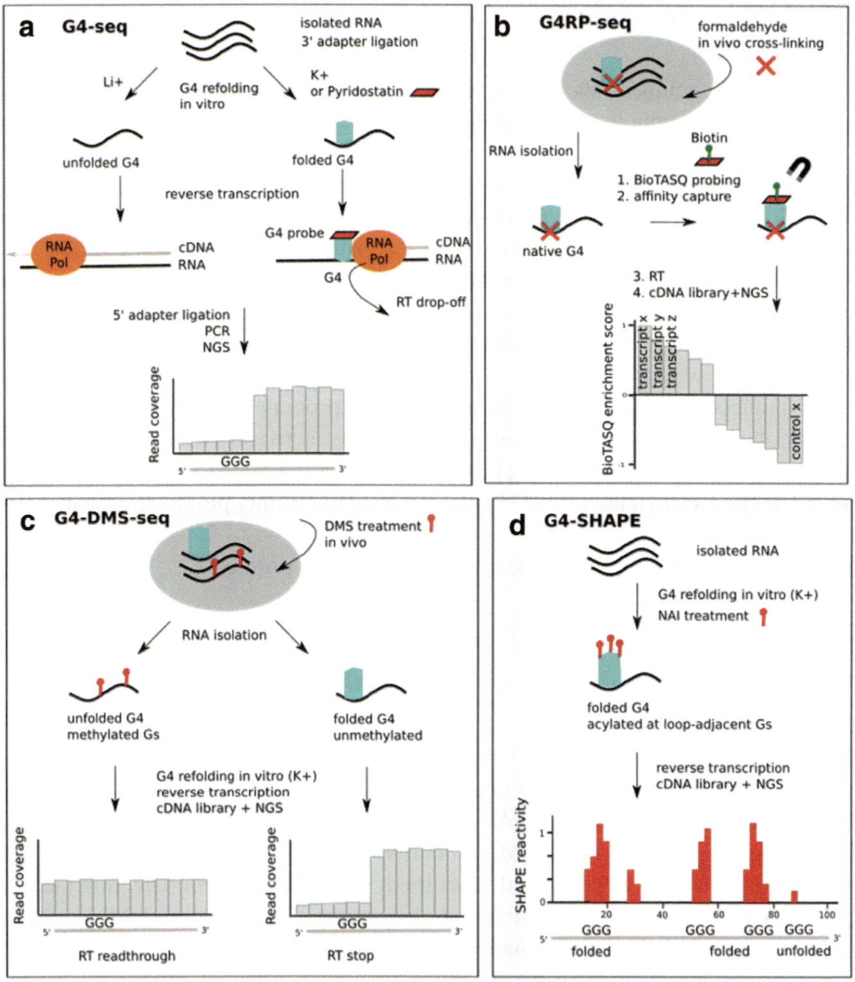

Fig. 4 Probing and visualization techniques for RNA G-quadruplex structures. **a** In G4-seq, total RNA is treated in vitro under G4-inducing conditions or with a stabilizing ligand. G4 sites are identified by a drop-off in RT reads at the site of G4 folding when compared to the untreated control under G4-disfavorable conditions. **b** G4RP-seq applies cross-linking in cells to freeze transiently folded G4s. Incubation in vitro with a biotinylated BioTASQ ligand and affinity capture with streptavidin beads is used to enrich and identify G4-containing transcripts by RT-qPCR or sequencing. **c** G4-DMS-seq makes use of specific methylation of G residues in unfolded G4 structures. RNA from DMS-treated cells is allowed to refold to G4s in vitro only if the G residues were protected from DMS methylation in folded G4 structures. This allows to probe for G4 folding in vivo after RT stop and NGS analysis. **d** G4-SHAPE probes the flexible 2′ OH of loop-adjacent G residues in a G4 structure. High SHAPE reactivities indicate folded G4s, whereas low reactivities indicate unfolded G4s in vivo

4.4 Immunoprecipitation of RNA G4s

G4RP-seq. Techniques involving antibodies such as BG4 are equivalent to ChIP-seq experiments, which are commonly used to capture DNA G4 structures in the native chromatin state. iCLIP protocols coupled to RNA sequencing can be used to elucidate binding regions of G4-binding proteins, and thus indirectly assess potential G4 sites (Kharel et al. 2020). More directly, the method G4RP-seq similarly works with crosslinking and affinity purification and with a novel G4-specific probe. The group synthesized a biotinylated probe, BioTASQ, that selectively binds, pulls down, and enriches G4-containing transcripts on streptavidin beads. but transient formation of G4 structures (**Error! Reference source not found.**B) (Yang et al. 2018). Formaldehyde was used to covalently freeze transient G4s in vivo and to minimize ligand-induced stabilization of G4s in the in vitro probing steps with BioTASQ.

In an alternative approach, the intrinsic peroxidase activity of a G4-hemin complex can be exploited in a reaction with H_2O_2 and a biotin substrate to self-biotinylate the G4. Here, the biotinylated DNA G4 was then used for affinity pulldown, purification, and PCR (Einarson and Sen 2017). The self-biotinylation of G4-hemin might also be applied to RNA G4s for RT stop analysis and NGS protocols.

4.5 Visualisation of RNA G4s with Fluorescent Probes

Small molecules that are used for G4 probing can also be applied for G4 visualisation in living cells. RNA G4s can be detected in vivo with turn-on fluorophore probes such as QUMA-1 and Naphtho-TASQ (N-TASQ) (Chen et al. 2018; Laguerre et al. 2015). Treatment with both QUMA-1 and N-TASQ does not require cell fixation and permeabilization as compared to antibody-based fluorescent detection (see 0). QUMA-1 is used for real-time imaging of dynamic folding and unfolding of G4s by tracking the mobility, appearance/disappearance, and merging of fluorescent foci over time. In this way, even the assembly to higher-order G4 structures and the dynamic unfolding of G4s by the helicase DHX36 could be visualized.

To screen for new ligands that are selective for endogenous RNA G4s as opposed to DNA G4s, a click-chemistry approach was developed by Di Antonio et al. (Di Antonio et al. 2012). An alkynylated pyridostatin was incubated with a library of azides containing variable functional groups. In the presence of the G4-forming telomeric-repeat RNA TERRA, adducts between PDS and azide would form only if they were successfully interacting with and stabilized by the G4 structure. By mass spectrometry quantification and competition assays with the DNA G4-forming telomere H-Telo, a carboxy-terminal PDS derivative, carboxyPDS, was validated as a novel RNA-selective small molecule probe.

4.6 *Disruption of G4 Structures with Antisense Oligonucleotides*

G4s are a unique example of a structural motif as individual G4s can be distinguished by sequence identity. Antisense oligonucleotide (ASO) probes have the advantage that they selectively target individual G4s due to sequence-specific base-pairing. In a recent study, DNA probes that disrupt genomic G4s were designed and applied to G4-forming DNA promoters to relieve the secondary structure. The precise positioning of chemically locked nucleosides (LNAs) improved the G4 sequence-binding affinities of the ASOs. The LNA probes led to disruption of DNA G4 structures in a reporter gene promoter. By this, gene expression was activated by facilitating RNA polymerase read-through (Chowdhury et al. 2022).

This example shows that in parallel to examining the sites, quantity, and dynamics of secondary structures, the interference with motif-specific probes is equally important to expand the data on G4s on their biological functions. Here, disruption of individual G4s alleviates polymerase stalling at promoters and allows to study the effect on gene expression. In general, stabilising or destabilising probes can be used as chemical biology tools for switching on or off a motif selectively and to explore its function in cells.

5 Conclusion

The development of next-generation sequencing techniques has paved the way to a high-throughput readout of chemical probing data. Since a decade, SHAPE-seq and DMS-seq have served as models for several variations of probing techniques. Of high importance are the mutational mapping (MaP)-approaches to study several flexible nucleotides simultaneously on one transcript.

Third-generation sequencing such as Oxford Nanopore technology is rapidly improving. It already allows the mapping of RNA base modifications, e.g., of m^6A and ψ, based on current and dwell time (Leger et al. 2021). Recently, a novel SHAPE-MaP reagent, 1-acetylimidazole, has been demonstrated to generate RNA adducts which can be used for structural mapping in single-molecule sequencing (Stephenson et al. 2022). It will be exciting to apply this technique to lncRNAs and to study different mRNA isoforms. Importantly, NanoSHAPE opens an unprecedented advance to analyse modifications and structural mapping in parallel. This is central to RNA research because base modifications impose an immense impact on RNA secondary structure. For instance, a single m^6A in the lncRNA MALAT1 disrupts a duplex hairpin structure, thereby exposing the single-stranded U-tract for access to an m^6A reader protein hnRNP-C, with downstream effects on mRNA processing (Liu et al. 2015). m^6A can also alter the RNA structure to facilitate the binding of low-complexity RBPs (Liu et al. 2017a). On a transcriptome-wide level, it will be of interest to develop deconvolution techniques for conformational ensembles of e.g. m^6A-containing transcripts, to derive preferred conformations for m^6A-modified mRNAs from structural probing data.

A full picture of 3D RNA tertiary structure can be obtained by applying biophysical low-resolution techniques. Solution structures are studied with single-angle X-ray scattering (SAXS) and with advanced atomic force microscopy (AFM) approaches (Ding 2023; Fang et al. 2015b; Lee et al. 2023). While traditional structural analysis by X-ray crystallography or NMR spectroscopy provide a high-resolution tertiary structure, those are limited to short transcripts or isolated domains, and have only been applied successfully to a few RNA targets (Chillón and Marcia 2020). Electron microscopy, i.e., cryo-EM or negative staining EM, is of increasing importance to study RNA structural dynamics and conformational ensembles. Careful sample preparation of full-length transcripts can provide detailed structures of single RNAs or in complex with their cognate RBP (Bonilla and Kieft 2022; Ma et al. 2022).

A combination of secondary structure probing and biophysical techniques for tertiary structure and dynamics in solution is best suited to gain a comprehensive picture of an RNA target. Importantly, the integration of different experimental data into bioinformatic prediction tools is constantly advancing to obtain more accurate RNA structure models (Li et al. 2020). If no solution or crystal samples are available, Förster resonance energy transfer (FRET) can be applied to study distances and long-range interactions in fluorescently tagged RNA molecules. Single-molecule FRET (smFRET) probes folding dynamics of an immobilized RNA molecule and can deconvolute conformational changes in RNP assembly processes such as ribosome biogenesis and transcription (Duss et al. 2018; Feng et al. 2021).

Drug development is poorly established for cellular RNA targets, mainly owing to their conformational diversity and dynamics. In future, it will be of high biomedical interest to screen for small molecules that target specific disease-associated RNAs, as exemplified for the lncRNA Xist (Aguilar et al. 2022). The targeting of a stable secondary structure motif, such as an individual G4 of specific sequence, can guide the way to target an individual disease-related transcript. The LNA-modified DNA probes provide an important basis to use antisense oligonucleotides as tools to interfere with stable DNA structures. It will be exciting to see how LNA probes can be designed to selectively bind G4 structures formed in RNA. Further development of G4-disrupting molecules as opposed to G4-stabilising probes will be crucial, since studies have reported the uncontrolled accumulation of DNA G4s associated with diseases like ALS (Simone et al. 2018).

References

Aguilar R, Spencer KB, Kesner B et al (2022) Targeting Xist with compounds that disrupt RNA structure and X inactivation. Nature 604:160–166

Aw JGA, Lim SW, Wang JX et al (2021) Determination of isoform-specific RNA structure with nanopore long reads. Nat Biotechnol 39:336–346

Biffi G, Di Antonio M, Tannahill D et al (2014) Visualization and selective chemical targeting of RNA G-quadruplex structures in the cytoplasm of human cells. Nat Chem 6:75–80

Bonilla SL, Kieft JS (2022) The promise of cryo-EM to explore RNA structural dynamics. J Mol Biol 434:167802

Brion P, Westhof E (1997) Hierarchy and dynamics of RNA folding. Annu Rev Biophys Biomol Struct 26:113–137

Butcher SE, Pyle AM (2011) The molecular interactions that stabilize RNA tertiary structure: RNA motifs, patterns, and networks. Acc Chem Res 44:1302–1311

Chambers VS, Marsico G, Boutell JM et al (2015) High-throughput sequencing of DNA G-quadruplex structures in the human genome. Nat Biotechnol 33:877–881

Chen X-C, Chen S-B, Dai J et al (2018) Tracking the dynamic folding and unfolding of RNA G-quadruplexes in live cells. Angew Chem Int Ed Engl 57:4702–4706

Cheng CY, Chou F-C, Kladwang W et al (2015) Consistent global structures of complex RNA states through multidimensional chemical mapping. Elife 4:e07600

Chillón I, Marcia M (2020) The molecular structure of long non-coding RNAs: emerging patterns and functional implications. Crit Rev Biochem Mol Biol 55:662–690

Chowdhury S, Wang J, Nuccio SP et al (2022) Short LNA-modified oligonucleotide probes as efficient disruptors of DNA G-quadruplexes. Nucleic Acids Res 50:7247–7259

Collie GW, Haider SM, Neidle S et al (2010) A crystallographic and modelling study of a human telomeric RNA (TERRA) quadruplex. Nucleic Acids Res 38:5569–5580

Costa M, Monachello D (2014) Probing RNA folding by hydroxyl radical footprinting. Methods Mol Biol 1086:119–142

Das R, Laederach A, Pearlman SM et al (2005) SAFA: semi-automated footprinting analysis software for high-throughput quantification of nucleic acid footprinting experiments. RNA 11:344–354

Di Antonio M, Biffi G, Mariani A et al (2012) Selective RNA versus DNA G-quadruplex targeting by in situ click chemistry. Angew Chem Int Ed Engl 51:11073–11078

Ding J (2023) High-resolution atomic force microscopy imaging of RNA molecules in solution. Methods Mol Biol 2568:133–145

Ding Y, Tang Y, Kwok CK et al (2014) In vivo genome-wide profiling of RNA secondary structure reveals novel regulatory features. Nature 505:696–700

Duss O, Stepanyuk GA, Grot A et al (2018) Real-time assembly of ribonucleoprotein complexes on nascent RNA transcripts. Nat Commun 9:5087

Einarson OJ, Sen D (2017) Self-biotinylation of DNA G-quadruplexes via intrinsic peroxidase activity. Nucleic Acids Res 45:9813–9822

Fang R, Moss WN, Rutenberg-Schoenberg M et al (2015a) Probing Xist RNA structure in cells using targeted Structure-seq. PLoS Genet 11:e1005668

Fang X, Stagno JR, Bhandari YR et al (2015b) Small-angle X-ray scattering: a bridge between RNA secondary structures and three-dimensional topological structures. Curr Opin Struct Biol 30:147–160

Feng XA, Poyton MF, Ha T (2021) Multicolor single-molecule FRET for DNA and RNA processes. Curr Opin Struct Biol 70:26–33

Gherghe CM, Shajani Z, Wilkinson KA et al (2008) Strong correlation between SHAPE chemistry and the generalized NMR order parameter (S2) in RNA. J Am Chem Soc 130:12244–12245

Götte M, Marquet R, Isel C et al (1996) Probing the higher order structure of RNA with peroxonitrous acid. FEBS Lett 390:226–228

Guo JU, Bartel DP (2016) RNA G-quadruplexes are globally unfolded in eukaryotic cells and depleted in bacteria. Science 353:aaf5371

Hajdin CE, Bellaousov S, Huggins W et al (2013) Accurate SHAPE-directed RNA secondary structure modeling, including pseudoknots. Proc Natl Acad Sci USA 110:5498–5503

Harris KA, Crothers DM, Ullu E (1995) In vivo structural analysis of spliced leader RNAs in *Trypanosoma brucei* and *Leptomonas collosoma*: a flexible structure that is independent of cap4 methylations. RNA 1:351–362

Hector RD, Burlacu E, Aitken S et al (2014) Snapshots of pre-rRNA structural flexibility reveal eukaryotic 40S assembly dynamics at nucleotide resolution. Nucleic Acids Res 42:12138–12154

Heilman-Miller SL, Woodson SA (2003) Effect of transcription on folding of the *Tetrahymena* ribozyme. RNA 9:722–733

Herschlag D (1995) RNA chaperones and the RNA folding problem. J Biol Chem 270:20871–20874

Homan PJ, Favorov OV, Lavender CA et al (2014) Single-molecule correlated chemical probing of RNA. Proc Natl Acad Sci U S A 111:13858–13863

Huppert JL, Balasubramanian S (2005) Prevalence of quadruplexes in the human genome. Nucleic Acids Res 33:2908–2916

Incarnato D, Oliviero S (2017) The RNA epistructurome: uncovering RNA function by studying structure and post-transcriptional modifications. Trends Biotechnol 35:318–333

Incarnato D, Neri F, Anselmi F et al (2014) Genome-wide profiling of mouse RNA secondary structures reveals key features of the mammalian transcriptome. Genome Biol 15:491

Jaeger JA, Turner DH, Zuker M (1989) Improved predictions of secondary structures for RNA. Proc Natl Acad Sci USA 86:7706–7710

Keating KS, Toor N, Pyle AM (2008) The GANC tetraloop: a novel motif in the group IIC intron structure. J Mol Biol 383:475–481

Kertesz M, Wan Y, Mazor E et al (2010) Genome-wide measurement of RNA secondary structure in yeast. Nature 467:103–107

Kharel P, Becker G, Tsvetkov V et al (2020) Properties and biological impact of RNA G-quadruplexes: from order to turmoil and back. Nucleic Acids Res 48:12534–12555

Kielpinski LJ, Vinther J (2014) Massive parallel-sequencing-based hydroxyl radical probing of RNA accessibility. Nucleic Acids Res 42:e70

Klein DJ, Schmeing TM, Moore PB et al (2001) The kink-turn: a new RNA secondary structure motif. EMBO J 20:4214–4221

Kwok CK, Marsico G, Sahakyan AB et al (2016) rG4-seq reveals widespread formation of G-quadruplex structures in the human transcriptome. Nat Methods 13:841–844

Laguerre A, Hukezalie K, Winckler P et al (2015) Visualization of RNA-quadruplexes in live cells. J Am Chem Soc 137:8521–8525

Lai D, Proctor JR, Meyer IM (2013) On the importance of cotranscriptional RNA structure formation. RNA 19:1461–1473

Lee Y-T, Fan L, Ding J et al (2023) Combining biophysical methods for structure-function analyses of RNA in solution. Methods Mol Biol 2568:165–177

Leger A, Amaral PP, Pandolfini L et al (2021) RNA modifications detection by comparative Nanopore direct RNA sequencing. Nat Commun 12:7198

Li B, Cao Y, Westhof E et al (2020) Advances in RNA 3D structure modeling using experimental data. Front Genet 11:574485

Liu N, Dai Q, Zheng G et al (2015) N^6-methyladenosine-dependent RNA structural switches regulate RNA-protein interactions. Nature 518:560–564

Liu N, Zhou KI, Parisien M et al (2017a) N^6-methyladenosine alters RNA structure to regulate binding of a low-complexity protein. Nucleic Acids Res 45:6051–6063

Liu F, Somarowthu S, Pyle AM (2017b) Visualizing the secondary and tertiary architectural domains of lncRNA RepA. Nat Chem Biol 13:282–289

Lu ZJ, Gloor JW, Mathews DH (2009) Improved RNA secondary structure prediction by maximizing expected pair accuracy. RNA 15:1805–1813

Lucks JB, Mortimer SA, Trapnell C et al (2011) Multiplexed RNA structure characterization with selective 2'-hydroxyl acylation analyzed by primer extension sequencing (SHAPE-Seq). Proc Natl Acad Sci U S A 108:11063–11068

Ma H, Jia X, Zhang K et al (2022) Cryo-EM advances in RNA structure determination. Signal Transduct Target Ther 7:58

Mathews DH (2004) Using an RNA secondary structure partition function to determine confidence in base pairs predicted by free energy minimization. RNA 10:1178–1190

Mathews DH, Disney MD, Childs JL et al (2004) Incorporating chemical modification constraints into a dynamic programming algorithm for prediction of RNA secondary structure. Proc Natl Acad Sci USA 101:7287–7292
McGinnis JL, Weeks KM (2014) Ribosome RNA assembly intermediates visualized in living cells. Biochemistry 53:3237–3247
McGinnis JL, Dunkle JA, Cate JHD et al (2012) The mechanisms of RNA SHAPE chemistry. J Am Chem Soc 134:6617–6624
Merino EJ, Wilkinson KA, Coughlan JL et al (2005) RNA structure analysis at single nucleotide resolution by selective 2'-hydroxyl acylation and primer extension (SHAPE). J Am Chem Soc 127:4223–4231
Mestre-Fos S, Penev PI, Suttapitugsakul S et al (2019) G-quadruplexes in human ribosomal RNA. J Mol Biol 431:1940–1955
Mestre-Fos S, Ito C, Moore CM et al (2020) Human ribosomal G-quadruplexes regulate heme bioavailability. J Biol Chem 295:14855–14865
Mitchell D, Ritchey LE, Park H et al (2018) Glyoxals as in vivo RNA structural probes of guanine base-pairing. RNA 24:114–124
Mitra S, Shcherbakova IV, Altman RB et al (2008) High-throughput single-nucleotide structural mapping by capillary automated footprinting analysis. Nucleic Acids Res 36:e63
Morandi E, van Hemert MJ, Incarnato D (2022) SHAPE-guided RNA structure homology search and motif discovery. Nat Commun 13:1722
Mortimer SA, Weeks KM (2007) A fast-acting reagent for accurate analysis of RNA secondary and tertiary structure by SHAPE chemistry. J Am Chem Soc 129:4144–4145
Mortimer SA, Weeks KM (2008) Time-resolved RNA SHAPE chemistry. J Am Chem Soc 130:16178–16180
Mortimer SA, Weeks KM (2009) Time-resolved RNA SHAPE chemistry: quantitative RNA structure analysis in one-second snapshots and at single-nucleotide resolution. Nat Protoc 4:1413–1421
Novikova IV, Dharap A, Hennelly SP et al (2013) 3S: shotgun secondary structure determination of long non-coding RNAs. Methods 63:170–177
Piao M, Li P, Zeng X et al (2022) An ultra low-input method for global RNA structure probing uncovers Regnase-1-mediated regulation in macrophages. Fundam Res 2:2–13
Puig Lombardi E, Londoño-Vallejo A (2020) A guide to computational methods for G-quadruplex prediction. Nucleic Acids Res 48:1–15
Quigley GJ, Rich A (1976) Structural domains of transfer RNA molecules. Science 194:796–806
Rice GM, Leonard CW, Weeks KM (2014) RNA secondary structure modeling at consistent high accuracy using differential SHAPE. RNA 20:846–854
Ritchey LE, Su Z, Tang Y et al (2017) Structure-seq2: sensitive and accurate genome-wide profiling of RNA structure in vivo. Nucleic Acids Res 45:e135
Rouskin S, Zubradt M, Washietl S et al (2014) Genome-wide probing of RNA structure reveals active unfolding of mRNA structures in vivo. Nature 505:701–705
Russell R (2008) RNA misfolding and the action of chaperones. Front Biosci 13:1–20
Russell R (2013) Biophysics of RNA folding, 1st edn. Biophysics for the Life Sciences, vol 3. Springer New York, Imprint Springer, New York
Schroeder R, Grossberger R, Pichler A et al (2002) RNA folding in vivo. Curr Opin Struct Biol 12:296–300
Sclavi B, Woodson S, Sullivan M et al (1997) Time-resolved synchrotron X-ray footprinting, a new approach to the study of nucleic acid structure and function: application to protein-DNA interactions and RNA folding. J Mol Biol 266:144–159
Seetin MG, Kladwang W, Bida JP et al (2014) Massively parallel RNA chemical mapping with a reduced bias MAP-seq protocol. Methods Mol Biol 1086:95–117
Shcherbakova I, Mitra S, Laederach A et al (2008) Energy barriers, pathways, and dynamics during folding of large, multidomain RNAs. Curr Opin Chem Biol 12:655–666

Shi H, Moore PB (2000) The crystal structure of yeast phenylalanine tRNA at 1.93 A resolution: a classic structure revisited. RNA 6:1091–1105

Siegfried NA, Busan S, Rice GM et al (2014) RNA motif discovery by SHAPE and mutational profiling (SHAPE-MaP). Nat Methods 11:959–965

Simone R, Balendra R, Moens TG et al (2018) G-quadruplex-binding small molecules ameliorate C9orf72 FTD/ALS pathology in vitro and in vivo. EMBO Mol Med 10:22–31

Spitale RC, Crisalli P, Flynn RA et al (2013) RNA SHAPE analysis in living cells. Nat Chem Biol 9:18–20

Spitale RC, Flynn RA, Zhang QC et al (2015) Structural imprints in vivo decode RNA regulatory mechanisms. Nature 519:486–490

Stephenson W, Razaghi R, Busan S et al (2022) Direct detection of RNA modifications and structure using single-molecule nanopore sequencing. Cell Genom 2:100097

Sutton JL, Pollack L (2015) Tuning RNA flexibility with helix length and junction sequence. Biophys J 109:2644–2653

Tamura M, Holbrook SR (2002) Sequence and structural conservation in RNA ribose zippers. J Mol Biol 320:455–474

Turner DH (1989) Thermodynamics and kinetics of base-pairing and of DNA and RNA self-assembly and helix coil transition. Nucleic Acids

Tyrrell J, McGinnis JL, Weeks KM et al (2013) The cellular environment stabilizes adenine riboswitch RNA structure. Biochemistry 52:8777–8785

Underwood JG, Uzilov AV, Katzman S et al (2010) FragSeq: transcriptome-wide RNA structure probing using high-throughput sequencing. Nat Methods 7:995–1001

Uroda T, Anastasakou E, Rossi A et al (2019) Conserved pseudoknots in lncRNA MEG3 are essential for stimulation of the p53 pathway. Mol Cell 75:982-995.e9

Vasa SM, Guex N, Wilkinson KA et al (2008) ShapeFinder: a software system for high-throughput quantitative analysis of nucleic acid reactivity information resolved by capillary electrophoresis. RNA 14:1979–1990

Waduge P, Sakakibara Y, Chow CS (2019) Chemical probing for examining the structure of modified RNAs and ligand binding to RNA. Methods 156:110–120

Walter AE, Turner DH (1994) Sequence dependence of stability for coaxial stacking of RNA helixes with Watson-Crick base paired interfaces. Biochemistry 33:12715–12719

Walter AE, Turner DH, Kim J et al (1994) Coaxial stacking of helixes enhances binding of oligoribonucleotides and improves predictions of RNA folding. Proc Natl Acad Sci USA 91:9218–9222

Wan Y, Qu K, Ouyang Z et al (2012) Genome-wide measurement of RNA folding energies. Mol Cell 48:169–181

Weeks KM (1997) Protein-facilitated RNA folding. Curr Opin Struct Biol 7:336–342

Wells SE, Hughes JM, Haller Igel A et al (2000) Use of dimethyl sulfate to probe RNA structure in vivo. In: RNA-ligand interactions part B, vol 318. Elsevier, pp 479–493

Weng X, Gong J, Chen Y et al (2020) Keth-seq for transcriptome-wide RNA structure mapping. Nat Chem Biol 16:489–492

Westhof E, Masquida B, Jaeger L (1996) RNA tectonics: towards RNA design. Fold Des 1:R78–R88

Xia T, SantaLucia J, Burkard ME et al (1998) Thermodynamic parameters for an expanded nearest-neighbor model for formation of RNA duplexes with Watson-Crick base pairs. Biochemistry 37:14719–14735

Yang SY, Lejault P, Chevrier S et al (2018) Transcriptome-wide identification of transient RNA G-quadruplexes in human cells. Nat Commun 9:4730

Yesselman JD, Denny SK, Bisaria N et al (2019) Sequence-dependent RNA helix conformational preferences predictably impact tertiary structure formation. Proc Natl Acad Sci USA 116:16847–16855

Ziehler WA, Engelke DR (2001) Probing RNA structure with chemical reagents and enzymes. Curr Protoc Nucleic Acid Chem Chapter 6:Unit 6.1

Structure and Functions of RNA G-quadruplexes

Prakash Kharel and Pavel Ivanov

Contents

Abstract G-quadruplexes (G4s) are four-stranded nucleic acid secondary structures that are formed by the stacking of square planar guanine arrangements and stabilized by favorable cations. Potential G4-forming sequences are distributed in the regulatory regions of the genome and transcriptome. G4s are proposed to modulate various physiological and pathophysiological cellular processes. As such RNA G4s (rG4s) have been implicated in several key processes of gene regulation such as RNA maturation, mRNA translation, and RNA transport. rG4s often impact cellular biology by associating different RNA binding proteins, both of which could act as crucial therapeutic targets in the fight for developing novel therapeutics for the diseases associated with rG4-containing transcripts.

P. Kharel · P. Ivanov (✉)
Division of Rheumatology, Inflammation, and Immunity, Department of Medicine, Brigham and Women's Hospital and Harvard Medical School, Boston, MA 02115, USA

P. Kharel
e-mail: pkharel@bwh.harvard.edu

J. Barciszewski (ed.), *RNA Structure and Function*, RNA Technologies 14,
https://doi.org/10.1007/978-3-031-36390-0_9

Keywords G-quadruplex · RNA metabolism · Translation · G-quadruplex targeting

1 Introduction

Guanine (G)-rich nucleic acid sequences can fold into four-stranded secondary structures called G-quadruplexes (G4s) via the stacking of two or more square planar G arrangements known as G-quartets (Sen and Gilbert 1988). It was first noted in 1910 that high concentrations of impure guanylic acid formed a gel in an aqueous solution (Bang 1910). 50 years later, Khorana and co-workers found similar highly ordered aggregation with the first synthesized deoxyguanosine oligonucleotides. These earlier observations were structurally rationalized using X-ray diffraction studies that could be explained by a hydrogen-bonding arrangements of four G bases, thus proposing the G-quartet assembly (Gellert et al. 1962). More than a century after the initial discovery of Bang, the G4 field has moved quite a remarkable distance.

G-quartets (or G-tetrads), structural units of G4s, are formed when guanines are organized into square planar arrangements where each G base is connected to two other bases. The G-quartet involves two edges of each of the four G bases with Watson–Crick and Hoogsteen base pairings (Fig. 1a) (Gellert et al. 1962). Hydrogen bonds (H-bond) between each G pair involve four donor/acceptor atoms, the N1, N7, N2, and O6 atoms, such that a G-quartet has eight total hydrogen bonds (four N2−H...N7 and four N1−H...O6 bonds). Importantly, four carbonyl oxygen (O6) atoms form a negatively charged core in the center of the G-quartet (Fig. 1b). Under the favorable condition, two or more G-quartets stack onto one another to form a right-handed helical G4 structure (Fig. 1c). The central anionic core of a G-quartet or the central space between two quartets provides a perfect space for the binding of a cation, which in turn provides key stability to the quartets and G4s (Bhattacharyya et al. 2016). Because of the defined geometry and size of the central channel, only cations with an adequate charge, size, and dehydration energy can coordinate a G4. Of particular biological importance, Na^+, K^+, and $NH_4{}^+$ cations are most physiologically relevant and G4-stabilizing (Fig. 1c). Cations like K^+ and $NH_4{}^+$ are too large to fit into the plane of G-quartet, but readily accommodate into the space between two G-quartets and coordinate with eight O atoms. On the other hand, smaller Na^+ is embedded into the middle of a single G-quartet and coordinates only four O atoms, thus contributing less to G4 stability. In contrast, cations with very small ionic radii such as Li^+ do not favor G4 formation.

Generally speaking, potential intramolecular G4-assembling sequences can be formed by repetition of a G-tract sequence motif within a single run of sequence. In such repetitive motif $G_mX_nG_mX_oG_mX_pG_m$, m is the number of G bases in every short G-track, which are connected by intervening X_n, X_o, X_p sequences with any combination of bases including Gs (Puig Lombardi and Londoño-Vallejo 2020). Despite having a lot of similarities in their basic building units, in fact, G4s are a diverse family of nucleic acid structures that can fold into various topologies (Lightfoot et al. 2019;

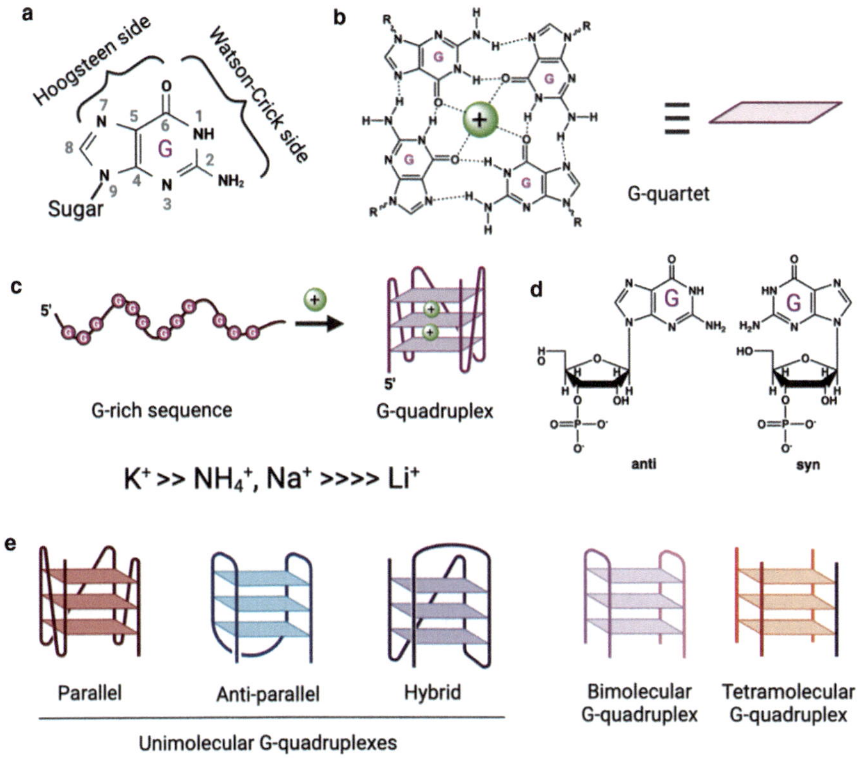

Fig. 1 Watson–Crick and Hoogsteen base pairing interaction sites in a guanine base. **b**. A square planar G-quartet arrangement stabilized by a centrally located cation. H-bonding between each pair of guanines involves four donor/acceptor atoms (the N1, N7, N2, and O6 atoms) resulting in 8 H-bonds per quartet. Four carbonyl oxygen (O6) atoms form a negatively charged core in the center of the G-quartet that favors the binding of monovalent cations (⊕). **c**. A G-rich sequence with at least 4 G-stretches with at least 2 Gs each can fold into G4 under favorable conditions (top); and Preferential binding of mostly used monovalent cations to G4s (bottom). **d**. syn and anti-conformation of glycosidic bond in Gs. In RNA, this remains almost exclusively in anti-conformation resulting in all parallel rG4 topology. **e**. G4s with different topologies and molecularities

Winnerdy and Phan 2020). In bona fide G4s, the G4 topologies are dictated by the pattern of strand polarities and the orientation of interconnecting loops. The G4s can have parallel (all backbones running in the same direction), anti-parallel (adjacent backbones run in the opposite direction), or mixed topologies (Fig. 1e). While the different topologies bring structural diversity, their influence on G4 formation and contributions to cellular functions is largely unknown. Another aspect of G4 structural diversity arises from the difference in the number of G-quartet stacks and the number of molecules involved. Based on the number of G-quartets, G4s can be 2-tier, 3-tier, 4-tier, and so on. Depending upon the number of nucleic acid molecules involved, besides unimolecular (intramolecular), G4s can also be bimolecular, or

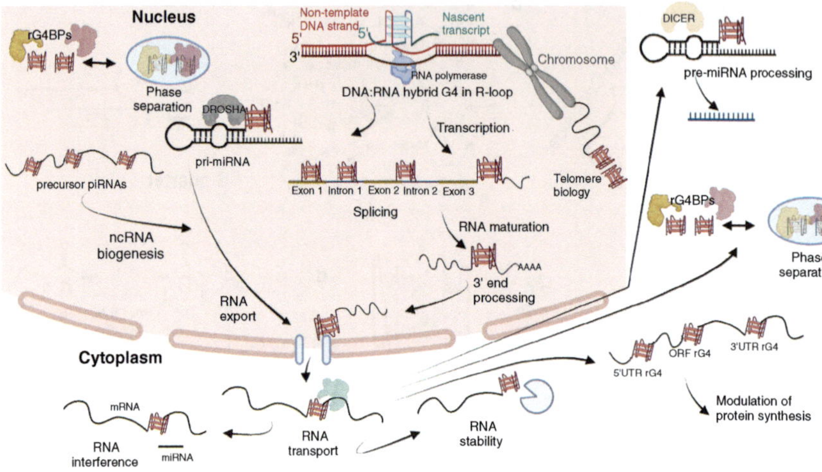

Fig. 2 **Role of RNA G-Quadruplexes (rG4s) in the nucleus and the cytoplasm.** rG4s have implications in almost every step of RNA life that ranges from the regulation of transcription, splicing, and 3′ end maturation in the nucleus to RNA transport, and the regulation of mRNA translation, ncRNA maturation, and RNA interference in the cytoplasm. Additionally, rG4s contribute to phase separation and/or aggregate formation in both the nucleus and cytoplasm

tetramolecular (Fig. 1e). Moreover, the nature of the flanking sequence can have a direct impact on the function of an rG4 (Zheng et al. 2022).

2 DNA G4s (dG4s) Versus RNA G4s (rG4s)

Both dG4s and rG4s look very similar at first glance. However, an assumption that rG4s are DNA counterparts is oversimplified. One of the key differences between dG4 and rG4 comes from the presence of a 2′-hydroxyl group (2′−OH) in the ribose sugar (Zaccaria and Fonseca Guerra 2018; Zhang et al. 2010). Not only, 2′−OH allows more intramolecular interactions within RNA G4s but also, they are favored to bring water molecules, making rG4s often more stable compared to their DNA counterparts. Additionally, the steric constraints posed by 2′-OH strongly favor the *anti*-conformation (via restrains on the glycosidic torsion angle), and imposition of additional constraints on sugar puckering (the ribose having a preference for C3′-*endo* puckering) (Fig. 1d). Consequently, the rG4 topology is almost always parallel where all four strands are oriented in the same direction (Fig. 1e). In contrast, dG4s are polymorphic and can adapt parallel, antiparallel or mixed conformations (Fig. 1e). rG4s also differ from dG4s in their cation interaction specificity. In a study based on a pair of G4 oligos, it was shown that while K^+ dramatically stabilizes both dG4s and rG4s, Na^+ only had a strong effect on dG4s. For divalent cations, only Sr^{2+} increases the stability of the rG4s. On the other hand, biologically relevant divalent cation Mg^{2+}

actually can destabilize rG4s (Balaratnam and Basu 2015). These unique features make rG4s more compact, less hydrated, and often more thermodynamically stable than dG4s. Furthermore, their presence in the cellular context makes the folding possibility of rG4s very different than that of dG4s; while the cellular DNA is almost always in a double-stranded form, cellular RNA is mostly in a single-stranded form.

3 Functions of rG4s in the Nucleus

3.1 *Transcriptional Regulation*

Putative G4s are commonly found in the genomic DNA, thus making it possible that corresponding rG4s are also formed upon transcription. In turn, the nascent RNA can base pair with the complimentary template DNA strand to form an RNA:DNA hybrid, which together with the displaced DNA strand, forms R-loop (Belotserkovskii et al. 2018). Bioinformatic analysis identified that such hybrid putative G4s (pG4s) are enriched downstream of the transcription start sites and are found in >97% of human genes, with an average of 73 hybrid pG4s per gene (Xiao et al. 2013). Indeed, the formation of R-loop G4s was confirmed using T7 RNA polymerase in vitro transcription. Such assay suggested that R-loop G4s inhibit transcription in vitro and represent *cis*-elements that are built into a gene and can be activated co-transcriptionally. The nascent RNA and non-template DNA strand of mitochondrial *CSBII*can co-transcriptionally form a stable DNA–RNA hybrid G4, which was suggested to promote transcription termination (Zheng et al. 2014). Furthermore, hybrid G4s formed by nascent transcript with DNA are shown to be dominating in number and thermodynamically more stable, which can help populate G4s in expense of duplex DNA (Shrestha et al. 2014). Furthermore, post-transcriptionally formed switch from rG4 to R-loop have been suggested to promote the class switch recombination (CSR) in the mouse immunoglobulin heavy chain (IHC) locus (Almeida et al. 2018). In mouse IHC, RNA helicase DDX1 directly binds to rG4s present in the intronic switch region, dissolving the structure thereby leading to a structural switch from rG4 to an R-loop form. R-loop formation results in a non-template single-strand DNA that could be a substrate for activation-induced cytidine deaminase (AID), the enzyme that initiates CSR by converting cytidines to uracils. Additionally, DNA:RNA hybrid G4s could contribute to transcription termination as potential G4s are proposed to act as terminator sequences that can stall RNA Polymerase II transcription. For example, R-loops formed behind elongating polymerase II are prevalent over G-rich sites located downstream of poly(A) signals, and are capable of G4 formation (Skourti-Stathaki et al. 2011). The DNA damage response protein Senataxin (SETX) is a DNA/RNA helicase which plays a key role in the resolving of R-loops thereby allowing $5' \rightarrow 3'$ exonuclease Xrn2 access to the 3' cleavage poly(A) sites causing nascent RNA release, 3′ cleavage product degradation and RNA polymerase II termination (Skourti-Stathaki et al. 2011). The depletion of SETX causes such pause-mediated transcription termination. RNA:DNA hybrid G4s can also contribute to transcription termination via coupling with 3′-end polyadenylation in association with heterogeneous nuclear ribonucleoprotein factors (HNRNP H/F) (Decorsière et al. 2011). For example, in tumor protein 53 (*TP53)* mRNA, rG4s interact with the splicing/polyadenylation factor HNRNP H/F to regulate polyadenylation. Under normal circumstances, mRNAs lacking rG4s at poly(A) signals are efficiently processed, whereas efficient 3′-end processing of *TP53* mRNA is inhibited presumably because of the rG4, resulting in the reduced gene expression. However, under genotoxic stress, there is global repression of mRNA 3′-end processing resulting in decreased mRNA maturation. In contrast, 3′-end processing of *TP53* mRNA is up-regulated to increase the expression of TP53. This anomalous mechanism is possible due to the recognition of rG4 in the *TP53* pre-mRNA by HNRNP H/F causing efficient recruitment 3'-end processing factors, which ultimately leads to an increased p53 expression. Another study showed that a G4 helicase DHX36 binds to TP53 rG4 under genotoxic condition and resolves the rG4 once the stress is removed thereby making TP53 mRNA available for immediate expression (Newman et al. 2017).

3.2 mRNA Maturation

Furthermore, rG4s also contribute to pre-mRNA splicing. Genome-wide analysis of alternatively spliced transcripts found over 3 million rG4 capable sites mapped to approximately 30,000 mammalian genes (Kikin et al. 2008). Alternative splicing is regulated by the synergic action of many RBPs with RNA elements that impact spliceosome assembly at neighboring splice sites (Wang et al. 2015), therefore rG4s assembled in the vicinity of splice sites may directly impact the binding of regulatory RBPs. For example, two rG4s are found within the FMRP-binding site (FBS) on its pre-mRNA (*FMR1*), which give rise to different FMRP isoforms (Didiot et al. 2008). As observed in a minigene system, the *FMR1* FBS can be a potent exonic splicing enhancer and acts as a control element that regulates alternative splicing in response to intracellular levels of FMRP isoforms. The binding of the longer FMRP isoform to FBS results in decreased synthesis of the longer FMRP isoforms (carrying a complete exon 15) concomitant with an increase of shorter isoforms. rG4 abrogating mutations in the FBS resulted in decreased FMRP binding, ablate exonic splicing enhancer activity and change the splicing pattern of *FMR1* pre-mRNA (Didiot et al. 2008). On the other hand, rG4s in intron 6 of the human telomerase reverse transcriptase (hTERT), the rate-limiting component of telomerase, can serve as an intronic splicing silencer as observed by G4-specific ligand-mediated impairment of hTERT splicing (Gomez et al. 2004). Additionally, an rG4 located in intron 3 of *TP53* pre-mRNA acts as an intronic splicing enhancer as it stimulates the splicing of intron 2 leading to a differential expression of transcripts encoding distinct p53 isoforms (Marcel et al. 2011). Furthermore, using a reporter system that consists of rG4 WT 3'UTR of *FXR1* mRNA, it has been shown that the presence of rG4 results in a more prominent shorter mRNA isoform while a G4 mutated version produced a prominent longer mRNA isoform, suggesting the role of 3′UTR mRNA rG4s in increasing alternative polyadenylation efficiency (Beaudoin and Perreault 2013).

3.3 Non-coding RNA Maturation in the Nucleus

In addition to mRNA maturation, rG4 structures can modulate the nuclear biology of noncoding RNAs, including both long non-coding RNA (lncRNAs) and short non-coding RNAs. There are relatively fewer studies in the role of rG4s in lncRNA. In the nucleus, nascent NEAT1 lncRNA binds to the non-POU domain-containing octamer-binding protein (NONO) through rG4 motifs (Simko et al. 2020). NONO plays an essential role in the initial paraspeckle formation stabilizing nascent NEAT1 transcript and providing the foundation necessary for the recruitment of the additional protein components needed for the subsequent steps of NEAT1 assembly and maturation (Clemson et al. 2009).

As such rG4s are also implicated in pre-miRNA maturation. Using computational analyses, two different groups proposed that 13–16% of pre-miRNAs harbor at least one putative rG4 motif in their sequence (Mirihana Arachchilage et al. 2015; Pandey et al. 2015). Based on in vitro data, rG4s in some pre-miRNAs exist in equilibrium with the canonical stem-loop structure such that their folding unwinds the stem-loop, thus hindering Dicer-mediated cleavage of the pre-miRNA and consequently affecting the pre-miRNA maturation process. First, it has been demonstrated that the maturation of the clinically relevant human *pre-miR92b* can be regulated by rG4 formation (Mirihana Arachchilage et al. 2015). Since the Dicer enzyme is stem-loop structure specific, disruption of the stem-loop because of the ion-dependent rG4 formation was found to inhibit Dicer-mediated maturation of *pre-miR-92b*, leading to reduction of mature *miR-92b* and de-repression of its targets. Similarly, it was found that rG4s in pre-miRNAs govern the biogenesis of mature miRNAs through a 'structural interference' mechanism (Pandey et al., 2015). A two-tier rG4 within *pre-let7e* interferes with Dicer-mediated processing, thus leading to a reduction of mature *miR-let7e* levels (Pandey et al. 2015). Furthermore, it has been proved that the formation of an rG4 structure in *pre-miR149* inhibits Dicer processing in vitro and this can be stabilized by the C8 acridine orange derivative and is used as a supramolecular carrier for the cancer-selective delivery of the ligand, considering the ability of such rG4 to bind to nucleolin (NCL) protein overexpressed on the surface of prostate cancer cells (Kwok et al. 2016). Interestingly, several pre-miRNA rG4s, such as *pre-miR-1229*, and *miR-1229-3p*, have been implicated in Alzheimer's disease, and *pre-miR-26a-1* rG4 has been linked to obesity regulation (Imperatore et al. 2020). Similarly, rG4s are implicated in Moloney leukemia virus 1 like (MOV10L1) mediated piRNA biogenesis (Zhang et al. 2019a).

4 RNA Transport

Subcellular RNA transport is a crucial post-transcriptional process that is key to spatiotemporal control of gene expression. RNA export from the nucleus to the cytoplasm is a ubiquitous phenomenon that is essential in the transport of a wider class of RNAs including mRNA, rRNA, tRNA, lncRNA, and miRNA. rG4s can play a crucial role in regulating the transport of G4-containing transcripts from the nucleus to the cytoplasm. In addition to nucleo-cytoplasmic export, the cytoplasmic mRNA transport mechanism is especially important in asymmetric cells such as neurons where transcribed mRNAs travel large distances to their sites of translation (Loya et al. 2010). It has been shown that 3′UTRs rG4s of PDS-95 (post-synaptic density protein 95; contains three G4s) and CaMKIIa (Ca2 + /calmodulin-dependent protein kinase II; contains one G4) mRNAs can regulate their dendritic localization (Subramanian et al. 2011). Furthermore, mRNA 3'UTR rG4s were shown to contribute to dendritic mRNA localization in an FMRP dependent manner (Goering et al. 2020).

5 Functions of rG4s in the Cytoplasm

5.1 Translation Regulation

Translation of mRNA to protein codes is one of the most important steps in RNA metabolism, and its regulation is tightly controlled. Secondary structures such as internal ribosome entry site (IRES)-like elements and rG4s in 5′UTR (untranslated regions) can significantly impact the translation efficiency (Georgakopoulos-Soares et al. 2022). Putative rG4s are overrepresented in the 5′UTRs of mRNAs implying important regulatory functions. When present, rG4s in mRNA 5′UTRs mostly inhibit translation (Kumari et al. 2007). However, 5′-UTR rG4s in the context of IRES-like elements, are known to augment the translation (Morris et al. 2010). mRNA 3'UTR rG4s also contribute to translation both negatively and positively (Arora and Suess 2011; Beaudoin and Perreault 2013; Thandapani et al. 2015).

Several cell-based reporter assays showed that rG4s in the mRNA 5'UTRs cause reduction in the efficiency of their translation (Kumari et al. 2007; Morris and Basu 2009). It has been shown that the rG4 density and position relative to the 5′ caps along with their stability contribute to their respective influence in translation (Kumari et al. 2008). Depletion or pharmacological inhibition of eukaryotic initiation factor 4A (eIF4A), a helicase that unwinds RNA secondary structures and facilitates the recruitment of the 43S preinitiation complex, generally reduces the translation efficiency of mRNAs. However, rG4-bearing transcripts are more sensitive to eIF4A depletion indicating that rG4s directly influence recruitment or scanning of preinitiation complexes/ribosome (Bordeleau et al. 2006; Wolfe et al. 2014). Unwinded rG4s in 5′UTRs can promote the formation of 80S ribosomes on alternative, upstream start codons, thus inhibiting the translation of the main open reading frame. rG4s in *FGF2*

(Bonnal et al. 2003), *α-Syn* (Koukouraki and Doxakis 2016), and *VEGF* (Morris et al. 2010) mRNAs are proposed to stimulate translation as a part of an internal ribosome entry site (IRES) or IRES-like elements, potentially by helping recruit the 40S ribosomal subunit (Bhattacharyya et al. 2015). Of note, rG4s in the mRNA open reading frame (ORF) have a much lower abundance than in the UTRs, and when present may act as translational repressors/ roadblocks for the elongating ribosomes (Mirihana Arachchilage et al. 2019). For example, rG4 within the ORF of *APP* mRNA inhibits its translation via association with FMRP, a known translational silencer (Westmark and Malter 2007). However, some rG4s, such as in *MLL1/4* mRNA ORF, can potentially enhance their translation. *MLL1/4* rG4 is recognized by the RGG-containing factor AVEN in a complex with rG4 helicase DHX36 (Thandapani et al. 2015). The binding of DHX36 stimulates *MLL1/4* mRNA translation presumably via its rG4-resolving activity, thus removing structure mediated blockade for elongating ribosomes. rG4s in the 3′ UTR of mRNA are shown to inhibit translation (e.g., *PIM1, APP*) (Arora and Suess 2011; Crenshaw et al. 2015)., however the molecular mechanism of such effects is unclear.

5.2 mRNA Stability

The stability of a given mRNA transcript is determined by the presence of sequence motifs (Koh et al. 2019; Siegel et al. 2021) and structures (Fischer et al. 2020), which can be bound by *trans*-acting RNA-binding proteins to inhibit or enhance mRNA decay. As such rG4s present in mRNA 3'UTRs can contribute to mRNA stability. Although the ubiquitous presence of rG4s in mRNA is clear and their role in the translation regulation has been a matter of several studies, their role in mRNA stability has only being recently being explored. We (Kharel et al. 2023) and others (Yang et al. 2022) have demonstrated that mRNA G4s are stress responsive elements such that rG4 folding is enhanced under different cellular stresses and 3'UTR mRNA rG4 folding contributes to mRNA stability. Additionally, 3'UTR rG4 folding has been shown to interfere with miRNA-mediated gene regulation (Rouleau et al. 2017).

5.3 ncRNA Biology

When present within transcripts, rG4s can directly influence RNA biogenesis and their downstream function in the cytoplasm as well. It has been independently reported by two laboratories that an rG4 present in pre-miRNAs can modulate their DICER-mediated maturation (Mirihana Arachchilage et al. 2015; Pandey et al. 2015). The formation of stable rG4s was reported in the ribosomal RNA as well (Mestre-Fos et al. 2019), although their functional roles in ribosome functions are still unclear. It has been reported that the formation of an rG4 in piRNA-48164 hinders PIWI protein binding thereby inhibiting the target reporter gene silencing in the cells (Balaratnam et al. 2019).

Furthermore, in response to various stresses cytosolic transfer RNAs are cleaved by ribonucleases in the anticodon loops (Akiyama et al. 2022; Yamasaki et al. 2009). Such stress-induced cleavage of tRNAs in the cytoplasm yields a novel class of small RNAs called tRNA-derived stress-induced tRNA fragments (tiRNAs), which represent tRNA 5′- and 3′-halves (Yamasaki et al. 2009). We have shown that 5′tiRNAs derived from $tRNA^{Ala}$ and $tRNA^{Cys}$ contain 5′G-rich motifs, which can adopt tetramolecular G4 structures that are functionally active and inhibit translation by directly interacting with eIF4G1 under stress (Ivanov et al. 2014; Lyons et al. 2017, 2020). As a consequence of such inhibition, cells undergo translational reprogramming, which aims on stress adaptation and cell survival. Partially, adaptation to stress can be explained by the abilities of G4-assembling tiRNAs to promote formation of stress granules (Emara et al. 2010), RNA granules with pro-survival roles in RNA metabolism (Ivanov et al. 2019).

6 rG4 Binding Proteins

While the dynamics of rG4 vs non-rG4 equilibrium is largely controlled by their ionic environment in vitro, within the cells, proteins potentially could solely (individually or as a part of protein–protein or RNP complexes) dictate or contribute to the cation-assisted G4 folding-unfolding dynamics (reviewed in (Kharel et al. 2020b)). A G4 binding protein can recognize and bind to a G4 in a multistep process involving main binding domains recognizing the G4 structure with the assistance of interactions from neighboring disordered regions (reviewed in (McRae et al. 2017)). In some cases, previously unstructured (intrinsically disordered) regions of rG4BPs become ordered upon canonical RNA binding to stabilize G4-interacting conformations. Several pull-down and cross-linking coupled with immunoprecipitation experiments show that many proteins can specifically bind to G4s in the cells. The analysis of reported rG4 interacting proteins reveals the presence of certain specific domains and motifs, or unstructured regions in the established or predicted binding regions of the rG4BPs (Kharel et al. 2020b). By virtue of their chemical nature and structural features, RRM (RNA-recognition motif) and RGG (Arginine-Glycine-Glycine) motifs within the RNA- binding proteins are mostly reported to be involved in the interaction with rG4s and hence are the most studied. In addition, some RBPs like Heterogeneous nuclear ribonucleoprotein H (HNRNPH1) and CCHC-type zinc finger nucleic acid binding protein (CNBP) bind to the G-rich motifs of RNA and actually prevent rG4 formation (Benhalevy et al. 2017a; Russo et al. 2010). Both CNBP and HNRNPH1 can also recognize, bind and destabilize the folded rG4s (Benhalevy et al. 2017a; Vo et al. 2022). On the other hand, helicase RBPs like DEAH-Box Helicase 36 (DHX36) bind to rG4s and actively resolve the structure (Booy et al. 2012). There are numerous other rG4 binding proteins that are recruited by rG4s to perform other cellular functions, such as splicing and translation regulation. Some rG4s might act as rG4BP sequestering elements thus preventing these factors from their other cellular functions (Conlon et al. 2016). Additionally, rG4-rG4BP interaction also contributes to RNA transport (discussed before) and RNA–protein condensate formation (discussed later). Some of the major rG4BPs and their known/ proposed functions are summarized in the Table 1 (also reviewed at (Fay et al. 2017b)).

Table 1 Representative list of rG4BPs and their functions

rG4BPs	Function
AFF3, AFF4	Modulate splicing by recognition of the exonic splicing enhancer Melko et al. (2011)
CNBP	Promotes the translation of G-rich mRNAs by preventing rG4 formation Benhalevy et al. (2017)
DHX36	Acts as dG4 and rG4 helicase Chen et al. (2018a); Tippana et al. (2019)
eIF4A	Alters translation efficiency of mRNAs with rG4 and other structural elements in the 5′ UTR (Schmidt et al. 2022; Wolfe et al. 2014)
eIF4G1	Under cellular stress, 5′tiRNA^{Ala} rG4 binds to HEAT1 domain of eIF4G1 thereby inhibiting translation Lyons et al. (2020)
FMR2	Modulates splicing by recognition of the exonic splicing enhancer Bensaid et al. (2009)
FMRP	Binds to rG4s and modulates the activity of microRNA (miRNA)-mediated silencing in the 3′ UTR of a subset of mRNAs through its interaction with RNA helicase Moloney leukemia virus 10 (MOV10) Kenny et al. (2019). Promotes mRNA localization in the neuronal cells Goering et al. (2020)
FUS/TLS	Binds rG4s and results in liquid–liquid phase separation and cellular condensate formation. Notably, ALS-linked mutations result in the dysregulation of liquid–liquid phase separation Ishiguro et al. (2021)
GRSF1	Melts mitochondrial rG4s and enhances degradosome-mediated degradation of G4 RNAs Pietras et al. (2018)
hnRNP A1	Regulates *MST1R* mRNA splicing and translation Cammas et al. (2016)
hnRNP A2	Promotes the translation of FMR1 by preventing G4 from forming; unfolds LTR promoter G4s Khateb et al. (2007)
hnRNP A3	Binds to G4C2 repeats and is a constituent of inclusions in the hippocampus of patients with C9orf72 mutations Mori et al. (2013)
hnRNP H1	Destabilizes rG4s and modulates splicing Vo et al., (2022). Sequestration of hnRNP H1 to G4C2 foci causes alterations in splicing Conlon et al. (2016)
Lin28	Remodels rG4s in its target mRNA and miRNA and affects mRNA stability and miRNA metabolism O'Day et al. (2015)
Nucleolin	Preferentially binds long-looped rG4s Lago et al. (2017). Binds to HCV viral core RNA G4 and suppresses its replication
TRF2	Binds TERRA rG4 and contributes to telomeric integrityMei et al. (2021)
YB1	Binds 5′tiRNA^{Ala} rG4s and contributes to stress granule formation and translation inhibition Lyons et al. (2016)

7 RG4s and Membrane-Less Biomolecular Condensates

Liquid–liquid phase separation (LLPS) is a biophysical phenomenon that contributes to the formation of membrane-less RNA–protein assemblies (or biocondensates) in the cells, such as cytoplasmic nuclear paraspeckles, cytosolic P bodies, and stress granules (Ivanov et al. 2019). Interestingly, some transcripts containing G-rich RNA repeat sequences can seed RNA only foci in vitro or RNA-containing protein complexes in lysates and live cells (Fay et al. 2017a; Ivanov et al. 2014; Yamasaki et al. 2009). rG4s have been heavily linked in the formation of such ribonucleoprotein (RNP) granules or RNA granules which are involved in various cellular processes and linked to several diseases including neurodegeneration and cancers (Wolozin and Ivanov 2019). Biophysical and structural gel-like features of poly(G) RNA assembly at higher concentrations could be the key contributing factor to LLPS and follows condensate formation. Furthermore, rG4s and their sequestered protein partners could assemble or aggregate to form RNP condensates. Several rG4 features qualify them as candidate contributors to LLPS (Asamitsu and Shioda 2021). First, at high concentrations, poly-guanosine can form gel-like structures in aqueous solutions. These largely static gel-like condensates might stimulate LLPS by increasing the local concentration of liquid phases. Second, RG4s formed *in cis* or *in trans* may promote the recruitment of multivalent protein factors leading to promotion of protein condensates that further contribute into LLPS-induced formation of RNA granules. Additionally, the arginine-/glycine-rich domains (RGG) of several rG4 binding proteins are structurally intrinsically disordered which brings conformational flexibility and degenerates specificity in RNA binding. Degeneracy in RNA binding could result in RBP oligomerization or multivalent interactions with other proteins or multiple RNAs/mRNPs at the same time, which is important to build RNP assemblies further promoting LLPS. Importantly, RGG domains mediate protein–protein interactions and can induce liquid–liquid phase separation even in the absence of RNA both in vitro and in live cells (Schuster et al. 2018).

We and others showed the role of rG4s in the formation of RNA granules in the transcripts generated by the GGGGCC hexanucleotide repeats (rG4C2) in the C9orf72 gene which is the most common genetic mutation associated with amyotrophic lateral sclerosis and frontotemporal dementia (C9-ALS/FTD) (Fay et al. 2017a). rG4C2 RG4s can also influence ALS/FTD-linked LLPS by modulating repeat-associated non-AUG (RAN) translation occurring at C9orf72 repeats, which generates toxic arginine-rich dipeptides that in turn can also promote LLPS. Similarly, short root mRNA rG4 was shown to induce a phase separation-like phenomenon in the plant cells (Zhang et al. 2019b). It has also been suggested that 3-tier rG4s are specially more efficient at inducing phase separation than their 2-tier counterparts (Zhang et al. 2019b). Contributions of rG4s into formation of biomolecular condensates is an area of active research currently and a matter of studies by several laboratories.

8 rG4s as Therapeutic Targets

Increasing evidence supporting the idea that rG4s can contribute or even regulate a variety of physiological and pathological processes has encouraged the design and development of rG4-interacting ligands that may act as therapeutic agents. Small molecule ligands have been the most investigated therapeutics and have been primarily used to stabilize the rG4s, enhancing their inherent repressive role in mRNA translation by obstructing ribosomal activity or interfering with translation machinery. G4 ligands often share common structural features such as an aromatic core, which permits stacking interactions with planar G-quartet, and one or more positive moieties that may interact with nucleic acids backbone phosphate groups in grooves and loops. Currently, several labs are working to develop ligands that are rG4-specific. There are numerous generic G4 targeting small molecule ligands that nonspecifically target and mostly stabilize both dG4s and rG4s (e.g., listed at http://www.g4ldb.com database). Nevertheless, a few of them show a low or high specificity toward rG4 (Select rG4 ligands in Fig. 3a) (reviewed in Kharel et al. 2020a; Santos et al. 2021).

It has been shown that the interaction of bisquinolinium ligands such as PhenDC3 with *TRF2* mRNA rG4 results in the suppression of its expression (Halder et al. 2011). The driving mechanism for the binding of bisquinolium ligands towards rG4s was proposed to be π–π stacking with square-planar G-quartets. Similarly, Miglietta et al. identified anthrafurandione derivatives as potential therapeutics that target 5'UTR *KRAS* mRNA rG4s to repress the mRNA translation in pancreatic cancer cells (Miglietta et al. 2017). The binding mechanism seems to involve the π–π stacking interactions of anthrafurandione core with G-quartets, whereas the cationic side chains bind to grooves and loops via electrostatic interactions. Despite of not being reported as a therapeutic, parallel G4 interacting porphyrin molecule, N-methyl mesoporphyrin-IX, brings selectivity towards rG4s and parallel dG4s (Sabharwal et al. 2014). This selectivity allows the use of NMM-IX as the rG4 trapping ligand under appropriate environmental conditions (Kharel et al. 2023). A polyaromatic molecule, RGB-1 has been shown to interact with TERRA and *NRAS* mRNA G4s where RGB-1 is proposed to selectively recognize rG4s due to the presence of H-bonding acceptors that interact with the 2′-OH group of the rG4s (Katsuda et al. 2016). However, a deeper structural analysis of the interacting RGB-1:rG4 complex is still lacking. Similarly, carboxy-pyridostatin has been used to selectively stabilize cytosolic G4s in the cells. c-PDS has been shown to establish π–π stacking interactions with TERRA G-quartets, and several hydrogen bonds with guanine residues (Rocca et al. 2017). Importantly, cPDS showed a stabilizing effect on TERRA rG4 ($\Delta T_m = 20.7$ °C), which was not affected by the addition of up to 100 equivalents of a dG4 competitor.

A small molecule library was used to screen ligands that could discerningly bind to the $(G_4C_2)_4$ rG4 formed by the mutagenic G_4C_2 repeats found in the first intron of the C9orf72, the most common genetic cause of C9-ALS/FTD (Simone et al. 2018). This repeat-associated intron sequence can be translated through the non-canonical

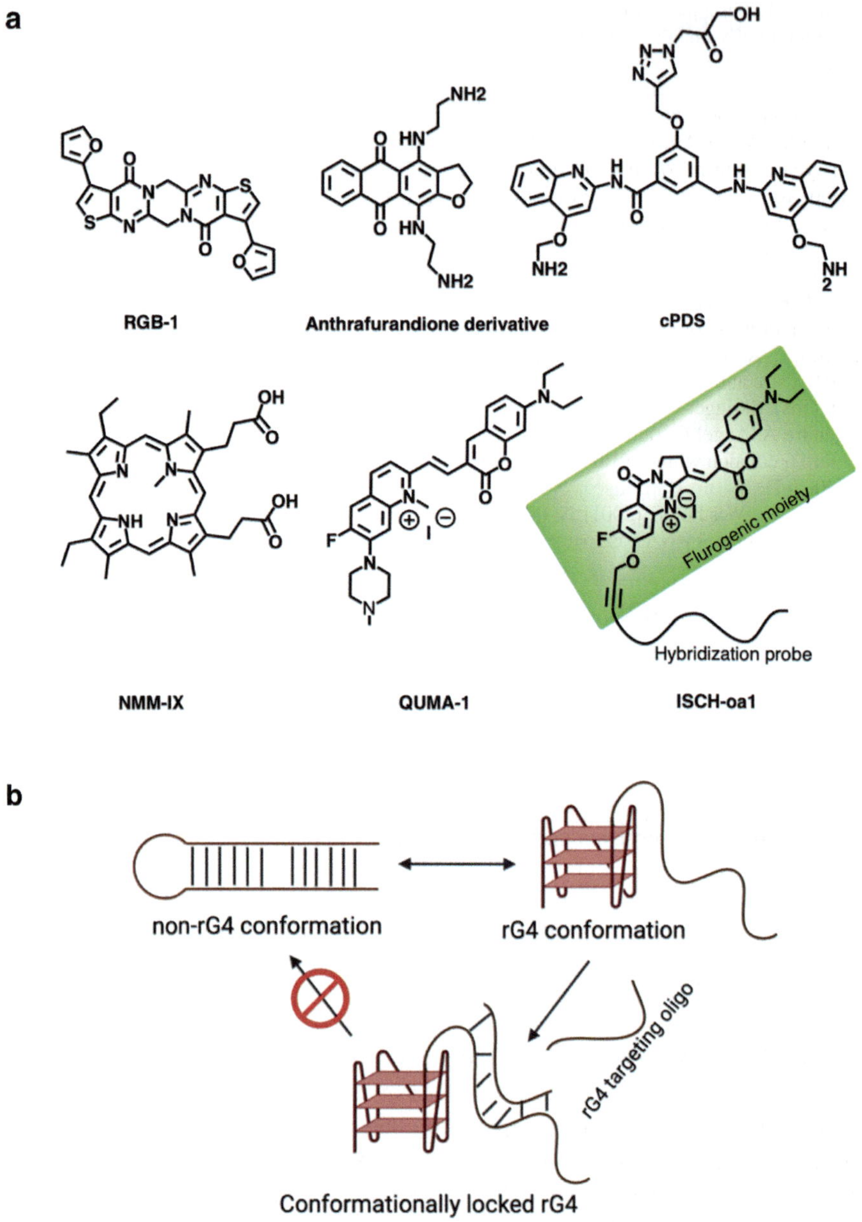

Fig. 3 rG4s targeting. a. Some of the representative small molecule ligands with a higher selectivity towards rG4s, and **b**. schematic of one of the strategies to target rG4 using engineered oligonucleotide therapeutic mechanism of translation (Repeat associated non-AUG translation), synthesizing toxic dipeptides. High quantities of dipeptides (poly-GA, GR, GP, PA, PR) can aggregate and induce pathogenic RNA granule formation, leading to cellular cytotoxicity. Targeting of such rG4s by small molecules in the iPSC-derived motor neurons and C9orf72 mutated Drosophila led to the reduction of dipeptide products, decreasing disease-related cytotoxicity. The small molecule ligands potentially destabilized the G4s formed by $(G_4C_2)_n$ sequence within the mutated intron and inhibit RNA foci formation, resulting in the inhibition of RAN translation of the dipeptide repeats.

Small molecules like QUMA-1 and ISCH-based fluorogenic probes provide a powerful platform to study rG4 folding dynamics inside the cells. A red-emitting fluorescent probe, QUMA-1, has been successfully developed for the selective, continuous, and real-time visualization of rG4s in both live and fixed cells (Chen et al. 2018b). Furthermore, G-quadruplex-triggered fluorogenic hybridization (GTFH) probes have been successfully engineered and synthesized, which are capable of specifically tracking a particular specific rG4 based on the covalently linked complimentary probe attached to the dye (ISCH) (Chen et al. 2016).

Engineered oligonucleotides that directly or indirectly target rG4 regions to stabilize the structure or to bring a molecular lock to a particular structure (Fig. 3b) further provide with a more specific rG4 targeting alternative over the ligand-based approaches (reviewed in Cadoni et al. 2021; Kharel et al. 2020a). However, nuclease-mediated cleavage of the phosphodiester bonds, unfavorable pharmacokinetics, and sub-optimal complex stability limit the direct clinical application of chemically unmodified nucleic-acids-based therapeutics. Chemical modifications, including the phosphate backbone (e.g., phosphorthioates), the sugar moiety (e.g., locked nucleic acids or 2′-O-*alkyl* ribonucleic acids), and the nucleobases (e.g., modified bases) have been designed to improve the stability and bioavailability of such therapeutics.

A strategy to lock miRNA-92b in the rG4 form using a complementary locked nucleic acid-based approach was used to minimize the canonical stem-loop structure resulting in blocked miRNA processing (Fig. 3b) (Mirihana Arachchilage et al. 2018). A rationally designed locked nucleic acid sequence that specifically binds to a region near the 3′-end of *pre-miR92b*, regulates the equilibrium between rG4 and stem-loop. Upon treatment, such equilibrium was shifted toward the G4 conformation. This, in turn, reduced the amount of mature miRNA-92, thus resulting in a therapeutic effect on its mRNA targets as demonstrated by the rescue of PTEN tumor suppressor gene expression in human non-small-cell lung cancer cells. Additionally, a strategy using complementary γ-peptide nucleic acid oligomers to invade an rG4 resulting in translation repression of a reporter gene was used (Oyaghire et al. 2016).

While nucleic acid-based therapeutic strategies for targeting rG4s possess incredible promise in terms of specificity and therapeutic output, challenges remain based on the poor pharmacodynamics of these larger therapeutic molecules. Cellular uptake of the larger, negatively charged molecules has poor efficiency in crossing the cell membrane or maintaining bio-stability, making use of co-delivery materials almost necessary. However, given the prevalence of rG4s in transcriptome and the lack of specificity of the currently used set of small molecules, finding ligands that will precisely bind to a particular rG4 remains immensely challenging. Thus, systematic efforts to identify and characterize unique rG4 features with clever drug engineering would be needed to develop effective structure–function-based rG4 drugs.

9 Concluding Remarks

Unfortunately, only few 3-D structures of rG4s have been determined to date. Recent advances in the G4 field clearly indicate the formation of rG4s in vivo and their broader role in RNA biogenesis, transport, stability, subcellular localization, and mRNA translation. Thus, understanding of structural features of rG4s in the context of endogenous transcripts is particularly important. Whenever present, rG4s and rG4BPs interact dynamically to guide RNA biology and cell biology. The dissecting of molecular details of such rG4-rG4BP interactions is particularly challenging. Nonetheless, recent advances in RNA biology fostered by cutting edge technologies at both proteome and transcriptome scales have already pushed rG4-rG4BP interaction studies to the nucleotide-amino acid resolutions. Furthermore, such approaches will allow detailed compositional characterization of dynamic rG4-RNP complexes in subcellular compartmentalization- and stimuli-dependent manners (e.g., under stress conditions). Expectedly, rG4s have the potential to serve as therapeutic targets. This partially stems from the fact that rG4-bearing RNA targets have only limited lifetime once transcribed, when compared to G4 DNA targets that embedded in the genomic context. We expect that future rG4 research will continue focusing on the atomic details of the molecular partnership between rG4s, rG4BPs, and rG4 ligands.

References

Akiyama Y, Lyons SM, Fay MM et al (2022) Selective cleavage at CCA ends and anticodon loops of tRNAs by stress-induced RNases. Front Mol Biosci 9:791094

Arora A, Suess B (2011) An RNA G-quadruplex in the 3' UTR of the proto-oncogene PIM1 represses translation. RNA Biol 8(5):802–805

Asamitsu S, Shioda N (2021) Potential roles of G-quadruplex structures in RNA granules for physiological and pathological phase separation. J Biochem 169(5):527–533

Balaratnam S, Basu S (2015) Divalent cation-aided identification of physico-chemical properties of metal ions that stabilize RNA g-quadruplexes. Biopolymers 103(7):376–386

Balaratnam S, Hettiarachchilage M, West N et al (2019) A secondary structure within a human piRNA modulates its functionality. Biochimie 157:72–80

Bang I (1910) Untersuchugen uber die Guanylsaure. Z Physiol Chem 31:407

Beaudoin JD, Perreault JP (2013) Exploring mRNA 3'-UTR G-quadruplexes: evidence of roles in both alternative polyadenylation and mRNA shortening. Nucleic Acids Res 41(11):5898–5911

Belotserkovskii BP, Tornaletti S, D'Souza AD et al (2018) R-loop generation during transcription: Formation, processing and cellular outcomes. DNA Repair 71:69–81

Benhalevy D, Gupta SK, Danan CH (2017) The human CCHC-type zinc finger nucleic acid-binding protein binds G-Rich elements in target mRNA coding sequences and promotes translation. Cell Rep 18(12):2979–2990

Bensaid M, Melko M, Bechara EG et al (2009) FRAXE-associated mental retardation protein (FMR2) is an RNA-binding protein with high affinity for G-quartet RNA forming structure. Nucleic Acids Res 37(4):1269–1279
Bhattacharyya D, Diamond P, Basu S (2015) An independently folding RNA G-quadruplex domain directly recruits the 40S ribosomal subunit. Biochemistry 54(10):1879–1885
Bhattacharyya D, Mirihana Arachchilage G, Basu S (2016) Metal cations in G-quadruplex folding and stability. Front Chem, 4(38)
Bonnal S, Schaeffer C, Créancier L et al (2003) A single internal ribosome entry site containing a G quartet RNA structure drives fibroblast growth factor 2 gene expression at four alternative translation initiation codons. J Biol Chem 278(41):39330–39336
Booy EP, Meier M, Okun N et al (2012) The RNA helicase RHAU (DHX36) unwinds a G4-quadruplex in human telomerase RNA and promotes the formation of the P1 helix template boundary. Nucleic Acids Res 40(9):4110–4124
Bordeleau ME, Cencic R, Lindqvist L et al (2006) RNA-mediated sequestration of the RNA helicase eIF4A by Pateamine A inhibits translation initiation. Chem Biol 13(12):1287–1295
Cadoni E, De Paepe L, Manicardi A et al (2021) Beyond small molecules: targeting G-quadruplex structures with oligonucleotides and their analogues. Nucleic Acids Res 49(12):6638–6659
Cammas A, Lacroix-Triki M, Pierredon S et al (2016) hnRNP A1-mediated translational regulation of the G quadruplex-containing RON receptor tyrosine kinase mRNA linked to tumor progression. Oncotarget 7(13):16793–16805
Chen SB, Hu MH, Liu GC et al (2016) Visualization of NRAS RNA G-quadruplex structures in cells with an engineered fluorogenic hybridization probe. J Am Chem Soc 138(33):10382–10385
Chen MC, Tippana R, Demeshkina NA et al (2018a) Structural basis of G-quadruplex unfolding by the DEAH/RHA helicase DHX36. Nature 558(7710):465–469
Chen XC, Chen SB, Dai J et al (2018b) Tracking the dynamic folding and unfolding of RNA G-quadruplexes in live cells. Ang Chem Int Ed 57(17):4702–4706
Clemson CM, Hutchinson JN, Sara SA et al (2009) An architectural role for a nuclear noncoding RNA: NEAT1 RNA is essential for the structure of paraspeckles. Mol Cell 33(6):717–726
Conlon EG, Lu L, Sharma A et al (2016) The C9ORF72 GGGGCC expansion forms RNA G-quadruplex inclusions and sequesters hnRNP H to disrupt splicing in ALS brains. eLife 5.
Crenshaw E, Leung BP, Kwok CK et al (2015) Amyloid precursor protein translation is regulated by a 3'UTR guanine quadruplex. PLoS ONE 10(11):e0143160
Decorsière A, Cayrel A, Vagner S et al (2011) Essential role for the interaction between hnRNP H/F and a G quadruplex in maintaining p53 pre-mRNA 3'-end processing and function during DNA damage. Genes Dev 25(3):220–225
Didiot MC, Tian Z, Schaeffer C et al (2008) The G-quartet containing FMRP binding site in FMR1 mRNA is a potent exonic splicing enhancer. Nucleic Acids Res 36(15):4902–4912
Emara MM, Ivanov P, Hickman T et al (2010) Angiogenin-induced tRNA-derived stress-induced RNAs promote stress-induced stress granule assembly. J Biol Chem 285(14):10959–10968
Fay MM, Anderson PJ, Ivanov P (2017a) ALS/FTD-Associated C9ORF72 repeat rna promotes phase transitions in vitro and in cells. Cell Rep 21(12):3573–3584
Fay MM, Lyons SM, Ivanov P (2017b) RNA G-quadruplexes in biology: principles and molecular Mmchanisms. J Mol Biol 429(14):2127–2147
Fischer JW, Busa VF, Shao Y et al (2020) Structure-mediated RNA decay by UPF1 and G3BP1. Mol Cell 78(1):70-84.e76
Gellert M, Lipsett MN, Davies DR (1962) Helix formation by guanylic acid. Proc Natl Acad Sci U S A 48(12):2013–2018
Georgakopoulos-Soares I, Parada GE, Hemberg M (2022) Secondary structures in RNA synthesis, splicing and translation. Comp Str Biotech J 20:2871–2884
Goering R, Hudish LI, Guzman BB et al (2020) FMRP promotes RNA localization to neuronal projections through interactions between its RGG domain and G-quadruplex RNA sequences. eLife 9, e52621

Gomez D, Lemarteleur T, Lacroix L et al (2004) Telomerase downregulation induced by the G-quadruplex ligand 12459 in A549 cells is mediated by hTERT RNA alternative splicing. Nucleic Acids Res 32(1):371–379

Halder K, Largy E, Benzler M et al (2011) Efficient suppression of gene expression by targeting 5′-UTR-based RNA quadruplexes with bisquinolinium compounds. ChemBioChem 12(11):1663–1668

Imperatore JA, Then ML, McDougal KB et al (2020) Characterization of a G-quadruplexsStructure in pre-miRNA-1229 and in its Alzheimer's Disease-associated variant rs2291418: implications for miRNA-1229 maturation. Int J Mol Sci 21(3)

Ishiguro A, Lu J, Ozawa D et al (2021) ALS-linked FUS mutations dysregulate G-quadruplex-dependent liquid-to-liquid phase separation and liquid-to-solid transition. J Biol Chem 297(5)

Ivanov P, O'Day E, Emara MM, Wagner, et al (2014) G-quadruplex structures contribute to the neuroprotective effects of angiogenin-induced tRNA fragments. Proc Natl Acad Sci U S A 111(51):18201–18206

Ivanov P, Kedersha N Anderson P (2019) Stress granules and processing bodies in translational control. Cold Spring Harb Perspect Biol 11(5).

Katsuda Y, Sato S, Asano L et al (2016) A small molecule that represses translation of G-quadruplex-containing mRNA. J Am Chem Soc 138(29):9037–9040

Kenny PJ, Kim M, Skariah G et al (2019) The FMRP–MOV10 complex: a translational regulatory switch modulated by G-quadruplexes. Nucleic Acids Res 48(2):862–878

Kharel P, Balaratnam S, Beals N et al (2020a) The role of RNA G-quadruplexes in human diseases and therapeutic strategies. Wires RNA 11(1):e1568

Kharel P, Becker G, Tsvetkov V et al (2020b) Properties and biological impact of RNA G-quadruplexes: from order to turmoil and back. Nucleic Acids Res 48(22):12534–12555

Kharel P, Fay M, Manasova EV et al (2023) Stress promotes RNA G-quadruplex folding in human cells. Nature Commun 14:205

Khateb S, Weisman-Shomer P, Hershco-Shani I et al (2007) The tetraplex (CGG)n destabilizing proteins hnRNP A2 and CBF-A enhance the in vivo translation of fragile X premutation mRNA. Nucleic Acids Res 35(17):5775–5788

Kikin O, Zappala Z, D'Antonio L et al (2008) GRSDB2 and GRS_UTRdb: databases of quadruplex forming G-rich sequences in pre-mRNAs and mRNAs. Nucleic Acids Res 36:D141-148

Koh WS, Porter JR, Batchelor E (2019) Tuning of mRNA stability through altering 3′-UTR sequences generates distinct output expression in a synthetic circuit driven by p53 oscillations. Sci Rep 9(1):5976

Koukouraki P, Doxakis E (2016) Constitutive translation of human α-synuclein is mediated by the 5'-untranslated region. Open Biol 6(4):160022

Kumari S, Bugaut A, Huppert JL et al (2007) An RNA G-quadruplex in the 5′ UTR of the NRAS proto-oncogene modulates translation. Nat Chem Biol 3(4):218–221

Kumari S, Bugaut A, Balasubramanian S (2008) Position and stability are determining factors for translation repression by an RNA G-quadruplex-forming sequence within the 5′ UTR of the NRAS proto-oncogene. Biochemistry 47(48):12664–12669

Kwok CK, Sahakyan AB, Balasubramanian S (2016) Structural analysis using SHALiPE to reveal RNA G-quadruplex formation in human precursor microRNA. Angew Chem Int Ed Engl 55(31):8958–8961

Lago S, Tosoni E, Nadai M, Palumbo M et al (2017) The cellular protein nucleolin preferentially binds long-looped G-quadruplex nucleic acids. Biochim Biophys Acta Gen Subj, 1861(5 Pt B):1371–1381

Lightfoot HL, Hagen T, Tatum NJ et al (2019) The diverse structural landscape of quadruplexes. FEBS Lett 593(16):2083–2102

Loya CM, Van Vactor D, Fulga TA (2010) Understanding neuronal connectivity through the post-transcriptional toolkit. Genes Dev 24(7):625–635

Lyons SM, Achorn C, Kedersha NL et al (2016) YB-1 regulates tiRNA-induced Stress Granule formation but not translational repression. Nucleic Acids Res 44(14):6949–6960

Lyons SM, Gudanis D, Coyne SM et al (2017) Identification of functional tetramolecular RNA G-quadruplexes derived from transfer RNAs. Nature Commun 8(1):1127
Lyons SM, Kharel P, Akiyama Y et al (2020) eIF4G has intrinsic G-quadruplex binding activity that is required for tiRNA function. Nucleic Acids Res 48(11):6223–6233
Marcel V, Tran PL, Sagne C et al (2011) G-quadruplex structures in TP53 intron 3: role in alternative splicing and in production of p53 mRNA isoforms. Carcinogenesis 32(3):271–278
McRae EKS, Booy EP, Padilla-Meier GP et al (2017) On characterizing the interactions between proteins and guanine quadruplex structures of nucleic acids. J Nucleic Acids 9675348
Mei Y, Deng Z, Vladimirova O et al (2021) TERRA G-quadruplex RNA interaction with TRF2 GAR domain is required for telomere integrity. Sci Rep 11(1):3509
Melko M, Douguet D, Bensaid M et al (2011) Functional characterization of the AFF (AF4/FMR2) family of RNA-binding proteins: insights into the molecular pathology of FRAXE intellectual disability. Human Mol Genet 20(10):1873–1885
Mestre-Fos S, Penev PI, Suttapitugsakul S et al (2019) G-quadruplexes in human ribosomal RNA. J Mol Biol 431(10):1940–1955
Miglietta G, Cogoi S, Marinello J et al (2017) RNA G-quadruplexes in Kirsten Ras (KRAS) oncogene as targets for small molecules inhibiting translation. J Med Chem 60(23):9448–9461
Mirihana Arachchilage G, Dassanayake Arosha C, Basu S (2015) A potassium ion-dependent RNA structural switch regulates human pre-miRNA 92b maturation. Chem Biol 22(2):262–272
Mirihana Arachchilage G, Kharel P, Reid J et al (2018) Targeting of G-quadruplex harboring pre-miRNA 92b by LNA rescues PTEN expression in NSCL cancer cells. ACS Chem Biol 13(4):909–914
Mirihana Arachchilage G, Hetti Arachchilage M, Venkataraman A et al (2019) Stable G-quadruplex enabling sequences are selected against by the context-dependent codon bias. Gene 696:149–161
Mori K, Lammich S, Mackenzie IR et al (2013) hnRNP A3 binds to GGGGCC repeats and is a constituent of p62-positive/TDP43-negative inclusions in the hippocampus of patients with C9orf72 mutations. Acta Neuropathol 125(3):413–423
Morris MJ, Basu S (2009) An unusually stable G-quadruplex within the 5'-UTR of the MT3 matrix metalloproteinase mRNA represses translation in eukaryotic cells. Biochemistry 48(23):5313–5319
Morris MJ, Negishi Y, Pazsint C et al (2010) An RNA G-quadruplex is essential for cap-independent translation initiation in human VEGF IRES. J Am Chem Soc 132(50):17831–17839
Newman M, Sfaxi R, Saha A et al (2017) The G-quadruplex-specific RNA helicase DHX36 regulates p53 pre-mRNA 3′-end processing following UV-induced DNA damage. J Mol Biol 429(21):3121–3131
O'Day E, Le MTN, Imai S et al (2015) An RNA-binding protein, Lin28, recognizes and remodels G-quartets in the microRNAs (miRNAs) and mRNAs it regulates. J Biol Chem 290(29):17909–17922
Oyaghire O, SN, Cherubim CJ, Telmer CA, et al (2016) RNA G-quadruplex invasion and translation inhibition by antisense γ-peptide nucleic acid oligomers. Biochemistry 55(13):1977–1988
Pandey S, Agarwala P, Jayaraj GG et al (2015) The RNA stem–loop to G-quadruplex equilibrium controls mature microRNA production inside the cell. Biochemistry 54(48):7067–7078
Pietras Z, Wojcik MA, Borowski LS et al (2018) Dedicated surveillance mechanism controls G-quadruplex forming non-coding RNAs in human mitochondria. Nat Commun 9(1):2558
Puig LE, Londoño-Vallejo A (2020) A guide to computational methods for G-quadruplex prediction. Nucleic Acids Res 48(1):1–15
Ribeiro de Almeida C, Dhir S, Dhir A (2018) RNA helicase DDX1 converts RNA G-quadruplex structures into R-loops to promote IgH class switch recombination. Mol Cell 70(4):650-662.e658
Rocca R, Talarico C, Moraca F et al (2017) Molecular recognition of a carboxy pyridostatin toward G-quadruplex structures: Why does it prefer RNA? Chem Biol Drug Des 90(5):919–925
Rouleau S, Glouzon JS, Brumwell A et al (2017) 3' UTR G-quadruplexes regulate miRNA binding. RNA 23(8):1172–1179

Russo A, Siciliano G, Catillo M et al (2010) hnRNP H1 and intronic G runs in the splicing control of the human rpL3 gene. Biochim Biophys Acta 1799(5):419–428

Sabharwal NC, Savikhin V, Turek-Herman JR et al (2014) N-methylmesoporphyrin IX fluorescence as a reporter of strand orientation in guanine quadruplexes. FEBS J 281(7):1726–1737

Santos T, Salgado GF, Cabrita EJ et al (2021). G-Quadruplexes and their ligands: biophysical methods to unravel G-quadruplex/ligand interactions. Pharmaceuticals (Basel) 14(8)

Schmidt T, Dabrowska A, Waldron JA et al (2022). Purine-rich RNA sequences in the 5'UTR site-specifically regulate eIF4A1-unwinding through eIF4A1-multimerisation to facilitate translation. bioRxiv 2022.2008.2008.503179.

Schuster BS, Reed EH, Parthasarathy R et al (2018) Controllable protein phase separation and modular recruitment to form responsive membraneless organelles. Nat Commun 9(1):2985

Sen D, Gilbert W (1988) Formation of parallel four-stranded complexes by guanine-rich motifs in DNA and its implications for meiosis. Nature 334(6180):364–366

Shrestha P, Xiao S, Dhakal S et al (2014) Nascent RNA transcripts facilitate the formation of G-quadruplexes. Nucleic Acids Res 42:7236–7246

Siegel DA, Le Tonqueze O, Biton A et al (2021). Massively parallel analysis of human 3′ UTRs reveals that AU-rich element length and registration predict mRNA destabilization. G3 Genes|Genomes|Genetics 12(1)

Simko EAJ, Liu H, Zhang T et al (2020) G-quadruplexes offer a conserved structural motif for NONO recruitment to NEAT1 architectural lncRNA. Nucleic Acids Res 48(13):7421–7438

Simone R, Balendra R, Moens TG et al (2018) G-quadruplex-binding small molecules ameliorate C9orf72 FTD/ALS pathology in vitro and in vivo. EMBO Mol Med 10(1):22–31

Skourti-Stathaki K, Proudfoot NJ, Gromak N (2011) Human senataxin resolves RNA/DNA hybrids formed at transcriptional pause sites to promote Xrn2-dependent termination. Mol Cell 42(6):794–805

Subramanian M, Rage F, Tabet R et al (2011) G–quadruplex RNA structure as a signal for neurite mRNA targeting. EMBO Rep 12(7):697–704

Thandapani P, Song J, Gandin V et al (2015) Aven recognition of RNA G-quadruplexes regulates translation of the mixed lineage leukemia protooncogenes. eLife 4, e06234

Tippana R, Chen MC, Demeshkina NA et al (2019) RNA G-quadruplex is resolved by repetitive and ATP-dependent mechanism of DHX36. Nat Commun 10(1):1855

Vo T, Brownmiller T, Hall K et al (2022) HNRNPH1 destabilizes the G-quadruplex structures formed by G-rich RNA sequences that regulate the alternative splicing of an oncogenic fusion transcript. Nucleic Acids Res 50(11):6474–6496

Wang Y, Liu J, Huang BO et al (2015) Mechanism of alternative splicing and its regulation. Biomed Rep 3(2):152–158

Westmark CJ, MalterJS, (2007) FMRP mediates mGluR5-dependent translation of amyloid precursor protein. PLoS Biol 5(3):e52

Winnerdy FR, & Phan AT (2020) Quadruplex structure and diversity. Ann Rep Med Chem (Vol. 54, pp. 45–73). Academic Press

Wolfe AL, Singh K, Zhong Y et al (2014) RNA G-quadruplexes cause eIF4A-dependent oncogene translation in cancer. Nature 513(7516):65–70

Wolozin B, Ivanov P (2019) Stress granules and neurodegeneration. Nat Rev Neurosci 20(11):649–666

Xiao S, Zhang JY, Zheng KW et al (2013) Bioinformatic analysis reveals an evolutional selection for DNA:RNA hybrid G-quadruplex structures as putative transcription regulatory elements in warm-blooded animals. Nucleic Acids Res 41(22):10379–10390

Yamasaki S, Ivanov P, Hu GF et al (2009) Angiogenin cleaves tRNA and promotes stress-induced translational repression. J Cell Biol 185(1):35–42

Yang X, Yu H, Duncan S et al (2022) RNA G-quadruplex structure contributes to cold adaptation in plants. Nat Commun 13(1):6224

Zaccaria F, Fonseca Guerra C (2018) RNA versus DNA G-quadruplex: the origin of increased stability. Chemistry 24(61):16315–16322

Zhang DH, Fujimoto T, Saxena S et al (2010) Monomorphic RNA G-quadruplex and polymorphic DNA G-quadruplex structures responding to cellular environmental factors. Biochemistry 49(21):4554–4563

Zhang Y, Yang M, Duncan S et al (2019b) G-quadruplex structures trigger RNA phase separation. Nucleic Acids Res 47(22):11746–11754

Zhang X, Yu L, Ye S et al (2019a) MOV10L1 binds RNA G-quadruplex in a structure-specific manner and resolves it more efficiently than MOV10. iScience 17:36–48.

Zheng KW, Wu RY, He YD et al (2014) A competitive formation of DNA:RNA hybrid G-quadruplex is responsible to the mitochondrial transcription termination at the DNA replication priming site. Nucleic Acids Res 42(16):10832–10844

Zheng AJL, Thermou A, Guixens P et al (2022) The different activities of RNA G-quadruplex structures are controlled by flanking sequences. Life Sci All 5(2):e202101232

Structure and Folding Patterns of RNA G-Quadruplexes

Patil Pranita Uttamrao, Sruthi Sundaresan, and Thenmalarchelvi Rathinavelan

Contents

Abstract G-quadruplexes are four-stranded nucleic acid structures, the bases of which are linked through Hoogsteen hydrogen bonding. RNA G-quadruplex is shown to take part in a wide range of cellular events such as telomere maintenance, gene expression mechanisms, etc. RNA G-quadruplex is found to be present in a significant portion of the non-coding transcriptome, thus acting as a biomarker for several diseases. RNA quadruplex is also shown to be associated with neurogenerative diseases and cancer; thus, it is a potential therapeutic target. For these reasons, it is necessary to understand the sequence, structure and functional relevance of RNA G-quadruplex. A comprehensive analysis of the structural and folding characteristics of RNA G-quadruplex structures deposited in the PDB has therefore been carried out. A total of 10 unique folds are found to be present in RNA quadruplex. Together with the existing algorithms for the transcriptome-wide prediction of G-quadruplexes, the structural features discussed here would help in the modeling of G-quadruplex structures.

P. P. Uttamrao · S. Sundaresan · T. Rathinavelan (✉)
Department of Biotechnology, Indian Institute of Technology Hyderabad, Kandi, Telangana State 502284, India
e-mail: tr@bt.iith.ac.in

J. Barciszewski (ed.), *RNA Structure and Function*, RNA Technologies 14,
https://doi.org/10.1007/978-3-031-36390-0_10

Keywords RNA G-quadruplex · Quadruplex fold · Quadruplex modeling · Quadruplex prediction · Quadruplex twist · Quadruplex rise · Quadruplex water interaction · Quadruplex ion interaction

1 Introduction

The inherent flexibility of nucleic acids facilitates the formation of several secondary structures other than duplexes. One of these is quadruplex which, as the name indicates, encompasses four nucleic strands. The formation of this structure is facilitated by guanine residues, whereby four guanines form a quartet/tetrad through Hoogsteen hydrogen bonding (Burge et al. 2006) (Fig. 1a). These quartets are stabilized by monovalent or divalent cations to circumvent the repulsion caused by O6 atoms of the guanines (Bhattacharyya et al. 2016; Milovanovic et al. 2020). The possibility of non-guanine quartets in the midst of G-quartets has also recently been discovered: C–C–C–C (PDB ID: 1EVO) (Fig. 1b), T-T-T-T (PDB ID: 6A85) (Fig. 1c), A-A-A-A (PDB ID:1J6S,1MDG, 1EVM) (Figs. 1d–f), I-I-I-I (PDB ID: 2GRB) (Fig. 1g), U-U-U-U (PDB ID: 6GE1) (Fig. 1h), G-A-G-A (PDB ID: 5M1L) (Fig. 1i), G-C-G-C (PDB ID:1A6H) (Fig. 1j), and A-T-A-T (PDB ID: 1JVC) (Fig. 1k). Nucleic acid quadruplexes have gained more attention due to their essential roles in replication (Siddiqui-Jain et al. 2002; Besnard et al. 2012), transcription (specifically, the oncogenes (Cogoi and Xodo 2006)) and telomeric modulations (Paeschke et al. 2005; Jansson et al. 2019). Thus, they play a major role in cancer (Hansel-Hertsch et al. 2020; Kosiol et al. 2021) and neurological diseases (Bernat and Disney 2015; Hanna et al. 2021).

The ability of guanine-rich nucleic acid to self-associate has a history of more than a century and pre-dates the discovery of the double helix itself by nearly 50 years. Both DNA (Burge et al. 2006), RNA (Fay et al. 2017) and their combination (Xiao et al. 2013) can lead to the formation of a quadruplex structure. It is now well understood that the nucleic acid strand(s) can form a quadruplex through a variety of intermolecular and intramolecular folds (Webba da Silva 2007; Wanrooij et al. 2012; Karsisiotis et al. 2013). Several studies have described different quadruplex folds (Burge et al. 2006; Malgowska et al. 2016; Patro et al. 2017; Dvorkin et al. 2018). Until recently, more emphasis has been given to the structural and functional understanding of DNA quadruplexes (Agarwala et al. 2015), while their RNA counterparts have received less attention as their existence in vivo is still a matter of debate. However, there is now convincing evidence of the existence of RNA quadruplexes in vivo (Laguerre et al. 2015; Yang et al. 2018).

While X-ray crystallography (Campbell et al. 2012) and nuclear magnetic resonance (NMR) (Adrian et al. 2012) are employed to derive the atomistic details of RNA G-quadruplexes, circular dichroism (CD) spectroscopy (Ajjugal et al. 2021a), ultraviolet (UV) spectroscopy (Mergny and Lacroix 2009) and electrophoretic mobility shift assay (EMSA) (Takahama et al. 2015) can, to a certain extent, provide the nature of RNA G-quadruplex folds. For instance, CD helps in distinguishing the parallel,

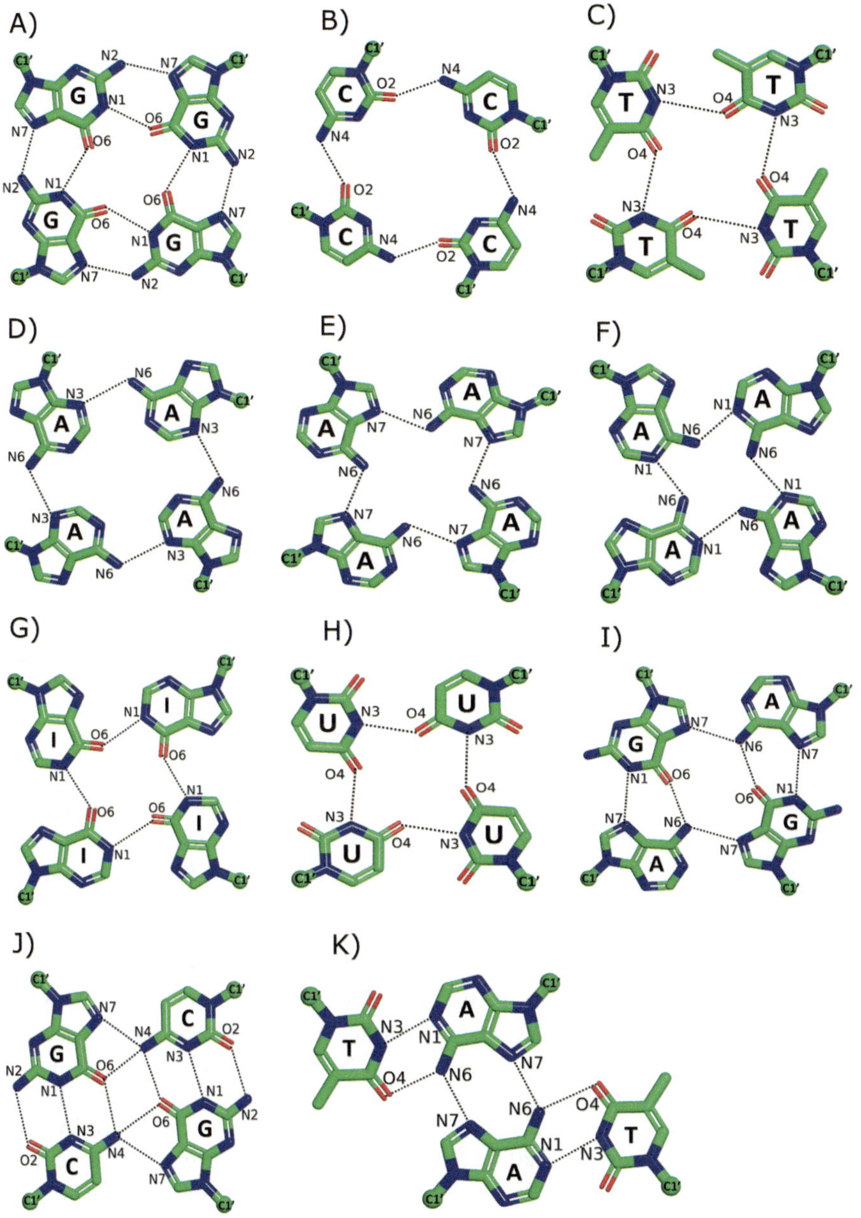

Fig. 1 Hydrogen-bonding patterns of G and non-G tetrads: A) G-G-G-G; **B)** C–C-C–C; **C)** T-T-T-T; **D)** A-A-A-A (N3…N6 hydrogen bond); **E)** A-A-A-A (N6…N7 hydrogen bond); **F)** A-A-A-A (N1…N6 hydrogen bond); **G)** I-I-I-I; **H)** U-U-U-U; **I)** G-A-G-A; **J)** G-C-G-C; and **K)** A-T-A-T. The green-colored sphere indicates the C1' atom of the sugar and the dotted line represents the hydrogen bond. For the sake of clarity, the sugar-phosphate backbone is not shown

antiparallel and hybrid quadruplex folds (Sathyaseelan et al. 2021). EMSA provides information about the intra- or inter-molecular nature of an RNA quadruplex. It is noteworthy that the Cartesian coordinates for RNA G-quadruplex structures are freely accessible from the protein databank (https://www.rcsb.org/).

Although the G-quartet acts as a repeating unit for DNA, RNA and RNA–DNA hybrid quadruplexes, one can expect a difference in their folding patterns due to the presence of a hydroxy group at the 2' position of the ribose sugar of RNA (Fay et al. 2017). Indeed, the presence of a hydroxy group is found to enhance the stability of RNA quadruplex compared to its DNA counterpart (Joachimi et al. 2009). This eventually exhibits a difference between the folding landscapes of DNA and RNA quadruplexes (Muller et al. 2021). Further, one can envisage that the single-stranded nature of RNA can facilitate the formation of a complex higher-order G-quadruplex architecture compared with the double-stranded DNA. Notably, the structural information of the RNA quadruplex is limited and significantly lower than the DNA quadruplex. Due to the limited availability of experimentally determined RNA G-quadruplex folds, computer-based modeling (Patro et al. 2017) and molecular dynamics simulations (Collie et al. 2010; Krepl et al. 2012) can aid in the understanding of their conformational dynamics and folding/unfolding mechanisms (Haldar et al. 2022), as well as in the modeling of their complex with the ligand and protein molecules. This review describes the RNA quadruplex folding architecture to assist in their modeling.

2 Biological Roles of RNA Quadruplex

The first evidence of the ability of polyinosinic acid and polyguanylic acid (which contain ribose sugar) to form four-stranded structures was discovered through fiber diffraction in 1962. The ability of RNA from an organism (5S RNA of *E. coli*) to form a G-quadruplex was first shown in 1991. Evidence is emerging of the inevitable role of RNA quadruplex (which mainly localizes in the cytoplasm (Biffi et al. 2014)), not only in regulating several biological phenomena (Fay et al. 2017; Liu et al. 2022), but also in many diseases (Reddy et al. 2013; Cammas and Millevoi 2017; Ajjugal et al. 2021a). Transcriptome-wide analyses further show the prevalence of transient RNA quadruplex in human cells (Kwok et al. 2016; Yang et al. 2018; Vannutelli et al. 2020), specifically their abundance in 5'UTR, 3'UTR, splicing junction of mRNA, microRNA (miRNA) (Chan et al. 2018) and long non-coding RNA (Lee et al. 2020; Vannutelli et al. 2020) across all the domains of life, from bacteria to humans (Kharel et al. 2020b). This implicates the evolutionary conservation of RNA quadruplex. Recent studies have indeed shown that RNA G-quadruplex is present in plants like *Arabidopsis thaliana* and rice (Yang et al. 2020) and regulates plant growth and development (Liu et al. 2022). RNA quadruplex is shown to be present in human mRNA and telomeric RNA, acting as a key regulator of biological processes (Tassinari et al. 2021). It plays a role in alternative splicing (Brown et al. 2001), 3'end processing (Decorsiere et al. 2011), translation regulation (Kumari et al. 2007), mitochondrial transcription termination (Kumari et al. 2008), mRNA transcription

(Fay et al. 2017), processing (Fay et al. 2017), translation (Fay et al. 2017) and localization (Subramanian et al. 2011), miRNA biogenesis/processing (Lyu et al. 2021; Tassinari et al. 2021) and their target recognition (Rouleau et al. 2017), telomere biological function (Cammas and Millevoi 2017; Fay et al. 2017; Shao et al. 2020) and post-translation modification (Song et al. 2016). Proteins that are shown to interact with RNA G-quadruplex also indicate its role in biological processes: FMRP (fragile X mental retardation protein) (Goering et al. 2020), RNA helicases (Caterino and Paeschke 2022) and HNRNPs (heterogeneous nuclear ribonucleoproteins) (Conlon et al. 2016; Huang et al. 2017). A study has shown that RNA G-quadruplex plays an important regulatory role in bacterial pathogenicity and metabolic pathways (Shao et al. 2020).

Besides their important biological roles, RNA quadruplex is found to play a major role in several human diseases such as cancer (Carvalho et al. 2020), neurodegenerative disorders (Fratta et al. 2012; Ajjugal et al. 2021a) and infectious diseases (Metifiot et al. 2014). For instance, G-quadruplex structure-forming regions are significant in the human non-coding transcriptome, thus, aberrations in the non-coding RNA lead to diseases (Tassinari et al. 2021). Most importantly, it leads to cytotoxicity in neurodegenerative diseases like FXTAS (Zhou et al. 2022) and amyotrophic lateral sclerosis (Xu et al. 2013), due to the formation of RNA granules by sequestering proteins (Cammas and Millevoi 2017) and the consequent aberrant phase separation (Tateishi-Karimata and Sugimoto 2021).

Since RNA quadruplex is significantly represented in viral genomes (for example, HIV, chikungunya, SARS-CoV-2, etc.), it regulates viral replication, host response to viral infection, integration of viral genome in the host telomere and cellular pathways, which are involved in viral particle maturation and immune modulation (Metifiot et al. 2014; Zhao et al. 2021). Similarly, RNA quadruplex structures play a major role in bacterial virulence regulation and pathogenesis (Miglietta et al. 2020; Shao et al. 2020). Thus, RNA quadruplex has the potential to be a novel therapeutic target (Miglietta et al. 2020; Miclot et al. 2021; Zhao et al. 2021). Additionally, the aptameric RNA G-quadruplex can also act as a potential drug molecule (Mashima et al. 2013; Wang et al. 2015; Fukunaga et al. 2020; Ni et al. 2021).

3 Topological Diversity of RNA Quadruplex

X-ray crystallography and NMR are the two important techniques to determine RNA quadruplex folds. Chemically synthesized short oligonucleotides can be subjected to X-ray crystallography and 2− and/or 3−dimensional NMR spectroscopy to obtain their fold information. These structures can be useful in identifying the folding patterns of RNA quadruplex. Indeed, these techniques are very useful in getting the structural details of higher-order RNA quadruplex structures that occur alongside other secondary structures such as duplex.

It has long been known that RNA quadruplexes are monomorphic parallel structures (Cammas and Millevoi 2017; Muller et al. 2021), with the exception of a few aptamer quadruplexes (Fernandez-Millan et al. 2017; Huang et al. 2017; Koirala et al. 2018; Jeng et al. 2021). RNA quadruplex formation, like its DNA counterpart, depends on several factors like the sequence length, number of Gs in the given sequence, length of the G stretch, type of nucleotide(s) present in the loop– and bulge-forming regions, and also environmental conditions such as salt. However, it is believed that the presence of a hydroxyl group at the C2' position of the ribose sugar restricts the guanines from taking a *syn* conformation, favoring the parallel G-quadruplex fold (Cammas and Millevoi 2017). Thus, it is believed that the topological diversity of the RNA G-quadruplex is lesser than that of its DNA counterpart, and the above-mentioned factors have less influence on the quadruplex fold (Cammas and Millevoi 2017). To shed more light on this, the RNA quadruplex folds are analyzed here through a careful inspection of RNA quadruplex structures determined using X-ray crystallography and NMR techniques by downloading them from PDB (Berman et al. 2000), NDB (Coimbatore Narayanan et al. 2014) and ONQUARDO (Zok et al. 2022). A total of 60 RNA G-quadruplex structures (Table 1), including 37 aptamer structures, formed by unique sequences, are downloaded and are inspected in detail to understand their topological diversity.

A total of 10 unique folds are found for RNA G-quadruplex (Fig. 2). This includes six monomeric (intramolecular; formed by a single strand, common to 19 unique sequences) (RNA monomeric fold (**RMF1–RMF6 in** Fig. 2)), 3 dimeric (RNA dimeric fold (**RDF1–RDF3 in** Fig. 2)) (intermolecular, formed by two strands, common to three unique sequences) and one tetrameric (intermolecular, formed by four strands, common to six unique sequences) (RNA tetrameric fold (**RTF in** Fig. 2)) folds (Table 1). While the majority of the monomeric folds are parallel ((RMF1-3) common to 11 unique sequences) (PDB IDs: 5BJO, 1MY9, 2RU7, 2RQJ, 2RSK, 6E 82, 6E80, 6E81, 6E84, 5BJP, 6K84, 6C63, 6C64, 6C65, 6V9D, 6V9B, 5V3F, 2LA5, 5DE5, 5DE8, 5DEA) in nature (Fig. 2), one antiparallel ((RMF4) common to four unique sequences) (PDB IDs: 4KZD, 4KZE, 4Q9Q, 4Q9R, 6B3K, 6B14, 5OB3, 7L0Z) and two hybrid (RMF5 and RMF6) (common to four unique sequences) (PDB IDs: 7OA3, 7OAV, 7OAW, 7OAX, 6UPO, 6E8S, 6E8T, 6E8U, 6PQ7) folds are found in the aptamer structures. Notably, a lateral loop having residues in the range of 3 (RMF6) to 18 (forms double helix) (RMF5) connects the quadruplex strands in the monomeric hybrid aptamer structures (Fig. 2). The dimeric (RDF1) and tetrameric (RTF) folds are parallel in nature with the exception of the RDF3 dimeric fold, which has an antiparallel topology (PDB IDs: 4TS0, 4TS2). In any case, the capability of the RNA quadruplex to take antiparallel and hybrid folds is clear from this analysis. Among these folds, parallel dimeric folds RDF1 and RDF2, and monomeric fold RMF3 are relevant to biological sequences. RDF1 and RDF2 correspond to pseudorabies virus and human telomeric RNA sequences respectively. Similarly, parallel RMF3 corresponds to sc1 RNA sequence. The remaining monomeric and dimeric folds belong to synthetic aptamer sequences, whereas the tetrameric fold sequences are simply short oligonucleotides.

In the monomeric and dimeric forms, the RNA quadruplex folds are principally governed by the propeller loop along with a lateral loop and/or a diagonal loop. The propeller and lateral loops facilitate the parallel and antiparallel orientations

Table 1 List of PDB IDs (column 2) corresponding to RNA quadruplex. Note that the first column indicates the nature of the fold (Refer Fig. 2). The third and fourth columns represent the nucleotide sequences and strand orientations (parallel, antiparallel and mixed/hybrid). The number of quartets (column 5) and the type of loops (column 6) are also given. The loop types P, B, L, D, and V in column 6 represent the propeller, bulge, lateral, diagonal, and V loop respectively. The number followed by these characters indicates the loop number (viz., P1 indicates the first propeller loop present in the quadruplex from the 5' direction). The parenthesis followed by this number indicates the number and type of nucleotides present in the corresponding loop. Column 7 names the ion(s) that coordinates within the quartet. Column 8 comments on the functional relevance of the structure, if any

Fold	PDB ID	Sequence	Strand polarity	No. of quartets	Type of loops	Ions	Remarks
RMF1	1MY9	GGAGGUUUUGGAGG	Parallel	2	P1(1-A), P2(4-UUUU), P3(1-A)	–	Aptamer
	2RQJ	GGAGGAGGAGGA			P1(1-A), P2(1-A), P3(1A)	–	
	2RSK					–	
	2RU7					–	
	5BJO	AGGAAGGAGG(IU)CUGAGGAGGU			P1(1-A), P2(5 -IUCUGA), P3(1-A)	K+	
	5BJP	GGAGGUCUGAGGAGGU			P1(1-A), P2(5-UCUGA), P3(1-A)		
	6E80					K+	
	6E81						
	6E84						
	6E82	GGUGGUCUGAGGAGG			B1(1-G), P1(0), V, P2(5-UCUGA), P3(1-A)		

(continued)

Table 1 (continued)

Fold	PDB ID	Sequence	Strand polarity	No. of quartets	Type of loops	Ions	Remarks
	6K84	GGAGGAGGAGGAAGGAGGAGGAGGA	Parallel	4	P1(1-A), P2(1-A), P3(1A), Lnk (2-AA), P1'(1-A), P2'(1-A), P3'(1A)	–	
RMF2	5V3F	GUGCGAAGGGACGGUGCGGA-GAGGGAGCAC	Parallel	3	B1(0), P1(2-AC), B2(1-U), P2(1-C), B3(1-A), P3(1-A), B4 (1-A)	K+	Aptamer
	6C63 6C64	GUACGAAGGAGAGGAGAGGAAGAGGA-GAGUAC			B1(1-A), P1(1-A), B2(1-A), P2(1-A), B3(2-AA), P3(1-A), B4(1-A)		
	6C65	GUACGAAGGAGAGGAGAGGUAGAGGA-GAGUAC			B1(1-A), P1(1-A), B2(1-A), P2(1-A), B3(2-UA), P3(1-A), B 4(1-A)		

(continued)

Table 1 (continued)

Fold	PDB ID	Sequence	Strand polarity	No. of quartets	Type of loops	Ions	Remarks
	6V9B	AUCGAGGGAGUGGUGAGGAUGAGGCGAU			B1(1-A), P1(1-U), B2(1-U), P2(1-A), B3(2-AU), P3(1-A), B4(1-C)		
	6V9D						
RMF3	2LA5	CGGUGUGGAAGGAGUGGCUGGGU	Parallel	3	L1(2-GU), B1(1-U), P1(2-AA), P2(1-A), B2(1-U), P3(2-CU)	–	Sc1 RNA
	5DE5					K+	
	5DE8						
	5DEA					K+ , Cs+	
RMF4	4KZD	GGACGCGACCGAAAUGGUGAAG-GACGGGUCCAGUGCGAAACACGCACUGU-UGAGUA-GAGUGUGAGCUCCGUAACUGGUCGCGUC	Anti-parallel	2	P1(2-AC), D1(28–52), B3(2-UA), P3(1-A), B4(1-U)	K +	Aptamer
	4KZE						
	4Q9Q						
	4Q9R						
	6B14	GACGCGACCGAAAUGGUGAAG-GACGGGUCCAGUGCGAAACACGCACUGU-UGAGUA-GAGUGUGAGCUCCGUAACUGGUCGCGUC			P1(2-AC), D1(28–53), B3(2-UA), P3(1-A), B4(1-U)	–	
	6B3K	GACGCGACCGAAAUGGUGAAG-GACGGGUCCAGUGCGAGACCCGCACUGU-UGAGUA-GAGUGUGAGCUCCGUAACUGGUCGCGUC					
	5OB3	GGGAGUACGGU-GAGGGUCGGGUCCAGUAGGUACGC-CUACUGUUGAGUA-GAGUGUGGGCUCCGUACUCCC			P1(2-UC), D1(21–45), B3(2-UA), P3(1-A), B4(1-U)	K+	
	7L0Z					K+	

(continued)

Table 1 (continued)

Fold	PDB ID	Sequence	Strand polarity	No. of quartets	Type of loops	Ions	Remarks
RMF5 RMF5	7OA3	GGCUAGCUGGAGGGGCGCCAGUUCGCUG-GUGGUUGGGUGCGGUCGGCUAGCC	Mixed	2	L1(2-GA), L2(14–31), P1(3-UUG), L3(3-UG	K+	Aptamer
	7OAV						
	7OAW						
	7OAX						
RMF6	6E8S	GCUACGAAGGAAGGAUUGGUAUGUG-GUAUAUUCGUAGC	Mixed	2	P1(2-AA), P2(3-AUU), L1(3-UAU), B1(1-U)	K+	Aptamer
	6UPO						
	6E8T	GUACGAAGGAAGGUUUGGUAUGUG-GUAUAUUCGUAC			P1(2-AA), P2(3-UUU), L1(3-UAU), B1(1-U)		
	6E8U	GUACGAAGGAAGGUUUG-GUAUGGGGUAGUUGUCGUAC			P1(2-AA), P2(3-UUU), L1(3-UAU), B1(1-G)		
	6PQ7						
RDF1	6JJH	GGCUCGGCGGCGGA	Parallel	4	P1(3-CUC), B1(1-C), P2(1-C)	K+	Pseudorabies virus RNA G-quadruplex
	6JJI						
	6XRQ						
RDF2	2KBP	UAGGGUUAGGGU	Parallel	3	P1(3-UUA)	–	Human telomeric RNA quadruplex
	2M18						
	3IBK					K+	
	3MIJ						

(continued)

Table 1 (continued)

Fold	PDB ID	Sequence	Strand polarity	No. of quartets	Type of loops	Ions	Remarks
RDF3	4TS0	Chain A: GACGCGACCGAAUGAAAUGGU-GAAGGACGGGUCCAGCCGGCUGC Chain B: GCAGCCGGCUUGUUGAGUA-GAGUGUGAGCUCCGUAACUGGUCGCGUC	Anti-parallel	2	P1A(2-AC) B1B(2-UA), P1B(1-A), B2B(1-U)	K+	Aptamer
	4TS2						
RTF	1J6S	(BRU)GAGGU (BRU)GAGGU	Parallel	5–10		Na+ , Ba2 +	RNA oligonucleotide
	1MDG					Na+	
	1J8G	UGGGGU				Na+ , Sr2 +	
	1RAU					–	
	4RJ1					Na+ , Sr2 +	
	4RKV						
	4RNE						
	4XK0					K+	
	2AWE	U(BGM)GUGU					
	2GRB	UG(I)GGU					
	6GE1	UGGUGGU				–	
	1N7A	U(BGM)AGGU				K +	
	1N7B						

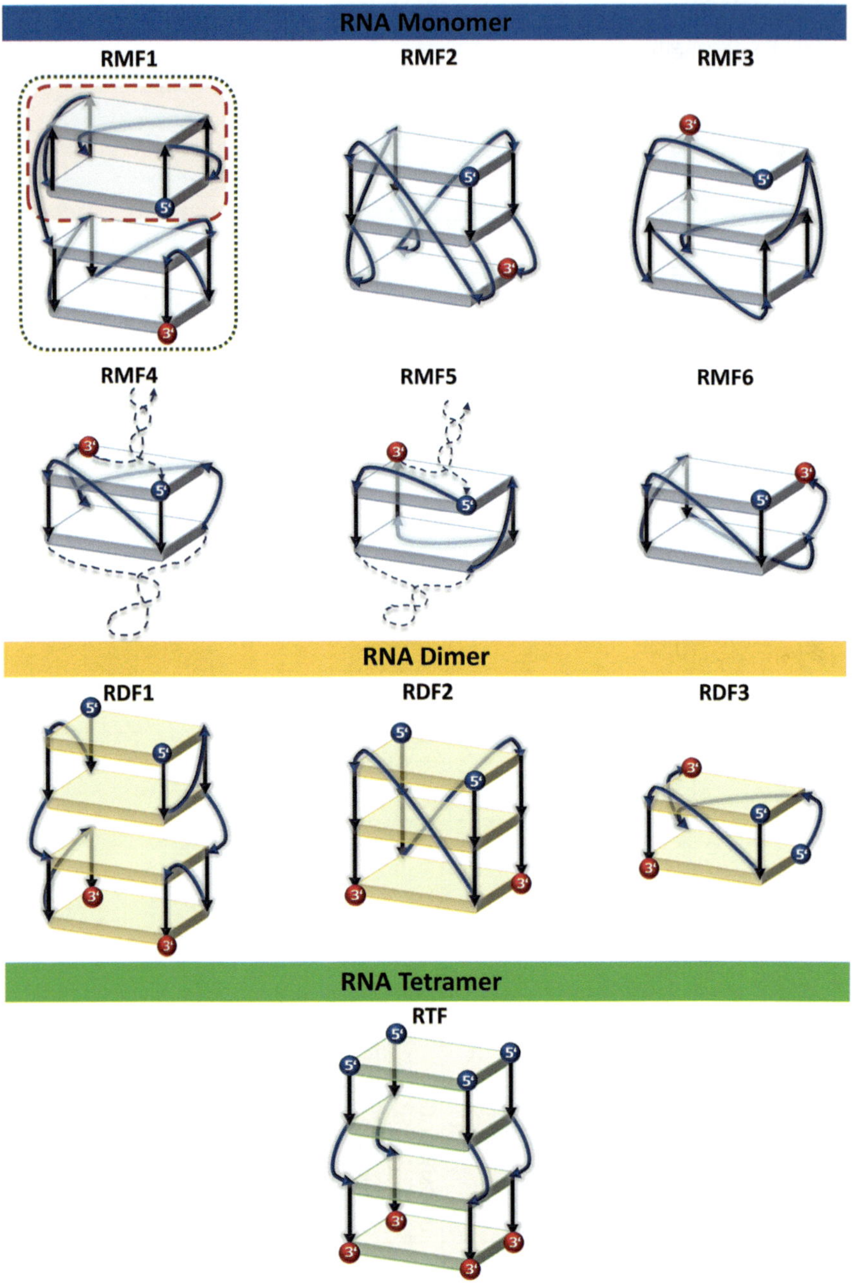

Fig. 2 Schematic illustration of 10 unique RNA quadruplex folds. Note that RMF1–RMF6 represent intramolecular RNA monomeric folds (RMF), RDF1–RDF3 represent the intermolecular RNA dimeric folds (RDF) and RTF represents intermolecular RNA tetrameric fold (RTF). The dotted lines in RMF4 and RMF5 represent the double helix formed by the loop residues. The polarity of the strands is indicated by the arrows

respectively of the contiguous strands (viz., ith and i + 1th strands). However, the diagonal loop facilitates the antiparallel orientations of ith and i + 2nd strands. Additionally, a V-loop mimicking the propeller loop with zero residues connects ith and i + 1st strands, leading to contiguous parallel strands. The number of residues in the propeller loop varies between 1 and 5 to connect two quartets (Table 1). For a lateral loop, two to three residues are found to be optimal. Exceptionally, a long lateral loop with 18 residues is shown to form a duplex in the aptamer structures (PDB IDs: 7OA3, 7OAV, 7OAW, 7OAX). For the diagonal loop, at least five residues are found to be necessary. A maximum of 26 residues are found in a diagonal loop in the aptamer structures which forms a duplex. Interestingly, higher-order quadruplexes are facilitated by the propeller or the bulge linkers in RMF1 (PDB ID: 6K84) and RDF1 (PDB IDs: 6XRQ, 6JJH) (which are parallel conformations), leading to the stacking of two quadruplexes. While a propeller linker in RMF1 connects two quadruplex fragments in an antiparallel fashion, the bulge linker in RDF1 facilitates the stacking of two quadruplex fragments in a parallel fashion. In the case of RTF (tetrameric fold), a higher-order octaplex structure results from a coaxial stacking of two tetraplexes (PDB ID: 1N7A).

It is quite interesting to note that among the 28 unique sequences (60 PDBs), there are only 10 unique RNA G-quadruplex folds. These folds can be used in the modeling of RNA G-quadruplexes if the sequence, oligomeric state, and strand polarity information are known.

3.1 Helical Twist and Rise of the RNA Quadruplex Folds

The helical twist and rise measured between the adjacent G-quartets using the scheme described in Fig. 3 exhibit average values of ~32° and 3.26 Å respectively. The helical rise is slightly higher for antiparallel quadruplex conformation (3.54 ± 0.24 Å) compared with parallel (3.3 ± 0.29 Å) and hybrid/mixed (3 ± 0.17 Å) conformations. The helical twist assumes values in the range of 21–37°, 32–35° and 29–34° respectively for parallel, antiparallel and hybrid conformations considering all the 10 folds. These helical parameters can be used in the modeling of RNA quadruplex folds.

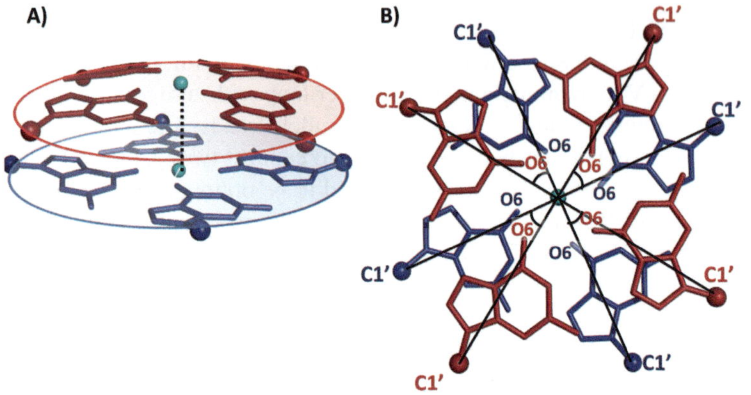

Fig. 3 **A)** The helical rise between the adjacent G-quartets (colored blue (first quartet) and red (second quartet)) is calculated by considering the center (cyan colored spheres, calculated using the four O6 atoms) of each quartet. Note that the distance between the two centers of the adjacent quartets are considered as the helical rise (solid vertical line). The C1' atoms are indicated in bigger spheres and are connected by a circle. **B)** The helical twist between the adjacent quartet is estimated by calculating the angle formed by the vectors (dotted line) connecting the C1' of the 1st quartet (colored blue) residue and the center and, C1' of the 2nd quartet (colored red) residue and the center after projecting the second quartet onto the first one. The average of the helical twists measured at the adjacent guanine residues of the individual quartet strands is considered as the quartet helical twist

3.2 Sugar-Phosphate Backbone and Glycosyl *Conformational Angle Preferences*

Intriguingly, the C2'-endo sugar pucker is equally preferred as the C3'-endo sugar pucker irrespective of the different RNA quadruplex folds (Fig. 4). It is also seen to be present in both G and non-G-quartets. A minor population of other sugar puckers is also seen. The C2'-endo sugar pucker does not exhibit any correlation with the *glycosyl* (χ) (C6–N9–C1'–O4') conformational angle of the corresponding nucleotide (Fig. 5f). As discussed in the previous studies (Malgowska et al. 2016; Fay et al. 2017), the Gs in the RNA G-quartets of the parallel fold predominantly assume *anti-glycosyl* conformation, whereas the Gs in the antiparallel and mixed folds assume both $\pm$ *syn* and *anti-glycosyl* conformations (Fig. 4f). Further, the A-A-A-A (**PDB IDs**: 1J6S, 1MDG) and U-U-U-U (**PDB IDs:6GE1**, 2AWE) quartets seen in the tetrameric folds prefer *high-anti* as well as *anti-glycosyl* conformations. Such *high-anti* and *anti-glycosyl* conformations facilitate N3…N6 (1MDG) (Fig. 1d) and N7…N6 (1J6S) (Fig. 1e) hydrogen bonds respectively between the adenines of the A-quartet. These hydrogen-bonding patterns mimic the A…A mismatch hydrogen-bonding patterns seen in the DNA duplex (Khan et al. 2015; Ajjugal and Rathinavelan 2021; Ajjugal et al. 2021b). Further, the Us prefer *anti-glycosyl* conformation and are stabilized by O4…N3 hydrogen bonds (Fig. 1h).

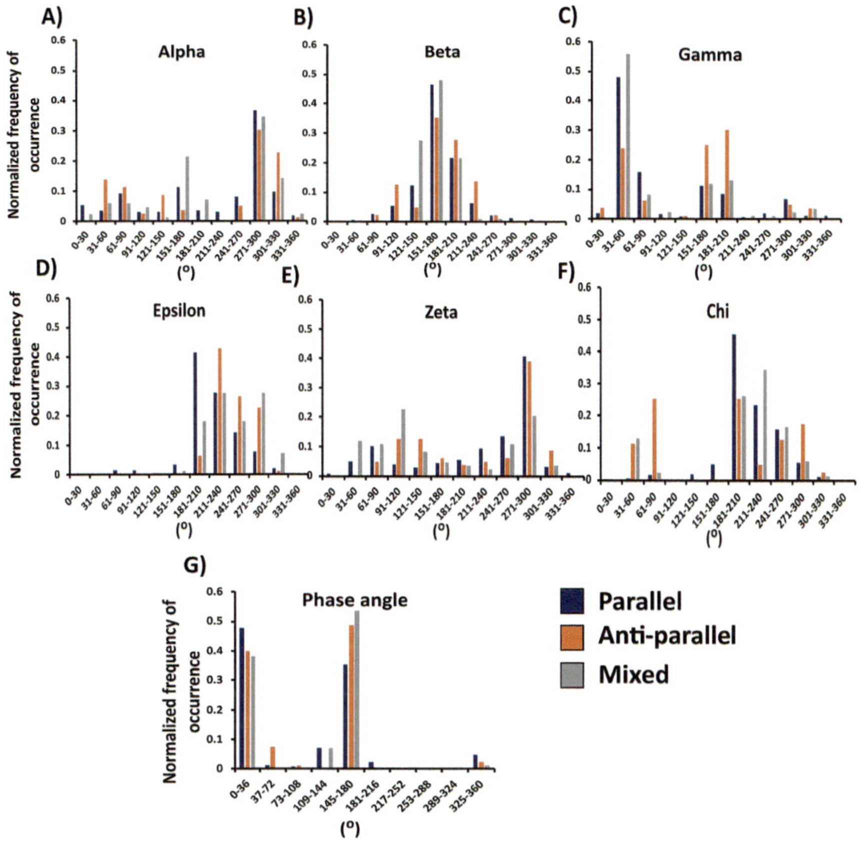

Fig. 4 Bar diagram illustrating the sugar-phosphate backbone and *glycosyl* conformational preferences for the parallel (colored blue), antiparallel (colored orange) and mixed/hybrid (colored grey) RNA quadruplex folds: **A)** α (C3'-P-O5'-C5'), **B)** β (P-O5'-C5'-C4'), **C)** γ (O5'-C5'-C4'-C3'), **D)** ε (C4'-O3'-C3'-P), **E)** ζ (O3'-C3'-P-O5') **F)** χ (O4'-C1'-N9/N1-C4/C2) and **G)** phase angle of pseudorotation (P). Note the predominance of ± *syn* conformation in antiparallel and mixed quadruplex folds and predominance of *anti* conformation in parallel quadruplex folds (F). The conformational angles (in degrees) are indicated in the X-axis and their normalized (with respect to each fold) frequency of occurrence is indicated in Y-axis

α (C3'–P–O5'–C5') (Fig. 4a) and ζ (O3'-C3'-P-O5') (Fig. 4e) take a wide range of conformational preferences ranging from *gauche* + to *trans* to *gauche*-. ε (C4'-O3'-C3'-P) assumes the conformational angles in the range of *trans* to *gauche*- (Fig. 4d). However, β (P–O5'–C5'–C4') (Fig. 4b) and γ (O5'–C5'–C4'–C3') (Fig. 4c) favor *trans* and, *gauche* + and *trans* respectively. Except ζ (O3'-C3'-P-O5'), the other conformational angles do not exhibit any direct correlation with the sugar-pucker preference (Fig. 5). ζ (O3'–C3'–P–O5') assumes a wide range of conformation (ranging from *gauche* + to *trans* to *gauche*-) in the case of C2'-endo, whereas it

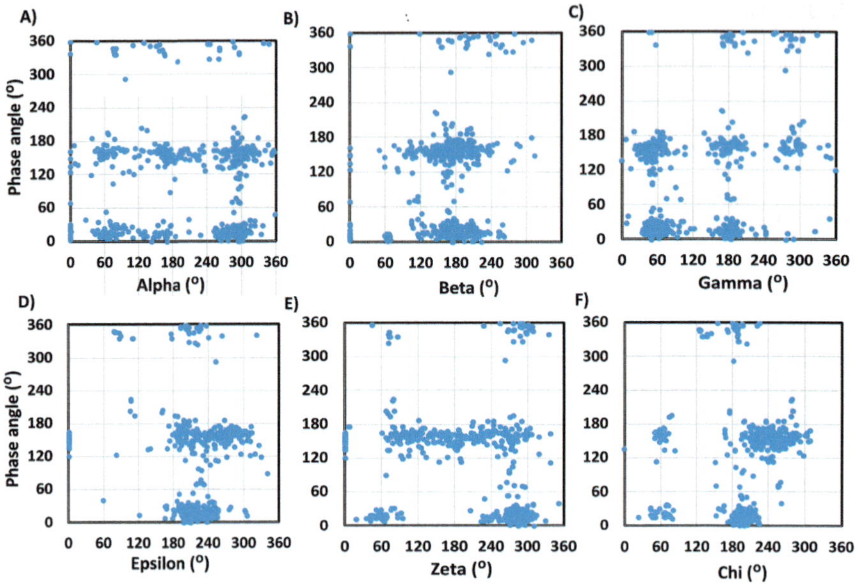

Fig. 5 2D Scatter plot illustrating the correlation between sugar pucker (Y-axis) and exocyclic and *glycosyl* torsion angles (X-axis)

is strictly confined to *gauche* + and *gauche*- in the case of C3'-endo sugar pucker (Fig. 5e).

Exploration of the impact of sugar-pucker preference on O2' and O4' hydrogen bond formation (which is seen in A-form RNA duplex) indicates that among all the folds, most of the structures under DMF2 fold and one structure under RTF fold exhibit such hydrogen-bonding pattern between the consecutive Gs (Fig. 6). For example, C3'endo sugar-pucker is favored by the Gs in PDB ID:2M18 and PDB ID:2GRB leading to consecutive O2'…O4' hydrogen bonding in all the four strands. Although such hydrogen bonding is also seen in the few other structures having DMF2 fold (PDB ID: 2KBP), most of the structures lack such consecutive hydrogen bonds, perhaps due to the presence of C2'endo sugar pucker, bulges, or loops. In other folds an O2'…O4' hydrogen-bonding pattern is also seen randomly. However, such instances are rarely seen in RMF4, RMF5 and RMF6 fold structures which have antiparallel or mixed conformations. Notably, the ± *syn glycosyl* conformation is seen in RMF4, RMF5 and RMF6 folds. In addition, the hydrogen bond formation between O2' and phosphate oxygen as well as the electronegative atoms of the bases in the same or adjacent strands are seen.

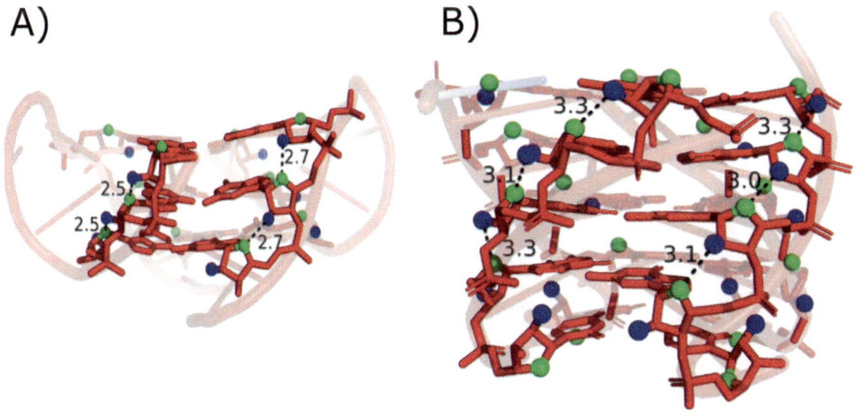

Fig. 6 Consecutive O2'…O4' hydrogen bonds seen in **A)** RDF2 fold (PDB ID: 2M18), and **B)** RTF fold (PDB ID: 2GRB). *Note* that O2' and O4' are indicated in blue and green colored spheres respectively. The dotted lines indicate the hydrogen bonds with the values indicated alongside. Note that the dark-shaded sticks represent the residues that participate in O2' and O4' hydrogen bond, whereas, the light-shaded sticks indicate the other residues

3.3 Fold-Dependent Water Network in RNA Quadruplex

Among the RNA quadruplex folds given in Fig. 2, RTF (parallel conformation) seems to be highly hydrated (Fig. 7a, b). In most of the structures, the RTF fold quadruplex grooves are filled up with short water wires involving N3 and N2 atoms of the G as well as the 2' hydroxyl group and anionic oxygen(s) of the phosphate. The possibility of ion-mediated water network is also seen. In the case of non-G tetrads, the electronegative atom in the groove participates in the water network. For example, O2 of U and N3 of I/A take part in the water network (Fig. 7b). Intriguingly, a continuous chain of water interaction is also seen in RTF fold (Fig. 7a). A similar continuous water network is also seen in RDF1 fold (Fig. 7c). In the other folds, random interactions of water molecules with the base and backbone electronegative atoms are seen. The bases in RMF1–RMF6 folds are found to be least hydrated due to their limited surface exposure, resulting from the presence of propeller loop(s) as well as narrow groove(s) (Fig. 7d). Better hydration is possible in RMF4, as seen in Fig. 7e. However, a continuous groove water network is absent in RMF4, unlike in the RTF and RDF1 folds.

3.4 Quartet–Ion Interaction in RNA Quadruplex

Not surprisingly, monovalent K^+ ion is found to predominantly stabilize the G-quartets of RNA quadruplex through coordinating with guanine's O6 atoms (Fig. 8a) (Table 1). Other monovalent ions like Na^+ and Cs^+ and divalent cations like Ba^{2+},

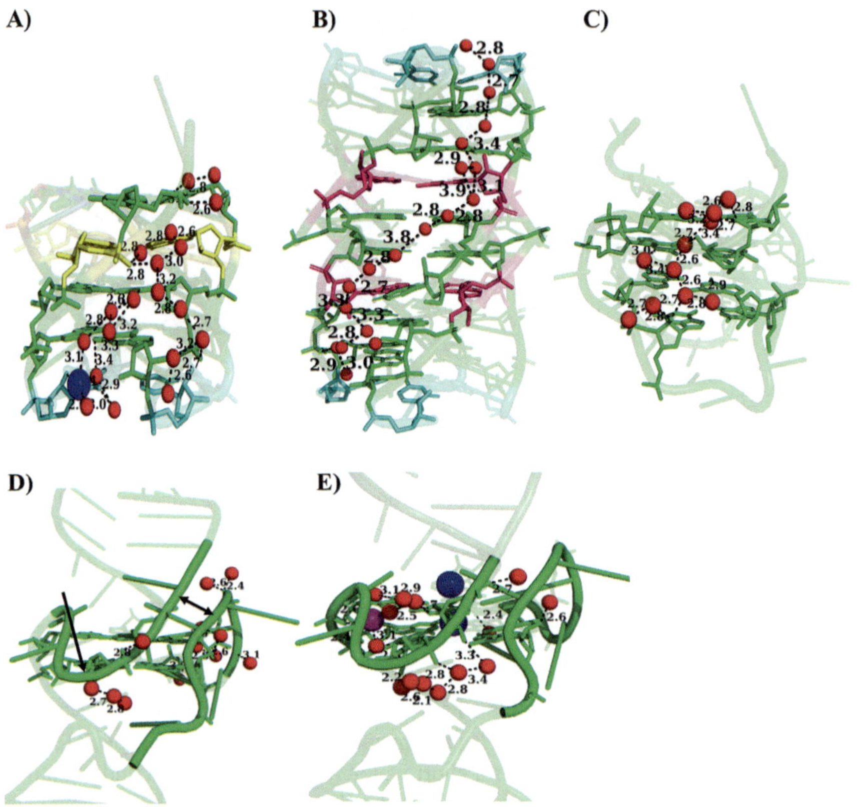

Fig. 7 **Water network around various RNA quadruplex folds**. A representative illustration of groove water (red colored sphere) network in **A, B)** RTF (PDB IDs:1N7A (A), 2GRB (B)), **C)** RDF1 (PDB ID:6XRQ), **D)** RMF4 (PDB ID: 5OB3) and **E)** RMF4 (PDB ID: 7LOZ) folds. The G, A, U and I residues are shown in green, pink, cyan and yellow colors respectively. Note that the dark-shaded sticks represent the water interacting groove residues, whereas, the light-shaded sticks indicate the other residues. Blue-colored spheres in C) and E) indicate K^+ ion and, the pink color sphere in E) indicates Na^+ ion. The double headed and single headed arrows in D) point out the narrow groove width and propeller loop respectively. The dotted lines indicate the hydrogen bonds with the values indicated alongside

Ca^{2+} and Sr^{2+} are also found to stabilize the G-quartets through O6 coordination. However, Cs^+ coordinates with O6 atoms along with K^+, in which both ions are present on either side of the G-quartet. Ba^{2+}, Ca^{2+}, and Sr^{2+} are mainly seen in the RTF structures. In the case of non-G-quartets, O4 atoms of U-quartet (Fig. 8b), N6 (Fig. 8c) or N3 atoms (Fig. 8d) of A-quartet, and O6 atoms of I-quartet (Fig. 8e) are involved in coordination with the ions. For the RNA quadruplex structural studies, the monovalent salt concentration in the range of 90–150 mM is used. A relatively lower divalent ion concentration is however used (in the range of 20–30 mM concentration)

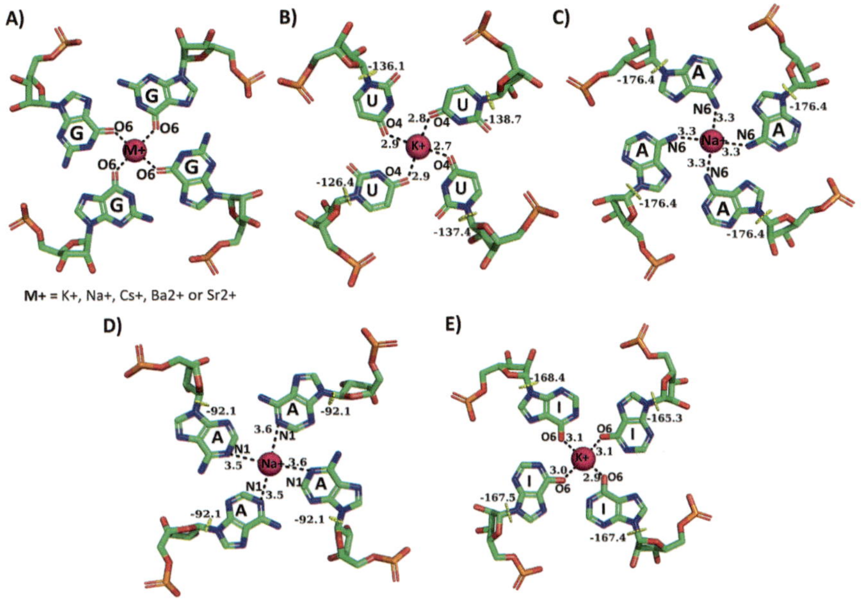

Fig. 8 Ion...quartet coordination in RNA quadruplex. Coordination of ions (indicated in sphere) with **A)** G-G-G-G, **B)** U-U-U-U, **C)** A-A-A-A, **D)** A-A-A-A and **E)** I-I-I-I quarts. The coordination distance is mentioned alongside the black colored dotted lines. The glycosyl conformational angles of the non-G quartets are marked adjacent to the yellow colored dotted rings.

along with the monovalent ions. The cations are also found to have coordination with the backbone and groove exposed base electronegative atoms.

4 Biophysical Techniques to Identify the Folding Nature of RNA Quadruplex

Circular dichroism, UV melting (Mergny and Lacroix 2009; Fay et al. 2017), nuclear magnetic resonance and EMSA may provide information about the strand orientation and oligomeric forms of quadruplexes. UV melting can provide information about the presence of a quadruplex (RNA or DNA) as it exhibits hyperchromicity at 295 nm against 260 nm in the case of the DNA or RNA duplex (Mergny and Lacroix 2009). Similarly, NMR can also be useful in differentiating the G-quadruplex forming RNA from the one that forms a double helix (Fay et al. 2017). Since the imino proton peaks of the quartet Gs fall in the range of 10–12 ppm (parts per million) in contrast to 12–15 ppm for the Gs that are engaged in Watson–Crick base pairs, NMR can be used in the initial differentiation of the A-form helix from the quadruplex (Fay et al. 2017). Although CD can mislead parallel RNA G-quadruplex conformation as it mimics the A-form duplex, the information from UV and 1D NMR could aid in distinguishing the RNA quadruplex from the duplex. However, CD spectra can be used to distinguish the antiparallel, parallel and hybrid (or mixed) conformations of RNA quadruplex based on the negative or positive peaks seen at different wavelengths (Kypr et al. 2009; Sathyaseelan et al. 2021). For instance, a negative peak around 260 nm and a positive peak around 295 nm are indicative of an antiparallel RNA quadruplex. The parallel RNA quadruplex can be identified with the help of the positive and negative peaks around 265 nm and 240 nm respectively. Similarly, mixed RNA quadruplex can be identified from the positive and negative peaks around 260 nm/295 nm and 240 nm respectively. Thus, one can obtain information about the nature of strand orientation from the CD. EMSA can help in differentiating the intramolecular monomeric quadruplex fold from the intermolecular dimeric and tetrameric folds (Ajjugal et al. 2021a). For instance, when a reference single-stranded RNA sequence having the same length as the quadruplex sequence is used in the EMSA, the quadruplex may run faster in the gel compared with the single-stranded RNA if it assumes a monomeric fold (Ajjugal et al. 2021a). Thus, to a certain extent, these experimental techniques provide information about the nature of quadruplex folds.

5 In Vitro and in Vivo Techniques to Identify RNA Quadruplex-Forming Regions

Several in vitro and in vivo experimental techniques such as rG4-seq, chemical profiling and transcriptome ligand binding can be used to detect the putative quadruplex-forming regions (Yu et al. 2022). Low-throughput experimental assays such as ligand-binding assays, reverse transcriptase (RT) footprinting assays, and biophysical assays are employed to determine the folding status of putative quadruplex-forming sequences (Paramasivan et al. 2007; Lyu et al. 2021). However, these techniques have limitations in detecting the transcriptome-wide potential RNA G-quadruplex (rG4)-forming sequences. The emergence of next-generation sequencing technologies has facilitated transcriptome-wide probing for quadruplex formation using rG4 assays such as rG4-seq, ligand-binding methods like G4RP-seq, and chemical probing methods such as Keth-seq, DMS probing, NAI probing, and SHALiPE-seq (Guo and Bartel 2016; Kwok et al. 2016; Yang et al. 2018, 2020; Yeung et al. 2019; Weng et al. 2020). Indeed, these studies have identified many RNA quadruplexes linked with critical biological processes across diverse transcriptomes (Kharel et al. 2020a; Dumas et al. 2021). G4Atlas is a newly developed database which has information about transcriptome-wide G-quadruplex-forming regions identified using these experimental techniques along with their functional information for 10 different species (Yu et al. 2022).

6 G–Quadruplex Prediction Tools

Several in silico prediction tools are available to detect the putative RNA G-quadruplex regions present in a sequence using conventional (Puig Lombardi and Londono-Vallejo 2020) and machine learning methods (Sahakyan et al. 2017; Klimentova et al. 2020). One can use G4Catchall (Doluca 2019), G4Hunter (Bedrat et al. 2016; Brazda et al. 2019), G4-Predictor V.2 (Mishra et al. 2016), G4RNA screener (Garant et al. 2018), ImGQfinder (Varizhuk et al. 2017), pqsfinder (Hon et al. 2017), QGRS Mapper (Kikin et al. 2006), and QPARSE (Berselli et al. 2020) tools to predict the potential of an RNA sequence to form intramolecular G-quadruplex. G4-iMGrinder, ImGQfinder and QPARSE tools can be used for predicting the regions that can form G-quadruplex with non-G quartets. Earlier studies estimated the quadruplex prediction accuracy of these tools to be in the range of 46–79%, within which G4CatchAll showed the highest prediction accuracy (Miskiewicz et al. 2021). For the 19 unique intramolecular RNA quadruplex sequences whose structures have already been solved experimentally (Table 1), these tools have exhibited a prediction accuracy in the range of 47–100% (Table 2) with G4Catchall and G4RNA screener having the highest, and G4Hunter and G4-Predictor V.2 having relatively lower prediction accuracy. While the majority of these tools fail to exactly pinpoint the quadruplex-forming regions, some of them have even failed to predict the ability of these sequences to form RNA quadruplex, especially for the aptamer sequences having long hairpin-like loop regions. However, none of these tools predict the nature of the RNA quadruplex fold. One can supplement 1D NMR, CD and EMSA data to reduce false predictions (Fay et al. 2017).

7 Modeling of RNA G-Quadruplex and Capturing Their Conformational Dynamics

After obtaining the potential of an oligoribonucleotide to form G-quadruplex, one can model the folds (Fig. 2) either manually or using existing tools (Patro et al. 2017). Subsequently, one can perform molecular dynamics (MD) simulation to capture the conformational dynamics of the RNA quadruplex, especially in the scenario of capturing the loop dynamics. AMBER (assisted model building and energy refinement) (Case et al. 2005), CHARMM (Chemistry at Harvard molecular mechanics) (Jo et al. 2008), NAMD (Phillips et al. 2020) and GROMACS (Berendsen 1995) are the MD tools commonly used for capturing the conformational dynamics of biomacromolecules. One can also use CHARMM-GUI (Jo et al. 2008) to generate the input scripts for the above programs.

Table 2 Accuracy of various in silico tools in predicting the ability of the sequences (whose structures are already known, see Table 1) to form RNA quadruplex*

	RMF1	RMF2	RMF3	RMF4	RMF5	RMF6	Total
Percentage accuracy in predicting the ability of the query sequences to form quadruplex (Correct prediction/total of unique sequences subjected for prediction). Note that NA stands for "not applicable"							
G4Catchall	100 (6/6)	100 (4/4)	100(1/1)	100(4/4)	100(1/1)	100(3/3)	100(19/19)
G4Hunter	50(3/6)	25(1/4)	100(1/1)	75(3/4)	100(1/1)	33 (1/3)	53 (10/19)
G4-Predictor V.2	33 (2/6)	100 (4/4)	100(1/1)	0	100(1/1)	33(1/3)	47 (9/19)
G4RNA screener	100(6/6)	100(4/4)	100(1/1)	100(4/4)	100(1/1)	100(3/3)	100 (19/19)
ImGQfinder	100(6/6)	100(4/4)	100(1/1)	100(4/4)	100(1/1)	100(3/3)	100(19/19)
Pqsfinder	100(6/6)	100(4/4)	100(1/1)	25 (1/4)	100(1/1)	100(3/3)	84 (16/19)
QGRS Mapper	100(6/6)	100(4/4)	100(1/1)	0.50 (2/4)	100(1/1)	100(3/3)	89 (17/19)
QPARSE	100(6/6)	100(4/4)	100(1/1)	0.50(2/4)	100(1/1)	100(3/3)	89 (17/19)

[*]Note that some of the folds have just one unique sequence, leading to 100% accuracy. Since none of the tools can predict the intermolecular RNA quadruplex formation, only the prediction accuracies of the monomeric folds are listed here

8 Challenges in the Prediction and Modeling of RNA G-Quadruplexes

Most of the existing algorithms only predict intramolecular (monomer) G-quadruplex-forming regions in a given sequence and not the intermolecular (dimer and tetramer) G-quadruplex-forming regions. Indeed, intramolecular G-quadruplex prediction is not 100% accurate as most of the tools simply use the conventional $G_{3+}N_{1\text{-}7}G_{3+}N_{1\text{-}7}G_{3+}N_{1\text{-}7}G_{3+}$ motif. However, non-canonical rG4 structures having quartets with 3 Gs and a non-G nucleotide and, 2 Gs and 2 non-Gs can also be formed under physiological ionic conditions (Kwok et al. 2016; Yu et al. 2022). Further, RNA sequences having nucleotides greater than seven loop residues are also shown to form quadruplexes (Pandey et al. 2013). Indeed, a long loop region can lead to the formation of secondary structures like duplexes as seen in the case of RNA aptamers (Huang et al. 2014; Fernandez-Millan et al. 2017; Koirala et al. 2018; Jeng et al. 2021; Mieczkowski et al. 2021). There are reports which state that sequences fulfilling the prerequisite (as per all the existing algorithms) of G-quadruplex formation are still unable to fold into a quadruplex (Beaudoin and Perreault 2010). When considering the whole transcriptome of an organism, there is a possibility of forming an intermolecular RNA quadruplex between remote regions, facilitating the formation of dimeric and tetrameric quadruplex folds. Additionally, the prediction of RNA G-quadruplex having non-G-quartets formed by As, Cs and Us poses a challenge. Yet another challenge is to predict the higher-order quadruplex structure formed in the transcriptome, in which two short quadruplexes can stack on top of each other and lead to a new fold as seen in RMF1, RDF1 and RTF folds (Fig. 2). While the machine learning approach may be useful in addressing such challenges in G-quadruplex prediction, obtaining a training set having all possible RNA folds is still a challenge due to the limited availability of the structures of RNA quadruplex folds. Yet another challenge is the functional relevance of these quadruplex folds. Unlike in the case of proteins, until now no clear RNA quadruplex fold-function information is available. Challenges also exist in the modeling of three-dimensional structures of RNA quadruplex folds. For instance, the 3D-NuS web server which models 17 different RNA G-quadruplex folds does not consider non-G-quartets (Patro et al. 2017). Similarly, it has a limitation in modeling higher-order quadruplex structures as well as quadruplex with overhangs. Such information is important for deriving the structural information of the biologically relevant RNA quadruplex-forming sequences and their functional relevance in vivo. Indeed, a handful of such information would facilitate RNA quadruplex sequence, structure, and function relationship. A recent initiative in this direction is the development of a manually curated database, G4Atlas, which provides experimentally identified transcriptome-wide G-quadruplex-forming regions along with their functional information in diverse biological processes for 10 different species (Yu et al. 2022). Although G4Atlas considers both canonical and non-canonical rG4 structures, it does not provide the corresponding folding pattern. Rigorous initiatives in this direction may pave the way for data-driven deep learning algorithms for the sequence, structure and the concomitant functional feature(s) of RNA quadruplex.

Acknowledgements The authors thank BIRAC-SRISTI GYTI (Grant/Award Number: PMU2019/007) for the funding and Chakkarai Sathyaseelan for the critical comments on the manuscript. PPU and SS acknowledge CSIR and MoE respectively for the fellowship.

Author Contributions PPU and SS analyzed the G-quadruplex folds. PPU analyzed the helical parameters, conformational angles and, water and ion network of the quadruplex. PPU and TR wrote the manuscript. TR designed and supervised the project.

References

Adrian M, Heddi B, Phan AT (2012) NMR spectroscopy of G-quadruplexes. Methods 57:11–24

Agarwala P, Pandey S, Maiti S (2015) The tale of RNA G-quadruplex. Org Biomol Chem 13:5570–5585

Ajjugal Y, Kolimi N, Rathinavelan T (2021a) Secondary structural choice of DNA and RNA associated with CGG/CCG trinucleotide repeat expansion rationalizes the RNA misprocessing in FXTAS. Sci Rep 11:8163

Ajjugal Y, Tomar K, Rao DK et al (2021b) Spontaneous and frequent conformational dynamics induced by A...A mismatch in d(CAA).d(TAG) duplex. Sci Rep 11:3689

Ajjugal Y, Rathinavelan T (2021) Sequence dependent influence of an A...A mismatch in a DNA duplex: An insight into the recognition by hZalphaADAR1 protein. J Struct Biol 213:107678

Beaudoin JD, Perreault JP (2010) 5'-UTR G-quadruplex structures acting as translational repressors. Nucleic Acids Res 38:7022–7036

Bedrat A, Lacroix L, Mergny JL (2016) Re-evaluation of G-quadruplex propensity with G4Hunter. Nucleic Acids Res 44:1746–1759

Berendsen HJC, van der Spoel D, van Drunen R (1995) GROMACS: A message-passing parallel molecular dynamics implementation. Computer Phys Commun 91:43–56

Berman HM, Westbrook J, Feng Z et al (2000) The Protein data bank. Nucleic Acids Res 28:235–242

Bernat V, Disney MD (2015) RNA structures as mediators of neurological diseases and as drug targets. Neuron 87:28–46

Berselli M, Lavezzo E, Toppo S (2020) QPARSE: searching for long-looped or multimeric G-quadruplexes potentially distinctive and druggable. Bioinformatics 36:393–399

Besnard E, Babled A, Lapasset L et al (2012) Unraveling cell type-specific and reprogrammable human replication origin signatures associated with G-quadruplex consensus motifs. Nat Struct Mol Biol 19:837–844

Bhattacharyya D, Mirihana Arachchilage G et al (2016) Metal cations in G-Quadruplex folding and stability. Front Chem 4:38

Biffi G, Di Antonio M, Tannahill D et al (2014) Visualization and selective chemical targeting of RNA G-quadruplex structures in the cytoplasm of human cells. Nat Chem 6:75–80

Brazda V, Kolomaznik J, Lysek J et al (2019) G4Hunter web application: a web server for G-quadruplex prediction. Bioinformatics 35:3493–3495

Brown V, Jin P, Ceman S et al (2001) Microarray identification of FMRP-associated brain mRNAs and altered mRNA translational profiles in fragile X syndrome. Cell 107:477–487

Burge S, Parkinson GN, Hazel P et al (2006) Quadruplex DNA: sequence, topology and structure. Nucleic Acids Res 34:5402–5415

Cammas A, Millevoi S (2017) RNA G-quadruplexes: emerging mechanisms in disease. Nucleic Acids Res 45:1584–1595

Campbell N, Collie GW, Neidle S (2012) Crystallography of DNA and RNA G-quadruplex nucleic acids and their ligand complexes. Curr Protoc Nucleic Acid Chem Chapter 17(Unit17):6

Carvalho J, Mergny JL, Salgado GF et al (2020) G-quadruplex, friend or foe: The role of the G-quartet in anticancer strategies. Trends Mol Med 26:848–861

Case DA, Cheatham TE 3rd, Darden T et al (2005) The Amber biomolecular simulation programs. J Comput Chem 26:1668–1688

Caterino M, Paeschke K (2022) Action and function of helicases on RNA G-quadruplexes. Methods 204:110–125

Chan KL, Peng B, Umar MI et al (2018) Structural analysis reveals the formation and role of RNA G-quadruplex structures in human mature microRNAs. Chem Commun (camb) 54:10878–10881

Cogoi S, Xodo LE (2006) G-quadruplex formation within the promoter of the KRAS proto-oncogene and its effect on transcription. Nucleic Acids Res 34:2536–2549

Coimbatore Narayanan B, Westbrook J, Ghosh S et al (2014) The nucleic acid database: new features and capabilities. Nucleic Acids Res 42:D114–D122

Collie GW, Haider SM, Neidle S et al (2010) A crystallographic and modelling study of a human telomeric RNA (TERRA) quadruplex. Nucleic Acids Res 38:5569–5580

Conlon EG, Lu L, Sharma A et al (2016) The C9ORF72 GGGGCC expansion forms RNA G-quadruplex inclusions and sequesters hnRNP H to disrupt splicing in ALS brains. Elife 5:e17820

Decorsiere A, Cayrel A, Vagner S et al (2011) Essential role for the interaction between hnRNP H/F and a G quadruplex in maintaining p53 pre-mRNA 3'-end processing and function during DNA damage. Genes Dev 25:220–225

Doluca O (2019) G4Catchall: A G-quadruplex prediction approach considering atypical features. J Theor Biol 463:92–98

Dumas L, Herviou P, Dassi E et al (2021) G-Quadruplexes in RNA Biology: Recent advances and future directions. Trends Biochem Sci 46:270–283

Dvorkin SA, Karsisiotis AI, Webba Da Silva M (2018) Encoding canonical DNA quadruplex structure. Sci Adv 4:eaat3007

Fay MM, Lyons SM, Ivanov P (2017) RNA G-Quadruplexes in biology: Principles and molecular mechanisms. J Mol Biol 429:2127–2147

Fernandez-Millan P, Autour A, Ennifar E et al (2017) Crystal structure and fluorescence properties of the iSpinach aptamer in complex with DFHBI. RNA 23:1788–1795

Fratta P, Mizielinska S, Nicoll AJ et al (2012) C9orf72 hexanucleotide repeat associated with amyotrophic lateral sclerosis and frontotemporal dementia forms RNA G-quadruplexes. Sci Rep 2:1016

Fukunaga J, Nomura Y, Tanaka Y et al (2020) A G-quadruplex-forming RNA aptamer binds to the MTG8 TAFH domain and dissociates the leukemic AML1-MTG8 fusion protein from DNA. FEBS Lett 594:3477–3489

Garant JM, Perreault JP, Scott MS (2018) G4RNA screener web server: User focused interface for RNA G-quadruplex prediction. Biochimie 151:115–118

Goering R, Hudish LI, Guzman BB et al (2020) FMRP promotes RNA localization to neuronal projections through interactions between its RGG domain and G-quadruplex RNA sequences. Elife 9:e52621

Guo JU, Bartel DP (2016) RNA G-quadruplexes are globally unfolded in eukaryotic cells and depleted in bacteria. Science 353:aaf5371

Haldar S, Zhang Y, Xia Y et al (2022) Mechanistic Insights into the ligand-induced unfolding of an RNA G-Quadruplex. J Am Chem Soc 144:935–950

Hanna R, Flamier A, Barabino A et al (2021) G-quadruplexes originating from evolutionary conserved L1 elements interfere with neuronal gene expression in Alzheimer's disease. Nat Commun 12:1828

Hansel-Hertsch R, Simeone A, Shea A et al (2020) Landscape of G-quadruplex DNA structural regions in breast cancer. Nat Genet 52:878–883

Hon J, Martinek T, Zendulka J et al (2017) pqsfinder: an exhaustive and imperfection-tolerant search tool for potential quadruplex-forming sequences in R. Bioinformatics 33:3373–3379

Huang H, Suslov NB, Li NS et al (2014) A G-quadruplex-containing RNA activates fluorescence in a GFP-like fluorophore. Nat Chem Biol 10:686–691

Huang H, Zhang J, Harvey SE et al (2017) RNA G-quadruplex secondary structure promotes alternative splicing via the RNA-binding protein hnRNPF. Genes Dev 31:2296–2309

Jansson LI, Hentschel J, Parks JW et al (2019) Telomere DNA G-quadruplex folding within actively extending human telomerase. Proc Natl Acad Sci USA 116:9350–9359

Jeng SCY, Trachman RJ 3rd, Weissenboeck F et al (2021) Fluorogenic aptamers resolve the flexibility of RNA junctions using orientation-dependent FRET. RNA 27:433–444

Jo S, Kim T, Iyer VG et al (2008) CHARMM-GUI: a web-based graphical user interface for CHARMM. J Comput Chem 29:1859–1865

Joachimi A, Benz A, Hartig JS (2009) A comparison of DNA and RNA quadruplex structures and stabilities. Bioorg Med Chem 17:6811–6815

Karsisiotis AI, O'Kane C, Webba Da Silva M (2013) DNA quadruplex folding formalism–a tutorial on quadruplex topologies. Methods 64:28–35

Khan N, Kolimi N, Rathinavelan T (2015) Twisting right to left: A...A mismatch in a CAG trinucleotide repeat overexpansion provokes left-handed Z-DNA conformation. PLoS Comput Biol 11:e1004162

Kharel P, Balaratnam S, Beals N et al (2020a) The role of RNA G-quadruplexes in human diseases and therapeutic strategies. Wiley Interdiscip Rev RNA 11:e1568

Kharel P, Becker G, Tsvetkov V et al (2020b) Properties and biological impact of RNA G-quadruplexes: from order to turmoil and back. Nucleic Acids Res 48:12534–12555

Kikin O, D'antonio L, Bagga PS (2006) QGRS Mapper: a web-based server for predicting G-quadruplexes in nucleotide sequences. Nucleic Acids Res 34:W676–W682

Klimentova E, Polacek J, Simecek P et al (2020) PENGUINN: Precise exploration of nuclear G-Quadruplexes using interpretable neural networks. Front Genet 11:568546

Koirala D, Shelke SA, Dupont M et al (2018) Affinity maturation of a portable Fab-RNA module for chaperone-assisted RNA crystallography. Nucleic Acids Res 46:2624–2635

Kosiol N, Juranek S, Brossart P et al (2021) G-quadruplexes: a promising target for cancer therapy. Mol Cancer 20:40

Krepl M, Zgarbova M, Stadlbauer P et al (2012) Reference simulations of noncanonical nucleic acids with different chi variants of the AMBER force field: quadruplex DNA, quadruplex RNA and Z-DNA. J Chem Theory Comput 8:2506–2520

Kumari S, Bugaut A, Huppert JL et al (2007) An RNA G-quadruplex in the 5' UTR of the NRAS proto-oncogene modulates translation. Nat Chem Biol 3:218–221

Kumari S, Bugaut A, Balasubramanian S (2008) Position and stability are determining factors for translation repression by an RNA G-quadruplex-forming sequence within the 5' UTR of the NRAS proto-oncogene. Biochemistry 47:12664–12669

Kwok CK, Marsico G, Sahakyan AB et al (2016) rG4-seq reveals widespread formation of G-quadruplex structures in the human transcriptome. Nat Methods 13:841–844

Kypr J, Kejnovska I, Renciuk D et al (2009) Circular dichroism and conformational polymorphism of DNA. Nucleic Acids Res 37:1713–1725

Laguerre A, Hukezalie K, Winckler P et al (2015) Visualization of RNA-Quadruplexes in Live Cells. J Am Chem Soc 137:8521–8525

Lee DSM, Ghanem LR, Barash Y (2020) Integrative analysis reveals RNA G-quadruplexes in UTRs are selectively constrained and enriched for functional associations. Nat Commun 11:527

Liu H, Chu Z, Yang X (2022) A Key Molecular Regulator, RNA G-quadruplex and Its Function in Plants. Front Plant Sci 13:926953

Lyu K, Chow EY, Mou X et al (2021) RNA G-quadruplexes (rG4s): genomics and biological functions. Nucleic Acids Res 49:5426–5450

Malgowska M, Czajczynska K, Gudanis D et al (2016) Overview of the RNA G-quadruplex structures. Acta Biochim Pol 63:609–621

Mashima T, Nishikawa F, Kamatari YO et al (2013) Anti-prion activity of an RNA aptamer and its structural basis. Nucleic Acids Res 41:1355–1362

Mergny JL, Lacroix L (2009) UV Melting of G-Quadruplexes. Curr Protoc Nucleic Acid Chem Chapter 17:Unit 17.1

Metifiot M, Amrane S, Litvak S et al (2014) G-quadruplexes in viruses: function and potential therapeutic applications. Nucleic Acids Res 42:12352–12366

Miclot T, Hognon C, Bignon E et al (2021) Structure and Dynamics of RNA Guanine Quadruplexes in SARS-CoV-2 genome. original strategies against emerging viruses. J Phys Chem Lett 12:10277–10283

Mieczkowski M, Steinmetzger C, Bessi I et al (2021) Large Stokes shift fluorescence activation in an RNA aptamer by intermolecular proton transfer to guanine. Nat Commun 12:3549

Miglietta G, Russo M, Capranico G (2020) G-quadruplex-R-loop interactions and the mechanism of anticancer G-quadruplex binders. Nucleic Acids Res 48:11942–11957

Milovanovic B, Stanojevic A, Etinski M et al (2020) Intriguing intermolecular interplay in guanine quartet complexes with alkali and alkaline earth cations. J Phys Chem B 124:3002–3014

Mishra SK, Tawani A, Mishra A et al (2016) G4IPDB: A database for G-quadruplex structure forming nucleic acid interacting proteins. Sci Rep 6:38144

Miskiewicz J, Sarzynska J, Szachniuk M (2021) How bioinformatics resources work with G4 RNAs. Brief Bioinform 22(3):bbaa201

Muller D, Bessi I, Richter C et al (2021) The folding landscapes of human telomeric RNA and DNA G-quadruplexes are markedly different. Angew Chem Int Ed Engl 60:10895–10901

Ni S, Zhuo Z, Pan Y et al (2021) Recent progress in aptamer discoveries and modifications for therapeutic applications. ACS Appl Mater Interfaces 13:9500–9519

Paeschke K, Simonsson T, Postberg J et al (2005) Telomere end-binding proteins control the formation of G-quadruplex DNA structures in vivo. Nat Struct Mol Biol 12:847–854

Pandey S, Agarwala P, Maiti S (2013) Effect of loops and G-quartets on the stability of RNA G-quadruplexes. J Phys Chem B 117:6896–6905

Paramasivan S, Rujan I, Bolton PH (2007) Circular dichroism of quadruplex DNAs: applications to structure, cation effects and ligand binding. Methods 43:324–331

Patro LPP, Kumar A, Kolimi N et al (2017) 3D-NuS: A Web server for automated modeling and visualization of non-canonical 3-dimensional nucleic acid structures. J Mol Biol 429:2438–2448
Phillips JC, Hardy DJ, Maia JDC et al (2020) Scalable molecular dynamics on CPU and GPU architectures with NAMD. J Chem Phys 153:044130
Puig Lombardi E, Londono-Vallejo A (2020) A guide to computational methods for G-quadruplex prediction. Nucleic Acids Res 48:1–15
Reddy K, Zamiri B, Stanley SYR et al (2013) The disease-associated r(GGGGCC)n repeat from the C9orf72 gene forms tract length-dependent uni- and multimolecular RNA G-quadruplex structures. J Biol Chem 288:9860–9866
Rouleau S, Glouzon JS, Brumwell A et al (2017) 3' UTR G-quadruplexes regulate miRNA binding. RNA 23:1172–1179
Sahakyan AB, Chambers VS, Marsico G et al (2017) Machine learning model for sequence-driven DNA G-quadruplex formation. Sci Rep 7:14535
Sathyaseelan C, Vijayakumar V, Rathinavelan T (2021) CD-NuSS: A web server for the automated secondary structural characterization of the nucleic acids from circular dichroism spectra using extreme gradient boosting decision-tree, neural network and kohonen algorithms. J Mol Biol 433:166629
Shao X, Zhang W, Umar MI et al (2020) RNA G-Quadruplex Structures Mediate Gene Regulation in Bacteria. mBio 11:e02926–19
Siddiqui-Jain A, Grand CL, Bearss DJ et al (2002) Direct evidence for a G-quadruplex in a promoter region and its targeting with a small molecule to repress c-MYC transcription. Proc Natl Acad Sci USA 99:11593–11598
Song J, Perreault JP, Topisirovic I et al (2016) RNA G-quadruplexes and their potential regulatory roles in translation. Translation (austin) 4:e1244031
Subramanian M, Rage F, Tabet R et al (2011) G-quadruplex RNA structure as a signal for neurite mRNA targeting. EMBO Rep 12:697–704
Takahama K, Miyawaki A, Shitara T et al (2015) G-Quadruplex DNA- and RNA-specific-binding proteins engineered from the RGG domain of TLS/FUS. ACS Chem Biol 10:2564–2569
Tassinari M, Richter SN, Gandellini P (2021) Biological relevance and therapeutic potential of G-quadruplex structures in the human noncoding transcriptome. Nucleic Acids Res 49:3617–3633
Tateishi-Karimata H, Sugimoto N (2021) Roles of non-canonical structures of nucleic acids in cancer and neurodegenerative diseases. Nucleic Acids Res 49:7839–7855
Vannutelli A, Belhamiti S, Garant JM et al (2020) Where are G-quadruplexes located in the human transcriptome? NAR Genom Bioinform 2:lqaa035
Varizhuk A, Ischenko D, Tsvetkov V et al (2017) The expanding repertoire of G4 DNA structures. Biochimie 135:54–62
Wang SK, Wu Y, Ou TM (2015) RNA G-quadruplex: The new potential targets for therapy. Curr Top Med Chem 15:1947–1956
Wanrooij PH, Uhler JP, Shi Y et al (2012) A hybrid G-quadruplex structure formed between RNA and DNA explains the extraordinary stability of the mitochondrial R-loop. Nucleic Acids Res 40:10334–10344
Webba Da Silva M (2007) Geometric formalism for DNA quadruplex folding. Chemistry 13:9738–9745
Weng X, Gong J, Chen Y et al (2020) Keth-seq for transcriptome-wide RNA structure mapping. Nat Chem Biol 16:489–492
Xiao S, Zhang J-Y, Zheng K-W et al (2013) Bioinformatic analysis reveals an evolutional selection for DNA:RNA hybrid G-quadruplex structures as putative transcription regulatory elements in warm-blooded animals. Nucleic Acids Res 41:10379–10390
Xu Z, Poidevin M, Li X et al (2013) Expanded GGGGCC repeat RNA associated with amyotrophic lateral sclerosis and frontotemporal dementia causes neurodegeneration. Proc Natl Acad Sci USA 110:7778–7783
Yang SY, Lejault P, Chevrier S et al (2018) Transcriptome-wide identification of transient RNA G-quadruplexes in human cells. Nat Commun 9:4730

Yang X, Cheema J, Zhang Y et al (2020) RNA G-quadruplex structures exist and function in vivo in plants. Genome Biol 21:226
Yeung PY, Zhao J, Chow EY et al (2019) Systematic evaluation and optimization of the experimental steps in RNA G-quadruplex structure sequencing. Sci Rep 9:8091
Yu H, Qi Y, Yang B et al (2022) G4Atlas: a comprehensive transcriptome-wide G-quadruplex database. Nucleic Acids Res Oct 16;gkac896
Zhao C, Qin G, Niu J et al (2021) Targeting RNA G-Quadruplex in SARS-CoV-2: A promising therapeutic target for COVID-19? Angew Chem Int Ed Engl 60:432–438
Zhou ZD, Jankovic J, Ashizawa T et al (2022) Neurodegenerative diseases associated with non-coding CGG tandem repeat expansions. Nat Rev Neurol 18:145–157
Zok T, Kraszewska N, Miskiewicz J et al (2022) ONQUADRO: a database of experimentally determined quadruplex structures. Nucleic Acids Res 50:D253–D258

Methods to Analyze Post-transcriptional Modifications Applied to Stable RNAs in *Staphylococcus aureus*

Roberto Bahena-Ceron, Jose Jaramillo-Ponce, Hiroki Kanazawa, Laura Antoine, Philippe Wolff, Virginie Marchand, Bruno P. Klaholz, Yuri Motorin, Pascale Romby, and Stefano Marzi

Contents

R. Bahena-Ceron and J. Jaramillo-Ponce contributed equally in this chapter.

R. Bahena-Ceron · J. Jaramillo-Ponce · H. Kanazawa · L. Antoine · P. Wolff · P. Romby · S. Marzi (✉)
Architecture Et Réactivité de L'ARN, CNRS, Université de Strasbourg, 9002 Strasbourg, France
e-mail: s.marzi@ibmc-cnrs.unistra.fr

V. Marchand · Y. Motorin
Université de Lorraine, CNRS, IMoPA (UMR7365), 54000 Nancy, France

Université de Lorraine, CNRS, INSERM, IBSLor (UAR2008/US40), F54000 Nancy, France

B. P. Klaholz
Centre for Integrative Biology (CBI), Department of Integrated Structural Biology, IGBMC (Institute of Genetics and of Molecular and Cellular Biology), 1 Rue Laurent Fries, Illkirch, France

Centre National de La Recherche Scientifique (CNRS) UMR 7104, Illkirch, France

Institut National de La Santé Et de La Recherche Médicale (Inserm) U964, Illkirch, France

Université de Strasbourg, Strasbourg, France

J. Barciszewski (ed.), *RNA Structure and Function*, RNA Technologies 14,
https://doi.org/10.1007/978-3-031-36390-0_11

Abstract RNA modifications contribute to the various functions of RNAs in all living organisms. Some of these modifications are dynamic and contribute to the regulation of gene expression. In bacteria, their roles in stress, environmental adaptation, and in infections caused by pathogens have been recently fully recognized. In this review, we describe several methodologies including mass spectrometry, next-generation RNA sequencing methods, biochemical approaches, and cryo-EM structural analysis that are used to detect and localize the modifications in tRNAs and rRNAs. We illustrate how the combination of methods was necessary to avoid technical biases for a successful mapping of the modifications in tRNAs and rRNAs in *Staphylococcus aureus*.

Keywords RNA modifications · Staphylococcus aureus · tRNA · rRNA · Mass Spectrometry · RNA-seq · Cryo-EM structure

1 Introduction

RNA molecules are chemically modified at different positions of the nitrogenous bases or the ribose cycle. These modifications are common to all domains of life and are present in all RNA classes, including stable non-coding RNAs (tRNAs, rRNAs, and some regulatory RNAs) and generally less stable mRNAs. They involve covalent alterations (i.e., methylations, acetylation, and deamination) or isomerization (i.e., pseudouridines) of nucleotides (Fig. 1a). To date, over 170 modifications have been found in RNAs from different organisms, with the most common modification being methylation (see MODOMICS database) (Boccaletto et al. 2022). Overall, they influence chemical and physical properties of RNAs with consequences on their stability, structure, recognition, transport, and cellular localization (Frye et al. 2016). Specific enzymes are responsible for their deposition and removal, providing a large degree of variability (de Crécy-Lagard and Jaroch 2020; Ishitani et al. 2008). These modulations add a sophisticated regulatory layer in gene expression and often connect external stimuli to various cellular processes. This is particularly the case for bacteria, which need to rapidly adapt their growth to environmental stresses, to the host responses, to nutrient availability, and to antimicrobials or dangerous chemical reactive species.

Epitranscriptomic regulation often targets the most modified RNA species, tRNAs and rRNAs, influencing the translation process. Both codon decoding and peptide bond formation are affected by specific modifications on tRNAs and rRNAs, which influence translation rates and accuracy. Consequently, they directly impact protein yields, their proper folding, and their activity (Samatova et al. 2020; Antoine et al.

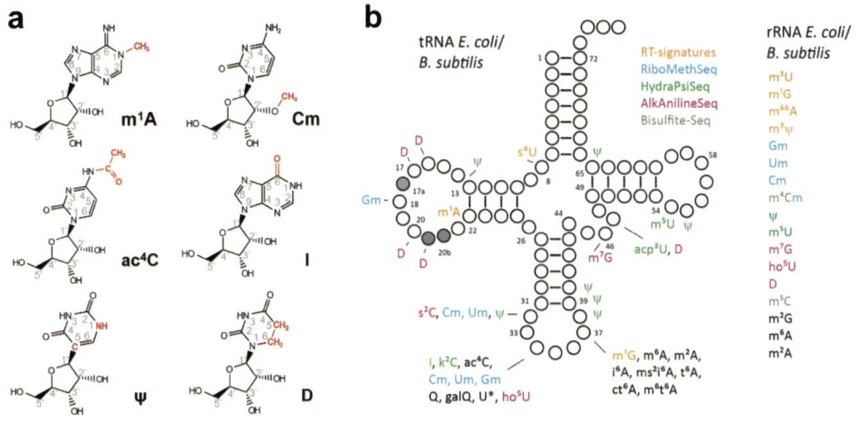

Fig. 1 Ribonucleoside modifications commonly found in bacteria. **a** Chemical structures of some modified ribonucleosides identified in *S. aureus* tRNAs using mass spectrometry (Antoine et al. 2019). The modification site is highlighted in red and the numbering systems for the nucleobases and ribose rings are indicated. m^1A: N^1-methyladenosine, Cm: 2′-O-methylcytidine, ac^4C: N^4-acetylcytidine, I: Inosine, ψ: Pseudouridine, D: Dihydrouridine. **b** Modified ribonucleosides in rRNAs and tRNAs from *E. coli* and *B. subtilis* identified by deep sequencing methods. The position of modifications are depicted in the tRNA cloverleaf structure, and modifications found in rRNAs are listed. Modified nucleosides are colored according to the RNA-seq protocol used for their detection (see Sect. 5). Modifications in black have been assigned with RNA-seq protocols not described. s^4U: 4-thiouridine, Gm: 2′-O-methylguanosine, Um: 2′-O-methyluridine, s^2C: 2-thiocytidine, k^2C: 2-lysidine, ho^5U: 5-hydroxyuridine, U*: hypermodified 2- and 5-substituted U34 residues, galQ: Galactosyl-queuosine, Q: Queuosine, ct^6A: cyclic N^6-threonylcarbamoyladenosine, m^6t^6A: N^6-methyl-threonyl-N^6-threonylcarbamoyladenosine, i^6A: N^6-isopentyladenosine, ms^2i^6A: 2-methylthio-N^6-isopentyladenosine, t^6A: N^6-threonylcarbamoyladenosine, m^1G: N^1-methylguanosine, m^6A: N^6-methyladenosine, m^2A: 2-methyladenosine, m^7G: N^7-methylguanosine, acp^3U: 3-(3-amino-3-carboxypropyl)uridine, m^5U: 5-methyluridine, m^3U: N^3-methyluridine, $m^3ψ$: N^3-methylpseudouridine, m^4Cm: N^4-, 2′-O-dimethylcytidine, m^5C: 5-methylcytidine, m^2G: N^2-methylguanosine

2021). Translation regulation fine tunes protein synthesis to balance the relative abundance of proteins in the same pathways or to provide the correct stoichiometry of complexes (Li et al. 2014). In addition, translational control is well appropriate when fast adaptation is required (Tollerson and Ibba 2020; Duval et al. 2015). In different stress conditions, modulation of RNA modifications can be rapidly achieved by changing the level of the modification enzymes or by affecting their activity.

Stable non-coding RNAs, rRNAs, and tRNAs, which represent in bacteria >95% of the total RNA fraction are extensively modified during their maturation. Despite decades of studies, the comprehensive and reliable profiling of tRNA and rRNA modifications was accomplished only for a few model species. For bacterial species, such comprehensive profiling was performed for *Escherichia coli* as a prominent representative of Gram-negative bacteria (see MODOMICS) and rather recently, for *Bacillus subtilis* (de Crecy-Lagard et al. 2020) and *Staphylococcus aureus* (Antoine et al.

2019) as representative Gram-positive species. Despite the fact that MODOMICS and tRNAdb (Jühling et al. 2009) databases show that all known bacterial tRNAs have numerous common modifications, they also revealed specificity among bacterial species. Dihydrouridine (D), pseudouridine (ψ), 5-methyluridine (m^5U/T), and 7-methylguanosine (m^7G) are highly common, while 2′-O-methylations (Nm), 4-thiouridine (s^4U), and anticodon loop modifications are rarer, and others are unique for specific tRNA species (Fig. 1b). Although Gram-negative and Gram-positive bacterial species share a number of common tRNA modifications, characteristics were identified in the D-arm, e.g., Gm18 is only found in Gram-negative bacteria, while m^1A22 is primarily observed in several Gram-positive bacteria. In contrast to tRNAs, rRNA modification profiles are less diverse among bacteria. However, the exact modification position is not necessarily conserved between different bacterial species but a similar modification is often present in the same rRNA structural domain (Piekna-Przybylska et al. 2008). Hence, it is important to gain more knowledge on the evolution of RNA modifications across bacterial phylogeny and analyze the structural environment of chemical modifications to obtain insights into their role and mode of interaction with neighboring residues.

Accurate detection of RNA modifications is still an extremely difficult task, as today there is still not a unique method that provides the whole pattern of RNA modifications. These approaches include various chromatographies (i.e., TLC, HPLC), reverse transcriptase analyses, affinity gel electrophoresis, and more sophisticated methods including liquid chromatography-tandem mass spectrometry analyses of nucleosides or RNA fragments (LC-MS/MS), and deep sequencing approaches (Schaefer et al. 2017; Wetzel and Limbach 2016; Motorin and Helm 2022; Yoluç et al. 2021). Recent nanopore technologies can potentially provide access to multiple modifications in native RNA molecules (Garalde et al. 2018; Begik et al. 2021). Even if these methods offer an extraordinary amount of information and depth, they nevertheless suffer from limitations and biases.

Here we describe the methodologies that were used to determine the modification patterns of stable RNAs (tRNAs and rRNAs) from *Staphylococcus aureus*, which is an opportunistic human bacterial pathogen causing a large variety of infections. We will especially underline the power of combining complementary methodologies.

2 RNA Purification

2.1 Isolation of Total RNA from S. Aureus

Sample requirements differ depending on the type of technique used for detecting RNA modifications. High amounts of homogenous RNAs are often required for successful analyses and appropriate methods for RNA purification should be optimized. A minimal number of purification steps following efficient cell lysis and RNA stabilization is crucial for correct RNA yield, purity, and integrity. The main challenge in RNA extraction from *S. aureus* is the disruption of the thick peptidoglycan cell wall. Successful approaches use enzymatic treatment with lysostaphin (Bastos et al. 2010), mechanical bead-beating, or a combination of both methods in the presence of protein denaturants to inhibit RNase activities and abolish RNA-protein interactions (i.e., phenol, guanidine isothiocyanate) (Atshan et al. 2012; Beltrame et al. 2015; França et al. 2011). *S. aureus* RNA samples are routinely prepared with the FastRNA® Pro blue kit (MP Biomedical), which uses mechanical bead-beating of bacteria suspended in the commercial solution designed to prevent RNA degradation. For large-scale preparations (up to 500 mL of culture in logarithmic phase), bead-beating lysis of *S. aureus*, suspended in a 1:1 mixture of aqueous buffer and acidic phenol/chloroform/isoamyl alcohol, is performed. RNA is further purified by chloroform extraction and ethanol precipitation.

2.2 *Purification of rRNAs and tRNAs*

As total RNA samples contain a mixture of different species (rRNAs, tRNAs, mRNAs, sRNAs, etc.), enrichment of the RNA molecule of interest is often beneficial for the analysis of modifications. Several approaches are based on size selection of RNAs (Poulson 1973). A rapid strategy is the use of specific filters (cutoff 50 kDa) to enrich for rRNAs (retentate) or tRNAs (filtrate) after ultrafiltration. Depending on the specific analysis to be performed on the RNAs, more laborious purification methods might be required. To purify small amount of individual 16S and 23S rRNAs as well as tRNAs, *S. aureus* RNA species can be separated by polyacrylamide gel electrophoresis (PAGE) and subsequently eluted from the excised bands (electroelution or "crush and soak") (Meyer and Masquida 2016; Petrov et al. 2013). To get larger amounts of RNAs and to avoid gel impurities, chromatographies performed under native conditions are preferred (Kanwal and Lu 2019). For instance, size-exclusion chromatography can be used to purify the different rRNAs (23S, 16S, and 5S) and the tRNA fractions (Chionh et al. 2013; McKenna et al. 2007). Alternatively, weak anion-exchange chromatography using DEAE (diethylaminoethyl) matrix is widely used to separate tRNAs and rRNAs, which elute at low and high salt concentration, respectively (Easton et al. 2010). Finally, to analyze modification profiles of rRNA incorporated in fully matured and actively translating ribosomes, rRNAs (23S, 16S, and 5S) can be recovered by phenol-chloroform extraction of purified *S. aureus* total ribosomes (Khusainov et al. 2016) or polysomes samples, obtained from sucrose gradients fractionation (Brielle et al. 2017).

2.3 *Purification of Individual tRNAs*

Individual tRNAs can be purified from total tRNA using biotinylated oligo DNA probes which specifically hybridize to the target tRNA molecule. Yokogawa and co-workers (Yokogawa et al. 2010) optimized this method for thermostable tRNAs and we adapted it for the purification of *S. aureus* tRNAs. The procedure consists of four steps: (i) design and preparation of the DNA probe, (ii) probe-resin immobilization, (iii) tRNA-probe hybridization, and (iv) tRNA isolation. The DNA probe is composed of 35–40 nucleotides and contains biotin at the 5′ or 3′ end. Its sequence is usually complementary to the region between the D-arm and the anticodon loop. The biotinylated probe is immobilized on a streptavidin Sepharose resin using standard procedures. Hybridization of target tRNA and DNA probe requires denaturation of tRNA under conditions that maintain tRNA:DNA interaction. The buffer containing tetra-ethylammonium chloride has been reported as the best to destabilize the tRNA tertiary structure and to enhance the interaction between tRNA and DNA probes (Yokogawa et al. 2010). After hybridization, the remaining non-target tRNAs are removed by washing the resin with a non-denaturing buffer. Then, the desired tRNA is eluted with the non-denaturing buffer pre-heated at 60 °C. The quality of the purification is assessed by denaturing PAGE and the identity of the tRNA is confirmed by mass spectrometry (MS). Depending on the yield of the purification, eventual contaminations can be removed by gel purification or ion-exchange chromatography. The purification method described above is efficient for most tRNAs, but further optimization is often required for tRNA isoacceptors, poorly transcribed tRNAs, and tRNAs with high GC content. Methods such as reciprocal circulating chromatography (RCC) (Miyauchi et al. 2007) and chaplet column chromatography (CCC) (Suzuki and Suzuki 2007) have been optimized to increase the yield of GC-rich tRNAs (Ohira et al. 2022).

3 Mass Spectrometry to Identify and Localize Modifications in RNAs

Liquid chromatography coupled to tandem mass spectrometry (LC-MS/MS) is the most trusted method for direct and unambiguous detection of RNA modifications, as it provides precise measurements of molecular masses (accuracy < 1 Da) even from very low amounts of material (Giessing and Kirpekar 2012; Jora et al. 2019). In this setup, the RNA sample is separated by reverse-phase high-performance liquid chromatography (RP-HPLC), and the eluting species are directly injected into a mass spectrometer equipped with two analyzers (Thomas and Akoulitchev 2006). An electrospray ionization (ESI) source (Kebarle and Tang 1993) evaporates and ionizes the analyte, and the formed (parent) ions are separated in the first analyzer according to their mass-to-charge (*m/z*) ratio. Ions with a defined *m/z* are selected for collision-induced dissociation (CID) (Ty et al. 2008) and the resulting fragments (daughter ions) are resolved in the second mass analyzer. The recorded MS and MS/MS spectra (plot of *m/z* values and their relative abundance) are used to depict mass shifts and fragmentation profiles that evidence the presence of additional chemical groups attached to the canonical nucleotides (Thakur et al. 2021).

Two types of complementary LC-MS/MS analyses are performed for the complete mapping of modifications in *S. aureus* RNAs. In the first analysis, the RNA of interest is hydrolyzed to its nucleoside (sugar+nucleobase) building blocks to identify all resident modifications (Pomerantz and McCloskey 1990). In the second, the RNA is digested with a nucleobase-specific RNase, and the sequences of the resulting oligonucleotides including modifications are determined and placed within the full-length RNA sequence (available from genomic data) (Kowalak et al. 1993). Both analyses require previous purification of the target RNA to obtain reliable data, limiting this "bottom up" mapping strategy to the analysis of one RNA at a time.

3.1 Analysis of Nucleosides

Samples for nucleoside analysis are prepared through enzymatic treatment of the RNA with nuclease P1, snake venom phosphodiesterase I, and bacterial alkaline phosphatase (Crain 1990). However, the reaction is performed under sub-optimal mildly acidic conditions (pH 5) (Wolff et al. 2020) to avoid alteration of labile modifications at higher pH (Jora et al. 2021; Miyauchi et al. 2013). Removal of phosphate groups is critical to make the nucleosides sufficiently hydrophobic to be separated on a reverse-phase column. Canonical nucleosides are the most abundant species and readily resolve into well-defined peaks, eluting according to their hydrophobicity (C < U < G < A) (Gehrke and Kuo 1990). Modified nucleosides are less abundant and they often appear earlier (e.g., D, ψ) or later (e.g., m^2G, t^6A) than their unmodified counterparts depending on the chemical nature of the modification (Su et al. 2014). Positional isomers (e.g., m^1A, m^6A, m^2A) have the same mass and cannot be distinguished by tandem MS analysis; however, they exhibit different retention times that facilitate their identification when standards are available (Jora et al. 2018).

RP-HPLC of nucleosides is performed with mobile phases (A: water, B: methanol) containing 0.1% formic acid to facilitate protonation of nucleobases during ESI (Cai et al. 2015). MS analysis is performed in positive mode using a triple quadrupole (QQQ) instrument consisting of two quadrupole mass analyzers separated by a collision cell. In most cases, CID of the nucleoside ion occurs at the N-glycosidic bond, resulting in a nucleobase ion and an uncharged ribose moiety (Ross et al. 2016). The recorded *m/z* values of parent (MS) and daughter ions (MS/MS) as well as the associated neutral loss of unmodified (132 Da) or 2′-O-methylated (146 Da) ribose enable identification of the modified nucleoside. Pseudouridine does not contain an N-glycosidic bond and shows a different but unique fragmentation profile allowing its identification (Dudley et al. 2005). Two strategies of data acquisition can be employed depending on the aim of the experiment. Dynamic multiple reaction monitoring

(DMRM) allows the identification and quantification of a defined set of modifications by restraining detection to only those specific *m/z* values (Thuring et al. 2016). In contrast, neutral loss scan (NLS) is used to identify the full range of modified nucleosides in the sample including novel modifications and consists in detecting all precursors ions that lose a mass equivalent to unmodified or methylated ribose (Kellner et al. 2014).

3.2 Analysis of Oligonucleotides

As RNase digestion of tRNA mixtures or long rRNAs often produces fragments with the same sequence, purification of the target molecule is critical to unambiguously map modifications by oligonucleotide analysis. Most of all individual tRNAs of *S. aureus* can be isolated using two-dimensional polyacrylamide gel electrophoresis (2D-PAGE), in which tRNAs are first separated by their length on a denaturing gel, and then by their conformation on a semi-denaturing gel (Fig. 2a) (Antoine and Wolff 2020; Antoine et al. 2019). For rRNAs, specific sequences can be obtained by RNase H cleavage using DNA probes flanking the region of interest followed by PAGE purification (Hansen et al. 2002; Nakai et al. 1994). In both cases, digestion is performed directly “in-gel” using RNase T1 (cleaving after G, m^2G and I leaving a 3’ phosphate end) and the resulting oligonucleotides elute passively from the gel (Taoka et al. 2010). Other RNases, such as RNase A (cleaving after pyrimidines and ψ and leaving a 3’ phosphate end) and RNase U2 (cleaving after purines, preferably A, leaving a 3’ phosphate end), are frequently used in separated analyses to increase the sequence coverage, particularly in G-rich regions (Houser et al. 2015).

The negatively charged backbone of oligonucleotides raises some challenges for LC-MS/MS analysis. A concern in sample preparation is the presence of alkali cations (Na^+, K^+), which must be replaced by “volatile” NH_4^+ cations (Stults et al. 1991). Otherwise, multiple salt adducts are formed upon ESI, leading to peak broadening and interferences that complicate the interpretation of spectra (Bleicher and Bayer 1994). In-gel digests are routinely desalted using “ZipTip” C18 reverse-phase pipette tips (Millipore), but other methods such as RNA precipitation with ethanol and ammonium acetate, cation exchange chromatography, and micro-dialysis can also be used (Castleberry et al. 2008). Due to the high hydrophilicity of phosphate groups, RP-HPLC of oligonucleotides requires a mobile phase containing ion-pairing agents such as triethylamine (TEA) and triethylammonium acetate (TEAA) (Apffel et al. 1997). The ammonium moiety of these molecules masks the charges in the oligonucleotide backbone, while the alkyl group provides hydrophobic surfaces for its interaction with the reverse-phase column. Hexafluoroisopropanol (HFIP) is also included in the mobile phase to improve the ionization of oligonucleotides (McGinnis et al. 2013), which occurs in negative mode (deprotonation). Because TEA/TEAA/HFIP contaminations are difficult to eliminate and interfere with proteomic analyses performed in positive mode (Wetzel and Limbach 2016), it is better to use a dedicated LC-MS/MS system for oligonucleotides analysis.

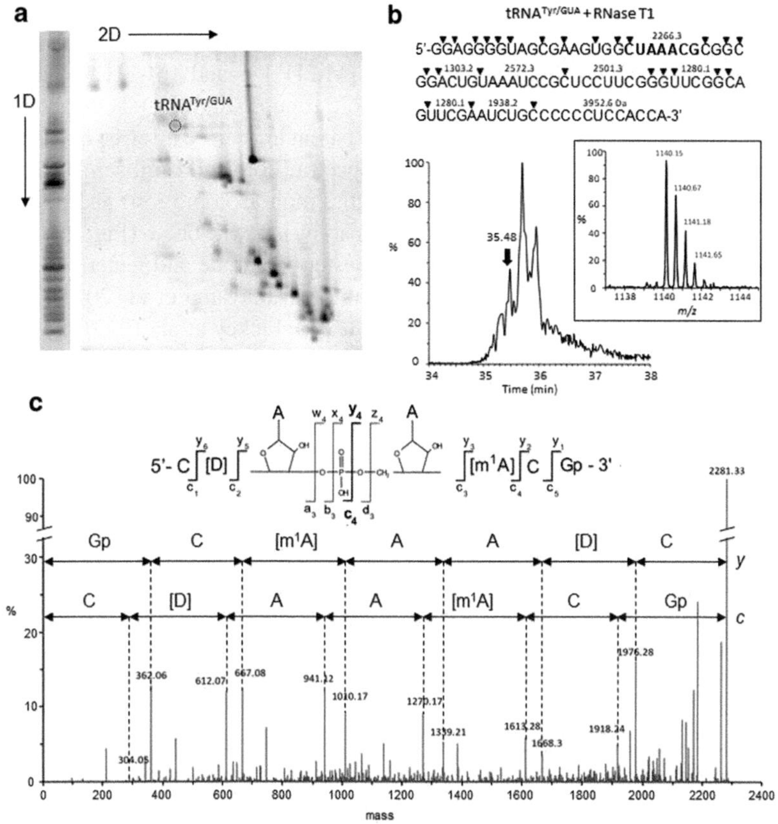

Fig. 2 Mapping of tRNA modifications by LC-MS/MS analysis of RNase digests. **a** Isolation of individual tRNAs by 2D-PAGE. Total tRNAs (10 μg) are first separated by size on a denaturing polyacrylamide gel (left). Then, the whole lane is excised and embedded on the top of a semi-denaturing gel. After electrophoretic separation, the different bands migrate into multiple spots, most of which contain individual tRNA species (right). Arrows indicate the direction of migration on the first and second dimensions. Both gels are stained with ethidium bromide. The position of $tRNA^{Tyr/GUA}$ in the 2D gel is indicated. **b** LC-MS analysis of in-gel T1 RNase digests of $tRNA^{Tyr/GUA}$. The unmodified sequence of the tRNA is shown and the cleavage sites of RNase T1 are labeled with arrowheads. Theoretical masses calculated with MongoOligo (http://rna.rega.kuleuven.be/masspec/mongo.htm) are displayed above the fragments of ≥4 nucleotides. The peak pointed on the elution profile corresponds likely to the oligonucleotide CUAAACG ($m = 2267.3$ Da) in bold. Indeed, the spacing of the isotopic peaks in its MS spectrum (inset) is 1/2, indicating thus a double-charged anion $[M-2H^+]^{2-}$ with $m = 2281.3$ Da. The discrepancy between the expected and measured mass already evidences the presence of modifications. **c** Deconvoluted CID spectrum of the parent ion with $m/z = 1140.15$. Manual interpretation of the MS/MS data yielded the sequence C[D]AA[m^1A]CGp containing two modified nucleotides: dihydrouridine and a N^1-methyladenosine. Although the position of the methyl group on the adenosine cannot be determined from this analysis, it is known that Gram-positive bacteria display m^1A at position 22 of tRNA. The phosphodiester bond between the two unmodified adenosines is shown explicitly to indicate the four possible cleavages during CID. Most abundant breakages occur at the 5′ P-O bond, generating *c*- and *y*- ion series (bold) that were used for sequencing

As ESI of oligonucleotides produces several multicharged anions (Potier et al. 1994), data-dependent acquisition (DDA) is used to select the most abundant *m/z* species for CID, which are typically $[M\text{-}2H^+]^{-2}$ and $[M\text{-}3H^+]^{-3}$ ions. On high-resolution instruments, as the quadrupole time-of-flight (Q-TOF, accuracy < 0.05 Da), the charge (and thus the mass) of the parent ion can be inferred by inspecting the spacing of isotope peaks in the MS/MS spectra (Polo and Limbach 2001). If the spacing between the selected peak (usually the monoisotopic ion) and its consecutive isotope peak is $1/x$, the charge of the parent ion will be $-x$ (Fig. 2b). CID of oligonucleotides occurs along the phosphodiester backbone and generates a ladder of fragments that allow sequence determination (Nakayama et al. 2011; Ni et al. 1996). Among the four possible cleavage sites (Mcluckey et al. 1992), fragmentation of the 5′ P-O bond is the most frequent and produces a series of complementary *c*- and *y*-type fragment ions (Fig. 2c). Deconvolution of the multicharge MS/MS spectrum (*m/z*) to mono-charge spectrum (mass) simplifies the sequencing process, which consists in finding the known terminus (e.g., 3’ Gp end for RNase T1 digests). Then the masses of individual nucleotides are added to match the different peaks until the mass of the parent ion is reached (Fig. 2c). The presence of a modified nucleotide is evidenced by a unique mass shift resulting from the attached chemical group. Only pseudouridine (same mass as U) cannot be mapped with this method. A specific chemical derivatization on pseudouridine with either acrylonitrile (Mengel-Jørgensen and Kirpekar 2002) or *N*-cyclohexyl-*N′*-(2-morpholinoehyl)carbodiimide metho-*p*-toluene-sulfonate (CMCT) (Patteson et al. 2001) produces a detectable mass shift. Although oligonucleotide analysis readily locates single-methylated nucleotides within the RNA sequence, the position of the methyl group on the ribose or the nucleobase cannot be determined, and further analyses are required (Wolff et al. 2020).

4 Biochemical Analyses to Easily Map Specific RNA Modifications

4.1 Primer Extension

RNA modifications induce changes in the physicochemical properties of the nucleobases and ribose rings, including changes in their polarity (Charette and Gray 2000), base stacking (Roost et al. 2015), and coordination with water and ions (Agris 1996). All these events have a direct impact on the activity of RNA-directed DNA polymerases, which are viral enzymes commonly known as reverse transcriptases (RTs) (Potapov et al. 2018). These effects can be exploited to assign modification sites using simple laboratory settings. RNA modifications at the Watson-Crick edge, (e.g., m^1A, m^1G, $m^2{}_2G$) readily interfere with cDNA synthesis, while other modifications (e.g., 2′O-methylation) are less problematic for primer extension, but can be detected with specific protocols. For example, under sub-optimal polymerization conditions, the presence of a modification at the nucleotide base or its ribose in the RNA template induces reverse transcription stop at the +1 position to the modification site (or, rarely, at the neighboring nucleotide) (Fig. 3a, left) (Rebane et al. 2002; Maden et al. 1995). Commonly used conditions include low deoxynucleotides (dNTPs) concentration (Motorin et al. 2007). This stop is readily detected after the separation of the cDNA products by PAGE, along with RNA sequencing profile. Most commonly, primer extension is performed with a radiolabeled DNA oligonucleotide and the radioactive products are detected by autoradiography (Fig. 3a, right). RTs from avian myeloblastosis virus (AMV) and Moloney murine leukemia virus (MMLV) are extensively used. Noteworthy, RT-signatures appear typical for a given modification, but they also depend on the reverse transcriptase used. A recent comparative study, performed on a similar RNA template using several RTs, revealed that RNA modifications can be distinguished by their RT-signature (Werner et al. 2020).

Although this technique is simple and inexpensive, it is sometimes difficult to distinguish between modifications and other possible causes of RT stops, such as

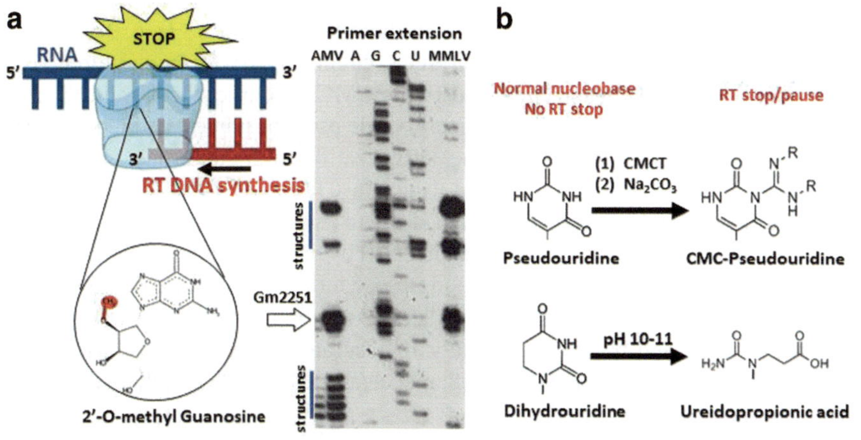

Fig. 3 Primer extension analysis to detect several RNA modifications. **a** On the right, a graphical representation explaining the principles of primer extension. Reverse transcriptase (RT, transparent light blue) stops DNA synthesis (red) at RNA modification sites, in this case at a 2′-O methyl guanosine in RNA (dark blue). On the left, an autoradiography film showing RT stop patterns induced by modifications or stable RNA structures, or unspecific cleavages. Avian myeloblastosis virus (AMV) and Moloney murine leukemia virus (MMLV) RT reactions were used at low nucleotides concentration (0.02 nM), to extend a radiolabeled DNA oligonucleotide complementary to *S. aureus* 23S rRNA. Using this procedure, the modification Gm 2251 (*E. coli* numbering) could be confirmed. **b** Chemical nucleobase derivatizations induced by the presence of pseudouridines and dihydrouridines to be detected by RT stops. Pseudouridines are modified by 1-cyclohexyl-(2-morpholinoethyl)carbodiimide metho-p-toluene sulfonate (CMCT) followed by alkali treatment with Na_2CO_3, which leaves only the CMC adduct attached to the N3 position. Dihydrouridines subjected to alkaline hydrolysis (pH 10–11) produce ureidopropionic acid after ring opening

RNA accidental breaks or presence of stable structures (Stern et al. 1988). To overcome misinterpretations, RT reactions are often conducted in parallel on in vitro transcribed RNAs, devoid of modifications, or on RNA purified from organisms in which specific RNA modification enzymes have been deleted or inactivated (Deryusheva et al. 2012).

Finally, it should be mentioned that many modifications cannot be directly detected by primer extension analysis, like pseudouridines, dihydrouridines, or 5-methylcytosines. For these modifications a specific pre-treatment of the sample prior to the primer extension is necessary (Fig. 3b).

4.2 Nucleotide Derivatization Coupled to Primer Extension Analysis

Given the specific chemical properties that modification brings to nucleotides, diverse options of derivatization inducing RT pauses/stops or nucleotide misincorporation have been developed for RT-silent modified nucleotides (Fig. 3b) (Behm-Ansmant et al. 2011). To map pseudouridines, chemical derivatization with CMCT is commonly used (Bakin and Ofengand 1993). CMCT modifies the N1 and N3 positions in guanosine and uridine, respectively, while pseudouridines are modified both at N1 and N3 positions. Removal of the derivatization from guanosine, uridine, and N1 of ψ is achieved by alkali treatment using Na_2CO_3 at pH 10 to 11. Only the N3 modification in pseudouridine is maintained due to its specific chemical environment (Ho and Gilham 1971) and can be easily detected by primer extension. Special attention is required at all steps of the labeling procedure to avoid false-positive stops. Therefore, it is important to run, in parallel, samples without treatment, or treated only with CMCT and Na_2CO_3 separately, to accurately discriminate stops induced by CMCT on pseudouridines from accidental cleavages of the RNA chain during the treatment (Adachi et al. 2019).

The main characteristic of dihydrouridine (D) is the absence of double bond between carbon 5 and 6 and the non-planar and non-aromatic nucleobase ring (Dalluge et al. 1996). The treatment under mild alkaline condition induces the disruption of the saturated nucleobase to produce beta-ureidopropionic acid (Xing et al. 2004) (Fig. 3b). This opened form of the nucleobase induces a specific RT stop during primer extension. Alternatively, the reduction with sodium borohydride (Cerutti and Miller 1967) causes also specific RT stops. Again, to avoid misinterpretation, it is always important to add controls with untreated samples (Igo-Kemenes and Zachau 1969; Xing et al. 2004).

4.3 Affinity Gels

Affinity electrophoretic methods rely on the strong interaction of specific functional groups present on proteins or nucleic acids with a specific reactant contained in the gel matrix (Nakamura and Takeo 1998). The displacement of the sample after gel migration provides signals about the presence or the absence of the RNA modification. For instance, boronate and organomercurial affinity gel electrophoresis have been widely used to identify specific tRNA modifications in various organisms (Tuorto et al. 2018).

Boronate affinity electrophoresis is based on the specific interaction between N-acryloyl-3-aminophenylboronic acid (APB) and free cis-hydroxyls of RNA. This is particularly the case of the queuosine (Q) modification, which possesses a *cis* diol group that delays the migration of the tRNAs during electrophoresis (Igloi and Kössel 1985). On the same principle, an interaction between the sulfur-modified (S) RNA with organomercurials leads to derivatized Hg-S adducts (Igloi 1988). The modified RNA with *cis* diol or mercury-sulfur can be visualized by different methods. For instance, ethidium bromide staining of the gel is often used for purified tRNA species while Northern blot analyses using 5'end labeled oligonucleotide allows the detection of one tRNA species among total tRNAs samples. Although boronate affinity gels are efficient to separate Q-containing RNA, it is noteworthy that the slower migration can also be observed for RNA species co-transcriptionally modified by NAD (Nübel et al. 2017). Nevertheless, this modification has not yet been observed in tRNAs. Finally, for organomercurial gels, different sensitivity has been reported for specific U-modified RNA, like s^4U, and s^2U (Zheng et al. 2017).

5 RNA-seq to Map Modifications on Single RNA and Complexed RNA Samples

Historically, profiling of *E. coli* tRNA modifications was first performed using classical methods of RNA modification analysis (i.e., semi-denaturing 2D RNA gels and RNA fingerprinting), and later combined with LC-MS analysis (Chakraburtty 1980; Kiesewetter et al. 1987; Maden et al. 1995; Mims et al. 1985). However, these analytical approaches are extremely laborious and require highly purified tRNAs, limiting their applicability to low-abundant tRNA species. Similarly, exhaustive profiling of rRNA modifications was hindered by the necessity of targeted rRNA fragmentation into short oligonucleotides of 50–100 nt long for further analysis (Yang et al. 2016). Although CMCT-RT mapping for pseudouridine (ψ) detection in the middle of 1980s (see Sect. 4.3) allowed precise profiling of this abundant modification in tRNAs and rRNAs, other RNA modifications remained understudied.

The development and implementation of deep sequencing mapping approaches as routine analytical protocols for RNA modification mapping was one of the major advances in the field. Major breakthroughs were made for comprehensive mRNA analysis, but also for rRNA and tRNA modification profiling (Motorin and Helm 2019; Motorin and Marchand 2021). Some adaptations had to be introduced for tRNAs characterized by stable tertiary structure and small size. These efforts allowed rapid and simultaneous profiling of different RNAs present in the same biological sample, but most of these methods are still able to detect only one given modification at a time. Analysis of low-abundant RNA species frequently requires enrichment of the RNA of interest, but this limitation does not hinder rRNA and tRNA analysis, since these species are very well represented in total RNA. In practice, fractionation of total RNA to rRNA and tRNA fractions may be beneficial to limit the required sequencing depth, but this step is optional and generally may be omitted. As already mentioned above, the presence of tRNA isodecoders and isoacceptors, which frequently differ only by one of few nucleotides, render the mapping difficult. Several approaches were suggested to alleviate this difficulty. For instance, only unambiguously mapped reads or "modification-aware" alignment algorithms can be retained (Arimbasseri et al. 2015; Behrens et al. 2021). Another possibility is to allow ambiguous or even multiple mapping for the same tRNA sequencing read. Finally, identical and highly similar tRNA sequences can be collapsed into one generic sequence used for reads' alignment (Pichot et al. 2021). This approach was proven to be efficient to reduce ambiguous mapping but showed modification biases in the case of rare tRNA species.

In addition to single-nucleotide resolution mapping, a comprehensive analysis of RNA modification profiles requires appreciation and/or quantification of the modification stoichiometry site-by-site. This goal can be achieved by LC-MS analysis (with careful calibration of appropriate standards) and by deep sequencing-based methods. Quantification with Ab-based enrichment protocols is relatively imprecise, but many chemicals-based NGS approaches determine modification stoichiometry with ~5–10% error, as it was demonstrated for the RNA Bisulfite sequencing (RBS), RiboMethSeq and HydraPsiSeq protocols. Hence, these methods are extensively applied to monitor RNA modification dynamics.

Considering the tRNA and rRNA modifications commonly found in bacterial species, the most relevant deep sequencing methods for analysis of these RNAs are AlkAnilineSeq to detect m^7G, D and ho^5C (Marchand et al. 2021a, 2018), HydraPsiSeq for pseudouridine (ψ), some 5-substituted U residues (namely, m^5U and wobble base modifications), and lysidine (k^2C) (Marchand et al. 2022, 2020) and RiboMethSeq for 2′-O-methylated (Nm) nucleotides (Marchand et al. 2016, 2017; Motorin and Marchand 2018). Other approaches include RNA bisulfite sequencing to identify m^5C (and also m^4C) and analysis of RT misincorporation signatures for Watson-Crick edge-modified nucleobases (m^1A, m^1G, etc.). These methods and their applications are discussed below.

5.1 AlkAnilineSeq

AlkAnilineSeq protocol (Marchand et al. 2018) takes advantage of RNA scission at the abasic site created by specific cleavage of the fragile N-glycosidic bond formed by certain RNA modifications. Exposure of RNA containing D, m^7G, m^3C, and ho^5C to elevated pH at high temperature leads to opening and subsequent loss of the modified nucleobase exposing RNA abasic site. Such chemical structures are known to be readily cleaved upon aniline treatment, leading to phosphodiester bond scission and liberating 5′-phosphate group at the N + 1 nucleotide relative to modification. This free 5′-phosphate is used for specific adapter ligation, providing unprecedented specificity of the detection. AlkAnilineSeq was successfully used for mapping all D and m^7G residues in tRNAs, and ho^5C which is a rather rare rRNA modification found in *E. coli* and closely related bacterial species. Specificity and sensitivity of AlkAnilineSeq are very high, low background allows detection of even substoichiometric modifications (minimum ~ 0.05 mol of modification/site) (Marchand et al. 2018, 2021a, b).

5.2 HydraPsiSeq

Detection of pseudouridines (ψ) and 5-substituted U residues is based on their resistance to hydrazine cleavage, while unmodified uridines (U) are efficiently cleaved under these conditions (Marchand et al. 2022; Peattie 1979). Thus, protected U residues in RNA may correspond to modified bases and the level of this protection is related to modification stoichiometry. Other bases in RNA (A, C, and G) are insensitive to hydrazine and cleaved only at the background level allowing the normalization of U signals. In addition, some other non-U-derived modified residues show a high sensitivity toward hydrazine cleavage, as shown for lysidine (k^2C) present in bacterial $tRNA^{Ile}$ (CAU). A similar observation was also made for m^7G, but this methylation can be also detected by AlkAnilineSeq. Since protection of the modified nucleotide may also be related to its accessibility/reactivity within the RNA structure, the proportion of false-positive hits is substantially higher for HydraPsiSeq compared to AlkAnilineSeq. Hence, it is highly recommended that pseudouridine (ψ) mapping should be further validated either by conventional CMCT-RT primer extension (Sect. 4.3) or by its high-throughput variant approach PseudoU-Seq (Schwartz et al. 2014). Recently, direct RNA sequencing by Oxford Nanopore Technology (ONT) have been proposed as a new means to detect different modifications including pseudouridine (ψ), which produce characteristic base-calling "error" signatures (Begik et al. 2021; Huang et al. 2021).

5.3 RiboMethSeq

RiboMethSeq protocol is extensively used for mapping and quantification of Nm residues in rRNAs, tRNAs, and mRNAs (Birkedal et al. 2015; Marchand et al. 2016). Detection of Nm residues is based on the resistance of the RNA phosphodiester bond following Nm residue, to cleavage induced by alkaline pH and high temperature.

These resistant phosphodiester bonds create "gaps" in random RNA cleavage profile. The depth of the "gap" corresponds to protection, which is related to the level of the 2′-O-methylation, allowing precise quantification.

As HydraPsiSeq, RiboMethSeq is measuring protection against RNA cleavage, the number of false-positive signals is relatively high. Additional validation of the modified candidates can be done using quantitative or semi-quantitative RTL-P or primer extension at low [dNTP] in standard or high-throughput version (2OMe-Seq) (Incarnato et al. 2017). When applied to *S. aureus* RNA analysis we have observed high rate of false-positive assignation for pseudouridines (ψ) by HydraPsiSeq, while Nm residues have been detected with high confidence.

5.4 RNA BisulfiteSeq

RNA bisulfite sequencing is derived from well-established DNA bisulfite sequencing and insures detection of m^5C and m^4C (m^4Cm in rRNA) in RNAs, together with precise quantification of the modification stoichiometry (Khoddami et al. 2019; Schaefer et al. 2009). Bisulfite treatment at moderate alkaline pH (to reduce RNA degradation) leads to chemical deamination of all C to U, while m^5C and m^4C are resistant to bisulfite, and thus can be detected as non-deaminated C residues. Application of the method is relatively straightforward for bacterial rRNAs, while m^5C was only rarely reported in bacterial tRNAs. Two major limitations have to be considered. First, only A, G, and U residues persist in RNAs after chemical deamination creating highly ambiguous alignment. Indeed, over 50% of RNA-derived reads may show multiple alignments to the reference and thus should be excluded. Second, deamination level of C residues in highly structured RNA regions may be incomplete and these non-deaminated Cs show up as false-positive hits (Squires et al. 2012).

5.5 Analysis of RT Signatures (RT-seq)

Analysis of RT-signatures, e.g., misincorporation/deletion and abortive cDNA synthesis was proposed for tentative mapping of RNA modifications having altered base-pairing properties, like I, m^1G, m^1A, m^3U, $m^1acp^3\psi$, and other modifications altering Watson-Crick edge. Since RNA structure or even specific RNA sequences perturb cDNA synthesis by RT, these data can be used only as tentative assignment of potential modifications. More reliable mapping data can be obtained using combinations of several RT enzymes with different properties (Werner et al. 2020), or using specific RT conditions, e.g., in the presence of Mn^{2+} ions (Kristen et al. 2020). Detection of misincorporation and RT-arrests require adapted protocols, but simple misincorporation/deletion can be directly extracted from RiboMethSeq, HydraPsiSeq, and AlkAnilineSeq raw sequencing data sets. Misincorporation and deletion profiles need to be analyzed in parallel since the signal will appear in one or another scoring system depending on the nature of RNA modification and of the RT enzyme used. For bacterial tRNAs, these protocols allow reliable detection of s^4U8, m^1A22, I34, m^1G37, and of many D and m^7G residues.

Combination of the deep sequencing protocols cited above allows, in principle, almost exhaustive mapping of bacterial rRNA and tRNA modifications. Few RNA modifications (i.e., m^2G, Q, m^6A, i^6A, and its derivatives), which are unreactive toward chemical reagents and/or do not lead to particular RT-signature, can be detected by LC-MSMS (Sect. 3) or by specific protocols adapted for them (i.e., NO-Seq (Werner et al. 2021) or specific RT for m^6A (Aschenbrenner et al. 2018)).

6 Analysis of rRNA Modifications Using Cryo-EM Structural Studies

In the past decade, cryo electron microscopy (cryo-EM) has emerged as an extraordinary tool to resolve ribosomal structures at high resolution. The tremendous advancements in the development of direct electron detectors and other technical improvements including advanced image processing and structure sorting (Kühlbrandt 2014; Klaholz 2019) have allowed cryo-EM reconstructions to reach resolution levels around 3 Å and better, which permits to visualize post-transcriptional modifications on bacterial and eukaryotic—including human—ribosomes (Cottilli et al. 2022; Natchiar et al. 2017; Stojković et al. 2020; Watson et al. 2020). Nevertheless, the presence of some rRNA elements that, more dynamic than others, limits the assignment to the most stable ribosomal regions. While structures cannot always identify the chemical nature of a modification, they provide key insights into the structural environment of a given chemical modification. For example, 2′-O-methylation can create a more hydrophobic environment that favors Van der Waals contacts with hydrophobic side chains of amino acids, pseudouridines provide an additional possibility for hydrogen bonds through their N1 position, that classical uridines do not have, and some base modifications can increase π-stacking between neighboring nucleotide bases (Natchiar et al. 2018) (Fig. 4). The structural analysis, combined with the various methods for chemical identification discussed above, thus provides important information on the structural role and function of chemical modifications.

Structures of several bacterial ribosomes in complex with mRNA and tRNAs have shown the importance of conserved tRNA and rRNA modifications in functional sites of the small (30S) and large ribosomal (50S) subunits. On the 30S ribosomal subunit, where the tRNA anticodon interacts with the mRNA codon in the decoding center, different rRNA modifications have been shown to kinetically regulate the initiation process and to be involved in proofreading (Demirci et al. 2010). For instance, m^2G966 and m^4Cm1402 (*E. coli* numbering) both contribute to stabilize the interactions taking place between the tRNA, mRNA, and rRNA in the P-site (Burakovsky et al. 2012; Jenner et al. 2010; Polikanov et al. 2015). On the 50S subunit, several conserved modifications are located at the peptide exit tunnel or are

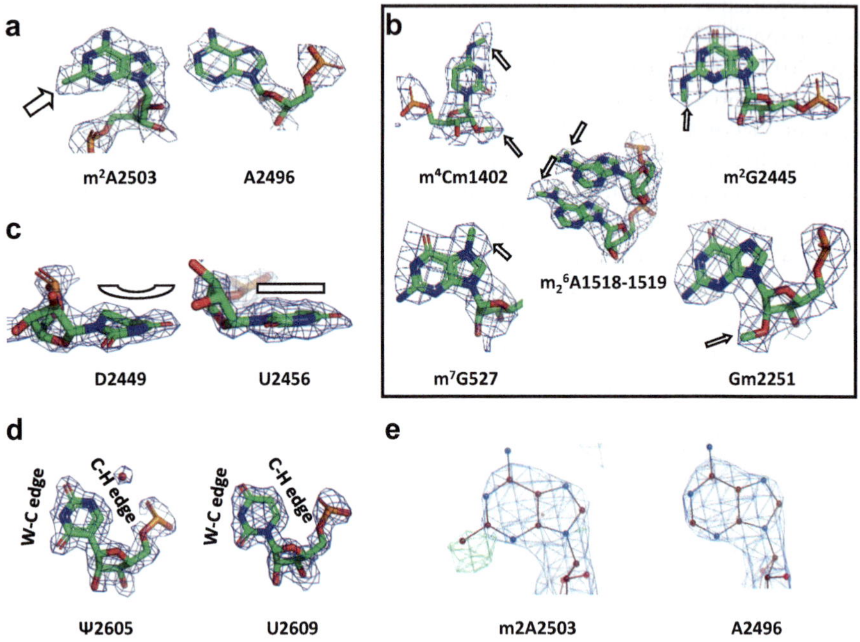

Fig. 4 rRNA modifications assigned by cryo-EM structure analyses. Nucleotides are numbered according to *E. coli* rRNA sequences. **a** *S. aureus* rRNA methylations can be assigned by cryo-EM looking at extra densities present at specific positions on the nucleotides. For example, an *e*lectron density shape difference between the two adenine residues A2496 and A2503 in 23S rRNA has assigned at position 2 a methyl group at the adenine nucleobase A2503 (model: PDB 6YEF; map: EMD-10791 at 3.2 Å, Wang et al. 2021). In this cryo-EM structure, several other *S. aureus* rRNA methylations have reported (**b**): m^4Cm: 4-methyl 2-O′ methyl cytidine; m^2G: 2-methyl guanosine; m^3U: 3-methyl uridine; m^7G: 7-methyl guanosine; $m^6{}_6$A: 6, 6-dimetyl guanosine, Gm: 2-O′ methyl guanosine. **c** Characteristic bent nucleobase density shape of Dihydrouridine (D) observed on *E. coli* cryo-EM structure (model: PDB 7K00; map: EMD- 22,586 at1.98 Å, Suzuki and Suzuki 2007). **d** Interaction between *E. coli* pseudouridine Ψ2605 with a possible water molecule and comparison with unmodified U2596 from the same 23S rRNA structure (model: PDB 7K00; map: EMD-22,586 at 1.98 Å). W-C edge: Watson-Crick edge; CH edge. **e** Example of the results obtained by using qptm software (Taoka et al. 2010). Calculated (blue) against difference (green) map from m^2A2503 (modified) against A2496 (non-modified)

clustered at the Peptidyl Transferase Center (PTC), which may indicate that they assist peptide bond formation. Interestingly, Gm2251 in the P-loop of the 23S rRNA (*E. coli* numbering) directly interacts with the CCA sequence of the tRNA and thus is expected to contribute to the correct positioning of the tRNA acceptor end (Fischer et al. 2015; Golubev et al. 2020; Polikanov et al. 2015; Watson et al. 2020).

Various strategies were used to analyze modifications from high-resolution structures of ribosomes. With X-ray structures, difference density mapping (between real densities and those artificially obtained from modification-less models) helps to indicate the position of modified nucleotides (Polikanov et al. 2015). With cryo-EM structures, the refinement of specific regions of the ribosomal subunits (focused refinement) allowed reaching better resolution to recognize rRNA modifications in combination with previous annotation using complementary methods (Watson et al. 2020; Golubev et al. 2020). Several cryo-EM structures of *S. aureus* ribosomes are now available (Belinite et al. 2021; Belousoff et al. 2017; Cimicata et al. 2022; Golubev et al. 2020; Halfon et al. 2019; Khusainov et al. 2020, 2017, 2016; Matzov et al. 2017; Wright et al. 2020), and 10 different methylation sites have been mapped (Golubev et al. 2020) (Fig. 4a, b), but not all modifications could be assigned. Dihydrouridines (D) and pseudouridines (ψ) can be mapped with good confidence at very high resolution, for example on the 2 Å structure of the *E. coli* ribosome (Watson et al. 2020). Dihydrouridines (D) impose a characteristic bent shape to their pyrimidine ring and can be easily discriminated from uridines (Fig. 4c). Pseudouridines (ψ) can be assigned indirectly, by analyzing in which manner they change the pattern of ions or water interactions. The isomerization of N1 and C5 in ψ maintains the same Watson-Crick base-pairing property (W-C edge) but includes an additional hydrogen bond donor (N1 imino proton) on the C-H edge, which is responsible for water or ions binding (Fig. 4d) (Watson et al. 2020). Assignations of ions, water molecules, and other interactions are still challenging but they will certainly contribute to alleviate better density interpretation (Wang et al. 2021).

To facilitate the analyses of thousands of nucleotides of the ribosome, dedicated software has been recently developed, which automatically calculates difference density maps. Positioning of the RNA modification signals, obtained at specific contour levels, offers now the possibility to restrict the analysis by other methods on only a subset of nucleotides (Fig. 4e) (Stojković et al. 2020). However, efforts are still needed to complete the mapping of all the modifications in *S. aureus* rRNAs and to gain an evolutionary analysis of the conserved and specific sites by comparing with the extensive analysis done for *E. coli* (Watson et al. 2020) and *T. thermophilus* (Polikanov et al. 2015). Structures at higher resolution will help overcoming uncertainty in extra density coming from methylations, to discriminate pseudouridines according to their specific interaction network, and to observe conformational changes induced by modified nucleobases.

7 Discussion

Global analysis of the bacterial epitranscriptome and its variations occurring in response to environmental changes and various stresses is still a major technical challenge. Combination of approaches involving high-resolution structures of RNA machineries, mass spectrometry, and next-generation sequencing is certainly one way to render this objective amenable. As discussed in the review, the choice of the approaches depends on the RNA species to be studied. Recent advances in the Nanopore technology allow now the direct sequencing of full-length native RNA molecules without the need of RT and PCR amplification, and the obtained data normally contain all the information to assign the modifications (reviewed in Begik et al. 2021). However, Nanopore sequencing is still in its infancy and further improvement in the direct RNA sequencing are required for routine application. In most cases, Nanopore data are used only for validation, but not for de novo mapping of RNA modifications. Moreover, application to non-polyadenylated RNA species, like rRNA and tRNA, requires the use of polyadenylation by polyA-polymerase or specific custom primers. The complexity of Nanopore ion current profiles also requires further development of appropriate interpretation informatics pipelines, frequently based on machine learning algorithms or artificial intelligence.

Another major issue is to gain more knowledge on the protein machineries that are required for the modification deposition (writer) or which prevents the modification to take place (eraser). Even if phylogeny can be used to predict the enzymes and pathways required for RNA modifications, little is known in *S. aureus*. Many questions still remained to be addressed: are modification enzymes regulated upon stress conditions? Are there specific modification enzymes for mRNAs, regulatory RNAs, tRNAs, and rRNAs? What are the enzymes required for biofilm formation and for virulence? Are the modified mRNAs conserved among bacteria? Previous sensitive LC-MS approach has revealed that m^6A/A ratio in mRNAs was higher in Gram-negative than in Gram-positive bacteria including *S. aureus* (Deng et al. 2015). Transcriptomic analysis performed in *E. coli* and *Pseudomonas aeruginosa* revealed that m^6A was enriched in mRNAs involved in energy metabolism, stress responses, and in several sRNAs (Deng et al. 2015). Due to the rather small genome size, the high diversity of species, and the existing tools to delete genes, the bacteria represent ideal model organisms to decipher the extent of the functions of RNA modifications and of the associated enzyme machineries. No doubt the development of innovative sequencing approaches will be essential to gain full understanding of the bacterial epitranscriptome and to monitor its impact in gene regulation.

Acknowledgements This work was supported by the Centre National de la Recherche Scientifique (CNRS), by the French National Research Agency ANR (ANR-21-CE12-0030-01 to [SM]), by the Region Grand Est (N°18P-09227- EpiRNA), and the previous French National Program Investissement d'Avenir (Labex NetRNA) (ANR-10-LABX-0036_NETRNA). This work of the Interdisciplinary Thematic Institute IMCBio, as part of the ITI 2021–2028 program of the University of Strasbourg, CNRS, and Inserm, was supported by IdEx Unistra (ANR-10-IDEX-0002), by SFRI-STRAT'US project (20-SFRI-0012), and EUR IMCBio (IMCBio ANR-17-EURE-0023) under the framework of the France 2030 Program. [HK] is supported by a fellowship from Fondation de la Recherche Médicale (FRM). [BPK] acknowledges support by CNRS, Association pour la Recherche sur le Cancer (ARC), Institut National du Cancer (INCa_16099), the Fondation pour la Recherche Médicale (FRM) and ANR.

References

Adachi H, DeZoysa MD, Yu YT (2019) Detection and Quantification of Pseudouridine in RNA. Methods Mol Biol 1870:219–235

Agris PF (1996) The importance of being modified: roles of modified nucleosides and Mg^{2+} in RNA structure and function. Prog Nucleic Acid Res Mol Biol 53:79–129

Antoine L, Bahena-Ceron R, Devi Bunwaree H et al (2021) RNA modifications in pathogenic bacteria: impact on host adaptation and virulence. Genes (basel) 12:1125
Antoine L, Wolff P (2020) Mapping of posttranscriptional tRNA modifications by two-dimensional gel electrophoresis mass spectrometry. Methods Mol Biol 2113:101–110
Antoine L, Wolff P, Westhof E et al (2019) Mapping post-transcriptional modifications in Staphylococcus aureus tRNAs by nanoLC/MSMS. Biochimie 164:60–69
Apffel A, Chakel JA, Fischer S et al (1997) Analysis of oligonucleotides by HPLC-electrospray ionization mass spectrometry. Anal Chem 69:1320–1325
Arimbasseri AG, Blewett NH, Iben JR et al (2015) RNA polymerase III output is functionally linked to tRNA dimethyl-G26 modification. PLoS Genet 11(12):e1005671
Aschenbrenner J, Werner S, Marchand V et al (2018) Engineering of a DNA polymerase for direct m(6) a sequencing. Angew Chem Int Ed Engl 57:417–421
Atshan SS, Shamsudin MN, Lung LTT et al (2012) Improved method for the isolation of RNA from bacteria refractory to disruption, including S. aureus producing biofilm. Gene 494:219–224
Bakin A, Ofengand J (1993) Four newly located pseudouridylate residues in Escherichia coli 23S ribosomal RNA are all at the peptidyltransferase center: analysis by the application of a new sequencing technique. Biochemistry 32:9754–9762
Bastos MdCdF, Coutinho BG, Coelho MLV (2010) Lysostaphin: a staphylococcal bacteriolysin with potential clinical applications. Pharmaceuticals 3:1139–1161
Begik O, Lucas MC, Pryszcz LP et al (2021) Quantitative profiling of pseudouridylation dynamics in native RNAs with nanopore sequencing. Nat Biotechnol 39:1278–1291
Behm-Ansmant I, Helm M, Motorin Y (2011) Use of specific chemical reagents for detection of modified nucleotides in RNA. J Nucleic Acids 2011:408053
Behrens A, Rodschinka G, Nedialkova DD (2021) High-resolution quantitative profiling of tRNA abundance and modification status in eukaryotes by mim-tRNAseq. Mol Cell 81:1802–1815
Belinite M, Khusainov I, Soufari H et al (2021) Stabilization of Ribosomal RNA of the Small Subunit by Spermidine in Staphylococcus aureus. Front Mol Biosci 8:738752
Belousoff MJ, Eyal Z, Radjainia M et al (2017) Structural Basis for Linezolid Binding Site Rearrangement in the Staphylococcus aureus Ribosome. mBio 8:e00395-17
Beltrame CO, Côrtes MF, Bandeira PT et al (2015) Optimization of the RNeasy mini kit to obtain high-quality total RNA from sessile cells of Staphylococcus aureus. Braz J Med Biol Res 48:1071–1076
Birkedal U, Christensen-Dalsgaard M, Krogh N et al (2015) Profiling of ribose methylations in RNA by high-throughput sequencing. Angew Chem Int Ed Engl 54:451–455
Bleicher K, Bayer E (1994) Various factors influencing the signal intensity of oligonucleotides in electrospray mass spectrometry. Biol Mass Spectrom 23:320–322
Boccaletto P, Stefaniak F, Ray A et al (2022) MODOMICS: a database of RNA modification pathways. 2021 update. Nucleic Acids Res 50:D231–D235
Brielle R, Pinel-Marie ML, Chat S et al (2017) Purification, identification, and functional analysis of polysomes from the human pathogen Staphylococcus aureus. Methods 117:59–66
Burakovsky DE, Prokhorova IV, Sergiev PV, a. (2012) Impact of methylations of m2G966/m5C967 in 16S rRNA on bacterial fitness and translation initiation. Nucleic Acids Res 40:7885–7895
Cai WM, Chionh YH, Hia F et al (2015) A platform for discovery and quantification of modified ribonucleosides in RNA: application to stress-induced reprogramming of tRNA modifications. Methods Enzymol 560:29–71
Castleberry CM, Rodicio LP, Limbach PA (2008) Electrospray ionization mass spectrometry of oligonucleotides. Curr Protoc Nucleic Acid Chem
Cerutti P, Miller N (1967) Selective reduction of yeast transfer ribonucleic acid with sodium borohydride. J Mol Biol 26:55–66
Chakraburtty K (1980) Recognition of E coli tRNAArg by arginyl tRNA synthetase. Nucleic Acids Res 8:4459–4472
Charette M, Gray MW (2000) Pseudouridine in RNA: what, where, how, and why. IUBMB Life 49:341–351

Chionh YH, Ho CH, Pruksakorn D et al (2013) A multidimensional platform for the purification of non-coding RNA species. Nucleic Acids Res 41:e168

Cimicata G, Fridkin G, Bose T et al (2022) Structural studies reveal the role of helix 68 in the elongation step of protein biosynthesis. mBio 13:e0030622

Cottilli P, Itoh Y, Nobe Y et al (2022) Cryo-EM structure and rRNA modification sites of a plant ribosome. Plant Commun 3:100342

Crain PF (1990) Preparation and enzymatic hydrolysis of DNA and RNA for mass spectrometry. Meth Enzymol 193:782–790

Dalluge JJ, Hashizume T, Sopchik AE, a. (1996) Conformational flexibility in RNA: the role of dihydrouridine. Nucleic Acids Res 24:1073–1079

de Crécy-Lagard V, Jaroch M (2020) Functions of bacterial tRNA modifications: from ubiquity to diversity. Trends Microbiol 29:41–53

de Crecy-Lagard V, Ross RL, Jaroch M et al (2020) Survey and validation of tRNA modifications and their corresponding genes in bacillus subtilis sp subtilis strain 168. Biomolecules 10:977

Demirci H, Ft M, Belardinelli R et al (2010) Modification of 16S ribosomal RNA by the KsgA methyltransferase restructures the 30S subunit to optimize ribosome function. RNA 16(12):2319–2324

Deng X, Chen K, Luo GZ et al (2015) Widespread occurrence of N6-methyladenosine in bacterial mRNA. Nucleic Acids Res 43:6557–6567

Deryusheva S, Choleza M, Barbarossa A et al (2012) Post-transcriptional modification of spliceosomal RNAs is normal in SMN-deficient cells. RNA 18:31–36

Dudley E, Tuytten R, Bond A et al (2005) Study of the mass spectrometric fragmentation of pseudouridine: comparison of fragmentation data obtained by matrix-assisted laser desorption/ionisation post-source decay, electrospray ion trap multistage mass spectrometry, and by a method utilising elect. Rapid Commun Mass Spectrom 19:3075–3085

Duval M, Simonetti A, Caldelari I et al (2015) Multiple ways to regulate translation initiation in bacteria: mechanisms, regulatory circuits, dynamics. Biochimie 114:18–29

Easton LE, Shibata Y, Lukavsky PJ (2010) Rapid, nondenaturing RNA purification using weak anion-exchange fast performance liquid chromatography. RNA 16:647–653

Fischer N, Neumann P, Konevega AL et al (2015) Structure of the E. coli ribosome-EF-Tu complex at <3 a resolution by Cs-corrected cryo-EM. Nature 520:567–570

França A, Melo LD, Cerca N (2011) Comparison of RNA extraction methods from biofilm samples of Staphylococcus epidermidis. BMC Res Notes 4:572

Frye M, Jaffrey SR, Pan T et al (2016) RNA modifications: what have we learned and where are we headed? Nat Rev Genet 17:365–372

Garalde DR, Snell EA, Jachimowicz D et al (2018) Highly parallel direct RNA sequencing on an array of nanopores. Nat Methods 15:201–206

Gehrke CW, Kuo KC (1990) Ribonucleoside analysis by reversed-phase high performance liquid chromatography. J Chromatogr Library 45(Part A):A3–A71

Giessing AMB, Kirpekar F (2012) Mass spectrometry in the biology of RNA and its modifications. J Proteomics 75:3434–3449. https://doi.org/10.1016/j.jprot.2012.01.032

Golubev A, Fatkhullin B, Khusainov I et al (2020) Cryo-EM structure of the ribosome functional complex of the human pathogen Staphylococcus aureus at 3.2 Å resolution. FEBS Lett 594:3551–3567

Halfon Y, Matzov D, Eyal Z et al (2019) Exit tunnel modulation as resistance mechanism of S. aureus erythromycin resistant mutant. Sci Rep 9:11460

Hansen MA, Kirpekar F, Ritterbusch W et al (2002) Posttranscriptional modifications in the A-loop of 23S rRNAs from selected archaea and eubacteria. RNA 8:202–213

Ho NW, Gilham PT (1971) Reaction of pseudouridine and inosine with N-cyclohexyl-N'-beta-(4-methylmorpholinium)ethylcarbodiimide. Biochemistry 10:3651–3657

Houser WM, Butterer A, Addepalli B, a. (2015) Combining recombinant ribonuclease U2 and protein phosphatase for RNA modification mapping by liquid chromatography-mass spectrometry. Anal Biochem 478:52–58

Huang S, Zhang W, Katanski CD et al (2021) Interferon inducible pseudouridine modification in human mRNA by quantitative nanopore profiling. Genome Biol 22:330
Igloi GL (1988) Interaction of tRNAs and of phosphorothioate-substituted nucleic acids with an organomercurial. Probing the chemical environment of thiolated residues by affinity electrophoresis. Biochemistry 27(10):3842–3849
Igloi GL, Kössel H (1985) Affinity electrophoresis for monitoring terminal phosphorylation and the presence of queuosine in RNA. Application of polyacrylamide containing a covalently bound boronic acid. Nucleic Acids Res 13:6881–6898
Igo-Kemenes T, Zachau HG (1969) On the specificity of the reduction of transfer ribonucleic acids with sodium borohydride. Eur J Biochem 10:549–556
Incarnato D, Anselmi F, Morandi E et al (2017) High-throughput single-base resolution mapping of RNA 2′-O-methylated residues. Nucleic Acids Res 45:1433–1441
Ishitani R, Yokoyama S, Nureki O (2008) Structure, dynamics, and function of RNA modification enzymes. Curr Opin Struct Biol 18:330–339
Jenner LB, Demeshkina N, Yusupova G et al (2010) Structural aspects of messenger RNA reading frame maintenance by the ribosome. Nat Struct Mol Biol 17:555–560
Jora M, Borland K, Abernathy S et al (2021) Chemical amination/imination of carbonothiolated nucleosides during RNA hydrolysis. Angew Chem Int Ed Engl 60:3961–3966
Jora M, Burns AP, Ross RL et al (2018) Differentiating positional isomers of nucleoside modifications by higher-energy collisional dissociation mass spectrometry (HCD MS). J Am Soc Mass Spectrom 29:1745–1756
Jora M, Lobue PA, Ross RL et al (2019) Detection of ribonucleoside modifications by liquid chromatography coupled with mass spectrometry. Biochim Biophys Acta Gene Regul Mech 1862:280–290
Jühling F, Mörl M, Hartmann RK et al (2009) tRNAdb 2009: compilation of tRNA sequences and tRNA genes. Nucleic Acids Res 37:D159-162
Kanwal F, Lu C (2019) A review on native and denaturing purification methods for non-coding RNA (ncRNA). J Chromatogr B Analyt Technol Biomed Life Sci 1120:71–79
Kebarle P, Tang L (1993) From ions in solution to ions in the gas phase. Anal Chem 65:972A-986A
Kellner S, Neumann J, Rosenkranz D et al (2014) Profiling of RNA modifications by multiplexed stable isotope labelling. Chem Commun 50:3516–3518
Khoddami V, Yerra A, Mosbruger TL et al (2019) Transcriptome-wide profiling of multiple RNA modifications simultaneously at single-base resolution. Proc Natl Acad Sci USA 116:6784–6789
Khusainov I, Fatkhullin B, Pellegrino S et al (2020) Mechanism of ribosome shutdown by RsfS in Staphylococcus aureus revealed by integrative structural biology approach. Nat Commun 11:1656
Khusainov I, Vicens Q, Ayupov R et al (2017) Structures and dynamics of hibernating ribosomes from Staphylococcus aureus mediated by intermolecular interactions of HPF. EMBO J 36:2073–2087
Khusainov I, Vicens Q, Bochler A et al (2016) Structure of the 70S ribosome from human pathogen Staphylococcus aureus. Nucleic Acids Res 44:10491–10504
Kiesewetter S, Fischer W, Sprinzl M (1987) Sequences of three minor tRNAsArg from E. coli. Nucleic Acids Res 15:3184
Klaholz BP (2019) Deriving and refining atomic models in crystallography and cryo-EM: the latest Phenix tools to facilitate structure analysis. Acta Crystallogr D Struct Biol 75:878–881
Kowalak JA, Pomerantz SC, Crain PF et al (1993) A novel method for the determination of post-transcriptional modification in RNA by mass spectrometry. Nucleic Acids Res 21:4577–4585
Kristen M, Plehn J, Marchand V et al (2020) Manganese Ions Individually Alter the Reverse Transcription Signature of Modified Ribonucleosides. Genes (basel) 11:950
Kühlbrandt W (2014) Biochemistry. The resolution revolution. Science 343:1443–1444
Li GW, Burkhardt D, Gross C et al (2014) Quantifying absolute protein synthesis rates reveals principles underlying allocation of cellular resources. Cell 157:624–635

Maden BE, Corbett ME, Heeney PA et al (1995) Classical and novel approaches to the detection and localization of the numerous modified nucleotides in eukaryotic ribosomal RNA. Biochimie 77:22–29

Marchand V, Blanloeil-Oillo F, Helm M et al (2016) Illumina-based RiboMethSeq approach for mapping of 2'-O-Me residues in RNA. Nucleic Acids Res 44:e135

Marchand V, Ayadi L, Ernst FGM et al (2018) AlkAniline-Seq: profiling of m(7) G and m(3) C RNA modifications at single nucleotide resolution. Angew Chem Int Ed Engl 57:16785–16790

Marchand V, Ayadi L, Bourguignon-Igel V et al (2021a) AlkAniline-Seq: a highly sensitive and specific method for simultaneous mapping of 7-methyl-guanosine (m(7)G) and 3-methyl-cytosine (m(3)C) in RNAs by high-throughput sequencing. Meth Mol Biol 2298:77–95

Marchand V, Bourguignon-Igel V, Helm M et al (2021b) Mapping of 7-methylguanosine (m(7)G), 3-methylcytidine (m(3)C), dihydrouridine (D) and 5-hydroxycytidine (ho(5)C) RNA modifications by AlkAniline-Seq. Methods Enzymol 658:25–47

Marchand V, Bourguignon-Igel V, Helm M et al (2022) Analysis of pseudouridines and other RNA modifications using HydraPsiSeq protocol. Methods 203:383–391

Marchand V, Pichot F, Neybecker P et al (2020) HydraPsiSeq: a method for systematic and quantitative mapping of pseudouridines in RNA. Nucleic Acids Res 48:e110

Marchand V, Pichot F, Thuring K et al (2017) Next-generation sequencing-based ribomethseq protocol for analysis of tRNA 2'-O-methylation. Biomolecules 7:13

Matzov D, Aibara S, Basu A et al (2017) The cryo-EM structure of hibernating 100S ribosome dimer from pathogenic Staphylococcus aureus. Nat Commun 8:723

McGinnis AC, Grubb EC, Bartlett MG (2013) Systematic optimization of ion-pairing agents and hexafluoroisopropanol for enhanced electrospray ionization mass spectrometry of oligonucleotides. Rapid Commun Mass Spectrom 27:2655–2664

McKenna SA, Kim I, Puglisi EV et al (2007) Purification and characterization of transcribed RNAs using gel filtration chromatography. Nat Protoc 2:3270–3277

Mcluckey SA, Van Berkel GJ, Glish GL (1992) Tandem mass spectrometry of small, multiply charged oligonucleotides. J Am Soc Mass Spectrom 3:60–70

Mengel-Jørgensen J, Kirpekar F (2002) Detection of pseudouridine and other modifications in tRNA by cyanoethylation and MALDI mass spectrometry. Nucleic Acids Res 30:e135

Meyer M, Masquida B (2016) Polyacrylamide gel electrophoresis for purification of large amounts of RNA. Methods Mol Biol 1320:59–65

Mims BH, Prather NE, Murgola EJ (1985) Isolation and nucleotide sequence analysis of tRNAAlaGGC from Escherichia coli K-12. J Bacteriol 162:837–839

Miyauchi K, Kimura S, Suzuki T (2013) A cyclic form of N 6-threonylcarbamoyladenosine as a widely distributed tRNA hypermodification. Nat Chem Biol 9:105–111

Miyauchi K, Ohara T, Suzuki T (2007) Automated parallel isolation of multiple species of non-coding RNAs by the reciprocal circulating chromatography method. Nucleic Acids Res 35:e24

Motorin Y, Helm M (2019) Methods for RNA modification mapping using deep sequencing: established and new emerging technologies. Genes (basel) 10:35

Motorin Y, Helm M (2022) RNA nucleotide methylation: 2021 update. Wiley Interdiscip Rev RNA 13:e1691

Motorin Y, Marchand V (2018) Detection and analysis of RNA ribose 2'-O-methylations: challenges and solutions. Genes (basel) 9:642

Motorin Y, Marchand V (2021) Analysis of RNA modifications by second- and third-generation deep sequencing: 2020 update. Genes (basel) 12:278

Motorin Y, Muller S, Behm-Ansmant I, a. (2007) Identification of modified residues in RNAs by reverse transcription-based methods. Meth Enzymol 425:21–53

Nakai C, Konishi A, Komatsu Y et al (1994) Sequence-specific cleavage of RNA by a hybrid ribonuclease H. FEBS Lett 339:67–72

Nakamura K, Takeo K (1998) Affinity electrophoresis and its applications to studies of immune response. J Chromatogr B Biomed Sci Appl 715:125–136

Nakayama H, Takahashi N, Isobe T (2011) Informatics for mass spectrometry-based RNA analysis. Mass Spectrom Rev 30:1000–1012

Natchiar SK, Myasnikov AG, Hazemann I et al (2018) Visualizing the role of 2'-OH rRNA methylations in the human ribosome structure. Biomolecules 8:125

Natchiar SK, Myasnikov AG, Kratzat H et al (2017) Visualization of chemical modifications in the human 80S ribosome structure. Nature 551:472–477

Ni J, Pomerantz C, Rozenski J et al (1996) Interpretation of oligonucleotide mass spectra for determination of sequence using electrospray ionization and tandem mass spectrometry. Anal Chem 68:1989–1999

Nübel G, Sorgenfrei FA, Jäschke A (2017) Boronate affinity electrophoresis for the purification and analysis of cofactor-modified RNAs. Methods 117:14–20

Ohira T, Minowa K, Sugiyama K et al (2022) Reversible RNA phosphorylation stabilizes tRNA for cellular thermotolerance. Nature 605:372–379

Patteson KG, Rodicio LP, Limbach PA (2001) Identification of the mass-silent post-transcriptionally modified nucleoside pseudouridine in RNA by matrix-assisted laser desorption/ionization mass spectrometry. Nucleic Acids Res 29:49–49

Peattie DA (1979) Direct chemical method for sequencing RNA. Proc Natl Acad Sci USA 76:1760–1764

Petrov A, Wu T, Puglisi EV et al (2013) RNA purification by preparative polyacrylamide gel electrophoresis. Meth Enzymol 530:315–330

Pichot F, Marchand V, Helm M et al (2021) Non-redundant tRNA reference sequences for deep sequencing analysis of tRNA abundance and epitranscriptomic RNA modifications. Genes (basel) 12:81

Piekna-Przybylska D, Decatur WA, Fournier MJ (2008) The 3D rRNA modification maps database: with interactive tools for ribosome analysis. Nucleic Acids Res 36:D178-183

Polikanov YS, Melnikov SV, Soll D et al (2015) Structural insights into the role of rRNA modifications in protein synthesis and ribosome assembly. Nat Struct Mol Biol 22:342–344

Polo LM, Limbach PA (2001) Analysis of Oligonucleotides by Electrospray Ionization Mass Spectrometry. Curr Protoc Nucleic Acid Chem

Pomerantz SC, McCloskey JA (1990) Analysis of RNA hydrolyzates by liquid chromatography-mass spectrometry. Methods Enzymol 193:796–824

Potapov V, Fu X, Dai N et al (2018) Base modifications affecting RNA polymerase and reverse transcriptase fidelity. Nucleic Acids Res 46:5753–5763

Potier N, van Dorsselaer A, Cordier Y et al (1994) Negative electrospray ionization mass spectrometry of synthetic and chemically modified oligonucleotides. Nucleic Acids Res 22:3895–3903

Poulson R (1973) Isolation, Purification and Fractionation of RNA. The ribonucleic acids, pp 243–261

Rebane A, Roomere H, Metspalu A (2002) Locations of several novel 2'-O-methylated nucleotides in human 28S rRNA. BMC Mol Biol 3:1

Roost C, Lynch SR, Batista PJ et al (2015) Structure and thermodynamics of N6-methyladenosine in RNA: a spring-loaded base modification. J Am Chem Soc 137:2107–2115

Ross R, Cao X, Yu N et al (2016) Sequence mapping of transfer RNA chemical modifications by liquid chromatography tandem mass spectrometry. Methods (san Diego, Calif) 107:73–78

Samatova E, Daberger J, Liutkute M et al (2020) Translational control by ribosome pausing in bacteria: how a non-uniform pace of translation affects protein production and folding. Front Microbiol 11:619430

Schaefer M, Kapoor U, Jantsch MF (2017) Understanding RNA modifications: the promises and technological bottlenecks of the 'epitranscriptome.' Open Biol 7:1–14

Schaefer M, Pollex T, Hanna K et al (2009) RNA cytosine methylation analysis by bisulfite sequencing. Nucleic Acids Res 37:e12

Schwartz S, Bernstein DA, Mumbach MR et al (2014) Transcriptome-wide mapping reveals widespread dynamic-regulated pseudouridylation of ncRNA and mRNA. Cell 159:148–162

Squires JE, Patel HR, Nousch M et al (2012) Widespread occurrence of 5-methylcytosine in human coding and non-coding RNA. Nucleic Acids Res 40:5023–5033

Stern S, Moazed D, Noller HF (1988) Structural analysis of RNA using chemical and enzymatic probing monitored by primer extension. Methods Enzymol 164:481–489

Stojković V, Myasnikov AG, Young ID et al (2020) Assessment of the nucleotide modifications in the high-resolution cryo-electron microscopy structure of the Escherichia coli 50S subunit. Nucleic Acids Res 48:2723–2732

Stults JT, Marsters JC, Carr SA (1991) Improved electrospray ionization of synthetic oligodeoxynucleotides. Rapid Commun Mass Spectrom 5:359–363

Su D, Chan CTY, Gu C et al (2014) Quantitative analysis of ribonucleoside modifications in tRNA by HPLC-coupled mass spectrometry. Nat Protoc 9:828–841

Suzuki T, Suzuki T (2007) Chaplet column chromatography: isolation of a large set of individual RNAs in a single step. Meth Enzymol 425:231–239

Taoka M, Ikumi M, Nakayama H et al (2010) In-gel digestion for mass spectrometric characterization of RNA from fluorescently stained polyacrylamide gels. Anal Chem 82:7795–7803

Thakur P, Jora M, Zhao R et al (2021) Mass spectrometry-based methods for characterization of hypomodifications in transfer RNA. Springer. Epitranscriptomics, pp 555–592

Thomas B, Akoulitchev AV (2006) Mass spectrometry of RNA. Trends Biochem Sci 31:173–181

Thuring K, Schmid K, Keller P et al (2016) Analysis of RNA modifications by liquid chromatography-tandem mass spectrometry. Methods 107:48–56

Tollerson R 2nd, Ibba M (2020) Translational regulation of environmental adaptation in bacteria. J Biol Chem 295:10434–10445

Tuorto F, Legrand C, Cirzi C et al (2018) Queuosine-modified tRNAs confer nutritional control of protein translation. EMBO J 37:e99777

Ty H, Kharlamova A, Liu J et al (2008) Ion trap collision-induced dissociation of multiply deprotonated RNA: c/y-Ions versus (a-B)/w-Ions. J Am Soc Mass Spectrom 19:1832–1840

Wang J, Natchiar SK, Moore PB et al (2021) Identification of $Mg(^{2+})$ ions next to nucleotides in cryo-EM maps using electrostatic potential maps. Acta Crystallogr D Struct Biol 77:534–539

Watson ZL, Ward FR, Méheust R et al (2020) Structure of the bacterial ribosome at 2 Å resolution. Elife 9:e60482

Werner S, Galliot A, Pichot F et al (2021) NOseq: amplicon sequencing evaluation method for RNA m6A sites after chemical deamination. Nucleic Acids Res 49:e23

Werner S, Schmidt L, Marchand V et al (2020) Machine learning of reverse transcription signatures of variegated polymerases allows mapping and discrimination of methylated purines in limited transcriptomes. Nucleic Acids Res 48:3734–3746

Wetzel C, Limbach PA (2016) Mass spectrometry of modified RNAs: recent developments. Analyst 141:16–23

Wolff P, Villette C, Zumsteg J et al (2020) Comparative patterns of modified nucleotides in individual tRNA species from a mesophilic and two thermophilic archaea. RNA 26:1957–1975

Wright A, Deane-Alder K, Marschall E et al (2020) Characterization of the core ribosomal binding region for the oxazolidone family of antibiotics using cryo-EM. ACS Pharmacol Transl Sci 3:425–432

Xing F, Hiley SL, Hughes TR et al (2004) The specificities of four yeast dihydrouridine synthases for cytoplasmic tRNAs. J Biol Chem 279:17850–17860

Yang J, Sharma S, Watzinger P et al (2016) Mapping of complete set of ribose and base modifications of yeast rRNA by RP-HPLC and mung bean nuclease assay. PLoS ONE 11:e0168873

Yokogawa T, Kitamura Y, Nakamura D et al (2010) Optimization of the hybridization-based method for purification of thermostable tRNAs in the presence of tetraalkylammonium salts. Nucleic Acids Res 38:e89

Yoluç Y, Ammann G, Barraud P et al (2021) Instrumental analysis of RNA modifications. Crit Rev Biochem Mol Biol 56:178–204

Zheng C, Black KA, Dos Santos PC (2017) Diverse mechanisms of sulfur decoration in bacterial tRNA and their cellular functions. Biomolecules 7:33

Bacterial Small RNAs: Diversity of Structure and Function

João Pedro Sousa, Alda Filipa Queirós Silva, Cecília Maria Arraiano, and José Marques Andrade

Contents

Abstract Small non-coding RNAs (sRNAs) are key post-transcriptional regulators of gene expression in bacteria, with a diversity of origins, sequences, structures, and modes of action among their members. The variety within these regulatory RNAs makes it difficult to unify this remarkable heterogeneous class of RNA molecules. Structural determinants along with nucleotide sequence play important roles in defining the sRNA ability to interact with their targets. Most sRNAs bind to mRNA targets, either acting as repressors or activators. Nevertheless, the sRNAs

J. P. Sousa and A. F. Q. Silva contributed equally in this chapter.

J. P. Sousa · A. F. Q. Silva · C. M. Arraiano · J. M. Andrade (✉)
Instituto de Tecnologia Química e Biológica António Xavier, Universidade Nova de Lisboa (ITQB NOVA), Avenida da República, 2780-901 Oeiras, Portugal
e-mail: andrade@itqb.unl.pt

J. P. Sousa
e-mail: joao.sousa@itqb.unl.pt

A. F. Q. Silva
e-mail: alda.silva@itqb.unl.pt

C. M. Arraiano
e-mail: cecilia@itqb.unl.pt

J. Barciszewski (ed.), *RNA Structure and Function*, RNA Technologies 14,
https://doi.org/10.1007/978-3-031-36390-0_12

themselves are subject to post-transcriptional regulation, either by ribonucleases, RNA chaperones, and other RNA-binding proteins, as well as RNA sponges. In this chapter we summarize the information on structure and function of bacterial sRNAs and provide several examples to better illustrate the broad diversity of this class of regulatory RNAs.

Keywords CRISPR · Hfq · Non-coding RNA · Regulatory RNA · RNA-RNA interaction · RNA sequence · RNA structure · Sponge RNA · sRNA

1 Introduction

Bacterial small non-coding RNAs (sRNAs) represent a class of regulatory RNAs, with sRNA-based networks virtually controlling all aspects of cell physiology. sRNAs are powerful regulators of gene expression that may bind to different macromolecules, either DNA, proteins, or other RNA molecules, but mRNAs are by far their most abundant targets (Quendera et al. 2020). Indeed, the vast majority of sRNAs act through an antisense mechanism, binding to their mRNA targets by complementary base pairing, resulting either in inhibition or activation of translation. sRNAs can be divided into two groups according to the location of the sRNA genes in relation to their targets: *cis*-acting sRNAs are expressed in the same location but in the opposite strand of the target whereas *trans*-encoded sRNAs are expressed from a different genomic region than their mRNA targets (Fig. 1). As consequence, *cis*-encoded sRNAs present a perfect base pairing with their mRNA targets, in contrast to the more numerous *trans*-encoded RNAs that establish short and imperfect antisense base pairing interactions with their targets, resembling the action of eukaryotic microRNAs. The base pairing of individual *trans*-encoded sRNAs can take place at different sites on the mRNA target, from the 5′ or 3′ untranslated regions (UTRs), as well as in the coding sequence. More often than not, *trans*-encoded sRNAs seem to be able to bind to more than one target mRNA and may use different regions for interactions with different targets (Andrade et al. 2013). This plasticity of sRNAs contributes to the rapid reprogramming of gene expression and helps to explain the regulatory success of sRNA-based pathways.

A plethora of sRNAs act as repressors, acting for example by sequestering the entry of ribosomes to mRNA (e.g., through occlusion of the ribosome binding site (RBS)) or affecting mRNA stability (e.g., by promoting the access to ribonucleases (RNases) that cleave mRNA leading to its inactivation), as it has been described for the sRNA MicA (Udekwu et al. 2005; Viegas et al. 2011). However, some other sRNAs have opposite effects and therefore act as activators of gene expression. For example, the sRNA SraL was shown to upregulate the expression of the transcription termination factor Rho, by interacting with the 5′UTR of *rho* mRNA and preventing its premature transcription termination (Silva et al. 2019). Broadly, sRNAs can vary in size from 50 to 500 nucleotides (nts), without a common sequence that can be used as signature, making this a very heterogeneous class of RNA molecules; nevertheless,

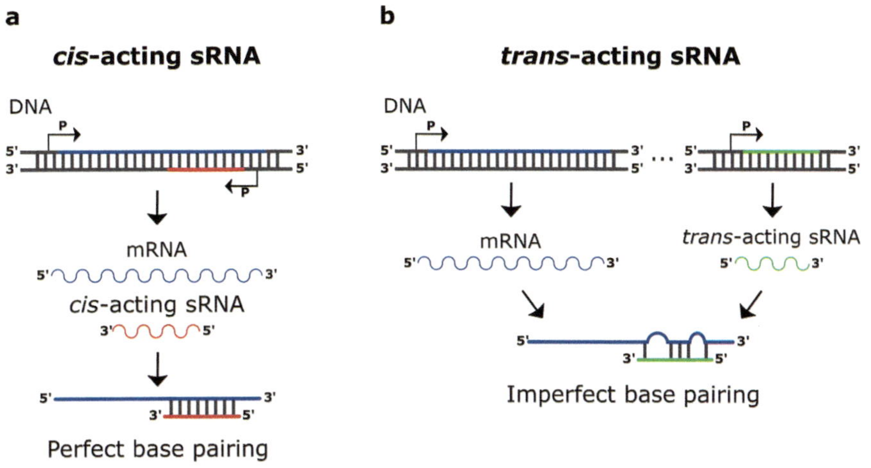

Fig. 1 Mechanisms of action for *cis*-acting and *trans*-acting sRNAs. **a** A *cis*-encoded sRNA and its target are located in the same genomic region but on different strands; as result the sRNA-mRNA antisense interaction exhibits a perfect base pairing region. **b** A *trans*-encoded sRNA is located in a different genomic region than its target; consequently, the sRNA establishes short and imperfect base pairing interactions with the mRNA target

a common feature of these regulatory RNAs is that they are highly structured, with the presence of hairpins and stem-loops. These structures serve as barriers against sRNA degradation by RNases, may act as anchor sites for RNA-binding proteins, and can also affect interaction with mRNA targets. In this chapter we present a comprehensive overview on bacterial sRNAs and how structure is linked to sRNA function, providing several examples to better illustrate the broad diversity of this class of regulatory RNAs.

2 sRNA that Use the 5′ Domain to Interact with mRNA Targets

sRNAs are usually structured molecules, and typically sequences enriched in GC are more structured. The presence of a stem-loop corresponding to the Rho-independent transcriptional terminator followed by a short U-rich sequence seems to be ubiquitous among bacterial small RNAs (Morita et al. 2017), and the presence of additional stems is frequent in many sRNAs (Fig. 2). Nevertheless, even such high structured molecules present linear regions, either located in the body of the sRNA or in the bulge loop position of hairpin stems. These linear domains are highly important since sRNA/mRNA base pairing generally occurs through interactions between these linear regions. Consequently, we can recognize different RNA structural modules within a sRNA (Andrade et al. 2013).

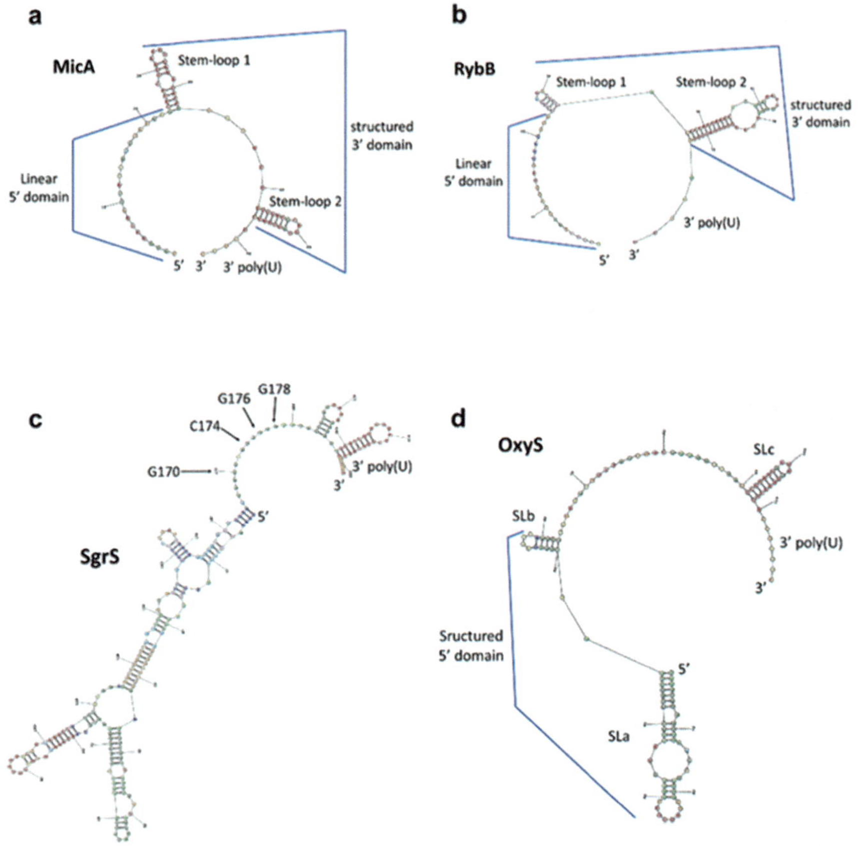

Fig. 2 Representative examples of sRNAs structural diversity. **a** MicA sRNA has a 5′-end linear sequence that is recognized as its main interaction region with mRNA targets. Two stem-loops present in the 3′-end of MicA and the short terminal 3′ poly(U) sequence are important elements for interaction with the RNA chaperone Hfq. **b** RybB sRNA also shows a 5′-end linear region for interaction with mRNA targets and a structured 3′-end with two stem-loops followed by a terminal poly(U) sequence. **c** SgrS sRNA is a highly structured sRNA that uses a linear sequence located near the 3′-end for interaction with its mRNA targets. The position of important nucleotides of SgrS for base pairing with *ptsG* mRNA is indicated with arrows. **d** OxyS sRNA displays a 5′-end structured region with two stem-loops (SLa and SLb), using a downstream linear region for interaction with its mRNA targets. A third stem-loop (SLc) that corresponds to the Rho-independent transcriptional terminator is located in the 3′-end, followed by the terminal 3′ poly(U) sequence. sRNAs structures were determined using the RNA structure webserver (Bellaousov et al. 2013)

The well-characterized small RNA MicA, firstly identified in *Escherichia coli*, is commonly used as model of study. MicA is a *trans*-encoded sRNA that represses several genes, including the expression of major outer membrane proteins, such as OmpA, LamB, Tsx, and EcnB (Gogol et al. 2011). The 5′ linear end of MicA consisting of ~20 nts, located before two strong stem-loops, was identified as the principal target recognition domain (Fig. 2a) (Udekwu et al. 2005; Andrade et al.

2013). Many other sRNAs that preferably use their 5′-end sequence for contact with their targets have also been identified. Interestingly, for a few it was even possible to define a short "seed" sequence responsible for the interaction with multiple targets. This is the case of *Salmonella Typhimurium* RybB (Fig. 2b) (Papenfort et al. 2010) whose conserved seed domain is also shared with *Vibrio cholerae* VrrA and MicV sRNAs (Peschek et al. 2019), and *E. coli* OmrA/B, two seemingly similar sRNAs that use their conserved 5′-end sequence to interact with multiple targets (Guillier and Gottesman 2008). Overall, several studies indicate the importance of the 5′-end domain of many sRNAs as the principal target interaction domain, with these few examples here depicted as illustrative ones. However, research on MicA further identified structural elements present in the 3′-end, namely stem-loops, which could also play a role in target recognition. While one of the stem-loops was more important for the in vivo repression of both *ompA* and *ecnB* mRNAs, the other stem-loops were found to be only critical for the regulation of *tsx* transcript levels (Andrade et al. 2013). This is probably consequence of conformational rearrangements of the sRNA as a result of target interactions with the linear sequence present in the bulge loop of the stems (Fig. 2a). Similar observations have been made in many studies such as the OxyS interaction with the *fhlA* mRNA (Altuvia et al. 1998).

3 sRNA that Use the 3′ Domain to Interact with mRNA Targets

SgrS is a well-characterized sRNA that is widespread in enteric bacteria (Rice and Vanderpool 2011). In *E. coli* SgrS is a 227-nt molecule (Fig. 2c) but the size is variable in other species. SgrS levels are upregulated when the balance between sugar uptake and its metabolism is disrupted; the intracellular accumulation of glucose-6-phosphate leads to activation of the transcription factor SgrR, which then induces SgrS expression (from the *sgrS* gene) to minimize sugar transport (Vanderpool and Gottesman 2007). In the case of glucose excess, SgrS binds reversibly to its main target, the *ptsG* mRNA encoding the EIICB domain of the glucose-specific phosphotransferase system (PTS), blocking its ribosome binding site and halting the translation of additional glucose importers; the SgrS-mRNA complex is subsequently degraded by RNase E (Rice and Vanderpool 2011). Strikingly, SgrS is also a dual-function molecule since it also encodes SgrT, a small 43-amino acid protein whose regulatory function mechanism acts independently from SgrS-mediated base pairing in response to phosphosugar stress (Wadler and Vanderpool 2007).

In contrast to MicA and other sRNAs, SgrS-mediated target recognition requires a conserved region near its 3′-end rather than a sequence in the 5′-end. SgrS nucleotides 168–187 are complementary to the 5′-UTR of the *ptsG* mRNA (Kawamoto et al. 2006). Particularly, nucleotides 168–181 have been shown to be enough to downregulate *ptsG* mRNA levels (Maki et al. 2010). Moreover, SgrS harbors a Hfq-binding motif at the 3′-end, which contains two adjacent stem-loops—the small stem-loop and the U-rich terminator stem-loop—followed by a poly(U) tail (Fig. 2c) (Kawamoto et al. 2006). The length of the poly(U) tail is a key feature of SgrS, not only for the correct biosynthesis of functional sRNAs but also for optimal Hfq binding to the two 3′ stem-loops, and target regulation (Morita et al. 2017).

The sequence and subsequential secondary structure of SgrS have been shown to be highly sensitive to changes, resulting in regulation defects in the cell. It has been shown that point mutations in the region spanning from U175 to G186, encompassed in the base pairing domain of SgrS to *ptsG* mRNA, led to negative functional consequences in the SgrS-mediated stress response (Poddar et al. 2021). Particularly, mutagenesis of G176 and G178 completely abolishes the ability of SgrS to inhibit the translation of *ptsG* mRNA (Kawamoto et al. 2006), while point mutations of C174 and G170 only weakly hinders SgrS base pairing (Fig. 2c) (Maki et al. 2008). Likewise, mutations in nucleotides 183–196, 199–219, and 220–227 corresponding to the small hairpin loop, the terminator stem-loop, and the poly(U) tail, respectively, also negatively impact SgrS function in the cell, mainly in the stem region of the terminator stem-loop, specifically in nucleotides C199 to G205, and C213 to G219 (Poddar et al. 2021).

A less specific target of SgrS is the *manXYZ* mRNA, encoding another PTS sugar transporter—involved mainly in mannose uptake, but also involved in the transport of glucose, glucosamine, and N-acetylglucosamine. SgrS inhibits the expression of this transport system by base pairing with low affinity to two distinct sites of the mRNA: within *manX* coding sequence and in an UTR between *manX* and *manY* (Rice and Vanderpool 2011). As it is the case with *ptsG*, SgrS annealing halts translation, recruits RNase E causing the degradation of *manXYZ* mRNA, and therefore minimizes phosphosugar toxicity.

4 sRNA Association with the RNA Chaperone Hfq

Many *trans*-encoded sRNAs associate with the RNA chaperone Hfq and this can affect their stability and their function (Quendera et al. 2020). Hfq forms an hexamer and it not only protects the sRNAs from degradation, namely from RNase E, RNase III and PNPase cleavage (Udekwu et al. 2005; Andrade and Arraiano 2008; Viegas et al. 2011; Andrade et al. 2012) but also stabilizes the imperfect base pairing between sRNA/mRNA pairs, functioning as an "RNA matchmaker" (Woodson et al. 2018). Indeed, the primary role of Hfq (at least in Gram-negative bacteria) is to promote the annealing of sRNA-mRNA duplexes (dos Santos et al. 2019). However, the role of Hfq in Gram-positive bacteria is more controversial even though extensive Hfq-dependent post-transcriptional regulation was recently identified in the Gram-positive human pathogen *Clostridioides difficile* (Fuchs et al. 2021). In addition, Hfq may play other functions independently of its association with sRNAs, as was observed with tRNA maturation, ribosome biogenesis and rRNA processing (Lee and Feig 2008; Andrade et al. 2018; dos Santos et al. 2020).

In *E. coli*, interactome studies identified thousands of mRNA-sRNA pairs exhibiting sequence complementarity (Melamed et al. 2016). Hfq promotes sRNA/ mRNA interaction as it is able to bind both molecules simultaneously making use of different protein surfaces; the Hfq distal side preferably binds to $(ARN)_n$ motifs frequently found on mRNA while the Hfq proximal side preferably binds the short and unstructured poly(U) tails present at the 3′-end of sRNAs (Sauer and Weichenrieder 2011). The basic patched rim surface of Hfq interacts with UA-rich sites present in both RNAs, accelerating the formation of the complex (Panja et al. 2013; Zhang et al. 2013). In addition, the position and structure of Rho-independent terminators located immediately before the 3′ poly(U) stretch of sRNAs are also important for Hfq interaction (Morita et al. 2017). It has been shown that mutations disrupting the terminator stem-loop structure or the poly(U) tail resulted in lower intracellular levels of sRNAs, like MicA and SgrS, due to higher levels of degradation related to defective Hfq interaction (Andrade et al. 2013; Poddar et al. 2021).

The binding and role of Hfq in sRNA function has been elucidated for many sRNAs, as it is the case of the well-studied OxyS, an important factor in the oxidative stress response (Altuvia et al. 1998; Seixas et al. 2022). Depletion of this sRNA results in considerably higher levels of both H_2O_2 and superoxide in *E. coli* (González-Flecha and Demple 1999). The overexpression of OxyS negatively affects the expression of the transcription factors RpoS and FlhA by binding to their respective mRNAs, blocking ribosome binding, and then recruiting RNase E for degradation (Fröhlich and Gottesman 2018). The OxyS sRNA is a 118 nt-long molecule, containing two stem-loops near its 5′-end (Fig. 2d): SLa which contains the largest hairpin structure of the molecule and a smaller stem-loop, SLb, immediately downstream. Moreover, this sRNA contains an additional stem-loop, SLc, on its 3′-end, followed by a poly(U) tail (Wang et al. 2015). The binding of Hfq is important for the binding of OxyS to its targets and OxyS contains in its sequence several Hfq-binding motifs: the $(AAN)_3$ motif downstream the second stem-loop (SLb) binds to the distal face of Hfq; the UUUU motif upstream of the third stem-loop (SLc) binds to the lateral surface of Hfq; and the 3′-end poly(U) motif binds to the proximal face of Hfq (Fig. 2d) (Cai et al. 2022). Through the multiple interactions with the RNA-binding surfaces of Hfq, OxyS is able to wrap around the Hfq hexamer, effectively distorting the overall conformation of OxyS from an extended configuration to a more packed and stable "wrapped" structure, which helps to stabilize sRNA regulation (Henderson et al. 2013; Cai et al. 2022). The structural changes of Hfq-bound OxyS better expose its base pairing region to bind to *fhlA* mRNA resulting in a more stable antisense-target RNA complex which prevents ribosome binding (Altuvia et al. 1998; Hoekzema et al. 2019). Interestingly, depletion of Hfq has not been shown to have a significant effect in OxyS-*rpoS* mRNA binding, but instead it seems to help to recruit RNase E (Henderson et al. 2013). Hfq binding can be observed in many other sRNAs, for instance, MicA: Hfq was found to bind an AU-rich single stranded region flanked by the two stem-loop structures and the 3′-end poly(U) stretch after the Rho-independent terminator of this sRNA (Andrade et al. 2013). Mutations of these Hfq-binding sequences were also found to disrupt target recognition highlighting the importance of Hfq/sRNA association for MicA function.

5 sRNAs that Bind to Proteins: The Example of CsrB

In addition to mRNA targets, sRNAs can also bind to and regulate the activity of RNA-binding proteins, as can be illustrated by the sRNA CsrB and the protein CsrA, the two main components of the Csr system that acts as the global regulator of carbon storage (Liu et al. 1997). CsrA is a conserved RNA-binding protein that was first discovered in *E. coli* (Romeo et al. 1993) and participates in the regulation of carbon metabolism, virulence, motility, and biofilm formation (Vakulskas et al. 2015). CsrA can act as a repressor when binding with its mRNA targets occludes the access to the ribosome binding sequence leading thus to translational arrest, as for example was shown to occur with the *cstA* mRNA (which encodes a transport protein) and *glgC* mRNA (which encodes glucose-1-phosphate adenylyltransferase) (Dubey et al. 2003). However, CsrA can also act as activator of gene expression, for example, by protecting transcripts from RNase E attack, as it was observed with the *flhDC* mRNA (which is involved in motility and chemotaxis) (Yakhnin et al. 2013).

Strikingly, CsrB sRNA inhibits the activity of CsrA by directly binding and sequestering the protein (Vakulskas et al. 2015). CsrB has 18 imperfect repeats of the 5′-CAGGA(U, C, A)G-3′ motif either in the loops or in the linear regions of the secondary structure of CsrB, which are the recognition elements for CsrA (Fig. 3). Thus, CsrA and CsrB assemble in a complex formed by 18 CsrA subunits and a single CsrB sRNA (Liu et al. 1997). As result, CsrA is sequestered and cannot participate in their normal post-transcriptional regulator activity. The sRNAs of the CsrB family participate in the regulation of several metabolic pathways and physiological functions in *Gammaproteobacteria* (Vakulskas et al. 2015). The expression of CsrB is highly controlled in *E. coli* by the two-component signal transduction system BarA-UvrY, stringent response factors, as ppGpp and DksA, and two DEAD-box RNA helicases, DeaD (CsdA) and SrmB (Suzuki et al. 2002; Vakulskas et al. 2014). CsrB levels are also controlled by the CsrD protein, which directs this sRNA for degradation by RNase E (Suzuki et al. 2006). In *Pseudomonas aeruginosa*, there is an homolog to the CsrA-CsrB system, the RsmA-RsmB system, repressor of secondary metabolism, which presents the same type of regulation and interaction (Vakulskas et al. 2015).

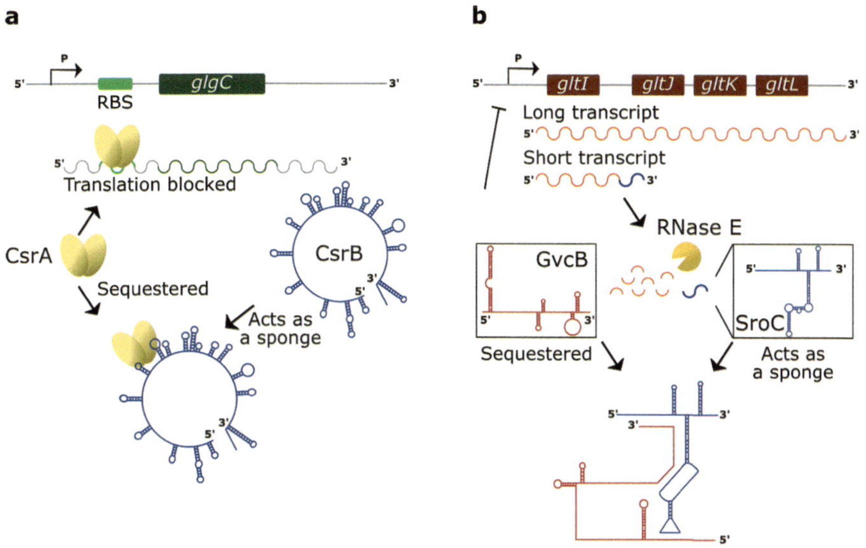

Fig. 3 sRNAs can bind to proteins and be sequestered by sponge RNAs. **a** The sRNA CsrB forms a complex with the protein CsrA. CsrB contains 18 repeated sequences that can bind to and sequester CsrA. **b** The regulatory RNA SroC derives from the processing of the intergenic region within the small transcript of the *gltIJKL* operon and acts as a sponge RNA that is able to bind to the sRNA GcvB

6 Sponge RNAs that Regulate sRNA Levels

sRNAs themselves can be post-transcriptionally regulated by the so-called sponge RNAs. These regulatory RNAs modulate sRNA levels through sequestration of the sRNA and/or promotion of degradation of the sRNA or the sponge RNA-sRNA complex (Denham 2020). The term sponge RNA was first used by Ebert and colleagues, when they engineered an RNA molecule capable of controlling microRNA activity by competing with its targets (Ebert et al. 2007). In bacteria, sponge RNAs can be divided into two groups, those that result from the processing of an mRNA and those that are independently transcribed. Most of the sponge RNAs already described were found in *Enterobacteriaceae*, in part due to their association with Hfq (Denham 2020). The best characterized sponge RNAs, which will be described in more detail, are the intergenic region between *chbB* and *chbC*, SroC, and the external transcribed spacer of the $tRNA^{Leu}$ ($3'ETS^{LeuZ}$).

6.1 Intergenic Region of the chbBC *Transcript*

The first sponge RNA described in bacteria was the intergenic region between *chbB* and *chbC*, in *Salmonella enterica* (Figueroa-Bossi et al. 2009). Immediately after the discovery of this mechanism in *Salmonella*, it was reported its existence in *E. coli* (Rasmussen et al. 2009). The translocation of chitin products to the cytosol (chitobiose and chitotriose that can be used as carbon and nitrogen sources) is carried out by the membrane protein porin ChiP which is repressed by the sRNA ChiX (Plumbridge et al. 2014). This sRNA ChiX also inhibits the expression of the chitin PTS encoded by the *chbBCARFG* operon. However, the *chbBCARFG* operon gives rise to a polycistronic mRNA that upon binding of ChiX is rapidly cleaved by RNase E, releasing an intercistronic spacer (around 400 nucleotides) between the *chbB* and *chbC* genes. This intergenic sequence will then act as sponge RNA capable of sequestering ChiX and directing it to degradation by RNase E (Overgaard et al. 2009; Figueroa-Bossi et al. 2009). This tight regulation allows the controlled expression of the chitin utilization operon: when chitosugars are present, the *chb* operon is transcriptionally upregulated and the repression of ChiP expression is relieved by the binding of the sRNA ChiX to the intercistronic spacer sequence between *chbB* and *chbC* (Denham 2020).

6.2 $3'ETS^{LeuZ}$

Bacterial tRNAs are derived from a polycistronic transcript, which contains external transcribed spacers (ETS), on the 5′ and 3′-ends, and sometimes internal transcribed spacers (ITS). For the maturation of tRNAs to occur, is necessary the action of endo and exoribonucleases separating the tRNAs and releasing the ITS and ETS, which were initially thought to be immediately degraded (Grüll and Massé 2019). Processing of the polycistronic tRNA transcript *glyW-cysT-leuZ* is done by RNase E, giving rise to three tRNA precursors and releasing a small fragment corresponding to the $3'ETS^{LeuZ}$, a sponge RNA with ~50 nts (Lalaouna et al. 2015). The $3'ETS^{LeuZ}$ is involved in the regulation of two sRNAs, RyhB and RybB, which are respectively related to the regulation of iron homeostasis and the integrity of the outer membrane (Massé et al. 2005; Salvail et al. 2010). In the absence of stress, $3'ETS^{LeuZ}$ sequesters RyhB and RybB, with no change in the expression of the targets of these sRNAs. However, in the presence of iron or envelope stress for RyhB or RybB, respectively, sRNA expression increases significantly. In this manner, the excess of sRNAs produced in comparison to the sponge RNA enables them to act on their target mRNAs, triggering the stress response. The regulation of RyhB and RybB by $3'ETS^{LeuZ}$ establishes a relationship between two essential metabolisms in the bacterial cell, iron homeostasis and the envelope stress response (Lalaouna et al. 2015).

For the interaction between the $3'ETS^{LeuZ}$ and the target sRNAs to occur, it is necessary the intervention of the RNA chaperone Hfq. $3'ETS^{LeuZ}$ binds to the distal and proximal faces of Hfq, although it is not yet known whether these interactions occur simultaneously. Since this sponge RNA has a shorter single stranded 3′ poly(U) tail, the interaction with the proximal face of the Hfq is not very strong, serving only to assist in binding to other regions of the protein. Thus, the proximal face of the

Hfq seems to have more importance in the stabilization of ternary complexes formed between the sponge RNA and the target, while the distal face binds to the sponge RNA to allow greater exposure of the sequence to base pair with the target. On the other hand, the interaction of RyhB and RybB with Hfq takes place in the proximal face and the rim of Hfq (Małecka et al. 2021). In silico predictions revealed that $3'ETS^{LeuZ}$ pairs with the central loop of RyhB and with the 5′-end of RybB, which are the regions of interaction between these sRNAs and the target mRNAs. After the discovery of this sponge RNA, it was already suggested that some 3′ETS of certain tRNAs may also perform sponge RNA functions, given their sequence conservation and size to pair with other RNAs (Lalaouna et al. 2015). However, up to now, this mechanism has not yet been found with other tRNA intergenic spacers nor in other species.

6.3 SroC

Two different transcripts are synthesized from the *gltIJKL* operon in *Salmonella*: (i) a larger transcript that results from the transcription of the entire operon; (ii) and a smaller one that is formed due to the presence of a Rho-dependent terminator between *gltI* and *gltJ*. The full-length transcript encodes for the glutamate/aspartate ATP-binding cassette transporter (Denham 2020). The 3′-end of the smaller transcript is cleaved by RNase E, originating SroC, a regulatory RNA fragment with 150 nts that binds to Hfq (Sittka et al. 2008; Chao et al. 2012). SroC is a sponge RNA which base pairs with the sRNA GcvB, leading to its inhibition and degradation by RNase E (Fig. 3). GcvB is master regulator, controlling the expression of 1% of *Salmonella* genes, especially in amino acid transport and biosynthesis genes (Sharma et al. 2011). GcvB is highly expressed during faster bacterial growth in a rich medium, suggesting that the main function of this sRNA is the optimization of energy consumption for the synthesis and transport of amino acids (Sharma et al. 2007). The SroC sponge RNA pairs with two distinct regions of GcvB, which are relatively distant from each other (Lalaouna et al. 2018). While the binding sites are only 14 nts apart in SroC, the distance between the binding sequences in GcvB is 137 nts (Sittka et al. 2008).

7 3′-UTR-Derived sRNAs

While the firstly identified sRNAs were monocistronic units transcribed from their own promoters, we now know that sRNAs can also derive from mRNA processing. Consequently, sRNAs can also originate from the processing of intergenic regions (Argaman et al. 2001), 5′-UTRs (Vanderpool and Gottesman 2004), 3′-UTRs (Chao et al. 2012), protein coding sequences (Dar and Sorek 2018) and pre-tRNAs (Lalaouna et al. 2015). An emerging class of regulatory RNAs includes those that come from fragments of the 3′-UTRs of bacterial mRNAs, and we will present this in more detail. The 3′-UTR-derived sRNAs are widely distributed in bacteria and control a broad variety of biological processes, they can act either in *trans*, directed to one or multiple mRNAs, or in *cis*, regulating the synthesis of the parental mRNA, usually blocking translation (Hoyos et al. 2020; Menendez-Gil and Toledo-Arana 2021).

The 3′-UTR-derived sRNAs represent an example of the differentiation between the regulatory and coding functions through the formation of two different RNA species (Mediati et al. 2021). sRNAs that derive from 3′-UTR regions can be classified into two categories based on their origin. Type I or monocistronic sRNAs are generated due to the presence of an internal promoter positioned in the coding sequence or immediately downstream. On the other hand, type II or polycistronic sRNAs result from the processing of the 3′-UTR region of an mRNA (Miyakoshi et al. 2015; Hoyos et al. 2020). Chemically, it is possible to distinguish the two types due to the presence of a 5′ triphosphate in type I sRNAs and a 5′ monophosphate in type II sRNAs (Miyakoshi et al. 2015). The majority of the 3′-UTR-derived sRNAs described were found in Gram-negative bacteria and belong to type II, reinforcing the important riboregulatory role of ribonucleases, for instance, RNase E in Gram-negative bacteria (Hoyos et al. 2020). However, RNase E is not conserved in all bacteria, being absent in most Gram-positive bacteria and, therefore, biogenesis of 3′-UTR-derived sRNAs may occur through a different mechanism in these bacteria (Ponath et al. 2022). A prominent endoribonuclease in Gram-positive bacteria that can process these sRNAs is RNase III. In fact, the role of RNase III was already described for the synthesis of RsaC in *Staphylococcus aureus*. This type II sRNA is generated from the *mntABC* operon mRNA, which encodes the largest manganese importer, hence it is only upregulated under manganese limiting conditions (Lioliou et al. 2012). One possible explanation for the smaller number of 3′-UTR-derived sRNAs in Gram-positive bacteria is that these bacteria have enzymes with 5′-3′ exoribonucleolytic activity, as RNase J1. Thus, these sRNAs are rapidly degraded unless a hairpin form at the 5′-end to stabilize it (Desgranges et al. 2022). In addition, there are also not as many processing sites near the stop codon described in Gram-positive bacteria when compared with Gram-negative bacteria (Mediati et al. 2021).

Despite the variability of 3′-UTR-derived sRNAs, it is possible to predict the formation of a sRNA from a 3′-UTR, due to some common characteristics. One of them is the formation of a stem-loop structure followed by a uridine tail, a preferred binding site of Hfq (Sauer and Weichenrieder 2011). Another is the identification of cleavage sites for RNases and conserved seed sequences that allow to base pair with the target mRNA (Miyakoshi et al. 2015). Besides, a differential expression pattern between the fragments produced from the 3′-UTRs and the corresponding mRNAs, also suggests different functions.

8 CRISPRs

A novel RNA-based system in prokaryotes has been discovered, revolutionizing gene editing systems (Jansen et al. 2002; Brouns et al. 2008). Clustered regularly interspaced short palindromic repeats (CRISPRs) constitute the key components of an adaptive immune system that can be found widespread in many archaea and bacteria (Jinek et al. 2012; Makarova et al. 2013). In association with specific Cas proteins, this system operates in a similar manner to RNA interference modules, using structured small RNAs to recognize and silence target nucleic acids, such as viral genetic material, plasmids and other mobile genetic elements (Barrangou et al. 2007). CRISPR-Cas systems have been divided into three major CRISPR-Cas types (Type I and Type III present in both archaea and bacteria and Type II just in the latter; then divided into additional subtypes) due to their variety in *cas* gene composition and consequent proteins necessary for the immune response to take effect (Makarova et al. 2011). Standard CRISPR *loci* are comprised of a CRISPR array, i.e., short direct repeats, interspersed with spacer sequences (invader-derived short variable DNA sequences) and a leader side directly upstream of the first repeat; these *loci* then tend to be flanked by *cas* genes (Jansen et al. 2002). All known CRISPR-Cas systems share a modular mechanism for gene silencing: mature CRISPR RNAs (crRNAs) that contain a unique spacer sequence responsible for recognizing complementary invading genetic material and guiding Cas proteins to degrade it (Chylinski et al. 2014; Charpentier et al. 2015).

Canonical CRISPR-Cas-based immunity consists of three distinctive stages: adaptation, expression, and interference. The adaptation stage is triggered by the presence of exogenous genetic material inside the bacteria. Viral or plasmid DNA is recognized, processed into small fragments known as protospacers and incorporated into the CRISPR array as new DNA spacers (Safari et al. 2019). In most cases, an initial viral attack results in the integration of a single 30 base pairs unique resistance-conferring spacer immediately downstream of the leader side of a CRISPR *locus*, followed by the duplication of the repeat in order to originate a new spacer–repeat unit (Makarova et al. 2011). The subsequent step is expression, and it consists in the expression and processing of the CRISPR RNA (crRNA or guide RNA), which is essential for the correct function of the CRISPR-Cas system. crRNA maturation can be divided into three distinct events (Safari et al. 2019): (i) an approximately 60–70-nucleotide long initial primary transcript of the CRISPR array is produced, named precursor crRNA (pre-crRNA). This RNA is transcribed from a promoter harbored in leader sequence upstream repeat-spacer array; (ii) pre-crRNA cleavage by nucleases at specific recognition sites in the repeats, resulting in a mature crRNA made up of the complete spacer sequence between two partial repeat sequences; (iii) occasionally, some transcripts undergo additional secondary processing to generate mature crRNA (Carte et al. 2008; Charpentier et al. 2015). While in CRISPR-Cas systems classified as Type I and III, a specific RNase belonging to the Cas6 family cleaves the pre-crRNA, in Type II systems an auxiliary *trans*-acting small RNA (tracrRNA) independently base pairs with each repeat sequence of the pre-crRNA and associates with the Cas9 protein, resulting in a dual-RNA which is subsequently cleaved by RNase III (Fig. 4) (Deltcheva et al. 2011). The third and final event of CRISPR-based immunity is interference. crRNA, in association with Cas effector endonucleases, recognizes and base pairs to cognate invader DNA sequences that match the prokaryote's spacers. This process is then followed by recognition of the target DNA by Cas effectors and cleaving of the nucleic acids within the protospacer sequence, leading to double-stranded DNA breaks and subsequent degradation of the invader DNA (Jinek et al. 2012). Different Cas proteins can have a role in one or several steps of CRISPR–Cas gene silencing, acting in most cases in association with other proteins.

CRISPR-based immunity is highly specific, as it has been demonstrated that a single mismatch between the spacer sequence and the target exogenous DNA completely nullified phage resistance in bacteria (Barrangou et al. 2007). Beyond primary nucleic acid sequence the repeat sequences in the flanking region usually

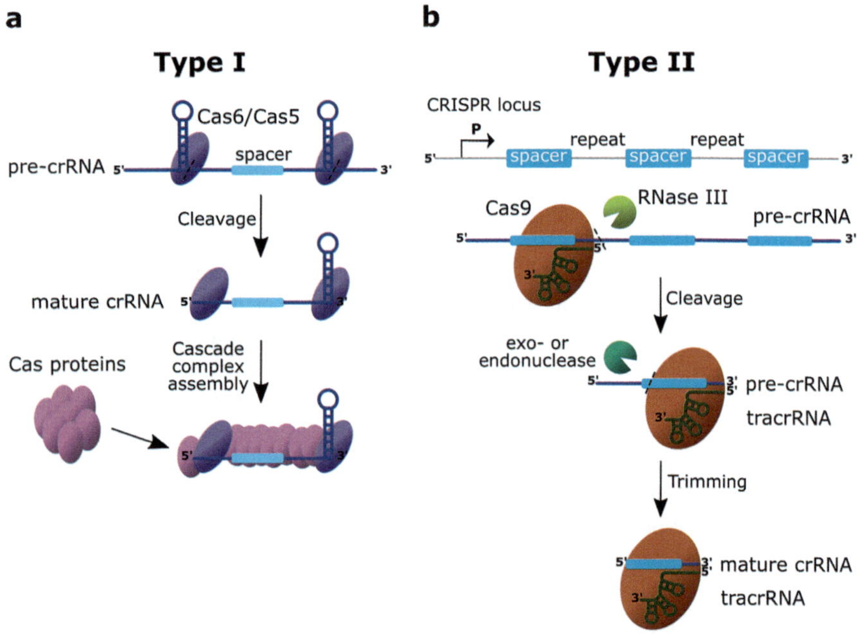

Fig. 4 CRISPR RNA processing mechanisms in Type I and Type II CRISPR-Cas systems. **a** In most Type I CRISPR-Cas systems, the short palindromic repeats form hairpin structures can be recognized and cleaved by nucleases Cas6 or Cas5, depending on the system subtype; the cleaved CRISPR RNA (crRNA) continues in association with the nuclease, and additional Cas subunits (forming the Cascade complex) bind to the 5′-end and spacer regions of the crRNA, which can consequently interact with and silence their specific target genetic element. **b** Type II CRISPR-Cas systems can be distinguished by the formation of duplexes with the pre-crRNA and the *trans*-acting tracrRNA; the latter helps in the recognition of the Cas9 protein, further stabilizing the RNA duplex which is then processed by RNase III cleavage. An additional processing step by still unknown endo- or exoribonucleases then generates the mature crRNAs (adapted from Charpentier et al. 2015)

adopt secondary structures, more specifically a partial duplex at the 5′-end and a hairpin structure at the 3′-end (Jore et al. 2011; Gu et al. 2019).

9 Conclusions

Small RNAs are an abundant and diverse group of non-coding RNAs present in bacteria. Diversity is the keyword that better describes these regulatory RNA molecules. As we have shown, sRNAs can act as repressors or activators of gene expression; they can be synthesized as independent transcriptional units or derive from mRNA processing; they can make use of different domains to interact with multiple targets; and they can associate with different RNA-binding proteins, among many other features. Importantly, sRNAs rely extensively on their sequence and structure for their function(s), as nicely illustrated in the various examples here provided. Overall, sRNA networks are widely generalized in bacteria. It is anticipated that research in this exciting field will continue to reveal additional players and new RNA features in the years to come.

Acknowledgements This work was supported by FCT—Fundação para a Ciência e a Tecnologia, I.P., through: MOSTMICRO-ITQB R&D Unit (UIDB/04612/2020, UIDP/04612/2020); LS4FUTURE Associated Laboratory (LA/P/0087/2020); Grant PTDC/BIA-MIC/32525/2017 to JMA; Doctoral Fellowships to AFQS (UI/BD/153390/2022) and JPS (2022.12475.BD).

References

Altuvia S, Zhang A, Argaman L et al (1998) The Escherichia coli OxyS regulatory RNA represses *fhlA* translation by blocking ribosome binding. EMBO J 17:6069–6075

Andrade JM, Arraiano CM (2008) PNPase is a key player in the regulation of small RNAs that control the expression of outer membrane proteins. RNA 14:543–551

Andrade JM, Pobre V, Arraiano CM (2013) Small RNA modules confer different stabilities and interact differently with multiple targets. PLoS ONE 8:e52866

Andrade JM, Pobre V, Matos AM et al (2012) The crucial role of PNPase in the degradation of small RNAs that are not associated with Hfq. RNA 18:844–855

Andrade JM, Santos RF, Chelysheva I et al (2018) The RNA-binding protein Hfq is important for ribosome biogenesis and affects translation fidelity. EMBO J 37:e97631

Argaman L, Hershberg R, Vogel J et al (2001) Novel small RNA-encoding genes in the intergenic regions of *Escherichia coli*. Curr Biol 11:941–950

Barrangou R, Fremaux C, Deveau H et al (2007) CRISPR provides acquired resistance against viruses in prokaryotes. Science 315:1709–1712

Bellaousov S, Reuter JS, Seetin MG et al (2013) RNAstructure: web servers for RNA secondary structure prediction and analysis. Nucleic Acids Res 41:W471–W474

Brouns SJJ, Jore MM, Lundgren M et al (2008) Small CRISPR RNAs guide antiviral defense in prokaryotes. Science 321:960–964

Cai H, Roca J, Zhao Y-F et al (2022) Dynamic refolding of OxyS sRNA by the Hfq RNA chaperone. J Mol Biol 434:167776

Carte J, Wang R, Li H et al (2008) Cas6 is an endoribonuclease that generates guide RNAs for invader defense in prokaryotes. Genes Dev 22:3489–3496
Chao Y, Papenfort K, Reinhardt R et al (2012) An atlas of Hfq-bound transcripts reveals 3′ UTRs as a genomic reservoir of regulatory small RNAs. EMBO J 31:4005–4019
Charpentier E, Richter H, van der Oost J et al (2015) Biogenesis pathways of RNA guides in archaeal and bacterial CRISPR-Cas adaptive immunity. FEMS Microbiol Rev 39:428–441
Chylinski K, Makarova KS, Charpentier E et al (2014) Classification and evolution of type II CRISPR-Cas systems. Nucleic Acids Res 42:6091–6105
Dar D, Sorek R (2018) High-resolution RNA 3′-ends mapping of bacterial Rho-dependent transcripts. Nucleic Acids Res 46:6797–6805
Deltcheva E, Chylinski K, Sharma CM et al (2011) CRISPR RNA maturation by trans-encoded small RNA and host factor RNase III. Nature 471:602–607
Denham EL (2020) The Sponge RNAs of bacteria – How to find them and their role in regulating the post-transcriptional network. Biochim Biophys Acta Gene Regul Mech 1863:194565
Desgranges E, Barrientos L, Herrgott L et al (2022) The 3′UTR-derived sRNA RsaG coordinates redox homeostasis and metabolism adaptation in response to glucose-6-phosphate uptake in *Staphylococcus aureus*. Mol Microbiol 117:193–214
dos Santos RF, Andrade JM, Pissarra J et al (2020) Hfq and RNase R mediate rRNA processing and degradation in a novel RNA quality control process. Mbio 11:e02398-e2420
dos Santos RF, Arraiano CM, Andrade JM (2019) New molecular interactions broaden the functions of the RNA chaperone Hfq. Curr Genet 65:1313–1319
Dubey AK, Baker CS, Suzuki K et al (2003) CsrA Regulates translation of the escherichia coli carbon starvation gene, cstA, by blocking ribosome access to the cstA transcript. J Bacteriol 185:4450–4460
Ebert MS, Neilson JR, Sharp PA (2007) MicroRNA sponges: competitive inhibitors of small RNAs in mammalian cells. Nat Meth 4:721–726
Figueroa-Bossi N, Valentini M, Malleret L et al (2009) Caught at its own game: regulatory small RNA inactivated by an inducible transcript mimicking its target. Genes Dev 23:2004–2015
Fröhlich KS, Gottesman S (2018) Small regulatory RNAs in the enterobacterial response to envelope damage and oxidative stress. Microbiol Spectr 6:211–228
Fuchs M, Lamm-Schmidt V, Sulzer J et al (2021) An RNA-centric global view of *Clostridioides difficile* reveals broad activity of Hfq in a clinically important gram-positive bacterium. Proc Natl Acad Sci USA 118:e2103579118
Gogol EB, Rhodius VA, Papenfort K et al (2011) Small RNAs endow a transcriptional activator with essential repressor functions for single-tier control of a global stress regulon. Proc Natl Acad Sci USA 108:12875–12880
González-Flecha B, Demple B (1999) Role for the *oxyS* Gene in Regulation of Intracellular Hydrogen Peroxide in *Escherichia coli*. J Bacteriol 181:3833–3836
Grüll MP, Massé E (2019) Mimicry, deception and competition: the life of competing endogenous RNAs. Wires RNA 10:e1525
Gu D-H, Ha SC, Kim J-S (2019) A CRISPR RNA is closely related with the size of the cascade nucleoprotein complex. Front Microbiol 10:2458
Guillier M, Gottesman S (2008) The 5′ end of two redundant sRNAs is involved in the regulation of multiple targets, including their own regulator. Nucleic Acids Res 36:6781–6794
Henderson CA, Vincent HA, Casamento A et al (2013) Hfq binding changes the structure of *Escherichia coli* small noncoding RNAs OxyS and RprA, which are involved in the riboregulation of *rpoS*. RNA 19:1089–1104
Hoekzema M, Romilly C, Holmqvist E et al (2019) Hfq-dependent mRNA unfolding promotes sRNA-based inhibition of translation. EMBO J 38:e101199
Hoyos M, Huber M, Förstner KU et al (2020) Gene autoregulation by 3' UTR-derived bacterial small RNAs. Elife 9:e58836
Jansen R, van Embden JDA, Gaastra W et al (2002) Identification of genes that are associated with DNA repeats in prokaryotes. Mol Microbiol 43:1565–1575

Jinek M, Chylinski K, Fonfara I et al (2012) A programmable dual-RNA–guided DNA endonuclease in adaptive bacterial immunity. Science (80-) 337:816–821

Jore MM, Lundgren M, van Duijn E et al (2011) Structural basis for CRISPR RNA-guided DNA recognition by Cascade. Nat Struct Mol Biol 18:529–536

Kawamoto H, Koide Y, Morita T et al (2006) Base-pairing requirement for RNA silencing by a bacterial small RNA and acceleration of duplex formation by Hfq. Mol Microbiol 61:1013–1022

Lalaouna D, Carrier M-C, Semsey S et al (2015) A 3′ external transcribed spacer in a tRNA transcript acts as a sponge for small RNAs to prevent transcriptional noise. Mol Cell 58:393–405

Lalaouna D, Eyraud A, Devinck A et al (2018) GcvB small RNA uses two distinct seed regions to regulate an extensive targetome. Mol Microbiol 111:303–551

Lee T, Feig AL (2008) The RNA binding protein Hfq interacts specifically with tRNAs. RNA 14:514–523

Lioliou E, Sharma CM, Caldelari I et al (2012) Global regulatory functions of the *Staphylococcus aureus* endoribonuclease III in gene expression. PLoS Genet 8:e1002782

Liu MY, Gui G, Wei B et al (1997) The RNA molecule CsrB binds to the global regulatory protein CsrA and antagonizes its activity in *Escherichia coli*. J Biol Chem 272:17502–17510

Makarova KS, Haft DH, Barrangou R et al (2011) Evolution and classification of the CRISPR–Cas systems. Nat Rev Microbiol 9:467–477

Makarova KS, Wolf YI, Koonin EV (2013) Comparative genomics of defense systems in archaea and bacteria. Nucleic Acids Res 41:4360–4377

Maki K, Morita T, Otaka H, Aiba H (2010) A minimal base-pairing region of a bacterial small RNA SgrS required for translational repression of *ptsG* mRNA. Mol Microbiol 76:782–792

Maki K, Uno K, Morita T et al (2008) RNA, but not protein partners, is directly responsible for translational silencing by a bacterial Hfq-binding small RNA. Proc Natl Acad Sci USA 105:10332–10337

Małecka EM, Sobańska D, Olejniczak M (2021) Bacterial chaperone protein Hfq facilitates the annealing of sponge RNAs to small regulatory RNAs. J Mol Biol 433:167291

Massé E, Vanderpool CK, Gottesman S (2005) Effect of RyhB small RNA on global iron use in *Escherichia coli*. J Bacteriol 187:6962–6971

Mediati DG, Lalaouna D, Tree JJ (2021) Burning the candle at both ends: have exoribonucleases driven divergence of regulatory RNA mechanisms in bacteria? Mbio 12:e01041-e1121

Melamed S, Peer A, Faigenbaum-Romm R et al (2016) Global mapping of small RNA-target interactions in bacteria. Mol Cell 63:884–897

Menendez-Gil P, Toledo-Arana A (2021) Bacterial 3′UTRs: a useful resource in post-transcriptional regulation. Front Mol Biosci 7:617633

Miyakoshi M, Chao Y, Vogel J (2015) Regulatory small RNAs from the 3′ regions of bacterial mRNAs. Curr Opin Microbiol 24:132–139

Morita T, Nishino R, Aiba H (2017) Role of the terminator hairpin in the biogenesis of functional Hfq-binding sRNAs. RNA 23:1419–1431

Overgaard M, Johansen J, Møller-Jensen J et al (2009) Switching off small RNA regulation with trap-mRNA. Mol Microbiol 73:790–800

Panja S, Schu DJ, Woodson SA (2013) Conserved arginines on the rim of Hfq catalyze base pair formation and exchange. Nucleic Acids Res 41:7536–7546

Papenfort K, Bouvier M, Mika F et al (2010) Evidence for an autonomous 5′ target recognition domain in an Hfq-associated small RNA. Proc Natl Acad Sci USA 107:20435–20440

Peschek N, Hoyos M, Herzog R et al (2019) A conserved RNA seed-pairing domain directs small RNA-mediated stress resistance in enterobacteria. EMBO J 38:e101650

Plumbridge J, Bossi L, Oberto J et al (2014) Interplay of transcriptional and small RNA-dependent control mechanisms regulates chitosugar uptake in *Escherichia coli* and *Salmonella*. Mol Microbiol 92:648–658

Poddar A, Azam MS, Kayikcioglu T et al (2021) Effects of individual base-pairs on in vivo target search and destruction kinetics of bacterial small RNA. Nat Commun 12:874

Ponath F, Hör J, Vogel J (2022) An overview of gene regulation in bacteria by small RNAs derived from mRNA 3′ ends. FEMS Microbiol Rev 46:fuac017

Quendera AP, Seixas AF, dos Santos RF et al (2020) RNA-binding proteins driving the regulatory activity of small non-coding RNAs in bacteria. Front Mol Biosci 7:78

Rasmussen AA, Johansen J, Nielsen JS et al (2009) A conserved small RNA promotes silencing of the outer membrane protein YbfM. Mol Microbiol 72:566–577

Rice JB, Vanderpool CK (2011) The small RNA SgrS controls sugar–phosphate accumulation by regulating multiple PTS genes. Nucleic Acids Res 39:3806–3819

Romeo T, Gong M, Liu MY et al (1993) Identification and molecular characterization of *csrA*, a pleiotropic gene from *Escherichia coli* that affects glycogen biosynthesis, gluconeogenesis, cell size, and surface properties. J Bacteriol 175:4744–4755

Safari F, Zare K, Negahdaripour M et al (2019) CRISPR Cpf1 proteins: structure, function and implications for genome editing. Cell Biosci 9:36

Salvail H, Lanthier-Bourbonnais P, Sobota JM et al (2010) A small RNA promotes siderophore production through transcriptional and metabolic remodeling. Proc Natl Acad Sci USA 107:15223–15228

Sauer E, Weichenrieder O (2011) Structural basis for RNA 3′-end recognition by Hfq. Proc Natl Acad Sci U S A 108:13065–13070

Seixas AF, Quendera AP, Sousa JP et al (2022) Bacterial response to oxidative stress and RNA oxidation. Front Genet 12:821535

Sharma CM, Darfeuille F, Plantinga TH et al (2007) A small RNA regulates multiple ABC transporter mRNAs by targeting C/A-rich elements inside and upstream of ribosome-binding sites. Genes Dev 21:2804–2817

Sharma CM, Papenfort K, Pernitzsch SR et al (2011) Pervasive post-transcriptional control of genes involved in amino acid metabolism by the Hfq-dependent GcvB small RNA. Mol Microbiol 81:1144–1165

Silva IJ, Barahona S, Eyraud A et al (2019) SraL sRNA interaction regulates the terminator by preventing premature transcription termination of rho mRNA. Proc Natl Acad Sci USA 116:3042–3051

Sittka A, Lucchini S, Papenfort K et al (2008) Deep sequencing analysis of small noncoding RNA and mRNA targets of the global post-transcriptional regulator, Hfq. Plos Genet 4:e1000163

Suzuki K, Babitzke P, Kushner SR et al (2006) Identification of a novel regulatory protein (CsrD) that targets the global regulatory RNAs CsrB and CsrC for degradation by RNase E. Genes Dev 20:2605–2617

Suzuki K, Wang X, Weilbacher T et al (2002) Regulatory circuitry of the CsrA/CsrB and BarA/UvrY systems of *Escherichia coli*. J Bacteriol 184:5130–5140

Udekwu KI, Darfeuille F, Vogel J et al (2005) Hfq-dependent regulation of OmpA synthesis is mediated by an antisense RNA. Genes Dev 19:2355–2366

Vakulskas CA, Pannuri A, Cortés-Selva D et al (2014) Global effects of the DEAD-box RNA helicase DeaD (CsdA) on gene expression over a broad range of temperatures. Mol Microbiol 92:945–958

Vakulskas CA, Potts AH, Babitzke P et al (2015) Regulation of bacterial virulence by Csr (Rsm) systems. Microbiol Mol Biol Rev 79:193–224

Vanderpool CK, Gottesman S (2004) Involvement of a novel transcriptional activator and small RNA in post-transcriptional regulation of the glucose phosphoenolpyruvate phosphotransferase system. Mol Microbiol 54:1076–1089

Vanderpool CK, Gottesman S (2007) The novel transcription factor SgrR coordinates the response to glucose-phosphate stress. J Bacteriol 189:2238–2248

Viegas SC, Silva IJ, Saramago M et al (2011) Regulation of the small regulatory RNA MicA by ribonuclease III: a target-dependent pathway. Nucleic Acids Res 39:2918–2930

Wadler CS, Vanderpool CK (2007) A dual function for a bacterial small RNA: SgrS performs base pairing-dependent regulation and encodes a functional polypeptide. Proc Natl Acad Sci U S A 104:20454–20459

Wang L, Wang W, Li F et al (2015) Structural insights into the recognition of the internal A-rich linker from OxyS sRNA by *Escherichia coli* Hfq. Nucleic Acids Res 43:2400–2411
Woodson SA, Panja S, Santiago-Frangos A (2018) Proteins that chaperone RNA regulation. Microbiol Spectr 6:21
Yakhnin AV, Baker CS, Vakulskas CA et al (2013) CsrA activates *flhDC* expression by protecting *flhDC* mRNA from RNase E-mediated cleavage. Mol Microbiol 87:851–866
Zhang A, Schu DJ, Tjaden BC et al (2013) Mutations in interaction surfaces differentially Impact *E. coli* Hfq association with small RNAs and their mRNA targets. J Mol Biol 425:3678–3697

A Moveable Feast. Molecular Modeling and Simulation Unraveling Cross-Talks Between RNA Structure and Its Biological Role

Aurane Froux, Emmanuelle Bignon, Guillaume Harlé, Stéphanie Grandemange, and Antonio Monari

Contents

Abstract We highlight the role played by molecular modeling and simulation to unravel, at an atomistic and even electronic scale, the complex structural dynamic equilibrium assumed by RNA sequences, either cellular or viral. After pointing out the role played by specific RNA structures in regulating key biological functions, or in assuring either viral replication or immune system activation, we will show how computationally efficient multiscale approaches lead to the understanding of the fundamental biological processes, in terms of nucleic acid structural dynamic

A. Froux
Department of Biological, Chemical and Pharmaceutical Sciences and Technologies, Università Degli Studi Di Palermo, Viale Delle Scienze, 90128 Palermo, Italy

E. Bignon (✉)
Université de Lorraine, CNRS, LPCT, UMR 7019, F-54000 Nancy, France
e-mail: Emmanuelle.bignon@univ-lorraine.fr

G. Harlé (✉) · S. Grandemange (✉)
Université de Lorraine, CNRS, CRAN, UMR 7039, F-54000 Nancy, France
e-mail: stephanie.grandemange@univ-lorraine.fr

A. Monari (✉)
Université Paris Cité and CNRS, ITODYS, F-75006 Paris, France
e-mail: Antonio.monari@u-paris.fr

J. Barciszewski (ed.), *RNA Structure and Function*, RNA Technologies 14,
https://doi.org/10.1007/978-3-031-36390-0_13

and their interaction with protein partners, and hence may be successfully used to rationally develop novel therapeutic approaches. By a selection of examples involving cellular and viral RNA, we will show that molecular modeling and simulation is nowadays assuming its role of a virtual microscope complementing and deepening the insight gained by structural and cellular biology.

Keywords RNA secondary structure · Cellular signaling pathways · RNA viruses · Immune system response · Classical molecular dynamics · Theoretical spectroscopy

1 Introduction

1.1 RNA Structures and Biological Role

Ribonucleic acid (RNA) is widely present in organisms (fungi, bacteria, eucaryotes) and viruses. RNAs present different structural similarities with DNA but are mostly present as a single-stranded nucleic acid. RNA backbone is constituted by a succession of phosphate group and ribose sugar in opposition to deoxyribose found in DNA. Different types of RNA are characterized such as messenger RNA (mRNA), ribosomal RNA (rRNA), transfer RNA (tRNA), and other non-coding RNA (long non-coding RNA, lncRNA; micro-RNA, miRNA, etc.).

Beside its nucleotide sequence, RNA structure is crucial for RNA functions, stability and regulation (Kamura et al. 2020; Xu et al. 2022). Indeed, RNA is capable of folding into complex higher-order structures, via base pairing across short or long interaction ranges (Xu et al. 2022). These structures include RNA helices or stem loops, bulges, Guanine-quadruplexes (G4), and define the RNA structure, whose evolution respond to environmental stimuli and stresses (Endoh and Sugimoto 2019). The RNA structural motifs are also influenced by the cellular environment including temperature, pH, ions concentration, as well as surrounding proteins, defined as *trans*-factors (Endoh and Sugimoto 2019). Multimorphism allows RNA to perform other diverse biological functions, besides the simple transfer of genetic information (Kamura et al. 2020). As a matter of fact, RNA structures are involved in the regulation of all the steps of the cell life events, particularly in transcription, splicing, nuclear export, trafficking and RNA stability.

RNA secondary structures are also present in bacteria and viruses and are particularly dense in the 5' and 3' untranslated regions (UTR) of the genome. These regions are important as *cis*-regulatory RNA elements, allowing the regulation of the gene expression in cells, via the modulation of translational component interactions and/or by regulating the stability of mRNA as shown in Fig. 1 (Xu et al. 2022). In RNA viruses, UTRs participate in the regulation of viral replication and are also associated with inflammatory capacity and with the recognition by the host innate

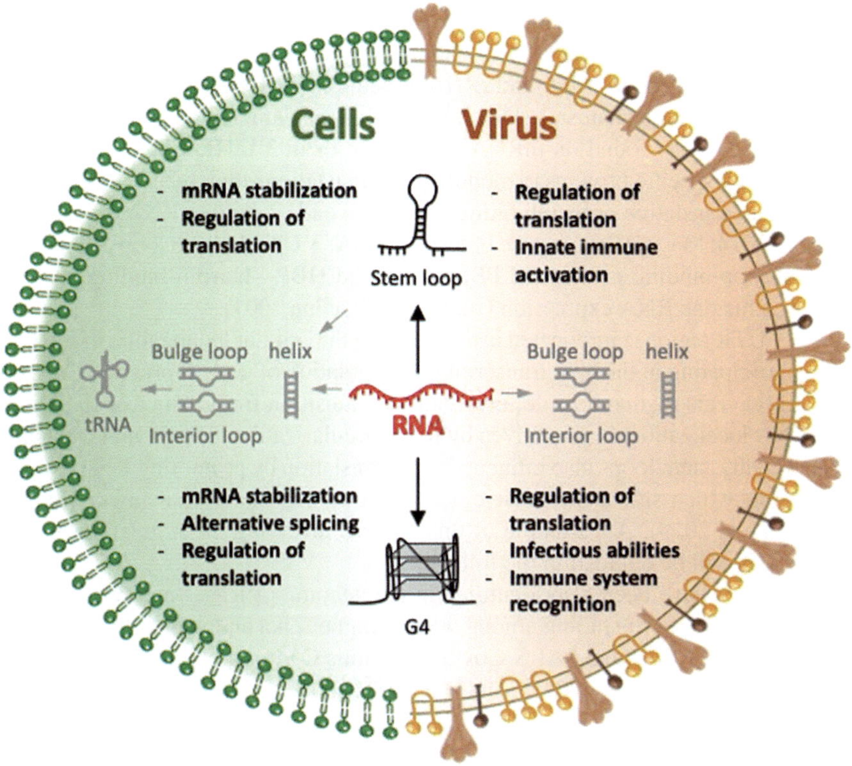

Fig. 1 RNA structures and functions in cells and viruses. Illustration of various RNA secondary structures focusing on stem-loop and G-quadruplex (G4) formation and their role

immune system (Fig. 1). Among the multiple secondary structures, we will focus our discussion on two major structures: stem loop and G4.

1.2 Stem Loops

Stem loops also known as hairpin loops can be formed by two complementary sequences of single-stranded RNA paired together. This matching leads to a double helical structure capped by a terminal loop composed by the connecting unpaired bases. Stem-loop stability depends on the length of the double helix, the specific sequence, and the presence of mismatches. Usually, smaller loops impart a higher structural stability than large loops, while cations, such as Na^+ and Mg^{2+}, stabilize the stem loop by electrostatically compensating the backbone negative charges (Tan and Chen 2008). Kinetic studies revealed that stem-loop structures fold rather rapidly with timescale between 0.4 and 8 microseconds depending on the size of the helix and loop (Kuznetsov et al. 2008). They play crucial roles in different biological functions especially by modulating RNAs stability, localization, translation mechanisms. For example, they shape tRNA and rRNA structure, which is crucial for their functions in translation mechanisms (Di Giulio 2004; Branciamore and Di Giulio 2011). Stem loops are also found on mRNA either in 5' or in 3'UTR, where they can be recognized by specific proteins to modulate the translation either by promoting RNA stability and translation or by inhibiting translation and promoting RNA degradation (Re et al. 2014; Mayr 2019). In case of histone RNA, 3'UTR hairpin loop is bound by the stem-loop binding protein (SLPB, also named HBP—hairpin-binding protein) leading to histone RNA expression (Battle and Doudna 2001).

The 3'UTR region is implicated in determining the cellular localization of mRNA, hence participating in the post-transcriptional regulation of gene expression (Yergert et al. 2021). This regulatory role, involving the transition from a nuclear to perinuclear RNA localization, is also driven by RNA circularization (Mayya and Duchaine 2019). Finally, stem loops also influence gene translation by promoting RNA/protein interactions which shield the nucleic acid from the ribosomal machinery (Mayya and Duchaine 2019). An example of this process involving the regulation of iron homeostasis will be detailed in the following.

Stem loops have been also identified in the genome of RNA- and retro-viruses such as Flaviviruses (including among other Dengue, Zika and yellow fever), Coronaviruses (SARS-CoV and SARS-Cov-2), Hepatitis C virus and human immunodeficiency virus (HIV) (Choi 2021; Vora et al. 2022). These structures shape virus life cycle by impacting viral RNA replication, translation, and immune system recognition (Yi and Lemon 2003; Dzananovic et al. 2013). Indeed, the recognition of stem loops in 5'UTR is fundamental to initiate viral expression and replication, for instance in coronaviruses binding of 5'UTR to RNA-binding motif 1 enhances replication.

1.3 Guanine Quadruplexes

Guanine-quadruplexes (G4) are non-canonical structures formed in a G-rich sequence environment. Under such conditions, four adjacent guanines stick together by Hoogsteen hydrogen bonds, leading to the formation of a planar motif, named G-tetrad, stabilized by the presence of a monovalent cation (mainly K^+). Although a large variability is found in the formation of G4s, they usually result from the stacking of at least 2 tetrads, joined by sugar-phosphate strands (loops) bearing free nitrogen bases (Kamura et al. 2020). Moreover, while DNA G4 can adopt different topologies depending on the strand direction, the RNA G4 mainly form compact parallel arrangements (Lyu et al. 2021).

Even if G4 formation is known to compete with other secondary structures (including stem loops) because of their slower kinetic of formation/dissociation, they are statistically present and highly conserved among mammalian species, as well as viruses (Kamura et al. 2020). Similarly to stem loops, RNA G4s are concentrated in 5' and 3'UTRs of mRNA, and are present in other non-coding RNA including lncRNA, tRNA, and rRNA (Tassinari et al. 2021; Varshney et al. 2021). From a biological perspective, G4s are highly involved in the regulation of translation, RNA metabolism, pre-mRNA processing, and immunity. In 5'UTR mRNA, G4s are acting as translation suppressor (Lyu et al. 2021), since the presence of a folded G4 will block the progression of the ribosome along the transcript and stop translation. In addition, 5'UTR mRNA constitutes a platform for several proteins and *trans*-factors involved in the cap-dependent translation regulation and efficiency, the presence of G4s influences the recruitment of these regulatory factors. However, RNA G4s can also lead to translation induction, especially for cap-independent translation (Kamura et al. 2020). Besides, G4s are involved in other RNA processes including transcript alternative polyadenylation, alternative splicing, miRNA targeting, and again mRNA localization, especially when present in the 3' UTR mRNA.

Viral RNA G4s have been described in all the groups of the Baltimore viral classification and are largely conserved (Ruggiero and Richter 2020) despite the high mutation rate of RNA viruses, pointing to their crucial role in viral evolution and replication. RNA G4s have been identified in regulatory regions, including the majority of the long terminal repeats (LTR) of the retrovirus family, and in part of other secondary structures, like stem loops of the 3'UTR. Furthermore, in the case of retroviruses, specific helicase unwinding the G4-folded region prior to reverse transcription is expressed, also suggesting the crucial role of these organized regions (Ruggiero and Richter 2020).

2 Molecular Modeling and Simulations

The recognition of the crucial functional roles of RNA in biological processes has fostered investigations aiming at better understanding the structure–function relationship underlying its mechanism of action. Defining the precise spatial organization assumed by a specific nucleic acids sequence can shed light on the presence of selective ligand-binding sites and RNA enzymatic activity. Advances in structural biology have allowed the release of 3D structures of different RNA folds, mainly by means of X-Ray crystallography, cryo-Electron Microscopy or Nuclear Magnetic Resonance spectroscopy (Kotar et al. 2020; Bonilla and Kieft 2022), sometimes assisted by computational modeling (Sripakdeevong et al. 2014; Kappel et al. 2020). In turn, these advances have offered new possibilities for rational RNA-targeting drug development.

Nevertheless, RNA exhibits a highly dynamic conformational behavior leading to specific folding patterns. This clearly poses supplementary burdens on the experimental structural determination, as attested by the small percentage (<1%) of RNA-only structures available on the Protein Data Bank. Even more fundamentally, such flexibility and conformational variability question the pertinence of a static picture, such as the one offered by crystallography, to catch the subtleties of the RNA biological role.

In this context, computational methods have proven their worth in the common effort to, not only palliate the lack of data related to RNA tertiary structures but also to elucidate their intrinsic dynamics (Schlick and Pyle 2017; Reinharz et al. 2021). A fair amount of algorithms have been developed to predict RNA secondary and tertiary structures from the primary sequence, often guided by integration of experimental data-based restraints and more recently by taking advantage of machine-learning approaches (Singh et al. 2019; Townshend et al. 2021; Yu et al. 2022).

As a matter of fact, the use of large-scale molecular dynamics (MD) simulations, which can nowadays reach, almost routinely, the μs time scale has paved the way to a totally different approach to the problem of RNA structure (Vangaveti et al. 2017; Šponer et al. 2018). Indeed, MD simulation may precisely describe complex and large-scale dynamic transitions with an atomistic resolution allowing the identification of specific hotspot for nucleic acid/protein interactions or drug design. The basis of MD simulations relies on the numerical resolution of Newton's equation of motion for each atom composing the systems, which should include the nucleic acid, its partners and the solvent, usually water. The progress of both MD algorithms and computer architecture, and in particular the use of graphical processing units (GPUs), has consistently pushed forward the limit of MD simulations, which today may be regarded as complementary experimental techniques offering a deeper and different view on crucial biological phenomena (Šponer et al. 2018).

Classical MD simulation relies on the description of the potential energy forces between the different atoms by an empirical function. Such empirical potential is known under the general name of force-field and its accuracy will clearly determine the quality, and the reliability, of the whole MD simulation. Specific force fields for nucleic acids, and DNA in particular are commonly developed and constantly refined, also taking profit from more precise structural determinations. Specific force fields for RNA structure have been developed, although they suffer from the additional complexity of having to accommodate different conformations and secondary structures. In particular, the reproduction of the correct behavior of the sugar and backbone dihedral angles is crucial to determine the accuracy of the force field. RNA force fields like those developed by Šponer and coworkers (Šponer et al. 2018) and integrated in the OL family have emerged as some of the more precise tools to correctly model the complex conformational behavior of RNA (Krepl et al. 2012). Undoubtedly, in the near future, the development of machine-learning-based force fields for RNA will constitute a game-changer in pushing forward the accuracy of MD simulations (Fröhlking et al. 2020).

Despite their increasing accuracy, classical force-fields are not adapted to describe bond formation and breaking nor electronically excited states. Hence, they cannot be used to understand spectroscopy or chemical reactivity, which would require the formalism of quantum chemistry. To cope with the increased computational cost of quantum chemistry approaches while preserving the description of the complex and inhomogeneous environment, the use of hybrid quantum mechanical/molecular mechanical (QM/MM) methods has emerged as the state-of-the-art technique (Boulanger and Harvey 2018). This approach exploits the local nature of chemical reactivity or electronic excitation to allow a partition between a relevant region, which will be treated at a higher quantum level of theory, and the larger environment which will be described following classical physics rules. QM/MM methods have allowed to correctly describe many enzymatic reactions, including those involving RNA, as well as to simulate crucial photochemical processes and optical response (Pokorná et al. 2018; Miclot et al. 2021; Giese et al. 2022; Bignon and Monari 2022).

Another intrinsic limitation common to either classic or QM/MM-based dynamics is due to the limited time scales which are accessible, which, in turn, impede the sampling of rare events necessitating to bypass significant free energy barriers. Thus, enhanced sampling algorithms have been proposed, which globally apply an external bias to the system, accelerating the sampling of the phase space and the occurrence of the rare events (Boulanger and Harvey 2018; Pokorná et al. 2018). However, the presence of the bias implies that the temporal resolution should be abandoned, at the profit of the reconstruction of precise free energy surfaces which characterize the studied phenomena. Different flavors of enhanced sampling methods exist, differing on the way the bias is applied and the free energy surface is statistically inferred. However, they usually require the definition of a reaction coordinate or collective variable. Thus, they require the reduction of the dimensionality of the complex free energy surface to a low-dimension space. Usually, only one or two coordinates are used to map the physical or chemical transformation under study, thus obtaining a reduced-dimension free energy profile (Bussi and Laio 2020). The choice of the reaction coordinate will strongly influence the quality of the final results and is hence crucial. While in some cases relatively simple geometrical parameters, such as bond distance, angles, or dihedrals, or even global descriptors such as differences in the root mean square deviation (RMSD) can be envisaged, in other situations, especially in the case of RNA/protein interactions, the definition of the collective variable is far from being trivial. Recently, advanced approaches such as string methods have been proposed for the description of more complex structural reorganization (Zinovjev and Tunon 2017; Chen et al. 2022b). Furthermore, the use of machine-learning analysis to infer relevant collective variables is also emerging as a way to a more efficient dimensionality reduction (Chen et al. 2022a; Giese et al. 2022). Despite their inherent difficulties, and the usual high computational cost, free energy methods are invaluable assets in modeling the complex conformational equilibrium of RNA, its reactivity, and its interactions with protein.

In the present section, we have provided a very general introduction to the principles of molecular modeling and their application to RNA. The interested readers are referred to more specialized reviews or book chapters such as one of Šponer and coworkers (Sponer et al. 2018).

3 Modeling RNA/Protein Interactions

3.1 Cellular Signal Pathways IRE/IRP

mRNA and its structural equilibrium may be exploited to regulate complex signaling pathways. The most paradigmatic example can be found in Fig. 2A and concerns the maintenance of iron homeostasis (Zhou and Tan 2017; Hognon et al. 2021). The concentration of the free iron pool can vary rapidly for instance through food intake. Consequently, its homeostasis should be regulated assuring a fast response of the cells to small variations in iron concentration. This is usually performed through the interaction between the Iron Response Element (IRE) mRNA and the Iron Response Protein 1 (IRP1). Indeed, free IRE promotes the translation of ferritin, i.e. the trans-membrane transporter eliminating free iron from the cellular medium. Sequestering of IRE by the formation of stable adducts with IRP precludes the translation, hence lowering the amount of ferritin and consequently increasing cellular iron concentration. However, the interaction between IRP1 and IRE should be specific and driven by peculiar structural and sequence motifs to avoid a global deregulation of the cellular mechanisms. From a structural point of view the IRP1-recognized RNA assumes a stem-loop secondary structure. Interestingly two unpaired extrahelical-extruded nucleobases can be evidenced (Dupuy et al. 2006). MD simulations of IRE have proven that the inherent structure of the IRE segment is rather stable as revealed for instance by the modest value of the RMSD assumed all along the simulation. Furthermore, the stem loop adopts a nearly ideal arrangement despite the presence of the extruded nucleobases (Hognon et al. 2021).

However, the formers are far from being innocent and may play an important role in the recognition between RNA and protein. Indeed, the simulations of the IRE/IRP1 complex (Fig. 2b, c) have shown the presence of specific interactions taking place with the extrahelical nucleobases. Differently from other similar protein/nucleic acid aggregates, in IRE/IRP1, the binding is mainly driven by hydrogen bonds with the exposed nucleobases, rather than nonspecific salt bridges between positive amino acids and the negative RNA phosphate backbone. As already pointed out, the interactions developing with the unpaired extrahelical nucleobases of IRE are of particular interest. Indeed, they point to the fact that the presence of interaction hot-spots, like exposed nucleobases, can enhance the selective recognition of RNA structures by regulating protein.

The IRP1 complex presents a rather hollow structure in which its two domains are widely open allowing access to the RNA binding site. This characteristic is maintained in the MD simulation of the unbound IRP1. However, the situation is dramatically changed when an iron-sulfur cluster is included in the protein structure (Hognon et al. 2021). In this case, a conformational transition takes place leading to a very compact structure, burying the RNA interaction region, and hence precluding the sequestration of the nucleic acid. The compact iron-sulfur containing structure is, actually, the catalytically active Aconitase enzyme (cAco) operating in the Krebs cycle. Interestingly, the use of enhanced sampling simulations has allowed to quantify

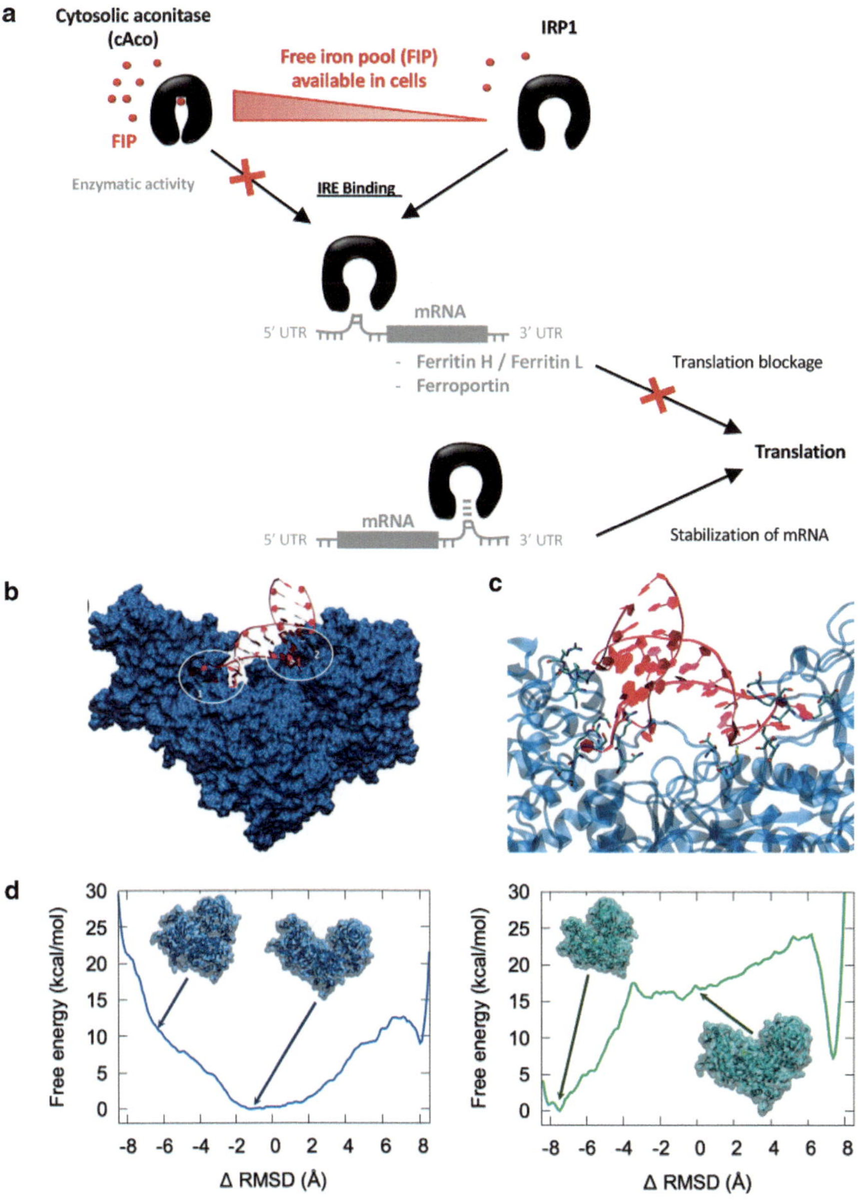

Fig. 2 Iron homeostasis regulatory mechanism based on the IRE/IRP1 interaction and on the IRP1/cAco transition (**a**). Representative snapshots showing the IRP1/IRE complex (**b**) and the specific interactions involving the extruded nucleobases (**c**). Free energy profile (**d**) for the transition between closed and open conformation for IRP1 (without Fe-S cluster, blue) and cAco (containing Fe-S cluster, green)

the free energy profiles related to the IRP1/cAco conformational transition (Fig. 2d). The free energy profile obtained as a function of RMSD clearly shows that in IRP1 (i.e. without Fe-S cluster) the open form, compatible with the IRE sequestration, is favored by more than 10 kcal/mol compared to the closed form (i.e. the structural signature of cAco. The presence of the Fe-S cluster, however, completely reverses this situation and now a penalty of about 17 kcal/mol should be overcome to obtain the open, IRE-interacting, conformation.

The combination of these results rationalizes the regulation of iron hemostasis. Indeed, at high iron concentration, the inclusion of the Fe-S cluster will be favored, which in turn will induce the conformational transition leading to the compact cAco and hence to the liberation of IRE. Consequently, the translation level of ferritin will increase, correlated to a more sustained expulsion of iron from the cytosol, thus decreasing its concentration. When a critical concentration level will be achieved, the chemical equilibrium will be reversed, the Fe-S cluster will be expulsed from the protein, favoring the open form of IRP1, inducing the formation of IRE/IRP1 complex, and, thus, lowering the ferritin expression limiting iron release and contributing to the increase its concentration.

3.2 Organization of Viral Genome in SARS-CoV-2

The role of specific, and non-canonical RNA sequences, is also fundamental in viral replication. In particular, the genome of diverse RNA viruses presents sequences, which are compatible with the folding into a guanine quadruplex (G4) motif. It has been pointed out that the interplay between G4 and viral proteins may decrease immune system response, as it is the case for the SARS Unique Domain (SUD) protein present in SARS-type coronaviruses (Hognon et al. 2020). Actually, the SUD non-structural motif is well conserved among SARS-type coronaviruses and could be related to their high escaping from the immune system. Nonetheless, the interaction between SUD and RNA fragments is crucial to define its mode of action. Indeed, the non-structural protein present a large and rather unspecific positive groove which may accommodate linear RNA. Thus, SUD is actually used by SARS viruses as a trap to sequester cellular mRNA coding for proapoptotic proteins. This action is in turn clearly related to the weakening of the immune response of the organism and favors the viral reproduction and spreading. Interestingly, it has been demonstrated that SUD protein should be activated through its self-dimerization. Such an activation is possible thanks to the interaction of SUD with viral or cellular RNA in G4 form (Hognon et al. 2020). Indeed, by using long-range MD simulation and free energy simulations, the binding modes and the interaction patterns between G4 and SUD have been evidenced. Two different binding modes have also been identified: the first one called monomeric developing only on one of the SUD monomers may be instrumental in recruiting the G4, while the second one (dimeric mode) bridges the two monomers locking SUD in its active form. Interestingly two-dimensional free energy profile has also confirmed that the dimeric form is favorable only in the presence of G4 in its dimeric binding mode, the free energy surface for the dimerization being relatively flat when the nucleic acid is non-interacting.

A part from the complex interplay between SUD and G4, the single-stranded RNA genome of SARS-CoV-2 also presents a number of putative G4-forming sequences, usually concentrated in untranslated and regulatory regions. In particular, it was surmised, also supported by circular dichroism spectroscopy measurement, that the RG-1 region of SARS-CoV-2 genome can fold into a G4 motif (Zhao et al. 2021). Although the spectroscopic analysis was convincing, the lack of a precise atomistically resolved structure of the RG-1 sequence was cruelly evident. More importantly, the absence of a clearly defined structure could hamper the development of specific drugs targeting the viral genome. To bridge this gap, a combination of multiscale modeling including MD simulations and QM/MM in silico spectroscopy has been performed to elucidate the structure of RG-1 (Miclot et al. 2021). Starting from homology models obtained on top of DNA G4, to palliate for the lack of RNA G4 resolved structures, an initial model of the RG-1 segment has been proposed.

The subsequent MD simulations have provided the first atomistic resolution of the structure of the viral RG-1 G4, proving that the it is stable and characterized by a very rigid guanine core, while presenting a large and flexible loop appended to the core, as can be observed from the superimposed snapshots in Fig. 3. However, to confirm the G4 theoretically-obtained model, the circular dichroism spectrum of the RNA arrangement has been calculated, using QM/MM approaches and on top of randomly selected structures issued from the MD. This strategy allows to take into account both the effects of the structural variability and of the thermal and vibrational motions, hence providing a one-to-one mapping with the experimental data. The striking agreement between the simulated and the experimental spectra (Fig. 3) confirms the soundness of the proposed G4 structure. Furthermore, the possible interaction between RG-1 and known G4 stabilizers, some of them already in clinical trials as anticancer agents, has also been analyzed, suggesting original therapeutic strategies. More generally, this protocol combining experimental circular dichroism and multiscale model represents a novel strategy allowing a straightforward and powerful determination of the precise arrangements of RNA and its temporal evolution.

3.3 Immune Response System and Selectivity of OAS

The OAS/RNaseL signaling pathway is one of the key-players in the immune response to viral infections (Sadler and Williams 2008). It relies on the sensing of exogenous double-stranded RNA by the oligoadenylate synthetase proteins (OAS), which induces the catalysis of short oligonucleotides called 2'-5'-oligoadenylates. These latter further activate the RNase L endoribonuclease leading to the destruction of all RNA within the infected cell. Investigating how OAS proteins recognize the viral RNA and the role of RNA secondary structures is crucial to bring insights into the molecular mechanisms driving the human immune response, and to set the framework for antiviral development targeting the OAS/RNaseL pathway. X-ray

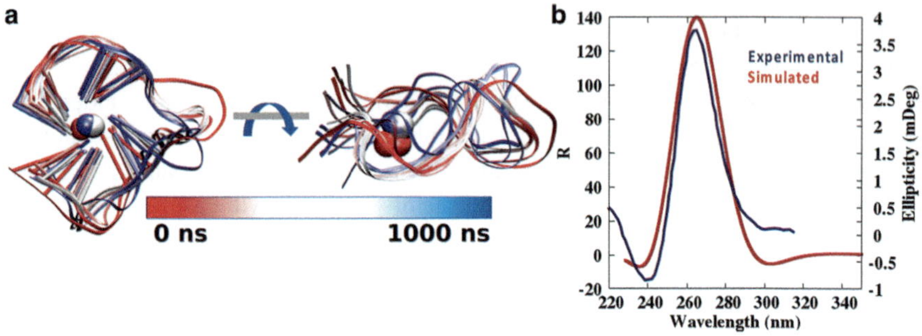

Fig. 3 Representative structures of the RG-1 sequence of the SARS-CoV-2 genome during the MD simulation. **b** Experimental and modeled circular dichroism spectra of the RG-1 RNA structure confirming the structural determination

structures of the human and porcine OAS1 have been released, in the apo form or in interaction with double-stranded RNA (Donovan et al. 2013; Lohöfener et al. 2015), shedding light on sequence preferences for OAS1 binding. Computational techniques have also been used to complement experimental methods in the urge to understand OAS1 activation mechanisms and to find potential inhibitors of OAS (Gonzalez et al. 2020). Some criteria facilitating the binding have been highlighted, such as the presence of a guanosine at the RNA double-strand termini, the helical stability or the length of the oligonucleotide, might facilitate the binding of specific OAS proteins (Schwartz et al. 2022). It has also been shown that OAS1 exhibits affinity for double-stranded regions in stem loops of the 5'-untranslated regions of SARS-CoV-2 (Wickenhagen et al. 2021) and West Nile (Deo et al. 2014) viruses. However, MD has been instrumental to reveal the binding mode of OAS1 to the 5'-UTR stem loops of these two viruses (Bignon et al. 2022a, b). Using protein-RNA docking, classical MD simulations and enhanced sampling approaches, it has been shown how OAS1 can bind to the first stem loop of SARS-CoV-2 genome through an interaction network more developed than for linear double-stranded RNA (Fig. 4). The atypical binding mode of OAS1 to the first stem loop in the 5'-UTR of West Nile Virus (WNV) genome has also been revealed through MD and characterized in terms of two distinct interaction areas (Fig. 4). This binding might be specific to the WNV 5'-UTR structure, which exhibits a bulge breaking the double-strand helix in two fractions, allowing a pronounced bending of the RNA. A similar behavior would be unlikely for more extended linear double-stranded RNA, however, long double-stranded RNA are actually more specifically recognized by the OAS3 member of the OAS family (Wang et al. 2020), which presents a more extended recognition region. Further computational studies, also involving long-range all-atoms MD simulations, are at present being conducted on the specific interaction modes of OAS3 to unravel its specific recognition of longer RNA segments.

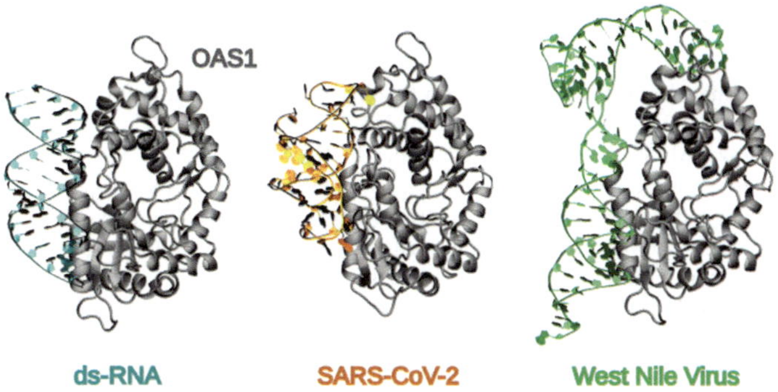

Fig. 4 Models of OAS1-dsRNA complexes as obtained by classical molecular dynamics simulations with a canonical RNA double-stranded helix (left), the stem loop 1 of SARS-CoV-2 (middle) and West Nile Virus (right) 5'-untranslated regions. OAS1 appears in gray and the RNA helices are colored differently for each sequence

4 Perspectives Towards Therapeutic Strategies

Despite recent fundamental progresses, such as cryo-electron microscopy, the determination of RNA structure is a complicated task, especially when the nucleic acid is interacting with proteins. The maturity of the molecular simulation techniques represents an efficient way to circumvent the lack of structural information available today. However, such a task requires on the on-side the development of precise potential and force fields, and on the other side the possibility to efficiently sample a rugged conformational landscape. The use of multiscale approaches combining atomistic MD simulations with QM and QM/MM approaches pushes further the limitations of the MD computational microscope providing one-to-one mapping with experimental spectroscopic observables such as circular dichroism. Importantly, such an approach will not only provide key insights on biological processes but it will also allow more powerful drug-design strategies. As an example, G4-ligands may be used as antivirals, blocking the action of helicases, or conversely avoiding the immune system escape exhibited by some SARS coronaviruses. Novel strategies for the control of homeostasis or cellular signaling may also be pursued by targeting specific structural features of RNA stem loops involved in post-transcriptional control of gene expression.

The study of the secondary structure of RNA, and its complex interaction network, is nowadays blooming like so many trees in the Spring. Molecular modeling and simulations are essential to bring out all the richness and the veritable essence of this bright and never-ending "Moveable Feast".

References

Battle DJ, Doudna JA (2001) The stem-loop binding protein forms a highly stable and specific complex with the 3' stem-loop of histone mRNAs. RNA 7:123–132

Bignon E, Marazzi M, Miclot T et al (2022a) Specific Recognition of the 5'-Untranslated Region of West Nile Virus Genome by Human Innate Immune System. Viruses 14:1282

Bignon E, Miclot T, Terenzi A et al (2022b) Structure of the 5' untranslated region in SARS-CoV-2 genome and its specific recognition by innate immune system via the human oligoadenylate synthase 1. Chem Commun (camb) 58:2176–2179

Bignon E, Monari A (2022) Modeling the Enzymatic Mechanism of the SARS-CoV-2 RNA-Dependent RNA Polymerase by DFT/MM-MD: An Unusual Active Site Leading to High Replication Rates. J Chem Inf Model 62:4261–4269

Bonilla SL, Kieft JS (2022) The promise of cryo-EM to explore RNA structural dynamics. J Mol Biol 434:167802

Boulanger E, Harvey JN (2018) QM/MM methods for free energies and photochemistry. Curr Opin Struct Biol 49:72–76

Branciamore S, Di Giulio M (2011) The presence in tRNA molecule sequences of the double hairpin, an evolutionary stage through which the origin of this molecule is thought to have passed. J Mol Evol 72:352–363

Bussi G, Laio A (2020) Using metadynamics to explore complex free-energy landscapes. Nat Rev Phys 2:200–212

Chen H, Liu H, Feng H et al (2022a) MLCV: Bridging Machine-Learning-Based Dimensionality Reduction and Free-Energy Calculation. J Chem Inf Model 62:1–8

Chen H, Ogden D, Pant S et al (2022b) A Companion Guide to the String Method with Swarms of Trajectories: Characterization, Performance, and Pitfalls. J Chem Theory Comput 18:1406–1422

Choi KH (2021) The Role of the Stem-Loop A RNA Promoter in Flavivirus Replication. Viruses 13:1107

Deo S, Patel TR, Dzananovic E et al (2014) Activation of 2' 5'-oligoadenylate synthetase by stem loops at the 5'-end of the West Nile virus genome. PLoS ONE 9:e92545

Di Giulio M (2004) The origin of the tRNA molecule: implications for the origin of protein synthesis. J Theor Biol 226:89–93

Donovan J, Dufner M, Korennykh A (2013) Structural basis for cytosolic double-stranded RNA surveillance by human oligoadenylate synthetase 1. Proc Natl Acad Sci USA 110:1652–1657

Dupuy J, Volbeda A, Carpentier P et al (2006) Crystal structure of human iron regulatory protein 1 as cytosolic aconitase. Structure 14:129–139

Dzananovic E, Patel TR, Deo S et al (2013) Recognition of viral RNA stem-loops by the tandem double-stranded RNA binding domains of PKR. RNA 19:333–344

Endoh T, Sugimoto N (2019) Conformational Dynamics of the RNA G-Quadruplex and its Effect on Translation Efficiency. Molecules 24:E1613

Fröhlking T, Bernetti M, Calonaci N et al (2020) Toward empirical force fields that match experimental observables. J Chem Phys 152:230902

Giese TJ, Zeng J, Ekesan Ş et al (2022) Combined QM/MM, Machine Learning Path Integral Approach to Compute Free Energy Profiles and Kinetic Isotope Effects in RNA Cleavage Reactions. J Chem Theory Comput 18:4304–4317

Gonzalez KJ, Moncada-Giraldo DM, Gutierrez JB (2020) In silico identification of potential inhibitors against human 2'-5'- oligoadenylate synthetase (OAS) proteins. Comput Biol Chem 85:107211

Hognon C, Bignon E, Harle G et al (2021) The Iron Maiden. Cytosolic Aconitase/IRP1 Conformational Transition in the Regulation of Ferritin Translation and Iron Hemostasis. Biomolecules 11:1329

Hognon C, Miclot T, García-Iriepa C, et al (2020) Role of RNA Guanine Quadruplexes in Favoring the Dimerization of SARS Unique Domain in Coronaviruses. J Phys Chem Lett 11:5661–6566

Kamura T, Katsuda Y, Kitamura Y et al (2020) G-quadruplexes in mRNA: A key structure for biological function. Biochem Biophys Res Commun 526:261–266

Kappel K, Zhang K, Su Z et al (2020) Accelerated cryo-EM-guided determination of three-dimensional RNA-only structures. Nat Methods 17:699–707

Kotar A, Foley HN, Baughman KM, Keane SC (2020) Advanced approaches for elucidating structures of large RNAs using NMR spectroscopy and complementary methods. Methods 183:93–107

Krepl M, Zgarbová M, Stadlbauer P et al (2012) Reference Simulations of Noncanonical Nucleic Acids with Different χ Variants of the AMBER Force Field: Quadruplex DNA, Quadruplex RNA, and Z-DNA. J Chem Theory Comput 8:2506–2520

Kuznetsov SV, Ren C-C, Woodson SA et al (2008) Loop dependence of the stability and dynamics of nucleic acid hairpins. Nucleic Acids Res 36:1098–1112

Lohöfener J, Steinke N, Kay-Fedorov P et al (2015) The Activation Mechanism of 2'-5'-Oligoadenylate Synthetase Gives New Insights Into OAS/cGAS Triggers of Innate Immunity. Structure 23:851–862

Lyu K, Chow EY-C, Mou X et al (2021) RNA G-quadruplexes (rG4s): genomics and biological functions. Nucleic Acids Res 49:5426–5450

Mayr C (2019) What Are 3' UTRs Doing? Cold Spring Harb Perspect Biol 11:a034728

Mayya VK, Duchaine TF (2019) Ciphers and Executioners: How 3′-Untranslated Regions Determine the Fate of Messenger RNAs. Front Genet 10:6

Miclot T, Hognon C, Bignon E et al (2021) Structure and Dynamics of RNA Guanine Quadruplexes in SARS-CoV-2 Genome. Original Strategies against Emerging Viruses. J Phys Chem Lett 12:10277–10283

Pokorná P, Kruse H, Krepl M et al (2018) QM/MM Calculations on Protein–RNA Complexes: Understanding Limitations of Classical MD Simulations and Search for Reliable Cost-Effective QM Methods. J Chem Theory Comput 14:5419–5433

Re A, Joshi T, Kulberkyte E et al (2014) RNA-protein interactions: an overview. Methods Mol Biol 1097:491–521

Reinharz V, Sarrazin-Gendron R, Waldispühl J (2021) Modeling and Predicting RNA Three-Dimensional Structures. Methods Mol Biol 2284:17–42

Ruggiero E, Richter SN (2020) Viral G-quadruplexes: New frontiers in virus pathogenesis and antiviral therapy. Annu Rep Med Chem 54:101–131

Sadler AJ, Williams BRG (2008) Interferon-inducible antiviral effectors. Nat Rev Immunol 8:559–568

Schlick T, Pyle AM (2017) Opportunities and Challenges in RNA Structural Modeling and Design. Biophys J 113:225–234

Schwartz SL, Dey D, Tanquary J et al (2022) Role of helical structure and dynamics in oligoadenylate synthetase 1 (OAS1) mismatch tolerance and activation by short dsRNAs. Proc Natl Acad Sci USA 119:e2107111119

Singh J, Hanson J, Paliwal K et al (2019) RNA secondary structure prediction using an ensemble of two-dimensional deep neural networks and transfer learning. Nat Commun 10:5407

Šponer J, Bussi G, Krepl M et al (2018) RNA Structural Dynamics As Captured by Molecular Simulations: A Comprehensive Overview. Chem Rev 118:4177–4338

Sripakdeevong P, Cevec M, Chang AT et al (2014) Structure determination of noncanonical RNA motifs guided by ^{1}H NMR chemical shifts. Nat Methods 11:413–416

Tan Z-J, Chen S-J (2008) Salt dependence of nucleic acid hairpin stability. Biophys J 95:738–752

Tassinari M, Richter SN, Gandellini P (2021) Biological relevance and therapeutic potential of G-quadruplex structures in the human noncoding transcriptome. Nucleic Acids Res 49:3617–3633

Townshend RJL, Eismann S, Watkins AM et al (2021) Geometric deep learning of RNA structure. Science 373:1047–1051

Vangaveti S, Ranganathan SV, Chen AA (2017) Advances in RNA molecular dynamics: a simulator's guide to RNA force fields: Advances in RNA molecular dynamics. Wires RNA 8:e1396

Varshney D, Cuesta SM, Herdy B et al (2021) RNA G-quadruplex structures control ribosomal protein production. Sci Rep 11:22735

Vora SM, Fontana P, Mao T et al (2022) Targeting stem-loop 1 of the SARS-CoV-2 5' UTR to suppress viral translation and Nsp1 evasion. Proc Natl Acad Sci U S A 119:e2117198119

Wang Y, Holleufer A, Gad HH et al (2020) Length dependent activation of OAS proteins by dsRNA. Cytokine 126:154867

Wickenhagen A, Sugrue E, Lytras S et al (2021) A prenylated dsRNA sensor protects against severe COVID-19. Science 374:eabj3624

Xu B, Zhu Y, Cao C et al (2022) Recent advances in RNA structurome. Sci China Life Sci 65:1285–1324

Yergert KM, Doll CA, O'Rouke R et al (2021) Identification of 3′ UTR motifs required for mRNA localization to myelin sheaths in vivo. PLoS Biol 19:e3001053

Yi M, Lemon SM (2003) Structure-function analysis of the 3' stem-loop of hepatitis C virus genomic RNA and its role in viral RNA replication. RNA 9:331–345

Yu H, Qi Y, Ding Y (2022) Deep Learning in RNA Structure Studies. Front Mol Biosci 9:869601

Zhao C, Qin G, Niu J et al (2021) Targeting RNA G-Quadruplex in SARS-CoV-2: A Promising Therapeutic Target for COVID-19? Angew Chem Int Ed 60:432–438

Zhou ZD, Tan E-K (2017) Iron regulatory protein (IRP)-iron responsive element (IRE) signaling pathway in human neurodegenerative diseases. Mol Neurodegener 12:75

Zinovjev K, Tunon I (2017) Adaptive Finite Temperature String Method in Collective Variables. J Phys Chem A 121:9764–9772

Viroids: Non-coding Circular RNAs Are Tiny Pathogens Provoking a Broad Response in Host Plants

Gerhard Steger, Kevin P. Wüsthoff, Jaroslav Matoušek, and Detlev Riesner

Contents

Abstract More than 50 years ago, viroids were firstly described as the smallest RNA molecules capable to infect certain plants and to autonomously self-replicate in host plants. Viroids are covalently closed circular single-stranded RNAs that are non-coding and depend for most of their infection cycle on host proteins. Today, viroids are subdivided into the two families *Avsunviroidae* and *Pospiviroidae*. Members of *Avsunviroidae* replicate in the chloroplast and have a highly bifurcated structure including hammerhead ribozymes, which cleave oligomeric replication intermediates into monomers and ligate them to mature circles. Members of *Pospiviroidae* accumulate in the nucleus, have a rod-like structure and depend on host proteins for cleavage and ligation. We will describe our present knowledge on sequence and structural elements of viroids in connection to their replication and trafficking.

G. Steger (✉) · K. P. Wüsthoff · D. Riesner
Institut für Physikalische Biologie, Heinrich-Heine-Universität Düsseldorf, 40204 Düsseldorf, Germany
e-mail: steger@biophys.uni-duesseldorf.de

K. P. Wüsthoff
e-mail: Kevin.Wuesthoff@hhu.de

D. Riesner
e-mail: Detlev.Riesner@hhu.de

J. Matoušek
Biology Centre of the Czech Academy of Sciences, Institute of Plant Molecular Biology, Czech Republic, 37005 Branišovská, Germany

J. Barciszewski (ed.), *RNA Structure and Function*, RNA Technologies 14,
https://doi.org/10.1007/978-3-031-36390-0_14

Keywords Potato spindle tuber viroid (PSTVd) · Peach latent mosaic viroid (PLMVd) · Thermodynamically metastable structures · Non-Watson–Crick basepairs

1 Introduction

In 1971, T. O. Diener coined the name "viroid" after showing that an unencapsidated, low-molecular weight RNA was sufficient to systemically infect, replicate, and raise symptoms in host plants without a helper virus (Diener 2003). At about the same time, Singh and Clark, Sänger, and Semancik and Weather confirmed the viroid concept. Viroids are single-stranded, non-coding, circular RNA pathogens of plants. The range of plant species infected by viroids includes vegetable and field crops (for example potato, tomato, cucumber, hop), fruit trees (for example peach) and grapevine, ornamental crops, and palm trees (Hadidi et al. 2017). Recently also fungi were experimentally shown to be hosts of viroids (Sun and Hadidi 2021). Despite not coding for any protein (Katsarou et al. 2022), viroids are able to induce pathogenic symptoms and complex changes at the molecular level; i.e., they incite changes in the host gene expression (for recent reviews, see Venkataraman et al. 2021; Flores et al. 2020; Ma et al. 2022b; Wang et al. 2022).

Today, viroids are subdivided into the two families *Avsunviroidae*, with five members and named after avocado sunblotch viroid (ASBVd), and *Pospiviroidae*, with 28 members and named after potato spindle tuber viroid (PSTVd) (Di Serio et al. 2020). Members of *Avsunviroidae* accumulate in chloroplasts and have a highly bifurcated structure (for example see peach latent mosaic viroid, PLMVd, Fig. 1) including, in the strands of both polarities, hammerhead ribozymes (HHR) as 'metastable' structural elements. A metastable structure is defined to form under particular conditions and the transition into the thermodynamically most stable structure is prevented or very slow; for example, RNA is synthesized at a rate of ~10 nt/s, and hairpin helices with small loops are formed in the microsecond to millisecond range in newly synthesized RNA segments, but their rearrangements into the thermodynamically preferred global structure, involving pairing of bases quite distant in the full-length RNA strand, may occur at a much slower rate (Repsilber et al. 1999). Members of *Pospiviroidae* have a rod-like structure (for example see PSTVd, Fig. 2) and accumulate in the nucleus.

Viroids have to rely only on their sequence and structural elements to perform all biological functions as done by viruses, but—as non-coding RNA—can not use viroid-derived proteins for these functions. Despite this, there is a tremendously rich response of plants to viroid infections at the molecular level. In the following, we will describe our present knowledge on sequence and structural elements of viroids in connection to biological function. For two of these functions—replication and systemic infection—we will extend to some detail. Finally, we will summarize present knowledge on the plant's response to viroid infection. For further topics, like

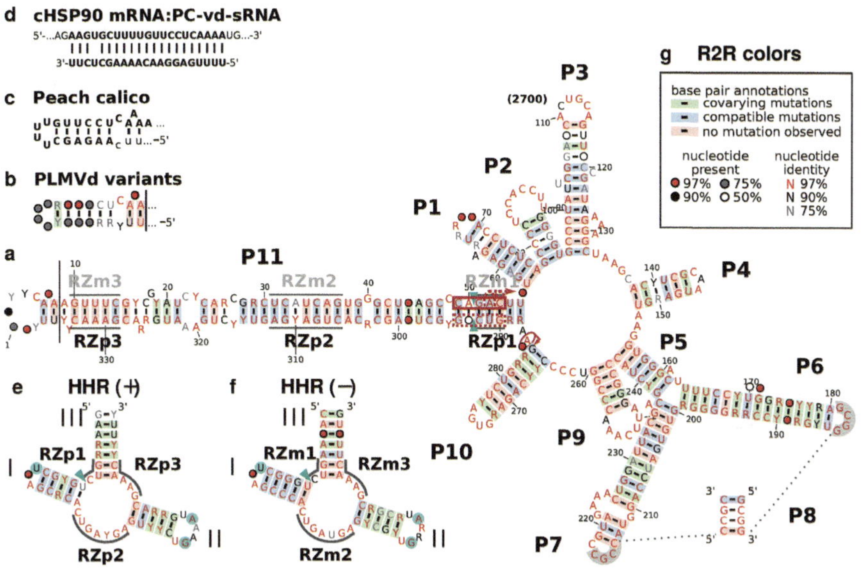

Fig. 1 Secondary structure schemes of PLMVd. The consensus sequences and structures presented were predicted for alignments with MAFFT/G-INS-I (Katoh et al. 2005); optimized in ConStruct (Wilm et al. 2008) at 37 °C, excluding lonely base pairs, and drawn with R2R (Weinberg and Breaker 2011); the number of unique sequence variants taken from GenBank was 2700. **a** The stem-loop structural elements of (+) PLMVd are labeled P1–P11; P8 is a highly conserved pseudoknot supported by in vitro probing. The green triangles indicate the self-cleavage sites of the hammerhead ribozymes in (+) (see e) and (−) (see f) sequences. The circled nucleotides A_{52} or C_{53} of (+) strand and U_{286} of (−) strand are replication start sites; the red boxes mark the sequence CAGAC in the vicinity of the replication initiation sites. **b** Of the 2700 PLMVd sequences, 134 had a sequence extension at the left hairpin P11. **c** From the PLMVd variant causing "Peach calico", which has a sequence extension at the left hairpin P11, a viroid-derived small RNA (vd-sRNA; (+)-strand sequence in bold) is expressed from the (−) strand. **d** The vd-sRNA derived from the (−) strand ((+) strand shown in c) leads to cleavage of the chloroplastic heat shock protein 90 (cHSP90). **e–f** The hammerhead ribozymes (HHR; e: (+) strand, 391 unique sequences; f: (−) strand, 310 unique sequences)—metastable structural elements—are based on three conserved sequence motifs (RZm1–3 and RZp1–3 for (−) and (+), respectively). **g** The color code used for the annotation of nucleotides and base pairs is given in the box. A co-varying position has Watson–Crick pairs that differ at both nucleotide positions among sequences; if only one position differs, the occurrence is classified as a compatible mutation (Weinberg et al. 2007). Note that no basepair annotated as "covarying" or "compatible" is statistically significant according to the analysis by R- scape (Rivas et al. 2017)

transmission, origin, evolution, and relation to other viroid-like RNAs, we refer to the literature (Hadidi and Randles 2021; Elena et al. 2009; Flores et al. 2022; Lee and Koonin 2022).

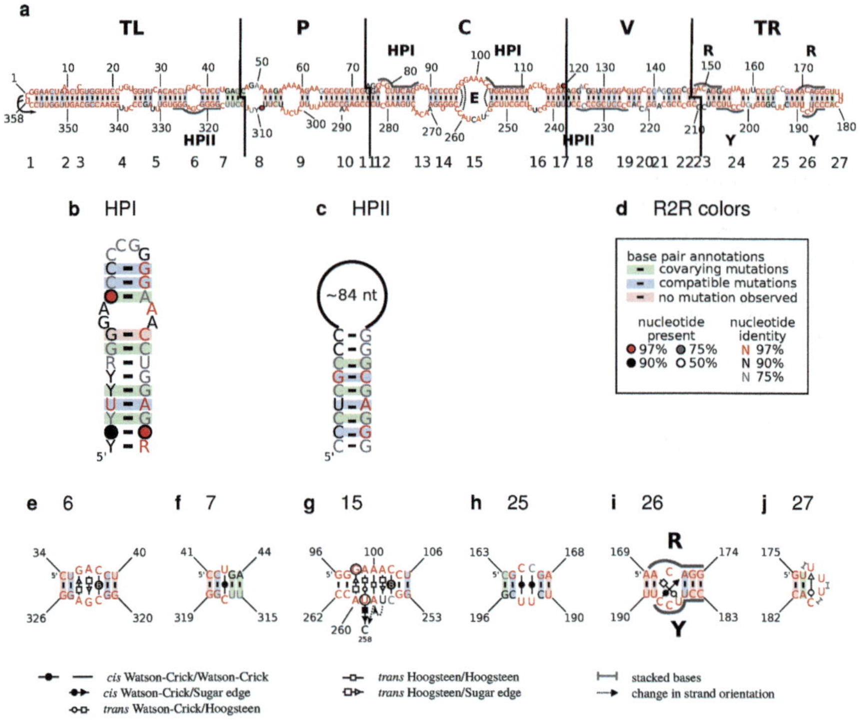

Fig. 2 Secondary structure schemes of PSTVd. The consensus sequences and structures presented were predicted for alignments with MAFFT/X-INS-I (Katoh and Toh 2008), for further details see Fig. 1. **a** Borders of the five domains of *Pospiviroidae* members (TL, terminal left; P, pathogenicity-related; C, central; V, variable; TR, terminal right) are marked by black lines. Nucleotides forming the extra-stable hairpin I (HPI; b) and hairpin II (HPII; c) in thermodynamically metastable structures are outlined. RY motifs (i) in the TR domain critical for binding of the viroid RNA-binding protein VirP1 are outlined and marked by R and Y. The replication start site for synthesis of (−) oligomers in the TL hairpin loop is marked by an arrow. Loops are numbered from 1 to 27; loop 15 is labeled with letter E. **e–j** Structural details of several loops shown in a. The symbols used for annotation of non-Watson–Crick pairs are given below the structures (Leontis and Westhof 2001). **g** A cross-link can be induced by UV light between the circled nucleotides G_{98} and U_{295} in loop E

2 Structure of Viroids

As examples for the structure of viroids, we will discuss in some detail those of PLMVd, a member of *Avsunviroidae*, and of PSTVd, a member of *Pospiviroidae*. For the structure of all other viroids, we refer to Steger and Perreault (2016). In general, members of *Avsunviroidae* possess a highly branched structure; the only exception is ASBVd with an elongated structure. In contrast, most if not all members of *Pospiviroidae* possess a rod-shaped secondary structure.

2.1 Structure of PLMVd

The secondary structure of (+)-stranded PLMVd is highly branched with at least 11 stem-loop elements P1–P11 (Fig. 1a). The shown structure is predicted based on sequence and structure conservation but is also compatible with RNA-selective 2'-hydroxyl acylation analyzed by primer extension (SHAPE) (Dubé et al. 2011). In addition to the secondary structure, (+)-stranded PLMVd contains a highly conserved pseudoknot P8, a kissing-loop interaction between the hairpin loops of P6 and P7 (Bussière et al. 2000). This pseudoknot seems to be essential for PLMVd accumulation in vivo (Dubé et al. 2010).

The long stem-loop P11 contains the sequences that are able to form the HHR in the (+) (Fig. 1e) as well as in the (−) sequence (Fig. 1f; De la Peña et al. 2017). Formation of HHR structures and cleavage by the HHR occurs only in multimeric strands, while the linear structure without hammerhead structures and cleavage capability is thermodynamically preferred in the monomeric, circular structure. The highly conserved sequence parts forming the core of a HHR are labeled RZp1–3 for (+) and RZm1–3 for (−) strands, respectively (Fig. 1a–f). Several nucleotides in the HHR core are involved in non-Watson–Crick pairs. Cleavage occurs after the $^{5'}$GUC$^{3'}$ in RZp1 as well as RZm1. A minimal HHR with only the catalytic core connected by three helices of any base composition are highly active in high Mg^{2+} concentrations in vitro, while activity in physiological, low Mg^{2+} concentration requires additional non-conserved sequence elements that stabilize the catalytically active conformation via tertiary interactions for example by interactions between the hairpin loops of helices I and II (not shown; Khvorova et al. 2003; Dufour et al. 2009).

The structures of (+)- and (−)-stranded PLMVd are similar but also exhibit several differences; for example, the pseudoknot P8, present in (+)-stranded PLMVd, is absent in (−)-stranded PLMVd in identical solvent conditions (Dubé et al. 2010).

In a SELEX (systematic evolution of ligands by exponential enrichment) experiment, the sequence 5'-CAGACG was enriched using a 93 nt long circular RNA as template and *Escherichia coli* DNA-dependent RNA polymerase (RNAP) to produce a 93 nt RNA resulting from self-cleavage (Motard et al. 2008); the template was composed of the complement of a hammerhead ribozyme and a stretch of 30 random nucleotides; RNAP was used as a model bacterial-like enzyme substituting for the nuclear-encoded polymerase (NEP) catalyzing the replication of PLMVd (see Sect. 3.1). Similar sequences, 5'-CAGACu in (+) strand and 5'-CAGACy in (−) strand, are present in PLMVd close to the initiation sites (Fig. 1a; Motard et al. 2008).

A severe albinism of peach is a disease called "peach calico" (PC) caused by PLMVd variants with an insertion of 12–14 nt in the hairpin loop of stem-loop P11 (Fig. 1c; Malfitano et al. 2003; Rodio et al. 2006, 2007); this insert is absent in latent and mosaic-inducing PLMVd variants. The PC-associated insertion—as well as the remaining PLMVd RNA—is a source of viroid-derived small RNAs (vd-sRNAs), which are produced by DICER-like (DCL) RNAses. A 21 nt vd-sRNA from the PC-associated insertion in the P11 hairpin in the (−) PLMVd RNA ((+)-strand sequence in bold in Fig. 1d) is complementary, except one nucleotide, to the mRNA of the nuclear-encoded chloroplastic heat-shock protein 90 (cHSP90; Navarro et al. 2012). The resulting duplex (Fig. 1d) is target of RISC-(RNA-induced silencing complex-) -mediated cHSP90 mRNA cleavage. The mRNA level of cHSP90 was substantially lower in albino leaf sectors infected with PC-inducing PMLVd than the steady-state level of the cHSP90 transcript in green tissues. The cHSP90 protein is known to participate in chloroplast biogenesis and plastid-to-nucleus signal transduction in *Arabidopsis thaliana* (Navarro et al. 2012). It should be noted that this is an example for the pathological actions of viroids via vd-sRNAs (see Sect. 5).

2.2 Structure of PSTVd

The native, rod-like structure is of high thermodynamic stability (Fig. 2a; for review see Steger and Riesner 2018). A comparison of in vitro and in vivo SHAPE-determined structures found only marginal differences, which supports PSTVd as a "naked" RNA not tightly bound by host proteins (López-Carrasco and Flores 2017). This rod-like structure does not include the most stable hairpin helices HPI (Fig. 2b) and HPII (Fig. 2c), which are present in thermodynamically stable structures at temperatures irrelevant for any in vivo function. Thus, at biologically relevant temperatures, the stable hairpins may have only a function as thermodynamically metastable structures. Synthetic mutations in HPII demonstrated that these are most critical during (−)-strand replication. HPII-containing structures are present in vivo in multimeric replication intermediates of PSTVd.

The rod-like secondary structure of *Pospiviroidae* members can be subdivided into five structural and functional domains (TL, terminal left; P, pathogenicity-related; C, central; V, variable; TR, terminal right; Fig. 2a) on the basis of sequence and structural alignment (Keese and Symons 1985; Wüsthoff and Steger 2022). The replication start site in (+)-stranded PSTVd is located in the TL domain (see Sect. 3.2). The P domain is associated with symptom severity, but mutations outside of the P domain may also influence virulence. The C domain is the most conserved part between different viroids; it harbors the processing site of multimeric (+) replication intermediates into monomeric circles (see Sects. 3.2 and 4). The V domain has the lowest sequence similarity even between closely related viroids. The TR domain contains up to two RY motifs that are the binding sites for viroid-binding protein 1 (VirP1), which is indispensable for replication and cell-to-cell transport (see Sect. 4). It is generally accepted that "viroids have evolved by rearrangement of domains between different viroids infecting the same cell and subsequent mutations within each domain" (Keese and Symons 1985). Further functional aspects of structural elements shown in Fig. 2 are discussed in Sects. 3.2 and 4.

3 Replication of Viroids

Viroids as non-coding RNAs have to rely on host proteins for their replication. The replication cycle of viroids, however, differs significantly between members of *Avsunviroidae* and *Pospiviroidae*. For an overview on viroid replication see Wang (2021).

3.1 *Replication of Members of* Avsunviroidae

Members of the family *Avsunviroidae* replicate in a symmetrical replication cycle in their host's chloroplast (Fig. 3a); that is, oligomeric replication intermediates as well as circular RNAs of both polarities are produced. Transcription is performed by NEP (Navarro et al. 2000) using viroid RNA as template despite it is a DNA-dependent RNA polymerase. Cleavage of the oligomeric intermediates into monomeric units is autocatalytic and co-transcriptionally mediated by HHR embedded in strands of both polarities. Interaction with a chloroplast RNA-binding protein (PARBP33), which functions as an RNA chaperone, increases cleaving efficiency (Daròs and Flores 2002). Ligation of monomeric to circular RNA is again mediated by the HHR and is supported by a chloroplastic tRNA ligase (Nohales et al. 2012b).

3.2 *Replication of Members of* Pospiviroidae

Members of the family *Pospiviroidae* replicate in an asymmetrical replication cycle inside their host's nucleus (Fig. 3b); that is, oligomeric replication intermediates of both polarities are produced but only (+) circles. Transcription is performed by the DNA-dependent RNA polymerase II (POLII; Dissanayaka Mudiyanselage and Wang 2020; Dissanayaka Mudiyanselage et al. 2022) using viroid RNA as template despite it is a DNA-dependent RNA polymerase. Transcription of (+) circles into (−) oligomers starts in the hairpin loop of TL in circular PSTVd's secondary structure (Kolonko et al. 2006). For transcription of oligomeric (−) strands into (+) strands, HP II is critical (see Sect. 2.2). The oligomeric (−) strand is then used as template for oligomeric (+) strand synthesis, which might be cleaved by an RNAse III-type enzyme (Gas et al 2008) into monomeric (+) strands and finally circularized by DNA ligase I (Nohales et al. 2012a) into circular monomeric (+) strands.

The canonical nine zinc finger (ZF) transcription factor IIIA (TFIIIA-9ZF) as well as its variant TFIIIA-7ZF with only seven ZF (Fu et al. 2009; Hammond et al. 2009; Layat et al. 2012) bind to (+) PSTVd; TFIIIA-7ZF binds to (−) PSTVd (Wang et al. 2016). Binding sites of TFIIIA-7ZF as well as of POLII are in the TL domain according to RNAse protection assays; note that the ${}_{7}U{:}G_{353}$ pair is critical for replication (Wu et al. 2020). TFIIIA-9ZF binds to the TR domain (Bojić et al. 2012; Wang

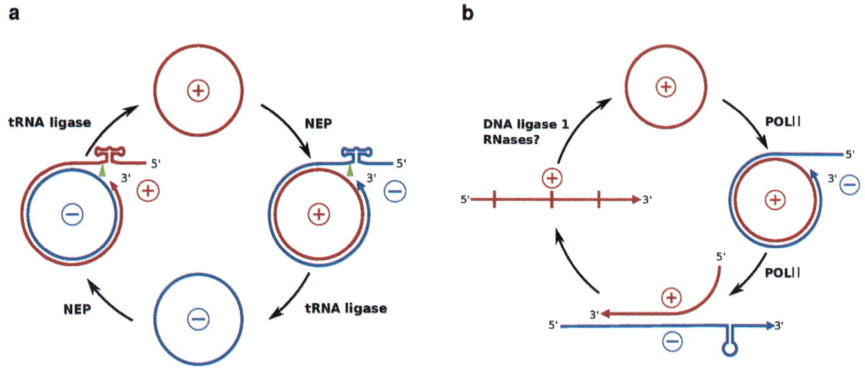

Fig. 3 Symmetrical replication of *Avsunviroidae* and asymmetrical replication of *Pospiviroidae*. **a** In case of members of the family *Avsunviroidae*, the nuclear-encoded chloroplastic DNA-dependent RNA polymerase (NEP) transcribes (+) as well (−) circles into oligomeric intermediates that are cleaved into monomers by viroid-encoded hammerhead ribozymes (Fig. 1). The cleavage sites are marked by green triangles. For the ligation of monomers, members of the *Avsunviroidae* family recruit a chloroplastic tRNA ligase or self-ligate. **b** In case of members of the *Pospiviroidae*, the host-encoded DNA-dependent RNA polymerase II (POLII) transcribes the mature circular viroid, defined as (+)-stranded, as well as the oligomeric (−)-stranded intermediates into oligomeric strands of opposite polarity. Cleavage of the oligomeric (+) strands by an RNAse III-type enzyme into monomers followed by their ligation by DNA ligase I results in (+) circular molecules. Note that the viroid structure is critical in all steps: the mature circle has a rod-shaped, thermodynamically stable structure; HPII (see Fig. 2c) is present in (−) strands due to sequential folding and is critical for (+) strand synthesis; structural elements of the CCR are critical for processing and ligation of (+) strands

et al. 2016). Suppression of TFIIIA-7ZF reduces PSTVd replication, and overexpression of TFIIIA-7ZF enhances PSTVd replication in plants (Wang et al. 2016; Matoušek and Steger 2022).

Transcription of the TFIIIA gene is dependent on the splicing regulator ribosomal protein L5 (RPL5). In the absence of this regulator, TFIIIA-7ZF is produced (Hammond et al. 2009). As members of the family *Pospiviroidae* possess a binding site for RPL5 in the C domain (Jiang et al. 2018), they are able to bind RPL5 and promote the transcription of TFIIIA-7ZF mRNA.

4 Trafficking of Viroids

PSTVd accumulates in the upper parts of tomato plants newly grown after inoculation and is predominantly found in association with the vascular tissue. Less than 20% of tomato cells are infected by PSTVd; up to 5×10^4 PSTVd molecules are present per nucleus with > 90% in nucleoli. That is, analysis of viroid-induced effects by experiments on total plants may underestimate viroid's influence (see Sect. 5).

In general, a viroid has to be replicated—either in the chloroplast or in the nucleus—, exported from its place of replication into the cytoplasm, has to traffic via plasmodesmata to neighboring cells, enter the vascular system for long-distance trafficking, enter a cell and therein its place of replication.

PSTVd's movement via plasmodesmata is mediated by a specific sequence or structural motif. According to a detailed analysis (Qi and Ding 2003), synthesis of PSTVd (+)- and (−)-strands occurs in the nucleoplasm, (−)-strand RNA is anchored in the nucleoplasm, and the (+)-strand RNA is transported selectively into the nucleolus. Responsible for the import into the nucleus is the upper part of the C domain (Zhao et al. 2001; Abraitiene et al. 2008).

In situ techniques suggested that synthetic PSTVd mutants showed specific blocks in trafficking and propagation in some floral organs (Zhong et al. 2008). For example, $_{76}$G:U$_{282}$, which is not present in Fig. 2a, and 155G:U$_{204}$ are required for PSTVd to enter vascular tissue; synthetic G$_{76}$A and G$_{156}$A mutants were able to traffic from phloem into bundle sheath cells, mesophyll cells, and epidermal cells but cannot transit in the reverse direction from bundle sheath into phloem cells (Wu et al. 2020).

The structure of most if not all loops in PSTVd's secondary structure is critical for trafficking and/or replication (Zhong et al. 2008; Ma and Wang 2022). In the following, we will give some examples on loop features and their biological function.

Loop 6 (Fig. 2e) with consensus sequence $^{5'}$G$_{36}$AC/$^{5'}$C$_{322}$GA forming three non-Watson–Crick (nWC) pairs is flanked by a U:G wobble pair and a *cis* Watson-Watson (cWW) G:C pair. It is critical for trafficking from palisade mesophyll to spongy mesophyll in *Nicotiana benthamiana* (Takeda et al. 2011).

Loop 7 (Fig. 2f) is an internal loop of nucleotides U/C that form a water-inserted cWW base pair flanked by cWW base pairs; it is required for PSTVd to traffic from nonvascular into the vascular tissue phloem to initiate systemic infection.

Loop E (or loop 15 in the standard numbering; Fig. 2g) is named according to its structural similarity to loop E of eukaryotic 5 S rRNA (Zhong et al. 2006; Freidhoff and Bruist 2019); for example, a crosslink can be induced by UV-light between nucleotides G$_{98}$ and U$_{295}$ in loop E (Fig. 2g; Wang et al. 2007; Eiras et al. 2007). Loop E is an asymmetric internal loop in which all nucleotides are involved in nWC basepairs including a triple basepair. Loop E is critical for cleavage of (+)-stranded replication intermediates into monomers (Gas et al. 2007). TFIIIA-9ZF, which is exported from the nucleus to the cytoplasm in a complex with 5 S rRNA, binds to loop E (and neighboring helices IV and V and hairpin loop D) of 5 S rRNA (Szymanski et al. 2003) but binds to the TR domain of PSTVd (Wang et al. 2016); the binding strengths of TFIIIA-9ZF to 5 S rRNA and PSTVd are similar (Eiras et al. 2010). RPL5 binds to 5 S rRNA at helix III and hairpin loop C, which are distant to loop E (Szymanski et al. 2003), but binds to loop E of PSTVd (Jiang et al. 2018). The binding strengths of RPL5 to 5 S rRNA and PSTVd are similar (Eiras et al. 2010). Overexpression of RPL5 leads to suppression of TFIIIA-7ZF expression that in turn leads to a decrease of PSTVd replication (Jiang et al. 2018).

Common to members of *Pospiviroidae* is the presence of one or two purine (R) and pyrimidine (Y) motifs in the upper and lower part of the TR's secondary structure; hence the name RY motif (Fig. 2a, i; Martínez de Alba et al. 2003). The VirP1 binds to the TR domain of PSTVd and about hop stunt viroid (HSVd) in vitro as well as in vivo. The distal RY motif has an fivefold stronger binding affinity than the more centrally located RY motif (Gozmanova et al. 2003; Martínez de Alba et al. 2003). This distal RY motif includes a C-loop motif (Ma et al. 2022a). Virp1 is a bromodomain-containing protein with an atypical RNA binding domain, a nuclear localization signal, and critical for PSTVd infectivity (Kalantidis et al. 2007). VIRP1 and IMPa-4 (importin α-4) likely form a complex for nuclear import of PSTVd (Ma et al. 2022a).

5 Plant Response to Viroid Infection

Viroids are targets of the host's RNA-induced silencing complex (RISC) in order to restrict viroid infection, but viroids resist, at least to some degree, to degradation by the RNA silencing machinery (Itaya et al. 2007). Anyway, RISC action results in viroid-derived small RNAs (vd-sRNA). Some of these vd-sRNA may target the host's DNA (Wassenegger and Dalakouras 2021; Gómez et al. 2022) or host's mRNAs in order to weaken the host's defense and/or induce symptoms (Flores et al. 2020; Navarro et al. 2021; Di Serio et al. 2022; Joubert et al. 2022; Matoušek and Steger 2022). A clear example is the case of PLMVd inducing PC outlined above (see Sect. 2.1); most other examples on viroid-induced silencing are more ambiguous. Clearly, some new approaches are necessary to fully understand these plant responses on cellular and intercellular levels.

There are major obstacles to the precise evaluation of gene regulation in multicellular plant organs; these include the monitoring of regulatory processes at levels other than steady-state transcript abundance, resolution of gene regulation in individual cells or cell types, and effective assessment of transient gene activity manifested during development or in response to external cues. The whole chain of molecular events from transcription to proteins including the expression of nuclear protein-coding genes is controlled by dynamic mechanisms ranging from DNA methylation, chromatin modification, and gene transcription to mRNA maturation, turnover, translation, and the posttranslational control of protein function (Bailey-Serres 2013). A genome-scale assessment of the spatiotemporal regulation of gene expression is essential for a comprehensive understanding of gene regulatory networks, therefore microgenomics is the inevitable way to push forward viroid research in the future. The methods for analysis of host-viroid interaction and host transcriptome changes upon viroid infection significantly improved with new techniques of sequencing starting from chemical fingerprinting in 1960s through two-dimensional fractionation of radiolabeled partial-digestion fragments in 1970s, Sanger's sequencing since in 1980, automated sequencing since 1990s, and finishing by NGS in 2005 representing a revolution in viroid research allowing for complex and comprehensive analysis of transcriptomes (Adkar-Purushothama and Perreault 2020). Besides sequencing of mRNA, NGS stimulated strong developments in analyses of other cellular RNA components and processes, in particular, degradomes, micro- and long RNAs (for methods and reviews see e.g. Addo-Quaye et al. 2008; German et al. 2008; Huesgen and Overall 2012; Mishra et al. 2015; Ma et al. 2015). Besides the improvement of analyses of plant transcriptomes, NGS and other techniques developed significantly to quantitatively and qualitatively assay the propagation and perform diagnostics of viroid-specific sequences and intermediates of different polarity (for review see e.g. Gucek et al. 2017; Pallás et al. 2018). For instance, single-strand-specific viroid reverse transcription-quantitative polymerase chain reaction (RT-qPCR) helped to analyze and quantify the unusual ratio of (−) and (+) intermediates of citrus bark cracking viroid (CBCVd) in hop and some *Solanaceae* plants (Matoušek et al 2017) or some preference in degradation of viroid (−) strands during elimination of seed non-transmissible viroids from pollen (Matoušek et al. 2020).

Acknowledgements Original work by the authors J. M., D. R., and G. S. was funded for many years by bilateral projects by the German Research Foundation (DFG) and the Czech Science Foundation (GACR). To all uncited authors, we apologize for not mentioning their work in this review for lack of space.

References

Abraitiene A, Zhao Y, Hammond R (2008) Nuclear targeting by fragmentation of the potato spindle tuber viroid genome. Biochem Biophys Res Commun 368:470–475

Addo-Quaye C, Miller W, Axtell M (2008) CleaveLand: a pipeline for using degradome data to find cleaved small RNA targets. Bioinformatics 25:130–131

Adkar-Purushothama C, Perreault J (2020) Impact of nucleic acid sequencing on viroid biology. Int J Mol Sci 21:5532

Bailey-Serres J (2013) Microgenomics: genome-scale, cell-specific monitoring of multiple gene regulation tiers. Annu Rev Plant Biol 64:293–325

Bojić T, Beeharry Y, Zhang D et al (2012) Tomato RNA polymerase II interacts with the rod-like conformation of the left terminal domain of the potato spindle tuber viroid positive RNA genome. J Gen Virol 93:1591–1600

Bussière F, Ouellet J, Côté F et al (2000) Mapping in solution shows the peach latent mosaic viroid to possess a new pseudoknot in a complex, branched secondary structure. J Virol 74:2647–2654

Daròs J, Flores R (2002) A chloroplast protein binds a viroid RNA in vivo and facilitates its hammerhead-mediated self-cleavage. EMBO J 21:749–759

Di Serio F, Owens R, Li SF et al (2020) Viroids. https://ictv.global/report_9th/subviral/Viroids, visited on November 2022

Di Serio F, Owens R, Navarro B et al (2022) Role of RNA silencing in plant-viroid interactions and in viroid pathogenesis. Virus Res 323:198964

Diener T (2003) Discovering viroids—a personal perspective. Nat Rev Microbiol 1:75–80

Dissanayaka Mudiyanselage S, Wang Y (2020) Evidence supporting that RNA polymerase II catalyzes de novo transcription using potato spindle tuber viroid circular RNA templates. Viruses 12:371

Dissanayaka Mudiyanselage S, Ma J, Pechan T et al (2022) A remodeled RNA polymerase II complex catalyzing viroid RNA-templated transcription. PLoS Pathog 18:e1010850

Dubé A, Baumstark T, Bisaillon M et al (2010) The RNA strands of the plus and minus polarities of peach latent mosaic viroid fold into different structures. RNA 16:463–473

Dubé A, Bolduc F, Bisaillon M et al (2011) Mapping studies of the Peach latent mosaic viroid reveal novel structural features. Mol Plant Pathol 12:688–701

Dufour D, De la Peña M, Gago S et al (2009) Structure-function analysis of the ribozymes of chrysanthemum chlorotic mottle viroid: a loop-loop interaction motif conserved in most natural hammerheads. Nucleic Acids Res 37:368–381
Eiras M, Kitajima E, Flores R et al (2007) Existence in vivo of the loop E motif in potato spindle tuber viroid RNA. Arch Virol 152:1389–1393
Eiras M, Nohales M, Kitajima E et al (2010) Ribosomal protein L5 and transcription factor IIIA from *Arabidopsis thaliana* bind in vitro specifically *Potato spindle tuber viroid* RNA. Arch Virol 156:529–533
Elena S, Gómez G, Daròs J (2009) Evolutionary Constraints to Viroid Evolution. Viruses 1:241–254
Flores R, Navarro B, Delgado S et al (2020) Viroid pathogenesis: a critical appraisal of the role of RNA silencing in triggering the initial molecular lesion. FEMS Microbiol Rev 44:386–398
Flores R, Navarro B, Serra P et al (2022) A scenario for the emergence of protoviroids in the RNA world and for their further evolution into viroids and viroid-like RNAs by modular recombinations and mutations. Virus Evol 8:veab107
Freidhoff P, Bruist M (2019) In silico survey of the central conserved regions in viroids of the *Pospiviroidae* family for conserved asymmetric loop structures. RNA 25:985–1003
Fu Y, Bannach O, Chen H et al (2009) Alternative splicing of anciently exonized 5S rRNA regulates plant transcription factor TFIIIA. Genome Res 19:913–921
Gas M, Hernández C, Flores R et al (2007) Processing of nuclear viroids in vivo: an interplay between RNA conformations. PLoS Pathog 3:e182
Gas M, Molina-Serrano D, Hernández C et al (2008) Monomeric linear RNA of citrus exocortis viroid resulting from processing in vivo has 5'-phosphomonoester and 3'-hydroxyl termini: implications for the RNase and RNA ligase involved in replication. J Virol 82:10321–10325
German M, Pillay M, Jeong D et al (2008) Global identification of microRNA-target RNA pairs by parallel analysis of RNA ends. Nat Biotechnol 26:941–946
Gómez G, Marquez-Molins J, Martinez G et al (2022) Plant epigenome alterations: an emergent player in viroid-host interactions. Virus Res 318:198844
Gozmanova M, Denti M, Minkov I et al (2003) Characterization of the RNA motif responsible for the specific interaction of potato spindle tuber viroid RNA (PSTVd) and the tomato protein Virp1. Nucleic Acids Res 31:5534–5543
Gucek T, Trdan S, Jaksé J et al (2017) Diagnostic techniques for viroids. Plant Pathol 66:339–358
Hadidi A, Randles J (2021) Viroids, and the legacy of Ricardo Flores (1947–2020). Cells 10:2570
Hadidi A, Randles J, Flores R, Palukaitis P (eds) (2017) Viroids and satellites. Academic Press, Elsevier
Hammond M, Wachter A, Breaker R (2009) A plant 5S ribosomal RNA mimic regulates alternative splicing of transcription factor IIIA pre-mRNAs. Nat Struct Mol Biol 16:541–549
Huesgen P, Overall C (2012) N- and C-terminal degradomics: new approaches to reveal biological roles for plant proteases from substrate identification. Physiol Plant 145:5–17
Itaya A, Zhong X, Bundschuh R et al (2007) A structured viroid RNA serves as substrate for Dicer-like cleavage to produce biologically active small RNAs but is resistant to RISC-mediated degradation. J Virol 81:2980–2994
Jiang J, Smith H, Ren D et al (2018) Potato spindle tuber viroid modulates its replication through a direct interaction with a splicing regulator. J Virol 92:e01004-e1018
Joubert M, van den Berg N, Theron J et al (2022) Transcriptomics advancement in the complex response of plants to viroid infection. Int J Mol Sci 23:7677
Kalantidis K, Denti M, Tzortzakaki S et al (2007) Virp1 is a host protein with a major role in *Potato spindle tuber viroid* infection in Nicotiana plants. J Virol 81:12872–12880
Katoh K, Toh H (2008) Improved accuracy of multiple ncRNA alignment by incorporating structural information into a MAFFT-based framework. BMC Bioinformatics 9:212
Katoh K, Kuma K, Miyata T et al (2005) Improvement in the accuracy of multiple sequence alignment program MAFFT. Genome Inform Ser 16:2233
Katsarou K, Adkar-Purushothama C, Tassios E et al (2022) Revisiting the noncoding nature of pospiviroids. Cells 11:265

Keese P, Symons R (1985) Domains in viroids: Evidence of intermolecular RNA rearrangement and their contribution to viroid evolution. Proc Natl Acad Sci U S A 82:4582–4586

Khvorova A, Lescoute A, Westhof E et al (2003) Sequence elements outside the hammerhead ribozyme catalytic core enable intracellular activity. Nat Struct Biol 10:708–712

Kolonko N, Bannach O, Aschermann K et al (2006) Transcription of potato spindle tuber viroid by RNA polymerase II starts in the left terminal loop. Virology 347:392–404

Layat E, Cotterell S, Vaillant I et al (2012) Transcript levels, alternative splicing and proteolytic cleavage of TFIIIA control 5S rRNA accumulation during Arabidopsis thaliana development. Plant J 71:35–44

Lee B, Koonin E (2022) Viroids and viroid-like circular RNAs: do they descend from primordial replicators? Life (basel) 12:103

Leontis N, Westhof E (2001) Geometric nomenclature and classification of RNA base pairs. RNA 7:499–512

López-Carrasco A, Flores R (2017) Dissecting the secondary structure of the circular RNA of a nuclear viroid *in vivo*: a "naked" rod-like conformation similar but not identical to that observed *in vitro*. RNA Biol 14:1046–1054

Martínez de Alba A, Sägesser R, Tabler M et al (2003) A bromodomain-containing protein from tomato specifically binds potato spindle tuber viroid RNA in vitro and in vivo. J Virol 77:9685–9694

Ma J, Wang Y (2022) Studies on viroid shed light on the role of RNA threedimensional structural motifs in RNA trafficking in plants. Front Plant Sci 13:836267

Ma J, Dissanayaka M, Park W et al (2022a) A nuclear import pathway exploited by pathogenic noncoding RNAs. Plant Cell 34:3543–3556

Ma J, Dissanayaka Mudiyanselage S, Wang Y (2022b) Emerging value of the viroid model in molecular biology and beyond. Virus Res 313:198730

Ma X, Tang Z, Qin J et al (2015) The use of high-throughput sequencing methods for plant microRNA research. RNA Biol 12:709–719

Malfitano M, Di Serio F, Covelli L et al (2003) Peach latent mosaic viroid variants inducing peach calico (extreme chlorosis) contain a characteristic insertion that is responsible for this symptomatology. Virology 313:492–501

Matoušek J, Steger G (2022) The splicing variant TFIIIA-7ZF of viroid-modulated transcription factor IIIA causes physiological irregularities in transgenic tobacco and transient somatic depression of "degradome" characteristic for developing pollen. Cells 11:784

Matoušek J, Siglová K, Jakse J et al (2017) Propagation and some physiological effects of *Citrus bark cracking viroid* and *Apple fruit crinkle viroid* in multiple infected hop (*Humulus lupulus* L.). J Plant Physiol 213

Matoušek J, Steinbachová L, Drábková L et al (2020) Elimination of viroids from tobacco pollen involves a decrease in propagation rate and an increase of the degradation processes. Int J Mol Sci 21:3029

Mishra A, Duraisamy G, Matoušek J (2015) Discovering microRNAs and their targets in plants. CRC Crit Rev Plant Sci 34:553–571

Motard J, Bolduc F, Thompson D et al (2008) The peach latent mosaic viroid replication initiation site is located at a universal position that appears to be defined by a conserved sequence. Virology 373:362–375

Navarro B, Gisel A, Rodio M et al (2012) Small RNAs containing the pathogenic determinant of a chloroplast-replicating viroid guide the degradation of a host mRNA as predicted by RNA silencing. Plant J 70:991–1003

Navarro B, Flores R, Di Serio F (2021) Advances in viroid-host interactions. Annu Rev Virol 8:305–325

Navarro J, Vera A, Flores R (2000) A chloroplastic RNA polymerase resistant to tagetitoxin is involved in replication of avocado sunblotch viroid. Virology 268:218–225

Nohales MÁ, Flores R, Darós J (2012a) Viroid RNA redirects host DNA ligase 1 to act as an RNA ligase. Proc Natl Acad Sci USA 109:1380513810

Nohales MÁ, Molina-Serrano D, Flores R et al (2012b) Involvement of the chloroplastic isoform of tRNA ligase in the replication of viroids belonging to the family *Avsunviroidae*. J Virol 86:8269–8276

Pallás V, Sánchez-Navarro J, James D (2018) Recent advances on the multiplex molecular detection of plant viruses and viroids. Front Microbiol 9:2087

De la Peña M, García-Robles I, Cervera A (2017) The hammerhead ribozyme: a long history for a short RNA. Molecules 22:78

Qi Y, Ding B (2003) Differential subnuclear localization of RNA strands of opposite polarity derived from an autonomously replicating viroid. Plant Cell 15:2566–2577

Repsilber D, Wiese U, Rachen M et al (1999) Formation of metastable RNA structures by sequential folding during transcription: Time-resolved structural analysis of potato spindle tuber viroid (−)-stranded RNA by temperature-gradient gel electrophoresis. RNA 5:574–584

Rivas E, Clements J, Eddy S (2017) A statistical test for conserved RNA structure shows lack of evidence for structure in lncRNAs. Nat Methods 14:45–48

Rodio M, Delgado S, Flores R et al (2006) Variants of Peach latent mosaic viroid inducing peach calico: uneven distribution in infected plants and requirements of the insertion containing the pathogenicity determinant. J Gen Virol 87:231–240

Rodio M, Delgado S, De Stradis A et al (2007) A viroid RNA with a specific structural motif inhibits chloroplast development. Plant Cell 19:3610–3626

Steger G, Perreault JP (2016) Structure and associated biological functions of viroids. Adv Virus Res 94:141–172

Steger G, Riesner D (2018) Viroid research and its significance for RNA technology and basic biochemistry. Nucleic Acids Res 46:10563–10576

Sun L, Hadidi A (2021) Mycoviroids: fungi as hosts and vectors of viroids. Cells 11:1335

Szymanski M, Barciszewska M, Erdmann V et al (2003) 5 S rRNA: structure and interactions. Biochem J 371:641–651

Takeda R, Petrov A, Leontis N et al (2011) A three-dimensional RNA motif in *Potato spindle tuber viroid* mediates trafficking from palisade mesophyll to spongy mesophyll in *Nicotiana benthamiana*. Plant Cell 23:258–272

Venkataraman S, Badar U, Shoeb E et al (2021) An inside look into biological miniatures: molecular mechanisms of viroids. Int J Mol Sci 22:2795

Wang C, Jiang F, Zhu S (2022) Complex small RNA-mediated regulatory networks between viruses/viroids/satellites and host plants. Virus Res 311:198704

Wang Y (2021) Current view and perspectives in viroid replication. Curr Opin Virol 47:32–37

Wang Y, Zhong X, Itaya A et al (2007) Evidence for the existence of the loop E motif of *Potato spindle tuber viroid* in vivo. J Virol 81:2074–2077

Wang Y, Qu J, Ji S et al (2016) A land plant-specific transcription factor directly enhances transcription of a pathogenic noncoding RNA template by DNA-dependent RNA polymerase II. Plant Cell 28:1094–1107

Wassenegger M, Dalakouras A (2021) Viroids as a tool to study RNA-directed DNA methylation in plants. Cells 10:1187

Weinberg Z, Breaker R (2011) R2R–software to speed the depiction of aesthetic consensus RNA secondary structures. BMC Bioinformatics 12:3

Weinberg Z, Barrick J, Yao Z et al (2007) Identification of 22 candidate structured RNAs in bacteria using the CMfinder comparative genomics pipeline. Nucleic Acids Res 35:4809–4819

Wilm A, Linnenbrink K, Steger G (2008) ConStruct: Improved construction of RNA consensus structures. BMC Bioinformatics 9:219

Wu J, Zhou C, Li J et al (2020) Functional analysis reveals G/U pairs critical for replication and trafficking of an infectious non-coding viroid RNA. Nucleic Acids Res 48:3134–3155

Wüsthoff K, Steger G (2022) Conserved motifs and domains in members of *Pospiviroidae*. Cells 11:230

Zhao Y, Owens R, Hammond R (2001) Use of a vector based on Potato virus X in a whole plant assay to demonstrate nuclear targeting of Potato spindle tuber viroid. J Gen Virology 82:1491–1497

Zhong X, Leontis N, Qian S et al (2006) Tertiary structural and functional analyses of a viroid RNA motif by isostericity matrix and mutagenesis reveal its essential role in replication. J Virol 80:8566–8581

Zhong X, Archual A, Amin A et al (2008) A genomic map of viroid RNA motifs critical for replication and systemic trafficking. Plant Cell 20:35–47

RNA (ecRNA)
Kenneth P, Tang L (1993) From ions in ...
Kellner S, Neumann J, Rosenkranz D ...
Khoddami V, Yerra A, ...
modifications ...

a different genomic region than its ...

Biology of Circular RNAs and Methodological Approaches to Their Study

Michaela Ruckova, Dagmar Al Tukmachi, and Ondrej Slaby

Contents

Abstract From initially overlooked as peculiarities of uncertain biological importance or rare isoforms generated as a result of splicing errors to commonly regarded as relevant regulatory molecules, circular RNAs (circRNAs) now represent an extensively studied subgroup of non-coding RNAs (ncRNAs). The structural uniqueness of these endogenous biomolecules confers resistance to exonuclease activity, resulting in higher stability than observed in their linear counterparts, which alongside their reported function of gene expression regulators, predisposes them to potential use as robust biomarkers and/or powerful therapeutic targets. During the last decade, we have witnessed a boom in circRNA research deciphering various mechanisms of their biogenesis and an in-depth description of their biological functions, including miRNA and protein sponging, yet regulation of these events remains unclear. Despite recent advances in high-throughput genomic technology and novel bioinformatic approaches allowing circRNAs characterization in detail, analyzing these molecules continues to be challenging, leaving multiple biological knowledge gaps unsolved.

M. Ruckova · D. Al Tukmachi · O. Slaby (✉)
Department of Biology, Faculty of Medicine, CEITEC Masaryk University, Masaryk University, Brno, Kamenice 753/5 625 00, Czech Republic
e-mail: oslaby@med.muni.cz

J. Barciszewski (ed.), *RNA Structure and Function*, RNA Technologies 14,
https://doi.org/10.1007/978-3-031-36390-0_15

Keywords Circular RNA · circRNAs · Function · Biogenesis · Biomarkers · Profiling · Analysis · Sequencing · Bioinformatic approach

1 Introduction

Circular RNAs (circRNAs) are long, covalently closed single-stranded molecules. Albeit still included in the sizeable non-coding RNA (ncRNA) family, their function as a template for peptides has also been reported multiple times to date. These RNAs, generated during an event called "backsplicing" and widely expressed across the eukaryotic tree of life, are increasingly acknowledged by the scientific community owing to their unique properties and powerful mode of action.

Although not generated during backsplicing, first circRNAs were identified in the middle of the 70 s as circular genomes in plant viroids. A few years later, the detection of first mammalian circRNAs extracted from cytoplasmic fraction of HeLa cells using electron microscopy was reported. In the breakthrough year, 2013, a sequence conservation study by Memczak et al. (2013) proved that circRNAs could act as post-transcriptional regulators. In the same month, other relevant studies were published, confirming the biological importance of circRNAs (Jeck et al. 2013). Hansen et al. (2013) were the first to report a functional analysis of naturally expressed circRNA and elucidated the now widely accepted role of circRNAs demonstrating that circRNAs could function as microRNA (miRNA) sponges to specifically bind a miRNA molecule and prevent it from binding to its target, thereby affecting the regulation of gene expression. Until the publication of crucial papers reporting circRNA regulatory properties, these molecules were considered functionless byproducts of aberrant splicing events.

In recent years, aberrant expression of circRNAs has been associated with a wide range of pathological conditions. According to various studies, it plays a crucial role in the pathogenesis and progression of various disorders, including for example cancer, neurodegenerative diseases, or cardiovascular diseases. Due to the circRNA loop structure with close ends unavailable for ribonucleases, these molecules are presumably more stable than their linear counterparts (Jeck et al. 2013), and, additionally, they are usually expressed in a tissue- or cell-type-specific manner (Memczak et al. 2013; Salzman et al. 2013). Therefore, they have been proposed as potential reliable prognostic or diagnostic biomarkers or therapeutic targets and have been extensively explored.

With the development of high-throughput sequencing technologies and dedicated bioinformatic approaches, the biology of circRNAs and their molecular and cellular functions have been gradually revealed, leading to the common understanding that these molecules represent key players in various molecular biological processes. However, primarily due to their unusual loop structure lacking free 5' and 3'-ends-ends, there are several reasons why the true nature of circRNA has long escaped the attention of scientists worldwide. When employing standard bioinformatic pipelines for high-throughput sequencing data analysis, any reads that span back-splice junctions (BSJs) will be discarded because they do not align to the linear reference genome. Nevertheless, there are now countless bioinformatics approaches for processing sequencing data (Szabo and Salzman 2016). Further, the lack of a poly(A) end must be considered when preparing sequencing complementary DNA (cDNA) libraries, as poly(A) selection is used in many protocols in the ribosomal RNA (rRNA) removal step (Pandey et al. 2019). Moreover, circRNAs are only a minority of total RNAs and must be purified before downstream analyses (Xiao and Wilusz 2019). Likewise, it is necessary to design modified primers when using RT-qPCR (reverse transcription followed by quantitative polymerase chain reaction) detection since primers designed according to the linear genome will not distinguish the linear RNA from the circRNA. CircRNAs need to be actively searched for, and standard molecular biology and bioinformatics approaches must be explicitly modified, thus, research on these unique molecules remains substantially challenging.

2 Biogenesis and CircRNA Subtypes

Although the complete mechanism of biogenesis and its regulation remain elusive, general knowledge of circRNA origins has improved considerably in recent years. During canonical splicing, a splice donor, an upstream 5' splice site, binds to a splice acceptor, a downstream 3' splice site, across an intron to discard it. Nevertheless, many pre-mRNAs can undergo an alternative splicing, such as backsplicing, during which circRNAs are generated (Fig. 1).

CircRNAs are derived from canonical splice sites (Jeck et al. 2013; Memczak et al. 2013) and are reported to be dependent on the splicing machinery, which usually is not efficient enough to form a linear RNA (Jeck et al. 2013; Ashwal-Fluss et al. 2014). Therefore, circRNAs transcription can compete with canonical pre-mRNA (precursor messenger RNA) splicing and affects its rate (Ashwal-Fluss et al. 2014). On the other hand, authors of a study on *Drosophila melanogaster* observed an increase in circRNA levels when depleted splicing factors (Kramer et al. 2015). Another study on the same organism cells demonstrated that depletion of U2 small nuclear ribonucleoprotein components leading to spliceosome inhibition increases the ratio of circRNAs to linear RNAs (Liang et al. 2017a). These findings suggest that when pre-mRNA processing is inhibited, emerging RNA can be guided into alternative pathways that promote backsplicing.

During spliceosome-mediated backsplicing, the downstream splice-donor site and the upstream splice-acceptor site on the same pre-mRNA must be brought into proximity by looping of flanking intron sequence to be ligated (Kramer et al. 2015). This event gives rise to the most abundant and well-studied circRNAs, which tend to have long introns flanking the exons involved in backsplicing, exonic circRNAs (ecircRNAs). Besides, these molecules are often derived from genes with highly active promoters. More seldom, if the circle retains introns, exon–intron circRNAs (EIciRNAs) are generated (Li et al. 2015). Loop structure formation may be enabled by various mechanisms, including base pairing in the presence of inverted repetitive

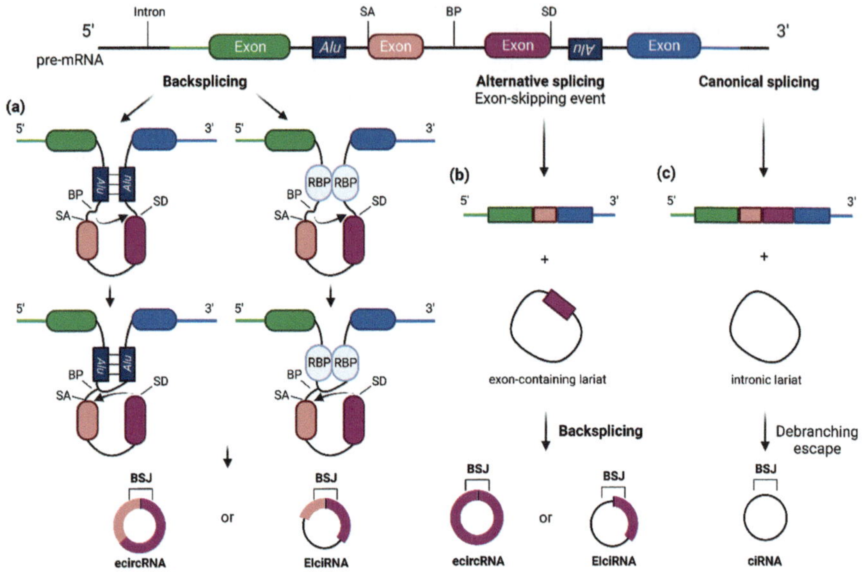

Fig. 1 Mechanisms of circRNA biogenesis

sequences (such as Alu elements) in flanking introns (Jeck et al. 2013), dimerization of RNA-binding proteins that bind to specific motifs located in flanking introns, such as quaking (Conn et al. 2015), muscleblind (Ashwal-Fluss et al. 2014), or FUS (Errichelli et al. 2017), or the presence of non-repetitive complementary sequences (Liang and Wilusz 2014). Furthermore, data from a study by Kramer et al. indicate that the production of circRNAs can be controlled by multiple hnRNP (heterogeneous nuclear ribonucleoprotein) and SR (serine-arginine) proteins, *cis*-acting elements and *trans*-acting splicing factors, acting in a combinatorial manner (Kramer et al. 2015).

Alternatively, if the lariat structure generated throughout pre-mRNA splicing avoids linearization, is cleaved from the nascent mRNA, maintains a circular form, and circumvents subsequent degradation, it can lead to the formation of circular intronic RNAs (ciRNAs) by so-called lariat-driven circularization. Above all, circRNAs composed of exons that are not surrounded by suitable complementary sequences in the pre-mRNA can arise by the process of exon skipping (Zhang et al. 2013). Additionally, tRNA intronic circular (tricRNAs) produced during metazoan tRNA splicing have been described (Lu et al. 2015). Chu et al. (2018) have even recently summarized 10 different types of circRNA.

During backsplicing (a), introns are excised from the pre-mRNA, and exons are covalently linked. The presence of long flanking introns, inverted repeat elements (such as Alu elements), or *trans*-acting RNA binding proteins (RBPs) supports the backsplicing. Base pairing between inverted Alu elements or the dimerization of RBPs guides a downstream splice-donor site (SD) into immediate proximity with an upstream splice-acceptor site (SA). An upstream branch point (BP) attacks a downstream SD leading to attacking an upstream SA, consequently forming exone-intron circRNAs (EIcircRNAs) or internally spliced exonic circRNAs (ecircRNAs). In the exon-skipping event (b), a lariat containing both introns and one or more exons forms during canonical splicing of the primary transcript and is cleaved from the molecule. The ends of the exons are brought to proximity allowing backsplicing to take place, giving rise to an ecircRNA or EIcircRNA. Moreover, circRNAs may arise from intronic lariat precursors escaping the debranching step of canonical linear splicing resulting in circular intronic RNA (ciRNA) (c). BSJ, back-splice junction; circRNA, circular RNA; pre-mRNA, precursor messenger RNA.

3 Export and Turnover

Following biogenesis, EcircRNAs are, for the most part, exported to the cytoplasm (Jeck et al. 2013; Memczak et al. 2013), whereas EIciRNAs and ciRNAs are found in the nucleus (Zhang et al. 2013; Li et al. 2015). Although the export is not fully understood, a recent study correlated the mechanism with the molecular length (Huang et al. 2018), providing the first findings on circRNAs export from the nucleus. The study indicated that ATP-dependent RNA helicase DDX39A and the spliceosome RNA helicase DDX39B are involved in this process. In addition to the cytoplasm, circRNAs can be exported into the extracellular space encapsulated in exosomes and enriched according to the cell of origin (Sun et al. 2021).

Albeit pathways behind circRNA turnover in vivo have just recently begun to be more perspicuous, much remains unknown. Due to the lack of free 5' and 3' ends and the resultant resistance to exonucleases, circRNAs are characterized as remarkably stable molecules. This fact allows some circRNAs to accumulate to levels that may exceed the linear form (Jeck et al. 2013; Memczak et al. 2013). Nevertheless, they are prone to degradation by circulating endonucleases (Park et al. 2019) or cleavage mediated by miRNAs, as shown in *Vitis vinifera* L. (Gao et al. 2019). Degradation in a manner dependent on a highly complementary miR-671 binding site has been observed in ciRS-7 (also known as CDR1as), a reported sponge for miR-7. The bond results in circRNA cleavage activation by the protein Argonaute 2 (AGO2). It has been shown that the regulatory circuit involves the long- intergenic ncRNA (lincRNA) *Cyrano* binding and targeting miR-7 for degradation leading to indirect modulation of the ciRS-7 degradation by miR-671. This event can be supported by miR-7 drafting the miR-67 silencing complex to the circRNA or by yet unexplained mechanism (Kleaveland et al. 2018).

As for endonucleases, which have been linked to circRNA turnover at a more general level, data indicate that ribonuclease P (RNase P) is able to degrade m^6A-modified circRNAs (Park et al. 2019) and, additionally, circRNAs are globally cleaved by 2-5A-dependent ribonuclease (Rnase L) during viral infection (Liu et al. 2019).

4 Biological Functions

Experimental evidence suggests that circRNAs function as significant cell biology and pathophysiology regulators. It has been reported that these molecules are able to modulate gene expression at both transcriptional and post-transcriptional levels (Fig. 2).

4.1 MiRNA Sponging

Although a myriad of circRNAs has been identified to date, only a minor fraction of this group has been assigned a specific biological function, with the majority being miRNA sponging in the cytoplasm (Hansen et al. 2013; Memczak et al. 2013). The process represents the inhibition of the miRNA function by binding target miRNA complementary sequences directly or indirectly to the specific circRNA "sponge", leading to increased expression of genes that would be silenced by the RNA interference mechanism in the absence of circRNA (Wang et al. 2016). Moreover, it has been reported that most of the circRNAs act as competitive endogenous RNAs (ceRNAs), which may modulate miRNA action by binding to miRNA response elements (MREs) (Mitra et al. 2018).

Probably the best-examined circRNA, ciRS-7, contains over 70 binding sites for miR-7 in humans and more than 60 conserved binding sites across 32 vertebrates and is abundantly expressed in various tissue types, suggesting its ability to regulate the expression of miR-7 target genes (Hansen et al. 2013; Memczak et al. 2013). An in vivo functional experiment conducted on mice demonstrated that removal of the ciRS-7 locus from the genome resulted in the downregulation of miR-7 expression (Piwecka et al. 2017), while, on the contrary, other findings indicated a negative correlation between ciRS-7 and miR-7 expression (Weng et al. 2017).

Another circRNA, circABCB1O, has been identified to sponge miR-1271 resulting in carcinogenesis promotion in breast cancer cells through the bioinformatically predicted circABCB10/miR-1271 axis (Liang et al. 2017b). CircHIPK3,

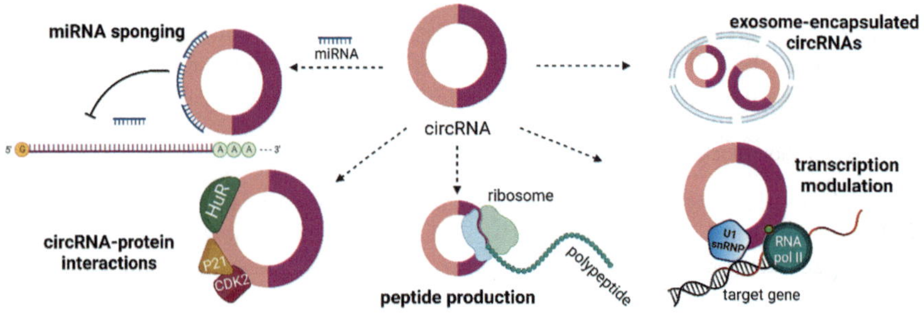

Fig. 2 Overview of circRNA functions

derived from the homeodomain interacting protein kinase 3 gene, has been reported multiple times to participate in tumorigenesis and cancer progression, and promoted, for example, gallbladder cancer cell growth by sponging miR-124 (Kai et al. 2018). Also, circ-Foxo3, derived from the forkhead box O3 gene, allegedly repressed cell proliferation and cell cycle progression by sponging specific miRNAs that regulate the production of Foxo3 mRNA (Du et al. 2016). Interestingly, it has been hypothesized that higher circRNA levels may work in cooperation to sponge many miRNAs (Bachmayr-Heyda et al. 2015). Nonetheless, this claim has not been verified experimentally.

However, this effect is presumably intrinsic only to some circRNAs. Militello et al. (2017) concluded that investigated circRNAs do not significantly bind miRNAs, so it is impossible to ascribe this function across the board to all circRNAs. For many circRNAs, their abundance is also limiting in this regard, i.e., the fact that there are not enough molecules present in the cell for efficient sponging of the relevant miRNAs.

4.2 *CircRNA–Protein Interaction*

Based on the available cross-linking immunoprecipitation datasets, some circRNAs are suggested to interact with RBPs in various manners. Besides the interaction with proteins during their own biogenesis, as mentioned previously, circRNAs have been investigated as protein sponges or decoys regulating RBP-dependent functions (Ashwal-Fluss et al. 2014; Abdelmohsen et al. 2017), protein scaffolds providing the formation of the enzyme–substrate complex (Du et al. 2016; Zeng et al. 2017), protein function enhancers (Zhang et al. 2013; Li et al. 2015), or recruiting proteins to a specific place of need (Chen et al. 2018). The emerging role of circRNA–protein interactions has been recently covered in a comprehensive review by Das et al. (2021).

In human cervical carcinoma HeLa cells, for example, circPABPN1 binds to RBP Hu-antigen R (HuR), a protein influencing gene expression programs and hence cellular phenotypes. This leads to the suppression of the nuclear poly(A) binding protein 1 (PABPN1) translation (Abdelmohsen et al. 2017). Holdt et al. (2016) demonstrated that circular antisense non-coding RNA in the INK4 locus (circANRIL), generally expressed at higher levels than the linear isoform, acts atheroprotectively by regulating rRNA maturation and atherogenesis pathways. The circRNA binds to pescadillo homolog 1 (PES1), an essential 60S-preribosomal assembly factor, effectuating the diminishment of pre-rRNA processing and ribosome biogenesis in vascular smooth muscle cells and macrophages. Consequently, circANRIL induces nucleolar stress and p53 activation, resulting in the possible removal of hyperproliferative cells from atherosclerotic plaques.

The previously acknowledged study of Du et al. (2016) reported, for the first time, a circRNA serving as a protein scaffold. In murine fibroblasts, circ-Foxo3 interacts with p-21- and cyclin-dependent kinase 2 (CDK2). The formation of the circ-FOxo3-p21-CDK2 ternary complex prevents the CDK2 from its function, subsequently resulting in cell cycle arrest. In another study, a circRNA highly expressed in neonatal human cardiac tissue, circ-Amotl1, binds to both 3-phosphoinositide-dependent protein kinase 1 (PDK1), leading to the PDK1-dependent phosphorylation of AKT1 and its subsequent nuclear translocation. In a mouse model, this cascade was reported to have a cardioprotective function (Zeng et al. 2017). Moreover, circRNAs can modulate the host genes transcription by controlling the initiation and elongation of RNA polymerase II (RNAP II) (Zhang et al. 2013). Further PAR-CLIP analysis showed that EIcircRNAs co-immunoprecipitate with RNAP II and can modulate transcription (Li et al. 2015).

CircRNAs have been hypothesized to have a protein-recruiting function. It was confirmed for circRNA FECR1, which coordinates the regulation of DNA methylating and demethylating enzymes and consequently represents an upstream controller of breast cancer growth. FECR1 recruits TET1 to the promoter region of *FLI1*, its host gene promoting tumor growth in solid tumors and acting as an oncogenic driver in hemato-oncological diseases, which leads to the demethylation of CpG sites and active transcription (Chen et al. 2018).

4.3 Translation of CircRNAs

Although circRNAs are commonly classified as ncRNAs, some of these molecules are confirmed and thousands are predicted to include a putative open reading frame (ORF) with an upstream IRES, thus, serving as a template for peptide synthesis. For linear mRNAs, translation depends on the presence of the 5' cap and poly(A) end, which are absent in circRNA. However, as demonstrated with engineered circRNAs with m^6A RNA modification incorporated (Wang and Wang 2015), or in the process of translation using the internal ribosome entry site (IRES), which allows binding of ribosome independently of the 5′ cap, has been observed. In human cells, the sequences identified as IRES contain a modified N6-methyladenosine (Yang et al. 2017), the most abundant base modification of RNA.

Albeit the functional relevance of most circRNA-originated peptides remains elusive, they are often shortened versions of the canonical protein lacking crucial domains to fulfill their original function and allegedly might compete with their full-length counterparts. They may naturally act in various manners, such as decoys, modulators of alternative protein complexes. For example, FBXW-185aa (Yang et al. 2018), PINT87aa (Zhang et al. 2018a, b), and SHPRH-146aa (Zhang et al. 2018a) (peptides derived namely from circ-FBXW7, circPINTexon2, and circ-SHPRH) are all suggested to work as tumor suppressors in glioblastoma. Moreover, recent findings of a novel study suggested a pervasive translation of circRNAs, providing profound implications in translation control (Fan et al. 2022).

Peptides of circRNA origin can be expressed under conditions unusual to the canonical protein, such as in times of cellular stress, or function in different cellular compartments to the canonical protein, therefore function as regulated product within the cell (Legnini et al. 2017; Pamudurti et al. 2017). The translation of circ-ZNF609, a circRNA derived from zinc finger protein 609 gene, or m6A-containing circRNAs were confirmed significantly increased upon heat shock proteins in human cells using advanced mass spectrometry-based analysis and overexpression of circRNA plasmids (Legnini et al. 2017; Yang et al. 2017). Circ-ZNF609 contains an ORF spanning from the start codon, as well as the linear counterpart, and terminating at an in-frame STOP codon, created upon circularization. The molecule was shown to control myoblast proliferation specifically (Legnini et al. 2017). Additionally, it was found that after starvation of *D. melanogaster* circ-Mbl-peptide was produced and stabilized (Pamudurti et al. 2017).

5 Profiling and Analysis of CircRNAs

As circRNAs have increasingly fallen under the radar of many researchers, the demand for molecular biology and bioinformatics methods to investigate these molecules with precision and depth has also increased. Current advances in high-throughput genomic technology and bioinformatic strategies allow circRNAs characterization in precise detail, yet there are many experimental hardships associated specifically with circRNA research. A set of guidelines for circRNA studies on the authors' experience in detail has been recently proposed (Nielsen et al. 2022).

5.1 Purification and Enrichment of CircRNAs

Expression of circRNAs has been observed in many cell types and tissues (Rahimi et al. 2021), and detectable levels have also been identified in various liquid biopsies (Wang et al. 2022). Besides, their higher stability compared to linear RNAs may favor them when analyzing partially degraded samples. These molecules can be isolated with various commercially available kits designed for total RNA isolation, nonetheless, their low abundance in total RNA, estimated at 1%, from biological material is problematic (Salzman et al. 2013). Apparently, this relates mainly to rapidly proliferating cells where circRNAs have less time to accumulate to the detection limit (Bachmayr-Heyda et al. 2015).

As the first step of circRNA profiling, total RNA isolation, quantity, and quality control are recommended (Nielsen et al. 2022). For relatively low-abundant circRNAs, a subsequent enrichment step prior to cDNA library preparation is crucial. The most widespread strategy is incorporating RNase R treatment into the procedure. This enzyme hydrolyzes linear RNA in the 3'-5' direction, leaving circular and lariat structures intact. However, Xiao and Wilusz (2019) showed that RNAs with highly structured 3' ends, such as snRNAs and histone mRNAs, resist RNase R treatment. Authors also found that RNase R stalls in the structure of many mRNAs, particularly in G-rich sequences that are referred to as G-quadruplexes. It was demonstrated that standard protocols involving RNase R could fail to digest > 20% of all highly expressed linear RNAs. An improved protocol for a more efficient method of RNase R treatment resolved these difficulties by adding a polyadenylation step, modifying the composition of the reaction buffer, and prolonging incubation time. In the same year, an approach called RPAD (RNase R Treatment, Polyadenylation, and Poly(A) + RNA Depletion) was reported, which includes an extra rRNA depletion step. It claims to drastically deplete linear RNAs leading to the isolation of highly pure circRNAs from total RNA pools (Pandey et al. 2019), ready as an input for the cDNA library preparation procedure and subsequent bioinformatic analysis.

5.2 Bioinformatic Approaches on CircRNAs and Their Global Profiling

Characterization and accurate quantification of circRNA is critical, but it is still a challenging research problem. Linear RNA and circRNA are co-expressed, and the presence of the BSJ is the only unique feature of circRNAs that can distinguish them from linear RNA. To precisely quantify circRNA expression levels using RNA-seq, it is crucial to perform deep sequencing with longer reads (>100 nucleotides) (Nielsen et al. 2022). Sequencing coverage not only increases sensitivity but also leads to higher false-positive rates. Fortunately, the false positive rate of BSJs can be reduced by adding the circRNA enrichment step during library preparation, determining the amount of circRNA in enriched vs. non-enriched samples, and setting a strict threshold for circRNA read counts (Szabo and Salzman 2016).

Most circRNA detection algorithms support both single-end (SE) and paired-end (PE) data analysis. Using PE sequencing leads to higher sensitivity and, in some cases, even higher specificity, but it is not always a possible approach due to its higher price (Szabo and Salzman 2016). CircRNA identification tools first recognize fusion junction sites and then use different filters to detect the corresponding circRNA. These tools can be simply classified as annotation-dependent and independent (Ye et al. 2017). Tools requiring gene annotation can only be used in an organism with well-annotated genes, such as a mouse or human (Memczak et al. 2013). De novo prediction algorithms are not dependent on gene annotation and show promising results but with a higher rate of false positives. Finally, combining several detection tools with different analytical approaches is strongly endorsed to diminish false positive results (Hansen et al. 2016).

Short-read sequencing platforms cannot detect the internal arrangement of exons located more than 100–150 nucleotides adjacent to the BSJ of circRNAs. In recent studies, Oxford Nanopore Technology long-read sequencing has been used to characterize circRNA isoforms and discovered novel exons and microexons. A relatively low read count limits the long-read sequencing methods compared to second-generation PE sequencing (Rahimi et al. 2021). According to the results, nanopore

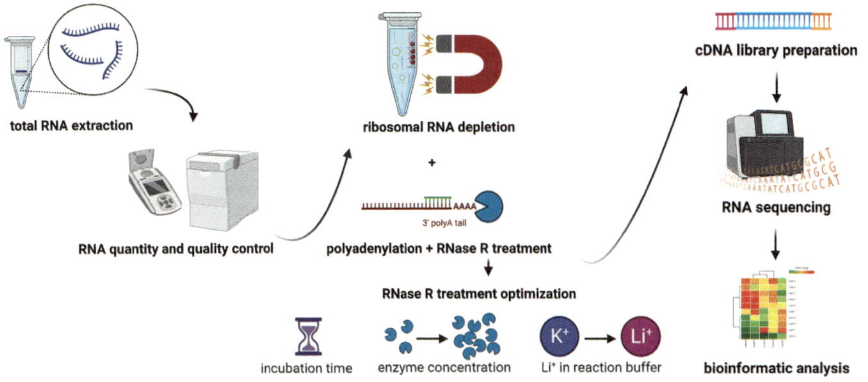

Fig. 3 Overview of circRNA profiling

long-read sequencing detected more circRNAs than short-read sequencing. Results suggest that long-read sequencing could determine expression levels of circRNAs that are not detected by short-read sequencing technology. The data show that highly expressed circRNAs are generally observed by both long-read and short-read sequencing methods. In addition, each method also detects many lower expressed circRNAs not discovered by the other technique (Rahimi et al. 2021). The length of the sequenced fragments was proved to be a significant factor affecting the efficiency of circRNA recognition by the CIRI-long tool (Zhang et al. 2021).

Interestingly, a study of scRNAseq (small conditional RNA sequencing) detected ~ 90% of specifically expressed circRNAs in fewer than ten cells in both human and mouse samples, suggesting further difficulties in detecting circRNAs by bulk RNA-seq techniques (Wu et al. 2022). Over of circRNA profiling is shown in Fig. 3.

5.3 *PCR-Based Analyses*

As there is a vast resemblance between circRNA and linear RNA sequences and, furthermore, enzymatic procedures used during library preparation for high-throughput sequencing may produce false-positive BSJs, it is necessary to perform the experimental validation (Nielsen et al. 2022). Since conventional RT-qPCR cannot distinguish precisely between circular and linear variants, the method must be modified. Among the most used approaches is RT-qPCR with divergent primers allowing the generation of BSJ amplicons (Fig. 4) followed or not by gel electrophoresis and/or Sanger sequencing (Abdelmohsen et al. 2017). Subsequent Sanger sequencing can further verify the junction sequence and may be able to reveal the complete sequence of the studied circRNA. Frequently, RT-qPCR must be preceded by RNase R treatment, especially when detecting low-abundant circRNAs. Recently,

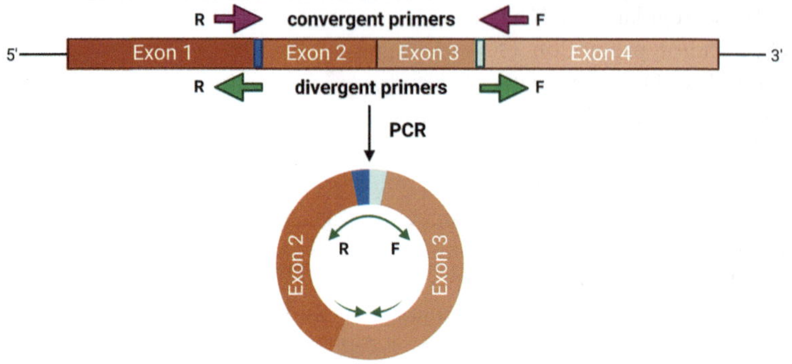

Fig. 4 Illustration of the complementary sequence of primers and the direction of replication on circular and linear molecules

a comprehensive protocol for circRNA quantification levels in cells using digital droplet PCR (dd-PCR) was reported (Das et al. 2022).

Particular attention should be paid to the RT step. Most RT enzymes exhibit template-switching activity and can probably contribute significantly to false positives in the detection of circRNAs (Tang et al. 2018). Other authors observed continuous circumnavigation of the single circRNA molecule, producing concatemeric cDNAs as a consequence of RT by either random primers or a gene-specific primer. Therefore, each of these cDNA molecules provided multiple priming targets for the RT-qPCR primer pairs, resulting in a false increase in circRNA levels. This difficulty can be minimized by combining PCR detection with hybridization-based methods such as, for example, northern blotting or microarrays, or in situ hybridization-based microscopy (Nielsen et al. 2022).

6 CircRNAs as Potential Biomarkers in Human Cancers

Since highly stable circRNAs are implicated in various diseases, including cancer, their biomarker potential in diagnosis or disease development monitoring has been considered. Additionally, they are currently being explored as potential therapeutic targets. Aberrant expression of some circRNAs in cells may be associated with their phenotypic changes, such as angiogenesis, reduced apoptosis, or the ability to invade and migrate, characteristics that can lead to their uncontrolled proliferation and/or metastasizing. Recently, a comprehensive review outlining circRNA modulation of cancer hallmarks and molecular pathways, resulting in cancer progression support and metastasizing, has been published (Yarmishyn et al. 2022). Although only several are highlighted in this chapter, numerous studies have focused on circRNA involvement in the pathophysiology of many diseases.

A focus on circRNA expression aberrations in papillary thyroid carcinoma has revealed several tissue circRNAs that could be exploited as biomarkers or therapeutic targets. Increased expression of circ_0058124 was observed, which silencing in vitro reduced the viability of tumor cells, their ability to generate colony formation and migration, and led to increased apoptosis. In the mouse model, silencing prevented circ_0058124 tumor growth (Liu et al. 2020). Similar properties were also revealed for circFOXM1, the blocking of which prevented tumor growth in vitro and in vivo. Its increased expression correlated with larger tumor size, higher grade, and the occurrence of lymph node metastases (Ye et al. 2020).

In circ_0016788, higher expression was observed in hepatocellular carcinoma tissue than in non-tumor tissue and was determined to be an independent factor predicting overall patient survival (Cheng et al. 2020). As an opposite example, lower expression of hsa_circ_0078602 in tumor tissue correlated with poor prognosis for patient survival, as demonstrated by Kou et al. (2019).

Hsa_circ_0004585 was reported to have significantly increased expression in tumor tissues of colorectal cancer patients. Its expression was positively correlated with tumor size, suggesting the molecule's involvement in the carcinogenesis process. In addition, hsa_circ_0004585 was also detected in peripheral blood and therefore has the potential to be used as a non-invasive diagnostic biomarker (Tian et al. 2019). Another promising diagnostic biomarker could be hsa_circ_0001696, expressed significantly higher in tumor tissue than in healthy tissue. The results of the research conducted by Li et al. (2021) suggest that decreased expression of hsa_circ_0001696 leads to an increase in the number of colonies and tumor cell migration promotion.

Zhou et al. (2021) found increased levels of circPARP4 in glioblastoma (GBM) tissue and showed that its presence promotes tumor cell division, migration, and invasiveness. An experiment in a mouse model showed that silencing circPARP4 led to significant tumor shrinkage. In contrast, lower expression of circ_0001946 in GBM cells led to increased invasiveness and proliferation of tumor cells and reduced apoptosis. This effect was further confirmed in an in vivo experiment in a mouse model, where increased expression of circ_0001946 reduced migration, invasiveness, and tumor proliferation (Li and Diao 2019). The authors of these studies suggest that the molecules under investigation have potential use in targeted therapy of GBM.

The involvement of specific circRNAs in pathophysiological cellular processes has been demonstrated not only in other cancers lung cancer but also in other types of diseases, such as Alzheimer's disease (Dube et al. 2019) or cardiovascular diseases (Wang et al. 2016).

7 Perspective

In the last decade, countless new insights have been gained about circRNA structure, biogenesis, properties, and functions. Moreover, the involvement of these unique molecules in the pathophysiology of many diseases, consisting of their ability to regulate gene expression by various mechanisms, has been described. Recent years have also seen significant improvements in the capabilities of molecular biological and bioinformatic approaches for circRNA study in detail. Despite all this, however, the field remains challenging. It awaits further remarkable discoveries that will move us closer to a complete understanding of circRNAs and, most importantly, harnessing their potential in clinical practice and improving the well-being of patients.

Acknowledgements The project National Institute for Cancer Research (Programme EXCELES, ID Project Number LX22NPO5102)—Funded by the European Union—Next Generation EU.

References

Abdelmohsen K, Panda AC, Munk R et al (2017) Identification of HuR target circular RNAs uncovers suppression of PABPN1 translation by CircPABPN1. RNA Biol 14:361–369

Ashwal-Fluss R, Meyer M, Pamudurti NR et al (2014) circRNA biogenesis competes with pre-mRNA splicing. Mol Cell 56:55–66

Bachmayr-Heyda A, Reiner AT, Auer K et al (2015) Correlation of circular RNA abundance with proliferation–exemplified with colorectal and ovarian cancer, idiopathic lung fibrosis, and normal human tissues. Sci Rep 5:8057

Chen N, Zhao G, Yan X et al (2018) A novel FLI1 exonic circular RNA promotes metastasis in breast cancer by coordinately regulating TET1 and DNMT1. Genome Biol 19:218

Cheng F, Wang L, Zhang J (2020) Circular RNA 0016788 displays as a biomarker for tumor progression and poor prognosis in surgical hepatocellular carcinoma patients. J Clin Lab Anal 34:e23300

Chu Q, Bai P, Zhu X et al (2018) Characteristics of plant circular RNAs. Brief Bioinform 21:135–143

Conn SJ, Pillman KA, Toubia J et al (2015) The RNA binding protein quaking regulates formation of circRNAs. Cell 160:1125–1134

Das A, Das D, Panda AC (2022) Quantification of Circular RNAs Using Digital Droplet PCR. J Vis Exp 187

Das A, Sinha T, Shyamal S, Panda AC (2021) Emerging Role of Circular RNA–Protein Interactions. Noncoding RNA 7:48

Du WW, Yang W, Liu E et al (2016) Foxo3 circular RNA retards cell cycle progression via forming ternary complexes with p21 and CDK2. Nucleic Acids Res 44:2846–2858

Dube U, Del-Aguila JL, Li Z et al (2019) An atlas of cortical circular RNA expression in Alzheimer disease brains demonstrates clinical and pathological associations. Nat Neurosci 22:1903–1912

Errichelli L, Dini Modigliani S, Laneve P et al (2017) FUS affects circular RNA expression in murine embryonic stem cell-derived motor neurons. Nat Commun 8:14741

Fan X, Yang Y, Chen C et al (2022) Pervasive translation of circular RNAs driven by short IRES-like elements. Nat Commun 13:1–15

Gao Z, Li J, Luo M et al (2019) Characterization and Cloning of Grape Circular RNAs Identified the Cold Resistance-Related Vv-circATS1. Plant Physiol 180:966–985

Hansen TB, Jensen TI, Clausen BH et al (2013) Natural RNA circles function as efficient microRNA sponges. Nature 495:384–388

Hansen TB, Venø MT, Damgaard CK et al (2016) Comparison of circular RNA prediction tools. Nucleic Acids Res 44:e58

Huang C, Liang D, Tatomer DC et al (2018) A length-dependent evolutionarily conserved pathway controls nuclear export of circular RNAs. Genes Dev 32:639–644

Jeck WR, Sorrentino JA, Wang K et al (2013) Circular RNAs are abundant, conserved, and associated with ALU repeats. RNA 19:141–157

Kai D, Yannian L, Yitian C et al (2018) Circular RNA HIPK3 promotes gallbladder cancer cell growth by sponging microRNA-124. Biochem Biophys Res Commun 503:863–869

Kleaveland B, Shi CY, Stefano J et al (2018) A Network of Noncoding Regulatory RNAs Acts in the Mammalian Brain. Cell 174:350-362.e17

Kou P, Zhang C, Lin J et al (2019) Circular RNA hsa_circ_0078602 may have potential as a prognostic biomarker for patients with hepatocellular carcinoma. Oncol Lett 17:2091–2098

Kramer MC, Liang D, Tatomer DC et al (2015) Combinatorial control of Drosophila circular RNA expression by intronic repeats, hnRNPs, and SR proteins. Genes Dev 29:2168–2182

Legnini I, di Timoteo G, Rossi F et al (2017) Circ-ZNF609 Is a Circular RNA that Can Be Translated and Functions in Myogenesis. Mol Cell 66:22-37.e9

Li PF, Zhang ZX, Yuan X et al (2021) Hsa_circ_0001696 modulates cell proliferation and migration in colorectal cancer. Oncol Lett 21:154

Li X, Diao H (2019) Circular RNA circ_0001946 acts as a competing endogenous RNA to inhibit glioblastoma progression by modulating miR-671-5p and CDR1. J Cell Physiol 234:13807–13819

Li Z, Huang C, Bao C et al (2015) Exon-intron circular RNAs regulate transcription in the nucleus. Nat Struct Mol Biol 22:256–264

Liang D, Tatomer DC, Luo Z et al (2017a) The output of protein-coding genes shifts to circular RNAs when the pre-mRNA processing machinery is limiting. Mol Cell 68:940

Liang D, Wilusz JE (2014) Short intronic repeat sequences facilitate circular RNA production. Genes Dev 28:2233–2247

Liang HF, Zhang XZ, Liu BG et al (2017b) Circular RNA circ-ABCB10 promotes breast cancer proliferation and progression through sponging miR-1271. Am J Cancer Res 7:1566

Liu CX, Li X, Nan F et al (2019) Structure and Degradation of Circular RNAs Regulate PKR Activation in Innate Immunity. Cell 177:865-880.e21

Liu L, Yan C, Tao S et al (2020) Circ_0058124 Aggravates the Progression of Papillary Thyroid Carcinoma by Activating LMO4 Expression via Targeting miR-370-3p. Cancer Manag Res 12:9459–9470

Lu Z, Filonov GS, Noto JJ et al (2015) Metazoan tRNA introns generate stable circular RNAs in vivo. RNA 21:1554–1565

Memczak S, Jens M, Elefsinioti A et al (2013) Circular RNAs are a large class of animal RNAs with regulatory potency. Nature 495:333–338

Militello G, Weirick T, John D et al (2017) Screening and validation of lncRNAs and circRNAs as miRNA sponges. Brief Bioinform 18:780–788

Mitra A, Pfeifer K, Park KS (2018) Circular RNAs and competing endogenous RNA (ceRNA) networks. Transl Cancer Res 7:S624–S628

Nielsen AF, Bindereif A, Bozzoni I et al (2022) Best practice standards for circular RNA research. Nat Methods 19:1208–1220

Pamudurti NR, Bartok O, Jens M et al (2017) Translation of CircRNAs. Mol Cell 66:9-21.e7

Pandey PR, Rout PK, Das A et al (2019) RPAD (RNase R treatment, polyadenylation, and poly(A)+ RNA depletion) method to isolate highly pure circular RNA. Methods 155:41–48

Park OH, Ha H, Lee Y et al (2019) Endoribonucleolytic Cleavage of m6A-Containing RNAs by RNase P/MRP Complex. Mol Cell 74:494-507.e8

Piwecka M, Glažar P, Hernandez-Miranda LR et al (2017) Loss of a mammalian circular RNA locus causes miRNA deregulation and affects brain function. Science 357:eaam8526

Rahimi K, Venø MT, Dupont DM et al (2021) Nanopore sequencing of brain-derived full-length circRNAs reveals circRNA-specific exon usage, intron retention and microexons. Nat Commun 12:4825

Salzman J, Chen RE, Olsen MN et al (2013) Cell-type specific features of circular RNA expression. PLoS Genet 9:e100377

Sun R, Liu W, Zhao Y et al (2021) Exosomal circRNA as a novel potential therapeutic target for multiple myeloma-related myocardial damage. Cancer Cell Int 21:311

Szabo L, Salzman J (2016) Detecting circular RNAs: bioinformatic and experimental challenges. Nat Rev Genet 17:679–692

Tang C, Yu T, Xie Y et al (2018) Template switching causes artificial junction formation and false identification of circular RNAs. bioRxiv 259556

Tian J, Xi X, Wang J et al (2019) CircRNA hsa_circ_0004585 as a potential biomarker for colorectal cancer. Cancer Manag Res 11:5413–5423

Wang K, Long B, Liu F et al (2016) A circular RNA protects the heart from pathological hypertrophy and heart failure by targeting miR-223. Eur Heart J 37:2602a–2611a

Wang Y, Wang Z (2015) Efficient backsplicing produces translatable circular mRNAs. RNA 21:172–179

Wang Z, Yu R, Chen X et al (2022) Clinical utility of cerebrospinal fluid-derived circular RNAs in lung adenocarcinoma patients with brain metastases. J Transl Med 20:74

Weng W, Wei Q, Toden S et al (2017) Circular RNA ciRS-7-A Promising Prognostic Biomarker and a Potential Therapeutic Target in Colorectal Cancer. Clin Cancer Res 23:3918–3928

Wu W, Zhang J, Cao X et al (2022) Exploring the cellular landscape of circular RNAs using full-length single-cell RNA sequencing. Nat Commun 13:3242

Xiao MS, Wilusz JE (2019) An improved method for circular RNA purification using RNase R that efficiently removes linear RNAs containing G-quadruplexes or structured 3' ends. Nucleic Acids Res 47:8755–8769. https://doi.org/10.1093/NAR/GKZ576

Yang Y, Fan X, Mao M et al (2017) Extensive translation of circular RNAs driven by N6-methyladenosine. Cell Res 27:626–641

Yang Y, Gao X, Zhang M et al (2018) Novel Role of FBXW7 Circular RNA in Repressing Glioma Tumorigenesis. J Natl Cancer Inst 110:304–315

Yarmishyn AA, Ishola AA, Chen CY et al (2022) Circular RNAs Modulate Cancer Hallmark and Molecular Pathways to Support Cancer Progression and Metastasis. Cancers (basel) 14:862

Ye CY, Zhang X, Chu Q et al (2017) Full-length sequence assembly reveals circular RNAs with diverse non-GT/AG splicing signals in rice. RNA Biol 14:1055–1063

Ye M, Hou H, Shen M et al (2020) Circular RNA circFOXM1 Plays a Role in Papillary Thyroid Carcinoma by Sponging miR-1179 and Regulating HMGB1 Expression. Mol Ther Nucleic Acids 19:741–750

Zeng Y, Du WW, Wu Y et al (2017) A Circular RNA Binds To and Activates AKT Phosphorylation and Nuclear Localization Reducing Apoptosis and Enhancing Cardiac Repair. Theranostics 7:3842–3855

Zhang J, Hou L, Zuo Z et al (2021) Comprehensive profiling of circular RNAs with nanopore sequencing and CIRI-long. Nat Biotechnol 39:836–845

Zhang M, Huang N, Yang X et al (2018a) A novel protein encoded by the circular form of the SHPRH gene suppresses glioma tumorigenesis. Oncogene 37:1805–1814

Zhang M, Zhao K, Xu X et al (2018b) A peptide encoded by circular form of LINC-PINT suppresses oncogenic transcriptional elongation in glioblastoma. Nat Commun 9:4475

Zhang Y, Zhang XO, Chen T et al (2013) Circular intronic long noncoding RNAs. Mol Cell 51:792–806

Zhou J, Wang H, Hong F et al (2021) CircularRNA circPARP4 promotes glioblastoma progression through sponging miR-125a-5p and regulating FUT4. Am J Cancer Res 11:138

Computational Tools for Functional Analysis of Circular RNAs

Tanvi Sinha, Sharmishtha Shyamal, and Amaresh C. Panda

Contents

Abstract The development of high-throughput total RNA sequencing technologies and novel computational tools have discovered the ubiquitous expression of circular RNAs (circRNAs) in eukaryotes across yeast to humans. Hundreds of studies demonstrated that circRNAs are abundant, conserved, stable, and expressed in a tissue-specific manner. Although a huge number of circRNAs are known to be expressed in various organisms, the cellular functions of most circRNAs remain to be explored. Several computational pipelines were developed for the systematic analysis of circRNA functions. This chapter describes an easy-to-use workflow for computational analysis of circRNA functions prior to experimental analysis. Here, we illustrate the main steps of circRNA functional analysis, including prediction of mature circRNA sequence, circRNA-RBP interaction, circRNA-miRNA-mRNA regulatory network, and protein-coding ability.

Keywords Circular RNA · miRNA · RNA-binding protein · Pathway · Translation

T. Sinha · S. Shyamal · A. C. Panda (✉)
Institute of Life Sciences, Nalco Square, Bhubaneswar, Odisha, India
e-mail: amaresh.panda@ils.res.in

J. Barciszewski (ed.), *RNA Structure and Function*, RNA Technologies 14,
https://doi.org/10.1007/978-3-031-36390-0_16

1 Introduction

Development of high-throughput transcriptome sequencing techniques discovered an intriguing class of circular RNA (circRNA) molecules without free ends (Jeck et al. 2013). Several reports established that circRNAs are an abundant and ubiquitously expressed RNA species in eukaryotes with regulatory potency (Hansen et al. 2013; Jeck et al. 2013). Furthermore, it has been well established that most circRNAs are generated via backsplicing of canonical mRNA precursors where one or multiple exons undergo head-to-tail joining to produce circRNA (Jeck et al. 2013). The unique chimeric backsplice junction sequence helps in circRNA identification, while the lack of a free end makes them resistant to exonucleases. Over the last few years, thousands of circRNAs have been identified in diverse biological samples (Vromman et al. 2021).

Identification of the huge number of circRNAs in recent years raised questions on the biological relevance of circRNAs in healthy and disease conditions. Over the years, it has been demonstrated that circRNAs play a critical role in gene regulation involving different mechanisms of action, including circRNA acting as miRNA sponges, association with RNA-binding proteins (RBPs), and serving as templates for protein translation (Hansen et al. 2013; Panda 2018; Das et al. 2021; Sinha et al. 2022b). Therefore, various computational tools have been developed to systematically explore possible functions of circRNAs before experimental validations. This chapter describes the detailed analysis of circRNA functions using computational tools (Fig. 1).

2 Tools

- Computer with Windows/Mac/Linux operating system and Internet browser such as Internet Explorer/Mozilla Firefox/Google Chrome
- Microsoft Excel to plot the bar graphs for data visualization
- BEDtools
- Python
- Cytoscape
- IRESfinder

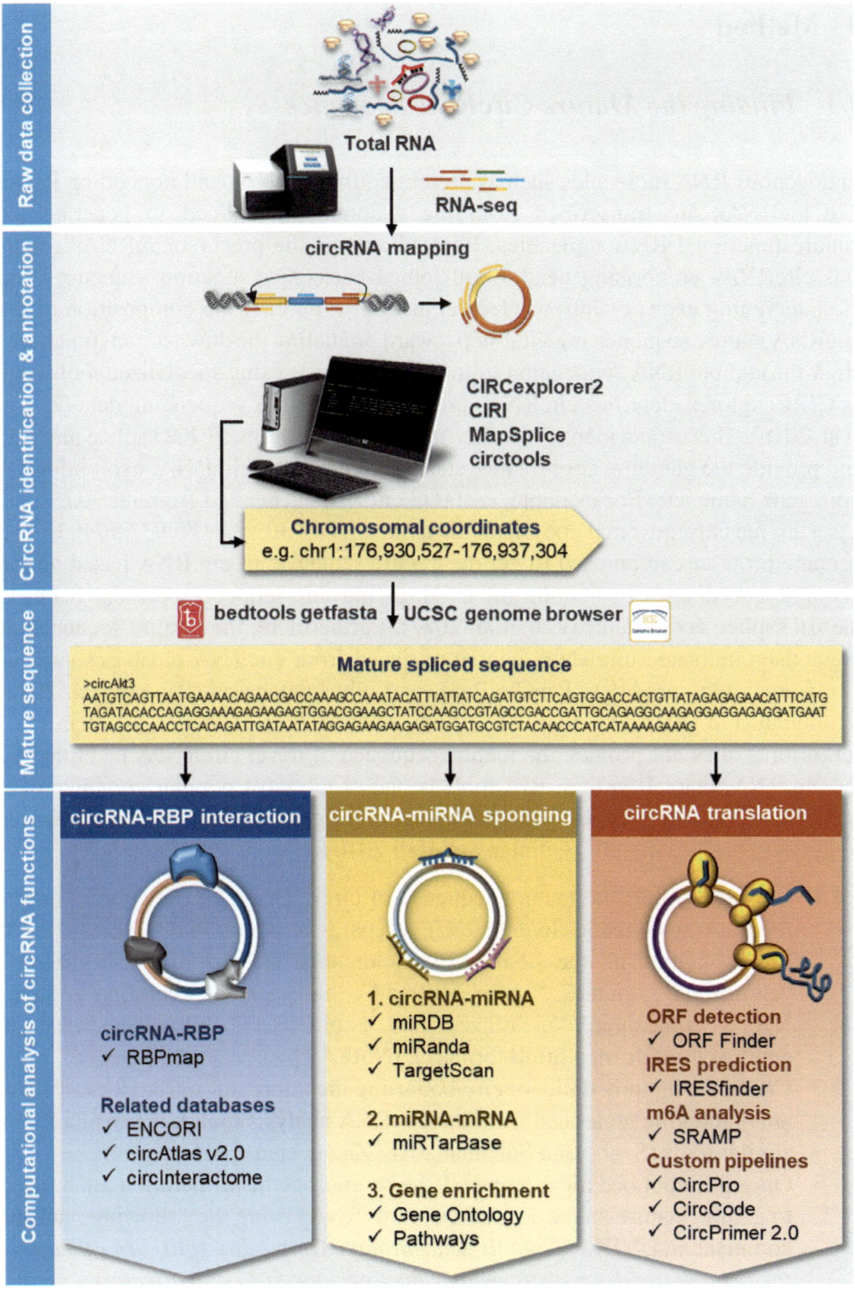

Fig. 1 Flowchart depicting computational tools for the analysis of circRNA functions

3 Method

3.1 Finding the Mature CircRNA Sequence

Endogenous RNA molecules such as protein-coding mRNAs and noncoding RNAs like lncRNAs are transcribed from the genome and spliced to generate the mature functional RNA molecules. Backsplicing of the precursor mRNAs generates circRNAs comprising head-to-tail joined backsplice junction sequences and the intervening exons or introns (Jeck et al. 2013). Finding the composition of the circRNA mature sequence is a vital step toward predicting the downstream functions. High-throughput RNA sequencing followed by analysis using specialized tools such as CIRCexplorer identifies circRNAs present in total RNA sequencing data (Zhang et al. 2016). These tools identify circRNAs based on the chimeric backsplice junction and provide the genomic coordinates, start, and end of the circRNA, exon information, gene name, etc. For example, *circAkt3* circRNA generated from the *Akt3* gene has a unique chromosomal coordinate ID as *chr1|176,930,527|176,937,304|*. UCSC genome browser can be used to get the mature sequence of circRNA based on the genomic coordinates by joining the exon, or intronic sequences present between the backsplice coordinates (Lee et al. 2022). Furthermore, the mature sequence of previously annotated circRNAs can be retrieved from circRNA databases such as circInteractome, circAtlas, PanCircBase, and other databases (Panda et al. 2018; Vromman et al. 2021; Wu et al. 2020; Sinha et al. 2022a). However, the mentioned set of tools does not provide the mature sequence of novel circRNAs identified by the circRNA annotation tool. BEDtools is one of the most popular computational software used for the retrieval and manipulation of genomic and spliced sequences of genes expressed in cells (Quinlan and Hall 2010).

3.1.1 In order to find the mature sequence of circRNAs using BEDtools (https://bedtools.readthedocs.io/en/latest/), prepare an appropriate BED12 input file containing all the 12 requisite fields such as "chrom", "chromStart", "chromEnd", "name", "score", "strand", "thickStart", "thickEnd", "itemRgb", "exonCount", "exonSizes" and "exonOffsets" (https://genome.ucsc.edu/FAQ/FAQformat.html#format1). [**Note** 4.1]

3.1.2 Create the input.bed file for *circAkt3* using the information from the circRNA annotation file generated from the circRNA analysis tool, CIRCexplorer2 or similar tools (Szabo and Salzman 2016; Zhang et al. 2016).

3.1.3 Once the input.bed file is prepared, the bedtools getfasta operation can be used to get the mature splice sequences of circRNAs using the following one-line command in UNIX—*bedtools getfasta-nameOnly-tab-s-split -bed <filedirectory> input.bed-fi <path to genome file> genome.fa-fo <filedirectory> prefix_ output.txt*.

3.1.4 Open the tab-delimited *getfasta output.txt* file using Microsoft Excel. The mature sequence of mouse *circAkt3* is 257 nucleotides long as shown below

>mmu_circAkt3_257
AATGTCAGTTAATGAAAACAGAACGACCAAAGCCAAATACATTTAT TATCAGATGTCTTCAGTGGACCACTGTTATAGAGAGAACATTTCATGT AGATACACCAGAGGAAAGAGAAGAGTGGACGGAAGCTATCCAAGCC GTAGCCGACCGATTGCAGAGGCAAGAGGAGGAGAGGATGAATTGTA GCCCAACCTCACAGATTGATAATATAGGAGAAGAAGAGATGGATGCG TCTACAACCCATCATAAAAGAAAG.

3.2 Analysis of RBPs Interacting with circAkt3

Several reports highlight the importance of the circRNA-protein interaction in cell physiology and diseases, leading to the differential regulation of the target gene expression by circRNAs acting as a protein decoy, protein sponge, and scaffold. Thus, it is paramount to discover the proteins interacting with circRNAs and study the functional outcome of such association in cells. There are a few web servers to predict the RBP binding to circRNAs. For example, the circInteractome (https://circinteractome.nia.nih.gov/index.html) database predicts RBP binding to human circRNAs with circBase IDs and their flanking sequences spanning the circularizing exons (Glazar et al. 2014; Panda et al. 2018). Another database namely ENCORI/starBase v2.0 (https://starbase.sysu.edu.cn/index.php) is useful for identifying circRNA-RBP interactions in human samples based on large-scale CLIP-Seq datasets (Li et al. 2014). circAtlas v2.0 (http://159.226.67.237:8080/new/index.php) database lists interacting RBPs of vertebrate circRNAs along with binding site counts in the upstream and downstream flanking sequences (Wu et al. 2020). However, these databases only predict the circRNA-RBP interaction for circRNAs present in that database. Here, we describe the prediction of RBP binding sites on the circRNA sequence using the RBPmap web server (Paz et al. 2014).

3.2.1 Open the RBPmap home page http://rbpmap.technion.ac.il/index.html.

3.2.2 Under the Input (mandatory) tab, select the "Genome" type and "Database assembly" version from the dropdown list. Select "Mouse" and "Dec. 2011 (GRCm38/mm10)" for mouse *circAkt3* (Fig. 2).

3.2.3 Paste the mature splice sequence of the *circAkt3* in the "Query sequences/ coordinates" box. [**Note** 4.2]

3.2.4 Under the Motif selection (mandatory) tab, click "Click here to select motifs from RBPmap full list" and select the "Human/Mouse motifs" in the pop-up window.

3.2.5 As shown in Fig. 2, clicking the "Submit" button will open a result page. The result page appears as a prediction summary list showing interacting protein names along with Position, Motif, Occurrence, Z-score, and P-value. The full result list can be downloaded as a text file.

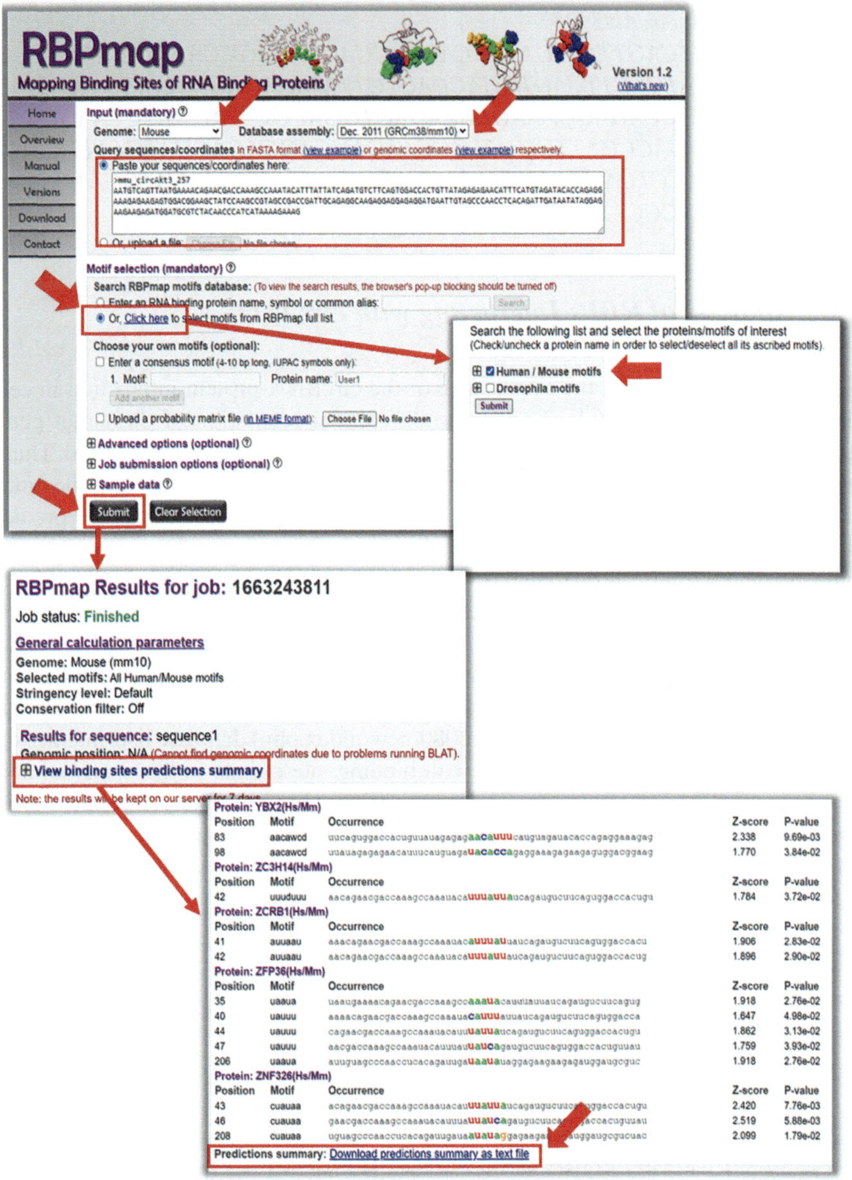

Fig. 2 Screenshots of RBPmap tool for predicting RBP binding to *circAkt3*

3.3 *Analysis of circRNA-miRNA-mRNA Regulatory Networks*

3.3.1 Identification of miRNAs Interacting with circAkt3

As mentioned in the previous sections, circRNA regulates both transcription and translation by binding RNA-binding proteins and miRNAs. miRNAs are known to suppress target mRNA stability and translation by binding to the 3'UTR of the target gene. CircRNAs with miRNA binding sites in the cytoplasm serve as miRNA sponges to inhibit miRNA activity through competing endogenous RNA (ceRNA) network, thereby regulating downstream target gene expression (Hansen et al. 2013; Panda 2018). Several databases and standalone tools have been developed to predict miRNA target sites, such as miRDB, TargetScan, and miRanda among many others (Enright et al. 2003; Agarwal et al. 2015; Riffo-Campos et al. 2016; Chen and Wang 2020). This section describes the method to predict miRNA binding sites on *circAkt3* using miRDB (Fig. 3).

3.3.1.1 Open the miRDB web tool (http://www.mirdb.org/). [**Note** 4.3]
3.3.1.2 Click the "custom prediction" tab and select "mouse" as the species and "mRNA target sequence" as the submission type.
3.3.1.3 Paste the *circAkt3* sequence in the sequence input box and click "go" for custom miRNA prediction.
3.3.1.4 After the completion of the prediction, click "Retrieve prediction results" to find the final list of miRNA binding sites in *circAkt3* sequence. Here, miRDB predicted 11 miRNAs that bind to *circAkt3* (Fig. 3).

3.3.2 Identification of Target Genes of miRNAs Associated with circAkt3

Hundreds of studies have highlighted the role of circRNA-miRNA-mRNA regulatory axis in gene regulation (Panda 2018). Numerous tools and web servers have been developed to predict the miRNA target genes computationally. However, since all miRNA targets may not be functionally validated, we used the mouse (mmu) miRTarBase 8.0 to find the experimentally validated mRNA targets of the miRNAs associated with *circAkt3* as follows (Huang et al. 2022).

3.3.2.1 Download the complete list of experimentally validated mouse miRNAs and their targets in miRTarBase-Release 8.0.
3.3.2.2 Find the common miRNAs targeted by *circAkt3* and present in miRTarBase. For *circAkt3*, four miRNAs were found to have experimentally validated mRNA targets in miRTarBase (Fig. 4).
3.3.2.3 Extract the target mRNAs of these four miRNAs in miRTarBase.
3.3.2.4 Construct the *circAkt3*-miRNA-mRNA regulatory network using Cytoscape software.

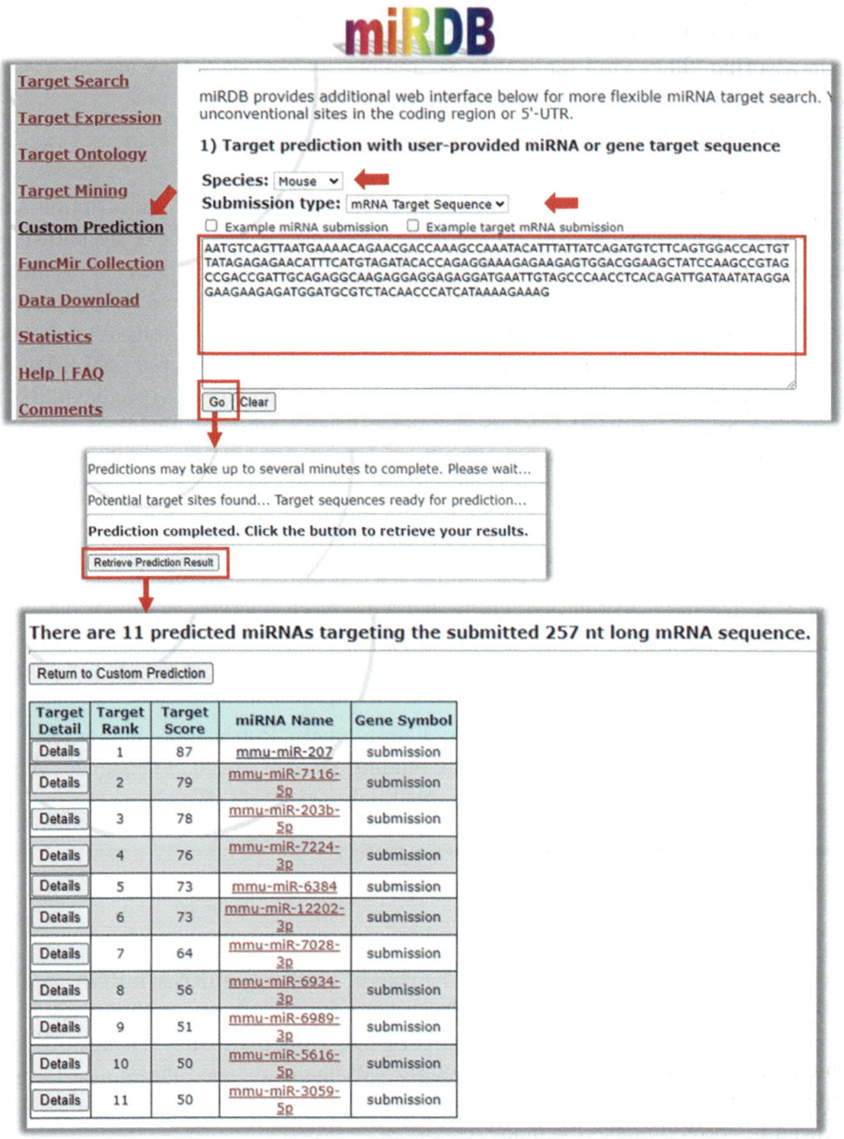

Fig. 3 Screenshots of miRDB web tool for predicting miRNA binding sites on *circAkt3*

3.3.3 Analysis of Gene Ontology and Pathways Regulated by circAkt3-miRNA-mRNA Network

As shown in Fig. 4, *circAkt3* may act as a sponge for four miRNAs that have experimentally validated downstream targets in miRTarBase. Therefore, the mRNA targets from the *circAkt3*-miRNA-mRNA regulatory network can be analyzed to predict the

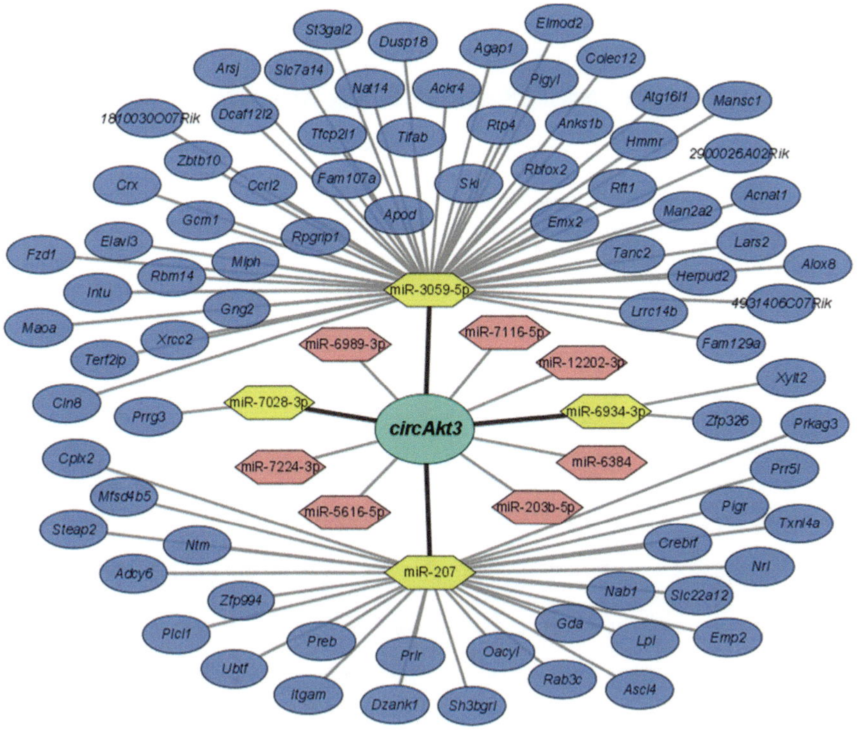

Fig. 4 The *circAkt3*-miRNA-mRNA regulatory network was constructed on Cytoscape. The hexagon nodes represent miRNAs associated with *circAkt3*. The yellow hexagons represent the miRTarBase miRNAs, and the blue circle nodes represent experimentally validated target genes in miRTarBase

biological functions of *circAkt3*. The association of target genes with different gene ontology terms and pathways can be predicted with various databases and web tools, including Ingenuity Pathway Analysis (IPA), Genomatix Software Suite (GSS), Gene Ontology database (GO) and KEGG pathways, PANTHER, DAVID, KOBAS, and STRING (Dennis et al. 2003; Kanehisa et al. 2016; The Gene Ontology 2019; Bu et al. 2021; Szklarczyk et al. 2021; Thomas et al. 2022). Statistical significance was calculated by student's t-test and considered significant with a p-value of <0.05. Here, we used the PANTHER database (v17) for the analysis for *circAkt3* target genes (Fig. 5) (Thomas et al. 2022).

3.3.3.1 To perform the gene enrichment analysis, open http://www.pantherdb.org/.

3.3.3.2 Paste the mRNA target gene list in the input box and select "Mus musculus" in the dropdown menu for "Organism".

3.3.3.3 Select the "Functional classification viewed in graphic charts" and "Pie chart".

3.3.3.4 Click the "submit" tab.

3.3.3.5 A result page will open with the pie chart results.

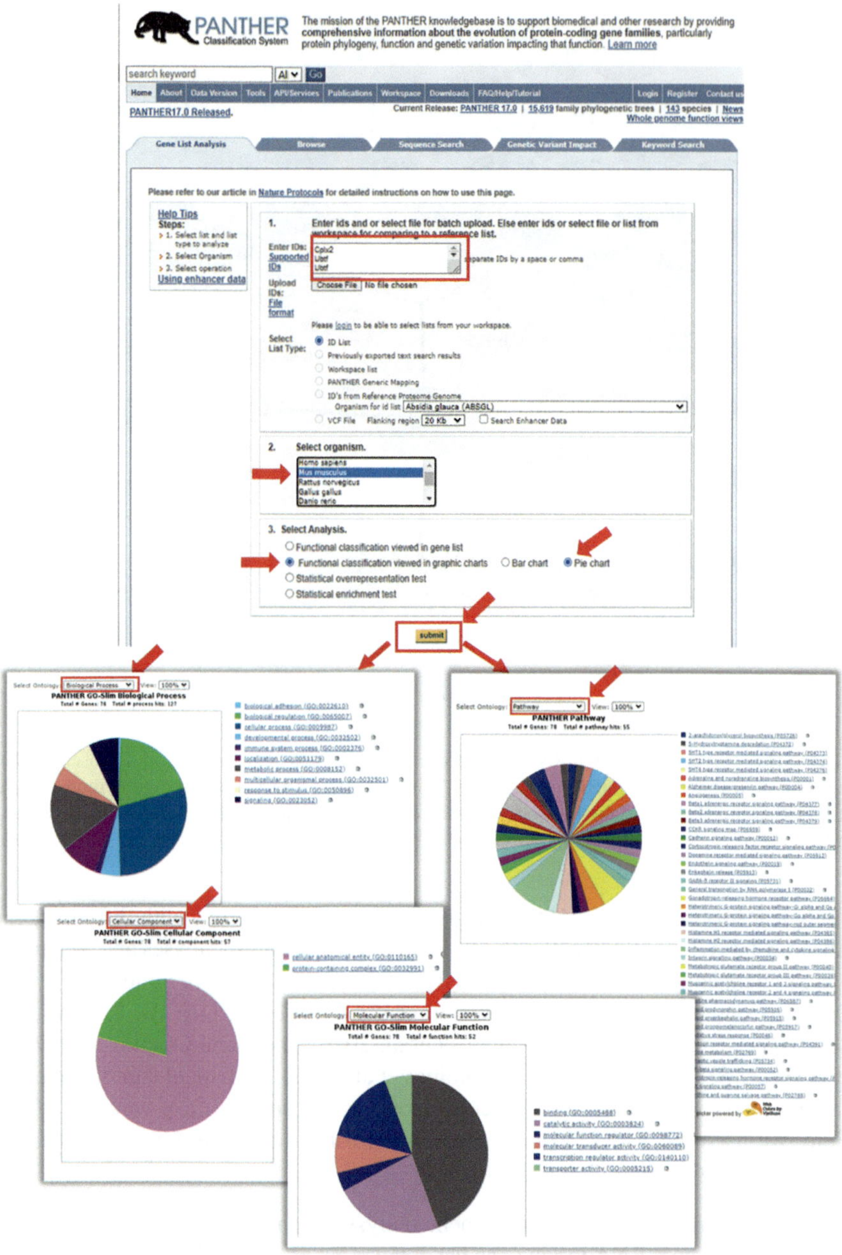

Fig. 5 Enrichment analysis of mRNA targets *circAkt3*-miRNA-mRNA regulatory network

3.3.3.6 Select the Gene Ontology terms in the dropdown menu to show the classification of genes under different GO terms, including Biological Process, Molecular Function, Cellular Component, and Panther Pathways (Fig. 5). [**Note** 4.4]

3.4 Analysis of the Protein-Coding Potential of circAkt3

Recent years witnessed a rise in research investigating the protein-coding potential of circRNAs. Only a handful of reports are available about the functional role of translating circRNAs and their peptide products (Fan et al. 2022; Sinha et al. 2022b). Recently, a few tools and databases have been developed to access the coding potential of circRNAs such as CPAT, CircPro, CircCode, CircPrimer, riboCIRC, and TransCirc among many others (Wang et al. 2013; Meng et al. 2017; Sun and Li 2019; Huang et al. 2021; Li et al. 2021; Zhong and Feng 2022). Nevertheless, the presence of open reading frames (ORFs), internal ribosome entry sites (IRES), m6A modification, and the association of circRNAs with translating polyribosomes suggested the possible cap-independent translation of circRNAs into peptides (Yang et al. 2017; Ye et al. 2021; Fan et al. 2022). Here, we used *circAkt3* sequence to study its protein-coding potential by analyzing the presence of ORF, IRES, and m6A sites.

3.4.1 Analysis of ORFs Spanning Backsplice Junction

We used the NCBI ORFfinder to identify the ORFs in *circAkt3* (Fig. 6). [**Note** 4.5]

3.4.1.1 Open the https://www.ncbi.nlm.nih.gov/orffinder/ and paste the *circAkt3* sequence three times to get ORFs spanning the junction.
3.4.1.2 Click “ATG and alternative initiation codons” and submit the form.
3.4.1.3 The new page will open with a list of ORFs present on *circAkt3*.
3.4.1.4 The ORFs spanning the backsplice junction should be identified manually. As shown in Fig. 6, the selected ORF1 spans the backsplice junction sequence.

3.4.2 Analysis of IRES

3.4.2.1 Reportedly presence of IRES in circRNAs may help drive protein translation. IRESfinder (https://github.com/xiaofengsong/IRESfinder) is a standalone Python script (*IRESfinder/IRESfinder.py-fcircRNA_2x_sequences.fa-ocircRNA_IRESfinder.result)* for screening of IRES sequences in eukaryotic cells based on logit model and k-mer features (Zhao et al. 2018).
3.4.2.2 IRESfinder results will score circRNAs into IRES and non-IRES sequences.

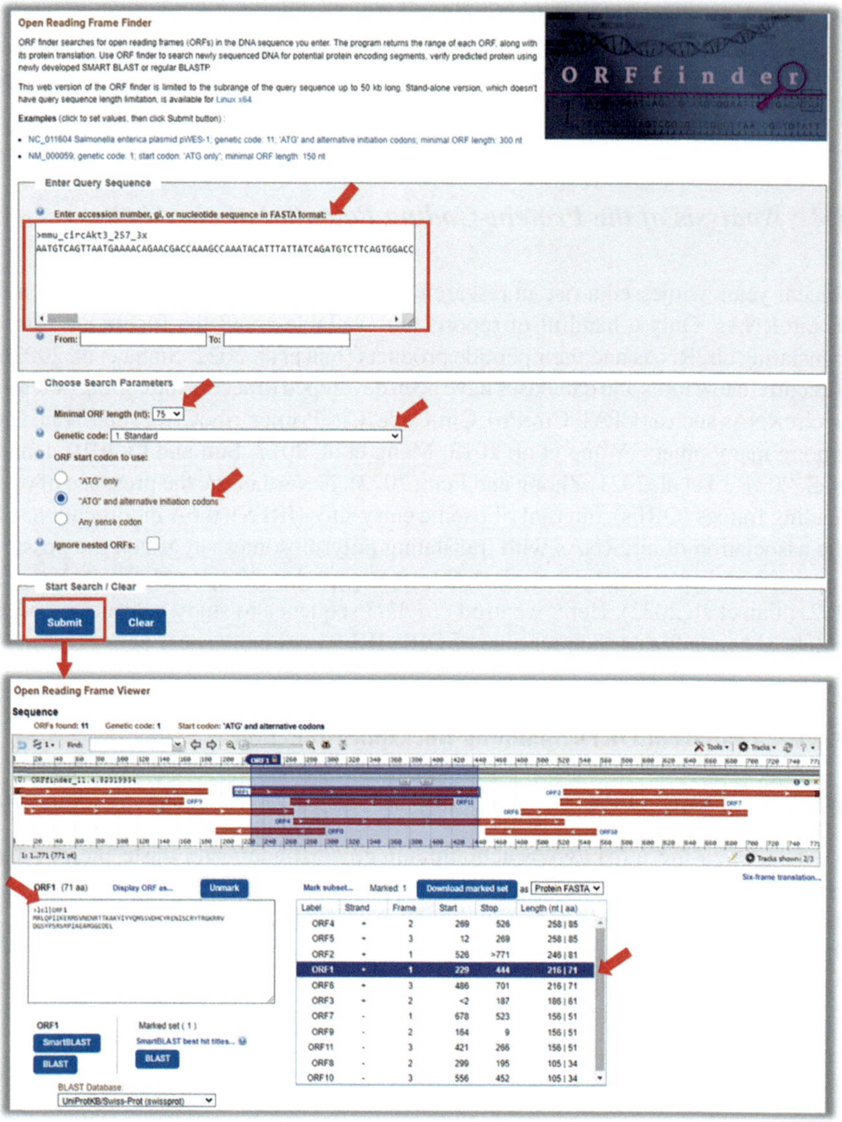

Fig. 6 ORFfinder interface showing ORF search result for 3 × sequence of *circAkt3*

3.4.2.3 Here, *circAkt3* was predicted as a non-IRES sequence suggesting that this may not be translated into protein.

3.4.3 Analysis of m6A Sites

3.4.3.1 The m6A sites on the circRNA sequence can be found by SRAMP web tool (Zhou et al. 2016).
3.4.3.2 Open the "prediction" tab on the https://www.cuilab.cn/sramp webpage.
3.4.3.3 Paste the *circAkt3* sequence in the input box for "Mature mRNA mode".
3.4.3.4 The new window shows the results of m6A prediction sites
3.4.3.5 As shown in Fig. 7, *circAkt3* does not have any m6A sites.

In addition to the above analysis, the association of *circAkt3* with polyribosomes must be tested to study its translation potential (Panda et al. 2017; Ye et al. 2021). Since *circAkt3* does not contain IRES or m6A sites, we assume it may not be translated into proteins.

4 Technical Notes

4.1 The sequence of multiple circRNAs can be extracted at once using the BED12 information for those circRNAs. The input.bed file used in BEDtools does not contain the headers.
4.2 The circRNA sequences in Fasta format can be uploaded as input for predicting RBPs for multiple circRNAs.
4.3 Any other sequence-based miRNA prediction webserver or standalone tool can be used to predict miRNAs targeting circRNAs. Examples include miRanda, TargetScan, miRWalk, and miRNet.
4.4 Statistical overrepresentation test or gene set enrichment test can also be performed to check specific enrichment of genes for specific GO terms and pathways.
4.5 The standalone ORFfinder can predict the ORFs present in multiple circRNAs using the sequences in Fasta format with the following command line in UNIX.
ORFfinder-in circRNA_3x_sequences.fa-s 1-strand plus-outfmt 1-out circRNA_ORF.txt

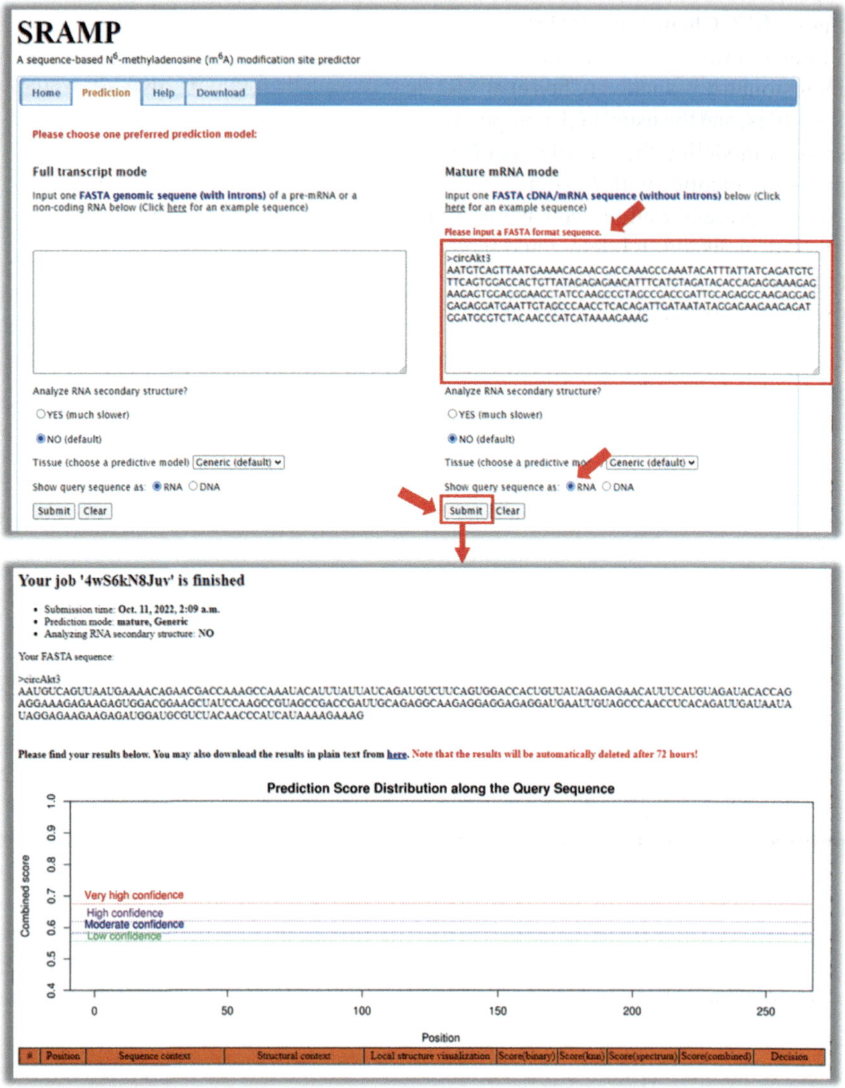

Fig. 7 Prediction of m6A sites on *circAkt3* using SRAMP web server

Acknowledgements This work was supported by intramural funding from the Institute of Life Sciences. The DST-Inspire JRF fellowship supported Tanvi Sinha. We thank Suman Singh for proofreading the article.

Conflicts of Interest The authors declare no conflict of interest.

References

Agarwal V, Bell GW, Nam JW et al (2015) Predicting effective microRNA target sites in mammalian mRNAs. Elife 4

Bu D, Luo H, Huo P et al (2021) KOBAS-i: intelligent prioritization and exploratory visualization of biological functions for gene enrichment analysis. Nucl Acids Res 49:W317–W325

Chen Y, Wang X (2020) miRDB: an online database for prediction of functional microRNA targets. Nucl Acids Res 48:D127–D131

Das A, Sinha T, Shyamal S et al (2021) Emerging role of circular RNA-protein interactions. Noncoding RNA 7

Dennis G Jr, Sherman BT, Hosack DA et al (2003) DAVID: database for annotation, visualization, and integrated discovery. Genome Biol 4:P3

Enright AJ, John B, Gaul U et al (2003) MicroRNA targets in drosophila. Genome Biol 5:R1

Fan X, Yang Y, Chen C et al (2022) Pervasive translation of circular RNAs driven by short IRES-like elements. Nat Commun 13:3751

Glazar P, Papavasileiou P, Rajewsky N (2014) Circbase: a database for circular RNAs. RNA 20:1666–1670

Hansen TB, Jensen TI, Clausen BH et al (2013) Natural RNA circles function as efficient microRNA sponges. Nature 495:384–388

Huang HY, Lin YC, Cui S et al (2022) miRTarBase update 2022: an informative resource for experimentally validated miRNA-target interactions. Nucl. Acids Res 50:D222–D230

Huang W, Ling Y, Zhang S et al (2021) TransCirc: an interactive database for translatable circular RNAs based on multi-omics evidence. Nucl Acids Res 49:D236–D242

Jeck WR, Sorrentino JA, Wang K et al (2013) Circular RNAs are abundant, conserved, and associated with ALU repeats. RNA 19:141–157

Kanehisa M, Sato Y, Kawashima M et al (2016) KEGG as a reference resource for gene and protein annotation. Nucl Acids Res 44:D457-462

Lee BT, Barber GP, Benet-Pages A et al (2022) The UCSC genome browser database: 2022 update. Nucl Acids Res 50:D1115–D1122

Li H, Xie M, Wang Y et al (2021) riboCIRC: a comprehensive database of translatable circRNAs. Genome Biol 22:79

Li JH, Liu S, Zhou H et al (2014) starBase v2.0: decoding miRNA-ceRNA, miRNA-ncRNA and protein-RNA interaction networks from large-scale CLIP-Seq data. Nucl Acids Res 42:D92-97

Meng X, Chen Q, Zhang P et al (2017) CircPro: an integrated tool for the identification of circRNAs with protein-coding potential. Bioinformatics 33:3314–3316

Panda AC (2018) Circular RNAs Act as miRNA Sponges. In: Xiao J (ed) Circular RNAs: biogenesis and functions. Springer, Singapore, pp 67–79

Panda AC, Dudekula DB, Abdelmohsen K et al (2018) Analysis of circular RNAs using the web tool circinteractome. Methods Mol Biol 1724:43–56

Panda AC, Martindale JL, Gorospe M (2017) Polysome fractionation to analyze mRNA distribution profiles. Bio Protoc 7

Paz I, Kosti I, Ares M Jr et al (2014) RBPmap: a web server for mapping binding sites of RNA-binding proteins. Nucl Acids Res 42:W361-367

Quinlan AR, Hall IM (2010) BEDTools: a flexible suite of utilities for comparing genomic features. Bioinformatics 26:841–842

Riffo-Campos AL, Riquelme I, Brebi-Mieville P (2016) Tools for sequence-based miRNA target prediction: what to choose? Int J Mol Sci 17

Sinha T, Mishra SS, Singh S et al (2022a) PanCircBase: an online resource for the exploration of circular RNAs in pancreatic islets. Front Cell Dev Biol 10:942762

Sinha T, Panigrahi C, Das D et al (2022b) Circular RNA translation, a path to hidden proteome. Wiley Interdiscip Rev RNA 13:e1685

Sun P, Li G (2019) CircCode: a powerful tool for identifying circrna coding ability. Front Genet 10:981

Szabo L, Salzman J (2016) Detecting circular RNAs: bioinformatic and experimental challenges. Nat Rev Genet 17:679–692

Szklarczyk D, Gable AL, Nastou KC et al (2021) The STRING database in 2021: customizable protein-protein networks, and functional characterization of user-uploaded gene/measurement sets. Nucl Acids Res 49:D605–D612

The Gene Ontology C (2019) The gene ontology resource: 20 years and still GOing strong. Nucl Acids Res 47:D330–D338

Thomas PD, Ebert D, Muruganujan A et al (2022) PANTHER: making genome-scale phylogenetics accessible to all. Protein Sci 31:8–22

Vromman M, Vandesompele J, Volders PJ (2021) Closing the circle: current state and perspectives of circular RNA databases. Brief Bioinform 22:288–297

Wang L, Park HJ, Dasari S et al (2013) CPAT: coding-potential assessment tool using an alignment-free logistic regression model. Nucl Acids Res 41:e74

Wu W, Ji P, Zhao F (2020) CircAtlas: an integrated resource of one million highly accurate circular RNAs from 1070 vertebrate transcriptomes. Genome Biol 21:101

Yang Y, Fan X, Mao M et al (2017) Extensive translation of circular RNAs driven by N(6)-methyladenosine. Cell Res 27:626–641

Ye Y, Wang Z, Yang Y (2021) Comprehensive identification of translatable circular RNAs using polysome profiling. Bio-Protoc 11:e4167

Zhang XO, Dong R, Zhang Y et al (2016) Diverse alternative back-splicing and alternative splicing landscape of circular RNAs. Genome Res 26:1277–1287

Zhao J, Wu J, Xu T et al (2018) IRESfinder: identifying RNA internal ribosome entry site in eukaryotic cell using framed k-mer features. J Genet Genomics 45:403–406

Zhong S, Feng J (2022) CircPrimer 2.0: a software for annotating circRNAs and predicting translation potential of circRNAs. BMC Bioinform 23:215

Zhou Y, Zeng P, Li YH et al (2016) SRAMP: prediction of mammalian N6-methyladenosine (m6A) sites based on sequence-derived features. Nucl Acids Res 44:e91

The Hidden Layer of RNA Variants

Kenzui Taniue and Nobuyoshi Akimitsu

Contents

Abstract The genome is pervasively transcribed and produces various messenger RNAs and noncoding RNAs. Defective transcripts are also produced, which are cleared by RNA surveillance and quality control systems. Nuclear RNA degradation pathways play important roles in these systems and in shaping the transcriptome and preventing diseases. In this review, we summarize current knowledge of nuclear noncoding RNAs. We then discuss nuclear RNA degradation factors involved in RNA surveillance and RNA quality control systems.

K. Taniue (✉) · N. Akimitsu (✉)
Isotope Science Center, The University of Tokyo, 2-11-16, Yayoi, Bunkyo-Ku, Tokyo 113-0032, Japan
e-mail: kenzui@ric.u-tokyo.ac.jp

N. Akimitsu
e-mail: akimitsu@ric.u-tokyo.ac.jp

K. Taniue
Department of Medicine, Asahikawa Medical University, 2-1 Midorigaoka Higashi, Asahikawa 078-8510, Hokkaido, Japan

J. Barciszewski (ed.), *RNA Structure and Function*, RNA Technologies 14,
https://doi.org/10.1007/978-3-031-36390-0_17

Keywords RNA surveillance · lncRNAs · eRNA · Gene region edge-associated transcript · RNA quality control · Nuclear RNA degradation · RNA exosome

Abbreviations

antisense CUT	Antisense cryptic unstable transcript
antisense SUT	Antisense stable unannotated transcript
ARS2	Arsenite-resistance 2
aTASR	Antisense TASR
CBC	Cap-binding complex
CBP	CREB-binding protein
CBP	Cap-binding protein
CCAT1	Colon cancer-associated transcript 1
CUT	Cryptic unstable transcript
ER	Estrogen receptor
GREATs	Gene region edge-associated transcripts
GRO-seq	Global run-on sequencing
H3K27ac	Acetylation on histone H3 lysine 27
H3K4me1	Mono methylation on histone H3 lysine 4
H3K4me2	Di-methylation on histone H3 lysine 4
HiNoCo body	Heat-inducible noncoding RNA-containing nuclear body
hnRNP	Heterogeneous nuclear ribonucleoprotein
KH	S1/K homology
LED	LncRNA activator of Enhancer Domain
lncRNA	Long ncRNA
MALAT1	Metastasis-associated lung adenocarcinoma transcript 1
mRNA	Messenger RNA
miRNA	MicroRNA
ncRNA	Noncoding RNA
NEAT1	Nuclear-enriched abundant transcript 1
NEXT	Nuclear exosome targeting
p53BER	P53-bound enhancer region
PALR	Promoter-associated long RNA
pancRNA	Promoter-associated noncoding RNA
PAR	Promoter-associated RNA
PASR	Promoter-associated short RNA
PH	Pleckstrin homology
piRNA	Piwi-interacting RNA
Pol II	RNA polymerase II
PRC2	Polycomb repressive complex 2
PROMPT	Promoter upstream transcript
RNP	Ribonucleoprotein
RRM	RNA-recognition motif

Rrp	RRNA-processing protein
SE	Super-enhancer
SF2	Superfamily 2
siRNA	Small interfering RNAs
snRNP	Small nuclear ribonucleoprotein
SUT	Stable unannotated transcript
TAR	Terminus-associated RNA
TASR	Terminus-associated small RNA
TCGA	The Cancer Genome Atlas
tiRNA	Tiny RNA
TPR	Translocated promoter region
TRAMP	TRF4-2-ZCCHC7-MTR4 polyadenylation
TSS	Transcription start site
TSSaRNA	TSS-associated RNA
TTF2	Transcription termination factor 2
TTSa RNA	Transcription termination site-associated RNA
uaRNA	Upstream antisense RNA
UaRNA	3′-UTR-derived RNA
UNT	Upstream noncoding transcript
XRN2	5′-To-3′ exoribonuclease

1 Introduction

Recent whole genome and transcriptome analyses have revealed that only 1% of the human genome encodes proteins and that most of the genome is actively transcribed into noncoding RNAs (ncRNAs) (Lander et al. 2001). Transcripts from noncoding regions of the genome are classified as either housekeeping or functional ncRNAs. Housekeeping ncRNAs, such as ribosomal RNAs (rRNAs), transfer RNAs (tRNAs), small nuclear RNAs (snRNAs) and small nucleolar RNAs (snoRNAs), are constitutively expressed independently of tissue and developmental stage. Functional ncRNAs, however, are composed of microRNAs (miRNAs), small interfering RNAs (siRNA), Piwi-associated RNAs (piRNAs), long ncRNAs (lncRNAs) and transcripts from gene edge regions (Carthew and Sontheimer 2009; Ponting et al. 2009). ncRNAs engage in diverse biological processes and some are related to pathological conditions, including cancer (Schmitt and Chang 2016; Anastasiadou et al. 2017; Taniue and Akimitsu 2021a). However, the roles of most ncRNAs are still unclear. In this chapter, we describe the biogenesis and characteristics of ncRNAs produced from intergenic regions and edge-of-gene regions and assess current knowledge regarding their roles in cancer (Fig. 1). Messenger RNAs (mRNAs) are primarily found in the cytoplasm, where they are translated, whereas ncRNAs are more frequently found in the nucleus. Many aberrant transcripts are produced by disrupted transcription and splicing and are detected and cleared by RNA surveillance and quality control systems in both the cytoplasm and nucleus. Nuclear RNA degradation pathways play important roles in these systems, and in this chapter we address the RNA surveillance system targeting ncRNAs and mRNAs in the nucleus. We also present an overview of nuclear RNA decay systems that are part of the RNA surveillance system in the nucleus.

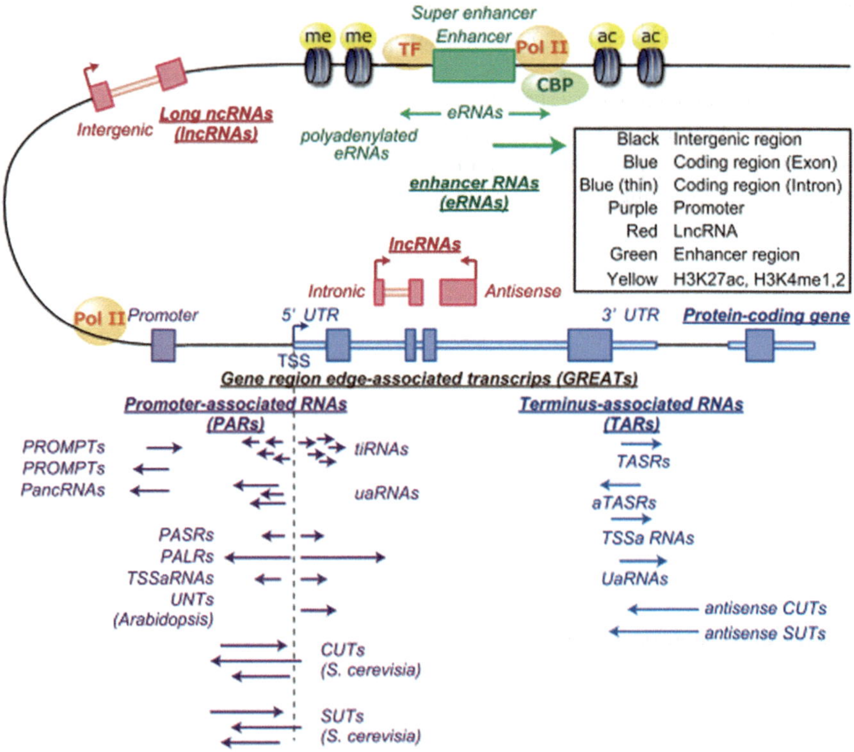

Fig. 1 Schematic representation of nuclear lncRNAs, eRNAs and gene region edge-associated transcripts (GREATs). Blue and pink bars represent coding region and lncRNAs, respectively. Intergenic lncRNAs do not intersect with any protein-coding genes. Antisense lncRNAs are transcribed from the opposite strand of protein-coding genes and intronic lncRNAs partially or entirely reside within protein-coding regions. Green bar represents enhancer region and eRNAs with or without polyadenylation are transcribed bi-directionally from the region. Active TFs are recruited into the enhancer region to bind to specific DNA sequences and promote transcription factor assembly, allowing them to recruit transcriptional co-activators, such as p300/CBP. CBP or other histone modification factors can acetylate H3K27 or methylate H3K4 to enhance open chromatin states, thereby engaging Pol II to promote eRNA transcription. The lower part of the figure shows the GREATs. GREATs are divided into promoter-associated RNA (PARs, Purple) and terminus-associated RNA (TARs, dark blue). Both types of RNAs are transcribed bi-directionally and produce various types of transcripts. See also list of abbreviations

2 Noncoding RNAs

Approximately 20,000 protein-coding genes are estimated to constitute a mere 1% of the human genome (Lander et al. 2001). For many decades, the majority of the non-protein-coding human genome was considered to be junk DNA (Elgar and Vavouri 2008; Kopp and Mendell 2018). However, over the past decade, massive parallel sequencing technology has revealed that the majority of the human genome is dynamically and differentially transcribed into ncRNAs (Djebali et al. 2012). Long-read sequencing has strengths in determining full-length RNA molecules. For instance, RNA Capture Long Seq, which combines targeted RNA capture with long-read sequencing, has identified novel noncoding RNAs in human and mouse genomes (Lagarde et al. 2017). These ncRNAs, although not all, are involved in almost all physiological processes.

ncRNAs are classified as short ncRNAs of less than 200 nucleotides and lncRNAs of more than 200 nucleotides. Short ncRNAs are composed of miRNAs, siRNAs, snoRNAs, rRNAs, tRNAs and piRNAs (Chan and Tay 2018), whereas lncRNAs include diverse functional RNAs, such as *XIST*, *NEAT1*, and *MALAT1*. Many of them play a role in physiological and pathological phenomena (Shirahama et al. 2020; Taniue and Akimitsu 2021a; Takahashi et al. 2021; Onoguchi-Mizutani and Akimitsu 2022). Several ncRNAs are present in the nucleus. Here we address typical nuclear ncRNAs (Fig. 1).

2.1 Nuclear LncRNAs

lncRNAs are characterized as RNAs longer than 200 bp with little or no coding potential (Anastasiadou et al. 2017; Taniue and Akimitsu 2021a). Accumulating evidence indicates that lncRNAs are essential for a variety of biological processes, including proliferation, differentiation, embryogenesis, neurogenesis, stem cell pluripotency, pathogenic infection, and cancer (Batista and Chang 2013; Huarte 2015; Anastasiadou et al. 2017; Kopp and Mendell 2018; Shirahama et al. 2020, 2021). lncRNAs also contribute to chromatin and genomic structural remodeling, RNA stabilization, RNA localization, and transcriptional regulation (Mercer et al. 2009; Yanagida et al. 2013; Taniue et al. 2016b; Ransohoff et al. 2018). Moreover, lncRNAs can regulate protein stability by inhibiting post-translational modifications that cause protein degradation (Taniue et al. 2016a). Mature mRNAs are mostly transported to the cytoplasm where they are translated, whereas lncRNAs are more often found in the nucleus than in the cytoplasm (Taniue and Akimitsu 2021a). Most nuclear-enriched ncRNAs are unstable, although certain lncRNAs have increased stability and function in the nucleus (Tano et al. 2010; Tani et al. 2012).

A well-studied nuclear lncRNA is human nuclear-enriched abundant transcript 1 (*NEAT1*), which is widely expressed in different tissues (Fox and Lamond 2010). *NEAT1* is transcribed from human chromosome 11 to generate two major isoforms,

NEAT1v1 (3.7 kb) and *NEAT1v2* (23 kb), by alternate 3′ end processing (Naganuma et al. 2012). *NEAT1v2*, but not *NEAT1v1*, is required for the formation and maintenance of paraspeckles (Clemson et al. 2009; Nakagawa et al. 2011), which are membrane-less compartments in the nuclear interchromatin space that play a variety of gene regulatory functions. Paraspeckles are involved in transcription and RNA processing and are stimulated to develop by p53 activation under stress conditions (Adriaens et al. 2016). Paraspeckles comprise more than 60 proteins, the majority of which have RNA-binding activity, and include NONO, SFPQ, and FUS (Yamazaki and Hirose 2015). In addition to p53 activation, *NEAT1* expression is influenced by a variety of factors, including influenza and herpes simplex virus infection, bacterial infection, lipopolysaccharide induction, and inflammasome activation (Imamura et al. 2014, 2018; Zhang et al. 2016, 2019). *NEAT1* also regulates the expression of a number of chemokines and cytokines, such as *IL6*, *IL8*, and *CXCL10*, after virus infection or lipopolysaccharide induction (Imamura et al. 2014, 2018; Zhang et al. 2016). Moreover, *NEAT1* is upregulated in several types of cancer and regulates cell growth, viability, and morphology of cancer cells (Blume et al. 2015; Adriaens et al. 2016).

Metastasis-associated lung adenocarcinoma transcript 1 (*MALAT1*) is a highly expressed lncRNA that was discovered as a prognostic factor for lung cancer survival (Ji et al. 2003; Tano et al. 2010, 2018). *MALAT1* is initially transcribed by RNA polymerase II (Pol II) as an approximately 7 kb polyadenylated RNA (Wilusz et al. 2008; Anastasiadou et al. 2017). To prevent endonuclease activity, mature *MALAT1* has a triple-helix structure composed of multiple U · A-U base triples and a single C · G-C base triple at its 3′ end instead of being polyadenylated (Wilusz et al. 2012; Brown et al. 2012). RNase P cleavage of 62 bp from the 3′ end is required for formation of the triple-helix structure. *MALAT1* is localized in nuclear speckles and is conserved across vertebrates (Ji et al. 2003; Hutchinson et al. 2007; Kopp and Mendell 2018). Nuclear speckles are dynamic membrane-less compartments that contain many splicing machinery components, including spliceosomal subunits, small nuclear ribonucleoproteins (snRNPs), and serine/arginine-rich proteins (Spector and Lamond 2011). *MALAT1* is recruited to speckles via direct interactions with many splicing-associated proteins and is involved in regulating alternative splicing but not speckle formation (Tano et al. 2010; Tripathi et al. 2010; Miyagawa et al. 2012). *MALAT1* interacts with actively transcribed gene bodies through associations with proteins that indirectly lead the lncRNA to nascent pre-mRNAs. *MALAT1*, therefore, functions as a scaffold or linker to promote the positioning of nuclear speckles at active gene loci (Engreitz et al. 2014). *MALAT1* has also recently been discovered to translocate from nuclear speckles to a distinct nuclear granule, termed the heat-inducible noncoding RNA-containing nuclear body (HiNoCo body), in response to heat shock through liquid–liquid phase separation (Onoguchi-Mizutani et al. 2021). *MALAT1* expression is associated with a variety of cancers, including colon, liver, and breast. *MALAT1*-deficient mice have no detectable abnormalities in splicing or in normal tissue homeostasis during development (Nakagawa et al. 2012; Zhang et al. 2012); however, its depletion in lung cancer cells results in a marked reduction in cell motility (Gutschner et al. 2013). The precise molecular roles of *MALAT1* are still unknown, but its neoplastic activity may be relevant to new therapeutic strategies targeting metastasis.

2.2 *Enhancer RNAs*

2.2.1 Enhancers and Super-Enhancers

Promoters and enhancers are two major cis-acting elements found in genomes that are involved in regulating spatiotemporal gene expression, a very accurate process affected by many factors (Smale and Kadonaga 2003; Shlyueva et al. 2014; Ye et al. 2020). Promoters are DNA sequences located at the transcription start site (TSS) of a gene that regulate transcription in a position- and direction-dependent manner (Smale and Kadonaga 2003). Enhancers are distal regulatory DNA elements that promote the expression of targeted genes (Shlyueva et al. 2014). Their activity is very specific for different cell types and stimuli (Harrison and Bose 2022). The first enhancer element, a 72-bp DNA repeat from the simian SV40 virus, the SV40 enhancer, was discovered in 1981 (Banerji et al. 1981; Moreau et al. 1981). This element can upregulate the expression of a heterogeneous-globin gene by more than 200-fold in transfected HeLa cells, regardless of genomic distance from the gene and orientation (Banerji et al. 1981). The first mammalian enhancer element was discovered in the intronic region of the immunoglobulin-heavy chain gene in 1983. This Ig enhancer element is necessary for effective Ig gene expression and has tissue and cell-type specificity (Banerji et al. 1983; Gillies et al. 1983; Neuberger 1983). Enhancers possess the following distinguishing characteristics (Table 1). (1) They can enhance the transcriptional activity of the promoter-determined transcription site of affected genes. (2) The orientation and chromosomal distance to the target-gene promoter have no influence on enhancer function (Banerji et al. 1981; Moreau et al. 1981). (3) Enhancers often have high levels of mono- or di-methylation on histone H3 lysine 4 (H3K4me1/me2) as well as acetylation on histone H3 lysine 27 (H3K27ac) but low levels or an absence of H3K4me3 (Heintzman et al. 2007, 2009; Creyghton et al. 2010). H3K4me1 linked with enhancers is generally independent of their activities (Heintzman et al. 2007), whereas H3K27ac relates to active enhancer elements (Creyghton et al. 2010; Rada-Iglesias et al. 2011). In addition, enhancers that are inactive or poised exhibit H3K27me3 enrichment (Rada-Iglesias et al. 2011). (4) An accessible chromatin architecture and enhanced DNase I hypersensitivity are characteristics of an active enhancer (Heintzman et al. 2009). (5) Enhancers, which contain specific DNA sequences that promote transcription factor assembly, are also enriched by transcriptional co-activators, such as BRD4, and by the histone acetyltransferases, CREB-binding protein (CBP) and p300, which control H3K27ac and have been employed for the genome-wide annotation of enhancer elements (Visel et al. 2009; Heintzman et al. 2009; Kim et al. 2010; Rada-Iglesias et al. 2011). (6) Pol II can occupy enhancers acting like promoter regions (Pulakanti et al. 2013).

Table 1 The feature of enhancers

1	Enhancers can enhance the transcriptional activity of target genes
2	Enhancer function is NOT affected by its orientation or chromosomal distance from the target gene promoter
3	Enhancer are enriched in H3K4mel/me2 and H3K27ac but not H3K4me3 mark
4	An active enhancer has an accessible chromatin architecture and DNase I hypersensitivity
5	Enhancers are enriched by transcriptional co-activators, such as CBP, p300 and BRD4
6	Pol II can occupy anhancers

In 2013, super-enhancers (SEs) were identified as a set of clustered enhancers with extensive occupancy by mediator and master transcription factors (Whyte et al. 2013). Compared with a typical enhancer, SEs have a dense assembly of Pol II, cofactors, chromatin regulators and typical enhancer histone modifications (H3K4me and H3K27ac) (Xiao et al. 2021). Moreover, in SE regions, the biomarkers of an active enhancer, such as H3K27ac, DNaseI hypersensitivity and p300 are extensively enriched, indicating that SEs have the potential to substantially enhance transcription activity (Whyte et al. 2013; Hnisz et al. 2013).

2.2.2 Enhancer RNA Features

It has been predicted that direct Pol II binding at enhancers may result in the generation of noncoding transcripts because enhancer regions are typically nucleosome-free with open DNA sequence, allowing Pol II enrichment (Heintzman et al. 2007; Ye et al. 2020). Several studies have revealed diverse and extensive transcription from active enhancers in neurons, macrophages, and cancer cells. The RNAs generated from these enhancers were termed enhancer RNAs (eRNAs) and are regarded as a key element in enhancer activity (Fig. 1) (Kim et al. 2010; de Santa et al. 2010; Hah et al. 2011; Li et al. 2013; Kaikkonen et al. 2013). Most 'classical' eRNAs are short (200–2000 nt), bi-directionally transcribed transcripts that are 5′ m^7G capped, non-polyadenylated, and unspliced (Fig. 2) (Li et al. 2013; Andersson et al. 2014a, b; Schwalb et al. 2016). eRNAs are generally relatively short-lived, transcribed at high rates but rapidly degraded, with half-lives in the range of minutes (Fig. 2) (Rabani et al. 2014; Schwalb et al. 2016). Only a few eRNAs are produced that are long (almost 5 kb), polyadenylated, spliced and unidirectionally transcribed from active enhancers, such as SEs (Fig. 2) (de Santa et al. 2010; Koch et al. 2011; Djebali et al. 2012; Natoli and Andrau 2012). The majority of polyadenylated eRNAs generated from SEs are more stable and have a longer half-life than non-polyadenylated eRNAs (Kowalczyk et al. 2012). The absence of polyadenylated eRNAs may result from degradation by the nuclear RNA exosome complex, which can degrade eRNAs from their 3′ end (Andersson et al. 2014a; Pefanis et al. 2015). Recent findings indicate that eRNAs produced from SEs function as lncRNAs, whereas enhancer-associated lncRNAs may behave similarly to eRNAs; therefore, the categorization of eRNAs

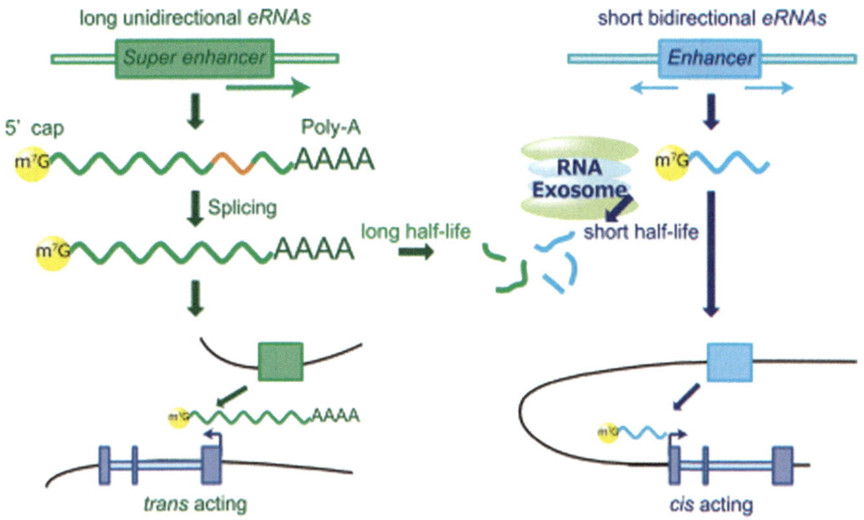

Fig. 2 Schematic representation of two types of eRNAs. The classical eRNAs are bidirectional, short, non-polyadenylated, capped, and unstable, and act *in cis*. Others are unidirectional, long, polyadenylated, capped, spliced, and more stable, and act *in trans*

and lncRNAs is not mutually exclusive (Alvarez-Dominguez et al. 2017; Kopp and Mendell 2018; Han and Li 2022).

2.2.3 Enhancer RNA Functions

eRNAs act as an activator of transcription in the enhancer region to promote expression of the target gene (Natoli and Andrau 2012; Lai et al. 2013). Specific stimuli can initiate eRNA transcription, which is dynamically controlled by transcription factors, and eRNAs can be a sign of enhancer activity (Melo et al. 2013; Mousavi et al. 2013). The control of eRNAs can be regulated by mRNA levels at proximal genes, whereas eRNAs can regulate their target genes (Wang et al. 2011; Li et al. 2013). eRNAs also perform critical roles in genomic stability (Ye et al. 2020). In mouse embryonic stem cells and B cells, RNA exosome-mediated eRNA degradation protects genomic integrity by preventing R-loop formation (Pefanis et al. 2015).

The majority of typical eRNAs are non-polyadenylated, short, and unstable; therefore, they act *in cis*, that is intrachromosomally (Fig. 2) (Andersson et al. 2014a). In contrast, long, stable, polyadenylated eRNAs can regulate the expression of many genes by relocating to other genomic regions and functioning *in trans*, that is interchromosomally (Fig. 2) (Tsai et al. 2018). eRNAs may have several functions, including chromatin loop (that is, enhancer-promoter loop) formation, chromatin accessibility, regulation of histone modification, and condensate formation via liquid–liquid phase separation. The estrogen receptor (ER) is a transcription factor that may bind with cohesin subunits, connecting an enhancer to its target-gene promoter to form the chromatin loop, which is assembled by the interactions between enhancer-bound transcription factors, mediators, cohesin and promoter-bound Pol II (Kagey et al. 2010). In addition, several estrogen-induced eRNA transcripts in breast cancer cells are required to reinforce chromatin looping, which might be further stabilized by ERα and cohesion interactions (Fullwood et al. 2009; Li et al. 2013). Moreover, eRNAs such as *CARMEN* play an important role in cardiovascular cell fate determination and differentiation, and interact with polycomb repressive complex 2 (PRC2), composed of the EZH2 and SUZ12 subunits, to inhibit its nucleosome-binding activity and repress the methyltransferase activity of PRC2. This lowers H3K27me3 levels and promotes gene transcription (Cifuentes-Rojas et al. 2014; Ounzain et al. 2015). These H3K27ac and H3K27me3 findings show the potential for eRNAs to associate with enzyme complexes and produce both activating and repressive histone modifications to establish a chromatin environment that is favorable for active transcription.

MYC, located at 8q24, is a proto-oncogene that promotes tumor formation in a variety of tumor types (Meyer and Penn 2008; Dang 2012). eRNAs are transcribed from SE regions of 8q24 (Pomerantz et al. 2009; Walavalkar et al. 2020). Long-distance physical connections between eRNAs from 8q24 and *MYC* have been demonstrated. Their dysregulation is a prominent marker of several cancer types (Jia et al. 2009; Pomerantz et al. 2009; Sotelo et al. 2010). Colon cancer-associated transcript 1 (*CCAT1*), a well-studied eRNA transcribed from the 8q24 locus with a length of 2628 nucleotides, was first reported as upregulated in colon cancer (Nissan et al. 2012). *CCAT1* is activated through long-range interaction with *MYC* in many cancers, such as colorectal (Pomerantz et al. 2009), prostate (You et al. 2019), squamous (Jiang et al. 2018), breast (Zhang et al. 2015) and esophageal (Hu et al. 2019) cancer. *CCAT1* binds directly to Asp-Glu-Ala-Asp (DEAD) box helicase 5 (DDX5) to function as a scaffold for DDX5 and the AR transcriptional complex, allowing AR-regulated genes to be expressed and tumor development to occur (You et al. 2019). *CCAT1-L*, a long isoform of *CCAT1*, is transcribed from a locus 515 kb upstream towards the SE of *MYC*. *CCAT1L* overexpression enhances *MYC* expression and tumor growth, whereas *CCAT1* knockdown suppresses *MYC* expression and tumor growth (Xiang et al. 2014). *CCAT1-L* can control *MYC* expression by interacting with CCCTC-binding factor and promoting long-range chromatin looping (Xiang et al. 2014). These findings are compelling evidence that eRNAs can control tumor development by regulating the expression of proto-oncogenes.

The transcription factor, p53, is a tumor suppressor (Taniue and Akimitsu 2021b) that interacts with and activates enhancers, termed p53-bound enhancer regions (p53BERs), resulting in the promotion of transcription from adjacent genes and the production of eRNAs (Melo et al. 2013). The p53-dependent eRNAs may increase gene transcription by binding to form a complex (Melo et al. 2013). p53 also controls chromatin accessibility and eRNA transcription in response to DNA damage (Léveillé et al. 2015; Younger and Rinn 2017). Interestingly, the majority of p53-activated enhancers lack canonical p53-binding sites, and some of them are activated by an lncRNA called LncRNA activator of enhancer domains (*LED*) (Léveillé et al.

2015). Knockdown of *LED*, located near the promoter of the *CDKN1A* gene, a p53-responsive cell-cycle inhibitor, resulted in suppression of *CDKN1A* enhancer activity in tumor cells after p53 activation. Moreover, *LED* is repressed in most human tumors, which decreases p53 activation (Léveillé et al. 2015).

2.3 Gene Region Edge-Associated Transcripts

Although the majority of ncRNAs are found in intergenic areas, some are found within protein-coding genes, such as antisense ncRNAs (He et al. 2008) and intronic ncRNAs (Brown et al. 2008). Furthermore, several ncRNAs have been discovered around the transcription demarcations of protein-coding gene regions in eukaryotes, bacteria, and archaea (Yu et al. 2018; Ni et al. 2020). These ncRNAs are classified into two types based on their genomic location: promoter-associated RNAs (PARs) and terminus-associated RNAs (TARs). The biogenesis mechanisms of PARs and TARs have not been fully elucidated, and their definition is relatively ambiguous. In this section, we present a brief description and taxonomy of these gene region edge-associated transcripts (GREATs) (Fig. 1).

2.3.1 Promoter-Associated RNAs

Consistent with the existence of bidirectional promoters, one additional lncRNA and many short ncRNAs have been discovered at corresponding mRNA TSSs. Various classes of PAR have been reported in several organisms, including cryptic unstable transcripts (CUTs) and stable unannotated transcripts (SUTs) (Wyers et al. 2005; Fejes-Toth et al. 2009; Neil et al. 2009; Xu et al. 2009), promoter-associated long RNAs (PALRs) and promoter-associated short RNAs (PASRs) (Kapranov et al. 2007; Ma et al. 2017), upstream noncoding transcripts (UNTs) (Chekanova et al. 2007), promoter upstream transcripts (PROMPTs) (Preker et al. 2008, 2011), TSS-associated RNAs (TSSaRNAs) (Seila et al. 2008; Yus et al. 2012; Zaramela et al. 2014), tiny RNAs (tiRNAs) (Taft et al. 2009), upstream antisense RNAs (uaRNAs) (Flynn et al. 2011), promoter-associated noncoding RNAs (pancRNAs) (Yamamoto et al. 2016; Uesaka et al. 2017), and other PARs (Davis and Ares 2006; Han et al. 2007; Core et al. 2008). Several PARs are short-lived RNA exosome-specific substrates (Wyers et al. 2005; Davis and Ares 2006; Chekanova et al. 2007; Preker et al. 2008). Moreover, the bidirectional transcriptional activity of Pol II is associated with the divergent distribution of most PARs around TSSs, and some PARs have been linked to transcriptional activation or repression of their target genes, demonstrating that transcription can be regulated not only during elongation but also during initiation (Saunders et al. 2006; Janowski et al. 2007; Morris et al. 2008). A novel type of PAR was discovered by investigating the density of transcriptionally active RNA polymerase by global run-on sequencing (GRO-Seq). These PARs are enriched within −250 to +50 bp of the TSSs of specific protein-coding genes in human lung fibroblasts

(Core et al. 2008). PARs are produced through Pol II-dependent divergent transcription and promoter-proximal pausing (Core et al. 2008). However, their biogenesis and mechanisms of action are poorly understood and it is uncertain whether these PARs with different names belong to the same or distinct ncRNA types (Yu et al. 2018). Among PARs, PROMPTs have been the most investigated, particularly regarding their sequence features, biogenesis, and biological functions.

2.3.2 Biogenesis and Function of PROMPTs

Promoter-associated RNAs have short half-lives because of fast exosome-mediated destruction; therefore, Preker et al. searched for short-lived PARs using oligodT primers in RNA exosome-depleted human HeLa cells. Their tiling microarray analysis showed 5′ capped and 3′ adenylated transcripts of over several hundred nucleotides transcribed in the upstream direction from TSSs of transcriptionally active genes. These were termed PROMPTs (Preker et al. 2008, 2011). PROMPTs, unlike the PARs found by Core et al. (Core et al. 2008), are transcribed from a region −2,500 to −500 bp upstream of a TSS (Preker et al. 2008). In addition, PROMPT transcription can occur in both directions and bidirectional transcription is a hallmark of most active promoters (Neil et al. 2009). PROMPTs are also rapidly degraded by the RNA exosome (Preker et al. 2008; Ntini et al. 2013). Furthermore, PROMPT synthesis is highly dependent on the activity of surrounding gene promoters (Preker et al. 2008).

Some PROMPTs are abundant in CpG-rich promoter regions, which influence the DNA methylation content of the promoters (Preker et al. 2008). The phosphorylation of RBM7 in response to DNA damage, such as UV irradiation, induces the stabilization of certain PROMPTs, possibly resulting in increased DNA methylation caused by the recruitment of the DNA methyltransferase, DNMT3B (Lloret-Llinares et al. 2016). Moreover, PROMPTs often originate in the antisense direction, close to the TSSs of the genes to which they are related. According to sequence motif analysis, 3′ poly(A) signals are more prevalent upstream of the promoter than downstream of the promoter. These functional poly(A) signals are implicated in the fast degradation of PROMPTs transcribed upstream of associated genes in the antisense direction (Ntini et al. 2013).

2.3.3 Terminus-Associated RNAs

In addition to PARs, diverse ncRNAs were also discovered at the 3′ ends of protein-coding genes (Ni et al. 2020): terminus-associated small RNAs (TASRs) (Kapranov et al. 2007), antisense TASRs (aTASRs) (Kapranov et al. 2010), transcription termination site-associated RNAs (TTSa RNAs) (Valen et al. 2011), 3′ noncoding transcript (Yue et al. 2010), 3′-UTR-derived RNAs (UaRNAs) (Mercer et al. 2011), antisense cryptic unstable transcripts (antisense CUTs), and antisense stable unannotated transcripts (antisense SUTs) (Neil et al. 2009).

TARs were first reported in mouse and human genomes in 2007 as TASRs ranging from 22 to 200 nucleotides in length (Kapranov et al. 2007). These RNAs are generally transcribed in various positions from both strands of protein-coding genes and lack distinctive features, such as specific lengths, unique base compositions or conventional secondary structures (Kapranov et al. 2007). The antisense RNA to 3′-UTRs, termed aTASRs, contain polyU tails at their 5′ ends (Kapranov et al. 2010). aTASRs lack poly(A) at their 3′ ends and possess a sequence of U residues at their 5′ ends; therefore, they are not detected by a conventional transcriptome analysis or library preparation.

TTSa RNAs, a different class of TASR, were found in humans by Argonaute 1/2 immunoprecipitated sequencing (Valen et al. 2011). TTSa RNAs are generally between 22 and 24 nucleotides in length and are particularly clustered close to the 3′ termination sites of mRNAs (Valen et al. 2011). TTSa RNAs feature an unusual oligo(A) tail at their 3′ end and a high proportion of G residues at their 5′ end and are derived from genes that regulate cell-cycle progression and DNA integrity checkpoints (Laudadio et al. 2018). Moreover, their expression is upregulated in tumor tissue compared with non-tumor tissues, indicating that TTSa RNAs are potential biomarkers for the diagnosis and prognosis of human cancers (Laudadio et al. 2018).

The lack of specific expression patterns and regular length, as well as their particular position inside 3′-UTRs and the presence of poly(A) tails at their 3′ ends, led to the consideration that TARs were degradation products from mRNAs or alternative 3′-UTRs. However, accumulating data indicates that they are essential molecules for a variety of cellular processes. A growing number of TARs have been identified; however, their biogenesis and functions are mostly unknown. Therefore, understanding their biological roles and modes of action has become a new frontier in RNA research.

3 RNA Surveillance and RNA Quality Control

During RNA biogenesis, various aberrant RNAs and RNP complexes are produced. RNA surveillance and RNA quality control systems ensure the integrity of RNAs and RNPs and prevent the production of potentially harmful proteins that may cause disease (Wolin and Maquat 2019). The RNA surveillance and RNA quality control systems in the cytoplasm have been extensively investigated and it has been discovered that nonsense-mediated RNA decay eliminates abnormal transcripts harboring premature termination, nonstop decay degrades transcripts lacking a termination codon, and no-go decay degrades transcripts stalled in translation elongation (Akimitsu 2008; Kurosaki et al. 2019). In the nucleus, these pathways monitor the biogenesis of both mRNAs and noncoding RNAs and the recognition of defective or nonfunctional RNAs leads to their rapid decay (Schmid and Jensen 2018). Many transcripts are pervasively produced from the genome and are also surveyed because their excess production causes activation of cytoplasmic innate immune sensors, the normal role of which is to recognize viral RNAs and trigger an antiviral response (Wolin and Maquat 2019).

Significant fractions of precursor and mature mRNAs and lncRNAs are degraded by nuclear RNA exosomes, indicating that cells do not distinguish exosome targets and non-targets by simply separating them into normal and aberrant RNA pools (Fan et al. 2017). The kinetic competition model is the currently accepted explanation of RNA surveillance in the nucleus. In this model, kinetic competition between RNA degradation pathways and RNA processing processes occurs with (1) assembly of nascent RNAs with RBPs to form distinct RNPs, (2) the processing enzymes for 5′ or 3′ ends of nascent RNAs and (3) binding of nuclear export factors for the matured RNPs (Wang and Cheng 2020). That is, the processing/export and degradation machinery binds in an unbiased manner to nascent RNAs, and the RNA fate is determined by the kinetic competition of these processes. Consistent with the kinetic competition model, the nuclear export process is part of the RNA surveillance system that reduces the export of nonfunctional and incompletely processed mRNAs.

Two nuclear RNA decay pathways mainly involved in nuclear RNA surveillance have been identified in eukaryotes: the 5′-to-3′ exoribonuclease (XRN2) pathway that degrades target RNAs from the 5′ end; and the RNA exosome that degrades RNAs from the 3′ end with cooperation from various adaptor complexes.

4 Nuclear RNA Decay Factors

4.1 XRN2

Distinct nuclear RNA surveillance mechanisms monitor the production of aberrant Pol II-transcribed RNAs, including ncRNAs and pre-mRNAs. During pre-mRNA processing, such as capping, splicing, and formation of 3′ ends, surveillance mechanisms sense the absence of one or more RNA-binding proteins during a critical time window, and degrade the aberrant RNAs or RNPs.

The formation of a cap structure occurs concomitantly with the release of Pol II from promoter-proximal pausing and the first critical step for mRNA biogenesis and RNP formation is the binding of the cap-binding complex (CBC), composed with cap-binding protein 20 (CBP20) and CBP80 (Gonatopoulos-Pournatzis 2014), to the cap structure. Transcripts that fail to bind the CBC undergo a decapping process by decapping enzymes and degradation by the exoribonuclease, XRN2, which is predominantly localized in the nucleus (Jiao et al. 2013). XRN2 is also involved in the elimination of prematurely terminated transcripts by cooperating with decapping proteins and transcription termination factor 2 (TTF2) (Brannan et al. 2012). XRN2 participates in Pol II pausing and premature transcription termination in a subset of human genes by cooperating with the microprocessor and RRP6, a component of the RNA exosome (Wagschal et al. 2012). Human endogenous pre-mRNAs are also co-transcriptionally degraded by XRN2 when their processing is inhibited (Davidson et al. 2012). Upon heat shock stress, XRN2 translocates to the nucleolus by cooperating with NF-κB repressing factor (NKRF), thereby controlling nucleolar 5′-to-3′ exoribonuclease levels and degradation of aberrant pre-rRNA (Coccia et al. 2017). These findings indicate that XRN2 participates in the regulation and quality control of transcription elongation. A recent study proposes that XRN2 associates with Pol II and engages on genes at all phases of the transcription cycle from before initiation to termination, supporting an important role of XRN2 in nuclear RNA surveillance (Cortazar et al. 2022).

4.2 RNA Exosome

RNA exosome is responsible for general RNA turnover in both the nucleus and cytoplasm. RNA exosome processes nuclear RNA precursors, such as pre-ribosomal RNAs, tRNAs, snoRNAs, snRNAs, and telomerase RNA (Zinder and Lima 2017). RNA exosome also plays a central role in RNA surveillance by degrading many unstable RNAs and misprocessed pre-mRNAs in the nucleus (Schmid and Jensen 2018; Ogami et al. 2018). Promoter-proximal short RNAs are degraded by RNA exosome, which acts as a transcriptional elongation checkpoint (Chiu et al. 2018).

The RNA exosome is a ribonuclease complex whose structure and composition are highly conserved among eukaryotes and archaea (Januszyk and Lima 2014). The human nuclear RNA exosome consists of a catalytically inactive barrel-shaped core complex (EXO9) with nine proteins and two catalytic subunits, EXOSC10 (also known as Rrp6 or PM/Scl-100) and DIS3 (also known as Rrp44 or EXOSC11) (Januszyk and Lima 2014). EXO9 consists of two sub-structures, a cap and ring structure, respectively. The cap sub-structure contains three S1/K homology (KH) proteins (EXOSC1–3, also known as Rrp4, Rrp40, and Csl4). The ring sub-structure contains six RNase pleckstrin homology (PH)-like proteins (EXOSC4–9, also known as Rrp41, Rrp42, Rrp43, Rrp45, Rrp46, and Mtr3) (Liu et al. 2006; Drązkowska et al. 2013; Makino et al. 2015). The cap sub-structure has an RNA entry pore, and the ring sub-structure has an RNA exit pore. EXOSC10 and DIS3 are located near the RNA entry and exit pores, respectively. The substrate RNA is threaded into the central channel of the core complex and degraded by DIS3 at the exit pore. Cryogenic electron microscopy has captured the structure of the threading human nuclear exosome (Weick et al. 2018; Gerlach et al. 2018).

The archaeal RNA exosome has active phosphorolytic exonuclease sites positioned in the central cavity of a characteristic barrel-shaped structure (Hartung and Hopfner 2009); however, human EXO9 has altered active site residues, which results in a catalytically inactive exosome core (Liu et al. 2006).

RNA exosome participates in diverse biological functions. For instance, RNA exosome activity involves strands-specific DNA mutations at the immunoglobulin-heavy chain locus and some other regions in mature B cells by resolution of RNA/DNA hybrid structures (Lim et al. 2017). RNA exosome is involved in processing and quality control of precursor telomerase RNA.

Mutations in human RNA exosome subunits cause diseases, such as cancer. It has been estimated that approximately 10% of all multiple myeloma cases are caused by mutations clustered in the DIS3 exonuclease domain (Tomecki et al. 2014). Integrated genomic and clinicopathological analyses across 32 tumor types using The Cancer Genome Atlas (TCGA) datasets showed that EXOSC4 amplification causes pancreatic cancer with poor survival (Taniue et al. 2022). In addition, several Mendelian diseases are associated with mutations in subunits of the human exosome complex (Morton et al. 2018; Fasken et al. 2020). For instance, *EXOSC3* mutations are causative in pontocerebellar hypoplasia and spinal muscular atrophy (Wan et al. 2012). *EXOSC8* mutations are responsible for the development of overlapping symptoms of pontocerebellar hypoplasia, spinal muscular atrophy and central nervous system demyelination (Boczonadi et al. 2014). Mutations in *EXOSC2* found in two unrelated German families cause neurodevelopmental defects of mild cerebellar atrophy and cerebellar hypoplasia, diffuse demyelination, hearing loss, mild intellectual disability, facial anomalies and premature aging (Di Donato et al. 2016). Therefore, exosome complex dysfunction is a major cause of severe complex childhood onset inherited neurological disorders.

4.3 Nuclear Cofactors of RNA Exosome

The RNA exosome can degrade diverse RNAs; however, it does not possess inherent substrate selectivity. Multiple distinct adaptor complexes are required to direct the RNA exosome to distinct RNA substrates (Schmid and Jensen 2019; Garland and Jensen 2020). RNA helicases belonging to superfamily 2 (SF2) play a role in the hub to connect RNA exosome and adaptor complexes. SKI2 and MTR4 connect the exosome to substrate RNAs by interacting with adaptor complexes in the cytoplasm and nucleus, respectively.

MTR4 is an SF2 DExH-box 3′–5′ RNA helicase intimately tied to nuclear exosome function (Weick and Lima 2021) and is a central component of several RNA adaptor complexes, such as TRAMP (LaCava et al. 2005), NEXT (Lubas et al. 2011), and PAXT (Meola et al. 2016; Ogami et al. 2017). MTR4 has a helicase core characterized by a ring-like arrangement of two RecA-like ATPase modules (RecA1 and RecA2) with winged helix and helical bundle domains (Weir et al. 2010; Jackson et al. 2010). Uniquely, MTR4 contains an arch-like protrusion with a KOW domain connected to the helicase core by an elongated anti-parallel coiled-coil motif. Binding of MTR4 with an RNA exosome enhances the unwinding activity of MTR4, allowing MTR4 to deliver single-stranded RNA to the exosome for degradation (Weick et al. 2018). The RNA exosome adapter complexes and their components are summarized below.

4.4 *TRF4-2-ZCCHC7-MTR4 Polyadenylation (TRAMP) Complex*

The first RNA exosome adaptor complex discovered was the TRAMP complex in yeast (LaCava et al. 2005). The TRAMP complex is trimeric, composed of Trf4/5p, Air1/2p, and Mtr4p. Trf4/5p and Air1/2p are non-canonical poly(A) polymerases and zinc knuckle RNA-binding proteins, respectively. TRf4-Air2 form a poly(A) polymerase sub-complex. In *Saccharomyces cerevisiae*, TRAMP substrates are precursors of rRNAs, snRNAs, and snoRNAs. Components of the TRAMP complex are conserved among a wide range of eukaryote, including yeast, flies, mouse, and human.

Human cells contain orthologs of all TRAMP subunits; however, whether these proteins actually form a complex that is analogous to yeast TRAMP remains under investigation. Human homologs of the TRAMP components are hTRF4-1 (POLS), hTRF4-2 (PAPD5) (Lubas et al. 2011), and ZCCHC7 (Air2) (Schneider and Tollervey 2013). PAPD5 and ZCCHC7 along with MTR4 add a short A tail to the 3′ ends of premature-rRNA and participate in the turnover of 5′ ETS fragments excised from pre-rRNAs during rRNA biosynthesis (Sudo et al. 2016). Therefore, to date, the best documented function for putative human TRAMP is the polyadenylation and degradation of prematurely terminated pre-rRNA.

4.5 *Nuclear Exosome-Targeting (NEXT) Complex*

The NEXT complex is composed of MTR4, the zinc finger protein ZCCHC8, and the RNA-recognition motif (RRM)-containing protein RBM7, which were identified through protein–protein interaction profiling using MTR4 as bait (Lubas et al. 2011). Initial studies revealed that the NEXT complex directs degradation of PROMPTs. NEXT-mediated degradation of PROMPTs ensures unidirectional mRNA output from bidirectional promoters (Almada et al. 2013; Ntini et al. 2013). Subsequently, many nuclear RNAs were shown to be degraded by the NEXT-RNA exosome pathway. For instance, NEXT directs degradation of enhancer RNAs (Lubas et al. 2015), 3′-end extended forms of snRNAs (Hrossova et al. 2015), snoRNA-encoding introns (Lubas et al. 2015), and many lncRNAs (Lubas et al. 2015). The diversity of NEXT targets highlights its importance in RNA metabolism.

The helicase activity of human MTR4 is comparatively weak but is enhanced by binding to nuclear RNA exosome or to complexes with ZCCHC8 (Puno and Lima 2018). RBM7 preferentially binds single-stranded uridine-rich motifs of approximately 20 nucleotides upstream of the non-polyadenylated or short polyadenylated 3′ ends of NEXT substrates (Hrossova et al. 2015; Lubas et al. 2015). ZCCHC8 recruits MTR4 onto RBM7 (Falk et al. 2016). It has been suggested that the CBC plays a crucial role in the recognition of the RNA targets by NEXT (Andersen et al. 2013).

Recently, cryogenic electron microscopy revealed a structural basis for RNA surveillance by the NEXT complex. ZCCHC8 works as a scaffold to anchor RBM7 to MTR4 and mediates NEXT homodimerization. RBM7 and ZCCHC8 survey upstream of RNA motifs to provide specificity and to facilitate capture of free 3′ ends by MTR4 (Puno and Lima 2022; Gerlach et al. 2022).

ZCCHC8 is involved in processing and quality control of precursor telomerase RNA, and its mutation causes short telomere syndrome-familial pulmonary fibrosis in humans (Gable et al. 2019). In addition, ZCCHC8 knockout mice show accumulation of retrotransposon element, LINE1, in oocytes and embryos and defects in early embryogenesis (Wu et al. 2019). Defective RBM7 causes spinal muscular atrophy-like symptoms (Giunta et al. 2016). These observations highlight the physiological importance of nuclear RNA degradation.

4.6 Poly(A) Exosome-Targeting (PAXT) Connection

While the NEXT complex and PAXT connection target similar substrates, PAXT prefers polyadenylated RNA, whereas NEXT generally targets non-polyadenylated transcripts to the exosome for decay (Wu et al. 2020). PAXT also degrades lncRNAs, such as NEAT1 and TUG1, and spliced transcripts of noncoding snoRNA host genes (Meola et al. 2016; Ogami et al. 2017).

During elongation, the CBC associates with arsenite-resistance 2 (ARS2), which serves as an interaction hub for the formation of mutually exclusive complexes that either promote pre-mRNA splicing, 3′ end formation, and nuclear export, or target incomplete and unprocessed transcripts for degradation by NEXT and the PAXT connection (Schulze et al. 2018).

The core factors of the PAXT connection are MTR4 and the zinc finger protein, ZFC3H1 (Meola et al. 2016; Ogami et al. 2017). The PAXT connection targets polyadenylated Pol II transcripts that are prematurely terminated or generated by bidirectional promoter activity for degradation, through interaction with PABPN1, a nuclear poly(A)-binding protein (Meola et al. 2016; Ogami et al. 2017). PABPN1 associates with ZFC3H1 through interactions that are partly RNase resistant, and ZFC3H1 binds MTR4. Depletion of the PAXT connection causes excess export of PAXT connection-target noncoding RNAs to the cytoplasm and inhibition of global translation (Ogami et al. 2017), indicating that controlling the amount of nuclear RNAs is important for correct gene regulation.

4.7 HNRNPH1

Heterogeneous nuclear ribonucleoproteins (hnRNPs) are polyvalent RNA-binding proteins with crucial roles in multiple aspects of RNA metabolism, including alternative splicing, mRNA decay, the packaging of nascent transcripts, and regulation of translation (Han et al. 2010). Among the members of the hnRNP family, hnRNPH1 has attracted increasing attention based on its essential role in neurological diseases and cancers (Uren et al. 2016). We recently reported that HNRNPH1 is a novel MTR4-interacting partner involved in the degradation of nuclear lncRNA (Tanu et al. 2021).

5 Concluding Remarks

RNA-seq technology for detecting several types of ncRNAs can be divided into three platforms: short-read cDNA sequencing (Illumina, MGI), long-read cDNA sequencing (Pacific Biosciences, Oxford Nanopore), and long-read direct RNA sequencing (Oxford Nanopore) (Stark et al. 2019; Onoguchi-Mizutani et al. 2022). Short-read sequencing technology is now the standard method for detecting and quantifying transcriptomes because it is less expensive and easier to conduct than microarrays or long-read sequencing technologies, and delivers high-quality transcriptome data. However, short-read approaches encounter some problems, such as difficulty in accurate identification and quantification of isoform diversity, because of ambiguity in mapping sequence reads. However, long-read approaches reduce and more correctly detect false-positive splice-junctions by short-read RNA-seq computational algorithms. In addition, direct long-read sequencing technology can read RNA sequence directly and detect RNA modification without cDNA synthesis during library preparation (Garalde et al. 2018; Onoguchi-Mizutani et al. 2022). Moreover, long-read sequencing technology will allow us to identify various types of aberrant RNAs and understand their relevance to cancer heterogeneity (Taniue and Akimitsu 2021c). Future integration of these approaches is expected to reveal more complex and diverse RNA worlds in cells, further demonstrating the crucial roles of RNA surveillance systems.

Acknowledgements This work was supported by the Japan Society for the Promotion of Science (JSPS) KAKENHI (Grant Numbers: 17KK0163, 20H04838, 21H04792, 21H00243, 21H02758, 21K19402 and 22KK0285). N.A. and K.T. were supported by the Takeda Science Foundation. N.A. was supported by the Mitsubishi Foundation and the Naito Foundation. K.T. was supported by the Uehara Memorial Foundation, the MSD Life Science Foundation, the Japan Foundation for Applied Enzymology, the ONO Medical Research Foundation and the Kobayashi Foundation. We thank Jeremy Allen, PhD, from Edanz (https://jp.edanz.com/ac) for editing a draft of this manuscript.

References

Adriaens C, Standaert L, Barra J et al (2016) P53 induces formation of NEAT1 lncRNA-containing paraspeckles that modulate replication stress response and chemosensitivity. Nat Med 22:861–868

Akimitsu N (2008) Messenger RNA surveillance systems monitoring proper translation termination. J Biochem 143:1–8

Almada AE, Wu X, Kriz AJ et al (2013) Promoter directionality is controlled by U1 snRNP and polyadenylation signals. Nature 499:360–363

Alvarez-Dominguez JR, Knoll M, Gromatzky AA et al (2017) The super-enhancer-derived alncRNA-EC7/bloodlinc potentiates red blood cell development in trans. Cell Rep 19:2503–2514

Anastasiadou E, Jacob LS, Slack FJ (2017) Non-coding RNA networks in cancer. Nat Rev Cancer 18:5–18

Andersen PR, Domanski M, Kristiansen MS et al (2013) The human cap-binding complex is functionally connected to the nuclear RNA exosome HHS public access author manuscript. Nat Struct Mol Biol 20:1367–1376

Andersson R, Gebhard C, Miguel-Escalada I et al (2014a) An atlas of active enhancers across human cell types and tissues. Nature 507:455–461

Andersson R, Refsing Andersen P, Valen E et al (2014b) Nuclear stability and transcriptional directionality separate functionally distinct RNA species. Nat Commun 5:1–10

Banerji J, Olson L, Schaffner W (1983) A lymphocyte-specific cellular enhancer is located downstream of the joining region in immunoglobulin heavy chain genes. Cell 33:729–740

Banerji J, Rusconi S, Schaffner W (1981) Expression of a beta-globin gene is enhanced by remote SV40 DNA sequences. Cell 27:299–308

Batista PJ, Chang HY (2013) Long noncoding RNAs: cellular address codes in development and disease. Cell 152:1298–1307

Blume CJ, Hotz-Wagenblatt A, Hüllein J et al (2015) P53-dependent non-coding RNA networks in chronic lymphocytic leukemia. Leukemia 29:2015–2023

Boczonadi V, Müller JS, Pyle A et al (2014) EXOSC8 mutations alter mRNA metabolism and cause hypomyelination with spinal muscular atrophy and cerebellar hypoplasia. Nat Commun 5:4287

Brannan K, Kim H, Erickson B et al (2012) mRNA decapping factors and the exonuclease Xrn2 function in widespread premature termination of RNA polymerase II transcription. Mol Cell 46:311–324

Brown JA, Valenstein ML, Yario TA et al (2012) Formation of triple-helical structures by the 3′-end sequences of MALAT1 and MENβ noncoding RNAs. Proc Natl Acad Sci USA 109:19202–19207

Brown JWS, Marshall DF, Echeverria M (2008) Intronic noncoding RNAs and splicing. Trends Plant Sci 13:335–342

Carthew RW, Sontheimer EJ (2009) Origins and mechanisms of miRNAs and siRNAs. Cell 136:642–655

Chan JJ, Tay Y (2018) Noncoding RNA:RNA regulatory networks in cancer. Int J Mol Sci 19:1310

Chekanova JA, Gregory BD, Reverdatto SV et al (2007) Genome-wide high-resolution mapping of exosome substrates reveals hidden features in the Arabidopsis transcriptome. Cell 131:1340–1353

Chiu AC, Suzuki HI, Wu X et al (2018) Transcriptional pause sites delineate stable nucleosome-associated premature polyadenylation suppressed by U1 snRNP. Mol Cell 69:648-663.e7

Cifuentes-Rojas C, Hernandez AJ, Sarma K et al (2014) Regulatory interactions between RNA and polycomb repressive complex 2. Mol Cell 55:171–185

Clemson CM, Hutchinson JN, Sara SA et al (2009) An Architectural role for a nuclear noncoding RNA: NEAT1 RNA is essential for the structure of paraspeckles. Mol Cell 33:717–726

Coccia M, Rossi A, Riccio A et al (2017) Human NF-κB repressing factor acts as a stress-regulated switch for ribosomal RNA processing and nucleolar homeostasis surveillance. Proc Natl Acad Sci USA 114:1045–1050

Core LJ, Waterfall JJ, Lis JT (2008) Nascent RNA sequencing reveals widespread pausing and divergent initiation at human promoters. Science 322:1845–1848

Cortazar MA, Erickson B, Fong N et al (2022) Xrn2 substrate mapping identifies torpedo loading sites and extensive premature termination of RNA pol II transcription. Genes Dev 36:1062–1078

Creyghton MP, Cheng AW, Welstead GG et al (2010) Histone H3K27ac separates active from poised enhancers and predicts developmental state. Proc Natl Acad Sci USA 107:21931–21936
Dang CV (2012) MYC on the path to cancer. Cell 149:22–35
Davidson L, Kerr A, West S (2012) Co-transcriptional degradation of aberrant pre-mRNA by Xrn2. EMBO J 31:2566–2578
Davis CA, Ares M (2006) Accumulation of unstable promoter-associated transcripts upon loss of the nuclear exosome subunit Rrp6p in Saccharomyces cerevisiae. Proc Natl Acad Sci USA 103:3262–3267
de Santa F, Barozzi I, Mietton F et al (2010) A large fraction of extragenic RNA pol II transcription sites overlap enhancers. Plos Biol 8:e1000384
Di Donato N, Neuhann T, Kahlert AK et al (2016) Mutations in EXOSC2 are associated with a novel syndrome characterised by retinitis pigmentosa, progressive hearing loss, premature ageing, short stature, mild intellectual disability and distinctive gestalt. J Med Genet 53:419–425
Djebali S, Davis CA, Merkel A et al (2012) Landscape of transcription in human cells. Nature 489:101–108
Drążkowska K, Tomecki R, Stoduś K et al (2013) The RNA exosome complex central channel controls both exonuclease and endonuclease Dis3 activities in vivo and in vitro. Nucl Acids Res 41:3845–3858
Elgar G, Vavouri T (2008) Tuning in to the signals: noncoding sequence conservation in vertebrate genomes. Trends Genet 24:344–352
Engreitz JM, Sirokman K, McDonel P et al (2014) RNA-RNA interactions enable specific targeting of noncoding RNAs to nascent pre-mRNAs and chromatin sites. Cell 159:188–199
Falk S, Finogenova K, Melko M et al (2016) Structure of the RBM7-ZCCHC8 core of the NEXT complex reveals connections to splicing factors. Nat Commun 7:13573
Fan J, Kuai B, Wu G et al (2017) Exosome cofactor hMTR4 competes with export adaptor ALYREF to ensure balanced nuclear RNA pools for degradation and export. EMBO J 36:2870–2886
Fasken MB, Morton DJ, Kuiper EG et al (2020) The RNA exosome and human disease. Methods Mol Biol 2062:3–33
Fejes-Toth K, Sotirova V, Sachidanandam R et al (2009) Post-transcriptional processing generates a diversity of 5′-modified long and short RNAs. Nature 457:1028–1032
Flynn RA, Almada AE, Zamudio JR et al (2011) Antisense RNA polymerase II divergent transcripts are P-TEFb dependent and substrates for the RNA exosome. Proc Natl Acad Sci USA 108:10460–10465
Fox AH, Lamond AI (2010) Paraspeckles. Cold Spring Harb Perspect Biol 2:1–15
Fullwood MJ, Liu MH, Pan YF et al (2009) An oestrogen-receptor-alpha-bound human chromatin interactome. Nature 462:58–64
Gable DL, Gaysinskaya V, Atik CC et al (2019) ZCCHC8, the nuclear exosome targeting component, is mutated in familial pulmonary fibrosis and is required for telomerase RNA maturation. Genes Dev 33:1381–1396
Garalde DR, Snell EA, Jachimowicz D et al (2018) Highly parallel direct RNA sequencing on an array of nanopores. Nat Methods 15:201–206
Garland W, Jensen TH (2020) Nuclear sorting of RNA. Wiley Interdiscip Rev RNA 11
Gerlach P, Schuller JM, Bonneau F et al (2018) Distinct and evolutionary conserved structural features of the human nuclear exosome complex. Elife 7:e38686
Gerlach P, Garland W, Lingaraju M et al (2022) Structure and regulation of the nuclear exosome targeting complex guides RNA substrates to the exosome. Mol Cell 82:2505-2518.e7
Gillies SD, Morrison SL, Oi VT, Tonegawa S (1983) A tissue-specific transcription enhancer element is located in the major intron of a rearranged immunoglobulin heavy chain gene. Cell 33:717–728
Giunta M, Edvardson S, Xu Y et al (2016) Altered RNA metabolism due to a homozygous RBM7 mutation in a patient with spinal motor neuropathy. Hum Mol Genet 25:2985–2996
Gutschner T, Hämmerle M, Eißmann M et al (2013) The noncoding RNA MALAT1 is a critical regulator of the metastasis phenotype of lung cancer cells. Cancer Res 73:1180–1189

Hah N, Danko CG, Core L et al (2011) A rapid, extensive, and transient transcriptional response to estrogen signaling in breast cancer cells. Cell 145:622–634

Han J, Kim D, Morris KV (2007) Promoter-associated RNA is required for RNA-directed transcriptional gene silencing in human cells. Proc Natl Acad Sci USA 104:12422–12427

Han SP, Tang YH, Smith R (2010) Functional diversity of the hnRNPs: past, present and perspectives. Biochem J 430:379–392

Han Z, Li W (2022) Enhancer RNA: what we know and what we can achieve. Cell Prolif 55

Harrison LJ, Bose D (2022) Enhancer RNAs step forward: new insights into enhancer function. Development 149:dev200398

Hartung S, Hopfner KP (2009) Lessons from structural and biochemical studies on the archaeal exosome. Biochem Soc Trans 37:83–87

He Y, Vogelstein B, Velculescu VE et al (2008) The antisense transcriptomes of human cells. Science 322:1855–1857

Heintzman ND, Hon GC, Hawkins RD et al (2009) Histone modifications at human enhancers reflect global cell-type-specific gene expression. Nature 459:108–112

Heintzman ND, Stuart RK, Hon G et al (2007) Distinct and predictive chromatin signatures of transcriptional promoters and enhancers in the human genome. Nat Genet 39:311–318

Hnisz D, Abraham BJ, Lee TI et al (2013) Super-enhancers in the control of cell identity and disease. Cell 155:934

Hrossova D, Sikorsky T, Potesil D et al (2015) RBM7 subunit of the NEXT complex binds U-rich sequences and targets 3′-end extended forms of snRNAs. Nucl Acids Res 43:4236–4248

Hu M, Zhang Q, Tian XH et al (2019) lncRNA CCAT1 is a biomarker for the proliferation and drug resistance of esophageal cancer via the miR-143/PLK1/BUBR1 axis. Mol Carcinog 58:2207–2217

Huarte M (2015) The emerging role of lncRNAs in cancer. Nat Med 21:1253–1261

Hutchinson JN, Ensminger AW, Clemson CM et al (2007) A screen for nuclear transcripts identifies two linked noncoding RNAs associated with SC35 splicing domains. BMC Genomics 8:1–16

Imamura K, Imamachi N, Akizuki G et al (2014) Long noncoding RNA NEAT1-dependent SFPQ relocation from promoter region to paraspeckle mediates IL8 expression upon immune stimuli. Mol Cell 53:393–406

Imamura K, Takaya A, Ishida Y et al (2018) Diminished nuclear RNA decay upon Salmonella infection upregulates antibacterial noncoding RNAs. EMBO J 37:1–15

Jackson RN, Klauer AA, Hintze BJ et al (2010) The crystal structure of Mtr4 reveals a novel arch domain required for rRNA processing. EMBO J 29:2205–2216

Janowski BA, Younger ST, Hardy DB et al (2007) Activating gene expression in mammalian cells with promoter-targeted duplex RNAs. Nat Chem Biol 3:166–173

Januszyk K, Lima CD (2014) The eukaryotic RNA exosome. Curr Opin Struct Biol 24:132–140

Ji P, Diederichs S, Wang W et al (2003) MALAT-1, a novel noncoding RNA, and thymosin β4 predict metastasis and survival in early-stage non-small cell lung cancer. Oncogene 22:8031–8041

Jia L, Landan G, Pomerantz M et al (2009) Functional enhancers at the gene-poor 8q24 cancer-linked locus. Plos Genet 5:e1000597

Jiang Y, Jiang YY, Xie JJ et al (2018) Co-activation of super-enhancer-driven CCAT1 by TP63 and SOX2 promotes squamous cancer progression. Nat Commun 9:3619

Jiao X, Chang JH, Kilic T et al (2013) A mammalian pre-mRNA 5′ end capping quality control mechanism and an unexpected link of capping to pre-mRNA processing. Mol Cell 50:104–115

Kagey MH, Newman JJ, Bilodeau S et al (2010) Mediator and cohesin connect gene expression and chromatin architecture. Nature 467:430–435

Kaikkonen MU, Spann NJ, Heinz S et al (2013) Remodeling of the enhancer landscape during macrophage activation is coupled to enhancer transcription. Mol Cell 51:310–325

Kapranov P, Cheng J, Dike S et al (2007) RNA maps reveal new RNA classes and a possible function for pervasive transcription. Science 316:1484–1488

Kapranov P, Ozsolak F, Kim SW et al (2010) New class of gene-termini-associated human RNAs suggests a novel RNA copying mechanism. Nature 466:642–646

Kim TK, Hemberg M, Gray JM et al (2010) Widespread transcription at neuronal activity-regulated enhancers. Nature 465:182–187
Koch F, Fenouil R, Gut M et al (2011) Transcription initiation platforms and GTF recruitment at tissue-specific enhancers and promoters. Nat Struct Mol Biol 18:956–963
Kopp F, Mendell JT (2018) Functional classification and experimental dissection of long noncoding RNAs. Cell 172:393–407
Kowalczyk MS, Hughes JR, Garrick D et al (2012) Intragenic enhancers act as alternative promoters. Mol Cell 45:447–458
Kurosaki T, Popp MW, Maquat LE (2019) Quality and quantity control of gene expression by nonsense-mediated mRNA decay. Nat Rev Mol Cell Biol 20:406–420
LaCava J, Houseley J, Saveanu C et al (2005) RNA degradation by the exosome is promoted by a nuclear polyadenylation complex. Cell 121:713–724
Lagarde J, Uszczynska-Ratajczak B, Carbonell S et al (2017) High-throughput annotation of full-length long noncoding RNAs with capture long-read sequencing. Nat Genet 49:1731–1740
Lai F, Orom UA, Cesaroni M et al (2013) Activating RNAs associate with mediator to enhance chromatin architecture and transcription. Nature 494:497–501
Lander ES, Linton LM, Birren B et al (2001) Initial sequencing and analysis of the human genome. Nature 409:860–921
Laudadio I, Formichetti S, Gioiosa S et al (2018) Characterization of transcription termination-associated RNAs: new insights into their biogenesis, tailing, and expression in primary tumors. Int J Genomics 2018:1243858
Léveillé N, Melo CA, Rooijers K et al (2015) Genome-wide profiling of p53-regulated enhancer RNAs uncovers a subset of enhancers controlled by a lncRNA. Nat Commun 6:6520
Li W, Notani D, Ma Q et al (2013) Functional roles of enhancer RNAs for oestrogen-dependent transcriptional activation. Nature 498:516–520
Lim J, Giri PK, Kazadi D et al (2017) Nuclear proximity of Mtr4 to RNA exosome restricts DNA mutational asymmetry. Cell 169:523-537.e15
Liu Q, Greimann JC, Lima CD (2006) Reconstitution, activities, and structure of the eukaryotic RNA exosome. Cell 127:1223–1237
Lloret-Llinares M, Mapendano CK, Martlev LH et al (2016) Relationships between PROMPT and gene expression. RNA Biol 13:6–14
Lubas M, Andersen PR, Schein A et al (2015) The human nuclear exosome targeting complex is loaded onto newly synthesized RNA to direct early ribonucleolysis. Cell Rep 10:178–192
Lubas M, Christensen MS, Kristiansen MS et al (2011) Interaction profiling identifies the human nuclear exosome targeting complex. Mol Cell 43:624–637
Ma X, Han N, Shao C et al (2017) Transcriptome-wide discovery of PASRs (Promoter-Associated Small RNAs) and TASRs (Terminus-Associated Small RNAs) in Arabidopsis thaliana. Plos One 12:e0169212
Makino DL, Schuch B, Stegmann E et al (2015) RNA degradation paths in a 12-subunit nuclear exosome complex. Nature 524:54–58
Melo CA, Drost J, Wijchers PJ et al (2013) eRNAs are required for p53-dependent enhancer activity and gene transcription. Mol Cell 49:524–535
Meola N, Domanski M, Karadoulama E et al (2016) Identification of a nuclear exosome decay pathway for processed transcripts. Mol Cell 64:520–533
Mercer TR, Dinger ME, Mattick JS (2009) Long non-coding RNAs: insights into functions. Nat Rev Genet 10:155–159
Mercer TR, Wilhelm D, Dinger ME et al (2011) Expression of distinct RNAs from 3' untranslated regions. Nucl Acids Res 39:2393–2403
Meyer N, Penn LZ (2008) Reflecting on 25 years with MYC. Nat Rev Cancer 8:976–990
Miyagawa R, Tano K, Mizuno R et al (2012) Identification of CIS- and trans-acting factors involved in the localization of MALAT-1 noncoding RNA to nuclear speckles. RNA 18:738–751
Moreau P, Hen R, Wasylyk B et al (1981) The SV40 72 base repair repeat has a striking effect on gene expression both in SV40 and other chimeric recombinants. Nucl Acids Res 9:6047–6068

Morris KV, Santoso S, Turner AM et al (2008) Bidirectional transcription directs both transcriptional gene activation and suppression in human cells. Plos Genet 4:e1000258
Morton DJ, Kuiper EG, Jones SK et al (2018) The RNA exosome and RNA exosome-linked disease. RNA 24:127–142
Mousavi K, Zare H, Dell'Orso S et al (2013) eRNAs promote transcription by establishing chromatin accessibility at defined genomic loci. Mol Cell 51:606–617
Naganuma T, Nakagawa S, Tanigawa A et al (2012) Alternative 3′-end processing of long noncoding RNA initiates construction of nuclear paraspeckles. EMBO J 31:4020–4034
Nakagawa S, Ip JY, Shioi G et al (2012) Malat1 is not an essential component of nuclear speckles in mice. RNA 18:1487–1499
Nakagawa S, Naganuma T, Shioi G et al (2011) Paraspeckles are subpopulation-specific nuclear bodies that are not essential in mice. J Cell Biol 193:31–39
Natoli G, Andrau JC (2012) Noncoding transcription at enhancers: general principles and functional models. Annu Rev Genet 46:1–19
Neil H, Malabat C, D'Aubenton-Carafa Y et al (2009) Widespread bidirectional promoters are the major source of cryptic transcripts in yeast. Nature 457:1038–1042
Neuberger MS (1983) Expression and regulation of immunoglobulin heavy chain gene transfected into lymphoid cells. EMBO J 2:1373–1378
Ni WJ, Xie F, Leng XM (2020) Terminus-associated non-coding RNAs: trash or treasure? Front Genet 11:552444
Nissan A, Stojadinovic A, Mitrani-Rosenbaum S et al (2012) Colon cancer associated transcript-1: a novel RNA expressed in malignant and pre-malignant human tissues. Int J Cancer 130:1598–1606
Ntini E, Järvelin AI, Bornholdt J et al (2013) Polyadenylation site-induced decay of upstream transcripts enforces promoter directionality. Nat Struct Mol Biol 20:923–928
Ogami K, Chen Y, Manley JL (2018) RNA surveillance by the nuclear RNA exosome: mechanisms and significance. Non-Coding RNA 4:8
Ogami K, Richard P, Chen Y et al (2017) An Mtr4/ZFC3H1 complex facilitates turnover of unstable nuclear RNAs to prevent their cytoplasmic transport and global translational repression. Genes Dev 31:1257–1271
Onoguchi-Mizutani R, Akimitsu N (2022) Long noncoding RNA and phase separation in cellular stress response. J Biochem 171:269–276
Onoguchi-Mizutani R, Kirikae Y, Ogura Y et al (2021) Identification of a heat-inducible novel nuclear body containing the long noncoding RNA MALAT1. J Cell Sci 134:jcs253559
Onoguchi-Mizutani R, Taniue K, Kawata K et al. (2022) Chapter 10. Techniques for analyzing genome-wide expression of non-coding RNA. In: Handbook of epigenetics: the new molecular and medical genetics, 3rd edn, pp 163–184
Ounzain S, Micheletti R, Arnan C et al (2015) CARMEN, a human super enhancer-associated long noncoding RNA controlling cardiac specification, differentiation and homeostasis. J Mol Cell Cardiol 89:98–112
Pefanis E, Wang J, Rothschild G et al (2015) RNA exosome-regulated long non-coding RNA transcription controls super-enhancer activity. Cell 161:774–789
Pomerantz MM, Ahmadiyeh N, Jia L et al (2009) The 8q24 cancer risk variant rs6983267 shows long-range interaction with MYC in colorectal cancer. Nat Genet 41:882–884
Ponting CP, Oliver PL, Reik W (2009) Evolution and functions of long noncoding RNAs. Cell 136:629–641
Preker P, Almvig K, Christensen MS et al (2011) PROMoter uPstream transcripts share characteristics with mRNAs and are produced upstream of all three major types of mammalian promoters. Nucl Acids Res 39:7179–7193
Preker P, Nielsen J, Kammler S et al (2008) RNA exosome depletion reveals transcription upstream of active human promoters. Science 322:1851–1854
Pulakanti K, Pinello L, Stelloh C et al (2013) Enhancer transcribed RNAs arise from hypomethylated, Tet-occupied genomic regions. Epigenetics 8:1303–1320

Puno MR, Lima CD (2018) Structural basis for MTR4-ZCCHC8 interactions that stimulate the MTR4 helicase in the nuclear exosome-targeting complex. Proc Natl Acad Sci USA 115:E5506–E5515
Puno MR, Lima CD (2022) Structural basis for RNA surveillance by the human nuclear exosome targeting (NEXT) complex. Cell 185:2132-2147.e26
Rabani M, Raychowdhury R, Jovanovic M et al (2014) High-resolution sequencing and modeling identifies distinct dynamic RNA regulatory strategies. Cell 159:1698–1710
Rada-Iglesias A, Bajpai R, Swigut T et al (2011) A unique chromatin signature uncovers early developmental enhancers in humans. Nature 470:279–285
Ransohoff JD, Wei Y, Khavari PA (2018) The functions and unique features of long intergenic non-coding RNA. Nat Rev Mol Cell Biol 19:143–157
Saunders A, Core LJ, Lis JT (2006) Breaking barriers to transcription elongation. Nat Rev Mol Cell Biol 7:557–567
Schmid M, Jensen TH (2018) Controlling nuclear RNA levels. Nat Rev Genet 19:518–529
Schmid M, Jensen TH (2019) The nuclear RNA exosome and its cofactors. Advances in experimental medicine and biology. Springer, Cham, pp 113–132
Schmitt AM, Chang HY (2016) Long noncoding RNAs in cancer pathways. Cancer Cell 29:452–463
Schneider C, Tollervey D (2013) Threading the barrel of the RNA exosome. Trends Biochem Sci 38:485–493
Schulze WM, Stein F, Rettel M et al (2018) Structural analysis of human ARS2 as a platform for co-transcriptional RNA sorting. Nat Commun 9:1701
Schwalb B, Michel M, Zacher B et al (2016) TT-seq maps the human transient transcriptome. Science 352:1225–1228
Seila AC, Calabrese JM, Levine SS et al (2008) Divergent transcription from active promoters. Science 322:1849–1851
Shirahama S, Miki A, Kaburaki T et al (2020) Long non-coding RNAs involved in pathogenic infection. Front Genet 11:454
Shirahama S, Taniue K, Mitsutomi S et al (2021) Human U90926 orthologous long non-coding RNA as a novel biomarker for visual prognosis in herpes simplex virus type-1 induced acute retinal necrosis. Sci Rep 11:12164
Shlyueva D, Stampfel G, Stark A (2014) Transcriptional enhancers: from properties to genome-wide predictions. Nat Rev Genet 15:272–286
Smale ST, Kadonaga JT (2003) The RNA polymerase II core promoter. Annu Rev Biochem 72:449–479
Sotelo J, Esposito D, Duhagon MA et al (2010) Long-range enhancers on 8q24 regulate c-Myc. Proc Natl Acad Sci USA 107:3001–3005
Spector DL, Lamond AI (2011) Nuclear speckles. Cold Spring Harb Perspect Biol 3:1–12
Stark R, Grzelak M, Hadfield J (2019) RNA sequencing: the teenage years. Nat Rev Genet 20:631–656
Sudo H, Nozaki A, Uno H et al (2016) Interaction properties of human TRAMP-like proteins and their role in pre-rRNA 5'ETS turnover. FEBS Lett 590:2963–2972
Taft RJ, Glazov EA, Cloonan N et al (2009) Tiny RNAs associated with transcription start sites in animals. Nat Genet 41:572–578
Takahashi K, Taniue K, Ono Y et al (2021) Long non-coding RNAs in epithelial-mesenchymal transition of pancreatic cancer. Front Mol Biosci 8:717890
Tani H, Mizutani R, Salam KA et al (2012) Genome-wide determination of RNA stability reveals hundreds of short-lived noncoding transcripts in mammals. Genome Res 22:947–956
Taniue K, Kurimoto A, Sugimasa H et al (2016a) Long noncoding RNA *UPAT* promotes colon tumorigenesis by inhibiting degradation of UHRF1. Proc Natl Acad Sci 113:1273–1278
Taniue K, Kurimoto A, Takeda Y et al (2016b) ASBEL–TCF3 complex is required for the tumorigenicity of colorectal cancer cells. Proc Natl Acad Sci 113:201605938
Taniue K, Akimitsu N (2021a) The Functions and Unique Features of LncRNAs in Cancer Development and Tumorigenesis. Int J Mol Sci 22:1–20

Taniue K, Akimitsu N (2021b) Aberrant phase separation and cancer. FEBS J 289:17–39

Taniue K, Akimitsu N (2021c) Fusion genes and RNAs in cancer development. Non-Coding RNA 7:1–14

Taniue K, Tanu T, Shimoura Y et al (2022) RNA exosome component EXOSC4 amplified in multiple cancer types is required for the cancer cell survival. Int J Mol Sci 23:496

Tano K, Mizuno R, Okada T et al (2010) MALAT-1 enhances cell motility of lung adenocarcinoma cells by influencing the expression of motility-related genes. FEBS Lett 584:4575–4580

Tano K, Onoguchi-Mizutani R, Yeasmin F et al (2018) Identification of minimal p53 promoter region regulated by MALAT1 in human lung adenocarcinoma cells. Front Genet 9:1–10

Tanu T, Taniue K, Imamura K et al (2021) hnRNPH1-MTR4 complex-mediated regulation of NEAT1v2 stability is critical for IL8 expression. RNA Biol 18:537–547

Tomecki R, Drazkowska K, Kucinski I et al (2014) Multiple myeloma-associated hDIS3 mutations cause perturbations in cellular RNA metabolism and suggest hDIS3 PIN domain as a potential drug target. Nucl Acids Res 42:1270–1290

Tripathi V, Ellis JD, Shen Z et al (2010) The nuclear-retained noncoding RNA MALAT1 regulates alternative splicing by modulating SR splicing factor phosphorylation. Mol Cell 39:925–938

Tsai PF, Dell'Orso S, Rodriguez J et al (2018) A muscle-specific enhancer RNA mediates Cohesin recruitment and regulates transcription in trans. Mol Cell 71:129-141.e8

Uesaka M, Agata K, Oishi T et al (2017) Evolutionary acquisition of promoter-associated non-coding RNA (pancRNA) repertoires diversifies species-dependent gene activation mechanisms in mammals. BMC Genomics 18:285

Uren PJ, Bahrami-Samani E, de Araujo PR et al (2016) High-throughput analyses of hnRNP H1 dissects its multi-functional aspect. RNA Biol 13:400–411

Valen E, Preker P, Andersen PR et al (2011) Biogenic mechanisms and utilization of small RNAs derived from human protein-coding genes. Nat Struct Mol Biol 18:1075–1082

Visel A, Blow MJ, Li Z et al (2009) ChIP-seq accurately predicts tissue-specific activity of enhancers. Nature 457:854–858

Wagschal A, Rousset E, Basavarajaiah P et al (2012) Microprocessor, Setx, Xrn2, and Rrp6 co-operate to induce premature termination of transcription by RNAPII. Cell 150:1147–1157

Walavalkar K, Saravanan B, Singh AK et al (2020) A rare variant of African ancestry activates 8q24 lncRNA hub by modulating cancer associated enhancer. Nat Commun 11:3598

Wan J, Yourshaw M, Mamsa H et al (2012) Mutations in the RNA exosome component gene EXOSC3 cause pontocerebellar hypoplasia and spinal motor neuron degeneration. Nat Genet 44:704–708

Wang D, Garcia-Bassets I, Benner C et al (2011) Reprogramming transcription by distinct classes of enhancers functionally defined by eRNA. Nature 474:390–397

Wang J, Cheng H (2020) Out or decay: fate determination of nuclear RNAs. Essays Biochem 64:895–905

Weick EM, Lima CD (2021) RNA helicases are hubs that orchestrate exosome-dependent 3′–5′ decay. Curr Opin Struct Biol 67:86–94

Weick EM, Puno MR, Januszyk K et al (2018) Helicase-dependent RNA decay illuminated by a Cryo-EM structure of a human nuclear RNA exosome-MTR4 complex. Cell 173:1663-1677.e21

Weir JR, Bonneau F, Hentschel J et al (2010) Structural analysis reveals the characteristic features of Mtr4, a DExH helicase involved in nuclear RNA processing and surveillance. Proc Natl Acad Sci USA 107:12139–12144

Whyte WA, Orlando DA, Hnisz D et al (2013) Master transcription factors and mediator establish super-enhancers at key cell identity genes. Cell 153:307–319

Wilusz JE, Freier SM, Spector DL (2008) 3′ end processing of a long nuclear-retained noncoding RNA yields a tRNA-like cytoplasmic RNA. Cell 135:919–932

Wilusz JE, JnBaptiste CK, Lu LY et al (2012) A triple helix stabilizes the 3′ ends of long noncoding RNAs that lack poly(A) tails. Genes Dev 26:2392–2407

Wolin SL, Maquat LE (2019) Cellular RNA surveillance in health and disease. Science 366:822–827

Wu G, Schmid M, Rib L et al (2020) A Two-Layered Targeting Mechanism Underlies Nuclear RNA Sorting by the Human Exosome. Cell Rep 30:2387-2401.e5
Wu Y, Liu W, Chen J et al (2019) Nuclear Exosome Targeting Complex Core Factor Zcchc8 Regulates the Degradation of LINE1 RNA in Early Embryos and Embryonic Stem Cells. Cell Rep 29:2461-2472.e6
Wyers F, Rougemaille M, Badis G et al (2005) Cryptic pol II transcripts are degraded by a nuclear quality control pathway involving a new poly(A) polymerase. Cell 121:725–737
Xiang JF, Yin QF, Chen T et al (2014) Human colorectal cancer-specific CCAT1-L lncRNA regulates long-range chromatin interactions at the MYC locus. Cell Res 24:513–531
Xiao S, Huang Q, Ren H et al (2021) The mechanism and function of super enhancer RNA. Genesis 59:e23422
Xu Z, Wei W, Gagneur J et al (2009) Bidirectional promoters generate pervasive transcription in yeast. Nature 457:1033–1037
Yamamoto N, Agata K, Nakashima K et al (2016) Bidirectional promoters link cAMP signaling with irreversible differentiation through promoter-associated non-coding RNA (pancRNA) expression in PC12 cells. Nucl Acids Res 44:5105–5122
Yamazaki T, Hirose T (2015) The building process of the functional paraspeckle with long non-coding RNAs. Front Biosci (Elite ed) 7:1–41
Yanagida S, Taniue K, Sugimasa H et al (2013) ASBEL, an ANA/BTG3 antisense transcript required for tumorigenicity of ovarian carcinoma. Sci Rep 3:1305
Ye R, Cao C, Xue Y (2020) Enhancer RNA: biogenesis, function, and regulation. Essays Biochem 64:883–894
You Z, Liu C, Wang C et al (2019) LncRNA CCAT1 promotes prostate cancer cell proliferation by interacting with DDX5 and MIR-28-5P. Mol Cancer Ther 18:2469–2479
Younger ST, Rinn JL (2017) p53 regulates enhancer accessibility and activity in response to DNA damage. Nucl Acids Res 45:9889–9900
Yu D, Ma X, Zuo Z et al (2018) Classification of transcription boundary-associated RNAs (TBARs) in animals and plants. Front Genet 9:168
Yue X, Schwartz JC, Chu Y et al (2010) Transcriptional regulation by small RNAs at sequences downstream from 3′ gene termini. Nat Chem Biol 6:621–629
Yus E, Güell M, Vivancos AP et al (2012) Transcription start site associated RNAs in bacteria. Mol Syst Biol 8:585
Zaramela LS, Vêncio RZN, Ten-Caten F et al (2014) Transcription start site associated RNAs (TSSaRNAs) are ubiquitous in all domains of life. Plos One 9:e107680
Zhang B, Arun G, Mao YS et al (2012) The lncRNA malat1 is dispensable for mouse development but its transcription plays a cis-regulatory role in the adult. Cell Rep 2:111–123
Zhang F, Wu L, Qian J et al (2016) Identification of the long noncoding RNA NEAT1 as a novel inflammatory regulator acting through MAPK pathway in human lupus. J Autoimmun 75:96–104
Zhang P, Cao L, Zhou R et al (2019) The lncRNA Neat1 promotes activation of inflammasomes in macrophages. Nat Commun 10:1–17
Zhang XF, Liu T, Li Y et al (2015) Overexpression of long non-coding RNA CCAT1 is a novel biomarker of poor prognosis in patients with breast cancer. Int J Clin Exp Pathol 8:9440
Zinder JC, Lima CD (2017) Targeting RNA for processing or destruction by the eukaryotic RNA exosome and its cofactors. Genes Dev 31:88–100

Functional Role of Non-coding RNAs in Prostate Cancer: From Biomarker to Therapeutic Targets

Dhirodatta Senapati, Vikas Sharma, and Snehasis Tripathy

Contents

Abstract Prostate cancer (CaP) is associated with considerably reduced overall survival with high mortality rates due to the development of resistance to both standard and novel therapies. This acquired resistance is a multifactorial problem involving the interplay of several genetic and epigenetic alterations that contribute significantly to the growth, invasiveness, and metastasis of CaP. A deeper understanding may lead to novel therapeutic interventions and ultimately prolong patient survival. The recent progress in high-throughput RNA sequencing (RNA-seq) techniques has identified several non-coding RNAs (ncRNAs) that play important roles in the progression of different diseases including cancer. ncRNAs are accumulated at

D. Senapati (✉) · S. Tripathy
GITAM (Deemed to Be University), GITAM School of Pharmacy, Visakhapatnam, Andhra Pradesh, India
e-mail: dhirusenapati63@gmail.com

V. Sharma
Department of Pharmacology, School of Medicine, Case Western Reserve University, Cleveland, OH, USA

J. Barciszewski (ed.), *RNA Structure and Function*, RNA Technologies 14,
https://doi.org/10.1007/978-3-031-36390-0_18

higher concentrations in serum, plasma, and/or urine samples of CaP patients. Thus, differential involvement of ncRNA from normal patients to localised and metastatic cancers, makes it an acceptable biomarker for early diagnosis of metastatic disease. In this chapter, we highlight the emerging impacts and the translational applications of non-coding RNA in CaP progression for developing noble therapeutic strategies.

Keywords Non-coding RNA · Prostate cancer · Biomarker · Drug resistance

Abbreviations

ADT	Androgen deprivation therapy
AR	Androgen receptor
ARSVs	AR splice variants
ASO	Antisense oligonucleotides
BPH	Benign prostatic hyperplasia
circRNAs	Circular RNAs
CaP	Prostate cancer
CT	Computed tomography
CTCs	Circulating tumour cells
CRPC	Castration-resistant prostate cancer
DRE	Digital rectal examination
ENZ	Enzalutamide
EMT	Epithelial-mesenchymal transition
FDA	Food and drug administration
lncRNAs	Long non-coding RNAs
miRNAs	MicroRNAs
ncRNAs	Non-coding RNAs
MRI	Magnetic resonance imaging
NEPC	Neuroendocrine prostate cancer
PAP	Prostatic acid phosphatase
PHI	Prostate Health Index
PSMA	Prostate-specific membrane antigen
PCA3	Prostate cancer antigen 3
PSA	Prostate-specific antigen
RNA-seq	RNA sequencing
snRNA	Small nuclear RNA
siRNA	Small interfering RNA
shRNAs	Short hairpin RNAs

1 Introduction

Despite the approval of several new agents to treat prostate cancer (CaP), the disease remains incurable, and CaP is still the most common cause of cancer death among men. These deaths occur because acquired resistance develops to the available therapies for CaP (Attard and Antonarakis 2016). Because of a greater awareness and the use of PSA for screening, the prevalence of CaP diagnosis has increased over the past few years. The results of digital rectal examination (DRE) and serum PSA estimate are abnormal, hence the diagnosis of CaP is frequently made by pathological examination of transrectal ultrasound-guided prostate needle biopsies. Sometimes, transrectal ultrasound-guided prostate needle biopsies provide inaccurate results or a diagnosis of the tumours with little significance (Andriole et al. 2009). The clinical behaviour of CaP can range from slow-growing tumours to more aggressive, lethal type neuroendocrine prostate cancer (NEPC) (Hamid et al. 2019). CaP that is clinically insignificant does not increase the risk of CaP mortality and treating such type of refractory CaP can have undesirable side effects. Identifying a new molecular target that works through a different mechanism than current therapies and controls cell biology that is relevant to lethal CaP progression has the potential to greatly improve the disease burden. Researchers are paying more attention to non-coding RNAs (ncRNAs) because they can target multiple signalling axes that are involved in tumour growth, invasion, metastasis, and chemoresistance (Ming et al. 2021). ncRNAs, which participate in a variety of molecular processes such as post-translational, posttranscriptional and epigenetic modifications, have been shown to be an essential molecular component in the pathogenic mechanisms of castration-resistant prostate cancer (CRPC) (Shih et al. 2015). ncRNAs are mainly transcripted by genomic sequences which are not involved to encode protein. Depending on the numberof nucleotide sequences, ncRNAs are divided into two main categories such as small ncRNAs (<200 bp) and long ncRNAs (lncRNA, > 200 bp). MicroRNA (miRNA), transfer RNA (tRNA), ribosomal RNA (rRNA), small nuclear RNA (snRNA), and small interfering RNA (siRNA) are included in the group of small ncRNAs. According to their positions in the genome in relation to protein-coding genes, lncRNAs are categorised into intergenic and intragenic types. A form of covalently closed ncRNA called circular RNAs (circRNA) that are expressed differentially in different cell types and is associated with both physiological development and various diseases (Anastasiadou et al. 2018). Classification of different types of ncRNAs based on size, regulatory role and biogenesis pathways are illustrated in Table 1. ncRNAs are considered as promising biomarkers for cancer diagnosis because it could expedite the diagnosis of the cancer, especially if the disease is diagnosed early or treated with a range of therapeutic alternatives. ncRNAs contribute a significant role in the development of drug resistance to chemo and anti-androgen therapy (Ding et al. 2021). As ncRNAs either directly or indirectly control the expression of other genes, this makes them prospective therapeutic targets for multiple

Table 1 ncRNAs classification into different type depending on size and regulatory roles

Type of ncRNAs	Size	Regulatory role	Biogenesis
tRNA	Small (15-50 nt)	Housekeeping	Synthesized as precursors which are converted to mature tRNA molecules by a sequence of events
snRNA	Small (<150 nt)	Housekeeping	These are transcribed by either RNA polymerase II or III and accumulated in nucleous
microRNA	Small (<22 nt)	Regulatory	Formed in nucleus by RNA polymerase II or III as long primary transcripts, which are then cleaved and processed to small nt
siRNAs	Small (<200 nt)	Regulatory	Generated by Dicer from longer precursors
CircRNA	medium	Regulatory	Failure of intronic lariat debranching during canonical splicing
lncRNA	Long (>200 nt)	Regulatory	Originate from pre-mRNAs by splicing

signalling pathways in CaP. Here, we provide a summary of the structure and clinical functions of ncRNAs in prostate cancer as biological markers, therapeutic targets, and mediators of treatment resistance.

2 Structural Variations in Different NcRNAs

ncRNAs are transcribed by the larger portion of the genome that does not encode proteins but generates non-coding transcripts that regulate gene expression and protein function. Although majority of ncRNAs have been identified in CaP, molecular mechanisms of few RNAs have been characterised so far due to insufficient structural information (Li et al. 2016). Therefore, investigations of individual ncRNAs structure and subsequent use of this structure in RNA-based therapeutics for the treatment of CaP are urgently required.

2.1 Structural Features of LncRNAs

lncRNAs are a group of RNAs that are always capped and polyadenylated and are typically longer than 200 nucleotides. The highly conserved secondary and tertiary structures of lncRNAs are intimately related to their biological functions. The majority of lncRNAs share many characteristics with mRNAs, including RNA polymerase II mediated transcription, frequent polyadenylation, splicing and 5′-cap formation. Indeed, except from not being translated, they frequently do not differ biologically from mRNAs. Nevertheless, lncRNAs can be distinguished from mRNAs by a few features. Some lncRNAs are non-poly-adenylated and transcribed by RNA polymerase III. Compared to mRNA, lncRNAs are generally shorter, less well conserved and expressed at lower levels (Quinn and Chang 2016). LncRNAs are divided into sense, antisense, bidirectional, intronic, and intergenic subgroups based on their genomic positions in relation to nearby protein-coding genes. Metastasis-associated lung adenocarcinoma transcript-1(MALAT1) is a type of long nuclear-retained transcript that was proposed as a biomarker for metastasis CaP (Helsmoortel et al. 2018). MALAT1 regulates alternative splicing within the cell by altering the amounts of serine/arginine (SR) factors. These lncRNAs' longer nuclear retention may be due to the presence of short poly(A)-rich tract at their 3′ ends. Given that the mRNA's poly(A) tail makes it more stable, it has been proposed that the short poly(A) tail-like moieties in MALAT1 may contribute to its stability and protect against exonuclease-mediated degradation. The 3′ end of MALAT1 is protected from exonuclease activity by base pairing of the highly conserved poly(A)- and its nearby U-rich motifs. In order to prevent exonuclease degradation and prolong the biological functions, it appears that nuclear-retained transcripts frequently form triple helixes on their 3′ ends. A p53-regulated long intragenic non-coding RNA called long intergenic non-coding RNA-p21 (lincRNA-p21) was to be downregulated in CaP (Huang et al. 2022). The name was derived due to its location near the cell cycle regulator gene, p21. It has a single exon with inverted repeat Alu elements (IRAlus), which fold independently of one another into a secondary structure. Additionally, the structures formed by IRAlus play a crucial role in lincRNA-p21 colocalisation in the nucleus with paraspeckles (D'Souza et al. 2022). Three-dimensional structures of sense and antisense lincRNA-p21 IRAlus contribute significantly to understand the role of lincRNA-p21 in cancer development and progression.

2.2 Structural Features of MiRNAs

Small ncRNAs known as microRNAs (miRNAs) control a variety of biological processes, including the progression of CaP. Through a number of processes, including miRNA amplification, deletion, improper transcriptional regulation, dysregulated epigenetic alterations, and defects in the miRNA biogenesis machinery, miRNA expression is dysregulated in human cancer. MiRNAs can act as tumour suppressors or oncogenes depending on the type of cells or under certain conditions. The miRNAs are non-coding single-stranded RNA molecules with short (about 18–22 nucleotides) sequences that regulate gene expression post-transcriptionally. These are formed in the nucleus by RNA polymerase II or III as long primary transcripts, which are then cleaved into 70–100 nucleotide precursor RNAs (pre-miRNAs) by the microprocessor complex, which is made up of the RNase III enzyme Drosha and its collaborating partner DGCR8. Pre-miRNA is exported to the cytoplasm by Exportin-5(XPO5)/RanGTP transporter system, where it is further processed by the RNase III endonucleases Dicer and TRBP into a 19–25-nucleotide duplex (one strand of this duplex is known as passenger strand and the other is referred to as mature or guide strand). The mature miRNA is incorporated into a protein complex known as the RNA-induced silencing complex (RISC) once the duplex has been unwound, and it directs RISC to the target mRNA. RISC complexed miRNAs negatively regulate the stability of target mRNAs by binding to their 3′ untranslated region (UTR), or, to the 5′ UTR or to the coding sequence. Perfect complementarity between miRNAs and target mRNA results in mRNA degradation or translational suppression through different mechanisms (Valinezhad Orang et al. 2014). Additionally, it has been shown that miRNAs can either directly or indirectly promote gene expression through interactions with micro-ribonucleoproteins like Ago2 and FXR1. It has been shown that each miRNA regulates the transcription of a number of mRNAs, and as a result, these mRNAs participate in critical cellular activities like proliferation, differentiation, migration and apoptosis. 20–40% of miRNAs, specifically in urologic disorders, are found close to CpG islands, supporting their potential involvement in epigenetic silencing (Rauhala et al. 2010). miRNAs are usually found in the chromosomal regions that are fragile and undergo frequent DNA amplifications, deletions, or translocations during the cancer progression.

2.3 Structural Features of CircRNAs

Different from linear RNAs, circRNAs form a covalently closed continuous loop and are most prevalent in eukaryotic transcriptomes. Thousands of endogenous circRNAs are produced by mammalian cells; the majority of these circRNAs are transcribed from reverse complementary, exonic or intronic sequences. CircRNAs are highly conserved across all species and resistant to RNA exonuclease or RNase R-mediated degradation. In contrast to linear RNAs, which are ended with 5′ caps and 3′ tails, circular RNAs form covalently closed loop structures without polyadenylated tails (Chen and Yang 2015). The majority of circRNAs come from nucleolar pre-mRNAs, but some also derive from mitochondria. There are two suggested models for circRNA biosynthesis. Exon skipping or "lariat-driven circularisation" is the name given to model 1, whereas "intron-pairing driven circularisation" or "direct backsplicing" is the name given to model 2. Growing evidence suggested that model 2 might occur more frequently than model 1. In addition to intron-pairing driven circularisation, reverse complementary sequences, including IRAlus, are crucial for circRNA formation. A consensus RNA motif with an 11-nucleotide C-rich portion close to the branch point and a 7-nucleotide GU-rich section adjacent to the 5′ splice site is required for ciRNAs biogenesis (Zhang et al. 2013). Extra-coding RNAs (ecRNAs), which make up the majority of circRNAs, are made up of one or more exons. However, certain circRNAs, such as exon–intron circRNAs (EIciRNAs) and circular intronic RNAs (ciRNAs), originate from the parent gene's intron. CircRNAs primarily function as regulatory ncRNAs, either directly by controlling gene transcription and splicing or indirectly by modifying other regulators such as proteins and miRNAs (Meng et al. 2017).

3 Clinical Implications of Non-coding RNAs in Prostate Cancer

CaP is highly heterogeneous in nature and metastatic CaP (mCaP) is associated with a number of genetic and epigenetic modifications that occur during cancer progression and treatment resistance. Novel disease-specific macromolecules are released from tumour tissue, and these molecules may later be used as novel biomarkers for cancer detection. Biomarkers are mostly detected in body fluids like urine, semen and blood, as well as tissue samples removed after a biopsy or surgical procedures. Analysis of these circulating materials in body fluid samples is generally termed as 'liquid biopsy'. Compared to invasive techniques like needle biopsies, computed tomography (CT), positron emission tomography (PET), and magnetic resonance imaging (MRI), liquid biopsy is the most effective method for the diagnosis of solid malignancies (Siravegna et al. 2017). When invasive biopsy is difficult to perform in metastatic tissue such as bones, liquid biopsy techniques are most frequently used to detect molecular heterogeneity in the tumour mass as well as to detect molecular signals in metastatic samples (Haffner et al. 2021). A promising strategy for treating several malignancies, including CaP, is the therapeutic targeting of ncRNAs such as miRNAs and lncRNAs. The development and clinical implementation of RNA-based therapies, primarily using antisense oligonucleotides (ASO) and siRNAs, have received significant attention from researchers during the past several years. It's interesting to note that some of them have recently received food and drug administration (FDA) approval, while others are still in the clinical trial phase. However, trial findings have been mixed thus far, with some research claiming considerable impacts and others merely demonstrating a limited efficacy or toxicity (Winkle et al. 2021). Figure 1, summarises the whole clinical relevance of ncRNAs' participation in CaP.

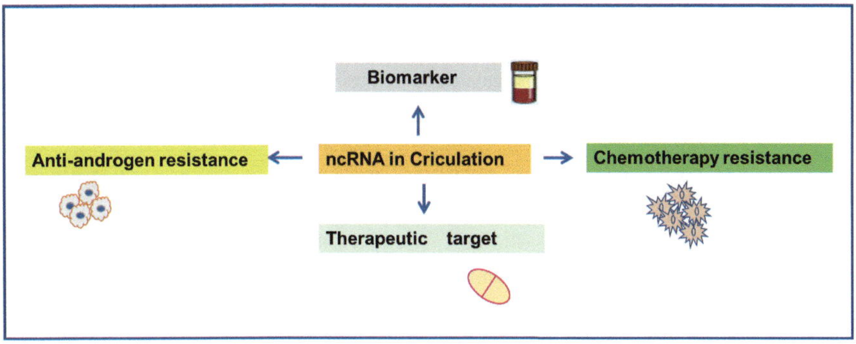

Fig. 1 Schematic presentation showing function of ncRNAs in prostate cancer

4 Non-coding RNA as Biomarker

In addition to their function in diagnosis, biomarkers may also be useful for assessing disease susceptibility, screening, prognosis, treatment response prediction, monitoring, and pharmacodynamic properties. Potential biomarkers include cells, proteins metabolites, DNA, RNA, and ncRNA (Sawyers 2008). The most utilised proteins as CaP biomarkers are prostate-specific membrane antigen (PSMA), caveolin-1, interleukin-6, CD147, TGF- 1, and human kallikrein-2. The wide range of protein concentrations still restricts the utility of proteins as biomarkers. The initial CaP biomarker identified was human prostatic acid phosphatase (PAP). ncRNAs have the ability to improve present diagnostic procedures or even surpass recognised biomarkers now employed in diagnostic procedures. When used alone or in combination with other potential markers, ncRNAs are promising diagnostic and prognostic markers (Das et al. 2019). ncRNAs that are most frequently altered in the tissue and fluid samples of CaP patients can be proposed as biomarkers for therapy response (Table 2).

Table 2 ncRNAs consistently altered in the fluid and tissue samples of prostate cancer patients

Noncoding RNAs	Type	Expression	Sample	References
PCA3	lncRNA	Upregulated	Urine, plasma	Hessels and Schalken (2009)
MALAT1	lncRNA	Upregulated	Plasma, Urine	Ren et al. (2012)
FR0348383	lncRNA	Upregulated	Urine	Zhang et al. (2015)
SChLAP1	lncRNA	Upregulated	Urine, plasma exosome	Prensner et al. (2014)
lincRNA-p21	lncRNA	Upregulated	Urine exosome	Işın et al. (2015)
miR-200b	miRNA	Upregulated	plasma	Souza et al. (2017)
miR-200c	miRNA	Upregulated	Plasma	Souza et al. (2017)
miR-20a	miRNA	Upregulated	Plasma	Shen et al. (2012)
miR-145	miRNA	downregulated	Urine, plasma exosome	Malla et al. (2018)
miR-221	miRNA	downregulated	Tissue	Spahn et al. (2010)
miR-141	miRNA	Upregulated	Urine, plasma exosome	Bryant et al. (2012)
miR-375	miRNA	Upregulated	Urine, plasma exosome	Fabris et al. (2015)

4.1 LncRNAs

PCA3 is one of the most explored prostatic lncRNAs. This is utilised specifically to develop a urine-based test that has already received FDA approval for use in clinics. Along with PCA3, other circulating lncRNAs related with CaP include MALAT1, FR0348383, **SCHLAP1**, and lincRNA-p21. MALT1 is predominantly present in plasma, while lincRNA-p21, FR0348383 and SCHLAP1 are isolated from post-DRE urine samples (Helsmoortel et al. 2018). It has been found that a high plasma concentration of MALAT1 is necessary to distinguish between BPH and CaP. After surgery, MALAT1 levels start to fall, which may indicate that MALT1 serves as a biomarker for progressive CaP. Additionally, the presence of MALAT1 RNA in urine samples revealed a potential relation between the levels of MALT1 and CaP (Ren et al. 2012). In order to minimise needless tissue biopsies, lncRNA FR0348383 level in the urine act as an acceptable biomarker for CaP diagnosis. However, FR0348383 needs to undergo comprehensive scientific and clinical validation before being considered as a biomarker (Zhang et al. 2015). Although lincRNA-p21 is markedly increased in CaP compared to BPH, it was not considered a biomarker due to its extremely low circulating levels (Işın et al. 2015). Similarly, SCHLAP1 concentrations in post-DRE urine samples also showed a good correlation with Gleason score.

4.2 MiRNAs

The ability to successfully isolate miRNAs from a range of biological samples makes them interesting candidates for biomarker markers. qRT-PCR, small RNA sequencing, microarray, and other widely used methods can all be used to identify and precisely quantify miRNAs because they are often stable. Multiple studies have revealed that miRNAs are significantly enriched in the serum, urine and plasma samples of CaP patients (Cortez et al. 2011). In silico expression of miRNA using different RNA data sets or qRT-PCR based experimental validation using RNA samples isolated from patients' body fluids may verify the role of miRNAs as possible biomarkers. Analysis of miRNA expression levels in TCGA datasets followed by gene expression profiles on plasma samples of untreated CaP patients have shown that expression levels of miR-200b and miR-200c are found to be CaP-specific in comparison to control. In contrast to miR-200c expression, which was linked to Gleason score, MiR-200b expression showed a correlation with bone metastases with PSA level above 10 ng/mL (Souza et al. 2017). Plasma levels of miR-221 and the Prostate Health Index (PHI) were provided in CaP as diagnostic and prognostic markers (Ibrahim et al. 2019). The decreased expression level of miR-221 in prostate tissues was directly associated with Gleason score, tumour growth, and chances of recurrence (Spahn et al. 2010). For solid tumours, such as CaP, miR-221 has been investigated as a potential circulating biomarker. It has been shown that there is an elevated level of circulating miR-20a level on serum samples of CaP patients compared to healthy controls. Therefore, the presence of miR-20a in CaP tissue and patient circulation suggests the potential use of miRNAs as biomarkers for CaP patients. Similar to this, miR-145 has been recommended as a circulating biomarker and was found to be altered in numerous types of malignancies, including CaP where low expression levels were proposed as a predictor of poor prognosis. Exosomes from urine samples of BPH patients showed elevated levels of circulating miR-145 compared to healthy control. CaP patients after radiation therapy also show the expression of miR-145 in the exosomes from blood samples (Malla et al. 2018). Interestingly, several studies have shown that elevated level of miR-141 and miR-375 are found in the exosomes derived from the urine and plasma samples of CaP patients (Bryant et al. 2012). Along with single oncogenic-miRNAs, groups of miRNAs have also been proposed as possible biomarkers in CaP. According to a recent report three miRNA profiles (miR-199a-5p, miR-511-5p, and miR-598-3p) have been found as a possible biomarker for early detection of low risk CaP (Gandellini et al. 2021). Both low- and high-risk CaP patients were found to have four miRNA signatures such as miR-20a, miR-21, miR-145, and miR-221. Similar to this, it was proposed that a four-miRNA profile consisting of miR-17, miR-20a, miR-20b, and miR-106a might be utilised to predict possibilities of recurrence following radical prostatectomy. A correlation between serum expression levels of four circulating miRNAs (miR-141, miR-182, miR-200b, and miR-375) and total PSA was found to help to distinguish between benign prostatic hyperplasia (BPH) and CaP. A recent study has shown that six elevated miRNAs level in urine (miR-234, miR-1238, miR-1913, miR-486-5p, miR-1825, and miR-484) can be used to distinguish between CaP and BPH patients (Haj-Ahmad et al. 2014).

4.3 CircRNAs

CircRNAs have potential characteristics that make them possible biomarkers for different cancers. CircRNAs can be accumulated in exosomes and released from the respective tissues into a variety of body fluids, such as plasma, saliva, and urine (Chao et al. 2021). The rupturing of cell membranes can also cause them to be liberated from dead or dying cells. CircRNAs have a covalently closed structure that makes them highly stable and resistant to RNase degradation. In cells, circRNA half-lives are approximately 2.5 times longer than linear RNA, and in exosomes, they are 6.3 times longer (Enuka et al. 2016). Expression level of circRNAs can also be easily measured using RNA-seq or qRT-PCR. Circulating exosomal RNA, circMYC was found in patients with nasopharyngeal carcinoma and was associated with clinical parameters such as tumour size, metastasis, and overall survival (Luo et al. 2020). In addition to nasopharyngeal carcinoma and other cancers, circRNAs were shown to be associated with clinicopathological features of CaP. The grade of the CaP's histology was correlated with the expression levels of circMBOAT2, circFoxo3, circCRKL, and circHIPK3. Circular RNAs may be able to predict the prognosis of CaP patients based on the expression levels of circFoxo3, circ-0016068, circ-MTO1, hsa circ

0,000,735 (hsa circ 0,000,735), Itchy E3 ubiquitin protein ligase (circ-ITCH), and ABCC4.

5 Non-coding RNA and Drug Resistance in Prostate Cancer

The androgen receptor (AR) has been the primary focus for treating mCaP for more than seven decades. Activation of AR is hindered by first-line androgen deprivation therapy (ADT). ADT provides remission initially, but later fails and gives rise to CRPC, which continues to grow by depending on AR (Formaggio et al. 2021). Abiraterone and enzalutamide are two potent FDA-approved drugs that generate brief remissions but do not cure CRPC and may promote the epithelial plasticity that results in NEPC. Chemo-, immuno-, or radiotherapy are further CRPC treatment options that can induce temporary remissions but are not curative (Kumari et al. 2022). A significant challenge to reducing CaP mortality is acquired treatment resistance. Therefore, understanding the underlying causes of resistance to anti-androgen therapies and chemotherapy is critical (Fig. 2).

5.1 *Non-coding RNAs and Anti-Androgen Resistance*

Numerous lncRNAs have been found to be highly expressed in mCaP. The emergence of resistance in metastatic CaP is driven by continued AR signalling after anti-androgen therapy. LncRNA has been thoroughly investigated in a number of cancers, such as colorectal, glioma, cervical, and liver cancer. LncRNA, HOXD-AS1 promotes the resistance of anti-androgen, bicalutamide, by recruiting WDR5 in

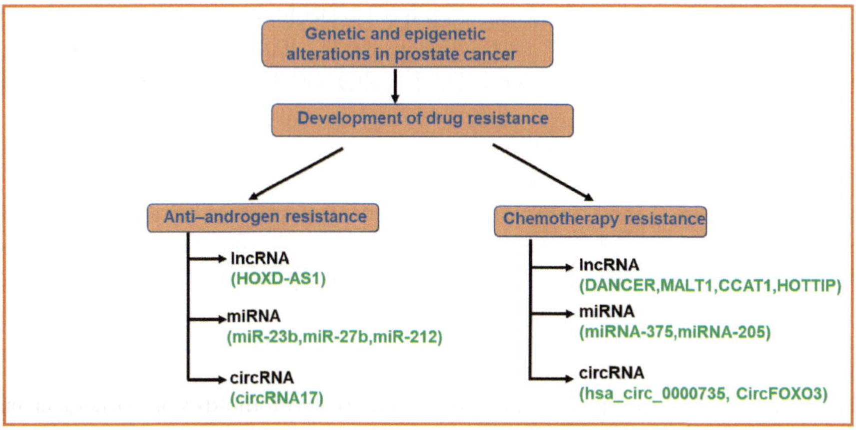

Fig. 2 ncRNAs which participate in different type of drug resistance in prostate cancer

CaP cells (Gu et al. 2017). miRNA, miR-23b and miR-27b can sensitize the effect of flutamide by decreasing the expression of CCNG1 in CRPC cell lines. Expression of AR mutants or AR splice variants (ARSVs) is primarily responsible for the development of resistance to anti-androgen drugs. Dysregulation of miR-212 promotes castration-resistance through hnRNPH1-mediated regulation of ARSVs. In addition to miR-212, miR361-3p was also found to target the AR-V7 and potentiate the sensitivity of enzalutamide in CaP. In hypoxic conditions, miR-361-3p directly recruits to the 3' UTR of AR-V7 and MKNK2 and increases the sensitivity of CaP cells to enzalutamide (Liu et al. 2020). circRNAs have been found to inhibit the resistance developed by anti-androgens such as enzalutamide (ENZ). In CaP specimens and ENZ-resistant CRPC cells, it was found that the expression level of circRNA, hsa circ 0,001,427(circRNA17), was downregulated. CircRNA17 suppression in ENZ-sensitive C4-2 cells leads to an increase in the expression of ARV7, which might be the cause of ENZ resistance and cell invasion. CircRNA17 could control the expression of ARV7 via modifying the expression of miR-181c-5p, which directly binds to the 3′UTR of ARV (Wu et al. 2019). These findings suggested that the signalling pathway involving circRNA17-miR-181c-5p-ARv7 axis makes a pivotal contribution to the development of a potential target for the intervention of ENZ-resistant CaP cells. Differential expression of circRNAs, such as hsa_circ_0001721 and hsa_circ_0004870 between ENZ-resistant and sensitive cell line indicating that these RNAs play a substantial role in the emergence of anti-androgen resistance (Greene et al. 2019).

5.2 Non-coding RNA's Impact in Chemotherapy Resistance

The first-line chemotherapy for mCaP is docetaxel, which significantly improves CRPC survival. Nevertheless, despite a positive response to the initial treatment, taxane therapy eventually develops resistance. Docetaxel-resistant prostate cancer exhibits a high upregulation of the lncRNA DANCR and silencing of DANCR enhances the proliferation of resistance-CaP cells to docetaxel therapy (Ma et al. 2019). Another lncRNA, MALAT1, which has been associated with the onset of CRPC, increases docetaxel resistance both in vivo and in vitro and promotes the disease's progression. Paclitaxel resistance in prostate cancer is facilitated by lncRNA CCAT1 via controlling the expression of miR-24-3p. Long intergenic nonprotein coding RNA 518 (Linc00518), which is frequent in prostate tumours, enhances paclitaxel resistance via increasing GATA6 expression by snubbing miRNA-216b-5p (He et al. 2019). The Lnc RNA, HOTTIP, maintains cisplatin resistance in CaP cells by modulating Wnt/β-catenin signalling pathways (Jiang et al. 2019). In paclitaxel-resistant cell lines, it has been observed that a number of miRNAs, including miRNA-200b3p, miRNA-375, and miRNA-34b-3p, are elevated. miRNA-205 impairs the CaP autophagic pathway and finally overcome cisplatin resistance by suppressing the protein RAB27A/LAMP3. MCaP tissues and cells have higher levels of Circ-'RNA, hsa_circ_0000735. Additionally, hsa_ circ _0000735 can increase

CaP chemoresistance by acting as an inhibitor of miRNA-7, which is downregulated in docitaxel-resistant CaP. CircFOXO3, a different circRNA, has been reported to slow down the tumour growth and chemoresistance through increased FOXO3 expression and inhibition of the epithelial-mesenchymal transition (EMT) (Shen et al. 2020).

6 Non-coding RNA as Therapeutic Target

A promising method for treating cancer and many other disorders is the pharmacological targeting of non-coding RNAs (ncRNAs), such as miRNAs and lncRNAs. Over the past ten years, a large number of ASO and siRNAs have been tested in clinical trials, and several of these have already been approved by the FDA. Numerous RNA-based treatments including ASOs, siRNAs, CRISPR-Cas9-based gene editing, miRNA mimics, short hairpin RNAs (shRNAs), ASO anti-miRNAs, miRNA sponges and therapeutic circRNAs have been designed (Rupaimoole and Slack 2017). Moreover, a large number of RNA treatments, including more recent ones like miRNA mimics and anti-miRNAs, are in phase II or III clinical development; however, no lncRNA-based therapeutics have reached the clinic. Intra-tumoural injection of MALAT-1 specific siRNA decreases tumour growth and metastasis in CaP (Ren et al. 2013). Subcutaneous delivery of ASO against ARLNC1-lncRNA has a strong anti-tumour effect in androgen-sensitive MDA-PCa-2b xenografts. Although siRNAs and ASOs that target coding transcripts are being explored in clinical trials for other cancer, RNAi-based techniques to target lncRNAs must be developed more for CaP therapy. Many lncRNAs execute their activity through interacting with proteins and/or protein complexes. Given that the pharmacological effects of small molecules outperform RNAi, focusing more emphasis on the lncRNA-interactome could increase specificity and minimise off-target effects. Through an epigenetic mechanism, Lnc RNA, CTBP1-AS inhibits the expression of the TP53 gene. It accomplishes this by recruiting the splicing factor proline and glutamine-rich (PSF/SFPQ) protein to the TP53 gene's promoter region. Therefore, pharmacological inhibition of PSF by small molecule inhibitors induces p53 mediated tumour suppressive action in CaP. Numerous compounds were discovered using chemical library screening, however, 7,8-dihydroxy-4-(4-methoxyphenyl) chromen-2-one, which disrupts CTBP1-AS-mediated p53 action, is the most potent inhibitor of castration-resistant tumour growth (Takayama et al. 2022). Recently, a special triple-helix structure, which was unique at the 3'end of MALAT1, was used effectively to discover small molecules for CaP therapy. MiR-34a is typically downregulated in a variety of cancer types by aberrant transcriptional regulation, chromosomal deletions, and promoter hypermethylation. An effective therapeutic strategy for the treatment of cancer is to increase the expression of miR-34a after the addition of miRNA mimics (Lodygin et al. 2008). miR-34a-mimics are tested in several preclinical animal models of cancer, including lung, breast, pancreatic, and CaP and all of these models have shown a successful increase in the expression of miR-34a as well as a subsequent decline in tumour growth (Rupaimoole and Slack 2017).

7 Future Perspectives

Although ncRNAs were only recently discovered, it has become clear that they are essential for CaP growth and development. However, the majority of ncRNAs' molecular roles in CaP are still poorly understood. The current challenge is to systematically explore these pathways in order to determine which ncRNAs are most important to CaP biology. Such ncRNAs may be targeted or altered for therapeutic purposes. The therapeutic application of the lncRNA PCA3 has already proved beneficial, particularly in predicting biopsy results for individuals with increased serum PSA. Other ncRNAs, particularly the miRNAs, have recently been shown to have potential as novel CaP biomarkers. It is becoming more important to identify and characterise circulating tumour cells (CTCs) in CaP patients. Patients with more than 5 CTCs in 7.5 mL of blood have a significantly poorer overall survival rate compared to patients with less than 5 CTCs in the same amount of blood. CaP patients receiving anti-androgens like abiraterone acetate or ENZ, the therapeutic value of monitoring CTCs is currently being studied in phase III trials (De Bono et al. 2008). Although CTCs from patients with metastatic breast cancer exhibit possible miRNAs like miR-183, it is unknown whether CTCs from patients with CaP exhibit particular ncRNAs. Numerous more potential miRNAs and other CaP biomarkers, like exosomes and CTCs, are in the pipeline, but they need to be confirmed by independent studies before being translated into clinical practise. There is growing evidence that different forms of ncRNA have contributed to chemoresistance by focusing on different signalling pathways. There is growing evidence that different forms of ncRNA have contributed to chemoresistance by focusing on different signalling pathways. NcRNA therapy is currently outperforming proteins to be utilised as a therapeutic target in tumour therapy since 80–85% of proteins lack the structural features needed to interact with drug molecules and almost 98% of the genome is composed of ncRNAs (Salami and Crews 2017). ncRNA therapy has not yet been associated to any drug resistance, but acquired resistance is a problem with ADT and chemotherapy. ncRNA also has a longer half-life than small molecules or antibodies due to chemical alteration or modifications. For instance, ASOs with the backbone modified with 2′-O-methoxyethyl may withstand degradation by nucleases better (Monia et al. 1993). Despite the optimistic outlook for tumour therapy, ncRNA therapy still faces a number of challenges. Different tissues may respond differently to nc RNA. As already mentioned, miR-375 may increase docetaxel resistance in prostate cancer. However, it has been noted that miR-375 promotes osteosarcoma growth, indicating that it functions differently in various tissue (Liu et al. 2018). Finding the greatest ncRNA candidates for cancer therapy is a top priority. It is very challenging to prevent an off-target effect in cells since ncRNAs often target a wide range of locations. There are still a number of obstacles to be overcome.

8 Conclusions

Together, the studies reviewed here show how ncRNAs can be used as biomarkers and for therapeutic purposes in CaP. A special interdisciplinary strategy involving pharmacology, chemistry, and nanotechnology is necessary for the practical applications of RNA-based medicines. An example of such an integrative strategy is the administration of a miRNA inhibitor to heart tissue utilising an ultrasound and microbubble-targeted delivery system, which reduced cardiac hypertrophy in a mouse model. We predict that genome sequencing of ncRNAs may become necessary for specific cancer therapy in the near future as a result of the advancement of ncRNA delivery technologies and a decrease in off-target effects. Additionally, we anticipate that ncRNA therapy might work well in combination with traditional therapies.

Disclosure Statement Authors declare no conflicts of interest for this article.

References

Anastasiadou E, Jacob LS, Slack FJ (2018) Non-coding RNA networks in cancer. Nat Rev Cancer 18:5–18

Andriole GL, Crawford ED, Grubb RL 3rd et al (2009) Mortality results from a randomized prostate-cancer screening trial. N Engl J Med 360:1310–1319

Attard G, Antonarakis ES (2016) AR aberrations and resistance to abiraterone or enzalutamide. Nat Rev Urol 13:697–698

Bryant RJ, Pawlowski T, Catto JW et al (2012) Changes in circulating microRNA levels associated with prostate cancer. Br J Cancer 106:768–774

Chao F, Wang S, Zhang C et al (2021) The emerging role of circular RNAs in prostate cancer: a systematic review. Front Cell Dev Biol 9

Chen LL, Yang L (2015) Regulation of circRNA biogenesis. RNA Biol 12:381–388

Cortez MA, Bueso-Ramos C, Ferdin J (2011) MicroRNAs in body fluids—The mix of hormones and biomarkers. Nat Rev Clin Oncol 8:467–477

Das R, Feng FY, Selth LA (2019) Long non-coding RNAs in prostate cancer: biological and clinical implications. Mol Cell Endocrinol 480:42–152

De Bono JS, Scher HI, Montgomery RB et al (2008) Circulating tumor cells predict survival benefit from treatment in metastatic castration-resistant prostate cancer. Clin Cancer Res 14:6302–6309

Ding L, Wang R, Shen D et al (2021) Role of noncoding RNA in drug resistance of prostate cancer. Cell Death Dis 12:1–10

D'Souza MH, Mrozowich T, Badmalia MD et al (2022) Biophysical characterisation of human LincRNA-p21 sense and antisense Alu inverted repeats. Nucl Acids Res 50:5881–5898

Enuka Y, Lauriola M, Feldman ME et al (2016) Circular RNAs are long-lived and display only minimal early alterations in response to a growth factor. Nucl Acids Res 44:1370–1383

Fabris L, Ceder Y, Chinnaiyan AM et al (2015) The potential of MicroRNAs as prostate cancer biomarkers. Eur Urol 70:312–322

Formaggio N, Rubin MA, Theurillat JP (2021) Loss and revival of androgen receptor signaling in advanced prostate cancer. Oncogene 40:1205–1216

Gandellini P, Ciniselli CM, Rancati T et al (2021) Prediction of grade reclassification of prostate cancer patients on active surveillance through the combination of a three-miRNA signature and selected clinical variables. Cancers (basel) 13:2433

Greene J, Baird AM, Casey O et al (2019) Circular RNAs are differentially expressed in prostate cancer and are potentially associated with resistance to enzalutamide. Sci Rep 9:1–11

Gu P, Chen X, Xie R et al (2017) lncRNA HOXD-AS1 regulates proliferation and chemo-resistance of castration-resistant prostate cancer via recruiting WDR5. Mol Ther 25:1959–1973

Haffner MC, Zwart W, Roudier MP et al (2021) Genomic and phenotypic heterogeneity in prostate cancer. Nat Rev Urol 18:79–92

Haj-Ahmad TA, Abdalla MA, Haj-Ahmad Y (2014) Potential urinary miRNA biomarker candidates for the accurate detection of prostate cancer among benign prostatic hyperplasia patients. J Cancer 5:182

Hamid AA, Gray KP, Shaw G et al (2019) Compound genomic alterations of TP53, PTEN, and RB1 tumor suppressors in localized and metastatic prostate cancer. Euro Urol 76:89–97

He J, Sun M, Geng H et al (2019) Long non-coding RNA Linc00518 promotes paclitaxel resistance of the human prostate cancer by sequestering miR-216b-5p. Biol Cell 111:39–50

Helsmoortel H, Everaert C, Lumen N et al (2018) Detecting long non-coding RNA biomarkers in prostate cancer liquid biopsies: hype or hope? Non-Coding RNA Res 3:64–74

Hessels D, Schalken JA (2009) The use of PCA3 in the diagnosis of prostate cancer. Nat Rev Urol 6:255–261

Huang Y, Yi Q, Feng J et al (2022) The role of lincRNA-p21 in regulating the biology of cancer cells. Hum Cell 35:1640–1649

Ibrahim NH, Abdellateif MS, Thabet G et al (2019) Combining PHI and miRNAs as biomarkers in prostate cancer diagnosis and prognosis. Clin Lab 65(10):7754

Işın M, Uysaler E, Özgür E et al (2015) Exosomal lncRNA-p21 levels may help to distinguish prostate cancer from benign disease. Front Genet 6:168

Jiang H, Xiong W, Chen L et al (2019) Knockdown of the long noncoding RNA HOTTIP inhibits cell proliferation and enhances cell sensitivity to cisplatin by suppressing the Wnt/β-catenin pathway in prostate cancer. J Cell Biochem 120:8965–8974

Kumari S, Sharma V, Tiwari R (2022) Therapeutic potential of p53 reactivation in prostate cancer: Strategies and opportunities. Euro J Pharmacol 174807

Li R, Zhu H, Luo Y (2016) Understanding the functions of long non-coding RNAs through their higher-order structures. Int J Mol Sci 17:702

Liu B, Sun Y, Tang M et al (2020) The miR-361-3p increases enzalutamide (Enz) sensitivity via targeting the ARv7 and MKNK2 to better suppress the Enz-resistant prostate cancer. Cell Death Dis 11:1–12

Liu G, Huang K, Jie Z et al (2018) CircFAT1 sponges miR-375 to promote the expression of Yes-associated protein 1 in osteosarcoma cells. Mol Cancer 17:1–16

Lodygin D, Tarasov V, Epanchintsev A et al (2008) Inactivation of miR-34a by aberrant CpG methylation in multiple types of cancer. Cell Cycle 7:2591–2600

Luo Y, Ma J, Liu F et al (2020) Diagnostic value of exosomal circMYC in radioresistant nasopharyngeal carcinoma. Head Neck 42:3702–3711

Ma Y, Fan B, Ren Z et al (2019) Long noncoding RNA DANCR contributes to docetaxel resistance in prostate cancer through targeting the miR-34a-5p/JAG1 pathway. Onco Targets Ther 12:5485

Malla B, Aebersold DM, Dal Pra A (2018) Protocol for serum exosomal miRNAs analysis in prostate cancer patients treated with radiotherapy. J Trans Med 16:1–13

Meng S, Zhou H, Feng Z et al (2017) CircRNA: functions and properties of a novel potential biomarker for cancer. Mol Cancer 16:94

Ming H, Li B, Zhou L et al (2021) Long non-coding RNAs and cancer metastasis: molecular basis and therapeutic implications. Biochim Biophysica Acta Rev Cancer 1875:188519

Monia BP, Lesnik E, Gonzalez C et al (1993) Evaluation of 2 '-modified oligonucleotides containing 2 '-deoxy gaps as antisense inhibitors of gene expression. J Biol Chem 268:14514–14522

Prensner JR, Zhao S, Erho N et al (2014) RNA biomarkers associated with metastatic progression in prostate cancer: a multi-institutional high-throughput analysis of SChLAP1. Lancet Oncol 15:1469–1480

Quinn JJ, Chang HY (2016) Unique features of long non-coding RNA biogenesis and function. Nat Rev Genet 17:47–62

Rauhala HE, Jalava SE, Isotalo J et al (2010) miR-193b is an epigenetically regulated putative tumor suppressor in prostate cancer. Int J Cancer 127:1363–1372

Ren S, Liu Y, Xu W et al (2013) Long noncoding RNA MALAT-1 is a new potential therapeutic target for castration resistant prostate cancer. J Urol 190:2278–2287

Ren S, Peng Z, Mao JH et al (2012) RNA-seq analysis of prostate cancer in the Chinese population identifies recurrent gene fusions, cancer-associated long noncoding RNAs and aberrant alternative splicings. Cell Res 22:806–821

Rupaimoole R, Slack FJ (2017) MicroRNA therapeutics: towards a new era for the management of cancer and other diseases. Nat Rev Drug Discov 16:203–222

Salami J, Crews CM (2017) Waste disposal—An attractive strategy for cancer therapy. Science 355:1163–1167

Sawyers CL (2008) The cancer biomarker problem. Nature 452:548–552

Shen J, Hruby GW, McKiernan JM, Gurvich I, Lipsky MJ, Benson MC, Santella RM (2012) Dysregulation of circulating microRNAs and prediction of aggressive prostate cancer. The Prostate 72:1469–1477

Shen Z, Zhou L, Zhang C et al (2020) Reduction of circular RNA Foxo3 promotes prostate cancer progression and chemoresistance to docetaxel. Cancer Lett 468:88–101

Shih JW, Wang LY, Hung CL et al (2015) Non-coding RNAs in castration-resistant prostate cancer: regulation of androgen receptor signaling and cancer metabolism. Int J Mol Sci 16:28943–28978

Siravegna G, Marsoni S, Siena S et al (2017) Integrating liquid biopsies into the management of cancer. Nat Rev Clin Oncol 14:531–548

Souza MF, Kuasne H, Barros-Filho MC et al (2017) Circulating mRNAs and miRNAs as candidate markers for the diagnosis and prognosis of prostate cancer. Plos One 12:e0184094

Spahn M, Kneitz S, Scholz CJ et al (2010) Expression of microRNA-221 is progressively reduced in aggressive prostate cancer and metastasis and predicts clinical recurrence. Int J Cancer 127:394–403

Takayama KI, Honma T, Suzuki T et al (2022) Targeting epigenetic and posttranscriptional gene regulation by PSF impairs hormone therapy-refractory cancer growth. Cancer Res 81:3495–3508

Valinezhad Orang A, Safaralizadeh R, Kazemzadeh-Bavili M (2014) Mechanisms of miRNA-mediated gene regulation from common downregulation to mRNA-specific upregulation. Int J Genomics 2014:970607

Winkle M, El-Daly SM, Fabbri M et al (2021) Noncoding RNA therapeutics—Challenges and potential solutions. Nat Rev Drug Discov 20:629–651

Wu G, Sun Y, Xiang Z et al (2019) Preclinical study using circular RNA 17 and micro RNA 181c–5p to suppress the enzalutamide-resistant prostate cancer progression. Cell Death Dis 10:1–14

Zhang W, Ren SC, Shi XL et al (2015) A novel urinary long non-coding RNA transcript improves diagnostic accuracy in patients undergoing prostate biopsy. Prostate 75:653–661

Zhang Y, Zhang XO, Chen T et al (2013) Circular intronic long noncoding RNAs. Mol Cell 51:792–806

3.3.1.4 After the completion of the prediction … to find the final list of miRNA binding … miRDB predicted 11 miRNAs that …

3.3.2 Identification of Target-site … with circRNA

Hundreds of studies …

4 … Notes

4.1 The sequence of multiple circRNA …

The Structure, Function, and Modification of Non-coding RNAs in Cardiovascular System

Xinxin Cui, Priyanka Gokulnath, Guoping Li, Lijun Wang, and Junjie Xiao

Contents

Abstract The occurrence, development and prognosis of cardiovascular disease is a multi-factor and multi-path pathological process. In addition to environmental factors, epigenetic regulation mechanisms also play an important role in the occurrence and development of cardiovascular disease. The most common and abundant internal modification of mRNA is m^{6}A. Together with RNA editing, which is an

X. Cui · L. Wang (✉) · J. Xiao (✉)
Cardiac Regeneration and Ageing Lab, Institute of Cardiovascular Sciences, Shanghai Engineering Research Center of Organ Repair, School of Medicine, Shanghai University, Shanghai 200444, China
e-mail: lijunwang@shu.edu.cn

J. Xiao
e-mail: junjiexiao@shu.edu.cn

P. Gokulnath · G. Li
Cardiovascular Division of the Massachusetts General Hospital and Harvard Medical School, Boston, MA 02114, USA

J. Barciszewski (ed.), *RNA Structure and Function*, RNA Technologies 14,
https://doi.org/10.1007/978-3-031-36390-0_19

alternative RNA modification, both play important roles in regulating gene expression and affect the fate of RNA molecules. In addition, with the advances in next-generation sequencing technology, non-coding RNAs such as microRNA, long non-coding RNA, and circular RNA which are usually not involved in protein synthesis, but can participate in cardiac homeostasis, cardiomyocyte growth, proliferation and apoptosis, endothelial cell function, cardiac remodeling and repair, and inflammatory response through various mechanisms. Grasping the cognition of RNA modifications and non-coding RNAs in cardiovascular disease may help us to better understand mechanisms and develop new biomarkers or therapeutic strategies in cardiovascular disease. This chapter summarizes the roles of long non-coding RNA, microRNA, circular RNA, and RNA modification in cardiovascular diseases.

Keywords Cardiovascular disease · miRNAs · lncRNAs · circRNAs · RNA modification

1 Introduction

Cardiovascular diseases (CVD) have become global burden which affects the population's life (Wang et al. 2020a, b, c). While remarkable progress has been made with drugs, interventional therapies, and surgery development, survival rates of CVD patients have improved substantially, efficacy is still far from satisfactory. With the development of next-generation sequencing, numerous non-coding RNAs (ncRNAs) have been reported to take critical roles in cardiovascular systems. Previous studies have suggested a tight correlation between ncRNAs and hypertension, heart failure, cardiac hypertrophy, and other CVD, and ncRNAs have also been proposed as the biomarkers and potential therapeutic targets in CVD (Poller et al. 2018). Thus, uncovering the mechanism of ncRNAs is important for investigating clinical strategies for CVD, and can provide promising targets for drug development. Here, we review the role of ncRNAs in CVD, and provide information for the diagnosis and treatment of CVD.

2 Overview of lncRNA, miRNA, circRNA, and RNA Modification

Long non-coding RNA (lncRNA) and circular RNA (circRNA) can interact with RNAs or proteins. As to RNA-RNA interactions, lncRNAs or circRNAs may directly affect microRNA (miRNAs) or mRNAs. Since lncRNAs or circRNAs and mRNA may share the same miRNA binding site, lncRNA can decrease the content of miRNA via competing with mRNA through binding to miRNA, and thus reduce the regulatory effect of miRNA on their target genes. On the other hand, miRNAs suppress target genes' function via directly binding to the 3′-UTR of mRNA, thereby promoting mRNA degradation or inhibiting translation to reduce the target genes expression for the purpose of regulating cell function (Ferragut Cardoso et al. 2021). A single miRNA has the capability of binding various mRNAs; therefore, the accumulation of synergistic regulation of multiple downstream mRNA transcripts significantly affects the characterization. RNA modification, especially m^6A modification, is involved in gene expression regulation by mediating RNA degradation, modifying target genes, and affecting the expression and function of modified-transcripts. Therefore, RNA network plays a critical role in the pathogenesis of CVD.

2.1 lncRNAs

LncRNA is characterized with non-coding transcripts above 200 nucleotides. Compared with mRNA, lncRNA is sometimes shorter and contains fewer exons. LncRNA can act on mRNA indirectly regulating miRNA through ceRNA mechanism, lncRNA could also bind to mRNA thus affecting the expression of downstream target genes and proteins, depending on the distance between them (Taniue and Akimitsu 2021). LncRNAs can be divided into five types according to their genomic locations: sense lncRNAs, sense long ncRNAs are transcribed on the same strand like mRNAs; Antisense lncRNAs, which are transcribed from the antisense strand of protein-coding genes, can complement and pair with partial or complete sequences of protein-coding genes; Intronic lncRNAs are transcribed from intronic sequences of protein-coding genes. Bidirectional lncRNAs are transcribed from the two opposite complementary strands of protein-coding gene; Intergenic lncRNAs are transcribed in the spacer sequence regions of protein-coding genes (Bridges et al. 2021).

In order to gain a more in-depth understanding of lncRNA functions, researchers have used RNA pull-down, RNA immunoprecipitation and other molecular techniques to explore the gene regulatory network formed by the interaction of lncRNAs with transcription factors, proteins, mRNAs, DNA and other biomolecules. Most lncRNAs are localized in the nucleus and have various expression patterns in different parts of tissues, but their secondary structure and subcellular localization are relatively conserved (García-Padilla et al. 2018). Because it lacks the complete open reading frame required for translation, most of the lncRNA transcript does not have the ability to encode proteins (Ransohoff et al. 2018). But lncRNAs have a specific and complex secondary spatial structure, which can provide multiple sites for binding with proteins and dynamically interact with DNA or RNA through the principle of base complementary pairing, forming a complex and subtle three-dimensional spatial morphology (Bridges et al. 2021). The open single-stranded RNA sequence allows lncRNAs to perform their basic functions by combining with other nucleotide sequences. Specifically, the action mode of lncRNAs mainly includes the following aspects: (1) Epigenetic regulation: certain specific lncRNAs participated in epigenetic regulation, which may regulate gene expression by recruiting some chromatin remodeling complexes, such as methylating histone modified proteins, thus transforming chromatin into heterochromatin state (Zhang et al. 2019). (2) Transcriptional regulation: some lncRNAs bind to transcription factors and affect the function of transcription factors, which can activate or inhibit the transcriptional activity of gene (Zhang et al. 2020a, b, c). (3) Post-transcriptional regulation: lncRNAs are directly involved in the regulation of mRNA stability. These processes influence gene functional polymorphisms (Ransohoff et al. 2018). In addition, lncRNA regulates the progression of CVD through the ceRNA mechanism via interacting with miRNA (Liao et al. 2021). Various mechanism of regulation mediated by lncRNA binding with RNA-binding protein (RBP) is also a hot topic of research.

The gene regulation and biological function of lncRNAs have been the focus of research in the field of life science in recent years. It is extremely important to identify specific lncRNAs for the diagnosis and therapeutic regimen of diseases. Then, the biological functions and mechanisms of lncRNAs were analyzed. So far, a variety of lncRNAs have been found to be linked to the physiologic and pathophysiologic processes of CVD such as heart failure, coronary artery disease, myocardial infarction, and arrhythmias (Gao et al. 2021; Li et al. 2021a, b, c; Lin et al. 2021; Viereck et al. 2020). Although only a few lncRNAs have been elucidated in current studies, lncRNAs can participate in the pathophysiological process of CVDs in various ways. It can be predicted that more and more lncRNAs will be discovered through further systematic and comprehensive studies. In summary, lncRNAs can attenuate or trigger gene transcriptional regulation through diversity mechanisms. Elucidating the complex regulatory mechanisms and functional targets of lncRNAs may provide information for the clinical diagnosis and cure of CVD.

2.2 *miRNAs*

MiRNAs are a kind of endogenous single-stranded RNAs with a length of approximately 20–25 nucleotides, showing the biological features as follows: Firstly, miRNA is highly conserved in different organisms (Liu et al. 2021). Secondly, miRNAs are widely distributed which may be related to their high conservatism. Thirdly, some miRNAs are specifically expressed in different tissues, for example, miR-208 is abundant in heart, while miR-205 mainly exists in breast (Nagalingam et al. 2021). In addition, miRNAs are stage-specific, with different expression levels in different developmental stages of organisms. The abundance of miR-199a is relatively low in testicular sperm, but it gradually accumulates in epididymal sperm as sperm matures (Sørensen et al. 2022). The basic synthesis and processing of animal-derived miRNAs is firstly to transcript and synthesize primary miRNA transcripts in the nucleus. After that, the product is cleaved to form precursor miRNA under the action of RNA hydrolase of Drosha, and then transferred to cytoplasm under the action of transporter, and further cleaved to produce mature miRNA under the action of Dicer hydrolase (Prasad et al. 2021). Mature miRNAs can regulate target genes by directly binding to the 3′-untranslated region (UTR), and regulate the expression of target genes via inducing degradation or inhibiting translation (Lerchenmüller et al. 2020).

The method to detect CVD in high-risk populations is to look for the presence of biomarkers, but the risk assessment in daily clinical practice lacks standard indicators and relies on highly skilled operators and relatively expensive equipment. Therefore, more direct, time-saving, operator-independent biomarkers are needed to correlate with these biomarkers as an alternative to the currently used biomarkers. Because miRNA is abundant in serum, has high stability, sensitivity, non-invasive, and easy to detect, miRNA can be used as a new biomarker to indicate CVD, and miRNAs have the potential to be used for early diagnosis and therapeutic schedule of CVD (Wang et al. 2021a, b, c, d). Previous studies have found that CVD exhibits specific miRNA expression profiles compared with normal tissues, providing new possibilities for potential target therapy (Zhou et al. 2018). These biomarkers not only provide indicative information for predicting disease, but their up-or-down-regulation may shed light on cardiovascular signaling networks.

In addition to being biomarkers, miRNAs also have a great regulatory effect in controlling the post-transcriptional expression of their target genes. The dysregulation of miRs is involved in the induction of vascular inflammation, apoptosis, cell proliferation, angiogenesis, and the development of CVD (Gabisonia et al. 2019; Ryu et al. 2021). Application of high-throughput sequencing techniques and bioinformatics has led to the discovery of numerous disease-related miRNAs and their target genes with the prospective to serve as therapeutic targets. For example, specific antisense oligonucleotide blocking miRs can be used to treat CVD (Ding et al. 2018). However, with the gradual deepening of the research on the function of miRNAs, researchers began to realize that there are still many problems to be solved before miRNAs can be used as targets for drug therapy. The distribution of most miRNAs has no significant tissue or organ specificity, so off-target problems of their acting organs and effects should be taken into account. That is, how to make the drug effective and specific to the heart is the first problem to be solved. Second, because the relationship between miRNA and target is not one-to-one, a target can be regulated by several miRNAs, and the same miRNA can also regulate different target genes, which leads to the effect of regulated miRNA may not be consistent with the expected one. To selectively exert the expected therapeutic effect through miRNA, large randomized controlled trials are further required.

2.3 *circRNAs*

The breakthrough in high-throughput deep sequencing technology has led to the discovery of plenty of circRNAs in humans, mice, nematodes, zebrafish, drosophila melanogaster, protozoa and plants. In mammalian cells, there are tens of thousands of different circRNAs, and the number continues to increase. CircRNA has a closed covalent ring structure and is not affected by RNA exonuclease. Consistently, actinomycin D assays suggested that circRNA's half-life was up to 23.7 h, which was significantly longer than that of its cognate mRNA (Wang et al. 2022a, b). Although circRNA has a relatively stable structure, the abundance of circRNA in cells and tissues is generally low compared with mRNAs (Chen et al. 2021). There are three classes of circRNAs, exonic circRNA (ecircRNAs), circular intronic RNA (ciRNA) and exon–intron circRNA (ElciRNA) according to the source. The majority of EcircRNAs are formed primarily by the reverse splicing of exons. The upstream 3′-region is inversely covalently connected to the downstream 5′ end. ciRNAs are RNAs with intron sequences linked by 2′-5′ phosphodiester bonds, elciRNAs' sources of sequence are exons and introns. Unlike EcircRNAs, these two types of circRNAs are usually detectable in the nucleus. Besides, due to the stable and tissue-specific character, circRNAs could be used as predictive biomarkers and therapeutic targets of diseases, such as CVD treatment (Ryu et al. 2021).

Most circRNAs are located in the cytoplasm. The most widely investigated mechanism of circRNA action is the "sponge mechanism". Studies have demonstrated that circRNA can block the degradation of downstream target genes by sponging miRNA (Li et al. 2020a, b). Similarly, circRNAs can interact with RNA binding proteins and act as protein antagonists to inhibit protein activity, thereby upregulating the expression level of target genes (Zhang et al. 2018). CircRNAs are not only sponge miRNAs, but can also directly bind to mRNA and coactivate transcription factors to affect translation (Liu and Chen 2022). In addition, some circRNAs have open reading frames with internal ribosome entry sites, which can bind to ribosomes to initiate translation, even if the efficiency of translation is low. CircRNA itself can also be translated, and m^6A modification can drive its translation (Di Timoteo et al. 2020).

2.4 RNA Modification

In recent years, greater than 170 different kinds of RNA chemical modifications have been involved in the modification process of coding and non-coding RNAs, which regulate gene expression to maintain cellular homeostasis. The dynamic feature of RNA modification allows cells to respond immediately to changes in the external condition, and the capacity to adapt to changing microenvironments, such as stress or drug stimulation, are critical for the survival of cardiomyocyte. Recent studies have demonstrated that RNA modification directly affects the chemical properties and metabolic processes of RNA, it affects base pairing, secondary structure and the ability to interact with proteins by regulating RNA splicing, localization and stabilization. RNA modification has opened a new chapter in cardiovascular disease research and has become an important emerging regulator of cardiovascular disease.

2.4.1 RNA m^6A Methylation

RNA m^6A methylation refers to the addition of A methyl group at the sixth nitrogenous base position of adenine residue in RNA, which can regulate the expression after transcription without sequence changing (Li et al. 2019). M^6A is by far the most plentiful RNA modification found in eukaryotic cells and can be detected in about 1/3 of mammalian mRNAs (Fu and Zhuang 2020). RNA m^6A exists in mRNA as well as in all categorizations of cellular RNA, including rRNA, tRNA and various non-coding RNAs, such as miRNA, lncRNA and circRNA. RNA m^6A has been widely researched for its important effect in regulating RNA splicing, translation, and stability (Di Timoteo et al. 2020).

There are key enzymes in the methylation process: methyltransferase, demethylase and m^6A binding protein. RNA m^6A is a dynamic and reversible modification that is catalyzed by RNA methyltransferase, removed by demethylase, and interacts with RNA m^6A-binding proteins, thereby affecting the splicing, translocation, stability, and translation of RNA, and thereby regulating various physiological processes of cells, especially in the development of CVD (Fig. 1). Methyltransferase-like 3 (METTL3)—Methyltransferase-like 14 (METTL14) complex is the primary enzyme that catalyzes the deposition of m^6A (Chen et al. 2019a, b). METTL3 catalyzes the conversion of adenosine to m^6A through its methyltransferase domain, and METTL14 is responsible for recognition of RNA substrates. Therefore, the RNA m^6A modification process requires the participation of METTL3-METTL14 complex. Wilms tumor 1-related protein (WTAP), as a cofactor of the above complex, has no methylation activity itself and can bind to the above complex to affect the efficiency of m^6A modification (Reichel et al. 2019). Unlike WTAP, methyltransferase-like 16 (METTL16) can exert the function of methyltransferase alone to affect RNA structure (Brown et al. 2016). Demethylases mainly include fat mass and obesity associated protein (FTO) and AlkB homolog 5 (ALKBH5) are primary demethylases which are part of the α-ketoglutarate-dependent dioxygenase family and can be catalyzed in a Fe^{2+}—and α-ketoglutarate-dependent manner (Wu et al. 2021). As the decrease of FTO and ALKBH5 expression, RNA m^6A modification in mRNA increased. The main function of RNA m^6A methylated reading protein is to specifically recognize and bind the m^6A site on RNA, and then participate in the modification and regulation of target RNA. So far, the discovered m^6A reading proteins include the YTHDF1/2/3, YTHDC1/2, HNRNPA2B1, HNRNPC and IGF2BP1/2/3 (Huang et al. 2018). YTHDF1, YTHDF2, and YTHDF3, regulate RNA metabolism, including RNA biosynthesis, translation, degradation, and cytosolic export, by recognizing the m^6A modification. YTHDF1 has been reported to promote m^6A modified-mRNA translation by increasing ribosomal assembly and interacting with initiation factors. YTHDF2 binds to m^6A-modified-mRNA and induces its degradation, while YTHDF3 regulates m^6A-modified RNA translation and other biogenesis (Chen et al. 2019a, b).

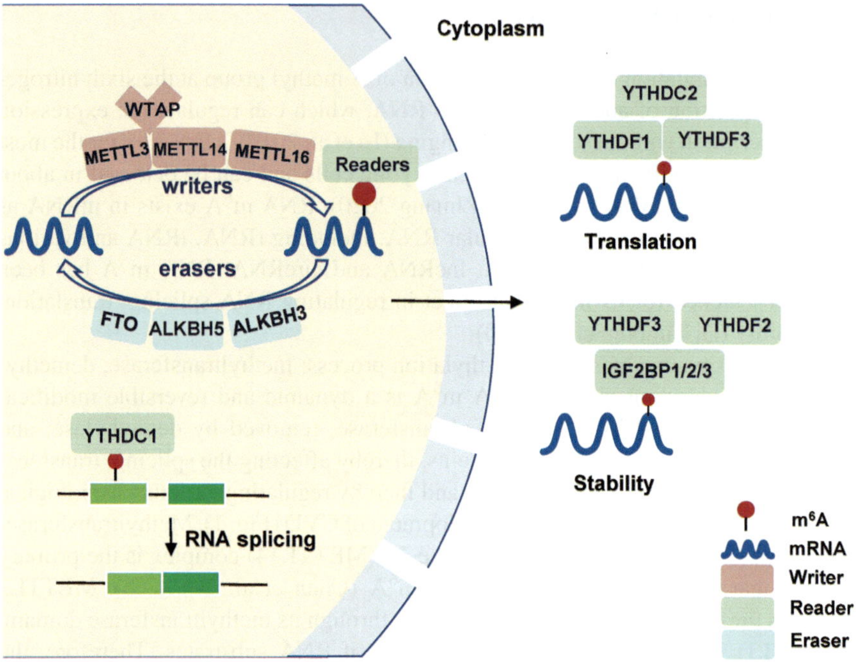

Fig. 1 The process and molecular function of m^{6}A RNA methylation. RNA m^{6}A methylation is dynamically regulated by "writers" (METTL3, METTL14, WTAP, and others) and "erasers" (FTO and ALKBH5). "readers" (YTHDF1, YTHDF2, YTHFDF3, YTHDC1, and YTHDC2) are binding proteins of RNA m^{6}A. Different regulators of m^{6}A have different functions, including RNA splicing, export, translation, and stability

2.4.2 RNA Editing

RNA editing is the processing modification of genetic information after transcription, including base replacement, insertion, deletion and so on. It provides a variety of information to RNA and increases the complexity of RNA regulation. So far, many different types of RNA editing phenomena have been discovered, including C-to-U, U-to-C, A-to-I, etc. Among them, the vital and common form of editing in mammals is A-to-I editing, which is a post-transcriptional modification process through adenosine deaminases acting on RNA (ADARs) mediating adenine to hypoxanthine. It's not only contributing to genome and proteome diversity, but also offering additional mechanisms of gene regulation (Xu et al. 2022). Although the biological meaning of RNA editing remains largely unclear, interestingly, researchers have found that stable expression of ADAR1 and ADAR2 as RNA editing enzymes is important for maintaining normal myocardial homeostasis. In fetal mammalian, ADAR1 knockdown can lead to myocardial apoptosis and even cardiac dysfunction. Other studies have shown that stable overexpression of ADAR2 attenuated cardiac damage caused by myocardial infarction and doxorubicin (DOX) (El Azzouzi et al. 2020; Wu et al.

2022). In addition, RNA editing also affects protein coding, alternative splicing, miRNA regulation and other biological processes, and acts an essential role in CVD (Landmesser et al. 2020).

3 Functions of lncRNA, miRNA, circRNA, and RNA Modification in Pathological Processes of CVDs

High-throughput sequencing revealed the abundance, evolution, conservation, and diversity of non-coding RNAs expressed in the heart according to developmental stage. There are differences in the expression levels of non-coding RNAs in healthy and injured hearts, which determines that non-coding RNA combined with myocardial injury marker detection can improve the accuracy of CVD diagnosis (Jusic and Devaux 2020). Non-coding RNA participated in the occurrence of CVD by mediating apoptosis, autophagy, oxidative stress, and other pathways, having a crucial effect on the pathophysiological process of CVD. LncRNA interacts with mRNA and miRNA-mRNA-lncRNA, forming a complex gene regulatory network with multiple pathways and layers (Fig. 2).

3.1 Cardiac Hypertrophy (CH) and Heart Failure (HF)

Pathological hypertrophy induced by stimuli such as chronic hypertension, mechanical pressure overload, or neurohumoral increase leads to pathological remodeling, often accompanied by abnormal expand in cardiomyocyte size, metabolic disorders, and subsequent ventricular dysfunction and heart failure (Zhang et al. 2020a, b, c). Currently, among CH and HF therapies, β-blockers, calcium antagonists, and angiotensin-converting enzyme inhibitors are the most common drug in clinical practice, which could only be used as symptomatic treatment, and patients still face a poor prognosis, which will develop into heart failure at the end stage, Therefore, new treatments are needed to delay or even reverse pathological cardiac hypertrophy. Ventricular remodeling, fibrosis and myocardial damage are the main pathologic basis of heart failure (Heidenreich et al. 2022).

One of the ways to delay the development of heart failure is to block myocardial remodeling. Non-coding RNAs are recognized as a suitable biomarker candidate in cardiac hypertrophy. Accumulating studies have reported that lncRNA is involved in this pathological process. The expression of lncRNA-H19 in the heart is relatively high and its regulation is dynamic. After birth, the expression level of H19 decreased gradually with age, but increased in the compensatory period of myocardial hypertrophy and decreased significantly in the decompensated period (Su et al.

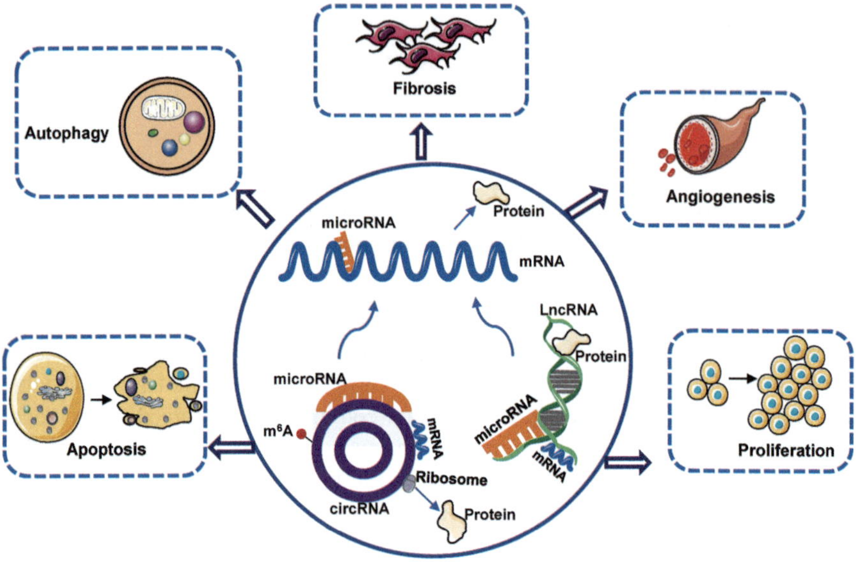

Fig. 2 The lncRNA/circRNA-miRNA-mRNA reciprocal regulatory network in cardiovascular system. miRNAs, lncRNAs and circRNAs in this regulatory network are inseparable and together play important roles in cardiovascular system. LncRNAs and circRNAs can act as competitive endogenous RNAs (ceRNAs) to bind miRNAs, and therefore, regulate the expression of miRNAs' target mRNAs. CircRNAs interaction with RBP to regulate its function and translocation. RNA m^6A modification can also regulate circRNA metabolism. In addition, lncRNAs can directly bind to mRNA or protein, thereby affecting the downstream gene expression and regulation

2021). As a negative regulator of CH, H19 interacted with PRC2 to attenuate pressure overload-induced cardiac hypertrophy by reducing the activity of the prohypertrophic nuclear factor of activated T cell signaling pathway (Viereck et al. 2020). In a similar study, cardiac hypertrophy-associated transcript (Chast) was found to be significantly increased in mice undergoing transverse aortic coarctation (TAC) surgery compared with control mice, restraining the expression of protein family M member 1 (Plekhm1) in the homeodomain of autophagy regulator Pleckstrin. Importantly, the circulating expression of Chast was significantly elevated in patients with early AMI cardiac systolic dysfunction and Chast inhibitor was found no toxicology side effects in mice and can be a therapeutic candidate for cardiac hypertrophy (Wang et al. 2020a, b, c).

Compared with mRNA, circulating miRNA is more stable, which arouses researchers' interest in miRNA as a biomarker. miR-208a could be released into plasma in response to myocardial stress, thus high expression of miR-208a could be detected in plasma after cardiac stress response. Plasma miR-208a and circulating cardiac troponin I, a generally acknowledged biomarker of cardiac injury, were simultaneously increased in isoproterenol-treated mouse and human patients with myocardial injury (Wang et al. 2010). In addition, some miRNA (such as miR-30c, miR-221) could combinate B-natriuretic peptide to distinguish different types of heart failure and improve the accuracy of diagnosis (Watson et al. 2015). Other miRNAs such as miR-27b-3p, miR-30d, miR-92b-3p, miR-142-3p, miR-199a are of great significance in the pathological process of cardiac hypertrophy (Table 1) (Hu et al. 2017; Huang et al. 2021; Lerchenmüller et al. 2020; Li et al. 2017, 2022; Seok et al. 2020). It is worth mentioning that the regulatory network of miRNA changes after the occurrence of CH, any single miRNA cannot fully explain the pathological process of CH. Most studies have only observed the function of miRNA with a regulatory role in cardiomyocytes or in fibroblasts, and whether it mediates in the crosstalk between cardiomyocytes and fibroblasts deserves further study.

Nowadays, circRNAs with potential roles in CVD have been identified by high-throughput sequencing methods such as RNA-seq and microarrays, which have been rapidly developed. circRNA can act as a miRNA sponge to regulate the expression of target genes. CircRNA_000203 was significantly up-regulated in the myocardium of HF patients and AngII-exposed cardiomyocytes. Bioinformatics showed that circRNA_000203 had potential binding sites of miR-26b-5p and miR-140-3p which can inhibit the increase of cell area and the expression of ANP and β-MHC in cells treated by AngII. CircRNA_000203 could eliminate the inhibition of GATA4 via sponge miR, this eventually led to the increase of GATA4 in cardiomyocyte and exacerbates pathological cardiac hypertrophy (Li et al. 2020a, b). Moreover, circRNA has the function of translating proteins. Expression of circNlgn was up-regulated during pathological remodeling, but its linear mRNA did not change. Because the post-splicing of circRNA changed the structure of the transcript, the translation product Nlgn173 was a novel protein isoform containing nuclear localization motif, allowing Nlgn173 to enter the nucleus and bind to downstream promoters, inhibited cell viability and increased fibrosis rates, ultimately leading to heart remodeling (Du et al. 2021).

The cardiac RNA m^6A transcriptome differs between healthy and failing cardiac tissues, and is dynamically regulated by modified "effectors", including "eraser", "writer" and "reader". RNA m^6A levels in cardiomyocytes are significantly up-regulated in response to pressure overload and stimulation with angiotensin. The expression of FTO and m^6A was negatively correlated, demethylation activity of FTO may ameliorate cardiac systolic dysfunction which was caused by abnormal increase of m^6A. Studies had shown that the reduction of FTO is one of the primary reasons for the increase of m^6A in failing hearts in humans and mice (Han et al. 2021). On the contrary, FTO overexpression in the heart of failing mice modulated the level of m^6A modification in the heart, enhanced the Ca^{2+} amplitude, prevented the degradation of sarcoplasmic reticulum calcium pump, and thus attenuated ischemia-induced myocardial contractile dysfunction (Mathiyalagan et al. 2019). In addition, METTL3 was a subunit with methyltransferase activity, which could promote the expression of mitogen-activated protein kinases, and regulate hypertrophy-related pathways, suggesting that "writers" also contribute to increased levels of m^6A in myocardial hypertrophy (Dorn et al. 2019). In other studies, as a "reader", YTHDF2's RNA and protein levels increased significantly during the development of HF, and it interacted with Myh7 mRNA, an important major marker of cardiac hypertrophy, and induced its expression in an RNA m^6A-dependent manner. Conversely, the development of

Table 1 The Role of ncRNA in the pathogenesis of CVD

ncRNAs	CVD	Model	Change	Target genes	Pathogenic mechanisms	Spices	References
LncRNA H19	CH	TAC	Down	NFAT	H19 reduced heart weight to tibia Length ratios and smaller cardiomyocyte sizes through suppressing NFAT	Pig/ human	Viereck et al. (2020)
LncRNA Chast	CH	TAC	Up	Plekhm1	Chast promotes hypertrophy by disrupting beneficial autophagic processes via down-regulation of Plekhm1, a regulator of autophagy located on the opposite strand of Chast	Mice/ human	Viereck et al. (2016)
miR-27b-3p	CH	TAC	Up	FGF	miR-27b-3p targeted FGF1 enhanced mitochondrial oxidative phosphorylation and attenuated CH	Mice	Li et al. (2022)
micro-208a	CH	Tg mice	Up	Thrap1 and myostatin 2	miR-208a affects contractility of the cardiac sarcomeres and fibrosis	Mice	Huang et al. (2021)

(continued)

Table 1 (continued)

ncRNAs	CVD	Model	Change	Target genes	Pathogenic mechanisms	Spices	References
miR-30d	CH	TAC	Down	CITED4/mTOR	miR-30d and CILTED depletion aggravated myocardial fibrosis	Mice	Lerchenmuller et al.(2020)
micro-92b-3p	CH	AngII infusion	Down	MEF2D	miR-92b-3p negatively regulated MEF2D expression	Mice	Hu et al. (2017)
miR-199a	CH	Tg mice	Up	GSK3β/mTOR	miR-199a inhibits cardiac autophagy and activates mTOR signaling	Mice	Li et al. (2017)
circRNA-000203	CH	AngII infusion	Up	miR-26b-5p, -140-3p/GATA	circRNA_000203 promote myocardium with pro-fibrotic effect by sponging microRNA-26b-5p	Mice	Li et al. (2020a, b)
lncRNA H19	MI	LAD-ligation	Down	miR-675	LncRNA H19 promoted angiogenesis and stimulated anti-apoptosis ability of ECs	Rats	Huang et al. (2020)
LncRNA LUCAT1	MI	H2O2	Down	miR- 181a-5p	LUCAT1 played a protective role on oxidative stress injury,inflammation and apoptosis of cardiomyocytes	H9c2/ human	Xiao et al. (2021)

(continued)

Table 1 (continued)

ncRNAs	CVD	Model	Change	Target genes	Pathogenic mechanisms	Spices	References
lncRNA MALAT1	IR	LAD-ligation	Up	miR-26b/PTGS2	MALAT1/miR26b/PTGS2 axis is probably an important mechanism for regulating inflammation in MI/R injury	Mice	Ruan et al. (2019)
Lnc NORAD	MI	LAD-ligation	Up	miR-577/COBLL1	Knockdown of NORAD alleviated cell injury by reducing apoptosis and decreasing expression levels of fibrogenic factors	Rats	Xiong et al. (2021)
lncRNA CAREL	MI	LAD-ligation	Up	miR-296/Trp53inp1, Itm2a	CAREL inhibited neonatal cardiomyocyte replication and cardiac regeneration after injury	Mice	Cai et al. (2018)
miR-19a/19b	MI	LAD-ligation	Down	PTEN	miR-19a/19b enhanced cardiomyocyte Proliferation and reduced MI-induced cell death	Mice	Gao et al. (2019)

(continued)

Table 1 (continued)

ncRNAs	CVD	Model	Change	Target genes	Pathogenic mechanisms	Spices	References
Lnc MIAT	MI	LAD-ligation	Up	miR-203/CF6 axis	Overexpression of Lnc MIAT and inhibited miR-203 promote the percentage of infarct area and the apoptosis of cardiomyocytes	Mice	Wang et al. (2021a, b, c, d)
CircUbe3a	MI	LAD-ligation	Up	miR-138-5p/RhoC	CircUbe3a promote the proliferation, migration, and phenotypic transformation of CFs	Mice	Wang et al. (2021a, b, c, d)
CircFndc3b	MI	LAD-ligation	Down	FUS/VEGF-A	Overexpression of circFndc3b increase angiogenic activity and reduces cardiomyocytes	Mice	Garikipati et al. (2019)
Circ-ZNF609	I/RI	LAD-ligate and loose	Up	Hippo-YAP/AKT	circ-ZNF609 can directly regulate the expression of YTHDF3 and promote the translation of YAP protein	Mice	Wang et al. (2022a, b)
Lnc Kcnq1ot1	DCM	Injection STZ	Up	miR-214-3p/caspase-1/TGF-β1	Silencing Kcnq1ot1 improved cardiac function and fibrosis, ameliorated pyroptosis	Mice	Yang et al. (2018)

(continued)

Table 1 (continued)

ncRNAs	CVD	Model	Change	Target genes	Pathogenic mechanisms	Spices	References
lncRNA MALAT1	DCM	Injection STZ	Up	miR-141	lncR-MALAT1/ miR-141 mediated inflammasome activation and fibrosis action	Mice	Che et al. (2020)
lncRNA DCRF	DCM	Injection STZ	Up	551b-5p/PCDH17	DCRF knockdown reduced cardiomyocyte autophagy, attenuated myocardial fibrosis	rats	Feng et al. (2019)
miR-30c	DCM	db/db	Down	PGC-1β	Reduce excessive ROS production and subsequently attenuated cardiomyocyte apoptosis	mice	Yin et al. (2019)
miR-21	DCM	Injection STZ	Up	SPRY1/ERK/mTOR	Overexpression of miR-21 aggravated fibrosis, reduced autophagy	Rats	Li et al. (2021a, b, c)
circRNA-000203	DCM	db/db	Up	miR-26b-5p/Col1a2, CTGF	circRNA-000203 contributing to the enhancement of the fibrotic phenotype	Mice	Tang et al. (2017)

(continued)

Table 1 (continued)

ncRNAs	CVD	Model	Change	Target genes	Pathogenic mechanisms	Spices	References
circHIPK3	DCM	Injection STZ	Up	miR-29b-3p-Col1a1-Col3a1	Overexpression of circHIPK3 aggravated fibrosis and progression of DCM	Mice	Wang et al. (2021a, b, c, d)
lncRNA MIAT	AS	ApoE −/−	Up	miR-149-5p/CD47	MIAT knockdown reduced necrotic core size, and promoted clearance of apoptotic cells	Mice	Ye et al. (2019)
lncRNA MALAT	AS	Injection STZ	Up	SAA3	MALAT1 regulates glucose-induced inflammatory changes, oxidative stress and influence endothelial stability	Mice	Puthanveetil et al. (2015)
Lnc-H19	AS	/	Up	MAPK/NF-kB	Overexpression of lncRNA H19 in HUVEC the proliferation ability was increased while apoptosis was suppressed	Human	Pan (2017)
miR-103	AS	ApoE −/− fed a HFD	Up	PTEN/MAPK	miR-103-induced inflammation and ERS	Mice	Jiang et al. (2020)

(continued)

Table 1 (continued)

ncRNAs	CVD	Model	Change	Target genes	Pathogenic mechanisms	Spices	References
miR-10a	AS	ApoE −/− fed a HFD	Down	GATA6/VCAM-1	miR-10 inhibiti inflammatory cell infiltration	Rats	Lee et al. (2018)
CircANRIL	AS	AS	Up	PES1	CircANRIL modulating apoptosis and cell proliferation	Human	Holdt et al. (2016)
Circ-000595	AS	AS	Up	miR-19a	Circ-000595 promote apoptosis	Human	Zheng et al. (2015)
LncRNA-AK098656	Hypertension	TgAK098656	Up	MYH11	AK098656 promotes cell Migration and increased proliferation in HASMCs	Human/ rats	Jin et al. (2018)
MALAT1	arrhythmia	LAD-ligation	Up	miR-200c/HMGB1	MALAT1 mediates cardiac Ito	Rats	Zhu et al. (2018)
miR-29	Hypertension	SHRs	Up	CTRP6/ERK/PPARγ	Inhibition of miR-29b improve proliferation and migration and suppression on production of ROS and LDH	Rats	Sun et al. (2019)

cardiac hypertrophy further enhanced the interaction between YTHDF2 and m^6A-methylated Myh7 mRNA (Xu et al. 2021). However, whether there are competing sites among "eraser", "writer" and "reader", or whether there are interactions among them to dynamically regulate m^6A level is worthy of further study.

3.2 *Myocardial Infarction (MI) and Ischemia–Reperfusion Injury (I/RI)*

MI is a kind of worsening cardiovascular disease due to its clinical symptom, with urgent onset, speedy development, poor prognosis, and high lethal (Gulati et al. 2020). The establishment of chest pain centers and the development of thrombolytic therapy and stenting have led to an observable reduction in the mortality rate of MI patients. Although modern reperfusion therapy can repair myocardial injury and reduce mortality in patients with acute myocardial infarction (AMI) to some extent, the mechanism is not clear. In fact, oxidative stress, inflammation, and cardiomyocyte apoptosis are considered as the potential mechanisms of MI and I/RI, because they are closely related to patient prognosis. Unfortunately, even with all the attention, there are currently no viable targeted therapies. It is urgent to find effective targets to prevent myocardial injury induced by MI and I/RI.

LncRNA H19 played a protective role not only in cardiac hypertrophy but also in MI. LncRNA H19 is involved in the regulatory role exertion of mesenchymal stem cells-derived exosomes on angiogenesis and endothelial survival by modulating pro-angiogenic factors (Huang et al. 2020). In addition, compared with healthy people, AMI patients lack lncRNA LUCAT1 expression, however, studies have found that lncRNA LUCAT1 plays a role as miRNA sponge and does not regulate protein expression through cis-regulatory mechanism. To further investigate the role of LUCAT1 in disease, proteins interacting with LUCAT1 were found by RNA pull-down approach. JMJD6 was screened out in several binding proteins, which were also located in the nucleus as LUCAT1. LUCAT1 bound to JMJD6 and directly recruited it to the FOXQ1 promoter and induced FOXQ1 demethylation, thereby downregulating Bax/Bcl-2 expression to alleviate apoptosis (Tao et al. 2022). In addition, lncRNAs also acted as miRNA sponges. The level of miR-181a-5p could be reduced by overexpressing LUCAT1, while down-regulation of LUCAT1 could up-regulate the expression of miR-181a-5p which was expected to become a potential way to against injury of AMI (Xiao et al. 2021). Another lncRNA MALAT1 induced inflammation by targeting miR-26b to regulate PTGS2 in myocardial I/R injury (Ruan et al. 2019). Similarly, as a risk factor for AMI, other lncRNAs such as NORAD played an important role in the lncRNA-miRNA-mRNA regulatory network in AMI (Table 1) (Cai et al. 2018; Xiong et al. 2021). However, in order to better elucidate the mechanism of NORAD, it is necessary to further explore its functional motifs and the proteins that interact with them.

MiRNAs are extensively distributed in biological tissues and fluids, which can be easily detected. MiR-21 and miR-30d-5p were found to be increased in plasma from patients after MI events (Yang et al. 2018). Similarly, miR-208a was detected in most of patients with AMI, but was not present in the plasma of healthy volunteers, which was remarkably linked with increased risk of cTnI, CK-MB and HF after AMI (Wang et al. 2010). In addition to being markers of disease, miRNAs can also be used as targets for disease treatment. The expression of miR-208a was increased in cardiomyocytes after MI, and it entered the peripheral blood circulation through exosomes and was eventually taken up by lung cells. Cardiomyocytes derived miR-208a in the lung could activate the NF-kB signaling pathway and up-regulated the expression of endothelial adhesion molecules. Pre-treatment of mice with miR-208a antagomir attenuated MI-induced target gene down-regulation of miR-208a in lungs. Moreover, Intramuscular injection of hsa-miR-199a-3p and has-miR-590-3p can promote cardiomyocyte proliferation and improve myocardial function. Intramuscular injection of miR-19a/19b mimics can also stimulate myocardial regeneration and further reduce MI-induced cardiac injury (Gao et al. 2019; Lesizza et al. 2017). Similarly, the area of infarct size and apoptosis of myocardial cells decreased by overexpressing miR-203 in the MI model. Consistently, overexpression of miR-203 in mice also protected against diabetic cardiomyopathy, which is related to the fact that miR-203 inhibits the pathological process of myocardial fibrosis and oxidative stress (Wang et al. 2021a, b, c, d; Yang et al. 2019a, b, c). There is no doubt that miRNA is a key regulator of MI combine these observations to identify several miRNAs as potential candidates could provide valuable information for clinical diagnosis and treatment. Although we have a profound understanding of the role of miRNA in MI, further studies are needed to verify its application as biomarkers in clinical diagnosis and treatment.

In addition to lncRNAs and miRNAs, circRNAs are also widely participated in the development of CVD. As mentioned above, exon-derived circRNAs are localized in the cytoplasm, circUbe3a was derived from exons of the *Ube3a* gene and was proved localized in the cytoplasm by fluorescent in situ hybridization. Pulldown assay and bioinformatic analysis further demonstrated that it was complementary to miR-138-5p. CircUbe3a targeted miR-138-5P/RhoC axis to regulate the proliferation, migration and myofibroblast transformation, leading to ventricular remodeling after MI (Wang et al. 2021a, b, c, d). Like other circRNAs, CircNfix also worked through a sponge mechanism. CircNfix was highly enriched in cardiomyocytes and acted as upstream of miR-214. Expression of CircNfix was elevated during heart development but significantly decreased after myocardial infarction. Mechanistically, CircNfix up-regulated the expression of GSK3β by targeting miR-214 and inhibiting β-catenin activity, leading to cardiomyocyte apoptosis and poor prognosis of MI (Huang et al. 2019). In contrast, circFndc3b was much lower in the MI of mouse, and the human circFndc3b homolog was also significantly down-regulated in cardiac tissue of patients with ischemic cardiomyopathy. CircFndc3b overexpression enhanced VEGF expression to promote angiogenesis and reduced left ventricular dysfunction after MI (Garikipati et al. 2019). In addition, myocardial infarct-associated circRNA MICRA was down-regulated in peripheral blood of patients with MI, resulting in left ventricular dysfunction (Salgado-Somoza et al. 2017).

The occurrence of cardiac injury is often closely related to RNA m^6A modification. m^6A-modified circ-ZNF609 (also named as MICRA) directly regulated YTHDF3 expression, which competitively regulated the binding of *Yap* mRNA to YTHDF1 and YTHDF2, thereby promoting the translation of YAP protein, inhibiting apoptosis, and improving cardiac function (Wang et al. 2022a, b). FTO and ALKBH5, as m^6A eraser, were significantly decreased in human and mouse MI, leading to the upregulation of m^6A level. The restoration of FTO level can obviously lessen the fibrosis and scar area in mouse with MI (Mathiyalagan et al. 2019). Overexpression of ALKBH5 also could significantly increase the proliferation of cardiomyocytes after MI, reduce the infarct area, and restore ventricle function. In terms of mechanism, ALKBH5-mediated RNA m^6A demethylation increases the level of YTHDF1 by improving its stability (Han et al. 2021). MI is also often accompanied by myocardial fibrosis, in which the fibrous heart tissue becomes stiffer and less compliant due to a lack of capacity for impulse and contraction, which can lead to decreased heart function. It has been shown that METTL3 is significantly up-regulated in cardiac tissues after TGFβ-treated cardiac fibroblasts and MI, leading to collagen deposition. In contrast, inhibition of METTL3 expression attenuated myocardial fibrosis and improves cardiac function (Li et al. 2021a, b, c). mRNA m^6A methylation was up-regulated in cardiomyocytes caused by hypoxia/reoxygenation and mice with myocardial ischemia. Silencing METTL3 attenuated I/R damage by enhancing autophagy flux and inhibiting apoptosis (Song et al. 2019). As a structural partner for m^6A substrate recognition, METTL14 activated METTL3 and mediated m^6A modification, which was important to cardiac homeostasis. Downregulation of METTL14 could inhibit *Phlpp2* mRNA levels, leading to dephosphorylation of Akt-s473, thus attenuating cardiac dysfunction caused by I/R injury (Wang et al. 2022a, b). Therefore, regulating m^6A is a promising therapeutic method for CVD, exploring the regulation complex mechanism of m^6A in CVD will help us understand the pathogenesis of CVD.

The RNA editing enzyme ADAR1 is essential for the survival and proliferation of embryonic cardiomyocytes. After ADAR1 knockdown, the apoptosis rate of cardiomyocytes was significantly increased, the proliferation of cardiomyocytes was inhibited, and finally, the embryo died (Moore et al. 2020). Studies have shown that ADAR2, another ubiquitously-expressed member of the ADAR family, is up-regulated in the heart during exercise training, and cardiac-specific overexpression of ADAR2 could attenuate cardiotoxicity caused by DOX. In addition, overexpression of ADAR2 inhibited DOX-induced cardiomyocyte apoptosis and promoted cardiomyocyte proliferation in neonatal rat cardiomyocytes. ADAR2 mediated pri-miR-34a editing, and overexpression miR-34a reversed the pro-proliferation and anti-apoptotic effects of ADAR2 on cardiomyocytes. Interestingly, ADAR2-N without deaminase activity had little ability to regulate miR-34 (Wu et al. 2022).

3.3 Diabetic Cardiomyopathy (DCM)

Diabetic cardiomyopathy (DCM) is often accompanied by cardiac structural disorders and cardiac dysfunction, resulting in imbalance of cardiac homeostasis (Dillmann 2019). Recently, more and more researchers have found that oxidative stress, inflammation, autophagy, activation of renin-angiotensin system, and abnormal myocardial metabolism are closely related to the pathogenesis of DCM.

Inflammation is a foremost risk element for the development of DCM, and inflammatory factors are usually elevated in DCM patients. The level of lncRNA MALAT1 was increased in diabetes mellitus mice heart and high glucose-treated cardiac fibroblasts. Classical inflammatory pathway NF-κB bound to the promoter region of lncRNA-MALAT1 and regulated its' expression. Overexpression of MALAT1 activated TGF-β1/Smads signaling pathway to promote cardiac fibrosis through interacting miR-141 (Che et al. 2020). LncRNA DCM associated factor (DCRF) could increase in autophagy of cardiomyocytes via sponging miR-551b-5p (Feng et al. 2019). In addition, silencing lncRNA Kcnq1ot1 alleviated DCM by inhibiting TGF-β1/SMAD pathway, which was a major mechanism of fibrosis. Mechanically, Kcnq1ot1 aggravated high glucose-induced myocardial inflammation and fibrosis via binding to endogenous miR-214-3p (Yang et al. 2018). Sequencing found that there were more than 100 differentially expressed miRNAs in vascular smooth muscle cells of db/db mice. Among them, significantly up-regulated miR-21 aggravated fibrosis, reduced autophagy, and impaired cardiac function (Li et al. 2021a, b, c). Overexpression of miR-30c increased glucose utilization, reduced excess reactive oxygen species production, cardiomyocyte apoptosis, and subsequently attenuated cardiac damage in db/db mice, which was related to the reduction of PPARα activity by miR-30c targeting PGC-1β (Yin et al. 2019).

Recently, emerging studies focused on the role of circRNA in DCM have been reported. MiR-26b-5p had an anti-fibrosis effect in cardiac fibroblasts, while circRNA_000203 could attenuate its effect and enhance the expression of fibrosis-related genes such as Col1a2, Col3a1 and α-SMA by sponging miR-26b-5p, and promote the expression of CTGF and other downstream targets. Therefore, circRNA000203 was likely to be a target against myocardial fibrosis in DCM (Tang et al. 2017). The difference was that circHIPK3 was involved in the pathological process of myocardial fibrosis. Overexpression of circHIPK3 up-regulated Col1a1 and Col3a1, and circHIPK3-miR-29b-3p-Col1a1 axis had a potential regulatory role in the pathogenesis of DCM (Wang et al. 2021a, b, c, d). CACR is differentially expressed in the serum of diabetic patients and healthy people, and the expression of CACR is higher under the condition of high glucose. Bioinformatics analysis showed that it had potential binding sites with miR-214-3p. Silencing CACR alleviated caspase-1 via miR-214-3p, and cell apoptosis was slightly inhibited. Luciferase detection and functional rescue assay further clarified ceRNA mechanism of CACR (Yang et al. 2019a, b, c). However, sequence analysis showed that there was no homologous CACR in mice, this study was not tested on animals and had certain limitations. Nonetheless, all the above studies provide insight into DCM and further studies are needed to better understand the regulatory role of circRNA.

In recent years, epigenetics has been reported to be associated with metabolic disorders, and glucose metabolism disorders are one of the key features of diabetes. MeRIP-seq results showed that the overall m^6A level in DCM was increased, while FTO was obviously down-regulated in type 2 diabetes. This result was also confirmed in another report, which may be related to the involvement of mRNA m^6A methylation participated in glucose metabolism via hepatic gluconeogenesis, and m^6A demethylase FTO regulated glucose metabolism through enhancing FOXO1 expression in an m^6A-dependent manner (Peng et al. 2019; Yang et al. 2019a, b, c). Therefore, overexpression of FTO could reduce myocardial fibrosis by selectively demethylating cardiac contractile transcripts. In addition, specific knockdown of METTL14 in mouse β-cells contributed to reduce B-cell proliferation and islet dysfunction, leading to worsening of diabetes (Wang et al. 2020a, b, c). These results suggested that RNA m^6A is a regulator of the development and progression of diabetes.

3.4 Atherosclerosis (AS)

Atherosclerosis is an arterial disease which is characterized by lipid metabolism disorder and inflammation (Libby 2021). lncRNA myocardial infarct-related transcript (MIAT) contents in serum of patients with coronary atherosclerosis were higher than those in healthy subjects, and were positively correlated with serum levels of proinflammatory cytokines IL6 and TNFα (Tan et al. 2019). To test the hypothesis that MIAT had a vital effect on atherosclerosis regulation, knockdown MIAT by a systematic delivery of MIAT shRNA adenoviral vectors in $ApoE^{-/-}$ mice reduced aortic atherosclerosis. Furthermore, it was found that MIAT acts as a sponge for miR-149 to prevent targeting to CD47 and prevent plaque progression (Ye et al. 2019). RNA sequence in human umbilical vein endothelial cells demonstrated that many lncRNAs, such as MEG3, TUG1, LINC00657 and MALAT1, were significantly up-regulated under high glucose environment or after oxidative stress. MALAT1 triggered an inflammatory cascade, and induced amyloid antigen expression in serum (Puthanveetil et al. 2015). Similarly, lncRNA H19 was highly expressed in the serum of patients with atherosclerosis. H19 could promote cell proliferation, reduce apoptosis of VSMCs, reduce vascular endothelial injury, and improve the formation of atherosclerosis through MAPK and NF-κB signaling pathways (Pan 2017). These results suggested that lncRNAs are key regulators of inflammatory response in atherosclerosis, and different lncRNAs regulate the expression of inflammation-related genes in different cell types.

In addition to lncRNAs, other ncRNAs are also implicated in the occurrence and development of atherosclerosis. miR-103 expression was significantly up-regulated in human aortic endothelial cells dealing with ox-LDL. miR-103 directly blocked PTEN expression and regulated MAPK signaling pathway, which has been shown to be participated in the inflammatory process and ER stress in atherosclerotic plaques (Jiang et al. 2020). Currently, the mainstream treatment for atherosclerosis was percutaneous coronary intervention or coronary artery bypass grafting. However, both methods have the risk of vascular injury, and a large proportion of patients need secondary intervention. The emergence of miRNA provides potential therapeutic targets for atherosclerosis. Researches have suggested that miR-10a has a protective role on atherosclerosis, and miR-10a could inhibit the inflammatory response in endothelial cells, which is related to the inhibition of GATA6/VCAM1 pathway (Lee et al. 2018). In addition, miR-222 also stabilized plaques by targeting p27 to inhibit the proliferation of vascular smooth muscle (Ding et al. 2017). MiRNAs were also regarded as biomarkers for disease occurrence. The expressions of miR-29, miR-100, miR-155, miR-199, miR-221, miR-363, miR-497 and miR-508 were up-regulated in early atherosclerotic lesions, on the contrary, the levels of miR-490 and miR-1273 were down-regulated (Wang et al. 2015). During plaque rupture, the level of miR-221 was down-regulated, while circRNA-284 was significantly increased, thus elevated circR-284/miR-221 ratio can be used as a biomarker for the carotid plaque rupture (Bazan et al. 2017). These data show that miRNAs play crucial roles in the pathological process of atherosclerosis.

Balanced expression of linear and circRNA transcripts is also important in atherosclerosis. It was found that the expression of lncANRIL and circANRIL changes in different directions in atherosclerosis. CircANRIL could disrupt ribosome biogenesis, induce P53 activation, and subsequently regulate cell apoptosis progression and proliferation. CircANRIL may be a potential therapeutic target for the treatment of atherosclerosis (Holdt et al. 2016). Furthermore, hypoxia is another pathogenesis of atherosclerotic. Circ-000595 is up-regulated under cobalt chloride-induced hypoxia in vitro. Knockdown of circ-000595 could reduce the apoptosis rate of aortic smooth muscle cells induced by hypoxia. In addition, knockdown of circ-000595 was associated with miR-19a expression elevation, which protected against atherosclerosis by regulating endothelial cell proliferation (Zheng et al. 2015). Therefore, it will be of great significance to further explore the impact of circRNA in atherogenesis and other disease progression.

In addition to the damaged endothelium being a pathological feature of atherosclerosis, it has been increasingly recognized that m^6A is involved in inflammatory and immune responses that have a major influence on the pathological development of atherosclerosis. METTL14 was significantly increased in inflammatory endothelial cells via enhancing FOXO1 translation by regulating Foxo1 mRNA m^6A modification. In this way, the expression of ICAM-1, VCAM-1 and E-selectin can be increased, which mediates endothelium-monocyte adhesion and participates in the development of atherosclerosis (Jian et al. 2020). In addition, METTL14 mediated the processing of primary miR-19a to mature miR-19a by promoting m^6A modification, thus promoting the migration and invasion of endothelial cells (Zhang et al. 2020a, b, c). Increased METTL3 expression and m^6A hypermethylation in oscillatory stress induced atherosclerosis. Inhibition of METTL3 can down-regulate NLRP1 and up-regulate KLF4, thereby inhibiting endothelial inflammation (Chien et al.

2021). Collectively, preventing abnormal RNA m^{6}A modification is expected to be an attractive treatment strategy for CVD.

3.5 Other Cardiovascular Diseases

Non-coding RNAs are participated in the occurrence and development of other cardiovascular related diseases, such as arrhythmias and hypertension. Hypertension is involved in genetic and environmental factors with high morbidity and mortality worldwide (Poller et al. 2018). The pathogenesis of hypertension has been extensively studied, but there are few studies about the impacts of ncRNA on hypertension. AngII-induced fibrosis was inhibited by miR-29b overexpression, suggesting a protective function for miR-29b in AngII-induced myocardial remodeling (Zhang et al. 2014). In contrast to these findings, miR-29 family was significantly up-regulated in aortic samples from both hypertensive patients and rats, and was positively correlated with blood pressure (Sun et al. 2019). Such contradictory results may be due to the fact that the aortic system and the peripheral vascular system are in different complex pathological microenvironments and the target genes of miR-29 were affected by different factors. It was demonstrated that the level of m^{6}A-modified circXpo6 and circTmtc3 are both changed in pulmonary hypertension, and m^{6}A relates to circRNA-miRNA-mRNA network, which affects the progression of pulmonary hypertension by accelerating the degradation of contractile proteins, increasing the synthesis of vascular smooth muscle cells and promoting hypertension (Jin et al. 2018; Su et al. 2020). In addition to m^{6}A modified-circRNA, RNA m^{6}A methylation may also appear on other RNAs and contribute to the pathogenesis of hypertension, however, the interaction between protein and RNA m^{6}A, as well as the specific biological process in hypertension still needs to be further studied.

At present, the drug prevention and treatment of arrhythmia is not satisfactory, which is still a difficult problem in the cardiovascular field. It is urgent to explore new drugs for the treatment of arrhythmia. Atrial fibrillation is the most ordinary atrial arrhythmia. miR-29b and miR-21 in plasma were significantly down-regulated in patients with atrial fibrillation, which were pro-fibrotic biomarkers and could regulate the participation of fibrotic proteins, while progressive fibrosis could promote the occurrence of spontaneous arrhythmia (Regouski et al. 2019). The occurrence of arrhythmia is usually along with unstable expression of non-coding RNA. Compared with the control group, the level of MALAT1 in the heart of AMI rats was significantly increased. Knocking down MALAT1 could delay the reduction of the peak of transient outward K^+ current in AMI rats. At the same time, it also revealed that MALAT1 may have a potential mechanism of inducing arrhythmia in the background of AMI (Zhu et al. 2018). These studies revealed that non-coding RNAs are highly associated with cardiovascular risk factors and are one of the key regulatory molecules during CVDs.

4 Conclusion and Perspective

The effect of non-coding RNA has attracted much attention in the field of CVD. The identification of specific non-coding RNA is extremely significant for the diagnosis and treatment of diseases. We mainly reviewed the role of non-coding RNA in CVD and the underlying regulatory role and mechanisms of some well-studied non-coding RNAs in CVD. The ability of non-coding RNA to bind RBP and miRNA, as well as the lncRNA/circRNA-miRNA-mRNA regulatory networks are widely involved in the development of CVD. Moreover, some ncRNAs are highly expressed in the serum or urine of patients and can be easily detected in body fluids with strong specificity, which can be used as diagnostic markers or prognostic indicators. Researchers have obtained extensive genome-wide transcriptome data from diverse cardiovascular diseases by making good use of next-generation sequencing technologies to identify ncRNAs in the CVD process.

Although a series of CVD related non-coding RNAs have been found, the research is still infancy and the deeper role of non-coding RNAs in CVD is far from elucidated. Firstly, with the explosion of new technologies and the rapid accumulation of exogenous transcriptomic studies, ncRNAs have added an additional layer of regulation at the post-transcriptional level. There are often multiple target genes for non-coding RNA, conversely, a target gene can be regulated by multiple non-coding RNA molecules. The complex regulation may lead to the non-coding RNA intervention not reaching the desired effect. Clarifying the complex regulatory mechanisms and functional targets of non-coding RNA will provide more effective clinical treatment for cardiovascular diseases. Secondly, by studying the role of non-coding RNA in CVD, many delivery approaches about non-coding RNA have been developed, such as lipid nanoparticle, polymer, and virus-based systems. In addition, miRNAs mimics, miRNAs inhibitors and small molecule inhibitors of m^6A have also received widespread attention due to their advantages such as low immunogenicity and convenient dose control (Erdos et al. 2021). These breakthroughs will help verify the therapeutic effect of many ncRNAs and promote the efficient and safe application of non-coding RNA therapy in clinical.

Last but not least, the efficacy of most proposed RNA therapies has not yet been tested in clinical trials and many non-coding RNAs have not yet been fully examined. It is still a need to conduct in-depth research to explore the molecular mechanisms of candidate ncRNAs which are closely associated with the diagnosis and treatment of CVD. Further efforts are required to accelerate the transformation of non-coding RNA from bench to clinical application.

Acknowledgements This work was supported by the grants from National Natural Science Foundation of China (82020108002 and 82225005 to JJ Xiao, 82270291 to LJ Wang), the grant from Science and Technology Commission of Shanghai Municipality (23410750100, 21XD1421300 and 20DZ2255400 to JJ Xiao), the "Dawn" Program of Shanghai Education Commission (19SG34 to JJ Xiao), the Natural Science Foundation of Shanghai, China (23ZR1423000 to LJ Wang).

References

Bazan HA, Hatfield SA, Brug A et al (2017) Carotid plaque rupture is accompanied by an increase in the ratio of serum circR-284 to miR-221 levels. Circ Cardiovasc Genet 10(4)
Bridges MC, Daulagala AC, Kourtidis A (2021) LNCcation: lncRNA localization and function. J Cell Biol 220(2)
Brown JA, Kinzig CG, DeGregorio SJ et al (2016) Methyltransferase-like protein 16 binds the 3′-terminal triple helix of MALAT1 long noncoding RNA. Proc Natl Acad Sci USA 113:14013–14018
Cai B, Ma W, Ding F et al (2018) The long noncoding RNA CAREL controls cardiac regeneration. J Am Coll Cardiol 72:534–550
Che H, Wang Y, Li H et al (2020) Melatonin alleviates cardiac fibrosis via inhibiting lncRNA MALAT1/miR-141-mediated NLRP3 inflammasome and TGF-β1/Smads signaling in diabetic cardiomyopathy. FASEB J 34:5282–5298
Chen XY, Zhang J, Zhu JS (2019a) The role of m(6)A RNA methylation in human cancer. Mol Cancer 18:103
Chen YG, Chen R, Ahmad S et al (2019b) N6-methyladenosine modification controls circular RNA immunity. Mol Cell 76:96–109
Chen L, Wang C, Sun H et al (2021) The bioinformatics toolbox for circRNA discovery and analysis. Brief Bioinform 22:1706–1728
Chien CS, Li JY, Chien Y et al (2021) METTL3-dependent N(6)-methyladenosine RNA modification mediates the atherogenic inflammatory cascades in vascular endothelium. Proc Natl Acad Sci USA 118:e2025070118
Di Timoteo G, Dattilo D, Centrón-Broco A et al (2020) Modulation of circRNA Metabolism by m(6)A Modification. Cell Rep 31:107641
Dillmann WH (2019) Diabetic cardiomyopathy. Circ Res 124:1160–1162
Ding S, Huang H, Xu Y et al (2017) MiR-222 in cardiovascular diseases: physiology and pathology. Biomed Res Int 2017:4962426
Ding Z, Wang X, Liu S et al (2018) PCSK9 expression in the ischaemic heart and its relationship to infarct size, cardiac function, and development of autophagy. Cardiovasc Res 114:1738–1751
Dorn LE, Lasman L, Chen J et al (2019) The N(6)-methyladenosine mRNA methylase METTL3 controls cardiac homeostasis and hypertrophy. Circulation 139:533–545
Du WW, Xu J, Yang W et al (2021) A neuroligin isoform translated by circNlgn contributes to cardiac remodeling. Circ Res 129:568–582
El Azzouzi H, Vilaça AP, Feyen DAM et al (2020) Cardiomyocyte specific deletion of ADAR1 causes severe cardiac dysfunction and increased lethality. Front Cardiovasc Med 7:30
Erdos MR, Cabral WA, Tavarez UL et al (2021) A targeted antisense therapeutic approach for Hutchinson-Gilford progeria syndrome. Nat Med 27:536–545
Feng Y, Xu W, Zhang W et al (2019) LncRNA DCRF regulates cardiomyocyte autophagy by targeting miR-551b-5p in diabetic cardiomyopathy. Theranostics 9:4558–4566
Ferragut Cardoso AP, Banerjee M, Nail AN et al (2021) miRNA dysregulation is an emerging modulator of genomic instability. Semin Cancer Biol 76:120–131
Fu Y, Zhuang X (2020) m(6)A-binding YTHDF proteins promote stress granule formation. Nat Chem Biol 16:955–963
Gabisonia K, Prosdocimo G, Aquaro GD et al (2019) MicroRNA therapy stimulates uncontrolled cardiac repair after myocardial infarction in pigs. Nature 569:418–422
Gao F, Kataoka M, Liu N et al (2019) Therapeutic role of miR-19a/19b in cardiac regeneration and protection from myocardial infarction. Nat Commun 10:1802
Gao R, Wang L, Bei Y et al (2021) Long noncoding RNA cardiac physiological hypertrophy-associated regulator induces cardiac physiological hypertrophy and promotes functional recovery after myocardial ischemia-reperfusion injury. Circulation 144:303–317
García-Padilla C, Aránega A, Franco D (2018) The role of long non-coding RNAs in cardiac development and disease. AIMS Genet 5:124–140

Garikipati VNS, Verma SK, Cheng Z et al (2019) Circular RNA CircFndc3b modulates cardiac repair after myocardial infarction via FUS/VEGF-A axis. Nat Commun 10:4317

Gulati R, Behfar A, Narula J et al (2020) Acute Myocardial Infarction in Young Individuals. Mayo Clin Proc 95:136–156

Han Z, Wang X, Xu Z et al (2021) ALKBH5 regulates cardiomyocyte proliferation and heart regeneration by demethylating the mRNA of YTHDF1. Theranostics 11:3000–3016

Heidenreich P, Bozkurt B, Aguilar D et al (2022) 2022 AHA/ACC/HFSA guideline for the management of heart failure: a report of the American college of cardiology/American heart association joint committee on clinical practice guidelines. Circulation 145:e895–e1032

Holdt LM, Stahringer A, Sass K et al (2016) Circular non-coding RNA ANRIL modulates ribosomal RNA maturation and atherosclerosis in humans. Nat Commun 7:12429

Hu ZQ, Luo JF, Yu XJ et al (2017) Targeting myocyte-specific enhancer factor 2D contributes to the suppression of cardiac hypertrophic growth by miR-92b-3p in mice. Oncotarget 8:92079–92089

Huang X, Zhang H, Guo X et al (2018) Insulin-like growth factor 2 mRNA-binding protein 1 (IGF2BP1) in cancer. J Hematol Oncol 11:88

Huang S, Li X, Zheng H et al (2019) Loss of super-enhancer-regulated circRNA Nfix induces cardiac regeneration after myocardial infarction in adult mice. Circulation 139:2857–2876

Huang P, Wang L, Li Q et al (2020) Atorvastatin enhances the therapeutic efficacy of mesenchymal stem cells-derived exosomes in acute myocardial infarction via up-regulating long non-coding RNA H19. Cardiovasc Res 116:353–367

Huang XH, Li JL, Li XY et al (2021) miR-208a in cardiac hypertrophy and remodeling. Front Cardiovasc Med 8:773314

Jian D, Wang Y, Jian L et al (2020) METTL14 aggravates endothelial inflammation and atherosclerosis by increasing FOXO1 N6-methyladeosine modifications. Theranostics 10:8939–8956

Jiang L, Qiao Y, Wang Z et al (2020) Inhibition of microRNA-103 attenuates inflammation and endoplasmic reticulum stress in atherosclerosis through disrupting the PTEN-mediated MAPK signaling. J Cell Physiol 235:380–393

Jin L, Lin X, Yang L et al (2018) AK098656, a novel vascular smooth muscle cell-dominant long noncoding rna, promotes hypertension. Hypertension 71:262–272

Jusic A, Devaux Y (2020) Mitochondrial noncoding RNA-regulatory network in cardiovascular disease. Basic Res Cardiol 115:23

Landmesser U, Poller W, Tsimikas S et al (2020) From traditional pharmacological towards nucleic acid-based therapies for cardiovascular diseases. Eur Heart J 41:3884–3899

Lee DY, Yang TL, Huang YH et al (2018) Induction of microRNA-10a using retinoic acid receptor-α and retinoid x receptor-α agonists inhibits atherosclerotic lesion formation. Atherosclerosis 271:36–44

Lerchenmüller C, Rabolli CP, Yeri A et al (2020) CITED4 protects against adverse remodeling in response to physiological and pathological stress. Circ Res 127:631–646

Lesizza P, Prosdocimo G, Martinelli V et al (2017) Single-dose intracardiac injection of pro-regenerative MicroRNAs improves cardiac function after myocardial infarction. Circ Res 120:1298–1304

Li Z, Song Y, Liu L et al (2017) miR-199a impairs autophagy and induces cardiac hypertrophy through mTOR activation. Cell Death Differ 24:1205–1213

Li Y, Wu K, Quan W et al (2019) The dynamics of FTO binding and demethylation from the m(6)A motifs. RNA Biol 16:1179–1189

Li H, Xu JD, Fang XH et al (2020a) Circular RNA circRNA_000203 aggravates cardiac hypertrophy via suppressing miR-26b-5p and miR-140-3p binding to Gata4. Cardiovasc Res 116:1323–1334

Li R, Jiang J, Shi H et al (2020b) CircRNA: a rising star in gastric cancer. Cell Mol Life Sci 77:1661–1680

Li M, Zheng H, Han Y et al (2021a) LncRNA Snhg1-driven self-reinforcing regulatory network promoted cardiac regeneration and repair after myocardial infarction. Theranostics 11:9397–9414

Li T, Zhuang Y, Yang W et al (2021b) Silencing of METTL3 attenuates cardiac fibrosis induced by myocardial infarction via inhibiting the activation of cardiac fibroblasts. FASEB J 35:e21162
Li X, Meng C, Han F et al (2021c) Vildagliptin attenuates myocardial dysfunction and restores autophagy via miR-21/SPRY1/ERK in diabetic mice heart. Front Pharmacol 12:634365
Li G, Shao Y, Guo HC et al (2022) MicroRNA-27b-3p down-regulates FGF1 and aggravates pathological cardiac remodelling. Cardiovasc Res 118:2139–2151
Liao Y, Cao W, Zhang K et al (2021) Bioinformatic and integrated analysis identifies an lncRNA-miRNA-mRNA interaction mechanism in gastric adenocarcinoma. Genes Genomics 43:613–622
Libby P (2021) The changing landscape of atherosclerosis. Nature 592:524–533
Lin H, Zhu Y, Zheng C et al (2021) Antihypertrophic memory after regression of exercise-induced physiological myocardial hypertrophy is mediated by the long noncoding RNA Mhrt779. Circulation 143:2277–2292
Liu CX, Chen LL (2022) Circular RNAs: characterization, cellular roles, and applications. Cell 185:2016–2034
Liu X, Chen D, Chen H et al (2021) YB1 regulates miR-205/200b-ZEB1 axis by inhibiting microRNA maturation in hepatocellular carcinoma. Cancer Commun (Lond) 41:576–595
Mathiyalagan P, Adamiak M, Mayourian J et al (2019) FTO-dependent N(6)-methyladenosine regulates cardiac function during remodeling and repair. Circulation 139:518–532
Moore JBT, Sadri G, Fischer AG et al (2020) The A-to-I RNA editing enzyme ADAR1 is essential for normal embryonic cardiac growth and development. Circ Res 127:550–552
Nagalingam A, Siddharth S, Parida S et al (2021) Hyperleptinemia in obese state renders luminal breast cancers refractory to tamoxifen by coordinating a crosstalk between Med1, miR205 and ErbB. NPJ Breast Cancer 7:105
Pan JX (2017) LncRNA H19 promotes atherosclerosis by regulating MAPK and NF-kB signaling pathway. Eur Rev Med Pharmacol Sci 21:322–328
Peng S, Xiao W, Ju D et al (2019) Identification of entacapone as a chemical inhibitor of FTO mediating metabolic regulation through FOXO1. Sci Transl Med 11(488)
Poller W, Dimmeler S, Heymans S et al (2018) Non-coding RNAs in cardiovascular diseases: diagnostic and therapeutic perspectives. Eur Heart J 39:2704–2716
Prasad A, Sharma N, Prasad M (2021) Noncoding but coding: Pri-miRNA into the action. Trends Plant Sci 26:204–206
Puthanveetil P, Chen S, Feng B et al (2015) Long non-coding RNA MALAT1 regulates hyperglycaemia induced inflammatory process in the endothelial cells. J Cell Mol Med 19:1418–1425
Ransohoff JD, Wei Y, Khavari PA (2018) The functions and unique features of long intergenic non-coding RNA. Nat Rev Mol Cell Biol 19:143–157
Regouski M, Galenko O, Doleac J et al (2019) Spontaneous atrial fibrillation in transgenic goats with TGF (Transforming Growth Factor)-β1 induced atrial myopathy with endurance exercise. Circ Arrhythm Electrophysiol 12:e007499
Reichel M, Köster T, Staiger D (2019) Marking RNA: m6A writers, readers, and functions in Arabidopsis. J Mol Cell Biol 11:899–910
Ruan Z, Wang S, Yu W et al (2019) LncRNA MALAT1 aggravates inflammation response through regulating PTGS2 by targeting miR-26b in myocardial ischemia-reperfusion injury. Int J Cardiol 288:122
Ryu J, Ahn Y, Kook H et al (2021) The roles of non-coding RNAs in vascular calcification and opportunities as therapeutic targets. Pharmacol Ther 218:107675
Salgado-Somoza A, Zhang L, Vausort M et al (2017) The circular RNA MICRA for risk stratification after myocardial infarction. Int J Cardiol Heart Vasc 17:33–36
Seok H, Lee H, Lee S et al (2020) Position-specific oxidation of miR-1 encodes cardiac hypertrophy. Nature 584:279–285
Song H, Feng X, Zhang H et al (2019) METTL3 and ALKBH5 oppositely regulate m(6)A modification of TFEB mRNA, which dictates the fate of hypoxia/reoxygenation-treated cardiomyocytes. Autophagy 15:1419–1437

Sørensen ST, Litman T, Gluud M et al (2022) miRNA Signature in Early-stage Mycosis Fungoides. Acta Derm Venereol 102:adv00785

Su H, Wang G, Wu L et al (2020) Transcriptome-wide map of m(6)A circRNAs identified in a rat model of hypoxia mediated pulmonary hypertension. BMC Genomics 21:39

Su W, Huo Q, Wu H et al (2021) The function of LncRNA-H19 in cardiac hypertrophy. Cell Biosc 11:153

Sun L, Zhang J, Li Y (2019) Chronic central miR-29b antagonism alleviates angiotensin II-induced hypertension and vascular endothelial dysfunction. Life Sci 235:116862

Tan J, Liu S, Jiang Q et al (2019) LncRNA-MIAT increased in patients with coronary atherosclerotic heart disease. Cardiol Res Pract 2019:6280194

Tang CM, Zhang M, Huang L et al (2017) CircRNA_000203 enhances the expression of fibrosis-associated genes by derepressing targets of miR-26b-5p, Col1a2 and CTGF, in cardiac fibroblasts. Sci Rep 7:40342

Taniue K, Akimitsu N (2021) The functions and unique features of LncRNAs in cancer development and tumorigenesis. Int J Mol Sci 22(2)

Tao Y, Liu Q, Wu R et al (2022) Long noncoding RNA LUCAT1 enhances the survival and therapeutic effects of mesenchymal stromal cells post-myocardial infarction. Mol Ther Nucleic Acids 27:412–426

Viereck J, Bührke A, Foinquinos A et al (2020) Targeting muscle-enriched long non-coding RNA H19 reverses pathological cardiac hypertrophy. Eur Heart J 41:3462–3474

Viereck J, Kumarswamy R, Foinquinos A et al (2016) Long noncoding RNA Chast promotes cardiac remodeling. Sci Transl Med 8:326ra22

Wang GK, Zhu JQ, Zhang JT et al (2010) Circulating microRNA: a novel potential biomarker for early diagnosis of acute myocardial infarction in humans. Eur Heart J 31:659–666

Wang R, Dong LD, Meng XB et al (2015) Unique MicroRNA signatures associated with early coronary atherosclerotic plaques. Biochem Biophys Res Commun 464:574–579

Wang C, Yuan Y, Zheng M et al (2020a) Association of age of onset of hypertension with cardiovascular diseases and mortality. J Am Coll Cardiol 75:2921–2930

Wang X, Wang L, Ma Z et al (2020b) Early expressed circulating long noncoding RNA CHAST is associated with cardiac contractile function in patients with acute myocardial infarction. Int J Cardiol 302:15–20

Wang Y, Sun J, Lin Z et al (2020c) m(6)A mRNA methylation controls functional maturation in neonatal murine β-cells. Diabetes 69:1708–1722

Wang F, Yu R, Wen S et al (2021a) Overexpressing microRNA-203 alleviates myocardial infarction via interacting with long non-coding RNA MIAT and mitochondrial coupling factor 6. Arch Pharm Res 44:525–535

Wang W, Zhang S, Xu L et al (2021b) Involvement of circHIPK3 in the pathogenesis of diabetic cardiomyopathy in mice. Diabetologia 64:681–692

Wang X, He Y, Mackowiak B et al (2021c) MicroRNAs as regulators, biomarkers and therapeutic targets in liver diseases. Gut 70:784–795

Wang Y, Li C, Zhao R et al (2021d) CircUbe3a from M2 macrophage-derived small extracellular vesicles mediates myocardial fibrosis after acute myocardial infarction. Theranostics 11:6315–6333

Wang L, Wang J, Yu P et al (2022a) METTL14 is required for exercise-induced cardiac hypertrophy and protects against myocardial ischemia-reperfusion injury. Nat Commun 13:6762

Wang L, Yu P, Wang J et al (2022b) Downregulation of circ-ZNF609 promotes heart repair by modulating RNA N(6)-methyladenosine-modified yap expression. Research (wash D C) 2022:9825916

Watson CJ, Gupta SK, O'Connell E et al (2015) MicroRNA signatures differentiate preserved from reduced ejection fraction heart failure. Eur J Heart Fail 17:405–415

Wu G, Yan Y, Cai Y et al (2021) ALKBH1-8 and FTO: potential therapeutic targets and prognostic biomarkers in lung adenocarcinoma pathogenesis. Front Cell Dev Biol 9:633927

Wu X, Wang L, Wang K et al (2022) ADAR2 increases in exercised heart and protects against myocardial infarction and doxorubicin-induced cardiotoxicity. Mol Ther 30:400–414

Xiao SH, Wang Y, Cao X et al (2021) Long non-coding RNA LUCAT1 inhibits myocardial oxidative stress and apoptosis after myocardial infarction via targeting microRNA-181a-5p. Bioengineered 12:4546–4555

Xiong X, Liu J, He Q et al (2021) Long non-coding RNA NORAD aggravates acute myocardial infarction by promoting fibrosis and apoptosis via miR-577/COBLL1 axis. Environ Toxicol 36:2256–2265

Xu H, Wang Z, Chen M et al (2021) YTHDF2 alleviates cardiac hypertrophy via regulating Myh7 mRNA decoy. Cell Biosci 11:132

Xu W, Biswas J, Singer RH et al (2022) Targeted RNA editing: novel tools to study post-transcriptional regulation. Mol Cell 82:389–403

Yang F, Qin Y, Lv J et al (2018) Silencing long non-coding RNA Kcnq1ot1 alleviates pyroptosis and fibrosis in diabetic cardiomyopathy. Cell Death Dis 9:1000

Yang F, Li A, Qin Y et al (2019a) A novel circular RNA mediates pyroptosis of diabetic cardiomyopathy by functioning as a competing endogenous RNA. Mol Ther Nucleic Acids 17:636–643

Yang X, Li X, Lin Q et al (2019b) Up-regulation of microRNA-203 inhibits myocardial fibrosis and oxidative stress in mice with diabetic cardiomyopathy through the inhibition of PI3K/Akt signaling pathway via PIK3CA. Gene 715:143995

Yang Y, Shen F, Huang W et al (2019c) Glucose is involved in the dynamic regulation of m6A in patients with type 2 diabetes. J Clin Endocrinol Metab 104:665–673

Ye ZM, Yang S, Xia YP et al (2019) LncRNA MIAT sponges miR-149-5p to inhibit efferocytosis in advanced atherosclerosis through CD47 upregulation. Cell Death Dis 10:138

Yin Z, Zhao Y, He M et al (2019) MiR-30c/PGC-1β protects against diabetic cardiomyopathy via PPARα. Cardiovasc Diabetol 18:7

Zhang Y, Huang XR, Wei LH et al (2014) miR-29b as a therapeutic agent for angiotensin II-induced cardiac fibrosis by targeting TGF-β/Smad3 signaling. Mol Ther 22:974–985

Zhang Z, Yang T, Xiao J (2018) Circular RNAs: promising biomarkers for human diseases. EBioMedicine 34:267–274

Zhang BY, Han L, Tang YF et al (2020a) METTL14 regulates M6A methylation-modified primary miR-19a to promote cardiovascular endothelial cell proliferation and invasion. Eur Rev Med Pharmacol Sci 24:7015–7023

Zhang M, Wang N, Song P et al (2020b) LncRNA GATA3-AS1 facilitates tumour progression and immune escape in triple-negative breast cancer through destabilization of GATA3 but stabilization of PD-L1. Cell Prolif 53:e12855

Zhang Y, Chen W, Wang Y (2020c) STING is an essential regulator of heart inflammation and fibrosis in mice with pathological cardiac hypertrophy via endoplasmic reticulum (ER) stress. Biomed Pharmacother 125:110022

Zhang X, Wang W, Zhu W et al (2019) Mechanisms and functions of long non-coding RNAs at multiple regulatory levels. Int J Mol Sci 20(22)

Zheng C, Niu H, Li M et al (2015) Cyclic RNA hsa-circ-000595 regulates apoptosis of aortic smooth muscle cells. Mol Med Rep 12:6656–6662

Zhou SS, Jin JP, Wang JQ et al (2018) miRNAS in cardiovascular diseases: potential biomarkers, therapeutic targets and challenges. Acta Pharmacol Sin 39:1073–1084

Zhu P, Yang M, Ren H et al (2018) Long noncoding RNA MALAT1 downregulates cardiac transient outward potassium current by regulating miR-200c/HMGB1 pathway. J Cell Biochem 119:10239–10249

Contribution of RNA Species in Sexually Transmitted Infections

Alexis Southwell, M. Neal Guentzel, and Rishein Gupta

Contents

Abstract Many studies have shown RNA modifications including deamination of cytosine to uracil (C-to-U), deamination of adenosine to inosine (A-to-I), N1-methyladenosine (m^1A), 5-methylcytosine (m^5C), N6-methyladenosine (m^6A), and pseudouridine (ψ), and ribose 2′-O-methyl to have key roles in regulating RNA stability, processing and gene expression. The manipulation and regulation of RNA modifications can be seen to create a trickling effect on biological processes. Here, we review the knowledge of RNA modifications of some of the most detrimental sexually transmitted infections and relate our current understanding of the regulatory roles behind such modifications in infections and their association with other diseases and cancer. With the increase in techniques to identify RNA modification, we anticipate further breakthrough of the RNA modifications' influence on bacterial biological processes and their link to various diseases.

Keywords RNA · Epitranscriptomics · Sexually transmitted infections · Diseases · Cancer

A. Southwell · M. N. Guentzel (✉) · R. Gupta (✉)
South Texas Center for Emerging Infectious Diseases, The University of Texas at San Antonio, San Antonio, Texas 78249, USA
e-mail: M.Guentzel@utsa.edu

R. Gupta
e-mail: Rishein.Gupta@utsa.edu

J. Barciszewski (ed.), *RNA Structure and Function*, RNA Technologies 14,
https://doi.org/10.1007/978-3-031-36390-0_20

1 Introduction

The World Health Organization has recently estimated that more than one million sexually transmitted infections (STIs) are acquired every day worldwide (WHO 2022). Still, control of the number of cases is difficult due to many STIs being asymptomatic, which in turn, causes their delayed treatment to often result in prominent sequelae (Low and Broutet 2017; Mabey et al. 2010). Cervical carcinoma, pelvic inflammatory disease, infertility or acquired immunodeficiencies, are only a few associated conditions affecting many individuals with persistent untreated STIs (Low and Broutet 2017). While bacterial STIs have curable capabilities, the opposite is known for viral infections, which are often of lifetime duration (Mabey et al. 2010). However, as more STIs are left untreated, an increase in antimicrobial resistance now poses an additional complication when managing STIs (Low and Broutet 2017). In search for new methods to manage the spread of STIs, observing transcriptional interactions can instead be useful biomarkers in regulating cellular responses upon infection with a STI (Kanyal et al. 2019).

The recent emergence of activity in ribonucleic acids (RNAs) modification research has sparked our desire in understanding the impact post-transcriptional modifications have on RNA behavior, gene expression and advanced biological processes (Seo and Kleiner 2021). RNA epigenetics, or epitranscriptomics is the thorough collection and analysis of RNA modifications to aim towards the identification of various RNA regulations and their functions. Since the reveal of modified nucleotides in mRNA and non-coding RNA (ncRNA), many questioned biological characteristics in epigenetic processes can be answered with RNA sequencing techniques analyzing these ncRNAs (Jonkhout et al. 2017; Marbaniang and Vogel 2016). It has been found that more than a hundred different associations between RNA modification enzyme mutations and human disease may exist, however, due to the greater proportion of associated studies being with tRNAs and rRNAs, additional relations have yet to be discovered (Jonkhout et al. 2017). Currently, the Next Generation Sequencing (NGS) technologies present the ability to sequence long reads from single RNA/DNA molecules and previous studies have proved their capabilities to quickly diagnose virus outbreaks in situ and soon the same can be expected of multiple RNA modification types (Jonkhout et al. 2017). Thus, identifying the location and abundance of these key modifications has the potential to be indicators for specific infections and/or diseases and further researched to determine their involvement in such biological processes (Jonkhout et al. 2017).

Additionally, more evidence is pointing towards the possibility that RNA may even mediate transgenerational inheritance after a study identified tRNA derived fragments in normal zygotes of mice displayed diet-induced metabolic disorders (Chen et al. 2016; Jonkhout et al. 2017; Sharma et al. 2016). As RNA modifications create a new world of uncharted genome function regulation, the possibilities of what these modifications are capable of in human diseases climb (Jonkhout et al. 2017). Though little is known on the influence of these modifications within microbial infection, the influx of research may soon provide us with answers on prevalent

STIs, other diseases, and certain cancers. In this book chapter, we assess the role of epitranscriptomics in various microbial STIs and study the shared modifications of subsequent diseases in order to highlight the potential of epitranscriptomics in perhaps lowering the severity of such diseases by regulating the expression of RNAs and changing its functionality.

2 Mode of Action

RNA is a single-stranded molecule composed of an alternating sugar (ribose) and phosphate group backbone where every sugar has one of four nitrogenous bases attached that together, ultimately form a nucleotide. The nitrogenous bases (nucleosides) in RNA are adenine (A), guanine (G), uracil (U), and cytosine (C). RNA nucleotides are usually synthesized in the nucleus but can mainly be found in the cytoplasm of the cell (Quin et al. 2021). Divided into two essential functional categories, RNA molecules can proceed towards the pathway of being *operational,* to assist with transcription/translation regulation, chromosomal structure and organization, and RNA processing and routing, or *informational,* to encode proteins and translation (Wu et al. 2014). The factors leading towards RNAs determining functions fall back on the RNA nucleosides that have been altered either during splicing and/or by RNA editing (Schaefer et al. 2017).

Non-coding RNAs (ncRNA) play a significant role in normal biological activity, viral infection, and tumorigenesis (Chen et al. 2018). These RNA species can be either for regular cellular functions including transfer RNAs (tRNAs) and ribosomal RNAs (rRNAs), or be of regulatory in nature (i.e., small or long regulatory RNA (Chen et al. 2018). The knowledge of RNA modifications is primarily based on our knowledge of transfer RNA (tRNA) and ribosomal RNA (rRNA) (Bohnsack et al. 2019; Harcourt et al. 2017; Schaefer et al. 2017). However, RNA modifications can also be found in messenger RNAs (mRNAs), enhancer RNAs (eRNAs), and various non-coding RNAs (ncRNAs) (Bohnsack et al. 2019). Thus, our current information on molecular functions is largely based on rRNA and tRNA modifications, and unfortunately, many functions are either unknown or less than the extent of modifications on messenger RNA (mRNA) (Harcourt et al. 2017; Seo and Kleiner 2021).

On the basis of length, ncRNAs are categorized into two groups: small ncRNAs such as microRNAs (miRNAs) or long non-coding RNAs (lncRNAs). LncRNAs contain transcripts greater than 200 nt while miRNAs are approximately 22 nt, both sharing the similarity of being engaged in regulating gene expression via transcriptional regulation (Bartel 2004; Chen et al. 2018).

Known base modifications in mRNA include adenosine to inosine (A-to-I) deamination and cytosine to uracil (C-to-U) deamination, N6-methyladenosine (m^6A), N1-methyladenosine (m^1A) among others (Quin et al. 2021). Ultimately controlled by chemical modifications ("markers"), understanding of cellular pathways continue to be enhanced as more focus is observed on three identifiers that cave way for post-transcriptional regulations of RNAs (Meyer and Jaffrey 2014). These three types of

RNA-modifying protein identifiers (shown in Fig. 1) include enzymes called "writers" that act as methylase and add the modification, "easers" that removes modifications through demethylation and finally, "readers" that identify modifications, interpret and execute phenotypic outcomes (Bierne et al. 2012; Courtney, Tsai et al. 2019b; Zhou et al. 2011).

Within these RNA-modifying proteins, any dysregulation, mutations or modifications are what human diseases have been shown to be associated with (Esteve-Puig et al. 2020). Through thorough analysis of the influence such modifications have on cellular responses, we have the potential to interfere with its usual pathway in ways that could lessen the severity of diseases or even eliminate the possibility of such.

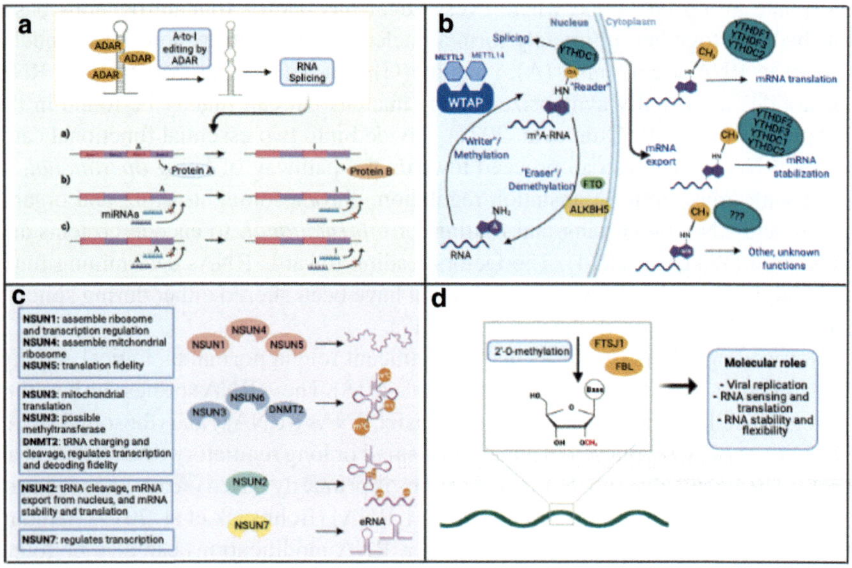

Fig. 1 Diagram of RNA modifications. **a** After A-to-I editing by ADAR, the pre-mRNA strand proceeds to remove introns and creates more opportunities for diversity. A-to-I editing by ADAR can lead to **(a)** codon change that causes alternative protein isoforms, **(b)** alter binding sites on pre-mRNAs for miRNAs to affect gene regulation, or **(c)** alter A-to-I expression on miRNAs instead, and also lead to changes in gene regulations (Eisenberg and Levanon 2018). **b** m^6A on RNA strands is involved in many cellular functions. "Readers" from both the YTHDF and YTHDC family of proteins contribute to RNA transcription and mRNA export to the cytoplasm where the modification plays a role in mRNA translation, stabilization, and other functions still unknown. **c** Cytosine-5-methylation utilizes multiple methyltransferases (NSUN1-7 and DNMT2) to influence a variety of roles in RNA processes. m^5C modifications are present in ribosomal RNAs (rRNAs), transfer RNAs (tRNAs), messenger RNAs (mRNAs), enhancer RNAs (eRNAs) and non-coding RNAs (ncRNAs). **d** 2′-O-methylation occurs on various types of RNAs by adding a methyl group to the 2′ hydroxyl on the ribose of any nucleoside. The expression of either methyltransferases, FTSJ1 or FBL, will initiate the methylation and have molecular roles that influence the stability and translation of RNAs

3 Sexually Transmitted Infections

3.1 Viral

3.1.1 Hepatitis

m^6A Modification in Hepatitis C Virus

Worldwide, approximately 520 million people are chronically infected with one of the five main hepatitis viruses, many of which being the cause behind cirrhosis, liver cancer and/or liver disease (Krajden 2001). A major RNA modification such as N6-methyladenosine (m^6A) methylation is increasingly being studied to understand its physiological function as more evidence reports m^6A modification in different pathogenic viruses with specific influence in the regulation of viral life cycles (Imam et al. 2020; Liu et al. 2021). The m^6A complex consists of two co-factors, METTL14 and WTAP, and a methyltransferase, METTL3 that contribute to the co-transcription of this modification (Courtney, Tsai et al. 2019b; Lichinchi et al. 2016; Meyer and Jaffrey 2014). m^6A has been found to interfere with gene expression and cellular function (Yang et al. 2020).

Hepatitis C Virus (HCV) is just one of the members of the Hepatitis family affected by m^6A modifications. During HCV infection, m^6A induced innate responses by affecting the abundancy of host RNAs, modifying it and causing destabilization in RNA that creates further disruption in later processes and ultimately decreases the viral particle production (Gokhale et al. 2016; Imam et al. 2020). A particular protein named YTHDF has this influence in negatively regulating HCV viral infection by contributing to the viral production and relocation of HCV genome causing difficulty in assembling viral particles (Gokhale et al. 2016; Imam et al. 2020).

A-to-I Editing in Hepatitis C Virus

It has been suggested that when infected with HCV, A-to-I editing by ADAR1 plays a role in preventing HCV replication by exhibiting antiviral activity (Netzband and Pager 2020; Pujantell et al. 2019; Taylor et al. 2005). RNA molecules called HCV replicons have been increasingly studied as it's been determined they have the capability to autonomously replicate hepatoma cells (Khan et al. 2020). Questions still remain on how the replicon systems contribute to HCV infection treatment, but what is known is that they've been responsible for encoding enzymes essential for viral replication (Khan et al. 2020). When these replicon cells receive IFN-α treatment, the result is a rapid loss of viral RNA and an increase in ADAR1 expression as the under expression of PKR leads to inhibited translation and HCV replicon synthesis was inhibited (Netzband and Pager 2020; Taylor et al. 2005; Vlachogiannis et al. 2021). RNA analyzed from these replicon cells showed enhanced mutative A-to-I editing with an increase in the amount of inosine present (Netzband and Pager 2020; Taylor et al. 2005). In addition, the presence of small inhibitory RNA (siRNA) transfection proved to influence ADAR1 expression and stimulated a ~40-fold increase of HCV replicon during ADAR1 knockdown while ADAR2 had no effect on HCV replicons during knockdown, indicating ADAR1 is responsible for instability of HCV replicon RNA and ADAR2 has no influence (Taylor et al. 2005; Vlachogiannis et al. 2021). Thus, A-to-I editing by ADAR1 in HCV infections contributes to the replication of the viral RNA (Taylor et al. 2005).

RNA Modifications in Hepatocellular Carcinoma

Driven by Hepatitis C, hepatocellular carcinoma (HCC), is the most frequent primary malignancy of the liver (Jemal et al. 2011). A-to-I editing of the antizyme inhibitor 1 (AZIN1) by ADAR1 modifications has a key role in the pathogenesis of HCC by promoting proliferation and correlates with tumor aggressiveness (Chen et al. 2013; Fumagalli et al. 2015; Slotkin and Nishikura 2013). In the progression of HCC, AZIN1 editing is increased causing a rise in the inhibition of antizyme tumor suppressor function (Chen et al. 2013; Slotkin and Nishikura 2013). Cell growth is regulated by antienzyme and involves degradation of growth promoting proteins like ornithine decarboxylase (ODC) and cyclin D1 (CCND1) but, when AZIN1 is elevated, degradation of these proteins is prevented so their upregulation increases proliferation, mobility, and colony formation (Chen et al. 2013; Slotkin and Nishikura 2013). In addition to HCC, ADAR enzymes have been found to be a target in several other cancers some of which being Esophageal cancer, Colorectal cancer, Breast cancer and Glioblastoma (Chen et al. 2013; Fumagalli et al. 2015; Han et al. 2015; Pujantell et al. 2019; Samuel 2019; Slotkin and Nishikura 2013). Further research on A-to-I editing of ADARs can be beneficial for not only cancers but other diseases as well, to increase the knowledge of the mechanisms behind such conditions. With a better understanding of said mechanisms, A-to-I editing can develop into therapeutic targets in the near future (Han et al. 2015).

3.1.2 Human Papillomavirus (HPV)

A-to-I Editing in HPV

Infection of Human Papillomavirus (HPV) has an estimated 266,000 cervical cancer deaths annually (Low and Broutet 2017). Recent theories suggest that A-to-I editing by ADAR1, an adenosine deaminase, has the capability of regulating innate immune activation by interferon-mediated immune responses to viral infection (Hartner et al. 2009; Lu et al. 2020; Pujantell et al. 2017). Though the exact role of ADAR1 on HPV infection has yet to be determined, studies have been able to recognize the modification to be key in marking double stranded RNAs (dsRNA) as self (cellular) or nonself (viral) causing an indirect response on HPV infection (O'Connell et al.

2015; Pujantell et al. 2019; Samuel 2019). dsRNA results in interferon (IFN) production and induced proteins required for its actions (Samuel 2019). A specific feature ADAR1 holds is that it expresses two isoforms—110 kDa (p110) and 150 kDa (p150) isoforms that are constitutive and predominately nuclear and interferon-induced and primarily cytoplasmic, respectively (O'Connell et al. 2015). When presented with ADAR1 p150, A-to-I editing proceeds to the inactivation of dsRNA-triggered innate immune responses (Samuel 2019). Viral dsRNAs activate interferon regulatory factor (IRF)-3 and IRF-7 after detection by cytoplasmic RLR and TLR3 which signals to produce IFN by the mitochondria adaptor MAVS (Pujantell et al. 2017; Samuel 2019). Both viral and cellular dsRNAs have also been found to lead to translation inhibition and/or RNA degradation from IFN-induced proteins (PKR protein kinase and OAS synthetases) when activated by dsRNA and experiences respectively, either autophosphorylation or cascades to activating the 2-5A-dependent RNase L (Samuel 2019). ADAR1 knockdown or deficiency has been found to both increase MDA5, PKR and OAS when cellular cytoplasmic dsRNAs sensors are triggered, while when presented with infected cells, also increases MDA5, RIG-I, and IRF inflicting deregulation of RLRs-MAVS signaling pathways and overproduction of IFN because of the increased levels of viral dsRNA (Pujantell et al. 2019, 2017; Samuel 2019). It has been reported that during ADAR1 knockdown, A-to-I editing by ADAR1 is thought to enhance the expression of HPV proteins through its influence on alternating innate immune functions/activation and play a role in HPV and alternative disease outcomes (Pujantell et al. 2019, 2017).

A-to-I Editing in HPV Related Cancers

ADAR1 influence has been reported to be linked to various cancer progression when its presence is upregulated (Samuel 2019). Specifically, ADAR1 has been present in the progression of esophageal cancers and myeloma which have been studied to possibly be associated with HPV infection status though the etiology of esophageal cancer is still unclear (Guo et al. 2016; Samuel 2019). Additionally, ADAR1 may be linked to cancer by its influence on programmed cell death-1 receptor (PD-1) (Lu et al. 2020). Lu et al. reported the loss of ADAR1 to suppress PD-1 immunotherapy resistance by increasing tumor inflammation which has been supported where Pujantell et al. confirmed ADAR1 knockdown to over-produce pro-inflammatory cytokines (Lu et al. 2020; Pujantell et al. 2019, 2017). An increased search for a better understanding of the ADAR1 modification of innate immune effectors has the potential to become a therapeutic strategy against HPV infection as a possibility to boost antiviral immune response (Pujantell et al. 2019, 2017).

3.1.3 Herpesviruses

m^{6}A Modification Among Herpesvirus Family

Herpesvirus family members such as Epstein-Barr Virus (EBV) and Kaposi's Sarcoma-Associated Herpesvirus (KSHV) both experience m^{6}A modifications playing a critical role in their life cycle, growth, and/or proliferation of infected cells (Imam et al. 2020). Though there is much to learn still, KSHV has been observed in various studies to significantly increase m^{6}A levels to promote lytic gene expression and replication of infected cells (Liu et al. 2021; Tan and Gao 2018; Tan et al. 2018; Ye 2017). When 3-deazaadenosine (DAA) is present, m^{6}A becomes blocked and inhibits the splicing of pre-mRNA encoding the replication transcription activator (RTA), and stops virion production in KSHV infected B cells (Liu et al. 2021; Tan and Gao 2018; Ye 2017). Additionally, KSHV lytic replication was shown to be hindered by m^{6}A 'reader' enzyme, YTHDF2, as it's seen to bind viral transcripts and promote the degradation and deadenylation of the viral mRNAs by mediating their stability (Tan and Gao 2018; Tan et al. 2018). The opposite was seen when cells were YTHDF2 deficient, in his case, there was an increase in virions production as well as in the half-life of viral transcripts by hindering the KSHV RNA degradation (Liu et al. 2021; Tan and Gao 2018; Tan et al. 2018). Interestingly, contrasting effects were seen during the knockdown of YTHDF3, where instead viral lytic replication can reduce by up to 40–70% (Tan et al. 2018). Other m^{6}A 'reader, 'writer', and 'eraser' enzymes also have an influence on the production and expression of viral mRNA. For example, the 'eraser', FTO, is a demethylase that, when inhibited, enhances the levels of m^{6}A and thus increases viral lytic replication (Tan and Gao 2018; Ye 2017). Yet, additional m^{6}A 'readers' (YTHDF1, YTHDC1 or YTHDC2) showed no influence or inconsistencies on viral lytic replication (Tan et al. 2018). Thus, the role of m^{6}A modifications can be ventured to impact the viral latency and replication of the KSHV infection.

A-to-I Editing in Herpesvirus Related Cancers

Commonly found in the vagina and cervix in women, individuals with herpes simplex viruses present an increased risk of potential cancerous progression as research suggests HSV to correlate with a greater incidence of cervical cancer (Jones 1995; Rafferty 1973). In-depth investigation is warranted to build on initial reports on A-to-I RNA editing in cervical carcinogenesis (Chen et al. 2007, 2017). Chen et al. reported the anti-tumorigenic role of bladder cancer-associated protein (BLCAP) was observed in conjunction with A-to-I editing by ADARs to gain an understanding of their relationship in cervical cancer cell lines (Chen et al. 2017). In which, A-to-I editing by ADAR1 was found to edit BLCAP towards a loss of function and potentially cause it to aid in initiating cervical cancer when normally unedited BLCAP would inhibit IL-6 induced signal transducer and activator of transcription 3 (STAT3) phosphorylation activation while the edited form did not (Chen et al. 2017). Further, elevated phosphorylation of STAT3 was detected in cervical and endometrial cancer cell lines in which STAT3 enhanced expression was associated with the increase of anti-apoptotic genes in the tissues and the development of tumors (Chen et al. 2007). Again, further observation of A-to-I editing of BLCAP is needed, but both it and the inhibition of STAT3 may have the potential to be a target for therapeutic intervention for endometrial and cervical cancer as well as various other STAT3-activated carcinomas (Chen et al. 2007, 2017). He et al. discovered lncRNA UCA1 was upregulated in cervical cancer tissues and cell lines (He et al. 2020). The mechanism of lncRNA sponging of miR-204 in turn promoted the invasion of cervical cancer cells through upregulating KIF20A expression (He et al. 2020). This interaction and others should also be further explored to access the potential for therapeutic use.

3.1.4 Human Immunodeficiency Virus I (HIV-1)

m^{6}A Modification in HIV-1

m^{6}A modifications of Human Immunodeficiency Virus (HIV-1) have been seen to share similarities as those seen in HCV. The methylases, demethylases and m^{6}A-binding proteins each play a significant role in the promotion and inhibition of HIV-1 (Tirumuru et al. 2017). When infected with HIV-1, m^{6}A modification levels presented an increase in host and viral mRNAs which in turn, also increased HIV-1 replication (Imam et al. 2020; Wang et al. 2022). Many proteins are seen to interfere with this enhancement of replication. Lichinchi et al. found a relationship between the silencing of m^{6}A erasers, like AlkBH5 (a protein highly expressed in T-cells), and an increase in HIV-1 replication (Lichinchi et al. 2016; Tirumuru et al. 2017; Wang et al. 2022). m^{6}A editing increased viral replication potential due to the enhancement of viral RNA nuclear export from the formation of Rev protein and Rev response element RNA complex (Imam et al. 2020; Kennedy et al. 2017; Lichinchi et al. 2016; Wang et al. 2022). This complex came about from the overexpression of the YTHDF proteins to mRNAs that had caused the enhancement of HIV-131 Rev mRNA (Imam et al. 2020; Kennedy et al. 2017; Lichinchi et al. 2016; Wang et al. 2022). In a couple of studies, researchers found the overexpression of YTHDF proteins to decrease HIV-1 reverse transcription when located in target cells, while the decrease of such proteins enhanced infection when located in CD4$^+$ T-cells (Kennedy et al. 2017; Tirumuru et al. 2017). Undoubtedly, additional research is necessary to improve the quality of information housed on these modifications.

m^{5}C Modification in HIV-1

In addition to m^{6}A, HIV-1 has high epitranscriptomic m^{5}C modification and may contribute to RNA splicing regulation and promote mRNAs translation (Courtney, Tsai et al. 2019b; Wang et al. 2022). Currently, limited knowledge is known of the function and role of m^{5}C modifications in mRNA metabolism (Yang et al. 2017). In

m^5C modifications, the primary writer, NSUN2, played a key role in HIV-1 infections by reducing HIV-1 protein expression, however, Wang et al. report this did not affect viral RNA levels (Courtney, Tsai et al. 2019b; Wang et al. 2022; Yang et al. 2017). During infection, the presence of m^5C ribonucleotides is approximately 14 times higher than cellular mRNA (Courtney, Chalem et al. 2019a; Courtney, Tsai et al. 2019b; Wang et al. 2022). When m^5C is lost, the alternative splicing of viral RNAs is dysregulated which leads to viral gene expression alterations and ultimately affects the promotion of mRNA export (Courtney, Tsai et al. 2019b; Yang et al. 2017). NSUN2 mediates m^5C methylation of HIV-1 RNA genome (Wang et al. 2022). The loss of NSUN2 disturbs HIV-1 transcript splicing and contributes to the reduction of HIV-1 protein expression and proceeds to inhibit the ribosome recruitment to bind to cellular mRNA and implies a reduction of translation has occurred (Courtney, Tsai et al. 2019b; Wang et al. 2022). m^5C is proven to promote viral gene expression, but still, the little knowledge of its functional capabilities limits our understanding of its roles in mRNA metabolism in both HIV infection as well as other conditions (Wang et al. 2022). Additional research can better explain these epitranscriptomic modifications that influence the expression of various diseases.

m^6A Modification in HIV-1 Related Diseases

If left untreated, HIV infections have the capability to progress to acquired immune deficiency syndrome (AIDS). A condition which causes people to be much more susceptible to a number of other diseases and places an individual at higher risk of cancer (Berkhout 2013). Leukemia is an AIDS associated malignancy that is considered to be induced by AIDS related viruses like EBV, HBV, HCV, and KSHV that have suggested roles in the development of some leukemia (Whitman et al. 2005). The most common of which being acute myeloid leukemia (AML), has seen to also be affected by RNA modifications that suggest playing a role in AML cell proliferation or growth, and with additional research, may promise the potential to be used to identify therapeutic strategies for leukemia treatment (Li et al. 2017; Xiao et al. 2019). In m^6A modifications, a dual role in cancer is witnessed as the m6A regulation abilities can both enhance the progression of cancer or assist in managing it (Zhang et al. 2022). The demethylase, FTO, in m^6A is seen to promote leukemogenesis and disrupt leukemia cell differentiation, thus, by targeting the enzyme for therapeutic purposes it has the capability to inhibit leukemia cell growth (Barbieri and Kouzarides 2020; Li et al. 2017). Additionally, the methylases, METTL14 and METTL3, have been present in AML whereas the latter's enhanced expression in AML has been noticed and identified as an essential key in AML cell proliferation, but together, this enzyme complex presents a negative correlation that supports the growth of leukemia cells (Barbieri and Kouzarides 2020). Thus, further understanding of the m^6A modification in AML presents a promising approach to leukemia treatments via mechanisms of immunotherapy (Zhang et al. 2022) (see Table 1).

Table 1 Modified nucleotides in RNA with notable STI and subsequent disease correlation

Nt	RNA modification	Enzymes	Molecular roles	Associated conditions	References
Editing	A-to-I	ADAR1 (W), ADAR2 (W), ADAR3 (W)	RNA stability, splicing, silencing, replication, translation efficiency, innate immune regulation	HCV, HPV, HIV, Esophageal cancer, Colorectal cancer, Hepatocellular carcinoma, Breast cancer, Glioblastoma, Cervical cancer	Chen et al. (2013, 2017), Fumagalli et al. (2015), Han et al. (2015), Pujantell et al. (2019), Samuel (2019), Slotkin and Nishikura (2013)
A	m^6A	METTL3 (W), METTL14 (W), WTAP (W), ALKBH5 (E), FTO (E), YTHDF1 (R), YTHDF2 (R), YTHDF3(R), YTHDC1 (R), YTHDC2 (R)	RNA stability, splicing, RNA folding, RNA degradation	HCV, HIV, KSHV, Myeloid leukemia, Breast cancer	Courtney, Tsai et al. (2019b), Gokhale et al. (2016), Imam et al. (2020), Kennedy et al. (2017), Meyer and Jaffrey (2014), Pujantell et al. (2019), Tan et al. (2018)
C	m^5C	NSUN1 to NSUN7 (W), DNMT2 (W) ALYREF (R)	RNA splicing, degradation, translation efficiency	HIV, Hepatocellular carcinoma, Glioblastoma multiforme, Leukemia	Courtney, Tsai et al. (2019b), Lu et al. (2020), Netzband and Pager (2020), Schaefer et al. (2017)
*N*m	2′-O-Me	FTSJ1 (W), FTSJ2 (W), FTSJ3 (W), FBL (W)	Viral replication, RNA sensing, translation, RNA stabilization	HIV, Flaviviruses, Ebola virus, Intellectual disabilities,	Dimitrova et al. (2019), Hyde and Diamond (2015), Ringeard et al. (2019)

3.2 Bacterial

With only approximately 60 rRNA and tRNA modifications known in bacteria, information on the mechanism and possibilities of alternation from RNA modifications is extremely limited (Marbaniang and Vogel 2016). Multiple investigations have been set in motion to uncover these RNA modifications but within each presents diversities among eukaryotic, bacteria, and archaea, and each group contains their own complexities in truly understanding how these modifications function (Schaefer et al. 2017). For bacteria, the characteristically short lifespan and lack of poly-A tail of bacterial mRNA create a great challenge in detecting modifications (Marbaniang and Vogel 2016). Since current information on bacterial RNA modifications is very limited, when searching for any sign of modifications and/or its influence in bacteria, little was found on their involvement, specifically in STIs. Instead, general information on proposed interferences between bacteria and modifications is largely introduced by a number of authors who have yet to determine a concrete cause of its functions.

It has been noted that when found with an in vitro bacterial system, modification introduced on mRNA showed altered amino acid incorporation due to the inhibition of translation (Marbaniang and Vogel 2016). As further information continues to be collected, recent studies have observed the presence of certain similar RNA modifications in bacteria as the ones in eukaryotic organisms, yet, still a significant number of modifications are expressed solely in bacteria or eukaryotes (Eigenbrod and Dalpke 2015). Of the known RNA modifications seen in eukaryotes such as m^6A, m^5C, 2′-O-methylation and ψ, many have been found to also play a role in bacterial species with the regulation of RNA recognition, cellular transcripts, translation, and various other bacterial pathogenesis functions, yet many of these modifications have not been investigated specifically for bacterial STIs (Deng et al. 2015; Eigenbrod and Dalpke 2015).

Interestingly, new transcriptome-wide modifications have been reported in bacteria that have never before been seen in other organisms, including 5′-NAD cap and those on a newly classed ncRNA, Y RNA. The former's functions have yet to be identified, while the latter, together with the Ro60 protein, have both been found to contribute towards structured RNA degradation (Chen et al. 2014; Marbaniang and Vogel 2016). Chen and team identified the bacterial Y RNAs to mimic tRNAs in such a way that the domain exhibited similar nucleotide modification characteristics and have become substrates for tRNA specific recognition enzymes (Chen et al. 2014). While additional information on this new class of ncRNA is highly needed, it confirms the potential of at least >250 bacteria and phages with Y RNAs, and the possibility of this RNA being present in other species is highly likely after observing RNAs eluding bioinformatic discovery via immunoprecipitation experiments (Chen et al. 2014).

Of the alternative RNA modifications first found in eukaryotes, many are happening to also make an appearance in bacterial biological processes and are increasingly being more thoroughly studied. m^6A profiling has been detected in all bacterial species, but particularly Escherichia coli and Pseudomonas aeruginosa displayed a distant m^6A pattern as from eukaryotes (Deng et al. 2015; Marbaniang and Vogel 2016). Further analysis from Deng and team, identified bacterial m^6A peaks and its modified genes, and proposed these genes operate in conjunction with functional roles in pathways for respiration, amino acid metabolism, and stress response (Deng et al. 2015). Further evidence of the similar aceA encoding isocitrate lyase gene (known typically to carry m^6A modifications) present on both *E. coli* and *P. aeruginosa* supports m^6A's stable involvement in bacteria RNA during bacterial growth, insinuating a connection between RNA modification and bacterial pathogenesis (Deng et al. 2015).

ψ and 2′-O-Me have both been identified in bacterial RNA. Ψ and dihydrouridine, also known as tRNA mimics, when present in the non-coding cytoplasmic Y RNAs of Salmonella Typhimurium, formulate a complex with Rsr protein and polynucleotide phosphorylase to assist in the degradation of structured cellular transcripts (Marbaniang and Vogel 2016). 2′-O-methylation in bacterial RNA creates a potent immunosuppressive effect that may pose the potential for its usage in some bacteria as an immune-evasion mechanism, similar to that seen with antiviral immunity breakthrough (Eigenbrod and Dalpke 2015).

From what is currently known, host lncRNAs can be affected by bacteria by using lncRNAs to (1) activate downstream antibacterial signaling, or (2) act as a positive regulator to contribute to infections (Wen 2020). In a recent study by Liu et al., 2149 lncRNAs were found to be differentially expressed in *Chlamydia infections* (Liu 2022). In particular, miR-378b was found to be upregulated after *Chlamydia muridarum* infection in mice and contributes in regulating host immune response (Liu 2022). This could potentially be further looked at in humans for similar effects.

Bacterial-related lncRNAs involved in epigenetic regulation or mRNA stability have not been reported yet (Wen 2020). In order to determine the potential of epitranscriptomics in bacteria, the investigation of new technological methods to identify these chemical modifications, or "markers", is necessary to understand cellular responses and introduce potential therapeutic strategies by interfering with its usual pathway. Though many of these sequencing technologies like RNA-seq and RIP-seq are actively being utilized to reveal an excess of conserved modifications (A-to-I, m^6A and m^5C) carried by mRNAs in eukaryotes, they could have the capability to uncover features of RNA modifications within different kingdoms of life as well (Edelheit et al. 2013).

4 microRNA and lncRNA Interactions

The interrelations of miRNA and lncRNA have increasingly become the focus of interest and will revolutionize our understanding of the molecular and cellular effects of their interactions (Ballantyne et al. 2016; Saiyed et al. 2022). The two ncRNAs share the ability to regulate gene expression at both transcriptional and post-transcriptional among other miRNAs and/or mRNA. However, little is known about the thorough description of the mechanisms of lncRNA-miRNA interactions illustrated by L. Chen et al. in which it may undergo one of four possibilities: (1) miRNA in RNA-induced silencing complexes (RISCs) target mRNAs and regulate target lncRNAs to reduce their stability via imperfect pairing; (2) lncRNA's absorption of miRNAs (as known as sponging) using miRNA recognition elements (MREs) to suppress target mRNAs degradation mediated by miRNAs; (3) via intracellular RNA splicing, lncRNA can act as miRNA precursors to create specific miRNAs to enhance regulation of target mRNAs; (4) lncRNA vs miRNA competition for recognition via 3′ UTR binding of target mRNAs (Chen et al. 2018). It's also important to note that ncRNA with the same seed sequence, often undergo similar processes, and thus can be impacted similarly by itself or other ncRNAs, making it challenging to determine the specific difference in the encounters.

5 ncRNA Therapeutic Interventions

Regulatory mechanisms have been identified among miRNA and lncRNA, and in causation of their interactions. The mechanisms have since begun to pave the way towards their potential use as biomarkers for disease detection and developed as therapeutics (Chen et al. 2018; Kelemen et al. 2019; Saiyed et al. 2022). Currently, more information concerning the use of miRNAs as therapeutic agents has been identified and tested. Previous studies suggest miRNA therapies are broadly classified in two categories: miRNA inhibition and miRNA replacement, depending on the intended mode of action and provide the greatest efficacy (Monga and Kumar 2019; Saiyed et al. 2022; Tagliaferri et al. 2012). In relation to the beforementioned diseases in this chapter, miRNAs and/or their lncRNA interactions have already begun testing as therapeutic agents to be manipulated and target HCV, liver cancer, or a number of other diseases and cancers (Table 2) (Ballantyne et al. 2016; Chakraborty et al.

Table 2 miRNA based therapeutics in clinical trials (Chakraborty et al. 2021; Monga and Kumar 2019; Zhang et al. 2021)

Company	Name	Therapeutic agent	Condition or disease	Clinical phase
Mima Therapeutics Inc.	MRX34	miR-34 mimic	Melanoma; primary liver cancer; hematologic malignancies	Phase I
Santaris Pharma	Miravirsen	Anti-miR-122	Hepatitis C virus	Phase II-suspended
Regulus Therapeutics	RG-101	Anti-miR-122	Hepatitis C virus	Phase Ib-discontinued
miRagen Therapeutics, Inc.	RGLS4326	miR-17	Polycystic kidney disease	Phase I
miRagen Therapeutics, Inc.	MRG-106	miR-155	Lymphomas; leukemias	Phase I
miRagen Therapeutics, Inc.	Remlarsen	miR-29	Idiopathic pulmonary fibrosis and other fibrosis	Phase I

2021; Monga and Kumar 2019; Zhang et al. 2018, 2021, 2022). With many additional biomarkers identified in other diseases, it is only a matter of time before new therapeutic trials begin.

Unfortunately, the same cannot be said for the use of lncRNA targeted interventions, which instead contain a number of further challenges of its own (Boon et al. 2016; Schmerer and Schulte 2021). Boon et al. likely acknowledge possibly the two biggest challenges being the expression of lncRNAs in numerous transcript variants which may account for their even more so diverse functions and mechanisms of action (Boon et al. 2016). Present research continues to investigate lncRNA biomarkers, uncovering new ones and diving further into the actions of known ones, and via trials of their assessment in different diseased individuals (Zhao et al. 2016). It is no doubt that clinical lncRNA targeting may be possible in the near future. However, ultimately, additional understanding of all ncRNAs' is necessary to be capable of creating a network illustrating their interactional connections. Thus, providing a tremendous advantage for the future exploration of ncRNA's capabilities and impact on biological processes (Agliano et al. 2019; Di Martino et al. 2021; Kundu and Basu 2021).

6 Discussion

Similarly, to DNA, RNA modifications have been proven to display significant influence in the biological processes and disease developments of both viral and bacterial infections in mice and humans. In this review, we analyzed the roles of multiple RNA modifications in sexually transmitted infections and common subsequent diseases and/or cancers associated with the mentioned infections. While additional modifications involving cytosine to uracil (C-to-U) deamination, N1-methyladenosine (m^1A) and pseudouridine (ψ) have been introduced and continue to be further investigated, this book chapter focuses on those commonly found with one or more of the beforementioned infections/conditions.

From our current understanding, Adenosine to inosine (A-to-I) deamination, N6-methyladenosine (m^6A), 5-methylcytosine (m^5A), and ribose 2′O-methyl were seen to share molecular roles in the conditions we studied. Jonkhout et al. confirm these observations as he found more than 100 different associations between RNA modification enzyme mutations and human disease (Jonkhout et al. 2017). As research comes closer to identifying the location and abundance of these modifications' components such as methyltransferases, demethylases and "readers", they're used as indicators so we may gain the knowledge to predict where and/or when in the cycle, specific changes to RNA will take place (Jonkhout et al. 2017). These RNA modifications' involvement in biological processes and their roles in RNA functional regulation are key in uncovering the unknown influence they have on various viral and microbial associated human diseases.

Currently, we know that any dysregulation, mutations or modifications are what diseases have been shown to be associated with (Esteve-Puig et al. 2020). In our observations, the modifications' major impact on RNA stability, involvement in innate immune responses, and cell proliferation are significant in their capability to be manipulated and developed into a therapeutic target in the future. Recent advancements in sequencing technology for single RNA/DNA molecules provide hope in its use to also analyze multiple RNA modification types to deliver a quicker reveal of their impact and potential to promptly diagnose and aid various diseases or outbreaks.

Funding There are no funding sources to declare.

References

Agliano F, Rathinam VA, Medvedev AE et al (2019) Long noncoding RNAs in host-pathogen interactions. Trends Immunol 40:492–510

Ballantyne MD, McDonald RA, Baker AH (2016) lncRNA/MicroRNA interactions in the vasculature. Clin Pharmacol Ther 99:494–501

Barbieri I, Kouzarides T (2020) Role of RNA modifications in cancer. Nat Rev Cancer 20:303–322

Bartel DP (2004) MicroRNAs: genomics, biogenesis, mechanism, and function. Cell 116:281–297

Berkhout B (2013) HIV, leukemia, and new horizons in molecular therapy. J Formos Med Assoc 112:441–444

Bierne H, Hamon M, Cossart P (2012) Epigenetics and bacterial infections. Cold Spring Harb Perspect Med 2:a010272

Bohnsack KE, Hobartner C, Bohnsack MT (2019) Eukaryotic 5-methylcytosine (m(5)C) RNA methyltransferases: mechanisms, cellular functions, and links to disease. Genes (basel) 10(2):102

Boon RA, Jae N, Holdt L et al (2016) Long noncoding RNAs: from clinical genetics to therapeutic targets? J Am Coll Cardiol 67:1214–1226

Chakraborty C, Sharma AR, Sharma G et al (2021) Therapeutic advances of miRNAs: a preclinical and clinical update. J Adv Res 28:127–138

Chen CL, Hsieh FC, Lieblein JC et al (2007) Stat3 activation in human endometrial and cervical cancers. Br J Cancer 96:591–599

Chen L, Li Y, Lin CH et al (2013) Recoding RNA editing of AZIN1 predisposes to hepatocellular carcinoma. Nat Med 19:209–216

Chen X, Sim S, Wurtmann EJ et al (2014) Bacterial noncoding Y RNAs are widespread and mimic tRNAs. RNA 20:1715–1724

Chen Q, Yan M, Cao Z et al (2016) Sperm tsRNAs contribute to intergenerational inheritance of an acquired metabolic disorder. Science 351:397–400

Chen W, He W, Cai H et al (2017) A-to-I RNA editing of BLCAP lost the inhibition to STAT3 activation in cervical cancer. Oncotarget 8:39417–39429

Chen L, Zhou Y, Li H (2018) LncRNA, miRNA and lncRNA-miRNA interaction in viral infection. Virus Res 257:25–32

Courtney DG, Tsai K, Bogerd HP et al (2019b) Epitranscriptomic addition of m(5)C to HIV-1 transcripts regulates viral gene expression. Cell Host Microbe 26(217–227):e6

Courtney DG, Chalem A, Bogerd HP et al (2019a) Extensive epitranscriptomic methylation of A and C residues on murine leukemia virus transcripts enhances viral gene expression. mBio 10:e01209–19

Deng X, Chen K, Luo GZ et al (2015) Widespread occurrence of N6-methyladenosine in bacterial mRNA. Nucleic Acids Res 43:6557–6567

Di Martino MT, Riillo C, Scionti F et al (2021) miRNAs and lncRNAs as novel therapeutic targets to improve cancer immunotherapy. Cancers (basel) 13:1587

Dimitrova DG, Teysset L, Carre C (2019) RNA 2′-O-methylation (Nm) modification in human diseases. Genes (basel) 10:117
Edelheit S, Schwartz S, Mumbach MR et al (2013) Transcriptome-wide mapping of 5-methylcytidine RNA modifications in bacteria, archaea, and yeast reveals m5C within archaeal mRNAs. PLoS Genet 9:e1003602
Eigenbrod T, Dalpke AH (2015) Bacterial RNA: an underestimated stimulus for innate immune responses. J Immunol 195:411–418
Eisenberg E, Levanon EY (2018) A-to-I RNA editing-immune protector and transcriptome diversifier. Nat Rev Genet 19:473–490
Esteve-Puig R, Bueno-Costa A, Esteller M (2020) Writers, readers and erasers of RNA modifications in cancer. Cancer Lett 474:127–137
Fumagalli D, Gacquer D, Rothe F et al (2015) Principles governing A-to-I RNA editing in the breast cancer transcriptome. Cell Rep 13:277–289
Gokhale NS, McIntyre ABR, McFadden MJ et al (2016) N6-methyladenosine in Flaviviridae viral RNA genomes regulates infection. Cell Host Microbe 20:654–665
Guo L, Liu S, Zhang S et al (2016) Human papillomavirus-related esophageal cancer survival: A systematic review and meta-analysis. Medicine (baltimore) 95:e5318
Han L, Diao L, Yu S et al (2015) The genomic landscape and clinical relevance of A-to-I RNA editing in human cancers. Cancer Cell 28:515–528
Harcourt EM, Kietrys AM, Kool ET (2017) Chemical and structural effects of base modifications in messenger RNA. Nature 541:339–346
He Q, Meng J, Liu S et al (2020) Long non-coding RNA UCA1 upregulates KIF20A expression to promote cell proliferation and invasion via sponging miR-204 in cervical cancer. Cell Cycle 19:2486–2495
Hyde JL, Diamond MS (2015) Innate immune restriction and antagonism of viral RNA lacking 2-O methylation. Virology 479–480:66–74
Imam H, Kim GW, Siddiqui A (2020) Epitranscriptomic(N6-methyladenosine) modification of viral RNA and Virus-host interactions. Front Cell Infect Microbiol 10:584283
Jones C (1995) Cervical cancer: is herpes simplex virus type II a cofactor? Clin Microbiol Rev 8:549–556
Jonkhout N, Tran J, Smith MA et al (2017) The RNA modification landscape in human disease. RNA 23:1754–1769
Kanyal A, Nahata S and Karmodiya K (2019) Epigenetics in infectious disease. In: Sharma S (ed), Prognostic Epigenetics (Academic Press) 15:171–201
Kelemen E, Danis J, Goblos A et al (2019) Exosomal long non-coding RNAs as biomarkers in human diseases. EJIFCC 30:224–236
Kennedy EM, Bogerd HP, Kornepati AVR et al (2017) Posttranscriptional m(6)A editing of HIV-1 mRNAs enhances viral gene expression. Cell Host Microbe 22:830
Khan S, Soni S, Veerapu NS (2020) HCV replicon systems: workhorses of drug discovery and resistance. Front Cell Infect Microbiol 10:325
Krajden M (2001) Hepatitis. Can J Infect Dis 12:329–331
Kundu M, Basu J (2021) The role of microRNAs and long non-coding RNAs in the regulation of the immune response to mycobacterium tuberculosis infection. Front Immunol 12:687962
Li Z, Weng H, Su R et al (2017) FTO plays an oncogenic role in acute myeloid leukemia as a N(6)-methyladenosine RNA demethylase. Cancer Cell 31:127–141
Lichinchi G, Zhao BS, Wu Y et al (2016) Dynamics of human and viral RNA methylation during zika virus infection. Cell Host Microbe 20:666–673
Liu C, Yang Z, Li R et al (2021) Potential roles of N6-methyladenosine (m6A) in immune cells. J Transl Med 19:251
Low N, Broutet NJ (2017) Sexually transmitted infections-research priorities for new challenges. PLoS Med 14:e1002481

Lu L, Gaffney SG, Cannataro VL et al (2020) Transfer RNA methyltransferase gene NSUN2 mRNA expression modifies the effect of T cell activation score on patient survival in head and neck squamous carcinoma. Oral Oncol 101:104554

Mabey D, Ndowa F, Latif A (2010) What have we learned from sexually transmitted infection research in sub-Saharan Africa? Sex Transm Infect 86:488–492

Marbaniang CN, Vogel J (2016) Emerging roles of RNA modifications in bacteria. Curr Opin Microbiol 30:50–57

Meyer KD, Jaffrey SR (2014) The dynamic epitranscriptome: N6-methyladenosine and gene expression control. Nat Rev Mol Cell Biol 15:313–326

Monga I, Kumar M (2019) Computational resources for prediction and analysis of functional miRNA and their targetome. Methods Mol Biol 1912:215–250

Netzband R, Pager CT (2020) Epitranscriptomic marks: emerging modulators of RNA virus gene expression. Wiley Interdiscip Rev RNA 11:e1576

O'Connell MA, Mannion NM, Keegan LP (2015) The epitranscriptome and innate immunity. PLoS Genet 11:e1005687

Pujantell M, Riveira-Munoz E, Badia R et al (2017) RNA editing by ADAR1 regulates innate and antiviral immune functions in primary macrophages. Sci Rep 7:13339

Pujantell M, Badia R, Galvan-Femenia I et al (2019) ADAR1 function affects HPV replication and is associated to recurrent human papillomavirus-induced dysplasia in HIV coinfected individuals. Sci Rep 9:19848

Quin J, Sedmik J, Vukic D et al (2021) ADAR RNA modifications, the epitranscriptome and innate immunity. Trends Biochem Sci 46:758–771

Rafferty KA Jr (1973) Herpes viruses and cancer. Sci Am 229:26–33

Ringeard M, Marchand V, Decroly E et al (2019) FTSJ3 is an RNA 2′-O-methyltransferase recruited by HIV to avoid innate immune sensing. Nature 565:500–504

Saiyed AN, Vasavada AR, Johar SRK (2022) Recent trends in miRNA therapeutics and the application of plant miRNA for prevention and treatment of human diseases. Futur J Pharm Sci 8:24

Samuel CE (2019) Adenosine deaminase acting on RNA (ADAR1), a suppressor of double-stranded RNA-triggered innate immune responses. J Biol Chem 294:1710–1720

Schaefer M, Kapoor U, Jantsch MF (2017) Understanding RNA modifications: the promises and technological bottlenecks of the 'epitranscriptome.' Open Biol 7:170077

Schmerer N, Schulte LN (2021) Long noncoding RNAs in bacterial infection. Wiley Interdiscip Rev RNA 12:e1664

Seo KW, Kleiner RE (2021) Mechanisms of epitranscriptomic gene regulation. Biopolymers 112:e23403

Sharma U, Conine CC, Shea JM et al (2016) Biogenesis and function of tRNA fragments during sperm maturation and fertilization in mammals. Science 351:391–396

Slotkin W, Nishikura K (2013) Adenosine-to-inosine RNA editing and human disease. Genome Med 5:105

Tagliaferri P, Rossi M, Di Martino MT et al (2012) Promises and challenges of MicroRNA-based treatment of multiple myeloma. Curr Cancer Drug Targets 12:838–846

Tan B, Gao SJ (2018) RNA epitranscriptomics: regulation of infection of RNA and DNA viruses by N(6)-methyladenosine (m(6) A). Rev Med Virol 28:e1983

Tan B, Liu H, Zhang S et al (2018) Viral and cellular N(6)-methyladenosine and N(6),2′-O-dimethyladenosine epitranscriptomes in the KSHV life cycle. Nat Microbiol 3:108–120

Taylor DR, Puig M, Darnell ME et al (2005) New antiviral pathway that mediates hepatitis C virus replicon interferon sensitivity through ADAR1. J Virol 79:6291–6298

Tirumuru N, Zhao BS, Lu W et al (2017) Correction: N(6)-methyladenosine of HIV-1 RNA regulates viral infection and HIV-1 Gag protein expression. Elife 6:e31482

Vlachogiannis NI, Verrou KM, Stellos K et al (2021) The role of A-to-I RNA editing in infections by RNA viruses: possible implications for SARS-CoV-2 infection. Clin Immunol 226:108699

Wang S, Li H, Lian Z et al (2022) The role of RNA modification in HIV-1 infection. Int J Mol Sci 23:7571

Whitman AG, Bryan BA, Dyson OF et al (2005) AIDS related viruses, their association with leukemia, and Raf signaling. Curr HIV Res 3:319–327

World Health Organization (2022) Sexually transmitted infections (STIs) https://www.who.int/en/news-room/fact-sheets/detail/sexually-transmitted-infections-(stis). Accessed 3 Sep 2022

Wu J, Xiao J, Zhang Z et al (2014) Ribogenomics: the science and knowledge of RNA. Genomics Proteomics Bioinform 12:57–63

Xiao H, Cheng Q, Wu X et al (2019) ADAR1 may be involved in the proliferation of acute myeloid leukemia cells via regulation of the Wnt pathway. Cancer Manag Res 11:8547–8555

Yang X, Yang Y, Sun BF et al (2017) 5-methylcytosine promotes mRNA export-NSUN2 as the methyltransferase and ALYREF as an m(5)C reader. Cell Res 27:606–625

Yang C, Hu Y, Zhou B et al (2020) The role of m(6)A modification in physiology and disease. Cell Death Dis 11:960

Ye F (2017) RNA N(6)-adenosine methylation (m(6)A) steers epitranscriptomic control of herpesvirus replication. Inflamm Cell Signal 4:e1604

Zhang G, Pian C, Chen Z et al (2018) Identification of cancer-related miRNA-lncRNA biomarkers using a basic miRNA-lncRNA network. PLoS ONE 13:e0196681

Zhang S, Cheng Z, Wang Y et al (2021) The Risks of miRNA Therapeutics: In a Drug Target Perspective. Drug Des Devel Ther 15:721–733

Zhang Z, Liu F, Chen W et al (2022) The importance of N6-methyladenosine modification in tumor immunity and immunotherapy. Exp Hematol Oncol 11:30

Zhao B, Lu M, Wang D et al (2016) Genome-wide identification of long noncoding RNAs in human intervertebral disc degeneration by RNA sequencing. Biomed Res Int 2016:3684875

Zhou Y, Kim J, Yuan X et al (2011) Epigenetic modifications of stem cells: a paradigm for the control of cardiac progenitor cells. Circ Res 109:1067–1081

Hypoxia and Epithelial-to-Mesenchymal Transition (EMT) in Cancer: A Non-coding RNA Perspective

Aastha Singh, Rahul Gupta, and Ritu Kulshreshtha

Contents

Abstract A recent focus on the regulatory roles of non-coding RNAs has widened our knowledge of their biological functions. These regulatory roles inherently exist owing to their unique structures. Cancer research is increasingly based on the roles of multiple non-coding RNAs. Crucial to explore are the major cancer hallmarks that combine to bring about an aggressive phenotype in cancer cells. Epithelial-to-mesenchymal transition (EMT) contributes towards an increasing metastatic potential in cancer cells, additionally leading to the domination of cancer stem cells (CSCs), which have a higher propensity to seed secondary tumors. An ever aggressively growing tumor gains a survival advantage by reprogramming its metabolism and becomes therapy resistant. Tumor hypoxia is centric to an aggressive cancer phenotype along with other physiological factors. Literature has shown how hypoxia could contribute towards EMT in cancer. This chapter focuses on the crucial role of several

Aastha Singh and Rahul Gupta have contributed equally.

A. Singh · R. Gupta · R. Kulshreshtha (✉)
Department of Biochemical Engineering and Biotechnology, Indian Institute of Technology Delhi, New Delhi 110016, India
e-mail: drritukulshreshtha@gmail.com; ritu@dbeb.iitd.ac.in

J. Barciszewski (ed.), *RNA Structure and Function*, RNA Technologies 14,
https://doi.org/10.1007/978-3-031-36390-0_21

non-coding RNAs that mediate hypoxia-driven EMT in multiple cancers. We aim to discuss the structural and functional aspects of these regulatory non-coding RNAs. Importantly, we focus on the cell signaling pathway crosstalk, mediated by ncRNAs in multiple cancers, and describe how different pathways that are crucial to both EMT and hypoxia converge onto some important molecular players. Interestingly, several ncRNAs have been implicated as novel biomarkers and therapeutic targets. A detailed understanding of the ncRNAs regulating hypoxia-driven EMT would help enable future studies of these phenomena.

Graphical Abstract

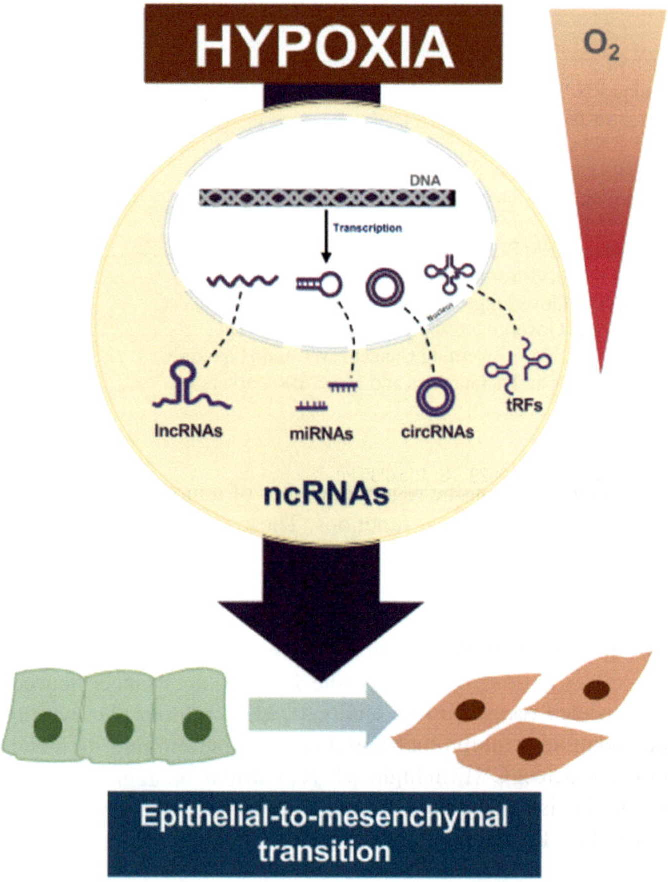

Keywords Cancer · hypoxia · Epithelial-to-mesenchymal transition (EMT) · Non-coding RNAs

List of Abbreviations

AGO	Argonaute
BSJ	Back-splicing junction
ceRNA	Competitive endogenous RNA
circRNAs	Circular RNAs
CSCs	Cancer stem cells
ECM	Extracellular matrix
EGF	Epithelial growth factor
EMT	Epithelial-to-mesenchymal transition
EMT-TFs	EMT transcriptional factors
EV	Extracellular vesicle
HCC	Hepatocellular carcinoma
HIF	Hypoxia-inducible factor
IRES	Internal ribosomal entry site
lncRNAs	Long non-coding RNAs
miRNAs	MicroRNAs
NAT	Natural antisense transcripts
ncRNAs	Non-coding RNAs
NSCLC	Non-small cell lung cancer
ODDD	Oxygen-dependent degradation domain
ORF	Open reading frames
PDGF	Platelet-derived growth factor
pVHL	Von Hippel-Lindau protein
PTGS	Post-transcriptional gene silencing
RBP	RNA-binding protein
RISC	RNA induced silencing complex
RCC	Renal cell carcinoma
TGF-β	Transforming growth factor beta
TME	Tumor microenvironment
tRFs	Transfer RNA fragments
VEGF	Vascular endothelial growth factor
UTR	Untranslated region

1 Introduction

The central tenet of molecular biology depicts the flow of genetic information from DNA to proteins, via RNA intermediates. Historically, proteins have remained in the spotlight of medical research, due to their varied structural and functional roles in health and disease. RNA, while central to biology, was assumed to simply convey the information from a gene's sequence to its coded protein. The completion of the Human Genome Project in 2003 signaled a major breakthrough in the field of genomics. An astonishing revelation was the fact that only about 2% of the human genome (around 20,000 genes) was found to code for proteins. The remaining 98% of the genome was believed to be non-coding with no apparent function, and was disappointingly termed "junk" DNA. Later in 2012, pioneering data from the ENCODE project revealed that the "dark" areas of the human genome were not static—they were extensively and widely transcribed into the bulk of RNA (Kapranov et al. 2007; Djebali et al. 2012). These RNA molecules, apart from their generic roles as messengers, ribosomal RNAs, ribozymes and tRNAs, were now found to also operate as major regulators of gene expression.

As of today, in addition to 19,370 protein-coding genes, we have annotated 19,095 long non-coding RNA genes and 7,566 small non-coding RNA genes (GENCODE v41). It is now widely accepted that ncRNA-mediated gene regulation via RNA/mRNA/protein interactions, both in cis and in trans, is a fundamental principle governing all cellular functions and occurs in all domains of life—from the simplistic prokaryotes to complex eukaryotes. The term ncRNA broadly refers to any transcript that does not translate into a protein. This includes, but is not limited to, snRNAs, snoRNAs, rRNAs, tRNAs, miRNAs, piRNAs, siRNAs, and lncRNAs. All of these ncRNAs differ in their structure and length, and exhibit varied functions. The levels of ncRNAs have been found to be highly dysregulated in diseases like cancer, pointing to their instrumental role in maintaining cell homeostasis.

It is widely agreed that global changes in the transcriptome and proteome of a cell lead to various diseases including cancer. Decades of research have still not yielded a cure for cancer, and it remains the biggest cause of deaths in human populations worldwide. Metastasis is the most formidable feature of carcinomas. This process, which refers to the dispersal of cancer cells from their primary site of origin, begins the process of establishment of secondary tumors across multiple tissues of the human body. It is estimated that 70–90% of deaths from cancer occur due to metastasis (Dillekås et al. 2019). Initiation of metastasis depends on the activation of complex molecular programs, like the epithelial-to-mesenchymal transition (EMT), inside the cancer cell. This trans-differentiation phenomenon, which occurs naturally during wound healing and early embryogenesis, alters the immobile, anchored epithelial cells into motile fibroblast-like mesenchymal cells which are more poised to escape the primary solid tumor. This escape involves multiple steps: the cells must lose their intercellular adhesions, reorganize their cytoskeleton, secrete ECM-degrading proteinases, break through the basement membrane, and intravasate into the nearby capillaries to utilize blood flow as a means of transport to distant sites. Hence, EMT relies on massive transcriptomic and proteomic changes, such that EMT-activated cells have been found to display enhanced immunosuppression, chemoresistance, stemness, and resistance to apoptosis and anoikis (Kurrey et al. 2009; Frisch et al. 2013). Certain features of the tumor microenvironment, like hypoxia, have been found to compound and hasten this process. Hypoxia, a condition of low oxygen, is a well-known physiological condition of solid carcinomas. Hypoxia causes a marked alteration to cancer cells' metabolism and their phenotype. Moreover, hypoxia is known to impact multiple hallmarks of cancer, making it an important driver of therapy resistance and tumor cell invasion.

Hence, the partnership between hypoxia and EMT presents a formidable challenge to cancer therapy. Owing to their myriad effects on the cell's transcriptome, ncRNAs have been found to be centrally involved in regulating cancer hallmarks (Slack and Chinnaiyan 2019), including metastasis and hypoxia. Deregulated ncRNA networks and their downstream signaling pathways in different types of cancers have shed light on the mechanism of tumor growth and progression. Since ncRNAs have also been shown to be secreted into exosomes and are found in blood, numerous ncRNAs have been identified as biomarkers for diagnosing cancers. They also serve as attractive targets for therapy. A diagrammatic representation of the mechanisms governing gene regulation in cancer is depicted in Fig. 1.

This chapter focuses on the major ncRNA entities such as miRNAs, lncRNAs, circRNA, transfer RNA fragments (tRFs) and how the understanding of their regulatory functions could intertwine EMT and hypoxia across multiple cancers. Importantly, ncRNAs seem to regulate multiple signaling pathways in a cancer cell that could crosstalk and further aid in tumor progression. Finally, an insight into the therapeutic aspect of ncRNAs and their role as potential disease biomarkers is provided.

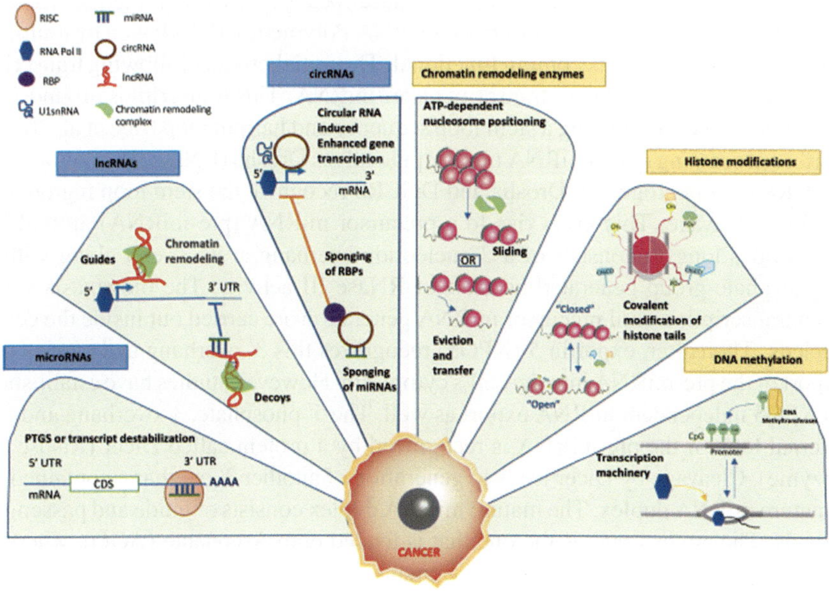

Fig. 1 A brief overview of the mechanisms of gene regulation in cancer

2 Non-coding RNAs: Structure and Function

2.1 microRNAs

MicroRNAs (miRNAs) are a class of small non-coding RNAs with a length of ~18–22 nucleotides. The discovery of miRNAs can be traced back to a report on lin-4 which was discovered as a non-coding RNA entity in *C. elegans*. lin-4 was shown to have antisense complementarity to the lin-14 gene. Investigation of its functional aspect unveiled that lin-4 could regulate lin-14 translation via the RNA–RNA antisense mechanism, thereby controlling the development of larval forms in *C. elegans* (Lee et al. 1993). It was in the early 2000s that miRNAs were discovered as distinct entities. let-7 was perhaps the first miRNA that was discovered to be functional and conserved across invertebrates and vertebrates including humans (Pasquinelli et al. 2000). After almost two decades of extensive work, miRNAs have been found to be deregulated in multiple diseases including cancers. Moreover, circulating miRNAs are being investigated as promising biomarkers for many diseases. Various miRNA clusters are now known to regulate multiple signaling pathways within a cell and are thereby instrumental in disease biology (Lu et al. 2005).

Except for a few miRNAs that are transcribed by RNA Polymerase III, the majority of the miRNAs are transcribed by RNA Polymerase II, followed by multiple processing steps before becoming functional. The initial product following transcription is known as the primary transcript or pri-miRNA. This transcript is around 200 nucleotides long comprising a stem loop sequence and hairpin loop (Lee et al. 2004). Further processing of pri-miRNA transcript involves Drosha (RNase III enzyme) and DGCR8 protein. Together, Drosha and DGCR8 recognize the stem loop region and perform cleavage. This gives rise to a precursor miRNA (pre-miRNA) around 70 nucleotides long. It consists of a 2-nucleotide overhang at its 3′ end along with a 5′ phosphate group generated because of RNase III activity. The biogenesis steps from transcription until precursor miRNA generation are carried out inside the cell's nucleus. Thereafter, exportin 5 (XPO5) recognizes this 3′ overhang and carries out export of the pre-miRNA into the cell's cytoplasm. However, studies have established an XPO5 independent miRNA export as well. The 5′ phosphate, 3′ overhang and the internal loop of the pre-miRNA is recognized by a protein called Dicer (RNase III enzyme). Cleavage by Dicer leads to generation of another 3′ overhang contained in a mature miRNA duplex. The mature miRNA duplex consists of guide and passenger strands. The guide strand is the one that is loaded onto Argonaute (AGO), whereas the passenger strand is discarded. In humans, AGO2 is the most highly expressed of all the Argonaute proteins. AGO2 prefers loading the strand whose 5′ end is less stably paired, with a preference for A or U at its 5′end. A mature miRNA consists of a −5p or −3p strand based on their origin from the pre-miR hairpin loop (Liu et al. 2004).

microRNAs mostly target mRNAs in their 3′UTR, to bring about gene silencing. This mechanism involves the AGO protein forming a complex called the RNA induced silencing complex (RISC). Nucleotides 2–8 in a mature miRNA comprise the 'seed' region. This region has nearly perfect complementarity with its mRNA target partner. The interaction between the seed region and target site on mRNA forms the basis of multiple target prediction algorithms for miRNAs. Such algorithms have varied research-based applications especially in human disease networks research. In addition to the seed region, there also exists a 'supplemental' region between nucleotides 13–16 where additional base-pairing is possible between the miRNA and its target mRNA (Schirle et al. 2014). To bring about gene silencing, AGO2 protein recognises the 5′ nucleotide of the miRNA along with a preference for an adenine (A) also known as the 't1A', on the target mRNA strand. This 't1A' is the complementary nucleotide to the first miRNA nucleotide, present on mRNA at the latter's 3′ end. Moreover, the AGO2 protein binds to a 3′ nucleotide on the miRNA to hold the miRNA–mRNA complex together. AGO2-mediated binding of miRNA with its target mRNA initially proceeds transiently between bases 2–6 of the miRNA and further binding is promoted only if there exists complementarity between bases 7 and 8 of the miRNA guide strand. In this manner, selectivity of target site is maintained (Chandradoss et al. 2015; Schirle et al. 2015). miRNA-mediated gene silencing may occur through two mechanisms, mRNA decay or translational repression. mRNA decay involves the removal of poly-adenylation signal (poly-A) or 5′decapping. This reduces the half-life of the mRNA transcript, making it susceptible to degradation. Translational decay, on the other hand, is mediated by release of eukaryotic initiation factor 4A-I (e-IF4A-I) or e-IF4A-II. Multiple miRNAs could silence one transcript by binding at different locations in its 3′UTR. Also, a single miRNA could target mRNA transcripts and bring about a repression of many targets (Guo et al. 2010; Fukao et al. 2014).

Multiple techniques have been employed to characterize miRNAs. The miRNA expression levels have been studied using techniques such as qRT-PCR, DNA microarrays and RNA sequencing. In silico databases such as "miRBase" contain all the information about miRNA structures and their genomic loci (Kozomara and Griffiths-Jones 2014). Crucial is the miRNA–mRNA target study, which ultimately helps in understanding miRNA-mediated gene silencing. Multiple bioinformatic tools help predict mRNA targets for miRNAs using certain parameters such as base-pairing interactions, thermodynamic stability of the miRNA–mRNA hybrid, sequence conservation and checking for multiple target regions in a 3′UTR (Agarwal et al. 2015). Experimental approaches such as AGO-CLIP sequencing is employed to discover biologically relevant miRNA–mRNA pairs. AGO pull-downs give an insight into the AGO complexed miRNA–mRNA pairs that could possibly interact to bring about a functional impact. Also, miRNA-induced target mRNA cleavage is identified using degradome sequencing, which helps understand the exact cleavage site in the mRNA (Chi et al. 2009).

Studied for their vast regulatory roles, miRNAs could serve as novel biomarkers for multiple diseases. Some of their functions have been thoroughly studied and hence certain miRNAs can be described as either oncogenic or tumor-suppressor miRNAs in different cancers. However, such regulatory roles have been established after thorough studies. Hence, the regulatory roles of miRNAs could be universal or highly context dependent. Usually oncogenic microRNAs silence tumor-suppressor genes in cancer cells. This miRNA activity leading to tumor formation is possible if their inherent expression levels are higher in cells. This would enable sustained oncogenic activity by repressing tumor-suppressor target genes. "OncomiR addiction" is the term used for the reliance of tumor cells on these highly expressed miRNAs. On the other hand, tumor-suppressor miRNAs effect their functions by repressing oncogenic players (Kinoshita et al. 2017).

2.2 *lncRNAs*

Long non-coding RNAs have lately emerged as vital mediators of health and disease. lncRNAs are transcripts over 200 nucleotides and may encompass hundreds to thousands of bases. Typically, lncRNAs are assumed to have little to no protein-coding potential, however, a minority of protein- or micropeptide-coding lncRNAs have also been discovered (Hartford and Lal 2020). In the 1990s, the lncRNA H19 and XIST were among the first mammalian lncRNAs to be discovered, initiating the dive into the vast unexplored world of the non-coding genome. As of 2022, GENCODE release v41 lists 19,095 lncRNA genes and 54,291 transcripts (GENCODE). Depending on their genomic location of origin, lncRNAs may be intergenic, intronic, sense, antisense, enhancer or promoter associated. They can also arise from pseudogenes. Except for a handful of lncRNAs transcribed from RNA polymerase III, the majority of lncRNAs are transcribed by RNA Polymerase II (Jarroux et al. 2017). Like mRNAs, they further undergo 5′ capping and 3′ polyadenylation. LncRNAs with introns may not always be properly spliced, as most lncRNAs harbor inefficient splicing sequences (Krchňáková et al. 2019). Some of the imperfectly spliced lncRNAs, and those transcribed from deregulated Polymerase II remain bound to the chromatin and can eventually become degraded by nuclear RNA exosome complexes. After processing, the lncRNA can be retained in the nucleus, or shipped to the cytosol, mimicking the same export pathway utilized by mRNAs (such as NFX1-mediated export). lncRNAs transported to the cytoplasm are sorted to their respective organelles utilizing cis-sequence motifs. lncRNAs have been found to localize to the mitochondria and are also secreted into exosomes (Statello et al. 2020).

Functionally, lncRNAs are very diverse, but are mostly involved in regulating gene expression. lncRNAs present in the nucleus are found to directly interact with DNA and chromatin elements. Owing to their negative charges, lncRNAs may bind to the positively charged histone complexes, thereby affecting chromatin remodeling and gene accessibility. They also assist in recruiting histone-modifying enzymes such as methyltransferases, demethylases, Polycomb repressive complex (PRC), proteins like GADD45a, by direct interaction, thereby guiding them to the site of action (Han and Chang 2015). The reverse is also observed: lncRNAs may prevent histone modifications by sequestering the modifying enzymes, hence acting as decoys (e.g., LncRNA LncPRESS1 prevents SIRT6 histone deacetylase from inactivating pluripotency-related genes). Interestingly, lncRNAs such as MEG3, HOTAIR, and Fendrr have been found to form triple-helical structures (triplexes) and "R-loops" by binding to double-stranded DNA inside the cell. lncRNAs may form triplexes and R-loops in cis, on site while being transcribed, or can work in trans for other protein-coding genes. These structures, while thought to be detrimental to the genome, have been found to be important for transcription regulation and also assist in DNA repair (Niehrs and Luke 2020). Another mechanism of lncRNA-mediated gene regulation is direct interruption of transcription, by means of steric occlusion of the site to be transcribed, rendering it inaccessible to RNA Polymerase II. The lncRNA B2 has been found to directly interact with RNA Polymerase II, and hinder the formation of the preinitiation complex. XIST is a well-characterized example of widespread gene silencing mediated by a single lncRNA. It is important to note that X-chromosome inactivation mediated by XIST relies on its interaction with multi-protein complexes, eventually leading to recruitment of PRC2 and histone methylation of silenced genes (Long et al. 2017).

Conversely, lncRNAs are equally important as gene activators. Enhancer-associated lncRNAs are a class of lncRNAs that enhance gene expression by binding to promoter regions of a gene and recruiting transcription activators (such as p300-CBP coactivator complex). They may also assist in looping of chromatin regions. lncRNAs may also intervene in gene expression at the post-transcriptional level. In this respect, they may interfere with splicing, mediate mRNA degradation, or regulate translation. lncRNAs can directly interact with mRNAs by base-pairing, or guide and recruit protein complexes. Direct base-pairing with mRNA has been found to increase loading into ribosomes and facilitates translation. lncRNAs have also been found to mediate A to I editing of mRNAs. Another crucial role of lncRNA is in the regulation of microRNAs. Such lncRNAs harbor complementary sequences to the miRNA seed region, and "sponge" the miRNA, thereby restraining it from inhibiting its target mRNAs. The competitive endogenous RNA theory is central to the function of both miRNAs and lncRNAs. lncRNAs can also act as ribozymes, or function as integral components of RNPs. Some lncRNAs have also been found to originate from mitochondrial DNA, and in turn affect key mitochondrial pathways such as respiration, metabolism, and apoptosis (Liu and Shan 2021).

Research on lncRNAs has revealed them to be complex, structured molecules. lncRNAs are dynamic and may adopt secondary or tertiary structures. These higher-order structures help lncRNAs to act as scaffolds and assemble multimolecular complexes with proteins. Deducing the structure of lncRNA has been challenging owing to their low abundance in cells, wide variability in lengths, low sequence conservation, poor annotation, exhibition of heterogeneity with sub-domains in their structure, strong association with multi-protein complexes, and challenges in purification (Zampetaki et al. 2018; Chillón and Marcia 2020; Sanbonmatsu 2021). Methods for structure determination of lncRNAs include enzymatic foot-printing, chemical probing, small-angle X-ray scattering, atomic force microscopy, NMR and cryo-EM (Graf and Kretz 2020).

2.3 Other Non-coding RNAs

2.3.1 Circular RNAs (circRNAs)

Circular RNAs (circRNAs) are yet another class of non-coding RNAs, first discovered in 1976. An investigation into the structure and function of viroids led to the serendipitous discovery of covalently closed circular molecules (Sanger et al. 1976). It was in 1979 that circular RNAs were reported in eukaryotic cells using electron microscopy (EM). In mouse testis, a sex-determining region Y (*Sry*)-specific circRNA was determined to have a biological function (Hsu and Coca-Prados 1979; Capel et al. 1993). Some reports have hinted at an epigenetic regulation such as methylation which could alter circRNA expression, independent of the corresponding linear mRNA expression (Rinaldi et al. 2016).

Upon biogenesis, circRNAs are exported out from the nucleus into the cell's cytoplasm. This export is known to be size dependent. UAP56 or DDX39B (DEAD box protein 56) facilitates export of longer circular RNAs usually >1000 nucleotides long. Nuclear RNA helicase URH49 can export small circular RNAs, usually <500 nucleotides long (Huang et al. 2018). circRNAs are products of different splicing events occurring on an mRNA molecule. The splicing event is mainly regulated by multiple small nuclear ribonuclear proteins (snRNPs) leading to exon junctions that ultimately give rise to a functional mRNA. In the case of a linear splicing event, exons are surrounded by introns and some introns flanked by ADAR1 and DHX9 that are RBPs, capable of disrupting association between the inverted repeat elements. This gives rise to linear mRNAs. However, in the case of circular RNA biogenesis, a back-splicing event has a mainstay. Back splicing can be facilitated by the presence of Alu elements having inverted repeat sequences or the presence of RNA-binding proteins (RBPs) such as FUS, HOK, NF90 and NH90/110, in the intronic regions. These elements can bring about internal proximity between the splice donor (SD) and splice acceptor (SA). The branchpoint (BP) and the SD are joined by means of the BP attacking the SD, and the SA is further attacked by the SD, leading to the formation of a circular RNA through a back-splicing junction (BSJ). Now, this circular RNA may have just exons, in which case it is called exonic circular RNA; sometimes circular RNAs having retained introns are also formed and are known as EIciRNAs. Exon skipping could lead to retention of an exon within the intronic lariat which is back spliced, giving rise to an exonic circRNA. Another mechanism through which circRNAs may be formed involves lariat precursors. Lariat precursors escaping the debranching steps of linear splicing may also form circRNAs, which, however, are intronic. Of notable importance here is the regulation of splicing machinery within a cell. This can be a deciding factor for the transcript to be able to form a mature mRNA or a circular RNA molecule. It has usually been reported that highly abundant circular RNAs are products of genes that consist of highly active promoters and possess long exons that flank introns (Zhang et al. 2014; Ivanov et al. 2015; Kelly et al. 2015; Rybak-Wolf et al. 2015).

circRNAs have varied biological roles. Of note is their ability to act as miRNA sponges. circRNAs may sponge multiple miRNAs and negate the miRNA-mediated silencing of their target transcripts. So the otherwise repressed mRNA transcripts can now undergo translation and demonstrate context-specific effects. For instance, a circRNA sponging a set of miRNAs that can repress oncogenes can now lead to their effective translation. Similarly, a circRNA sponging a set of miRNAs that can repress tumor-suppressor genes may lead to normal translation of those genes. However, not all circRNAs sponge tumor-suppressor/oncogenes in a cell at the same time. An important factor determining this sponging behavior of circRNAs is the turnover of miRNAs and circRNA present at the time. Additionally, for any complex biological regulation by circRNAs to be effective, there must exist a set of circRNA–miRNA interacting sites, arising from their abundant expression at that instance. circRNAs also directly interact with RBPs, which can lead to the regulation of the RBP domains of proteins. In this way, circRNAs act as protein sponges (Memczak et al. 2013). For instance, if a circRNA binds to AGO2 protein, it may stoichiometrically make AGO2 less available and impact the global miRNA-mediated silencing in that instance. Interestingly, a protein reservoir may also be created by the interaction of proteins with multiple circRNAs, and this may prompt a response to external stimuli. circRNAs may also recruit enzyme substrate together and act as molecular scaffolds to impact enzyme kinetics. U1snRNP's activity is known to be enhanced by interaction with circRNA. Occasionally, circRNAs undergo cap-independent translation owing to the presence of internal ribosome entry sites (IRES) followed by the start codon, AUG (Abe et al. 2015). Various methods are available for characterizing and studying circRNA expression, their interactions with proteins and miRNAs. Northern blotting, digestion with RNase R, and qPCR involving the use of BSJ-specific divergent primers are some of the common methods used in labs to characterize circRNAs. Multiple techniques are available for studying circRNA interactions. These include in silico methods which predict circRNA–miRNA interactions by scanning multiple miRNA binding sites in a circRNA. In lab, co-immunoprecipitation (co-IP) for AGO2 is employed to pull down bound RNAs to the AGO2 complex. This gives an insight into the bound circRNAs to the AGO2 complexed miRNAs. Through this approach, biologically relevant and miRNA-bound circRNAs are deciphered. The circRNA is further confirmed along with its interacting miRNA using qPCR with BSJ-specific divergent primers. On a similar principle, circRNA–protein/RBP interactions are investigated. However, linear mRNA depletion beforehand helps in reducing noise and enriches circRNA abundance upon pull-down. Investigation of circRNA translation through their IRES requires careful analysis of open reading frames (ORFs) which should be distinct from their mRNA counterpart. This is because a majority of circRNAs arise from coding gene exons. However, upon elimination of mRNA ORFs, it becomes convenient to predict the ORF and peptide sequence for a circRNA capable of undergoing translation. Experimental approaches such as cloning the circRNA IRES elements in luciferase-based reporters help predict the ability of the circRNA to translate into a peptide. Mass spectrometry and circRNA peptide or circRNA fusion tag-specific antibody pull-down can help decipher the translation of the candidate circRNA. circRNAs are relatively more stable than other RNA entities owing to their circular nature, which protects them from RNase cleavage. Their high expression levels have also been reported in multiple cancers such as breast, colon, lung, etc. circRNAs are enriched with extracellular vesicles, hence they are promising candidates for biomarker research in multiple cancers. Many cancer hallmarks such as angiogenesis, metabolic reprogramming, immune machinery evasion, metastasis, and cell cycle arrest have now been attributed to the presence of circRNAs (Li et al. 2015; Memczak et al. 2015; Szabo and Salzman 2016).

2.3.2 tRNA-Derived Fragments (tRFs)

Recently, research has focused on tRNA-derived fragments (tRFs) or tRNA-derived small RNAs (tsRNAs). These small RNAs, typically 14–32 nucleotides long, are formed after cleavage of mature or pre-tRNAs in the cell. tRFs are found to be very similar to miRNAs in terms of their biogenesis and function. The endonucleotic cleavage of tRNAs to generate tRFs has been found to involve Dicer, and other nucleases such as RNase Z. The cleaved tRFs may be 5′ (if cleaved from the TψC arm) or 3′ (if cleaved from the D arm) (Keam and Hutvagner 2015). Both 5′ and 3′ tRFs are found to bind to the 3′UTR of mRNAs and suppress translation. Analysis of PAR-CLIP and CLASH data identified the presence of tRFs bound to the Argonaute complex in cells. tRFs bind to their target mRNAs using a short sequence of 7–8 bases from their 5′ end (Kuscu et al. 2018).

In cancer, tRFs have been found to interact with RNA-binding proteins, and regulate the stability of mRNA transcripts. tRFs have also been shown to inhibit or promote the proliferation of cells, and hence affect their viability. Conventional approaches to detect tRFs in cellular fractions include deep sequencing and microarrays. However, these methods are complicated by the fact that tRFs, like tRNAs, are chemically modified, which might interfere with their detection. Newer methods include ARM-Seq (AlkB-facilitated RNA methylation sequencing), DM-tRNA-Seq (engineered demethylases-based tRNA sequencing), and dumbbell PCR (Yu et al. 2020).

3 EMT Signaling and Cancer

EMT was first discovered as a developmental process in avian embryos. In 1982, it was observed that adult corneal epithelial cells, when suspended in a collagen gel, acquired mesenchymal features and started to migrate (Greenburg and Hay 1982). In the early 2000s, research on EMT in cancer unveiled a set of "core" transcription factors—the EMT-TFs (proteins belonging to ZEB1/2, SNAIL1/2, TWIST families)—that were the main drivers of EMT and were found to be dysregulated in multiple cancers. Higher than normal expression of EMT-TFs was reported to increase the risk of tumor metastasis. Conversely, reduction in ZEB1, SNAIL1, and

TWIST1 was found to abrogate metastasis in breast cancer and pancreatic cancer models.

Traditionally, epithelial cells are characterized by the presence of strong adherens junctions (with a constitutive expression of E-cadherin), cytokeratin intermediary filaments and an apical-basal polarity—features which are lost, wholly or partially, during the conversion to mesenchymal states. EMT-TFs like ZEB1 and SNAIL1 bind to E-box elements in the promoter of E-cadherin, and repress its expression, resulting in the dissolution of cell junctions. The loss of E-cadherin expression has been found to initiate metastasis in some studies, and has widespread effects on the cellular transcriptome (Onder et al. 2008). The EMT-TFs are pleiotropic in nature, and are also known to activate each other. Apart from phenotypic conversion, EMT-TFs have been found to have other oncogenic roles: TWIST protein was shown to drive the oncogenic transformation of cells in vitro by destabilizing p53 (Piccinin et al. 2012); SNAIL1 and ZEB1 have been associated with increased resistance to chemotherapeutic drugs by averting DNA damage and associated apoptosis (Seo et al. 2021); and SNAIL1 and SLUG were found to be responsible for maintaining cancer stem cells (Stemmler et al. 2019). It is imperative to note that EMT-TFs work in coordination with microRNAs to bring about these vast changes in the cell's genetic program. MiRNAs of the miR-200 and miR-34 family, shown to be powerful suppressors of the EMT process, are in their turn repressed by ZEB1 and SNAIL1.

Heterotypic signaling cues from the tumor stroma have been shown to assist in EMT activation (de Wever and Mareel 2003). Signaling molecules released from cells present in the tumor microenvironment (fibroblasts, macrophages, endothelial cells) contribute to the formation of a "reactive" stroma—a tumor-sustaining milieu with altered ECM components, increased secretion of proteinases and growth factors, inflammation, and heightened angiogenesis (Frankenstein et al. 2020). TGF-β is one crucial molecule that has established roles in promoting invasion and migration by stimulating the expression of EMT-TFs. Downstream of TGF-β receptor activation, the Smad proteins convey signals to the nucleus and activate the canonical targets of TGF-β signaling, EMT-TFs, Fibronectin, N-cadherin, etc. TGF-β also induces the expression of matrix-metalloproteinases (MMPs 2 and 9), which are crucial for degrading ECM/basement membrane components to make way for the cancer cells to invade (Wendt et al. 2009). TNF-α and NF-κB signaling pathways are also known to act in coordination with TGF-β to initiate the EMT program. Notch signaling pathways activate Snail to initiate EMT in gastric cancer. FGF, HGF, and IGF act via RTK signaling to converge on the activation of EMT-associated genes. The Wnt signaling pathway is a major upstream activator of EMT. It is known to directly suppress E-cadherin, and induces the expression of EMT-TFs. In breast cancer, Wnt signaling also represses BRCA1, leading to severe forms of the disease (Gonzalez and Medici 2014). Members of the Hedgehog signaling pathway also induce EMT, in parallel activating the Wnt and TGF-β signaling cascade.

Lately, much research has focused on the plastic nature of the EMT phenomenon—the fact that EMT is transient and reversible makes it difficult to design molecular therapies against EMT-undergoing cells. While initially considered to be a binary switch, recent evidence points to the acquisition of an intermediate/hybrid EMT state

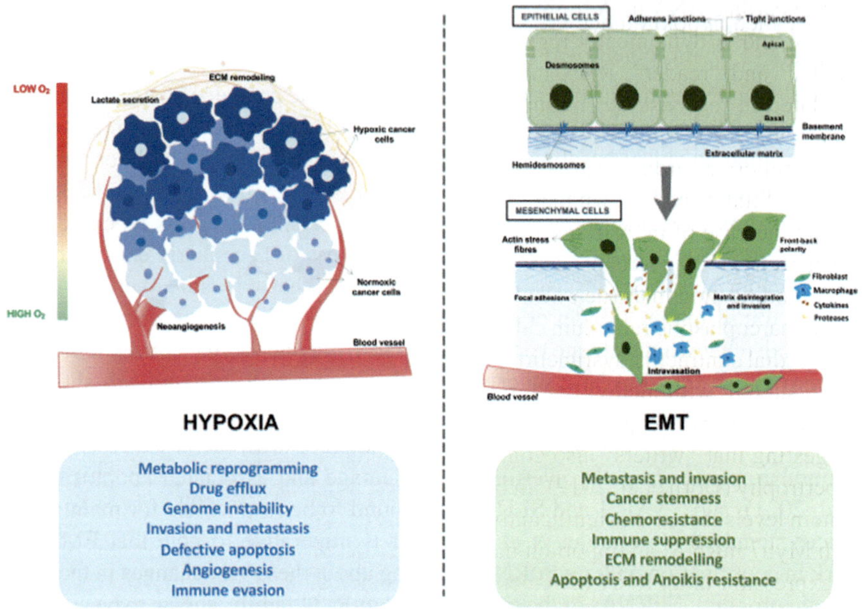

Fig. 2 A representation of the altered cellular phenotypes within cancer cells upon hypoxia and EMT

(retains characteristics common to both epithelial and mesenchymal phenotypes) by tumor cells in circulation. This has been hypothesized to assist in collective migration of conjoined clusters, dramatically increasing their survival and seeding probabilities. As cells with activated EMT cease to proliferate and acquire stem cell-like properties, they are also invisible to conventional chemotherapeutic drugs. It is also known that patient prognosis dips severely as soon as metastases are detected in the body. Hence, preventing the dispersal of cells from the primary tumor remains the best bet to keep cancers under control. Since the activation of EMT relies heavily on the crosstalk between the components and environment of the tumor-associated stroma, it is of the utmost necessity to obtain a complete molecular picture of all triggers that initiate EMT in the tumor microenvironment. Figure 2 is a diagrammatic representation of the altered phenotypes upon the activation of EMT in cancer cells.

4 Hypoxia Signaling and Cancer

Hypoxia refers to conditions of low oxygen concentration within the cells. Varied oxygen levels are observed in hypoxic tissues and their values typically vary from 0.2 to 1% O_2 as against the physiological or "physoxia" levels of 3–7% O_2 (Begg and Tavassoli 2020). Almost three decades of extensive work has been invested into hypoxia signaling research. Under ischemic or hypoxic conditions, a nuclear factor called hypoxia-inducible factor (HIF) was discovered bound at the enhancer elements of erythropoietin (EPO) for promoting expression. It was the first report to open avenues of hypoxia research (Semenza et al. 1991). The role of HIF was further investigated, and a mammalian oxygen-sensing mechanism was discovered. Glycolysis was perhaps the first pathway that was directly shown to be regulated by HIF1. Later in 1995 it was discovered that HIF exists as a heterodimer consisting of two sub-units, HIF-1α and HIF-1β (Wang et al. 1995). The discovery of HIF-1α-mediated activation of vascular endothelial growth factor (VEGF) was a milestone that continues to guide cancer and disease research to this day. Angiogenesis and vascular proliferation were discovered to be crucial hypoxic responses (Forsythe et al. 1996; Iyer et al. 1998). The exact regulatory mechanism of HIF-1α was unveiled and it was reported that von Hippel-Lindau (pVHL) protein targets HIF-1α in its oxygen-dependent degradation domain (ODDD), which consists of a crucial proline residue. This showed how cells respond to hypoxic conditions by stabilizing HIF-1α, which otherwise is degraded under normoxia (Iliopoulos et al. 1996; Maxwell et al. 1999; Jaakkola et al. 2001).

Hypoxia signaling stems from an array of HIFs which are crucial, primarily transcriptional, regulators. HIF-1α is expressed widely across body tissues, while HIF-2α and HIF-3α are found expressed in specific tissues (Wiesener et al. 2003; Semenza 2012). It is these HIFs whose levels are modulated upon cellular stress or hypoxia, whereas HIF-1β is constitutively expressed (Kaelin and Ratcliffe 2008). Upon forming a heterodimer, HIF enters the nucleus where it binds to the hypoxia response elements (HREs) present in the promoters of multiple gene targets. The transcriptional activation of the downstream gene targets brings about modulation of a variety of cell signaling pathways (Wu et al. 2015; Bao and Wong 2021). Interestingly, multiple pathways such as mTOR, JAK-STAT, T-cell receptor signaling, NF-κB pathway, Wnt/β-catenin and Notch signaling have some crosstalk via hypoxia signaling (Luo et al. 2022). Hypoxia-induced crosstalk or hypoxia-perturbed crosstalk of multiple cellular signaling elements is an illustration of the widespread regulatory possibilities within cells exhibiting hypoxic stress. This is of great significance in cancer research.

Solid tumors exhibit hypoxia. The tumor microenvironment (TME) is a niche comprising multiple elements such as proliferating tumor cells, stromal cells, immune cells, and a hypoxic core. The sub-populations also encompass cancer-associated fibroblasts (CAFs), immune cells, lymphatic cells, and various other blood cells. Hypoxia invariably brings about tumor growth aggression by impacting all the components of TME. Tumor hypoxia can be chronic or acute, severe or mild, perfusion or diffusion limited. The latter two terms describe lack of oxygen availability in cells, either on the basis of blood vessel occlusion or as a function of distance from the blood vessel, respectively. The other terminologies are based on either the duration of hypoxia or the levels of O_2. However, chronic hypoxia lasting for more than 24 h and severe hypoxia ($<1\%$ O_2 levels) are those which present therapeutic challenges (Hammond et al. 2014). Although hypoxic stress would trigger stress response mechanisms in cancer cells, microvasculature formation along with angiogenesis are also accelerated. In this manner, a tumor becomes capable of rapid proliferation and enhanced survival by increasing its nutrient supply even further. Yet another effect of an aggressively growing tumor is the exacerbation of the hypoxic tumor which expands upon rapid growth and the formation of leaky blood vessels that facilitate nutrient supply to the tumor core (Krock et al. 2011; Carmeliet and Jain 2011). Cellular metabolism is accordingly reprogrammed due to pH, and nutrient levels change which alters various metabolic pathways. Hypoxia triggers immune resistance, therapy resistance, and phenotype switching over to increased cellular motility and metastasis. Overall, hypoxia poses a great therapeutic challenge on account of cell cycle arrest, inhibition of apoptosis, or drug efflux in both radio-treated and chemo-treated cancer cells (Höckel and Vaupel 2001). A diagrammatic representation of the altered phenotypes in hypoxic cancer cells is given in Fig. 2.

5 Hypoxia-Induced EMT and the Cancer Pathway Crosstalk

As we have seen, hypoxia-driven induction of HIFs leads to overexpression of nearly 200 genes. Important among these are growth factors, such as VEGF, transforming growth factor-β (TGF-β), platelet-derived growth factors (PDGF) and epidermal growth factor (EGF). Moreover, hypoxia influences glycolytic enzymes and anti-apoptotic factors (Carmeliet et al. 1998; Jung-Whan et al. 2007). Increasing number of studies have highlighted the role of hypoxia in promoting cancer stem cell (CSC) formation and maintenance. CSCs maintain stem cell-like abilities and have the potential to infiltrate nearby cells, leading to invasion by the tumor cells to new sites. This markedly enhances the metastatic potential of the cancer cells, inter alia. Hypoxia can induce an EMT program through various approaches: regulation of EMT signaling pathway elements or the EMT-TFs can be acted upon by hypoxic stress.

Both EMT and hypoxia regulate two major cancer hallmarks, metastasis and angiogenesis, through multiple mechanisms that were initially studied in isolation. However, emerging literature strongly suggests that both EMT and hypoxia converge on some important cancer-related pathways. The TGF-β signaling pathway is known to recruit EMT-TFs via Smad2, Smad3, and Smad4 which can activate EMT-TFs. Alternatively, Smad-independent recruitment of EMT-TFs also occurs via PI3K and ERK pathways (Xu et al. 2009; Lamouille et al. 2014). TGF-β has been shown to stabilize HIF-1 under hypoxic conditions by inhibiting PHD2 expression, whereas EMT-TFs may even be recruited during this process (McMahon et al. 2006). Studies probing the levels of HIF-1 and TGF-β have been performed in different cancers. In lung cancer, TGF-β levels were found to be elevated under chronic hypoxia (Furuta et al. 2015). Interestingly, in gastric cancers there exists an autocrine signaling that maintains inherently higher levels of TGF-β in hypoxic conditions, further leading to EMT (Matsuoka et al. 2013). Similarly, the NF-κB pathway coincides with both hypoxia and EMT. This pathway is activated upon cellular stress in multiple cancers

(Xia et al. 2014). Hypoxic stress activates NF-κB, which in turn recruits EMT-TFs such as TWIST1 to activate EMT (Li et al. 2012; D'Ignazio et al. 2016). The Notch pathway is crucial in multiple cancers where it may have either an oncogenic or tumor-suppressor role (Aster et al. 2017). For instance, Notch1 through HIF1 recruitment to LOX promoter enhances the latter's expression. SNAIL1 stability results from LOX overexpression, ultimately leading to EMT (Sahlgren et al. 2008). Additionally, Notch1 may directly promote SNAIL2 activity (Shao et al. 2015).

Wnt signaling is known to be deregulated in various cancers such as breast and prostate cancers and glioblastoma, and plays an important role in their progression. There is increasing evidence that Wnt-induced stability of HIF-1α could regulate downstream HIF targets such as EMT-TFs, leading to promotion of EMT in cancer cells. Notably, this mechanism is mediated via PI3K/mTOR pathways that are switched "on" upon cellular stress. This way, different pathways including Notch, Wnt, mTOR, ERK and NF-κB crosstalk with each other at multiple points. Ultimately, they can help stabilize HIF. These are some mechanisms by which HIF1, through different pathways, leads to EMT. Hypoxia-induced EMT typically leads to a decrease in epithelial phenotype-specific markers such as E-cadherin and an increase in mesenchymal phenotype markers such as N-cadherin and Vimentin. Apart from indirect regulation of EMT-TFs through multiple pathways, HIF1 is known to directly upregulate TWIST1, SNAIL1 and ZEB1 by binding to HREs in their promoters. Additionally, a deubiquitinating enzyme, USP47, is elevated under hypoxic conditions. USP47, which is a direct target of SOX9, leads to SNAIL1 deubiquitination, making it more resistant to proteasomal activity. Hence, upon HIF-1 binding, ZEB1 and TWIST1 are recruited (Yang et al. 2008; Chen et al. 2010; Zhang et al. 2013; Choi et al. 2017).

6 Non-coding RNAs Governing EMT and Hypoxia

The control of hypoxia-driven EMT in cancer cells lies significantly with non-coding RNAs. Analysis of gene and small-RNA expression data from cancer cells has led to the identification of potential oncogenic and tumor-suppressive miRNAs, lncRNAs and circRNAs. Functional studies on cancer cells grown under hypoxia have further revealed important genes and signaling pathways targeted by these ncRNAs. Here, we list and describe known ncRNAs that work to advance or suppress hypoxia-driven EMT in cancer, and the cellular networks that are involved in the same.

6.1 miRNAs Governing Cancer EMT and Hypoxia

6.1.1 Oncogenic miRNAs that Promote Hypoxia-Induced EMT

MiR-210-3p is a well-characterized miRNA that is highly induced under hypoxia, and exhibits elevated levels in multiple cancers. In gliomas, hypoxic conditions stimulate the expression of miR-210-3p, which in turn increases the transcript levels of TGF-β, independent of HIF-1α. This stabilization of TGF-β has been found to promote EMT, accompanied by a change in morphology and expression of mesenchymal-specific markers. Additionally, this overexpression of miR-210-3p reinforced resistance to the commonly used drug Temozolomide in glioma cells. These oncogenic features were attributed to the activation of NF-κB signaling; however, the targets of the miRNA in this respect are yet to be discovered (Liu et al. 2021a, b). The oncogenic role of miR-210-3p has also been established for other cancers. In pancreatic cancer, it was found to be overexpressed in bone metastatic tissues. On inhibiting miR-210-3p, the size and proportion of tumors metastasizing to the bone was significantly reduced in mouse models in vivo. Furthermore, its inhibition also interfered with the accumulation of NF-κB/p65 in the nucleus. TNIP1 and SOCS1 were reported to be the direct targets of miR-210-3p, which are well-known negative regulators of NF-κB signaling (Ren et al. 2017). miR-210-3p was also found to be enriched in $CD44^{+}/CD24^{-}$ breast cancer stem cells (BCSCs) and MCF7 spheroids generated in vitro, under hypoxia treatment. E-cadherin was found to be directly suppressed by miR-210-3p at the transcript as well as protein levels (Tang et al. 2018). Overexpression of miR-210-3p in hypoxia was found to be associated with a decrease in the expression of HECTD1. HECTD1 is an E3 ubiquitin ligase, associates with SNAIL1, and mediates its degradation. The downregulation of HECTD1 was found to heighten EMT in HeLa cells (Wang et al. 2020). Therefore, hypoxia-induced miR-210-3p directly affects tumor metastasis.

Studies have unraveled multiple molecular players involved in the regulation of HIF-1α protein stability. FBXL16 is one such protein that was shown to interact with HIF-1α and induce the latter's ubiquitination-mediated degradation, regardless of the oxygen concentration in the cell. FBXL16 is lowly expressed in the triple-negative breast cancer (TNBC) subtype of breast cancer. FBXL16 upregulation curbed the expression of EMT and angiogenesis markers in TNBC. Investigation of the mechanism for FBXL16 downregulation led to activated p38-mediated increase in miR-135b-3p being identified as responsible for directly suppressing FBXL16. p38 is a member of the MAPK family of kinases, which is found to be activated in TNBC and promotes migration and proliferation (Kim et al. 2021).

miR-574-3p is found to be highly expressed in gastric cancer cell lines, which correlates with a poorer prognosis and also indicates advanced stages and metastasis. MiR-574-3p brings about EMT by diminishing E-cadherin levels and inducing Snail and Vimentin, both in vitro and in vivo, and also induces HIF-1α expression. Examining the predictive targetome of miR-574-3p led to the identification of CUL2 as a target. CUL2, when overexpressed, conferred protection against miR-574-3p-induced EMT, and also suppressed HIF-1α by direct binding (Ji et al. 2022).

Breast cancer stem cells have been demonstrated to display activated EMT, as well as overexpression of HIF-1α. miR-21 is also highly expressed in breast cancer stem cells. In order to discern the involvement of miR-21 in BCSC formation and sustenance, miR-21 was knocked down in these cells. Strikingly, antagonizing miR-21 led to a reversal of EMT, as observed by a decrease in Vimentin, N-cadherin and α-SMA. HIF-1α was also consequently decreased. While this demonstrates the close interconnection between hypoxia and EMT in driving and maintaining cancer stem cells, deciphering the targetome and mechanism of miR-21 action may be helpful to fully understand this process (Han et al. 2012).

miR-130b, a highly expressed miRNA in hepatocellular carcinoma (HCC), has been validated to target the tumor-suppressor PTEN. While it is correlated with low survival in HCC patients, it has been found to promote cell proliferation and metastasis in vitro and in mouse models. Phosphorylated Akt and HIF-1α were found to be stabilized in miR-130b overexpressing cells, hinting at a putative role in hypoxia induction (Chang et al. 2016).

Hypoxia in pancreatic cancer cells drives the conversion to mesenchymal phenotype, and also induces expression of miR-301a. The latter, in turn, potentiates the expression of EMT markers, and increases the expression and nuclear accumulation of HIF-1α. The overexpression of miR-301a in hypoxia was found to be HIF-2α dependent. TP63 was confirmed to be a direct target of the miRNA. Notably, TP63 has been found to inhibit EMT and metastasis (Zhang et al. 2020a, b, c). Hypoxia also induces miR-205 expression in cervical and lung cancers. The mechanism of suppression of apoptosis-stimulating protein of p53-2 (ASPP2) in hypoxia was attributed to miR-205. ASPP2 was shown to curb hypoxia-induced EMT and migration, although the mechanism remains unknown (Wang et al. 2016). However, in prostate cancer, miR-205 was shown to have opposite roles. MiR-205 expression was suppressed transcriptionally by HIF-1α, and forced overexpression was found to reverse EMT induced by cancer-associated fibroblasts (CAFs) (Gandellini et al. 2014). CAFs, key members of the tumor-associated reactive stroma, have also been studied for their role in EMT/hypoxia progression in other cancers. Co-culturing CAFs with non-small cell lung cancer (NSCLC) cells induced mesenchymal markers, angiogenesis and enhanced proliferation of the tumor cells. Interestingly, miR-224, which was initially found to be overexpressed only in CAFs, was later induced in co-cultured NSCLC cells. It was found to interact with and dampen Sirt3 mRNA. Consequently, it was found to activate the mTOR/HIF-1α signaling pathway, and stimulate the expression of VEGFA (Zhang et al. 2021).

6.1.2 Tumor-Suppressive miRNAs that Suppress Hypoxia-Induced EMT

Vasodilator-stimulated phosphoprotein (VASP) plays an important role in cell migration and cell adhesion, via regulation of cytoskeleton dynamics. In hepatocellular carcinoma, multiple mechanisms are undertaken by cancer cells during hypoxia to upregulate VASP, which in turn increases their metastatic potential and induces the expression of EMT-TFs and MMPs. VASP was shown to be a transcriptional target of HIF-1α, which binds to HREs in the promoter of the gene. VASP was also aberrantly expressed on activation of TGF-β signaling, which occurred downstream of hypoxia induction. Lastly, HIF-1α suppressed miR-204, which directly targets VASP by binding to its 3′ UTR. VASP protein was found to activate the Akt and ERK signaling in cells (Liu et al. 2018). The migration-suppressive role of miR-204 has also been documented in NSCLC, where Kruppel-like factor 7 (KLF7) was reported to be its direct target. Interestingly, miR-204 was found to be shuttled via exosomes derived from the mesenchymal stem cells into the NSCLC cells, where it suppressed the expression of mesenchymal markers and dampened the HIF1/Akt signaling (Liu et al. 2021a, b).

In gastric cancer, miR-4521 is under-expressed, which is in turn associated with a poorer prognosis. This miRNA is shown to be inhibitory to metastasis, both in cell culture and in vivo. Its suppression under hypoxia was attributed to the induction of ETS1, a well-characterized hypoxia-inducible transcription factor. Mechanistically, miR-4521 is involved in the suppression of IGF2, which leads to lower levels of phosphorylated Akt and IGF1R, the receptor for IGF2. Since the Akt/GSK3b signaling directly activates the transcription of EMT-TF SNAIL1, miR-4521 directly assists in inhibiting the spurious activation of Snail-activated EMT pathway. FOXM1 is another validated target of this miRNA (Xing et al. 2021).

miR-137 is another tumor-suppressive miRNA downregulated under hypoxia in multiple cancers. Forced expression of this miRNA reduces cell viability and migration, and promotes apoptosis in prostate cancer. miR-137 abrogates EGFR-mediated ERK signaling, thereby reducing the expression of N-cadherin, Vimentin and MMP2. LGR4, which binds to EGFR and stabilizes the receptor, was verified to be the suppressed target of miR-137 (Zhang et al. 2020a, b, c). Another mechanism for miR-137-based migration inhibition has been discovered in colorectal carcinoma, where the miRNA suppresses KDM1A. KDM1A is a histone demethylase, and forms an integral part of SNAI1-associated corepressor complex, that leads to transcriptional repression of the E-cadherin gene (Ding et al. 2021).

miR-200a, the EMT-suppressive miRNA that binds to and suppresses ZEB1, has also been found to target and decrease STAT4 in ovarian cancer. STAT4 is induced under hypoxia, and is associated with ovarian cancer, EMT and metastasis. The signaling partners of STAT4 in promoting hypoxia-induced EMT are yet to be determined for this cancer model (Li et al. 2021). miR-200c, another member of the miR-200 family that downregulates ZEB1, also suppresses HIF-1α, and VEGF in lung cancer. miR-200c suppresses EMT as well as stemness in breast cancer cells (Mansoori et al. 2021). In prostate cancer cell lines, miR-200c could undo hypoxia-induced mesenchymal features and inhibited their migration (Basu et al. 2020). In colorectal cancer, HIF-1α mediates hypoxia-induced EMT via Ascl2 (Achaete-Scute Family bHLH Transcription Factor 2). HIF-1α promotes the downregulation of miR-200b (another member of the EMT-suppressive miR-200 family) by Ascl2-mediated binding of E-boxes in the promoter of the miRNAs leading to their repression. In a feedback loop, miR-200b suppresses HIF-1α by targeting its 3′UTR (Shang et al. 2017).

miR-34a is a transcriptional target of p53, which binds to its promoter and activates its expression. Upstream of the transcription start site of miR-34a, three HREs have been discovered. In colorectal cancer, HIF-1α has been found to occupy the second HRE and suppress the expression of miR-34a. Remarkably, this inhibition is more pronounced in CRC cells with a mutant p53. miR-34a, when overexpressed, rescues the hypoxia-induced EMT phenotype by dampening the expression of SNAIL1, SNAIL2, ZEB1 and Vimentin. It also decreases the levels of phosphorylated STAT3. Inh3/PPP1R11 was found to be the direct target of miR-34a. Inh3 is a protein-phosphatase inhibitor, and likely stabilizes the phosphorylated STAT3. Inh3 was also found to have multiple HREs, indicating its direct activation by HIF-1α as another mechanism for sustaining STAT3 signaling and activating EMT. However, cells with wild-type p53 do not undergo hypoxia-induced EMT via the Inh/pSTAT3 axis, notably due to the high expression of miR-34a. The presence of a wild-type p53, hence, is considered a major threat to the induction of hypoxia-associated aggressive hallmarks (Li et al. 2017a, b).

In a study on breast cancer, miR-338-3p was found to be downregulated in breast cancer cell lines as compared to normal breast epithelial cell lines. Forceful overexpression of this miRNA interfered with the cells' proliferation and migration abilities, and augmented the epithelial phenotype. These findings were recapitulated in vivo. ZEB2, another major EMT-TF, was found to be a direct target of miR-338-3p. In this study, ZEB2 was shown to advance breast cancer progression by activation of the NF-κB and PI3K/Akt pathway. Interestingly, miR-338-3p was found to be suppressed in breast cancer cells when subjected to hypoxia, and this decrease was HIF-1α dependent via direct recruitment to the miRNA promoter (He et al. 2020).

TUFT1 protein has been previously shown to be induced via hypoxia and Hedgehog signaling. TUFT1 transcript and protein was found to be overexpressed in hepatocellular carcinoma (HCC) and corresponded with a bleak prognosis. TUFT1 promoted EMT and metastasis both in vitro and in vivo. Hypoxia was found to induce the expression of TUFT1, via HIF-1α-mediated suppression of miR-671-5p (which targets TUFT1) in HCC cells. TUFT1 was found to be central to hypoxia-mediated HCC proliferation and invasion, as gauged by experimental knockdown of TUFT1. TUFT1 activated the PI3K/Akt pathway through regulation of intracellular calcium levels (Dou et al. 2018).

miR-1224-5p is found to be significantly downregulated in colorectal cancer (CRC), and its expression was inversely associated with tumor metastasis. In CRC cell lines, miR-1224-5p suppressed invasion and migration, and increased the expression of E-cadherin, thereby abolishing EMT. The SP1 transcription factor was found to be a direct target of miR-1224-5p. The levels of miR-1224-5p were found to decrease under hypoxia, and enforcing its expression abolished hypoxia-driven EMT in CRC cells. miR-1224-5p and SP1 were found to be involved in NF-κB signaling, as overexpression of miR-1224-5p led to lower levels of phosphorylated p65 (Li et al. 2020).

Hypoxia has been shown to downregulate the expression of miR-1236. It was found that the EMT-TF, TWIST, which is expressed in hypoxia, binds to the promoter region of miR-1236 and suppresses its expression. Conversely, miR-1236 directly targets SENP1, which is involved in mediating the downstream effects of hypoxia-driven EMT (Chen et al. 2016). Among miRNAs found to be downregulated by hypoxia is miR-30c in renal cell carcinoma. The expression of miR-30c was directly linked to the VHL status in cells, and was shown to directly target the EMT-TF, SLUG (Huang et al. 2013).

In hepatocellular carcinoma, miR-1296, via downregulating SRPK1, decreases the levels of phosphorylated Akt, thereby interfering with PI3K/Akt signaling. miR-1296 was found to be antagonistic to EMT activation and cell migration. Consequently, its expression was suppressed under hypoxia. The mechanism behind this downregulation has not been explored further (Xu et al. 2017).

MicroRNAs that target and downregulate HIF1 directly abrogate the hypoxia pathway. MiR-143-5p suppresses HIF-1α through binding to the 3′UTR in gall bladder cancer. When miR-143-5p was overexpressed artificially in gall bladder cancer cells, a considerable decrease in proliferation and migration, as gauged by wound healing and trans-well assay, was observed. In parallel, it decreased TWIST1, Vimentin and β-catenin transcripts and proteins (He et al. 2017). miR-142 is another miRNA that targets HIF-1α in pancreatic cancer, and consequently is downregulated in cancer cells, especially under hypoxia (Lu et al. 2017). miR-622 has also been found to bind to the 3′UTR of HIF-1α mRNA and mediate its degradation, in lung cancer. Overexpression of miR-622 was shown to decrease SNAIL1, Vimentin and β-catenin protein levels, with an increase in E-cadherin. FOXO3a was found to bind to the promoter of miR-622 and promote its expression (Cheng et al. 2015). Another miRNA, miR-182-5p, has been shown to directly target HIF-2α in NSCLC (Yang et al. 2021).

Novel compounds and drugs have also been tested to interfere with hypoxia-induced EMT. Treatment of gastric cancer cells with Crocin, the active compound present in saffron, was found to decrease HIF-1α and KLF5 expression, thereby inhibiting EMT and migration. Crocin treatment was found to increase the expression of miR-320, which directly suppresses KLF5 (Zhou et al. 2019). Another alkaloid compound, Matrine, obtained from Chinese medicinal plants, has been found to inhibit EMT. Matrine was found to drive the expression of miR-199a-5p, which directly suppresses HIF-1α (Dai et al. 2021).

6.1.3 Exosomal miRNAs that Regulate Hypoxia-Induced EMT

Exosomal miRNAs are prominent mediators of intercellular communication and signaling, which is a vital component of the tumor tissue. Bone marrow-derived mesenchymal stem cells (BMSCs) are stable components of the tumor microenvironment, and directly assist in tumor progression. Exosomal miRNAs profiled from hypoxic BMSCs identified miR-193a, miR-210-3p and miR-5100 as transferring to lung cancer cells and initiating their invasion. This invasion was brought about by

STAT3-mediated EMT activation (Zhang et al. 2019a, b). Hypoxic mesenchymal stem cells were also found to secrete miR-21-5p through their exosomes. These exosomes supported growth, viability, M2 macrophage differentiation and EMT of NSCLC cells (Ren et al. 2019). In hepatocellular carcinoma cells exposed to hypoxia, secretory exosomes were found to promote EMT. MiR-1273f was found to be the key component of hypoxic exosomes responsible for the downstream effects by targeting LHX6. LHX6 is a negative regulator of the WNT/β-catenin signaling, hence hypoxic exosomes promote Wnt signaling activation in nearby cells under hypoxia to drive their invasion (Yu et al. 2019). In lung carcinoma, miR-31-5p was an active member of hypoxic exosomes, and could be internalized in normoxic cancer cells. MiR-31-5p was validated to target SATB2, and activated the MEK/ERK signaling to induce EMT (Yu et al. 2021). On the other hand, low expression of tumor-suppressive miRNAs has also been detected in hypoxic exosomes. In CRC, miR-1255b-5p was reduced in exosomes under hypoxia, and hTERT was postulated to be targeted by this miRNA (Zhang et al. 2020a, b, c).

6.1.4 Regulation of Dicer

Nur77, a type of nuclear receptor, is transcriptionally activated by HIF-1α. Under hypoxia, Nur77 was shown to activate Akt to stimulate β-catenin signaling in CRC cells. Nur77 overexpression could potentiate the spheroid-forming ability of cells, resulting in increased stemness. Nur77 was also found to suppress the levels of Dicer in the cell, indicating a global impairment of miRNA biogenesis. Expression of mature let-7i-5p was found to be directly affected by this process (Shi et al. 2021). Dicer protein has been found to be suppressed in other cancers as well during the onset of hypoxia-induced EMT. In hepatocellular carcinoma, Dicer levels were severely downregulated under hypoxia. Overexpressing Dicer abrogated the expression of HIF-1α and associated EMT and migration (Ibrahim et al. 2017). Interestingly, HIF-1α has been found to directly interact with Dicer, and initiate its ubiquitination, resulting in diminished levels of Dicer and inefficient miRNA processing (Lai et al. 2018). The major miRNAs governing cancer EMT and hypoxia are listed in Table 1, and their network of interactions is shown in Fig. 3.

6.2 lncRNAs Governing Cancer EMT and Hypoxia

We have seen how tumor hypoxia may lead to EMT, exacerbating the tumor invasiveness. For instance, crosstalk at the level of different pathways can inter-regulate both hypoxia and EMT. One such pathway is Wnt signaling, which is mostly activated in different cancers. It is already known that Wnt signaling is central to promoting EMT in cancers. A study in breast cancer showed that hypoxia may lead to EMT via lncRNA RBM5-AS1/β-catenin axis. Additionally, it was shown that $CD44^{+}$/$CD24^{-}$ positive breast cancer stem cells (BCSCs) had higher expression levels of

Table 1 List of miRNAs governing cancer hypoxia and EMT

miRNA	Role in hypoxia-induced EMT	Mechanism of action	Cancer type	References
miR-210-3p	Activates	Activates NF-κB signaling, TGF-β signaling	Glioma, pancreatic cancer, breast cancer	Ren et al. (2017), Tang et al. (2018), Liu et al. (2021a, b)
miR-135b-3p	Activates	Activated by p38 MAP Kinase, suppresses FBXL16-mediated degradation of HIF-1α	Triple-negative breast cancer	Kim et al. (2021)
miR-574-3p	Activates	Induces HIF1, decreases E-cadherin, targets CUL2	Gastric cancer	Ji et al. (2022)
miR-21	Activates	Induces EMT, induces HIF-1α	Breast cancer stem cells	Han et al. (2012)
miR-130b	Activates	Targets PTEN, stabilizes HIF-1α and phosphorylated Akt	Hepatocellular carcinoma	Chang et al. (2016)
miR-301a	Activates	Increases nuclear accumulation of HIF-1α, targets p63 that suppresses EMT	Pancreatic cancer	Zhang et al. (2020a, b, c)
miR-205	Activates Inhibits	Induced by hypoxia, suppresses ASPP2 Suppressed by HIF-1α, reverses EMT	Cervical cancer, lung cancer Prostate cancer	Wang et al. (2016), Zhang et al. (2020a, b, c) Gandellini et al. (2014)
miR-224	Activates	Activates mTOR/HIF-1α signaling	Non-small cell lung cancer	Zhang et al. (2021)
miR-1273f	Activates	Promotes Wnt signaling	Hepatocellular carcinoma hypoxic exosomes	Yu et al. (2019)
miR-31-5p	Activates	Activates MEK/ERK signaling	Lung cancer hypoxic exosomes	Yu et al. (2021)
mir-204	Inhibits	Suppresses VASP, suppresses HIF1/Akt signaling	Hepatocellular carcinoma, NSCLC	

(continued)

Table 1 (continued)

miRNA	Role in hypoxia-induced EMT	Mechanism of action	Cancer type	References
miR-4521	Inhibits	Inhibits Akt/GSK3-β signaling via suppression of IGF2	Gastric cancer	Xing et al. (2021)
miR-137	Inhibits	Inhibits EGFR-ERK signaling, suppresses KDM1A	Prostate cancer, colorectal cancer	
miR-200a	Inhibits	Represses STAT4	Ovarian cancer	Li et al. (2021)
miR-200c	Inhibits	Downregulates ZEB1, HIF-1α, VEGFA	Lung cancer, prostate cancer	Mansoori et al. (2021), Basu et al. (2020)
miR-200b	Inhibits	Represses HIF-1α	Colorectal cancer	Shang et al. (2017)
miR-34a	Inhibits	Repressed by HIF-1α, inhibits expression of SNAIL1, SNAIL2, ZEB1, decreases phosphorylated STAT3	Colorectal cancer	Li et al. (2017a, b)
miR-338-3p	Inhibits	Inhibits ZEB2 consequently suppressing PI3K/Akt signaling, suppressed by HIF-1α	Breast cancer	He et al. (2020)
miR-671-5p	Inhibits	Targets TUFT1 to inhibit PI3K/Akt signaling, repressed by HIF-1α	Hepatocellular carcinoma	Dou et al. (2018)
miR-1224-5p	Inhibits	Decreases NF-κB signaling	Colorectal cancer	Li et al. (2020)
miR-1236	Inhibits	Decreases hypoxia-driven EMT by targeting SENP1, suppressed by TWIST1	Multiple cancers	Chen et al. (2016)
miR-30c	Inhibits	Downregulated in hypoxia, downregulates SLUG	Renal cell carcinoma	Huang et al. (2013)

(continued)

Table 1 (continued)

miRNA	Role in hypoxia-induced EMT	Mechanism of action	Cancer type	References
miR-1296	Inhibits	Decreases PI3K/Akt signaling, expression is suppressed in hypoxia	Hepatocellular carcinoma	Xu et al. (2017)
miR-143-5p	Inhibits	Suppresses HIF-1α, decreases TWIST1, Vimentin	Gallbladder cancer	He et al. (2017)
miR-142	Inhibits	Suppresses HIF-1α	Pancreatic cancer	Lu et al. (2017)
miR-622	Inhibits	Suppresses HIF-1α, decreases SNAIL1, Vimentin	Lung cancer	Cheng et al. (2015)
miR-199a-5p	Inhibits	Suppresses HIF-1α, inhibits EMT	Hepatocellular carcinoma	Dai et al. (2021)

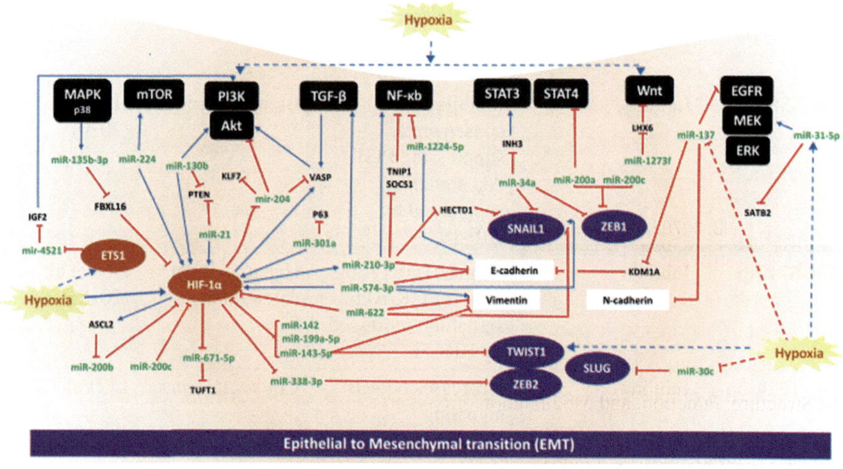

Fig. 3 An overview of regulatory network involving miRNAs in cancer EMT and hypoxia. The black boxes indicate important signaling pathways, and microRNAs are colored green. Red lines indicate negative regulation. Blue arrows indicate positive regulation or activation. Orange circles depict important hypoxia-signaling elements, while purple circles indicate EMT players including EMT-TFs

RBM5-AS1. It was reported that hypoxia-induced RBM5-AS1 could upregulate β-catenin via RUNX2 leading to the overexpression of RBM5-AS1, thus making it an HIF-independent hypoxia-responsive lncRNA in breast cancer (Li et al. 2022). In lung cancer, hypoxia-induced EMT was reported via HIF-2α-dependent induction of lncRNA NEAT1, sponging miR-101-3p. Moreover, SOX9 that activated the Wnt pathway was reported to be upregulated upon hypoxic stress (Kong et al. 2019). In gastric cancer tissues tumor hypoxia is exhibited by the gastric mucosal cells. A recent study showed that hypoxia-induced lncRNA HCP5 led to EMT by sponging miR-186-5p which targeted WNT5A. WNT5A has been previously known to be a crucial regulator of EMT in prostate, lung, breast, and skin cancers (Gao et al. 2021). Hypoxia-induced lncRNA KDM4A-AS1 was reported to sponge miR-411-5p in hepatocellular carcinoma (HCC), the downstream effect being upregulation of KPNA2 leading to activated Akt signaling. KDM4A-AS1 levels were higher in HCC patients, favoring a poor prognosis (Chen et al. 2021). In the same cancer, yet another hypoxia-regulated lncRNA, MAPKAPK5-AS1, was discovered to have an oncogenic role. Here, miR-154-5p was shown to be sponged by MAPKAPK5-AS1, leading to the expression of PLAGL2, which was known to promote EMT via ZEB1 upregulation. Moreover, PLAGL2 could activate EGFR/PI3K/Akt signaling pathways that could sustain higher HIF-1α levels. This positive feedback loop could promote EMT in HCC patients (Wang et al. 2021). In colorectal cancer, it was reported that lncRNA XIST could sponge miR-93-5p and in turn upregulate HIF-1α/AXL signaling, leading to EMT (Yang et al. 2020). Similarly, the lncRNA HOTAIR was reported as a sponge for miR-217 in renal cell carcinoma, thereby promoting HIF-1α/AXL signaling. AXL signaling was known to further activate PI3K/Akt, NF-κB and JAK-STAT pathways, leading to EMT (Hong et al. 2017). In ovarian cancer, the lncRNA DSCR8 was reported to sponge a tumor-suppressor miR-98-5p, causing HIF-1α and STAT3 upregulation, contributing towards EMT (Dong et al. 2020). The same axis HIF-1α/STAT3 was reported in breast cancer, where expression of lncRNA VCAN-AS1 sponged miR-106a-5p, contributing to EMT (Du et al. 2021). In HCC, it was reported that AGAP2-AS1 was upregulated by hypoxia, and it sponged miR-16-5p, further impacting the ANXA11/Akt axis. Further downstream, ANXA11 can recruit the Akt pathway to sustain EMT (Liu et al. 2019). Hypoxia, by directly impacting EMT-TFs, can also promote EMT. In pancreatic cancer, it was found that hypoxic conditions led to the overexpression of lncRNA BX111. This overexpression had significant correlation with patient survival data, indicating that high BX111 levels corresponded with a poor prognosis in pancreatic cancer patients. It was shown that HIF-1α-mediated induction of BX111 led to the recruitment of ZEB1 via YB-1, where the latter promoted ZEB1 upregulation leading to EMT. Abrogation of BX111 crucially impacted the levels of YB-1 and in turn ZEB1 overexpression. This showed how BX111 could serve as a crucial EMT-promoting lncRNA in pancreatic cancers (Deng et al. 2018). Hypoxia-regulated lncRNA HOTTIP was reported to induce EMT in glioma cells. This regulation occurred through the miR-101-3p/ZEB1 axis where ZEB1 expression was promoted. HOTTIP could sponge miR-101-3p in glioma. Interestingly, HOTTIP and HIF-1α levels hold prognostic significance in glioblastoma (Zhang et al. 2017). In gastric cancer, it was reported that hypoxia-regulated lncRNA PCGEM1 could recruit SNAIL1 and promote EMT. Interestingly, PCGEM1 induction was through HIF-1α, and this lncRNA, probably by acting as scaffold between HIF-1α and SNAIL1, indirectly led to EMT (Zhang et al. 2019a, b). In different cancer cell lines, it was reported that hypoxia-regulated lncRNA lnc-RP11-390F4.3 was overexpressed upon HIF-1α activity. This was shown to recruit multiple regulators of EMT such as ZEB1, SNAIL, ZEB2 and TWIST1 (Peng et al. 2020). Oncofetal lncRNA H19 was discovered to be induced by both hypoxia and TGF-β signaling in a mouse breast cancer model. This paved the way for the understanding of a convergence between hypoxia and the TGF-β pathway. TGF-β-induced EMT via H19 is due to the PI3K-Akt signaling pathway. Moreover, SLUG expression could positively sustain H19 levels (Matouk et al. 2014). H19 was reported to regulate miR-675, modulating SLUG expression (Zhu et al. 2014). H19 could sponge miR-138 and induce ZEB1/ZEB2 expression, leading to EMT. In gall bladder cancer, LINC00152 was reported to sponge miR-138-5p, leading to upregulation of HIF-1α. The sustained higher HIF-1α levels led to EMT via SLUG upregulation (Cai et al. 2022). lncRNA NORAD was reported to be elevated under hypoxia, and importantly, its higher levels corresponded to a worse prognosis in pancreatic cancer patients. Upon hypoxia induction, NORAD could sponge miR-125a-3p and upregulate RhoA, ultimately leading to EMT. Additionally, it was reported that miR-125a-3p could participate in MAPK/PI3K/mTOR pathways, which indicated the ability of NORAD to promote EMT via convergence of hypoxia and other signaling pathways (H. Li et al. 2017a, b). In another study it was found that NORAD could sponge miR-495-3p in colorectal cancer cells. Interestingly, HIF-1α and NORAD levels were found to be positively correlated in colorectal cancers. HIF-1α/EMT signaling was shown to be regulated upon NORAD knockdown in cells. This hinted at its role in regulating hypoxia-induced EMT, although the exact mechanisms remain unclear (Zhang et al. 2022). The major lncRNAs governing cancer EMT and hypoxia are listed in Table 2, and their network of interactions is shown in Fig. 4.

6.3 *Other ncRNAs Governing Cancer EMT and Hypoxia*

Recent studies have revealed the role of other ncRNAs, such as circRNAs and tRFs, in regulation of EMT and hypoxia. CircRNAs were discovered earlier than miRNAs and lncRNAs. However, it is only during the last decade that research on their specific roles in cancers has gathered pace. In a study on cervical cancer, circ-HIPK3 levels were found to be elevated in tissues and cancer cells. circ-HIPK3 could sponge miR-338-3p leading to the upregulation of HIF-1α. This indirectly hinted at its role in promoting EMT. However, the exact EMT-promoting mechanism remains elusive (Qian et al. 2020). circ-0001875 was found to be significantly upregulated in lung cancer cells. It was reported to sponge miR-31-5p by upregulating its target SP1. Hence, HIF-1α-induced circ-0001875 could promote EMT via the TGF-β/Smad2 pathway (Wu et al. 2022). In gastric cancer cells, circ-0081143 was reported to upregulate EGFR by sponging miR-497-5p, contributing towards EMT in hypoxic gastric cancer cells (Tang et al. 2020). In lung cancer, SAE2-mediated SUMOylation of an EMT marker such as Vimentin was promoted upon circ-CCDC66 and FAK

Table 2 List of lncRNAs governing cancer EMT and hypoxia

lncRNA	Role in Hypoxia-induced EMT	Mechanism of action	Cancer type	References
RBM5-AS1	Activates	Activation of Wnt/β-catenin signaling	Breast cancer	Li et al. (2022)
HCP5	Activates	Upregulation of WNT5 by sponging miR-186-5p	Gastric cancer	Gao et al. (2021)
MAPKAPK5-AS1	Activates	Activation of EGFR/Akt pathway by upregulation of PLAGL2 via sponging miR-154-5p leading to ZEB1 activation	Hepatocellular carcinoma	Wang et al. (2021)
HOTAIR	Activates	Activation of AXL, NF-κB and JAK-STAT signaling by sponging miR-217	Renal cell carcinoma	Hong et al. (2017)
DSCR8	Activates	Activation of STAT3 by sponging miR-98-5p	Ovarian cancer	Dong et al. (2020)
BX111	Activates	ZEB1 activation by recruitment of YB1	Pancreatic cancer	Deng et al. (2018)
HOTTIP	Activates	Sponges miR-101-3p and upregulating ZEB1	Glioma	Zhang et al. (2017)
PCGEM1	Activates	SNAIL1 activation by its scaffolding with HIF-1α	Gastric cancer	Zhang et al. (2019a, b)
lnc-RP11-390F4.3	Activates	Recruitment of ZEB1, SNAIL1, ZEB2, TWIST1	Multiple	Peng et al. (2020)
H19	Activates	Induction of TGF-β signaling via Akt pathway and SLUG expression	Breast cancer	Zhu et al. (2014)
LINC00152	Activates	SLUG upregulation by sponging miR-138-5p	Gall bladder cancer	Cai et al. (2022)
NORAD	Activates	RhoA upregulation by sponging miR-125a-3p	Pancreatic cancer	Li et al. (2017a, b)
		Activation of mTOR/Akt pathways by sponging miR-125a-3p		Tang et al. (2015)

(continued)

Table 2 (continued)

lncRNA	Role in Hypoxia-induced EMT	Mechanism of action	Cancer type	References
VCAN-AS1	Activates	Upregulation of STAT3 by sponging miR-106a-5p	Breast cancer	Du et al. (2021)

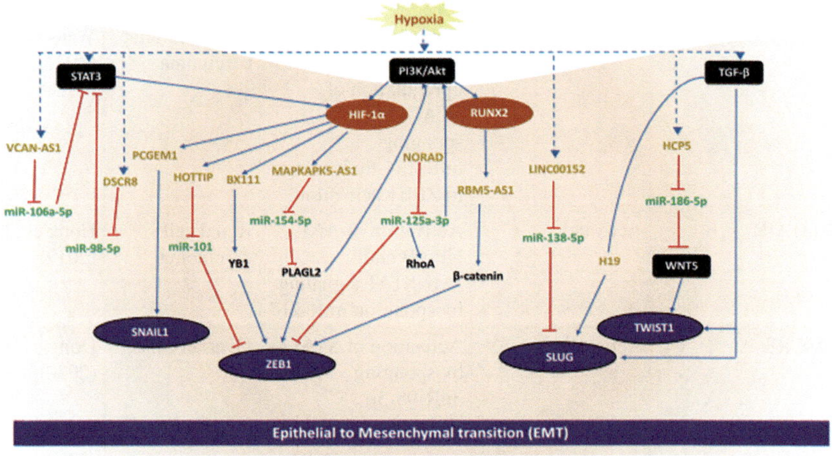

Fig. 4 An overview of the regulatory network involving lncRNAs in cancer EMT and hypoxia. The black boxes indicate important signaling pathways, microRNAs are colored green, and lncRNAs are colored dark yellow. Red lines indicate negative regulation and sponging of microRNAs by lncRNAs. Blue arrows indicate positive regulation or activation. Orange circles depict important hypoxia-signaling elements, while purple circles indicate EMT players including EMT-TFs

expression, under hypoxic conditions. This highlighted the crucial role of a circRNA in mediating HGF-MET pathway-induced EGFR expression (Joseph et al. 2018).

The tRNA-derived fragment, tRF-20-M0NK5Y93, was found to be downregulated under hypoxia, in colon cancer cells. Suppression of tRF-20-M0NK5Y93 was found to increase the transcript and protein levels of MMP2 and MMP7, which are important for EMT progression, as well as HIF-1α. tRF-20-M0NK5Y93 was found to directly bind to the 3′UTR of the Claudin1 gene and mediate its downregulation (Luan et al. 2021). A list of circRNAs and tRFs regulating cancer EMT and hypoxia is provided in Table 3.

Table 3 List of circRNAs and tRFs regulating cancer EMT and hypoxia

Name of ncRNA	Role in hypoxia-induced EMT	Mechanism of action	Cancer type	References
circ-HIPK3	Activates	Sponges miR-338-3p to upregulate HIF-1α	Cervical cancer	Qian et al. (2020)
circ-0001875	Activates	Sponges miR-31-5p leading to EMT via TGF-β/Smad2 pathway, induced by HIF-1α	Lung cancer	Wu et al. (2022)
circ-0081143	Activates	Upregulates EGFR by sponging miR-497-5p leading to EMT	Gastric cancer	Tang et al. (2020)
circ-CCDC66	Activates	Induces EGFR expression through HGF-MET pathway	Lung cancer	Joseph et al. (2018)
tRF-20-M0NK5Y93	Inhibits	Decreases MMP2, MMP7, HIF-1α, targets Claudin1 gene	Colon cancer	Luan et al. (2021)

7 Non-coding RNAs as Therapeutics and Novel Biomarkers

Ideal biomarkers must possess certain characteristics such as high serum stability, presence in different body fluids, ease of analysis by existing technological regimes, high specificity for a disease condition, and high level of sensitivity that correlates with the disease progression or burden. ncRNAs, in addition to their widely studied regulatory roles, have recently been exciting candidates for disease biomarker research. Earlier attempts at studying this aspect of ncRNAs involved large-scale profiling across tumor and normal tissues. ncRNAs could be assayed from different body fluids such as blood, serum, urine, saliva. In the context of cancer, biopsies taken from tumor sites could help decipher disease progression based on ncRNA profiles. However, tissue biopsies are highly temporal and may not convey the precise transcriptomic profile of a tumor. There are multiple confounding factors and the presence of tumor heterogeneity that might interfere in understanding of the ncRNA profile of a cancer. More recently, research has shown that ncRNAs can be extracted from the extracellular vesicle (EV) secreted by tumor cells. Freely circulating RNA entities are largely amenable to RNase activity, hence the RNA entities that can be profiled from such EVs may be stable enough to act as biomarkers under a specific context. ncRNAs can be assayed by simple techniques such as cDNA microarray profiles, RNA sequencing data or qRT-PCR. Hence, ncRNAs as potential biomarkers appear to be highly promising. To judge by recent statistics, lncRNA PCA3 appears to be the only ncRNA candidate receiving FDA approval. PCA3 levels are highly elevated in prostate cancer cases and studies have shown that it could be used as a prostate cancer biomarker in conjunction with other existing diagnostic tests. However, this field is surely not limited by the infancy of FDA-approved ncRNA biomarkers. Their inherent biogenesis and functional roles facilitate biomarker research in cancers. For instance, lncRNAs are highly stable in the bloodstream owing to their association with exosomes and their ability to form secondary structures. Many lncRNAs are currently being explored as cancer biomarkers: MALAT1 for HCC, prostate, NSCLC; LIN00152 for gastric cancer, HCC; H19 for breast cancer; and HOTAIR for colorectal and cervical cancer. Additionally, a panel of lncRNAs could also be used as novel biomarkers (Badowski et al. 2022). Various cancer studies have drawn attention to circRNAs. circRNA abundance has been reported in body fluids like blood and urine. circRNAs are highly stable owing to their single-stranded circular structures which shield them from RNase digestion. They need to complex themselves with other proteins or biomolecules to be stable. Although no circRNA has yet entered clinical stages, emerging studies seem promising. circRNAs exhibit tissue specificity, among other advantages, such as ease of detection and stability. hsa_circ_0000190, hsa_circ_002059 have been implicated in gastric cancer. circ-FARSA is being studied in NSCLC, hsa_circ_0027089 has been implicated in HCC. circ-AR has been studied in differentiating metastatic castration-resistant prostate (mCRPC) cancer from primary tumor. ciRS-7 has been implicated in colorectal cancer. The potential of circRNAs as biomarkers is still in its infancy owing to doubts regarding the regulation of their biogenesis and their exact role in normal, well-studied cellular pathways (Verduci et al. 2021). miRNAs have been found associated with EVs, especially exosomes. Such miRNAs could play a role as disease biomarkers. Circulating miRNAs hold promising potential for biomarker development. Many tumor-suppressor and oncogenic miRNAs and their regulatory roles have been studied across different cancers and under different contexts. miRNAs have been studied as potential biomarkers from body fluids such as blood and urine. miR-10b has been reported in pancreatic cancer, and miR-21 in cancers of breast, colorectal, lung, prostate, and leukemia. miR-16 and miR-34b have been implicated in lung cancer; miR-155 in leukemia; miR-221 and miR-375 in prostate cancer. miR-1290 has been implicated in colorectal and prostate cancers. However, no miRNA candidate biomarkers have yet been approved by the FDA.

The past decade has witnessed the rapid development of RNA-based therapeutics, culminating in an approved list of eleven RNA-based drugs. Notably, all these drugs are antisense oligonucleotides (ASO), however lncRNA/miRNA/circRNA-based drugs have not yet been approved. This has not discouraged emerging research which has led to some miRNA-based drug candidates entering phase II/phase III trials. In Huntington's disease, a pri-miR-451 backbone that targets Huntingtin (HTT) protein, is currently being tested under the name rAAV5-miHTT. Yet another miRNA drug candidate is RG-125, which is an anti-miR to miR-103/107, currently under regulatory trials for non-alcoholic fatty liver disease (NAFLD) and type II diabetes (T2D). Previously, it was shown that miR-103/107 targets and regulates insulin sensitivity. Hence, the anti-miR was shown to increase insulin sensitivity in diabetes. Remlarsen, an miR-29 mimic, has been developed for the treatment of a skin disorder, keloid fibrosis. The anti-miR-122 Miravirsen has been developed to treat hepatitis C infection. miRNA-based drugs offer some advantages. Owing to their widespread regulatory effects, a cascade of specific players may be regulated by a single miRNA, offering precision. Also, they are naturally synthesized in the body, hence their immunogenicity could be lower than that of chemically synthesized siRNA/ASOs. lncRNAs, on the other hand, can act as natural antisense transcripts (NATs), unlike siRNAs which are synthesized in vitro. Antisense-lncRNAs can bring about silencing in cis to their host gene (Winkle et al. 2021). As far as cancer is concerned, some miRNA-based therapeutics have entered clinical trials. For instance, miR-16 mimic for NSCLC is under phase I trials in the US. Similarly, anti-miR-155 has nearly reached phase II trials for cutaneous T-cell lymphoma (CTCL). Apart from these two promising candidates, miRNA clusters such as the miR-17/92, miR-200 family, miR-21, miR-210, miR-21, and miR-155 have been implicated in multiple cancers with their expression levels highly dysregulated. Further investigations into their biological roles may enable their development as potential drug candidates (Forterre et al. 2020).

However, there are some challenges associated with ncRNA-based drug candidate delivery, specificity, and their tolerability. These challenges can be tackled through appropriate inter-disciplinary research. To summarize, it may be possible to turn ncRNA-based therapeutics into a reality.

8 Conclusions

Cancer is a heterogeneous disease that involves massive deregulation of both protein-coding genes and ncRNAs. Lately, emerging studies have uncovered the relevance of miRNAs, lncRNAs and circRNAs in regulating each hallmark of cancer. Hypoxia in the tumor microenvironment has been reported to aggravate tumor growth and metastasis and contribute to changes in cell phenotype and metabolism. The mechanism behind hypoxia-induced EMT has been found to rely heavily on regulatory ncRNA-based interactions. Here, we have obtained a comprehensive overview of such mediators across a variety of cancers. Hypoxia-induced miRNAs and lncRNAs have been found to sustain TGF-β, PI3K/Akt, Wnt, NF-κB, EGFR and STAT signaling pathways, and directly or indirectly activate EMT transcription factors. They are also reported to suppress the expression of E-cadherin and induce mesenchymal markers. Conversely, ncRNAs downregulated under hypoxia have been reported to negatively regulate EMT-TFs and associated markers. LncRNAs and circRNAs majorly mediate these functions by sponging miRNAs. More functional and mechanistic studies are required to unravel the complicated networks involving various ncRNAs, hypoxia and EMT and identify the hub networks with the potential to be targeted for efficient therapy. Such ncRNAs could be important candidates for further investigations and may ultimately lead to the development of novel biomarkers linking the EMT and hypoxia in cancers.

References

Abe N, Matsumoto K, Nishihara M et al (2015) Rolling circle translation of circular RNA in living human cells. Sci Rep 5:16435

Agarwal V, Bell GW, Nam J-W et al (2015) Predicting effective microRNA target sites in mammalian mRNAs. Elife 4:e05005

Aster JC, Pear WS, Blacklow SC (2017) The varied roles of notch in cancer. Annu Rev Pathol Mech Dis 12:245–275

Badowski C, He B, Garmire LX (2022) Blood-derived lncRNAs as biomarkers for cancer diagnosis: the Good, the Bad and the Beauty. npj Precis Oncol 6:40

Bao MH, Wong CC (2021) Hypoxia, metabolic reprogramming, and drug resistance in liver cancer. Cells 10:1715

Basu S, Chaudhary A, Chowdhury P et al (2020) Evaluating the role of hsa-miR-200c in reversing the epithelial to mesenchymal transition in prostate cancer. Gene 730:144264

Begg K, Tavassoli M (2020) Inside the hypoxic tumour: reprogramming of the DDR and radioresistance. Cell Death Discov 6:77

Cai Q, Wang Z, Wang S et al (2022) Long non-coding RNA LINC00152 promotes gallbladder cancer metastasis and epithelial–mesenchymal transition by regulating HIF-1α via miR-138. Open Biol 7:160247

Capel B, Swain A, Nicolis S et al (1993) Circular transcripts of the testis-determining gene Sry in adult mouse testis. Cell 73:1019–1030

Carmeliet P, Jain RK (2011) Molecular mechanisms and clinical applications of angiogenesis. Nature 473:298–307

Carmeliet P, Dor Y, Herbert J-M et al (1998) Role of HIF-1α in hypoxia-mediated apoptosis, cell proliferation and tumour angiogenesis. Nature 394:485–490

Chandradoss SD, Schirle NT, Szczepaniak M et al (2015) A dynamic search process underlies MicroRNA targeting. Cell 162:96–107

Chang RM, Xu JF, Fang F et al (2016) MicroRNA-130b promotes proliferation and EMT-induced metastasis via PTEN/p-AKT/HIF-1α signalling. Tumour Biol 37:10609–10619

Chen J, Imanaka N, Chen J et al (2010) Hypoxia potentiates Notch signaling in breast cancer leading to decreased E-cadherin expression and increased cell migration and invasion. Br J Cancer 102:351–360

Chen SY, Teng SC, Cheng TH et al (2016) miR-1236 regulates hypoxia-induced epithelial–mesenchymal transition and cell migration/invasion through repressing SENP1 and HDAC3. Cancer Lett 378:59–67

Chen T, Liu R, Niu Y et al (2021) HIF-1α-activated long non-coding RNA KDM4A-AS1 promotes hepatocellular carcinoma progression via the miR-411-5p/KPNA2/AKT pathway. Cell Death Dis 12:1152

Cheng CW, Chen PM, Hsieh YH et al (2015) Foxo3a-mediated overexpression of microRNA-622 suppresses tumor metastasis by repressing hypoxia-inducible factor-1α in erk-responsive lung cancer. Oncotarget 6:44222

Chi SW, Zang JB, Mele A et al (2009) Argonaute HITS-CLIP decodes microRNA–mRNA interaction maps. Nature 460:479–486

Chillón I, Marcia M (2020) The molecular structure of long non-coding RNAs: emerging patterns and functional implications. Crit Rev Biochem Mol Biol 55:662–690

Choi B-J, Park S-A, Lee S-Y et al (2017) Hypoxia induces epithelial-mesenchymal transition in colorectal cancer cells through ubiquitin-specific protease 47-mediated stabilization of Snail: a potential role of Sox9. Sci Rep 7:15918

D'Ignazio L, Bandarra D, Rocha S (2016) NF-κB and HIF crosstalk in immune responses. FEBS J 283:413–424

Dai M, Chen N, Li J et al (2021) In vitro and in vivo anti-metastatic effect of the alkaliod matrine from Sophora flavecens on hepatocellular carcinoma and its mechanisms. Phytomedicine 87:153580

de Wever O, Mareel M (2003) Role of tissue stroma in cancer cell invasion. J Pathol 200:429–447

Deng S, Chen H, Ye Z et al (2018) Hypoxia-induced LncRNA-BX111 promotes metastasis and progression of pancreatic cancer through regulating ZEB1 transcription. Oncogene 37:5811–5828
Dillekås H, Rogers MS, Straume O (2019) Are 90% of deaths from cancer caused by metastases? Cancer Med 8:5574–5576
Ding X, Zhang J, Feng Z et al (2021) MiR-137-3p inhibits colorectal cancer cell migration by regulating a KDM1A-dependent epithelial-mesenchymal transition. Dig Dis Sci 66:2272–2282
Djebali S, Davis CA, Merkel A et al (2012) Landscape of transcription in human cells. Nature 2012 489:7414 489:101–108
Dong L, Cao X, Luo Y et al (2020) A positive feedback loop of lncRNA DSCR8/miR-98-5p/STAT3/HIF-1α plays a role in the progression of ovarian cancer. Front Oncol 10:1713
Dou C, Zhou Z, Xu Q et al (2018) Hypoxia-induced TUFT1 promotes the growth and metastasis of hepatocellular carcinoma by activating the Ca^{2+}/PI3K/AKT pathway. Oncogene 38:1239–1255
Du P, Luo K, Li G et al (2021) Long non-coding RNA VCAN-AS1 promotes the malignant behaviors of breast cancer by regulating the miR-106a-5p-mediated STAT3/HIF-1α pathway. Bioengineered 12:5028–5044
Forsythe JA, Jiang BH, Iyer NV et al (1996) Activation of vascular endothelial growth factor gene transcription by hypoxia-inducible factor 1. Mol Cell Biol 16:4604–4613
Forterre A, Komuro H, Aminova S et al (2020) A comprehensive review of cancer MicroRNA therapeutic delivery strategies. Cancers 12:1–21
Frankenstein Z, Basanta D, Franco OE et al (2020) Stromal reactivity differentially drives tumour cell evolution and prostate cancer progression. Nat Ecol Evol 4:870–884
Frisch SM, Schaller M, Cieply B (2013) Mechanisms that link the oncogenic epithelial-mesenchymal transition to suppression of anoikis. J Cell Sci 126:21–29
Fukao A, Mishima Y, Takizawa N et al (2014) MicroRNAs trigger dissociation of eIF4AI and eIF4AII from target mRNAs in humans. Mol Cell 56:79–89
Furuta C, Miyamoto T, Takagi T et al (2015) Transforming growth factor-β signaling enhancement by long-term exposure to hypoxia in a tumor microenvironment composed of Lewis lung carcinoma cells. Cancer Sci 106:1524–1533
Gandellini P, Giannoni E, Casamichele A et al (2014) miR-205 hinders the malignant interplay between prostate cancer cells and associated fibroblasts. Antioxid Redox Signal 20:1045
Gao M, Liu L, Yang Y et al (2021) LncRNA HCP5 induces gastric cancer cell proliferation, invasion, and EMT processes through the miR-186-5p/WNT5A axis under hypoxia. Front Cell Dev Biol 9:663654
Gonzalez DM, Medici D (2014) Signaling mechanisms of the epithelial-mesenchymal transition. Sci Signal 7:re8
Graf J, Kretz M (2020) From structure to function: Route to understanding lncRNA mechanism. BioEssays 42:2000027
Greenburg G, Hay ED (1982) Epithelia suspended in collagen gels can lose polarity and express characteristics of migrating mesenchymal cells. J Cell Biol 95:333
Guo H, Ingolia NT, Weissman JS et al (2010) Mammalian microRNAs predominantly act to decrease target mRNA levels. Nature 466:835–840
Hammond EM, Asselin M-C, Forster D et al (2014) The Meaning, Measurement and Modification of Hypoxia in the Laboratory and the Clinic. Clin Oncol 26:277–288
Han P, Chang CP (2015) Long non-coding RNA and chromatin remodeling. RNA Biol 12:1094–1098
Han M, Wang Y, Liu M et al (2012) MiR-21 regulates epithelial-mesenchymal transition phenotype and hypoxia-inducible factor-1α expression in third-sphere forming breast cancer stem cell-like cells. Cancer Sci 103:1058
Hartford CCR, Lal A (2020) When long noncoding becomes protein coding. Mol Cell Biol 40:e00528-e619
He M, Zhan M, Chen W et al (2017) MiR-143-5p deficiency triggers EMT and metastasis by targeting HIF-1α in gallbladder cancer. Cell Physiol Biochem 42:2078–2092

He J, Wang J, Li S et al (2020) Hypoxia-inhibited miR-338-3p suppresses breast cancer progression by directly targeting ZEB2. Cancer Sci 111:3550
Höckel M, Vaupel P (2001) Tumor hypoxia: definitions and current clinical, biologic, and molecular aspects. JNCI J Natl Cancer Inst 93:266–276
Hong Q, Li O, Zheng W et al (2017) LncRNA HOTAIR regulates HIF-1α/AXL signaling through inhibition of miR-217 in renal cell carcinoma. Cell Death Dis 8:e2772–e2772
Hsu M-T, Coca-Prados M (1979) Electron microscopic evidence for the circular form of RNA in the cytoplasm of eukaryotic cells. Nature 280:339–340
Huang J, Yao X, Zhang J et al (2013) Hypoxia-induced downregulation of miR-30c promotes epithelial-mesenchymal transition in human renal cell carcinoma. Cancer Sci 104:1609
Huang C, Liang D, Tatomer DC et al (2018) A length-dependent evolutionarily conserved pathway controls nuclear export of circular RNAs. Gen Develop 32:639–644
Ibrahim AA, Schmithals C, Kowarz E et al (2017) Hypoxia causes downregulation of dicer in hepatocellular carcinoma, which is required for upregulation of hypoxia-inducible factor 1α and epithelial–mesenchymal transition. Clin Cancer Res 23:3896–3905
Iliopoulos O, Levy AP, Jiang C et al (1996) Negative regulation of hypoxia-inducible genes by the von Hippel-Lindau protein. Proc Natl Acad Sci USA 93:10595–10599
Ivanov A, Memczak S, Wyler E et al (2015) Analysis of intron sequences reveals hallmarks of circular RNA biogenesis in animals. Cell Rep 10:170–177
Iyer NV, Kotch LE, Agani F et al (1998) Cellular and developmental control of O_2 homeostasis by hypoxia-inducible factor 1 alpha. Genes Dev 12:149–162
Jaakkola P, Mole DR, Tian YM et al (2001) Targeting of HIF-α to the von Hippel-Lindau ubiquitylation complex by O2-regulated prolyl hydroxylation. Science 292:468–472
Jarroux J, Morillon A, Pinskaya M (2017) History, discovery, and classification of lncRNAs. Adv Exp Med Biol 1008:1–46
Ji Z, Wang X, Liu Y et al (2022) MicroRNA-574-3p regulates HIF-α isoforms promoting gastric cancer epithelial-mesenchymal transition via targeting CUL2. Dig Dis Sci 67:3714–3724
Joseph NA, Chiou SH, Lung Z et al (2018) The role of HGF-MET pathway and CCDC66 cirRNA expression in EGFR resistance and epithelial-to-mesenchymal transition of lung adenocarcinoma cells J Hematol Oncol 11:74
Jung-Whan K, Ping G, Yen-Chun L et al (2007) Hypoxia-inducible factor 1 and dysregulated c-Myc cooperatively induce vascular endothelial growth factor and metabolic switches hexokinase 2 and pyruvate dehydrogenase kinase 1. Mol Cell Biol 27:7381–7393
Kaelin WG Jr, Ratcliffe PJ (2008) Oxygen sensing by metazoans: the central role of the HIF hydroxylase pathway. Mol Cell 30:393–402
Kapranov P, Cheng J, Dike S et al (2007) RNA maps reveal new RNA classes and a possible function for pervasive transcription. Science 316:1484–1488
Keam SP, Hutvagner G (2015) tRNA-derived fragments (tRFs): emerging new roles for an ancient RNA in the regulation of gene expression. Life 5:1638
Kelly S, Greenman C, Cook PR, Papantonis A (2015) Exon skipping is correlated with exon circularization. J Mol Biol 427:2414–2417
Kim YJ, Zhao Y, Myung JK et al (2021) Suppression of breast cancer progression by FBXL16 via oxygen-independent regulation of HIF1α stability. Cell Rep 37:109996
Kinoshita T, Yip KW, Spence T et al (2017) MicroRNAs in extracellular vesicles: potential cancer biomarkers. J Hum Genet 62:67–74
Kong X, Zhao Y, Li X et al (2019) Overexpression of HIF-2α-dependent NEAT1 promotes the progression of non-small cell lung cancer through miR-101-3p/SOX9/Wnt/β-catenin signal pathway. Cell Physiol Biochem 52:368–381
Kozomara A, Griffiths-Jones S (2014) miRBase: annotating high confidence microRNAs using deep sequencing data. Nucleic Acids Res 42(Database issue):D68–73
Krchňáková Z, Thakur PK, Krausová M et al (2019) Splicing of long non-coding RNAs primarily depends on polypyrimidine tract and 5′ splice-site sequences due to weak interactions with SR proteins. Nucleic Acids Res 47:911–928

Krock BL, Skuli N, Simon MC (2011) Hypoxia-induced angiogenesis: good and evil. Genes Cancer 2:1117–1133

Kurrey NK, Jalgaonkar SP, Joglekar AV et al (2009) Snail and slug mediate radioresistance and chemoresistance by antagonizing p53-mediated apoptosis and acquiring a stem-like phenotype in ovarian cancer cells. Stem Cells 27:2059–2068

Kuscu C, Kumar P, Kiran M et al (2018) tRNA fragments (tRFs) guide Ago to regulate gene expression post-transcriptionally in a Dicer-independent manner. RNA 24:1093–1105

Lai HH, Li JN, Wang MY et al (2018) HIF-1α promotes autophagic proteolysis of Dicer and enhances tumor metastasis. J Clin Invest 128:625

Lamouille S, Xu J, Derynck R (2014) Molecular mechanisms of epithelial–mesenchymal transition. Nat Rev Mol Cell Biol 15:178–196

Lee RC, Feinbaum RL, Ambros V (1993) The C. elegans heterochronic gene lin-4 encodes small RNAs with antisense complementarity to lin-14. Cell 75:843–854

Lee Y, Kim M, Han J et al (2004) MicroRNA genes are transcribed by RNA polymerase II. EMBO J 23:4051–4060

Li C-W, Xia W, Huo L et al (2012) Epithelial-mesenchymal transition induced by TNF-α requires NF-κB–mediated transcriptional upregulation of twist1. Cancer Res 72:1290–1300

Li Z, Huang C, Bao C, et al (2015) Exon-intron circular RNAs regulate transcription in the nucleus. Nat Struct Mol Biol 22:256–264. https://doi.org/10.1038/nsmb.2959

Li H, Rokavec M, Jiang L et al (2017a) Antagonistic effects of p53 and HIF1A on microRNA-34a regulation of PPP1R11 and STAT3 and hypoxia-induced epithelial to mesenchymal transition in colorectal cancer cells. Gastroenterology 153:505–520

Li H, Wang X, Wen C et al (2017b) Long noncoding RNA NORAD, a novel competing endogenous RNA, enhances the hypoxia-induced epithelial-mesenchymal transition to promote metastasis in pancreatic cancer. Mol Cancer 16:169

Li J, Peng W, Yang P et al (2020) MicroRNA-1224-5p inhibits metastasis and epithelial-mesenchymal transition in colorectal cancer by targeting SP1-mediated NF-κB signaling pathways. Front Oncol 10:294

Li Y, Wang J, Chen W et al (2021) Overexpression of STAT4 under hypoxia promotes EMT through miR-200a/STAT4 signal pathway. Life Sci 273:119263

Li X, Yang J, Ni R et al (2022) Hypoxia-induced lncRNA RBM5-AS1 promotes tumorigenesis via activating Wnt/β-catenin signaling in breast cancer. Cell Death Dis 13:95

Liu X, Shan G (2021) Mitochondria encoded non-coding RNAs in cell physiology. Front Cell Develop Biol 9:713729

Liu J, Carmell MA, Rivas FV et al (2004) Argonaute2 Is the catalytic engine of mammalian RNAi. Science 305:1437–1441

Liu Z, Wang Y, Dou C et al (2018) Hypoxia-induced up-regulation of VASP promotes invasiveness and metastasis of hepatocellular carcinoma. Theranostics 8:4649

Liu Z, Wang Y, Wang L et al (2019) Long non-coding RNA AGAP2-AS1, functioning as a competitive endogenous RNA, upregulates ANXA11 expression by sponging miR-16-5p and promotes proliferation and metastasis in hepatocellular carcinoma. J Exp Clin Cancer Res 38:194

Liu H, Chen C, Zeng J et al (2021a) MicroRNA-210-3p is transcriptionally upregulated by hypoxia induction and thus promoting EMT and chemoresistance in glioma cells. PLoS ONE 16:e0253522

Liu XN, Zhang CB, Lin H et al (2021b) microRNA-204 shuttled by mesenchymal stem cell-derived exosomes inhibits the migration and invasion of non-small-cell lung cancer cells via the KLF7/AKT/HIF-1α axis. Neoplasma 68:719–731

Long Y, Wang X, Youmans DT et al (2017) How do lncRNAs regulate transcription? Sci Adv 3:eaao2110

Lu J, Getz G, Miska EA et al (2005) MicroRNA expression profiles classify human cancers. Nature 435:834–838

Lu Y, Ji N, Wei W et al (2017) MiR-142 modulates human pancreatic cancer proliferation and invasion by targeting hypoxia-inducible factor 1 (HIF-1α) in the tumor microenvironments. Biol Open 6:252

Luan N, Chen Y, Li Q et al (2021) TRF-20-M0NK5Y93 suppresses the metastasis of colon cancer cells by impairing the epithelial-to-mesenchymal transition through targeting Claudin-1. Am J Transl Res 13:124–142

Luo Z, Tian M, Yang G et al (2022) Hypoxia signaling in human health and diseases: implications and prospects for therapeutics. Signal Transduct Target Ther 7:218

Mansoori B, Silvestris N, Mohammadi A et al (2021) miR-34a and miR-200c have an additive tumor-suppressive effect on breast cancer cells and patient prognosis. Genes (Basel) 12:1–17

Matouk IJ, Raveh E, Abu-lail R et al (2014) Oncofetal H19 RNA promotes tumor metastasis. Biochim Biophys Acta-Mol Cell Res 1843:1414–1426

Matsuoka J, Yashiro M, Doi Y et al (2013) Hypoxia stimulates the EMT of gastric cancer cells through autocrine TGFβ signaling. PLoS ONE 8:e62310

Maxwell PH, Wiesener MS, Chang G-W et al (1999) The tumour suppressor protein VHL targets hypoxia-inducible factors for oxygen-dependent proteolysis. Nature 399:271–275

McMahon S, Charbonneau M, Grandmont S et al (2006) Transforming growth factor beta1 induces hypoxia-inducible factor-1 stabilization through selective inhibition of PHD2 expression. J Biol Chem 281:24171–24181

Memczak S, Jens M, Elefsinioti A et al (2013) Circular RNAs are a large class of animal RNAs with regulatory potency. Nature 495:333–338

Memczak S, Papavasileiou P, Peters O et al (2015) Identification and characterization of circular RNAs as a new class of putative biomarkers in human blood. PLoS ONE 10:e0141214

Niehrs C, Luke B (2020) Regulatory R-loops as facilitators of gene expression and genome stability. Nat Rev Mol Cell Biol 21:167–178

Onder TT, Gupta PB, Mani SA et al (2008) Loss of E-cadherin promotes metastasis via multiple downstream transcriptional pathways. Cancer Res 68:3645–3654

Pasquinelli AE, Reinhart BJ, Slack F et al (2000) Conservation of the sequence and temporal expression of let-7 heterochronic regulatory RNA. Nature 408:86–89

Peng P-H, Chieh-Yu Lai J, Hsu K-W et al (2020) Hypoxia-induced lncRNA RP11-390F4.3 promotes epithelial-mesenchymal transition (EMT) and metastasis through upregulating EMT regulators. Cancer Lett 483:35–45

Piccinin S, Tonin E, Sessa S et al (2012) A "Twist box" code of p53 inactivation: twist box:p53 interaction promotes p53 degradation. Cancer Cell 22:404–415

Qian W, Huang T, Feng W (2020) Circular RNA HIPK3 promotes EMT of cervical cancer through sponging miR-338-3p to Up-regulate HIF-1α. Cancer Manag Res 12:177–187

Ren D, Yang Q, Dai Y et al (2017) Oncogenic miR-210-3p promotes prostate cancer cell EMT and bone metastasis via NF-κB signaling pathway. Mol Cancer 16:117

Ren W, Hou J, Yang C et al (2019) Extracellular vesicles secreted by hypoxia pre-challenged mesenchymal stem cells promote non-small cell lung cancer cell growth and mobility as well as macrophage M2 polarization via miR-21-5p delivery. J Exp Clin Cancer Res 38:62

Rinaldi L, Datta D, Serrat J et al (2016) Dnmt3a and Dnmt3b associate with enhancers to regulate human epidermal stem cell homeostasis. Cell Stem Cell 19:491–501

Rybak-Wolf A, Stottmeister C, Glažar P et al (2015) Circular RNAs in the mammalian brain are highly abundant, conserved, and dynamically expressed. Mol Cell 58:870–885

Sahlgren C, Gustafsson MV, Jin S et al (2008) Notch signaling mediates hypoxia-induced tumor cell migration and invasion. Proc Natl Acad Sci USA 105:6392–6397

Sanbonmatsu K (2021) Getting to the bottom of lncRNA mechanism: structure–function relationships. Mamm Genome 33:343–353

Sanger HL, Klotz G, Riesner D et al (1976) Viroids are single-stranded covalently closed circular RNA molecules existing as highly base-paired rod-like structures. Proc Natl Acad Sci USA 73:3852–3856

Schirle NT, Sheu-Gruttadauria J, MacRae IJ (2014) Structural basis for microRNA targeting. Science 346:608–613

Schirle NT, Sheu-Gruttadauria J, Chandradoss SD et al (2015) Water-mediated recognition of t1-adenosine anchors Argonaute2 to microRNA targets. Elife 4:e07646

Semenza GL (2012) Hypoxia-inducible factors in physiology and medicine. Cell 148:399–408

Semenza GL, Nejfelt MK, Chi SM et al (1991) Hypoxia-inducible nuclear factors bind to an enhancer element located 3′ to the human erythropoietin gene. Proc Natl Acad Sci USA 88:5680–5684

Seo J, Ha J, Kang E et al (2021) The role of epithelial–mesenchymal transition-regulating transcription factors in anti-cancer drug resistance. Arch Pharm Res 44:281–292

Shang Y, Chen H, Ye J et al (2017) HIF-1α/Ascl2/miR-200b regulatory feedback circuit modulated the epithelial-mesenchymal transition (EMT) in colorectal cancer cells. Exp Cell Res 360:243–256

Shao S, Zhao X, Zhang X et al (2015) Notch1 signaling regulates the epithelial–mesenchymal transition and invasion of breast cancer in a Slug-dependent manner. Mol Cancer 14:28

Shi Z, To SKY, Zhang S et al (2021) Hypoxia-induced Nur77 activates PI3K/Akt signaling via suppression of Dicer/let-7i-5p to induce epithelial-to-mesenchymal transition. Theranostics 11:3376

Slack FJ, Chinnaiyan AM (2019) The Role of Non-coding RNAs in Oncology. Cell 179:1033–1055

Statello L, Guo CJ, Chen LL et al (2020) Gene regulation by long non-coding RNAs and its biological functions. Nat Rev Mol Cell Biol 22:96–118

Stemmler MP, Eccles RL, Brabletz S et al (2019) Non-redundant functions of EMT transcription factors. Nat Cell Biol 21:102–112

Szabo L, Salzman J (2016) Detecting circular RNAs: bioinformatic and experimental challenges. Nat Rev Genet 17:679–692. https://doi.org/10.1038/nrg.2016.114

Tang H, Li R, Liang P et al (2015) miR-125a inhibits the migration and invasion of liver cancer cells via suppression of the PI3K/AKT/mTOR signaling pathway. Oncol Lett 10:681–686

Tang T, Yang Z, Zhu Q et al (2018) Up-regulation of miR-210 induced by a hypoxic microenvironment promotes breast cancer stem cell metastasis, proliferation, and self-renewal by targeting E-cadherin. FASEB J 32:6965–6981

Tang J, Zhu H, Lin J et al (2020) Knockdown of Circ_0081143 mitigates hypoxia-induced migration, invasion, and EMT in gastric cancer cells through the miR-497-5p/EGFR axis. Cancer Biother Radiopharm 36:333–346

Verduci L, Tarcitano E, Strano S et al (2021) CircRNAs: role in human diseases and potential use as biomarkers. Cell Death Dis 12:468

Wang GL, Jiang BH, Rue EA et al (1995) Hypoxia-inducible factor 1 is a basic-helix-loop-helix-PAS heterodimer regulated by cellular O_2 tension. Proc Natl Acad Sci USA 92:5510–5514

Wang X, Yu M, Zhao K et al (2016) Upregulation of MiR-205 under hypoxia promotes epithelial–mesenchymal transition by targeting ASPP2. Cell Death Dis 7:e2517–e2517

Wang X, de Geyter C, Jia Z et al (2020) HECTD1 regulates the expression of SNAIL: implications for epithelial-mesenchymal transition. Int J Oncol 56:1186

Wang L, Sun L, Liu R et al (2021) Long non-coding RNA MAPKAPK5-AS1/PLAGL2/HIF-1α signaling loop promotes hepatocellular carcinoma progression. J Exp Clin Cancer Res 40:72

Wendt MK, Allington TM, Schiemann WP (2009) Mechanisms of epithelial-mesenchymal transition by TGF-β. Future Oncol 5:1145

Wiesener MS, Jürgensen JS, Rosenberger C et al (2003) Widespread, hypoxia-inducible expression of HIF-2α in distinct cell populations of different organs. FASEB J 17:271–273

Winkle M, El-Daly SM, Fabbri M et al (2021) Noncoding RNA therapeutics—challenges and potential solutions. Nat Rev Drug Discov 20:629–651

Wu D, Potluri N, Lu J et al (2015) Structural integration in hypoxia-inducible factors. Nature 524:303–308

Wu D, Chen T, Zhao X et al (2022) HIF1α-SP1 interaction disrupts the circ-0001875/miR-31-5p/SP1 regulatory loop under a hypoxic microenvironment and promotes non-small cell lung cancer progression. J Exp Clin Cancer Res 41:156

Xia Y, Shen S, Verma IM (2014) NF-κB, an active player in human cancers. Cancer Immunol Res 2(9):823–830

Xing S, Tian Z, Zheng W et al (2021) Hypoxia downregulated miR-4521 suppresses gastric carcinoma progression through regulation of IGF2 and FOXM1. Mol Cancer 20:1–19

Xu J, Lamouille S, Derynck R (2009) TGF-β-induced epithelial to mesenchymal transition. Cell Res 19:156–172

Xu Q, Liu X, Liu Z et al (2017) MicroRNA-1296 inhibits metastasis and epithelial-mesenchymal transition of hepatocellular carcinoma by targeting SRPK1-mediated PI3K/AKT pathway. Mol Cancer 16:1–15

Yang W, Yan H-X, Chen L et al (2008) Wnt/β-catenin signaling contributes to activation of normal and tumorigenic liver progenitor cells. Cancer Res 68:4287–4295

Yang L, Cao M, Zhang J et al (2020) LncRNA XIST modulates HIF-1A/AXL signaling pathway by inhibiting miR-93-5p in colorectal cancer. Mol Genet Genomic Med 8:e1112

Yang W, Yin Y, Bi L et al (2021) MiR-182-5p promotes the metastasis and epithelial-mesenchymal transition in non-small cell lung cancer by targeting EPAS1. J Cancer 12:7120

Yu Y, Min Z, Zhou Z et al (2019) Hypoxia-induced exosomes promote hepatocellular carcinoma proliferation and metastasis via miR-1273f transfer. Exp Cell Res 385:111649

Yu M, Lu B, Zhang J et al (2020) TRNA-derived RNA fragments in cancer: Current status and future perspectives. J Hematol Oncol 13:1–14

Yu F, Liang M, Huang Y et al (2021) Hypoxic tumor-derived exosomal miR-31-5p promotes lung adenocarcinoma metastasis by negatively regulating SATB2-reversed EMT and activating MEK/ERK signaling. J Exp Clin Cancer Res 40:1–15

Zampetaki A, Albrecht A, Steinhofel K (2018) Long non-coding RNA structure and function: Is there a link? Front Physiol 9:1201

Zhang Q, Bai X, Chen W et al (2013) Wnt/β-catenin signaling enhances hypoxia-induced epithelial–mesenchymal transition in hepatocellular carcinoma via crosstalk with hif-1α signaling. Carcinogenesis 34:962–973

Zhang X-O, Wang H-B, Zhang Y et al (2014) Complementary sequence-mediated exon circularization. Cell 159:134–147

Zhang S, Wang W, Liu G et al (2017) Long non-coding RNA HOTTIP promotes hypoxia-induced epithelial-mesenchymal transition of malignant glioma by regulating the miR-101/ZEB1 axis. Biomed Pharmacother 95:711–720

Zhang J, Jin HY, Wu Y et al (2019a) Hypoxia-induced LncRNA PCGEM1 promotes invasion and metastasis of gastric cancer through regulating SNAI1. Clin Transl Oncol 21:1142–1151

Zhang X, Sai B, Wang F et al (2019b) Hypoxic BMSC-derived exosomal miRNAs promote metastasis of lung cancer cells via STAT3-induced EMT. Mol Cancer 18:1–15

Zhang H, Liang F, Yue J et al (2020a) MicroRNA-137 regulates hypoxia-mediated migration and epithelial.mesenchymal transition in prostate cancer by targeting LGR4 via the EGFR/ERK signaling pathway. Int J Oncol 57:540–549

Zhang KD, Hu B, Cen G et al (2020b) MiR-301a transcriptionally activated by HIF-2α promotes hypoxia-induced epithelial-mesenchymal transition by targeting TP63 in pancreatic cancer. World J Gastroenterol 26:2349

Zhang X, Bai J, Yin H et al (2020c) Exosomal miR-1255b-5p targets human telomerase reverse transcriptase in colorectal cancer cells to suppress epithelial-to-mesenchymal transition. Mol Oncol 14:2589

Zhang J, Han L, Yu J et al (2021) miR-224 aggravates cancer-associated fibroblast-induced progression of non-small cell lung cancer by modulating a positive loop of the SIRT3/AMPK/mTOR/HIF-1α axis. Aging (Albany NY) 13:10431

Zhang L, Wu H, Zhang Y et al (2022) Induction of lncRNA NORAD accounts for hypoxia-induced chemoresistance and vasculogenic mimicry in colorectal cancer by sponging the miR-495-3p/hypoxia-inducible factor-1α (HIF-1α). Bioengineered 13:950–962

Zhou Y, Xu Q, Shang J et al (2019) Crocin inhibits the migration, invasion, and epithelial-mesenchymal transition of gastric cancer cells via miR-320/KLF5/HIF-1α signaling. J Cell Physiol 234:17876–17885

Zhu M, Chen Q, Liu X et al (2014) lncRNA H19/miR-675 axis represses prostate cancer metastasis by targeting TGFBI. FEBS J 281:3766–3775

A Study on the Role of piRNAs in Cancer Epigenetics

Alagu Theivanai Ganesan, Subhamay Adhikary, Alakesh Das, Amit Dey, Antara Banerjee, and Surajit Pathak

Contents

Abstract Piwi-interacting RNAs (piRNAs) are single-stranded, small noncoding RNAs that are 24–30 nucleotides long and have a 5′-U end and a 2′-O-methylation at the 3′ end. piRNAs associate with Piwi proteins that cleave them then enters the nucleus to induce epigenetic modification, transcriptional and post-transcriptional gene silencing. The piRNA and PIWI proteins mediate various epigenetic changes, such as DNA methylation, histone modification via methylation, gene regulation via

A. T. Ganesan and S. Adhikary contributed equally in this chapter.

A. T. Ganesan
Department of Biotechnology, University College of Engineering, BIT Campus, Anna University, Tiruchirappalli 620024, India

A. T. Ganesan · S. Adhikary · A. Das · A. Dey · A. Banerjee · S. Pathak (✉)
Medical Biotechnology, Faculty of Allied Health Sciences, Chettinad Hospital and Research Institute (CHRI), Chettinad Academy of Research and Education (CARE), Kelambakkam 603103, India
e-mail: drsurajitpathak@care.edu.in

J. Barciszewski (ed.), *RNA Structure and Function*, RNA Technologies 14,
https://doi.org/10.1007/978-3-031-36390-0_22

transposon silencing, and chromatin remodelling. Various DNA methyltransferases, such as DNMT1, DNMT3A, and DNMT3B, catalyse the cytosine methylation of DNA molecules. Histone methylation on lysine and arginine residues of H3 and H4 histone proteins is done by various methyl transferases. Suv39H1 and SETDB1 methyltransferase enzymes are involved in lysine methylation on histone H3 and H4, and the protein arginine methyltransferase-5 (PRMT-5) enzyme causes the methylation of arginine residues on histone H3. Upregulation and downregulation of Piwi proteins and piRNAs may have the ability to suppress cancer growth and metastasis. Various studies have shown that piRNA-mediated modifications can be used as biomarkers for the diagnosis, prognosis of various diseases, such as different types of human malignancies, type-II diabetes mellitus, and Alzheimer's disease. In this chapter, the current findings and functions of piRNAs in cancer epigenetics are discussed.

Keywords piRNA · Cancer · Gene silencing · Epigenetic regulation · Argonaute · PIWI proteins.

1 Introduction

RNA is essential for gene expression in all life forms, serving as a messenger, carrying information from the nucleus towards the cytoplasm, where genes (mRNA) are translated into proteins on ribosomes with the help of tRNA and rRNA. Almost every biological action is regulated by RNA, and the structures that RNA molecules adopt are regarded as being essential to their activities. The rapid emergence of high-throughput sequencing tools for mapping RNA structure has enabled rapid in vivo structural probing of entire cellular transcriptomes (Spitale and Incarnato 2022). RNA is also the genetic material in some viruses and is implicated in various fundamental life processes, including metabolism, splicing, and biocatalysis. Cells contain several types of RNA that have different roles. Following the human genome sequencing, sources claim that about 20,000–25,000 genes, or around 2% of the genome, were likely encoded by human cells (International Human Genome Sequencing Consortium 2004). The remaining 98% are called noncoding genes since they do not code for proteins. Evolutionary and comparative genomic studies have uncovered the structure and functions of noncoding RNAs, either conserved or varied over millions of years of evolution (Bhartiya and Scaria 2016). Transfer RNAs (tRNAs) and ribosomal RNAs (rRNAs) are two examples of commonly known noncoding RNAs. Noncoding RNAs can be grouped into two classes based on their length: small noncoding RNAs (sncRNAs) (less than 200 nucleotides) and long noncoding RNAs (which contain over 200 nucleotides). MicroRNAs (miRNAs), small interfering RNAs (siRNAs), PIWI-interacting RNAs (piRNAs), small nuclear RNAs (snRNAs), and small nucleolar RNAs (snoRNAs) are all subclasses of small ncRNAs. The biological complexity of organisms can be correlated with a plethora of noncoding RNAs rather than the genes that encode proteins, as was assumed earlier. Noncoding RNAs are involved in various cellular activities (Lee et al. 2019), such as DNA replication, RNA splicing, translation, gene regulation at different levels (Morris 2012), cellular development, differentiation, and genomedefence, and contribute to various diseases, such as Alzheimer's, type II diabetes, and cancer, and clinical stages (Zhang et al. 2021; Chi et al. 2021; Kumar et al. 2020).

Originally thought to have been confined to the germline, about 20,000 piRNA genes have been identified, making it the largest group of sncRNA molecules found and expressed in the eukaryotic genome (Peng and Lin 2013). The length of a piRNA is about 24–31 nucleotides, depending on the species. The first PIWI-interacting RNAs (piRNAs) were found to function in *Drosophila melanogaster*, where they interacted with sub-members of the Argonaute proteins known as PIWI protein to carry out their functions (Huang et al. 2013a, b). They primarily function in the nucleus, where they are associated with transcriptional control of various genes, promotion of fertility in animal germlines, repression of transposable elements during the formation of germ cells, ensuring genome integrity during gametogenesis, as well as viral defence. Attributes that make them distinct from other noncoding silencing RNAs are their length (24–31 nucleotides), their production from single-stranded RNA precursors as opposed to double-stranded precursors, interaction with PIWI proteins, and a unique 2′-O-methylation at their 3′ end (Tian et al. 2011; Kirino and Mourelatos 2007). Numerous piRNAs and PIWI proteins may be linked to the emergence of various diseases, according to an increasing amount of research (Mani and Juliano 2013). Comprehensive knowledge about the modifications in the gene expression patterns in various tissues and organs is a key to understanding the disease pathogenesis and changes in host homeostasis.

2 Argonaute and PIWI Proteins

The Argonaute family of proteins plays a chief role in RNA-mediated gene silencing as a critical component of the RNA-induced silencing complex (RISC) (Bartel 2018; Bodak et al. 2017). Eight Argonaute proteins have been recognised, four of which are present in humans and mice with high structure similarity (Ago1-4). Structural studies of these proteins exposed four conserved domains: an N-terminal domain (N), the PIWI/Argonaute/Zwille (PAZ) domain, the MID domain, and the P-element induced wimpy testis (PIWI) domain (Cramer et al. 2018). Ago2 has mRNA slicing activity predominantly among the others due to its unique structural features (Liu et al. 2004). Ago2 has two motifs at its N-terminal domain (residues 44–48 and 134–166) and both are necessary for its catalytic activity. Ago2 belongs to the PAZ and Piwi domains (PPD) protein family, which is distinguished by the N-terminal PAZ domain (called after the Piwi, Argonaute, and Zwille/Pinhead proteins) and the C-terminal Piwi domain (Lingel et al. 2003). One study reported that the addition of external siRNAs was found to be associated with Ago2 to guide the target RNA cleavage. Independently, siRNA-based knockdown of individual Ago members in conjunction with a sensitive positive-readout reporter experiment showed Ago2's unique function in directing target RNA cleavage (Meister et al. 2004). Mutant motifs cause the failure of the processes of RISC activation and mRNA cleavage. These residues, located in the close vicinity of the PIWI domain, are necessary for proper guide-target positioning during protein folding. While Ago3 and Ago4 lack motifs, Ago1 exhibits complete catalytic activity with even one of these motifs. According to recent studies, Ago3 shows slicing activity because of a particular insertion (3SI) in the N-terminal domain, which is broader and less precise than Ago2′s nucleotide binding (Park et al. 2017).

Argonaute proteins evolved into two subgroups: AGO proteins and PIWI proteins; one binds with miRNAs and siRNAs and other binds with piRNAs, respectively. PIWI proteins are one type of animal-specific subclass of proteins that are mainly expressed in certain adult germ cells and germline stem cells (GSCs), primordial germ cells (PGCs), and in the germplasm, as well as in some larval somatic cells, such as disc cells, salivary gland cells, and terminal filament cells, and adult female gonadal cap cells (Cox 1998). When a guide piRNA interacts with PIWI proteins, ribonucleoprotein complexes are formed, called piRNA-induced silencing complexes (piRISCs), which is thought to be a highly conserved mechanism (Yamashiroet al. 2020). Humans have four isoforms of Piwi proteins: PIWI-like protein 1 (PIWIL1/HIWI), PIWI-like protein 2 (PIWIL2/HILI), PIWI-like protein 3 (PIWIL3/HIWI3), and PIWI-like protein 4 (PIWIL4/HIWI2) (Qiao et al. 2002). Mice have PIWIL1, PIWIL2, and PIWIL4,which are indispensable for spermatogenesis (Kuramochi-Miyagawa et al. 2004), and flies have PIWI, AUB, and AGO. The MID, PAZ, and PIWI domains of both clades of Argonaute proteins are highly conserved. Each Piwi-like protein has a large binding site within the PAZ domain that allows binding of the 3′ end of the guide RNA, while the 5′ end binds to the MID domain. The PIWI domain also contains an RNaseH fold domain. Through complementary base pair recognition, piRNAs bind to the PIWI domain and guide them to their targets.

3 piRNA Biogenesis

Extensive studies on piRNAs have led to their establishment as a diverse class of noncoding RNAs based on their production and interaction with PIWI proteins, but not with AGO and their Dicer-independent production (Czech et al. 2018)). Among the other small RNA molecules, piRNA is the longest and is produced from a long-strand precursor without Dicer and Drosha (Iwasaki et al. 2015). They were formerly known as repeat-associated small interfering RNAs (rasiRNAs), i.e., dsRNAs that can silence repeat elements; rasiRNAs are now considered a subclass of piRNAs. Although the functions and biogenesis of piRNAs have been studied to a greater extent in *Drosophila* (Khurana and Theurkauf 2010), they are thought to play similar roles in vertebrates as well.

3.1 Transcription of Uni-strand and Dual-Strand piRNA Cluster and Their Nuclear Processing

Based on their origin, piRNA may be broadly classified into five categories: mRNA-derived, transfer RNA (tRNA)-derived, long noncoding RNA (lncRNA), snoRNA-derived, and transposon-derived piRNAs. mRNA-derived piRNAs are mainly produced from the 3′ UTR region of mRNA and can bind with the same mRNA molecule after transcription and processing. PiRNAs derived from tRNA, primarily use the 5′ end of tRNA halves rather than mature tRNAs. The lncRNA-derived piRNAs originate from the exonic region of lncRNA. Some piRNAs generated from the middle part of the snoRNA, which has C/D and C′/D′ boxes, are known as sno-derived piRNAs. The piRNAs synthesised from transposons are both sense and antisense piRNAs, which are generated from single-stranded cluster transcription (Chen et al. 2021). There are two phases of piRNA biogenesis. Initially in the nucleus, long-stranded precursor RNA is generated by transcription and then transported to the cytoplasm, where maturation of precursor piRNA takes place and finally loaded into the PIWI-protein for function. There are also two conventional mechanisms involved in piRNA biogenesis: cluster transcripts and the ping-pong cycle (piRNA amplification loop) (Leva et al. 2014; Gunawardane et al. 2007). The piRNA clusters are of two types in *Drosophila*: uni-stranded (single-stranded) clusters and dual-stranded (double-stranded) clusters. Genomic regions called piRNA clusters that rarely carry transposable fragments are transcribed to produce piRNA precursors (pre-piRNAs). The precursors produced from single-stranded clusters map to only one genomic strand, whereas those from double-stranded clusters map to both genomic strands. Inside somatic follicle cells of the *Drosophila ovary*, single-stranded clusters are transcribed; inside germline-derived cells, double-stranded clusters are transcribed (Malone et al. 2009).

Until now, around 142 piRNA clusters have already been discovered in *Drosophila.* Most of them are found in heterochromatic regions (pericentromeric and sub-telomeric regions), and piRNA clusters are found in euchromatin regions (Yamanaka et al. 2014). Some well-known dual-stranded cluster examples are 42AB, 38C, and 80F and uni-stranded clusters examples are flamenco. Generally, the heterochromatin region is transcriptionally inactive, but transcription of these clusters takes place from this region (Huang et al. 2017).

The transcription of single-stranded clusters followed the same canonical mRNA transcription process. These single-stranded clusters have histone-3 lysine-4 demethylation (H3K4me2) marks in the promoter region, which is associated with transcription (Lim and Kai 2015; Mohn et al. 2014) and piRNA cluster body of both uni-stranded and double-stranded clusters have repressive histone-3 lysine-9 trimethylation (H3K9me3). The presence of H3K9me3 in the heterochromatin region is thought to be mark of transcriptional repression but this methylation does not hamper the transcription of piRNA precursor molecules (Mohn et al. 2014). *Flamenco,* a single-stranded piRNA cluster, has an initiator sequence and downstream promoter element (DPE) in its promoter region where RNAP-II can bind. So, similar to the canonical mRNA synthesis process, RNA is generated from single-stranded clusters (*flamenco*) with a 5′ cap and a 3′ poly (A) tail that is sometimes spliced using RNA polymerase II (RNAP-II). The transcription of these single-stranded piRNA clusters (*flamenco*) is a unidirectional transcription process, and the piRNAs generated from these transcripts are mostly antisense to transposon mRNA (Goriaux et al. 2014). Figure 1 shows that after transcription, *flamenco* RNA undergoes alternative splicing. For the release of *flamenco* RNA from transcription sites, the exon junction complex (EJC) proteins (Mago nashi (Mago) and Tsunagi (Tsu/Y14), as well as EJC accessory proteins (such as RNA-binding protein S1 (RnpS1) and Acinus (Acn)), are required. A dead box helicase U2AF65-associated protein (UAP56), an important EJC complex interactor protein, is also necessary for *flamenco* RNA export. Then the *flamenco* transcripts are exported to the cytoplasm with the help of the nuclear RNA export factor 1 (Nxf1)/Tap and nuclear transport factor 2 like export factor 1/p15 (Nxt1/p15) mRNA export complexes and stored in the flam bodies. In cytoplasm, flam transcripts are stored in the flam bodies (Dennis et al. 2016).

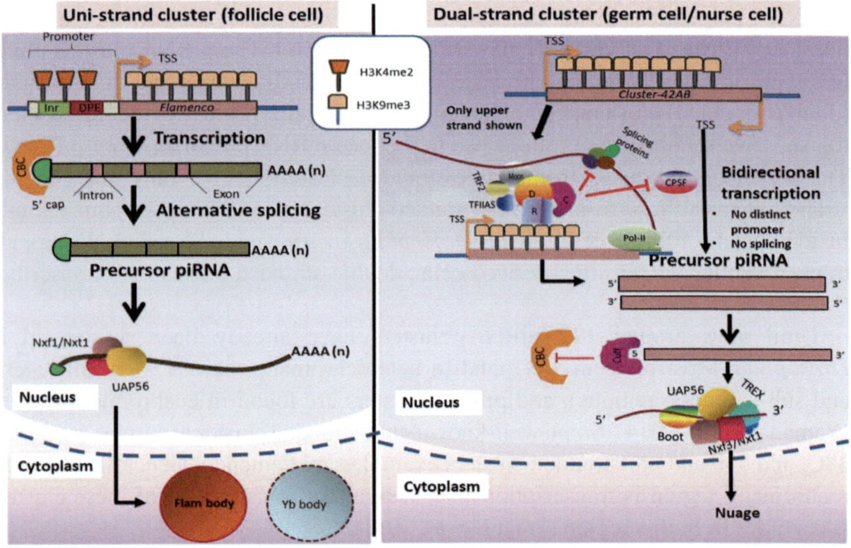

Fig. 1 Transcription of uni-strand cluster (*Flamenco)* and dual-strand cluster (Cluster 42 AB) and generation of precursor piRNA. Inr—Initiator, DPE—Downstream Promoter Element, CBC—Cap Binding Complex, Nxf1—Nuclear RNA export Factor 1, Nxt1—Nuclear transport factor 2 like export factor 1, UAP56—U2AF65-associated protein 56, H3K9me3—Histone-3 Lysine-9 Trimethylation, H3K4me2—Histone-3 Lysine-4 Dimethylation, Cuff—Cutoff, TREX—TRanscription Export, Boot—Bootlegger, Nxf3—Nuclear RNA export Factor 3, TRF2—Telomeric Repeat Factor 2, TFIIAS—Transcription Factor IIA-S, TSS—Transcription Start Site, Moon—Moonshiner, R—Rhino, D—Deadlock, C—Cutoff, PolII—RNA Polymerase II, and CPSF—Cleavage and Polyadenylation Specificity Factor

Figure 1 shows that RNA generated from double-stranded clusters is non-polyadenylated RNA, and this kind of cluster does not have a clear promoter for RNA polymerase II (Huang et al. 2017). Double-stranded clusters have no specific promoter region for RNAP-II binding, so the transcription starts using flanking protein-coding gene promoters. piRNA biogenesis from double-stranded clusters requires these proteins, such as Rhino (Rhi), Deadlock (Del), Cutoff (Cuff), Moonshiner (Moon), TFIIA-S, and TRF2, to drive RNAP-II-dependent transcription (Fig. 2) (Le Thomas et al. 2013; Mohn et al. 2014). Rhino is a paralog of heterochromatin protein-1d (HP-1d), which marks the H3K9me3 positions and determines the transcription initiation site (Zhang et al. 2014). Del is responsible for making a connection between Rhi and Cuff and is also responsible for recruiting the Moon, which is a protein that has similarity with transcription factor II A-L (TFIIA-L), which also has a co-factor such as transcription factor II A-S (TFIIA-S), and telomeric repeat binding factor (TRF2), a variant TATA box-binding protein related to transcription factor IID core. Cuff is a protein that has nuclease activity (homolog of yeast decapping nuclease) and is a transcription termination factor that is related to Rai1. At first, Rhi recruits these two co-factors, Del and Cuff, to make a trimeric complex known as Rhino-Deadlock-Cutoff (RDC) on the H3K9me3 region of the cluster, which then recruits the Moon, TFIIA-S, and TRF2 (the basic transcription initiation factor) to start transcription (Fig. 1). Cuff is thought to be involved in splicing inhibition and precursor-piRNA termination prevention by binding to the 5′ end of precursor-piRNA via their 5′ cap-binding pocket. It inhibits the termination by preventing the cleavage and polyadenylation specificity factor (CPSF)-mediated cleavage at the poly-A site (Fig. 1). Other required factors UAP56, a protein component of Transcription Export (TREX), also influences Cuff-mediated splicing inhibition. The TREX complex also helps in the export of precursor piRNA from the nucleus to the cytoplasm and helps in the accumulation of a perinuclear membraneless organelle known as Nuage, which is a well-known piRNA biogenesis centre in germ cells (Fig. 1) (Zhang et al. 2012; Hur et al. 2016).

Germ cell-specific Nxf3 and its partner Nxt1/p15 form a heterodimer complex and play a pivotal role in the nuclear export of piRNA precursors to nuage. This translocation mechanism is accompanied by the nucleo-cytoplasmic shuttling capability of Nxf3. Del and Bootlegger (Boot), with their physical interaction help in the recruitment of the UAP56 protein and the Nxf3-Nxt1/p15 complex to the transcription sites (Fig. 1) (ElMaghraby et al. 2019).

So, the transcription of single-stranded clusters resembles the well-known mRNA transcription due to unique promoters, a 5′ cap, and a 3′ poly-A terminal end and does not need Rhino. Mouse piRNA precursors have clear promoters and poly-A cleavage sites and are long, continuous transcripts. Reliable tools for cluster identification, visualization, and analysis of piRNAs are piClust, protract, PILFER, and piClusterBusteR (Jung et al. 2014; Rosenkranz and Zischler 2012; Ray and Pandey 2018; Schreiner and Atkinson 2017).

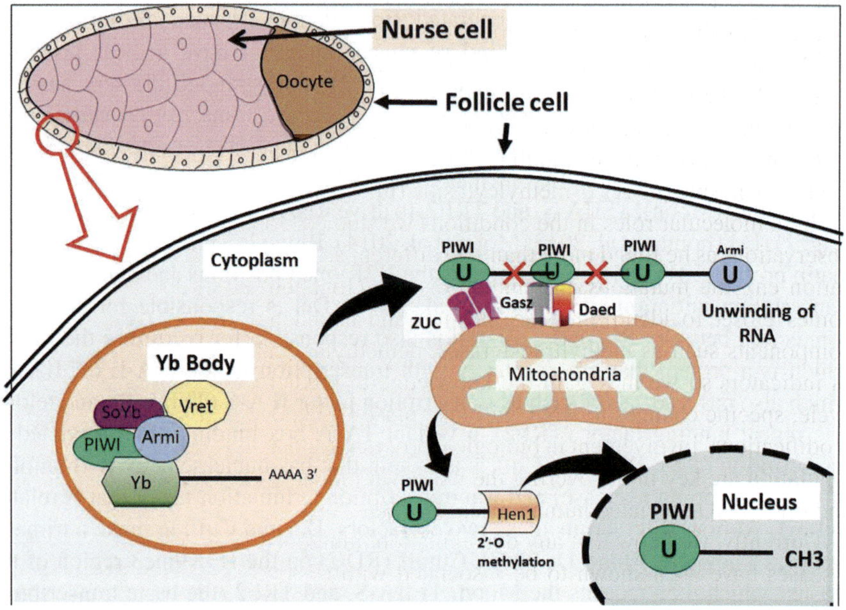

Fig. 2 Processing of piRNA-precursors in the follicle cells of *drosophila* ovary. PIWI—P-element Induced WImpy testis, Armi—Armitage, SoYb—Sister of Yb, Vret—Vreteno and ZUC—Zucchini

3.2 *Maturation of Precursor piRNA*

After completion of transcription, flam RNA (precursor-piRNA) is transported to the flam bodies located in the cytoplasm, and the processing of precursor-piRNA starts inside the Yb bodies, which are located close to the flam bodies. The entire piRNA processing factor is present in Yb bodies, and the assembly of Yb bodies is dependent on Yb-Yb (protein-protein) and Yb-flam RNA (protein-RNA) interactions. Yb body can directly interact with the precursor RNA via its N-terminal domain, which has DEAD-box RNA helicase activity (Murota et al. 2014). The factors involved in precursor-piRNA processing in the Yb bodies are Yb (also known as female sterile (1) Yb), which has a Tudor domain and helicase activity; Vreteno (Vret); Sister of Yb (SoYb); and Armitage (Armi), a dead-box protein. Other co-chaperone proteins, shutdown (Shu), and heat-shock protein-83 (HSP-83), are also found in the Yb body. The primary source of transposon targeting piRNAs in ovarian somatic cells is Armi-mediated interaction with flam RNA (Ishizu et al. 2019). When Armi binds to flam RNA, it aids in the translocation of this complex to the mitochondria. During this inter-organelle translocation, PIWI also binds to the armi-bound flam RNA complex (Ishizu et al. 2019). Maturation of piRNA from flam RNA depends on both the mutual interaction of PIWI and Armi (Ge et al. 2019). The mitochondrial outer membrane (MOM) contains Gasz, Minotaur (Mino), and Daedalus (Daed) factors required for piRNA maturation (Fig. 2). Gasz and Daed form a platform for Zucchini (Zuc)-mediated piRNA maturation upon interaction among them (Munafò et al. 2019). Zuc is a Drosophila-specific riboendonuclease that acts on Armi-bound PIWI-piRNA intermediates that have been stalled on MOM by Gasz and Daed. Mino is a member of the Glycerol-3-phosphate Acyltransferase (GPAT) family, which is essential for piRNA maturation, but its function is still unknown. The Zuc-dependent piRNA processing does not require any pre-existing piRNA, that's why it is called the primary biogenesis pathway (Vagin et al. 2013).

The 5′ end of piRNA can be processed using either endonuclease (slicer) activity or cleavage. The PIWI proteins, aubergine (Aub) and argonaute-3 (Ago3), of the cytoplasm have the slicer activity, and Zuc endonuclease has the cleavage property Zuc endonuclease has no sequence specificity and cleaves upstream of the U nucleotide of the U nucleotide containing RNA stretch to generate 1 U bias in downstream primary piRNA (Fig. 2). Zuc endonuclease forms the 5′ end of piRNA in somatic cells, resulting in phased piRNA loaded into the PIWI protein. This PIWI-bounded phased piRNA has a clear 1 uracil (1 U) nucleotide at the 5′ end, known as 1U bias at the 5′end (Nishimasu et al. 2012).

The 3′ end formation of piRNA involved three mechanisms: Zuc-mediated cleavage, slicer activity, and 3′–5′ exonuclease trimming activity by nibbler (Nbr) (or poly (A)-specific ribonuclease-like domain-containing 1 (PNLDC1) in mice). In the case of somatic cells, Zuc endonuclease causes the trimming of nucleotides from the 3′ to the 5′ direction to form the 3′ end of piRNA (Han et al. 2015). The piRNAs generated by Zuc endonuclease action have a phasing signature, which means that when one piRNA is mapped to the sequence of the precursor one piRNA, it is immediately followed by another piRNA, indicating that a single cleavage by Zuc can create the 3′ end of the upstream piRNA and the 5′ end of the followed downstream piRNA (Wang et al. 2015; Czech et al. 2016). At last, the transfer of one methyl group to the 2′-hydroxyl group of the 3′ end is catalysed by a methyltransferase enzyme known as Hen1 to generate mature piRNA associated with PIWI protein in the ovarian somatic cells or follicle cells of *Drosophila.* This modification enhances the stability of mature PIWI-associated piRNAs (Saito et al. 2007; Wang et al. 2016).

Maturation of piRNA in nurse cells (germ cells) of the drosophila ovary takes place in the nuage located in the cytoplasm and follows three distinct mechanisms: de novo biogenesis, Ping-Pong amplification loop, and phasing (Fig. 3) (Han et al. 2015). The piRNA biogenesis mechanism following the de novo pathway is still unknown. The piRNA generated by following the Ping-Pong cycle is known as the piRNA amplification pathway. Several proteins are required for the Ping-Pong pathway to function properly, including vasa (a dead box RNA helicase), Aub, AGO3 (Siwi, homolog of AGO3 found in Bombyx mori- silkworm ovaries), and several Tudor domain-containing proteins (such as Spindle-E, Krimper, Tejas, and Tapas in *drosophila* and TDRD1-9 in mice) (Fig. 3). Miwi2, MVH, TDRD1 (Tudor domain-containing protein 1), TDRD6, and Maelstrom (MAEL) are the mouse homologs required for the production of secondary piRNAs and are associated with piP-bodies. Following the de novo pathway, the generated Aub-loaded piRNAs are the starting complex for the Ping-Pong pathway (Kawaoka et al. 2009). Initially, it was reported that the

Tudor domain protein could only bind with symmetrical dimethylarginine residues loaded on PIWI proteins, and this modification was accomplished by transferring two methyl groups onto arginine residues using the methyltransferase enzyme Capsuleen (Csul) found in silkworm ovaries or the *Drosophila* homolog protein methyl transferase 5 (dPRMT5) (Nishida et al. 2009). Krimper plays a pivotal role in correctly attaching of AGO3 by interacting with Aub. With complementary base pairing, Aub protein binds to the precursor piRNA and forms an Aub-bound piRNA complex, which acts on transposon mRNA and cleaves it into two fragments. Vasa protein loads these two fragments onto AGO3, and a 3′–5′ exonuclease enzyme comes to play a trimming role from the 3′ end of the fragments to generate a new mature piRNA (Hayashi et al. 2016; Xiol et al. 2014). Simultaneously, AGO3 can cleave the complementary transposon transcripts (Nishida et al. 2015). The Vasa helicase loads the AGO3-mediated cleavage products, having 3′ fragments attached to the Aub to form the Aub-pi RNA-induced silencing complex (Aub-piRISC). AGO-3 also has been found in phased piRNA generation. The fragments generated by AGO-3 mediated cleavage of precursor-piRNA further targeted by Zuc and process to generate 3′ and 5′ end and finally load into PIWI to form piRISC (Mohn et al. 2015; Wang et al. 2015). Like Aub, PIWI has the ability to attach to piRNAs generated via the de novo pathway, but the abundance of PIWI-bound piRNA is found more in phased piRNAs than in de novo pathway piRNAs (Senti et al. 2015).

The Ping-Pong piRNA amplification pathway occurs with the help of the AGO3, Aub, and PIWI proteins, which have nuclease activity and are also involved in the 5′ end generation of new piRNAs. Aub proteins bind to antisense strand piRNAs, which guide Aub to initiate the cleavage of pre-piRNAs produced from sense strands,

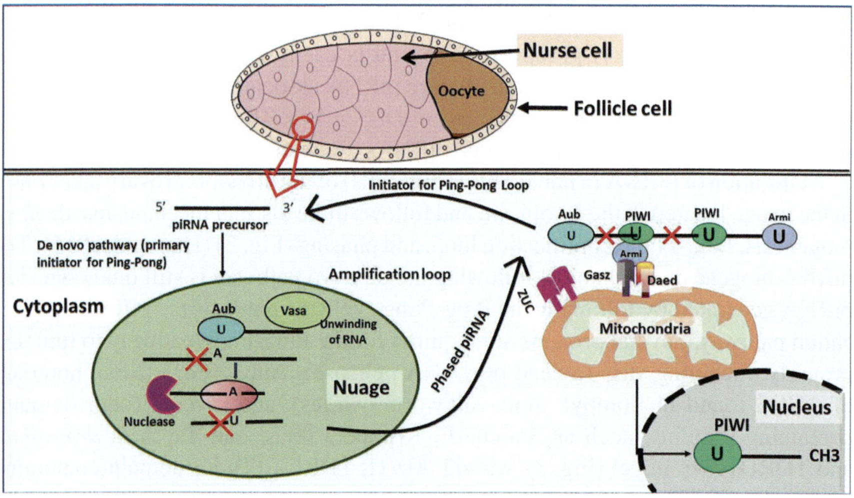

Fig. 3 Processing of piRNA precursor in the germ cell or nurse cell of *drosophila* ovary. PIWI—P-element Induced WImpy testis, Armi—Armitage, AGO3—Argonaute 3, Aub—Aubergine, and ZUC—Zucchini

10 nucleotides away from the 5′ end, creating the 5′ end of a new mature piRNA (Brennecke et al. 2007; Gunawardane et al. 2007) Shu and Hsp83 chaperone proteins help in the incorporation of this piRNA into Ago3, and trimming and methylation occurred on this piRNA as in the primary synthesis. Conversely, the sense-strand of piRNAs is the possible target of Ago3 binding, which guides and helps in the cleaving of antisense-strand pre-piRNAs, thus producing mature antisense piRNAs that incorporate into Aub, again with the help of Shu and Hsp83. The amplification loop repeats, producing a number of piRNAs in a short period. This loop has also been identified in mice and zebrafish (Spitale and Incarnato 2022; Houwing et al. 2007). The piRNAs, serving as guide RNAs, bind to the respective Piwi proteins and relocate back to the nucleus to carry out target gene silencing.

The processing of 5′ and 3′ ends inside the ovarian germ cells, or nurse cells, and is dependent on slicer activity and Zuc endonuclease activity. Zuc endonucleases cleave the precursor piRNA to generate both the 3′ and 5′ ends of mature piRNA without requiring a guide piRNA (Fig. 3). Armi is one type of RNA helicase that helps open precursor piRNA to enhance Zuc-mediated cleavage by using ATP (Ishizu et al. 2019). Slicer-dependent cleavage can generate new mature piRNA and load it into the Aub and AGO3 proteins, which can start a new piRNA-generating cycle known as the “Ping-Pong cycle” (Fig. 3). Aub-mediated cleavage products load into AGO3 and vice versa. Ping-Pong piRNA is produced by Aub and AGO3-mediated slicer cleavage, but the Zuc endonuclease-mediated piRNA processing complex may also target these cleavage products (downstream fragments) to produce PIWI-bound piRNA (Huang et al. 2017).

4 Functions of piRNAs

Due to the vast differences in sequences and modes of action between various animals, the roles of piRNAs have only been partially defined (Aravin et al. 2006; Brennecke et al. 2007). Based on their mode of action, piRNAs may be divided into transcriptional and post-transcriptional levels. Transcriptional activity includes regulating the expression of protein-coding genes and transposable elements, epigenetic modifications, genome defence, spermatogenesis, immune response, and associated human diseases. Epigenetic modifications, such as changes in chromatin structure and histone proteins, are carried out through the activity of DNA methyltransferase (DNMT) (Huang et al. 2013a, b; Watanabe et al. 2011a, b; Yin and Lin 2007). By targeting the coding region of mRNA, piRNAs play an important role in mRNA degradation. In the embryonic stage of *drosophila* melanogaster, Aub mediates the facilitation of maternal mRNA degradation in a piRNA-dependent manner because of its binding with the coding region and 3′ untranslated region (UTR) of mRNA (Barckmann et al. 2015). It has also been discovered that the Aub-piRNA complex can bind with the C-C motif chemokine receptor 4 (CCR4) in *Drosophila* germline stem cells, inhibiting Casitas B-cell lymphoma (Cbl) mRNA expression. This mRNA can also be targeted by transposon-derived piRNAs in their 5′ UTR and 3′ UTR regions

(Rojas-Ros et al. 2017). PIWI-piRNA complexes have been implicated in mRNA degradation through interaction with the mRNA decay machinery, that is, via deadenylation complexes. Inside the germline cell of *drosophila*, Aub and AGO3bind with the mRNA decay machinery complex, such as decapping protein 1/2 (DCP1/2), maternal expression at 31B (Me31B), and Pacman (PCM) in the nuage to form the complex and influence the piRNA-dependent mRNA degradation. Piwi, Aub, Ago3, Armi, Spn-E, and Squash (Squ) are important proteins of the piRNA biogenesis pathway, and they play a significant role in the deadenylation and decay of nanos mRNA by piRNA-dependent targeting of the 3′ UTR region during the early developmental axis determination of the *drosophila* embryo (Wang et al. 2021).

4.1 Silencing of Transposons

Several controls of gene expression and its regulation exist in all eukaryotic species to guard the genome against the harmful effects of external and endogenous genetic elements. One such intrusive genetic element is the transposon. Transposons, also known as jumping genes important in shaping the genome as they can move and relocate within the genome using DNA or RNA intermediates. They disrupt the genetic sequence and create potentially detrimental effects, causing diversity and instability of the genome, pathogenesis through chromosome rearrangement and gene deregulation, and cancers (Chénais 2013; Fedoroff 2012; Kazazian 2004; O'Donnell and Boeke 2007; Wicker et al. 2007). DNA transposons are mobilised without the need for transcription and propagate by a cut-and-paste method in which a transposase is produced that identifies the inverted terminal repeats of a target DNA and facilitates the excision of the element and insertion of the new dsDNA intermediate into a functional genomics location where the inserted DNA transposon can be duplicated along with the target DNA to produce more copy transposon elements, which code for the transposase enzyme. Helitrons are DNA transposons that spread through ssDNA intermediates. The encoded helitron transposase cleaves the donor element at one end and the target site at the other, ligating the target's 3′ end to its 5′ end while leaving the donor element unharmed. Retrotransposons propagate via a copy-paste mechanism, where a reverse transcriptase encoded by the retrotransposon carries out the reverse transcription of the transposon sequence. Then, a DNA copy is placed into a new location via RNA intermediates (Hancks and Kazazian 2012). Transposable elements that behave abnormally cause DNA, resulting in cell cycle blocking, insertional mutagenesis, and inappropriate recombination (Hancks and Kazazian 2012; Hsu et al. 2021).

To guard the genome against such invasive genetic elements, piRNAs play a role in suppressing transposable elements during gametogenesis via forming piRISCs (Spitale and Incarnato 2022; Aravin et al. 2007; Gainetdinov et al. 2017). The ping-pong loop is the defence mechanism in various species, where the pathway has a memory of the previous transposon movements and can produce acute responses to active transposons. piRNAs, complementary to transposable elements' transcripts, repress their functioning by binding to the PIWI proteins and directing them toward the target transposons. Various non-long terminal repeats (non-LTRs) such as SVA, L1, and Alu are involved in leukaemia and cancers of the breast, ovary, and colon. Additionally, L1 repeats in vitro show increased binding of RNA polymerase II to transposons compared to non-cancerous cell lines and in vivo disrupt commonly mutated genes in cancer (Konkel and Batzer 2010).

PIWI-mediated silencing occurs through the creation of a restrictive chromatin state at the transcriptional level by increasing the trimethylation of histone H3's 9th lysine residue (H3K9me3) and heterochromatin protein HP1 chromatin labels at the targeted loci, which further promotes interaction with the Piwi complex (Le Thomas et al. 2013, 2014). Overexpression of piRNAs results in the recruitment of epigenetic regulator HP1a and histone methyltransferase Su(var)3-9 by the piRNA-PIWI complex and HP1, respectively, as well as the reduction of RNA polymerase II and the enrichment of H3K9me2/3 (Giauque and Bickel 2016; Huang et al. 2013a, b). Heterochromatic loci targeted by PIWI for silencing could be identified by the existence of a transposon or its fragments in an intronic sequence or proximity to a gene (Sienski et al. 2012). As perceived in the ping-pong cycle in Drosophila, Aub, and Ago3 move towards the cytoplasm and function in the post-transcriptional repression of transposons through the cleavage of transcripts (Post et al. 2014). Also, the production of anti-sense piRNAs via Ago3-mediated cleavage drives transcriptional silencing by Piwi. In mice, Mili and Miwi2 have been characterised to affect repetitive elements' methylation to sustain stable transposons suppression (Aravin et al. 2007; Carmell et al. 2007). A derepression of piRNA clusters was implicated in increased transposable element activity. Piwi mutants revealed defects in the suppression of transposons, which could be seen in studies of piRNA loci. The clusters exerted control over specific transposons in flies. For instance, the *flamenco* locus, a significant piRNA cluster located within the pericentromeric region of heterochromatin on the X chromosome, repressed the movement of retrotransposons such as gypsy, ZAM, and Idefix (Desset et al. 2003; Prud'homme et al. 1995). In the *flamenco* mutants, gypsy was activated, and all the piRNAs generated from this cluster became lost. DNA methylation in the transposon silencing of piRNAs also leads to transcriptional gene silencing.

4.2 Gene Regulation and Development

piRNAs regulate functional and non-functional gene fragments, which are not extensively complementary to transposons. Prostate cancer is associated with altered expression of piR 015,520, located in the human melatonin receptor 1A gene, which negatively regulates the expression of MNTR1A by binding to its genomic site (Esposito et al. 2011). piR-39980 is one type of tumour suppressor that belongs to the fibrosarcoma family, represses the ribonucleotide reductase regulatory subunit M2 gene (RRM2) by binding to its 3′ UTR via sequence complementarity (Das et al. 2019). Thus, the piRNA expression levels could be said to influence gene expression at the loci of the piRNA. piR-1245, which facilitates tumour growth and invasion in colorectal cancer, can regulate gene expression by interacting with a group of target genes such as ATF3, BTG1, DUSP1, INP1, TP53, and many others and causes RNA suppression via abnormal base-pairing after the formation of specific RISC complexes (Ma et al. 2004). piR-36712 inhibits the expression of seleno- protein W1, SEPW1, by competing with SEPW1P (retro processed pseudogene of SEPW1) RNA for miR-7 and miR324 and affecting the SEPW1 mRNA. It has been observed that the upregulation of SEPW1 due to the downregulation of piR-36712 will inhibit P53 and P21, making piR-36712 a therapeutic target in breast cancer (Tan et al. 2019).

The piRNAs also control the stability of their associating proteins by binding to them. They regulate the ubiquitination of Miwi during the later stages of mouse spermatogenesis by binding to the anaphase-promoting complex (APC)/C substrate-binding subunit (Cox et al. 2000). Studies have revealed that piR-L-163 interacts with phosphorylated ERM proteins (p-ERM; ERM—ezrin, radixin, and moesin—essential in the control of signal transduction pathways by associating transmembrane proteins); the connection is required for p-binding ERM's to F-actin and ERM-binding phosphoprotein 50 (EBP50)(EBP50) (Mei et al. 2015). piR-36026 binds to tumour suppressors, Serpin peptidase inhibitor, clade A, member 1 (SERPINA1), and lecithin retinol acyltransferase (LRAT), which are target genes in breast cancer cells, and subsequent molecular activities of caspase-3 and phosphatidyl inositol were observed within the nucleus with piR-36026 as the oligonucleotide probe (Lee et al. 2016). The piRNA-PIWI pathway is also present subtly in somatic cells such as ovarian follicle cells, salivary glands, mouse hippocampal cells, pluripotent stem cells (PSCs), mesenchymal stem cells (MSCs), and hematopoietic stem cells (HSCs), but very infrequently in mature multipotent stem cells (Martinez et al. 2015; Sato and Siomi 2020; Liu et al. 2019). They appear to be associated with protein-coding genes as well. In *Drosophila*, the stellate protein-coding gene on the X chromosome was the target of the first piRNAs discovered and transcribed from the suppressor of the stellate locus located on the Y chromosome (Peng et al. 2013). piRNAs produced by mitochondrial DNA colocalize with PIWI proteins and have a role in the stress response. The nuage is one type of germline granules found in *Drosophila* melanogaster (Lim and Kai 2007), also seen in lower metazoan stem cells but confined to germline cells in higher metazoan stem cells. These are membrane-less perinuclear granules containing various tudor domain-bearing proteins, where piRNA maturation occurs. Nuage is also known as pi-body intermitochondrial cement or pi-body based on their localization and morphology (Watanabe et al. 2011a, b). As it is confined to germline cells of higher metazoans, suggesting that the germline genes have progenital significance in monitoring stem ness (Kwon et al. 2014).

4.3 Changes in Diseases

Early research on piRNAs concentrated on how they function in the germline, such as germline development, maintenance of genome integrity, and transposon silencing, among other cells, since many PIWI clade member proteins are predominantly limited to the germline. Investigation of the functional activities of piRNAs beyond the germline, which included binding to the certain genomic region in somatic cells and epigenetic targeting of DNA loci, proved to be an icebreaker. Instances include the identification of the involvement of HP1A in the methylation of H3K9 in somatic cells, the localization of hippocampal piRNAs in neurons and dendrites of mouse, the effects of piRNAs in spine morphogenesis, and many more (Giauque et al. 2016; Lee et al. 2011). piRNAs' biological roles have been greatly expanded in the studies of various diseases that have an impact on human health, including cancer, cardiology, and metabolic function (Ayarpadikannan and Kim 2014; Gao et al. 2020; Rajan et al. 2016). The increasing number of harmful substances in the environment has increased the number of diseases. Treating diseases such as cancer is often unsuccessful due to the reasonably late detection, high metastasis, and recurrence rates, which has led to exploring new biomarkers for the detection, prognosis, and efficient therapy of cancer. Numerous studies have shown that piRNAs have a variety of possible impacts on the control of illness that need further investigation (Wakisaka and Imai 2019; Kausar et al. 2019; Shaw and Catteruccia 2019; Sleiman et al. 2016).

Analysis has documented that after being treated with oxidised low-density lipoprotein, it was found that piR-41220 and piR-46956, were downregulated in retinal pigment epithelial cells around the globe all over the globe (oxLDL) (Zhao et al. 2019; Wu et al. 2020). In particular, they guarded themselves against the potentially harmful effects of oxidative stress by focusing their attention on genes linked with the operation of the eye. This allowed them to avoid any potential damage. Analyses of the expressions of piRNA were performed on cardiac progenitor cells (CPCs), which comprised cardio fibroblasts (CFs), cardiosphere-derived cells (CDCs), and cardiospheres. The upregulation of some piRNAs led to their targeting of transposons over the whole of the human genome, including LINE retrotransposons. The LINE-1 retrotransposon was the focus of most research conducted on the LINE family of retrotransposons (Vella et al. 2016; Wu et al. 2020). It was found that several Alzheimer's disease-associated piRNAs were discovered to be connected with genome-wide single-nucleotide polymorphisms, and around one hundred Alzheimer's disease (AD) patients were shown to have dysregulated expression of piRNAs (SNPs). Researchers were able to find that the microRNAs piR-38240, piR34393, and piR-40666 played a part in the process of maintaining hemostasis in the nervous system. This was one significant major discoveries made by these studies. In addition to influencing the trafficking mechanism that regulates antibody levels, their overexpression may also induce cell death and neurodegeneration. The study into the roles played by piRNA in non-cancerous illnesses is much less in-depth when compared to the research into the functions played by piRNA in cancerous disorders (Qiu et al. 2017; Wu et al. 2020). Due to the difficulty of most tissue samples, the vast majority of research is conducted on the tissues of model animals rather than on actual human tissue samples. The findings of a number of different studies have shown that piRNAs. Suppose it is possible to summarise the connections between the piRNAs that may be detected in physiological fluids and these diseases. In that case, physiological fluids and these diseases, then there will be new options for clinical diagnosis and treatment (Wu et al. 2020).

5 Epigenetic Mechanisms

Studies have suggested that piRNAs control genome integrity by suppressing mobile genetic transposable elements and silencing euchromatic genes to varying degrees when localized in close proximity to a heterochromatic region. For the preservation of chromatin structure and the progression of the cell cycle during early embryogenesis, maternal PIWI proteins play a vital role. Cancer stem cells (CSCs) are characterized by their unique self-renewal properties, absence of senescence, maintenance of an undifferentiated state, and rapid proliferation. These attributes are kept in check by epigenetic regulations that provoke the changes in expression profiling of various genes in tumour cells. Other non-ageing cells include germline and somatic stem cells. The genetic instability of these cells is triggered by the impaired repair mechanisms leaving mutations with a few or no limited number of genomic changes. Epigenetic changes influence the expression of specific genes rather than altering the genetic code and could result from environmental perturbations, DNA methylation, post-transcriptional histone modifications, and deregulation of noncoding RNAs. The human genome sequence does not explain transcriptional regulation. All species need complex coordination and communication between numerous well-conserved chromatin-modifying factor. Therefore histone post-translational modifications (HPTMs), DNA methylation, histone variants, remodelling enzymes, and effector proteins modify chromatin structure and function that affects DNA repair, replication, growth, and proliferation. Thus, histone modification is transmitted through the generations in the form of epialleles, while PIWI is directly linked with DNA methylation. (Morris 2012; Rajasethupathy et al. 2012). Histone modification is transmitted through the generation to generation in the form of epialleles, while DNA methylation is functionally related to PIWI (Morris 2012).

Methylation of DNA influences individual genes' expression as well as DNA domains through interaction with nucleosomes and hence modifies DNA packaging. DNA methyltransferases (DNMT1, DNMT3A, and DNMT3B) aid in the targeted regulation of modification of the structure of chromatin and histones. The CpG islands in the promoter regions are methylated by DNMT, resulting in the suppression of the initiation of transcription of the transposable elements and target genes. PIWI-piRNA complexes recognize the sequences and up-regulate the expression of DNMTs, which prevent the binding of transcription factors by methylating the promoter regions. The molecular basis of this process is unclear, and the evolutionary basis seems highly conservatory. Methylation of the promoter site methylation represses the gene's transcription, whereas that of the gene's coding region might result in post-transcriptional gene silencing. PIWIL1 stimulates the overexpression of DNMT1, which act on hemimethylated CpG islands and add more methyl groups to this island and DNMT3a, and PIWIL2 and PIWIL4 induce the overexpression of DNMT1, DNMT3a, and DNMT3b. DNMT3a and DNMT3b, which methylate new CpG sites, are upregulated by piR-823 (Yang et al. 2018). In primary testicular cancers, a 5′ methylation end in the CpG island promoters of the PIWIL1, PIWIL2, PIWIL4, and TDRD1 genes was discovered in conjunction with transcriptional silence. Loss of PIWIL2 and PIWIL4 results in the downregulation of DNA methylation in promoter regions (Kuramochi-Miyagawa et al. 2004). Global hypomethylation of DNA causes transposable gene reactivation, mitotic recombination, and enhanced expression of various oncogenes. With the development and progression of tumour malignancy, the extent of DNA hypomethylation becomes more severe. Local hypermethylation of DNA leads to the repression of tumour suppressor genes and leads to various types of epigenetic modification of histone proteins, resulting in the transformation of proliferating cells into highly malignant cells that differ from one another by a pro-metastatic condition, invasion, uncontrolled cell division, and epithelial mesencymal transition (EMT) (Zhang et al. 2013). piRNAs are also involved in the methylation of non-transposable sites of DNA, as seen in the Ras-specific guanine nucleotide-releasing factor 1 (Rasgrf1) locus of the mouse male germline, where piRNAs modulate genomic imprinting. To affect long-term memory plasticity, a similar event takes place in the promoter of CAMP response element-binding protein 2 (CREB2) in the neurons of Aplysia (Esposito et al. 2011). It is yet unknown how piRNAs and PIWI proteins contribute to aberrant DNA methylation's role in the epigenetic control of germline cells.

Another mechanism of epigenetic regulation in cancer is histone modification. Normal cells have stable levels of heterochromatin during the various phases of cell cycle, which are silenced during transcription, where euchromatic genes are actively transcribed. The methylation of lysine residue of histone H3 and H4 are carried out by recruiting histone methyltransferases (HMTs), Suv39H1 (a Histone Lysine Methyltransferase), and SET Domain Bifurcated Histone Lysine Methyltransferase 1 (SETDB1), which aid in the upregulation of H3K9 methylation to suppress transcription. Moreover, these complexes have the binding ability with the multiple isoforms of HP1, guiding them to combine with the H3K9me repressive marks in the target regions. As a result, the accumulation of H3K9me invokes a heterochromatin state during which chromosomes segregate at the time of cell division and genes are not accessible to transcription factors. Epigenetic alterations of Suv39 lead to genomic instability and increased risk of B-cell lymphoma and lung cancer. Epigenetic changes associated with cancers are shown in Table 1 Other epigenetic modifications of histone proteins include acetylation, phosphorylation, and ubiquitination, which are known to promote transcription, and biotinylation and sumoylation to downregulate gene expression (Martin and Zhang 2005; Lee et al. 2005; Larson and Narlikar 2018; Maksimov et al. 2018). Also, Piwi proteins in *Drosophila* germline stem cells upregulate the expression of chromosome 3R telomere-associated sequence (3R-TAS) (Yong et al. 2016). Negative effects in the absence of PIWI, such as enhanced telomeric position in the white eye gene, could be seen, indicating that Piwi was necessary for the expression of the reporter gene to promote the active epigenetic state of the gene 3R-TAS. Suppression or activation of transcription due to Piwi-binding depends on the elements that interact with Piwi and/or the chromatin microenvironment that influences it or its sublocalization within the nucleus (Yin and Lin 2007; Ng et al. 2016; Han et al. 2017; Ross et al. 2014).

Table 1 Epigenetic roles of piRNAs and PIWI proteins

Cancer	piRNA/PIWI	Function	Effect	References
Breast cancer	piR-021285	Suppresses expression of ARHGAP11A by methylating a 5′ CpG site	Inhibition of apoptosis	Han et al. (2017)
	piR-932	Binds to PIWIL2 to promote CpG island methylation of Latexin promoter	Suppressed Latexin expression	Yin et al. (2007)
Colorectal cancer	piR-823	Suppresses the ubiquitination of hypoxia-inducible factor 1α by elevating the expression of glucose-6-phosphate dehydrogenase	Increased glucose consumption and decreased content of intracellular reactive oxygen species	Han et al. (2017)
Multiple Myeloma	piR-823	Inhibits in vitro and in vivo tumour formation and induces the expression of cell cycle and apoptotic proteins	Downregulation of tumour growth	Han et al. (2017)
		Downregulates de novo methyltransferases DNMT3A and DNMT3B	Downregulation of DNMT3B causes re-expression of methylated p16	Yan et al. (2011)
Cervical cancer	PIWIL2	Involved in the transformation of the epithelial cells of the cervix into tumour-causing cells via epigenetic-based cell programming	Inhibition of apoptosis	Han et al. (2017)
Endometrial cancer	PIWIL1	Mediates PTEN hypermethylation via DNMT1	Loss of PTEN expression	Sun et al. (2011)

6 piRNAs in Cancer

piRNAs and sncRNAs play a pivotal role in the regulation of gene expression in tumour cells. Due to the similar properties of cancer and germ cells, several germline-specific factors are also involved in oncogenesis. miRNAs are extensively studied in cancers; some show tumour-suppressive function and some show oncogenic function. By playing this role they trigger tumour formation (Jothimani et al. 2018; Ling et al. 2013; Hayes et al. 2014; Di et al. 2014; Pathak et al. 2015). By acting at the transcriptional and post-transcriptional levels upstream or downstream of various oncogenes and tumour suppressor genes in different tumour tissues or cell lines, piRNAs play critical roles in gene silencing and alterations. PIWI-interacting RNAs has gain attention in recent times to be used as potential diagnostic biomarkers and therapeutic target in various diseases, including colorectal cancer (Fig. 4). They will aid in developing combination therapies that will target multiple pathways (Gibb et al. 2011; Wicker et al. 2007). It has been found that the levels of piRNAs and PIWI are notably altered between non-tumour tissues. Several piRNAs and PIWI proteins like PIWIL2 and HIWI are expressed in tumour cells in the cancers of the stomach, brain, bladder, breast, colon, and lungs (Liu et al. 2006). It is hypothesised that the piRNA pathway contributes to the incidence and progression of cancer. The potential roles of these regulating RNAs have just emerged and remain to be investigated. Oncogene inhibition and tumour suppressor gene degradation are thought to be regulated by aberrant piRNA expression. The disruption of piRNAs, on the other hand, may cause mutagenic retro transpositions and genomic instability which have an impacting impact on cancer progression. Epigenetic changes in cancers include gene-specific DNA hypermethylation, DNA hypomethylation, and histone hypoacetylation, which result in the activation of oncogenes such as Ras, cyclin D2 and the silencing of tumour suppressor genes RB1, and p16 that alters the balance between DNA methylation and remethylation (Sienski et al. 2012).

The abnormal expression of piRNAs has been implicated with cancer hallmarks containing proliferation, apoptosis, invasion, and metastasis. Accumulation of genetic and epigenetic alterations regulates carcinoma development. Lu et al.'s analysis of the whole small RNA library in HeLa cells revealed that breast cancer expresses more than 100 piRNAs. Cell cycle progression and the oncosuppressive oestrogen receptor (ER) have an impact on certain piRNAs. According to the results of in-situ-hybridization, piR-49322 was shown to be localised in HeLa cells' cytoplasm and nucleolus, particularly towards the nuclear membrane's edge (Lu et al. 2010). Since most piRNAs are estrogen-sensitive, a lack of oestrogen may cause changes in the levels of piRNA expression. The oestrogen receptor, or ER, regulates cancer spread by directly regulating piRNA production in breast cancer and responding to stimuli both within and outside the malignant cells. Hence, breast cancer proliferation can be controlled by the influence of ERβ and the deficiency of estrogen. Table 2 showed that piR-34736, piR-36249, piR-35407, and piR-34377 were among the downregulated piRNAs. PiR-36743, PiR-36026, and PiR-31106 levels were found to have noticeably increased (Hashim et al. 2014).

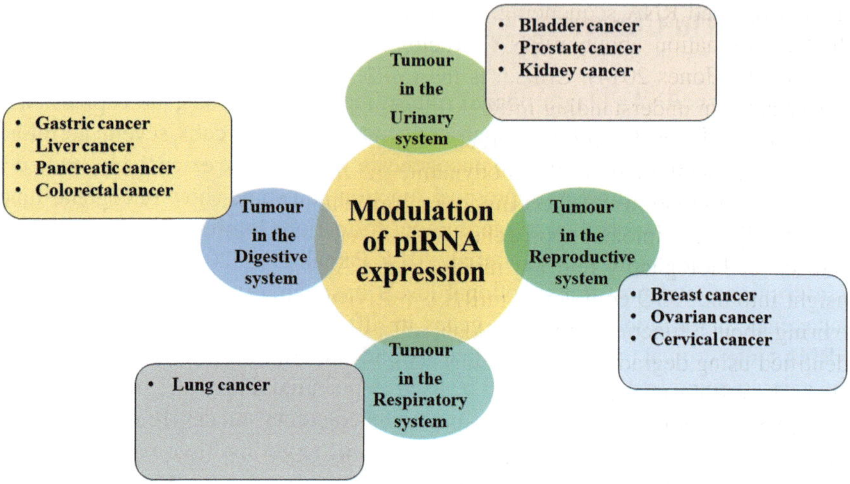

Fig. 4 Clinical conditions associated with the modulation of piRNA expression

It has been shown that the piRNA gene expression patterns of 6260 human tissue samples from benign and malignant tissues from diverse organs and examined the piRNA expression profile in twelve various cancer tissue types (Fig. 4) (Martinez et al. 2015). The observed patterns were unique to each malignancy and clinical category. Most cancer types showed varying degrees of piRNA expression, while some were specific to particular clinical traits and subgroups that were specific to each cancer type. Microarray data of piRNA from bladder cancer tissues and nearby normal tissues showed that. 106 piRNAs were found to be upregulated, and 91 piRNAs were downregulated (Chu et al. 2015). The most down-regulated piR was piR-DQ594040, whose overexpression can prevent cell division and colony formation and stimulate cell death by increasing the level of the TNFSF4 protein. Recent findings uncovered that piR-55490 was silenced in lung carcinoma samples and cell lines, in comparison with normal lung tissues, and that the suppression of piR-55490 resulted in an increased proliferation rate. Also, piR-55490 was found to bind to the 3′ UTR of mTOR mRNA, suppressing the activation of the Akt/mTOR pathway and inducing its degradation (Peng et al. 2016). Studies on the functions of piRNAs in breast cancer, gastric cancer, and hepatocellular carcinoma highlight the potential role of piRNAs in preventing tumour invasion and metastasis. piRNAs are known to block metastases via gene methylation and phosphorylation of the Akt pathway.50 breast cancer biopsies were used to confirm the upregulation of piR-4987, piR-20365, piR-20485, and piR-20582 by RT-PCR (Table 2). The upregulation of piR-4987 was positively correlated with lymph node metastasis. Deep sequencing was used to screen differentially expressed piRNAs in four breast cancer tissues and four corresponding normal tissues (Huang et al. 2013a, b). Analysis showed through their study that the piR-932/PIWIL2 complex may promote latex in methylation (a tumour suppressor), which would then promote the EMT, and that

Table 2 Function and expression of piRNAs in different cancers

Cancer	piRNA/PIWI	Function	Expression	References
Prostate cancer	piR_015520	Downregulates the expression of the MNTR1A gene	Upregulated	Esposito et al. (2011)
Fibrosarcoma	piR-39980	Tumour suppressor; binds to 3′ UTR of ribonucleotide reductase regulatory subunit M2 (RRM2)	Upregulated	Das et al. (2019)
Colorectal cancer	piR-1245	Accelerates tumour growth and invasion; regulates gene expression by binding to target genes such as ATF3, BTG1, DUSP1, INP1 and TP53	Upregulated	Weng et al. (2018)
Breast cancer	piR-36712	Inhibits the expression of selenoprotein W1; therapeutic target	Upregulated	Tan et al. (2019)
	piR-36026	Binds to breast cancer target genes such as SERPINA1 and LRAT	Upregulated	Lee et al. (2016)
	piR-36743, piR-36026, piR-31106		Upregulated	Hashim et al. (2014)
	piR-34736, piR-36249, piR-35407, piR-34377	Affected by the progression of cell cycle and on cosuppressive estrogen receptor β (ERβ); ERβ controls cancer metastasis	Downregulated	Hashim et al. (2014)
	piR-4987, piR-20365, piR-20485, piR-20582	piR-4987 was involved in lymph-node metastasis	Upregulated	Huang et al. (2013a, b)
	piR-932	Promotes methylation of Latexin and hence epithelial-mesenchymal transition (EMT)	Upregulated	Zhang et al. (2013)
Bladder cancer	piR-DQ594040	Involved in the regulation of TNFSF4	Downregulated	Chu et al. (2015)

(continued)

Table 2 (continued)

Cancer	piRNA/PIWI	Function	Expression	References
Lung cancer	piR-55490	Increased proliferation; binds to 3′ UTR of mTOR mRNA. Suppressing the activation of the Akt/mTOR pathway and inducing its degradation	Downregulated	Peng et al. (2016)
	piR-651	Promotes increased levels of cyclins and CDKs	Upregulated	Han et al. (2017)
	piR-L-163	Regulates signal transduction pathways by binding to ERM proteins	Upregulated	Han et al. (2017), Mei et al. (2015)
Gastric cancer	piR-823	Inhibits tumour growth, TNM, and distant metastasis	Downregulated	Han et al. (2017)
	piR-651	Metastasis	Upregulated	Han et al. (2017)
Hepatic carcinoma	piR-651	Metastasis	Upregulated	Han et al. (2017)
	piR-Hep1	Increased cell viability and invasiveness; increased levels of Akt phosphorylation	Upregulated	Law et al. (2013)

piR-932 and PIWIL2 may be useful targets for preventing the spread of breast cancer (Zhang et al. 2013).PiR-823 was downregulated in human gastric cancer, in contrast to piR-651, which was strongly associated with tumour-node-metastasis, according to microarray data (TNM). Further, piR-651 had also been reported to be increased in other types of cancer such as hepatocellular carcinoma (HCC), breast cancer, cervical cancer, and lung cancer to name a few. A tentative study was carried out to evaluate piRNAs as biomarkers in detecting circulating cancer cells in gastric cancer samples. The levels of piR-651 and piR-823 were considerably lower than those in the control samples, and the levels of piR-823 exhibited a strong positive correlation with distant metastases and TNM stages. piR-Hep1, a novel piRNA, was upregulated in hepatic carcinoma, compared to surrounding non-cancerous tissues, found in the deep sequencing study (Law et al. 2013). The expression of PIWIL2 correlated positively with piR-Hep1, indicating the role of the piR/PIWIL2 complex. With a decreased level of active AKT phosphorylation, piR-Hep1 knockdown inhibited cell viability, mobility, and invasiveness (Han et al. 2017; Law et al. 2013).

7 PIWI Expression in Cancer

Seminomas, a cancer of the male germ cell, were the first cancer type to exhibit the PIWI gene. Qiao et al. (2002) and Lee et al. (2006) studied the roles of HIWI and HILI respectively in seminomas and discovered that HIWI was found in seminomas, but not in those that developed from somatic cells such Sertoli cells or Leydig cells, and HILI has been linked to apoptosis, adhesion, and cell proliferation. This was followed by studies on different cancers. Upregulation of HILI inhibited apoptosis of cancer cells in breast cancer (Hashim et al. 2014). HIWI expression progressively increased from pre-neoplastic lesions to advanced malignancy, which suggests a relationship between HIWI and the growth of gastric cancer cells (Liu et al. 2006). Cell cycle arrest in the M/G2 phase was caused by HIWI inhibition, which suggests that HIWI could prove to be a promising gastric cancer therapy target. Increased levels of HIWI protein in early sarcoma inhibited cell differentiation, and decreased levels of HIWI protein inhibited cell growth. In cervical cancer cells, HIWI and HILI were expressed and showed an association with HPV infection, and HILI inhibited p53 and repressed the apoptosis of tumour cells. Immunohistochemical investigations of samples of different stages of cancer proposed PIWI as a potent biomarker for cervical and breast cancer (He et al. 2010; Liu et al. 2010,). Several studies followed that supported the potential of PIWI expression in cancer prognosis. The expression level of HIWI in the form of mRNA and protein in non-small cell lung cancer (SSCloAldebr cells) was found to be significantly high. Silenced HIWI lost the ability to form a sphere and the capacity to form a colony in the SSCloAldebr cells, and it suppressed the tumour growth in nude mice (Liang et al. 2013). Tudor domain-containing protein 9 (TDRD9) is one type of helicase that has been found to be involved in the piRNA biosynthesis in lung adenocarcinoma Although lung adenocarcinoma TDRD9 expression is associated with a poor prognosis, TDRD9 knockdown can reduce apoptosis, arrest the cell cycle, and promote proliferation (Guijo et al. 2017). It has been shown that altered HIWI expression in pancreatic cancer patients posed a high risk of cancer-related death. HILI can play an oncogenic role in the production of tumours by influencing the Stat3/Bcl-X and Stat3/cyclinD1 signaling cascades (Grochola et al. 2008). An in vitro and in vivo study revealed that HILI in its inhibited form has tumour growth suppressive activity (Lee et al. 2006). Silencing of HIWI expression in gliomas has been found to induce apoptosis, block the cell cycle, and thereby inhibit cellular proliferation. Therefore, it affects the expression of apoptotic and cell cycle-related proteins, such as cyclin D1, Bcl2, and Bax. Other investigations were carried out to study the roles of PIWI proteins like PIWIL4 in hepatic carcinoma, ovarian cancer, glioma, gastrointestinal cancer, and many more. Members of the PIWI clade have been observed to influence the metastasis and invasion in tumours via the regulation of MMP-2 and MMP-9, the mechanism of which is not yet understood in detail. Also, HIWI, PIWIL2 (especially in colon cancer), and PIWIL4 have diverse effects on metastasis along with piRNAs (Han et al. 2017).

8 Therapeutic Approach of piRNAs and PIWI Proteins

The use of artificial piRNAs as weapons to bind to mRNAs and prevent the production of oncoproteins and other proteins related to malignancy is intriguing, which would have the benefit of not using the Dicer enzyme, which is necessary for miRNAs. piRNAs may have other hypothetical advantages beyond miRNAs, such as the potential for targets with higher specificity since various miRNAs regulate numerous mRNAs and the capacity to access unwanted long non-coding miRNAs with potential repercussions. In addition, PIWI antibodies can be used for therapeutic purposes because of having an advantageous therapeutic effect on the growth of cancer cells and may be used in combination therapy for various tumours as a post-translational strategy. Certain synthetic piRNAs can be designed in such a way that they can bind to PIWI proteins, thereby silencing PIWI genes at the transcriptional level through genomic silencing, which is one of the potential approaches. These therapeutic interventions might potentially be employed in combination with other conventional therapies. For instance, PIWI antibodies may be used as a possible drug delivery approach against cancer cells, which has been used to deliver other antibodies (Liu et al. 2019).

9 Conclusion

The studies on PIWI-interacting RNAs since its discovery more than a decade ago have attracted significant attention due to the development of sRNA-seq technologies and bioinformatic analyses. Crosslinking immunoprecipitation sequencing (CLIP-seq) and RNA immunoprecipitation sequencing (RIP-seq) are the two frequently used methods of sequencing where piRNAs are crosslinked with PIWI and immunoprecipitated with antibodies targeting the proteins. Several other studies have been conducted to analyse piRNAs computationally, such as a Support Vector Machine (SVM), Fisher separator algorithm for piRNA identification, multiple kernel fusion and SVM, and a k-mer approach for piRNA prediction, and many more. Compared to existing algorithms, the precision and sensitivity of the piRNA prediction and detection tools still have room for improvement. The number of newly discovered piRNAs has increased dramatically; however, there is a lack of thorough and precise knowledge of their roles and mechanism of interactions. Although a numerous studies have detailed reports on the complex biogenesis and the expression patterns of piRNAs and PIWI proteins in various diseases, the underlying molecular mechanisms remain to be elucidated. How piRNAs regulate their targets through mRNA decay or methylation of promoters and the regulation of differentially expressed piRNAs are yet to be answered, and along with this, there are also still several unanswered problems regarding how PIWI proteins control tumour growth, apoptosis, metastasis, and invasion in the cytoplasm either independently or in conjunction with piRNAs. Synthetic piRNAs have been designed to bind the mRNAs of cancer-related proteins to block their synthesis. PIWI antibodies are a different hypothetical strategy that may impact tumour growth. The discovery of potential interactions would deepen the understanding of the therapeutic insights of piRNAs in cancer and other diseases, the link between PIWI and CSCs, and the influence of piRNAs secreted in one site on distant targets via exosomes like microRNAs.

Availability of data and materials: NA.

10 Competing Interests

The authors declare that they have no known competing financial interests or personal relationships that could have appeared to influence the literature survey work reported in this chapter.

Funding: NA.

Acknowledgements We would like to thank Chettinad Academy of Research and Education, Chettinad Hospital, and Research Institute for providing the facilities.

Authors' Contributions Conceptualization: **Surajit Pathak and Antara Banerjee**, Writing original draft: **Subhamay Adhikary, Alagu Theivanai Ganesan, Alakesh Das, Amit Dey, Editing: Surajit Pathak and Antara Banerjee**; the content of the submitted manuscript was approved by all authors.

References

AravinA A, Sachidanandam R, Girard A et al (2007) Developmentally regulated piRNA clusters implicate MILI in transposon control. Science 316:744–747

Aravin A, Gaidatzis D, Pfeffer S et al (2006) A novel class of small RNAs bind to MILI protein in mouse testes. Nature 442:203–207

Ayarpadikannan S, Kim HS (2014) The impact of transposable elements in genome evolution and genetic instability and their implications in various diseases. Genomics Inform 12:98–104

Barckmann B, Pierson S, Dufourt J et al (2015) Aubergine iCLIP reveals piRNA-dependent decay of mRNAs involved in germ cell development in the early embryo. Cell Rep 12:1205–1216

Bartel DP (2018) Metazoan MicroRNAs. Cell 173:20–51

Bhartiya D, Scaria V (2016) Genomic variations in non-coding RNAs: structure, function and regulation. Genomics 107:59–68

Bodak M, Cirera-Salinas D, Luitz J et al (2017) The role of RNA interference in stem cell biology: beyond the mutant phenotypes. J Mol Biol 429:1532–1543

Brennecke J, Aravin AA, Stark A et al (2007) Discrete small RNA-generating loci as master regulators of transposon activity in Drosophila. Cell 128:1089–1103

Carmell MA, Girard A, van de Kant HJ et al (2007) MIWI2 is essential for spermatogenesis and repression of transposons in the mouse male germline. Dev Cell 12:503–514

Chen S, Ben S, Xin J et al (2021) The biogenesis and biological function of PIWI-interacting RNA in cancer. J Hematol Oncol 14:1–18

Chénais B (2013) Transposable elements and human cancer: a causal relationship? Biochim Biophys Acta 1835:28–35
Chi T, Lin J, Wang M et al (2021) Non-coding RNA as biomarkers for Type 2 diabetes development and clinical management. Front Endocrinol 12:630032
Chu H, Hui G, Yuan L et al (2015) Identification of novel piRNAs in bladder cancer. Cancer Lett 356:561–567
Cox DN, Chao A, Lin H (2000) Piwi encodes a nucleoplasmic factor whose activity modulates the number and division rate of germline stem cells. Development 127:503–514
Cox DN, Chao A, Baker J et al (1998) A novel class of evolutionarily conserved genes defined by piwi are essential for stem cell self-renewal. Genes Dev 12:3715–3727
Cramer W, Guiot J, Fader M et al (2018) Climate change and interconnected risks to sustainable development in the Mediterranean. Nat Clim Change 8:972–980
Czech B, Hannon GJ (2016) A happy 3′ ending to the piRNA maturation story. Cell 164:838–840
Czech B, Munafò M, Ciabrelli F et al (2018) piRNA-guided genome defense: from biogenesis to silencing. Annu Rev Genet 52:131–157
Das B, Roy J, Jain N et al (2019) Tumor suppressive activity of PIWI-interacting RNA in human fibrosarcoma mediated through repression of RRM2. Mol Carcinog 58:344–357
Dennis C, Brasset E, Sarkar A et al (2016) Export of piRNA precursors by EJC triggers assembly of cytoplasmic Yb-body in Drosophila. Nat Commun 7:13739
Desset S, Meignin C, Dastugue B et al (2003) COM, a heterochromatic locus governing the control of independent endogenous retroviruses from Drosophila melanogaster. Genetics 164:501–509
Di Leva G, Garofalo M, Croce CM (2014) MicroRNAs in cancer. Annu Rev Pathol 9:287–314
ElMaghraby MF, Andersen PR, Pühringer F et al (2019) A heterochromatin-specific RNA export pathway facilitates piRNA production. Cell 178:964–979
Esposito T, Magliocca S, Formicola D et al (2011)piR_015520 belongs to Piwi-associated RNAs regulates expression of the human melatonin receptor 1A gene. PloS One 6:e22727
Fedoroff NV (2012) Transposable elements, epigenetics, and genome evolution. Science 338:758–767
Gainetdinov I, Skvortsova Y, Kondratieva S et al (2017) Two modes of targeting transposable elements by piRNA pathway in human testis. RNA 23:1614–1625
Gao XQ, Zhang YH, Liu F et al (2020) The piRNA CHAPIR regulates cardiac hypertrophy by controlling METTL3-dependent N^6-methyladenosine methylation of Parp10 mRNA. Nat Cell Biol 22:1319–1331
Ge DT, Wang W, Tipping C et al (2019) The RNA-binding ATPase, Armitage, couples piRNA amplification in nuage to phased piRNA production on mitochondria. Mol Cell 74:982–995
Giauque CC, Bickel SE (2016) Heterochromatin-associated proteins HP1a and Piwi collaborate to maintain the association of achiasmate homologs in drosophila oocytes. Genetics 203:173–189
Gibb EA, Brown CJ, Lam WL (2011) The functional role of long non-coding RNA in human carcinomas. Mol Cancer 10:38
Goriaux C, Desset S, Renaud Y et al (2014) Transcriptional properties and splicing of the flamenco pi RNA cluster. EMBO Rep 15:411–418
Grochola LF, Greither T, Taubert H et al (2008) The stem cell-associated Hiwi gene in human adenocarcinoma of the pancreas: expression and risk of tumour-related death. BJC 99:1083–1088
Guijo M, Ceballos-Chávez M, Gómez-Marín E et al (2017) Expression of TDRD9 in a subset of lung carcinomas by CpG island hypomethylation protects from DNA damage. Oncotarget 9:9618–9631
Gunawardane LS, Saito K, Nishida KM et al (2007) A slicer-mediated mechanism for repeat-associated siRNA 5' end formation in Drosophila. Science 315:1587–1590
Han BW, Wang W, Li C et al (2015) piRNA-guided transposon cleavage initiates Zucchini-dependent, phased piRNA production. Science 348:817–821
Han YN, Li Y, Xia SQ et al (2017) PIWI proteins and PIWI-interacting RNA: emerging roles in cancer. Cell Physiol Biochem 44:1–20

Hancks DC, Kazazian HH Jr (2012) Active human retrotransposons: variation and disease. Curr Opin Genet Dev 22:191–203
Hashim A, Rizzo F, Marchese G et al (2014) RNA sequencing identifies specific PIWI-interacting small non-coding RNA expression patterns in breast cancer. Oncotarget 5:9901–9910
Hayashi R, Schnabl J, Handler D et al (2016) Genetic and mechanistic diversity of piRNA 3′-end formation. Nature 539:588–592
Hayes J, Peruzzi PP, Lawler S (2014) MicroRNAs in cancer: biomarkers, functions and therapy. Trends Mol Med 20:460–469
He G, Chen L, Ye Y et al (2010) Piwil2 expressed in various stages of cervical neoplasia is a potential complementary marker for p16. Am J Transl Res 2:156–169
Houwing S, Kamminga LM, Berezikov E et al (2007) A role for Piwi and piRNAs in germ cell maintenance and transposon silencing in Zebrafish. Cell 129:69–82
Hsu PS, Yu SH, Tsai YT et al (2021) More than causing (epi) genomic instability: emerging physiological implications of transposable element modulation. J Biomed Sci 28:1–4
Huang G, Hu H, Xue X et al (2013a) Altered expression of piRNAs and their relation with clinicopathologic features of breast cancer. Clin Transl Oncol 15:563–568
Huang XA, Yin H, Sweeney S et al (2013b) A major epigenetic programming mechanism guided by piRNAs. Dev Cell 24:502–516
Huang X, Tóth KF, Aravin AA (2017) piRNA Biogenesis in Drosophila melanogaster. Trends Genet 33:882–894
Hur JK, Luo Y, Moon S et al (2016) Splicing-independent loading of TREX on nascent RNA is required for efficient expression of dual-strand piRNA clusters in Drosophila. Genes Dev 30:840–855
International Human Genome Sequencing Consortium (2004) Finishing the euchromatic sequence of the human genome. Nature 431:931–945
Ishizu H, Kinoshita T, Hirakata S et al (2019) Distinct and collaborative functions of Yb and Armitage in transposon-targeting piRNA biogenesis. Cell Rep 27:1822–1835
Iwasaki YW, Siomi MC, Siomi H (2015) PIWI-interacting RNA: its biogenesis and functions. Annu Rev Biochem 84:405–433
Jothimani G, Sriramulu S, Chabria Y et al (2018) A review on theragnostic applications of micrornas and long non-coding RNAs in colorectal cancer. Curr Top Med Chem 18:2614–2629
Jung I, Park JC, Kim S (2014) piClust: a density based piRNA clustering algorithm. Comput Biol Chem 50:60–67
Kausar S, Abbas MN, Zhao Y et al (2019) Immune strategies of silkworm, Bombyx mori against microbial infections. Invertebr Surviv J 1:130–140
Kawaoka S, Hayashi N, Suzuki Y et al (2009) The Bombyx ovary-derived cell line endogenously expresses PIWI/PIWI-interacting RNA complexes. RNA 15:1258–1264
Kazazian HH Jr (2004) Mobile elements: drivers of genome evolution. Science 303:1626–1632
Kirino Y, Mourelatos Z (2007) Mouse Piwi-interacting RNAs are 2'-O-methylated at their 3' termini. Nat Struct Mol Biol 14:347–348
Konkel MK, Batzer MA (2010) A mobile threat to genome stability: the impact of non-LTR retrotransposons upon the human genome. Semin Cancer Biol 20:211–221
Kumar S, Gonzalez EA, Rameshwar P et al (2020) Non-coding RNAs as mediators of epigenetic changes in malignancies. Cancers (basel) 12:3657
Kuramochi-Miyagawa S, Kimura T, Ijiri TW et al (2004) Mili, a mammalian member of piwi family gene, is essential for spermatogenesis. Development 131:839–849
Kwon C, Tak H, Rho M et al (2014) Detection of PIWI and piRNAs in the mitochondria of mammalian cancer cells. Biochem Biophys Res Commun 446:218–223
Larson AG, Narlikar GJ (2018) The role of phase separation in heterochromatin formation, function, and regulation. Biochemistry 57:2540–2548
Law PT, Qin H, Ching AK et al (2013) Deep sequencing of small RNA transcriptome reveals novel non-coding RNAs in hepatocellular carcinoma. J Hepatol 58:1165–1173

Le Thomas A, Tóth KF, Aravin AA (2014)To be or not to be a piRNA: genomic origin and processing of piRNAs. Genome Biol 15:1–6
Le Thomas A, Rogers AK, Webster A et al (2013) Piwi induces piRNA-guided transcriptional silencing and establishment of a repressive chromatin state. Genes Dev 27:390–399
Lee DY, Teyssier C, Strahl BD et al (2005) Role of protein methylation in regulation of transcription. Endocrine Rev 26:147–170
Lee EJ, Banerjee S, Zhou H et al (2011) Identification of piRNAs in the central nervous system. RNA 17:1090–1099
Lee H, Zhang Z, Krause HM (2019) Long noncoding RNAs and repetitive elements: junk or intimate evolutionary partners? Trends Genet 35:892–902
Lee JH, Schütte D, Wulf G et al (2006) Stem-cell protein Piwil2 is widely expressed in tumors and inhibits apoptosis through activation of Stat3/Bcl-XL pathway. Hum Mol Genet 15:201–211
Lee YJ, Moon SU, Park MG et al (2016) Multiplex bioimaging of piRNA molecular pathway-regulated theragnostic effects in a single breast cancer cell using a piRNA molecular beacon. Biomaterials 101:143–155
Liang D, Yang Y, Liu Y (2013) The role Hiwi gene in the maintenance of lung cancer stem cell populations. Neoplasma. https://doi.org/10.4149/neo_2014_022
Lim AK, Kai T (2007) Unique germ-line organelle, nuage, functions to repress selfish genetic elements in Drosophila melanogaster. Proc Natl Acad Sci USA 104:6714–6719
Lim RS, Kai T (2015) A piece of the pi(e): The diverse roles of animal piRNAs and their PIWI partners. Semin Cell Dev Biol 47–48:17–31
Ling H, Fabbri M, Calin GA (2013) MicroRNAs and other non-coding RNAs as targets for anticancer drug development. Nat Rev Drug Discov 12:847–865
Lingel A, Simon B, Izaurralde E et al (2003) Structure and nucleic-acid binding of the Drosophila Argonaute 2 PAZ domain. Nature 426:465–469
Liu Y, Dou M, Song X et al (2019) The emerging role of the piRNA/piwi complex in cancer. Mol Cancer 1:1–5
Liu JJ, Shen R, Chen L et al (2010) Piwil2 is expressed in various stages of breast cancers and has the potential to be used as a novel biomarker. Int J Clin Exp Pathol 3:328–337
Liu J, Carmell MA, Rivas FV et al (2004) Argonaute2 is the catalytic engine of mammalian RNAi. Science 305:1437–1441
Liu X, Sun Y, Guo J et al (2006) Expression of hiwi gene in human gastric cancer was associated with proliferation of cancer cells. Int J Cancer 118:1922–1929
Lu Y, Li C, Zhang K et al (2010) Identification of piRNAs in Hela cells by massive parallel sequencing. BMB Rep 43:635–641
Ma JB, Ye K, Patel DJ (2004) Structural basis for overhang-specific small interfering RNA recognition by the PAZ domain. Nature 429:318–322
Maksimov V, Oya E, Tanaka M et al (2018) The binding of Chp2's chromodomain to methylated H3K9 is essential for Chp2's role in heterochromatin assembly in fission yeast. PLoS ONE 13:e0201101
Malone CD, Brennecke J, Dus M et al (2009) Specialized piRNA pathways act in germline and somatic tissues of the Drosophila ovary. Cell 137:522–535
Mani SR, Juliano CE (2013) Untangling the web: the diverse functions of the PIWI/piRNA pathway. Mol Reprod Dev 80:632–664
Martin C, Zhang Y (2005) The diverse functions of histone lysine methylation. Nat Rev Mol Cell Biol 6:838–849
Martinez V, Vucic E, Thu K et al (2015) Unique somatic and malignant expression patterns implicate PIWI-interacting RNAs in cancer-type specific biology. Sci Rep 5:10423
Mei Y, Wang Y, Kumari P et al (2015) modulates p-ERM RNA interacts with and modulates p-ERM proteins in human somatic cells. Nat Commun 6:7316
Meister G, Landthaler M, Patkaniowska A et al (2004) Human Argonaute2 mediates RNA cleavage targeted by miRNAs and siRNAs. Mol Cell 15:185–197

Mohn F, Handler D, Brennecke J (2015) piRNA-guided slicing specifies transcripts for Zucchini-dependent, phased piRNA biogenesis. Science 348:812–817

Mohn F, Sienski G, Handler D et al (2014) The rhino-deadlock-cutoff complex licenses noncanonical transcription of dual-strand piRNA clusters in Drosophila. Cell 157:1364–1379

Morris KV (2012) Non-coding RNAs and epigenetic regulation of gene expression: drivers of natural selection. Caister Academic Press ISBN 978-1-904455-94-3

Munafò M, Manelli V, Falconio FA et al (2019) Daedalus and Gasz recruit Armitage to mitochondria, bringing piRNA precursors to the biogenesis machinery. Genes Dev 33:844–856

Murota Y, Ishizu H, Nakagawa S et al (2014) Yb integrates piRNA intermediates and processing factors into perinuclear bodies to enhance piRISC assembly. Cell Rep 8:103–113

Ng KW, Anderson C, Marshall EA et al (2016) Piwi-interacting RNAs in cancer: emerging functions and clinical utility. Mol Cancer 15:1–3

Nishida KM, Iwasaki YW, Murota Y et al (2015) Respective functions of two distinct Siwi complexes assembled during PIWI-interacting RNA biogenesis in Bombyx germ cells. Cell Rep 10:193–203

Nishida KM, Okada TN, Kawamura T et al (2009) Functional involvement of Tudor and dPRMT5 in the piRNA processing pathway in Drosophila germlines. EMBO J 28:3820–3831

Nishimasu H, Ishizu H, Saito K et al (2012) Structure and function of Zucchini endoribonuclease in piRNA biogenesis. Nature 491:284–287

O'Donnell KA, Boeke JD (2007) Mighty Piwis defend the germline against genome intruders. Cell 129:37–44

Park MS, Phan HD, Busch F et al (2017) Human Argonaute3 has slicer activity. Nucleic Acids Res 45:11867–11877

Pathak S, Grillo AR, Scarpa M et al (2015) MiR-155 modulates the inflammatory phenotype of intestinal myofibroblasts by targeting SOCS1 in ulcerative colitis. Exp Mol Med 47:e164

Peng JC, Lin H (2013) Beyond transposons: the epigenetic and somatic functions of the Piwi-piRNA mechanism. Curr Opin Cell Biol 25:190–194

Peng L, Song L, Liu C et al (2016) piR-55490 inhibits the growth of lung carcinoma by suppressing mTOR signaling. Tumour Biol 2:2749–2756

Post C, Clark JP, Sytnikova YA et al (2014) The capacity of target silencing by Drosophila PIWI and piRNAs. RNA 20:1977–1986

Prud'homme N, Gans M, Masson M et al (1995) Flamenco, a gene controlling the gypsy retrovirus of Drosophila melanogaster. Genetics 139:697–711

Qiao D, Zeeman AM, Deng W et al (2002) Molecular characterization of hiwi, a human member of the piwi gene family whose overexpression is correlated to seminomas. Oncogene 21:3988–3999

Qiu W, Guo X, Lin X et al (2017) Transcriptome-wide piRNA profiling in human brains of Alzheimer's disease. Neurobiol Aging 57:170–177

Rajan KS, Velmurugan G, Gopal P et al (2016) Abundant and altered expression of PIWI-interacting RNAs during cardiac hypertrophy. Heart Lung Circ 25:1013–1020

Rajasethupathy P, Antonov I, Sheridan R et al (2012) Role for neuronal piRNAs in the epigenetic control of memory-related synaptic plasticity. Cell 149:693–707

Ray R, Pandey P (2018) piRNA analysis framework from small RNA-Seq data by a novel cluster prediction tool - PILFER. Genomics 110:355–365

Rojas-Ríos P, Chartier A, Pierson S et al (2017) Aubergine and pi RNA s promote germline stem cell self-renewal by repressing the proto-oncogene Cbl. EMBO J 36:3194–3211

Rosenkranz D, Zischler H (2012) proTRAC–a software for probabilistic piRNA cluster detection, visualization, and analysis. BMC Bioinform 13:5

Ross RJ, Weiner MM, Lin H (2014) PIWI proteins and PIWI-interacting RNAs in the soma. Nature 505:353–359

Saito K, Sakaguchi Y, Suzuki T et al (2007) Pimet, the Drosophila homolog of HEN1, mediates 2′-O-methylation of Piwi-interacting RNAs at their 3′ ends. Genes Dev 21:1603–1608

Sato K, Siomi MC (2020) The piRNA pathway in Drosophila ovarian germ and somatic cells. Proc Jpn Acad Ser B Phys Biol Sci 96:32–42

Schreiner P, Atkinson PW (2017) piClusterBusteR: software for automated classification and characterization of piRNA cluster loci. BioRxiv 133009
Senti KA, Jurczak D, Sachidanandam R et al (2015) piRNA-guided slicing of transposon transcripts enforces their transcriptional silencing via specifying the nuclear piRNA repertoire. Genes Dev 29:1747–1762
Shaw WR, Catteruccia F (2019) Vector biology meets disease control: using basic research to fight vector-borne diseases. Nat Microbiol 4:20–34
Sienski G, Dönertas D, Brennecke J (2012) Transcriptional silencing of transposons by Piwi and maelstrom and its impact on chromatin state and gene expression. Cell 151:964–980
Sleiman M, Logue JM, Montesinos VN et al (2016) Emissions from electronic cigarettes: key parameters affecting the release of harmful chemicals. Environ Sci Technol 50:9644–9651
Spitale RC, Incarnato D (2022) Probing the dynamic RNA structurome and its functions. Nat Rev Genet 1–19
Sun G, Wang Y, Sun L et al (2011) Clinical significance of Hiwi gene expression in gliomas. Brain Res 1373:183–188
Tan L, Mai D, Zhang B et al (2019) PIWI-interacting RNA-36712 restrains breast cancer progression and chemoresistance by interaction with SEPW1 pseudogene SEPW1P RNA. Mol Cancer 18:9
Tian Y, Simanshu DK, Ma JB et al (2011) Structural basis for piRNA 2'-O-methylated 3'-end recognition by Piwi PAZ (Piwi/Argonaute/Zwille) domains. Proc Natl Acad Sci USA 108:903–910
Vagin VV, Yu Y, Jankowska A et al (2013) Minotaur is critical for primary piRNA biogenesis. RNA 19:1064–1077
Vella S, Gallo A, Nigro AL et al (2016) PIWI-interacting RNA (piRNA) signatures in human cardiac progenitor cells. Int J Biochem Cell Biol 76:1–1
Wakisaka KT, Imai Y (2019) The dawn of pirna research in various neuronal disorders. Front Biosci Landmark 24:1440–1451
Wang C, Lin H (2021) Roles of piRNAs in transposon and pseudogene regulation of germline mRNAs and lncRNAs. Genome Biol 22:1–21
Wang H, Ma Z, Niu K et al (2016) Antagonistic roles of Nibbler and Hen1 in modulating piRNA 3′ ends in Drosophila. Development 143:530–539
Wang W, Ha BW, Tipping C et al (2015) Slicing and binding by Ago3 or Aub trigger Piwi-bound piRNA production by distinct mechanisms. Mol Cell 59:819–830
Watanabe T, Chuma S, Yamamoto Y et al (2011a) MITOPLD is a mitochondrial protein essential for nuage formation and piRNA biogenesis in the mouse germline. Dev Cell 20:364–375
Watanabe T, Tomizawa S, Mitsuya K et al (2011b) Role for piRNAs and noncoding RNA in de novo DNA methylation of the imprinted mouse Rasgrf1 locus. Science 332:848–852
Weng W, Liu N, Toiyama Y et al (2018) Novel evidence for a PIWI-interacting RNA (piRNA) as an oncogenic mediator of disease progression, and a potential prognostic biomarker in colorectal cancer. Mol Cancer 17:16
Wicker T, Sabot F, Hua-Van A et al (2007) A unified classification system for eukaryotic transposable elements. Nat Rev Genet 8:973–982
Wu X, Pan Y, Fang Y et al (2020) The biogenesis and functions of piRNAs in human diseases. Mol Ther Nucleic Acids 21:108–120
Xiol J, Spinelli P, Laussmann MA et al (2014) RNA clamping by Vasa assembles a piRNA amplifier complex on transposon transcripts. Cell 157:1698–1711
Yamanaka S, Siomi MC, Siomi H (2014) piRNA clusters and open chromatin structure. Mob DNA 5:1–12
Yamashiro H, Negishi M, Kinoshita T et al (2020) Armitage determines Piwi– piRISC processing from precursor formation and quality control to inter-organelle translocation. EMBO Rep 21:e48769
Yan Z, Hu HY, Jiang X et al (2011) Widespread expression of piRNA-like molecules in somatic tissues. Nucleic Acids Res 39:6596–6607

Yang J, Xue FT, Li YY et al (2018) Exosomal piRNA sequencing reveals differences between heart failure and healthy patients. Eur Rev Med Pharmacol Sci 22:7952–7961

Yin H, Lin H (2007) An epigenetic activation role of Piwi and a Piwi-associated piRNA in Drosophila melanogaster. Nature 450:304–308

Yong WS, Hsu FM, Chen PY (2016) Profiling Genome-Wide DNA Methylation. Epigenetics Chromatin 9:26

Zhang F, Wang J, Xu J et al (2012) UAP56 couples piRNA clusters to the perinuclear transposon silencing machinery. Cell 151:871–884

Zhang H, Ren Y, Xu H et al (2013) The expression of stem cell protein Piwil2 and piR-932 in breast cancer. Surg Oncol 22:217–223

Zhang Y, Zhao Y, Ao X et al (2021) The role of non-coding RNAs in Alzheimer's disease: from regulated mechanism to therapeutic targets and diagnostic biomarkers. Front Aging Neurosci 13:654978

Zhang Z, Wang J, Schultz N et al (2014) The HP1 homolog rhino anchors a nuclear complex that suppresses piRNA precursor splicing. Cell 157:1353–1363

Zhao C, Tolkach Y, Schmidt D et al (2019) Mitochondrial PIWI-interacting RNAs are novel biomarkers for clear cell renal cell carcinoma. World J Urol 37:1639–1647

Modified Nucleosides as RNA Components. Structure, Biological Role and Drug Design

Mikhail S. Drenichev, Anastasia A. Zenchenko, and Cyril S. Alexeev

Contents

Abstract More than 100 modified nucleosides with different structures and functions are known as components of nucleic acids. Carbohydrate-modified (disaccharide) nucleosides are components of tRNA and poly(ADP-ribose) and also participate as second messengers in plants and animals. Base-modified nucleosides contain methylations, acylations, hydroxylations, amino acid and hydrocarbon functionalities, cyclic structure, sulfur or selenium. These modifications are represented in various types of RNA (transport—t, ribosomal—r, matrix—m, small-interfering—si, non-coding—nc RNAs) among all domains of life. They have many important biological implications: RNA splicing, protein biosynthesis, altering RNA structure and functional organization of ribosomes. Many reviews and books were devoted to this theme, but the problems accompanied by the structural diversity of ribonucleosides and their participation in the regulation of macromolecules' biosynthesis are updated with novel, complex data. In this chapter, general aspects of the structure and functions of modified nucleosides as minor RNA components are given in considering novel scientific achievements. This work highlights essential structural features of various general classes of naturally modified nucleosides and their biosynthetic formation and biological functions. A significant part of this work is devoted to

M. S. Drenichev (✉) · A. A. Zenchenko · C. S. Alexeev
Engelhardt Institute of Molecular Biology, Russian Academy of Sciences, Vavilova Street, 32, 119991 Moscow, Russian Federation
e-mail: room517@eimb.ru

J. Barciszewski (ed.), *RNA Structure and Function*, RNA Technologies 14,
https://doi.org/10.1007/978-3-031-36390-0_23

medicinal chemistry. Here we consider the mechanism of action of synthetic nucleosides and drugs on their basis, changing properties of viral RNAs and thus leading to inhibition of viral reproduction and application of nucleoside stable isotope labeled internal standards (SILIS) for analysis of RNA probes.

Keywords Nucleosides · RNA · Minor components · Disaccharides · Mutagenesis · Drugs

Abbreviations

ATP	Adenosine triphosphate
Tpt1	TRNA-2′-phosphotransferase 1 (archaeal)
NUDIX hydrolase	**Nu**cleoside **Di**phosphates linked to **X** hydrolase
NudC hydrolase	**Nu**cleoside **Di**phosphates in NAD^+ **C**ap hydrolase
snoRNA	Small nucleolar RNA
TARG1	Terminal (ADP-ribosyl)hydrolase
MACROD	Mono(ADP-ribosyl)hydrolase

1 Carbohydrate-Modified Nucleoside Derivatives

Disaccharide nucleosides represent an essential and diverse class of natural compounds containing an *extra* carbohydrate residue linked to one of the nucleoside hydroxyl groups via an *O*-glycosidic bond. A disaccharide residue and a heterocyclic base make their properties similar to those of carbohydrates and nucleosides. They can also contain various functional substituents, such as acyl and alkyl groups, phosphate and sulfate residues, fatty acids, amino acids, peptides and additional sugars (Efimtseva et al. 2007). Natural disaccharide nucleosides may be divided into three main groups according to their structure: (1) 2′ (3′)-*O*-Hexapyranosylribonucleosides; (2) nucleoside derivatives having two hexapyranosyl moieties (most of them have *N*-glycosidic linkage); (3) pentafuranosylnucleosides. Compounds belonging to the first and the second group are parts of nucleoside antibiotics; compounds belonging to the third group are constituents of biopolymers and participate in some biosynthetic pathways and regulation of metabolic processes. Cyclic adenosine diphosphate ribose, ADPR (Fig. 1, cADPR, **2**), represents a cyclic adenine nucleotide, where adenosine and ribofuranose residues are joined by a 5′-5″-pyrophosphate bond and ribofuranose residue, in turn, closes the cycle by glycosidic bonding to the nitrogen at position 1 (N^1) of the same adenine base (Moreau et al. 2011). cADPR is produced from nicotinamide adenine dinucleotide (NAD^+, Fig. 1, **1a**) by ADP-ribosyl cyclases (EC 3.2.2.5) as part of a second messenger system or by the bifunctional ectoenzymes of the CD38 family and potentiates intracellular Ca^{2+} release that in turn controls a diverse range of highly regulated cellular processes (Watt et al. 2018). N^1-Phosphoribosyl-AMP (Fig. 1, **3**), structurally similar to cADPR, is formed by an unusual reaction that joins ATP to 5′-phosphoribosyl-1-pyrophosphate (PRPP) and is a metabolic product in the biosynthesis of histidine (Mathews and Van Holde 1996). Analogously to cADPR, bacterial and plant Toll/interleukin-1 receptor (TIR) domains produce cyclic signaling nucleotides with immune and virulence functions, combining NAD^+-hydrolyzing ADPR-polymerase-like activities (Manik et al. 2022). These two TIR domain-produced cADPR isomers (v-cADPRs, Fig. 1), 3′-cADPR (**10**) and 2′-cADPR (**9**), contained 2′(3′)-*O*-glycosidic bond in β-configuration and were shown to be effector molecules associated with plant immunity suppression by phytopathogens such as *Pseudomonas syringae* (Manik et al. 2022). Structurally related compounds pRib-AMP/ADP (2′-(5″-phosphoribosyl)-5′-adenosine mono-/di-phosphate), which trigger immune signaling in plants (Huang et al. 2022; Jia et al. 2022) are metabolic precursors of various disaccharide nucleosides alternatively to NAD^+. Natural compounds (Fig. 1) Ar(p) **7** and Gr(p) **8** were firstly isolated from yeasts and contained an additional ribofuranosyl residue attached via an *O*-glycosidic bond to the adenosine or guanosine. These compounds are synthesized during post-synthetic modification of $tRNA^{Met}$ by ribosyl transferring (Rit1)-enzyme through the addition of a PRPP group directly onto A64 or G64 positions of the tRNA (Efimtseva et al. 2007; McCown et al. 2020). Such modification regulates the translation process in *Saccharomyces*, preventing the initiator $tRNA^{Met}$ from binding to eukaryotic elongation factor eEF1A (Kolitz and Lorsch 2010). Interestingly, the post-synthetic modification of tRNA by PRPP is known only in yeasts (McCown et al. 2020), while modifications associated with mono/poly(ADP-ribosyl)ation (MAR and PAR) bear more common character influencing regulator mechanisms mostly in all domains of life. PARylation is a posttranslational modification of proteins in eukaryotic cells, catalyzed by poly(ADP-ribose)polymerases (PARPs) (D'Amours et al. 1999; Meyer et al. 2006; Gagne et al. 2008). These enzymes promote the conversion of NAD^+ to poly(ADP-ribose) (PAR) with the release of nicotinamide (Fig. 1) (Efimtseva et al. 2007; Drenichev and Mikhailov 2015; Meyer et al. 2006). At present, about 17 enzymes of the PARP family are known; among them are PARP-1, PARP-2, PARP-3, tankyrases, etc. PARP-1 (EC 2.4.2.30) is responsible for the synthesis of 90% of PAR in cells (Efimtseva et al. 2007; Drenichev and Mikhailov 2015) and appears to be the major PARylating enzyme in higher eukaryotes after DNA-damage. PAR is a complex branched biopolymer in which 2′-*O*-α-D-ribofuranosyladenosine moieties **5** are bound by pyrophosphate linkages. PAR molecules, either isolated from natural sources or synthesized in vitro, have an irregular structure with more than 200 monomeric units, where linear segments of 20–50 units alternate with branched fragments (Ferro and Oppenheimer 1978; Miwa et al. 1981; Kistemaker et al. 2015a, b; Liu et al. 2017). The chemical structure of PAR was determined by enzymatic hydrolysis and further characterization of the products by NMR spectroscopy (Ferro and Oppenheimer 1978). Hydrolysis of the pyrophosphate bonds by snake venom phosphodiesterase yielded nucleotide as a major product. Subsequent enzymatic dephosphorylation by alkaline phosphatase gave disaccharide and trisaccharide nucleosides.

Although PAR exists in an extended conformation in low salt buffers, it may undergo structural changes in high salt buffers, showing cooperative helix-coil transitions with hypochromicity up to 40% for a long-chain molecule (Minaga and Kun 1983). The ability to form helicoidal secondary structure by long-chain PAR stabilized by base stacking and hydrogen bonding interactions has much in common with the structure of DNA and RNA (D'Amours et al. 1999; Minaga and Kun 1983). It was shown that protein binding depends on PAR's molecular weight. Several proteins like p53 manifest the ability to form complexes with long PAR chains (>40 units), while other proteins, for example, XPA, can bind to short (20–40 units) and long (>40 units) PAR chains (Fahrer et al. 2007). In solution, one protein can form several complexes with PAR of different molecular weights (Fahrer et al. 2007). PARP-1 catalyzes the synthesis of PAR chains with variable lengths. Understanding PAR functions in cells and opening the proper mechanism of association with proteins requires the chemical synthesis of regular PAR oligomers (Kistemaker et al. 2015a, b). Biochemical degradation of PAR is mediated by several enzymes (Fig. 1): ADP-ribosyl protein lyase (ADPRPL, splitting covalent bonds with proteins), poly(ADPribose)glycohydrolase (PARG, splitting glycosidic bonds in PAR), NUDIX enzyme (hydrolysis of pyrophosphate linkages) and ADPR-hydrolases (ARH, splitting single ADPR-moieties and short PAR fragments, not shown on Fig. 1) (Drenichev and Mikhailov 2015, 2016). PAR-degrading enzymes regulate complex biochemical events in a cell associated with PAR. If getting into the cytosol from the nucleus, PAR causes programmed cell death (parthanatos, from Greek Θάνατος, "Death") (Drenichev and Mikhailov 2016). Therefore, PARG also performs a cytoprotective function, hydrolyzing PAR in the cytosol.

Mono(ADP-ribosyl) moieties (MAR, ADPR) are covalently linked to proteins through backbone carboxyl groups of glutamic/aspartic acid residues (formation of ester linkages) (Meyer et al. 2006) or on lysine ε-amino groups (Messner et al. 2010) and can be considered as the first unit in assembling of long PAR chains and further interactions with proteins associated with PAR. The absence of an N^1-glycosidic bond to adenine distinguishes the non-cyclic ADP-ribose (ADPR) from cADPR. Besides proteins, nucleic acid substrates can also be ADP-ribosylated, extending the set of substrates for ADP-ribosylation (Fig. 2). Several mammalian PARPs were shown to catalyze this reaction. PARP10 catalytic domain can modify phosphorylated single-stranded (ss) RNA ends with a preference for 5′-terminal phosphate (Groslambert et al. 2021; Munnur et al. 2019). Additionally, PARP11 and PARP15 could also ADP-ribosylate 5′-phosphorylated ssRNA (Munnur et al. 2019). Full-length PARP10/PARP1 can catalyze the modifications less efficiently than their catalytic domains alone (Groslambert et al. 2021), thus suggesting several mechanisms of regulation of RNA stability and functioning on the ADPR level. An unusual RNA-cap at the 5′-end containing the entire NAD^+ molecule was identified in mammals, bacteria and yeasts when RNA polymerase initiates transcription with NAD^+ (Walters et al. 2017; Jiao et al. 2017). NAD^+ caps were shown to promote RNA decay in mammals by decapping exoribonuclease (DXO)-mediated deNADing process (Jiao et al. 2017). Under DXO-catalyzed removal of NAD^+ cap, the RNA-5′-phosphate is formed, which is sensitive to further enzymatic digestion by RNAses. It was shown

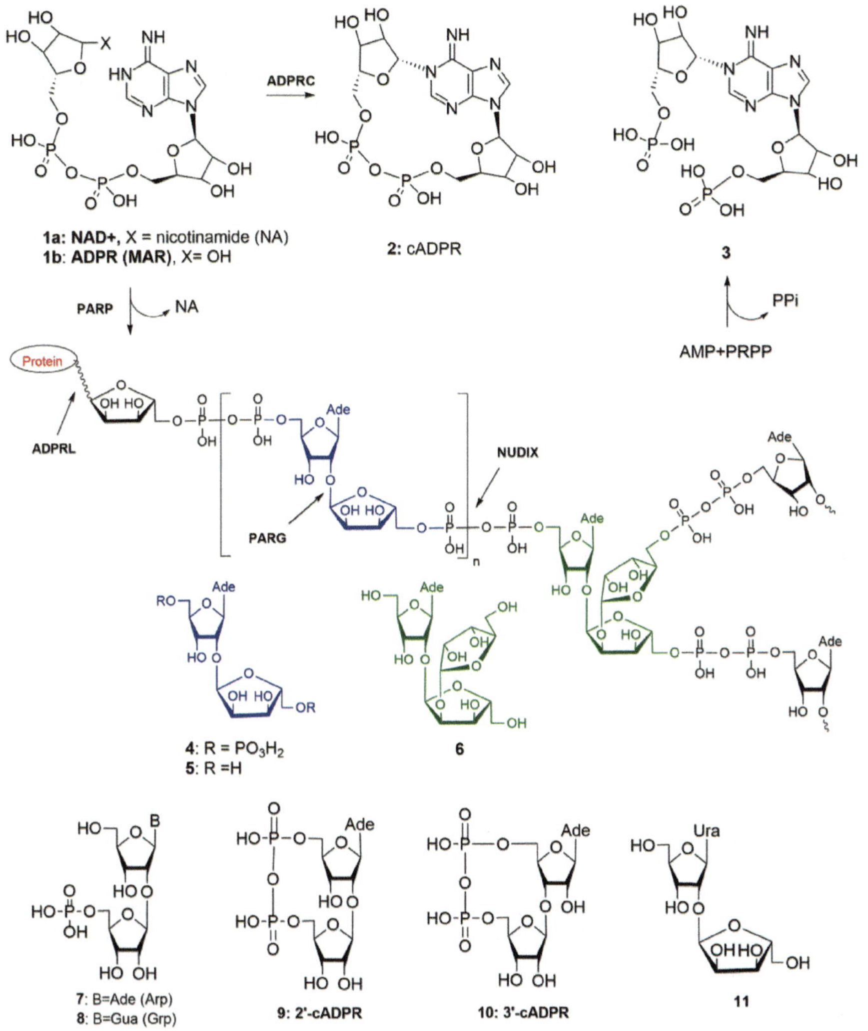

Fig. 1 Disaccharide nucleosides as components of biopolymers and signaling molecules

that some snoRNA containing NAD$^+$cap corresponded to intron sequences (Jiao et al. 2017). A non-canonical ADPR 5′-cap structure (Fig. 2) can be formed in 5′-phosphorylated ends of RNA by enzymatic transfer of ADPR from NAD$^+$ to RNA catalyzed by human tRNA-2′-phosphotransferase (TRPT1), PARPs and their homologues from fungi, archaea (enzyme Tpt1) and bacteria, suggesting RNA ADP-ribosylation as a universal splicing pathway for all domains of life (Munnur et al. 2019; Munir et al. 2018). ADPR modification of RNA by TRPT1 and PARP10 rendered the oligonucleotide substrates resistant to treatment with calf intestine alkaline phosphatase (CIP), which is indicative of RNA capping. The enzymatic mechanism of archaeal Tpt1 is associated with the transfer of an RNA 2′-monophosphate to NAD^+, yielding a 2′–OH RNA and cADPR (Munnur et al. 2019) and probably precedes the transfer of ADPR residue to RNA 5′-phosphorylated end. However, many species expressing a TRPT1 homologue do not possess intron-containing tRNAs for further splicing or mechanisms that would generate 2′-phosphate RNA, suggesting that RNA capping could be a primary activity of many TRPT1 homologues (Munnur et al. 2019). Similarly to NAD^+, RNA modification by ADPR is reversible and removed by hydrolases PARG, TARG1, MACROD1, MACROD2 and ARH3 (Munnur et al. 2019). TRPT1-mediated 5′-phosphate RNA modification can be reversed by PARG, TARG1, MACROD enzymes, ARH3 and NUDT16 (Groslambert et al. 2021). All these data indicate that RNA ADPR and NAD^+ capping may be a part of a delicate biochemical mechanism regulating transcription and translation in cells. PARP10-mediated modification of RNA ends could have an immune function as its expression inhibits Venezuelan equine encephalitis virus (VEEV) translation and participates in modulating inflammation. It was shown that VEEV and SARS-CoV macrodomains encoded within the non-structural proteins can remove ADP-ribosylation from PARP10-modified RNA, counteracting PARP10 activity (Groslambert et al. 2021). It is hypothesized that the PARP10 RNA recognition motif (RRM) specifically binds and ADP-ribosylates viral RNA, distinguishing it from a host RNA. According to the biological diversity of biochemical ribosylation, novel RNA components may exist (McCown et al. 2020). 2′-*O*-α-D-ribofuranosyluridine (RFU, Fig. 1, **11**), as a part of the previously unknown metabolite, Hellecaucaside A, was isolated from the hellebore plant *Helleborus caucasicus* (Sylla et al. 2014; Kulikova et al. 2015). As ADPR moiety can participate in regulation of RNA functions, PARylation of RNA 5′-ADPR cap can be alternatively assumed

as a possible regulator mechanism. Finding out of cyclic phosphate forms of 2′/3′-*O*-ribofuranosyladenosine (Fig. 1, **9–10**) may also indicate the existence of an RNA-capping mechanism by *β*-D-ribofuranosyladenosine in plants analogously to ADPR-capping and phosphate transfer associated with enzymatic activities similar to archaeal Tpt1.

Interestingly, 3′-*O*-β-D-ribofuranosyladenosine, which has an identical glycosidic linkage to 3′-cADPR, has been shown to accumulate in leaves infected with *Pseudomonas syringae* DC3000 (Drenichev et al. 2019; Bednarek et al. 2004). This fact suggests that cADPR isomers can be further modified in plants to form additional novel nucleosides associated with plant immunity.

2 Base-Modified Nucleoside Derivatives

Base-modified purine and pyrimidine nucleotides represent more structural diversity as RNA components than disaccharide compounds. To date, more than 160 modified ribonucleosides have been isolated from various RNA sources (transfer, ribosomal,

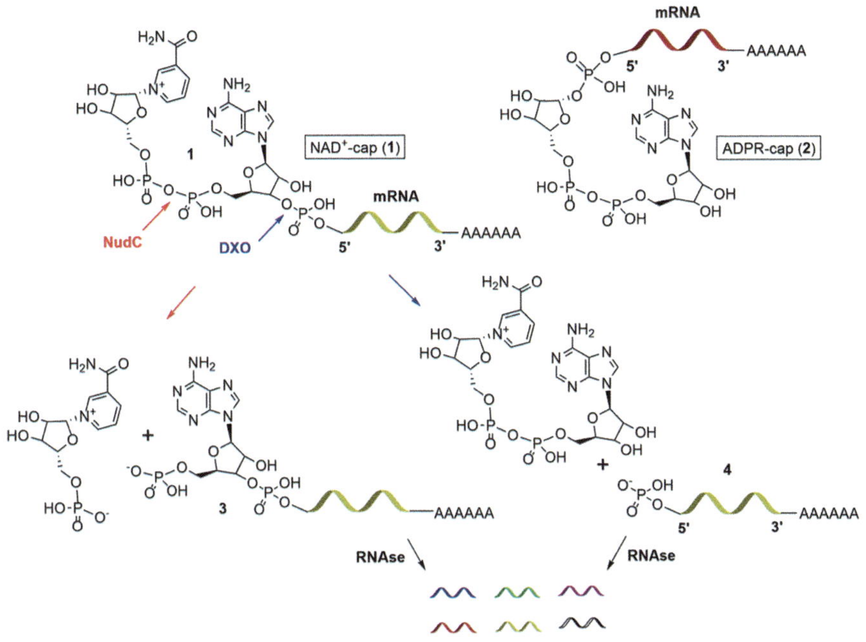

Fig. 2 mRNA modifications by ADPR- and NAD-caps

messenger, and non-coding RNA) (Efimtseva et al. 2007; McCown et al. 2020). Most minor RNA components are methylated compounds; some of them biochemically convert to more structurally complex compounds. There are also cross-over pathways in which hybrid products containing both nucleoside and amino acid residue are formed.

2.1 Purine-Modified Nucleosides

The most common purine modifications occurring in bacteria, archaea and eukaryotes include adenine with inosine replacement (A to I editing), adenosine, inosine, guanosine methylation and addition of amino acid residues to adenosine (Fig. 3).

Deamination of A-to-I has been reported in tRNAs as a post-synthetic modification at position I34, which is essential for translation in bacteria and eukaryotes (McCown et al. 2020). This process of A-to-I editing is catalyzed by ADARs, enzymatic regulators of protein biosynthesis.

m^1A was found in three domains of life and represented dynamic base modification in all types of RNA (mRNA, tRNA, snRNA). This modification disrupts A-U Watson–Crick base pairings, leading to various biologically important effects (Table 1).

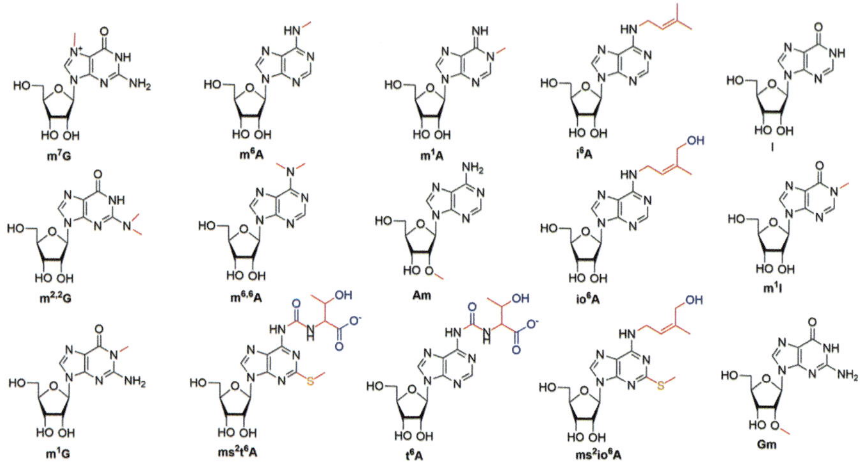

Fig. 3 Some of frequently occurring base-modified purine nucleosides as RNA components

Table 1 Biochemical effects of some frequently occurring base-modified purine ribonucleosides

Modification	Source	Functions (position in RNA)
I	tRNA	Recognition of I as G, altering the original codon and changing a protein sequence
m^1A	mRNA, tRNA, nsRNA	In tRNA: tRNA hairpin stabilization at A9; structural thermal stability of tRNAs and their maturation (yeasts) at A58; metabolic precursor for m^1I, at the position of 57 of TΨC-loop of tRNAs (archaea)
m^1G	tRNA	Hydrogen bond disrupting modification in the anticodon loop of tRNA
m^7G, $m^{2,2}G$	mRNA	mRNA capping and protection from hydrolytic cleavage
m^6A	mRNA	Destabilization of hairpins for interactions with proteins ($tRNA^{Val}$ *E. coli*), capping and regulation of mRNA hydrolysis
$m^6{}_2A$	mRNA	Small subunits of eukaryotic and bacterial ribosomes (*E. coli* and *T. Thermophilus*) at positions A1518 and A1519 are essential for their assembly and proper functioning
Am, Gm	siRNA	Resistance to alkaline degradation
t^6A/ms^2t^6A	tRNA	Position 37 of tRNA for recognizing and reading codons. t^6A decodes ANN codons (where N is any nucleotide), and ms^2t^6A decodes AAR codons (where R is purine, less specificity)
i^6A	tRNA	Position 37 of bacterial and eukaryotic tRNA for disruption of base pairings with the neighboring U33 and formation of tertiary structure. Codon recognition in yeasts and probably in bacteria. Metabolic conversion to ms^2i^6A and ms^2io^6A (yeasts)
ms^2i^6A	tRNA	Stabilization of tertiary tRNA structure by mispairing complementary bases. tRNA distribution in mitochondria (mammals)

Adenosine modifications by methyl group represent an extensive field of research in molecular biology due to their influence on many vital biochemical events in the cell. Methyl group is introduced into adenine residue of $tRNA^{Val}$ *Escherichia coli* or mRNAs using methyl transferases (m^6A writers). On the contrary, deleting m^6A marks by demethylases (m^6A erasers) stabilizes hairpins and slows down mRNA interactions with proteins.

N^6-isopentenyladenosine (i^6A) was found in bacterial and eukaryotic tRNAs (McCown et al. 2020). This mark is formed by the isopentenyl pyrophosphate (IPP) molecule at position A37 during post-synthetic of some tRNA modification (Table 1). This process is catalyzed by bacterial tRNA dimethylallyltransferase A (miaA), yeast tRNA isopentenyltransferase 1 (mod5) or human tRNA isopentenyltransferase TRIT1 enzyme. Hydroxylation of i^6A in tRNA by the bacterial enzyme (MiaE) forms zeatin (io^6A).

In mammalian cells and bacteria, a selective thiomethylation of i^6A/t^6A serves as the parallel biosynthetic pathway for the formation of ms^2i^6A/ms^2t^6A. ms^2i^6A can be further hydroxylated in bacteria to form ms^2io^6A. In thiomethylation reactions, sulfur can be transferred to i^6A by different pathways from methionine or other sources such as thiamine, iron-sulfur proteins, s^2U, s^4U or previously formed ms^2i^6A derivatives (McCown et al. 2020). In plants, i^6A and its hydroxylated derivatives are also synthesized by an independent pathway forming cytokinins, an important class of phytohormones, stimulating cell division and influencing plant growth and development. The key step is the enzymatic transfer of a hydrocarbon moiety from dimethylallyl diphosphate (DMAPP) to the adenine residue in adenosine 5′-phosphates (AMP, ADP or ATP). The next step is the phosphatase-catalyzed hydrolysis of phosphate groups of the resulting i^6A-5′-phosphates. The reversible enzymatic cleavage of the *N*-glycosidic bond in i^6A and hydroxylated derivatives with adenosine nucleosidase or nucleoside phosphorylase yields N^6-alkylated purine as cytokinin (Drenichev et al. 2016). Hydroxylated i^6A-derivatives *cis*/*trans*-zeatin ribosides (cZR/tZR) and dihydrozeatin riboside (DZR) are formed from i^6A as metabolic successors by enzymatic oxidation of i^6A with MAO on the level of 5′-phosphorylated i^6A derivatives. The formation of i^6A and io^6A as single compounds or components of tRNA indicates the similarity of their biochemical pathways in plants. Geranyl substituent, a longer terpenoid fragment, activates sulfur in s^2U to yield ges^2U (Fig. 4), a metabolic precursor to a range of 2-selenouridine products in bacteria (McCown et al. 2020). A bacterial enzyme SelU elucidates two biosynthetic steps starting from s^2U: subsequent geranylation and selenation to form se^2U. SelU also converts 5-carboxymethylaminomethyl-2-thiouridine ($cmnm^5s^2U$) and its analogues to the corresponding thiogeranyl and selenyl nucleosides ($cmnm^5se^2U$ or mnm^5se^2U). Isopentenyl transfer reactions propose that different organisms use other additional IPP derivatives to modify adenosine molecules. In this sense, a geranyl transfer on s^2U can assume the proceeding of analogous biochemical conversion for adenosine molecule as tRNA component or as ATP.

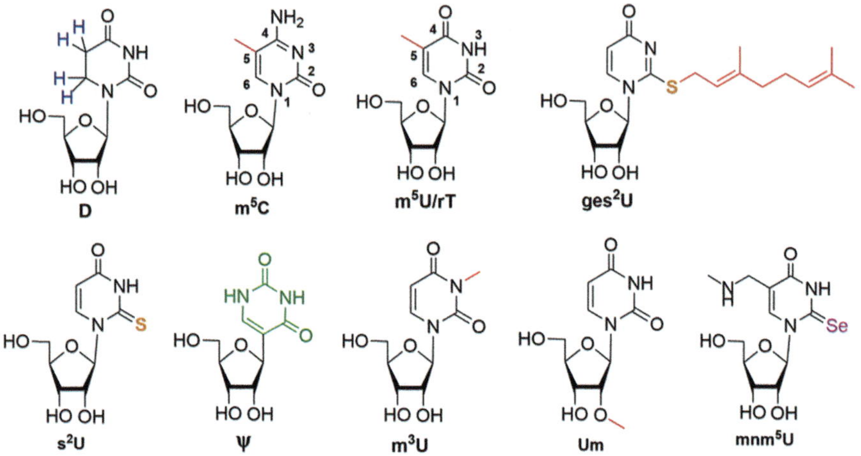

Fig. 4 Some of frequently occurring base-modified purine nucleosides as tRNA components

2.2 *Pyrimidine-Modified Nucleoside Derivatives*

Pseudouridine (Ψ) was the first discovered modified nucleoside found in tRNA in 1950s (Schultz and Kothe 2021). Ψ is an isomer of uridine, where the pyrimidine base is inverted to form a carbon–carbon bond and imine group (Fig. 4). Ψ is synthesized by uridine isomerization in post-translational RNA modification. This process is catalyzed by pseudouridine synthases (Psu, Ψ-writers), which are subdivided into several families specific to different sites in tRNA (McCown et al. 2020).

Ψ is the most widespread modification of RNA, found in tRNA, rRNA and small nuclear RNA (snRNA). This fact indicates its biological significance for cells (Table 2).

Another abundant tRNA modification is associated with dihydrouridine (D, Fig. 4). This is posttranslational modification of tRNA by enzymatic reduction of the C5–C6 double bond in uridine. A D-loop in the cloverleaf structure of tRNA is reach with dihydrouridine marks, which destabilize stacking interactions with other nucleosides and favor the formation of the double-helical structure by tRNA (McCown et al. 2020). Various pyrimidine base modifications of RNA are briefly described in Table 2. 5-Methylcytosine (m^5C) marks are present in the tRNA and rRNA of various organisms (McCown et al. 2020; Schultz and Kothe 2021). Being introduced into mRNA, m^5C can play diverse functions depending on the type of cells and organs. A general pathway for the formation of uridine minor nucleosides is enzymatic deamination of the corresponding cytidine precursors. Alternative biosynthetic pathways starting from uridine also exist (McCown et al. 2020). 5-Methyluridine (m^5U) or ribothymidine (rT) is a common modification in the tRNAs of bacteria and eukaryotes and a less common modification in archaea. Sulfur (S)- and selenium(Se)-containing nucleosides s^2U and mnm^5se^2U were found in the tRNA of all three domains of life (Schultz and Kothe 2021; McCown et al. 2020). The

Table 2 Biochemical effects of several frequently occurring base-modified pyrimidine nucleosides as RNA components

Modification	Source	Functions (position)
Ψ	tRNA, rRNA, snRNA	Thermo- and hydrolytic stability of RNA molecules, assembly of spliceosomes by intermolecular RNA–RNA and RNA–protein interactions. Enabling translation by pseudouridylation of mRNA (*E. coli*). Suppression of gene transcription by pseudouridinylation (mice). Ribosome association and disassembling of post-translational termination complex in *E. coli* ($m^3Ψ$ in bacterial 23S rRNA)
D	tRNA	Destabilization of stacking interactions and formation of the double helical structure by tRNA (D-loop in the cloverleaf structure of tRNA). Preferable C2′-endo conformation of D favors increased flexibility and dynamic motion of tRNA tertiary structure, which may be essential for translation
m^5U (rT)	tRNA	Stabilization of tRNA tertiary structure by incorporation into T54ΨC-loop. Mitochondrial and cytosolic rRNA in bacteria and eukaryotes (probably participates in translation)
Um	tRNA, mRNA, rRNA, ncRNA	Alkaline degradation, the disruption of protein-binding and DNA-binding activities
m^5C	tRNA, rRNA	Contribution to L-shape formation of tRNA at position 48. The presence of m^5C in 28S rRNAs stabilizes small and large subunits through a bridge between m^5C-rRNA and eL41 protein (yeast, human and mice). Metabolically oxidized to 5-hydroxymethylcytidine (hm^5C) and 5-formylcytidine (f^5C) which up-regulate mRNA translation in vitro (humans)
m^3U	tRNA, rRNA	May affect tRNA and rRNA functionalities
s^2U	tRNA	Maintenance of tertiary structural stability and modification by enzymes (location in T54ΨC-loop), regulation of translation and negative regulation of C34-to-U34 (location in anticodon loop at positions 32, 33, 34, 37). Discrimination of multiple codons by s^2U34 in $tRNA^{Glu}$, $tRNA^{Gln}$, $tRNA^{Lys}$
ges^2U	tRNA	A metabolic precursor for selenouridine. Terpene transfer reactions
mnm^5U	tRNA	Discrimination of the formation of G–U Wobble pairs in acceptor tRNAs, $tRNA^{Glu}$, $tRNA^{Gln}$, $tRNA^{Lys}$

formation of Wobble pairs by nucleosides s^2U and mnm^5se^2U in acceptor tRNA with mRNA prevents frameshifting during translation. Thus, s^2U and mnm^5se^2U modifications control the correct assembling of proteins during translation.

2.3 *Cellular Pathological States and Metabolic Disorders Associated with Minor Nucleosides*

Disruption of RNA modifications can drastically affect RNA functions in a cell, leading to metabolic and developmental disorders, which are summarized in Table 3.

Ψ can be a potential biomarker for cancer as the appearance of various types of cancer is likely to be associated with the shortening of telomeres encoding Ψ or defected pseudouridinylation of regulator proteins and receptors (McCown

Table 3 Human pathologies caused by disruptions of minor RNA components

Modification	Pathology	Cause of occurrence
m^1A	Gastrointestinal cancers, carcinomas, pancreatic adenocarcinoma	Up- or downregulation of m^1A-writers/erasers
m^1G	Neurodegenerative defects. Disorders of glucose metabolism	Mutations in human tRNA methyl transferase 10A (hTRMT10A)
m^6A	Carcinogenesis	Enhancement of translation of oncogenes (EGFR, TAZ, MK2, DNMT3A) by methyltransferase-like protein 3 (METTL3)
m^1I	Autoimmune diseases (polymyositis)	Targeting antibodies to m^1I-tRNA
I	Astrocytoma, glioblastoma, hepatocellular carcinoma, myeloid leukemia, neuronal disorders	Disruption of ADAR (adenosine deaminase acting on RNA) activity
m^5C	Mitochondrial-associated disease, Leber's hereditary optic neuropathy, myopathy, hypotonia	Lack of proper methylation or oxidation of m^5C. Knock-down of *nsun3* (encodes methyltransferase) и *abh1* (encodes protein of CAP-binding complex) genes
	William-Beuren syndrome (cardiac defects, lack of weight, mental deflects)	Deletion of *nsun5* gene (encodes methyltransferase)
m^3U	Obesity	Demethylation of m^3U by obesity-linked protein (FTO)
Ψ	Various types of cancer	Shortening telomeres encoding Ψ
	X-DC (diskeratosis), Hoyeraal–Hreidarsson syndrome (Leukoplakia)	Defects of Ψ formation
	Mitochondrial myopathy and anemia	Defected pseudouridinylation of steroid receptor RNA activator
D	Lung cancer	Increased activity of human dihydrouridine synthase protein (hDus2p) leading to greater amounts of D

et al. 2020; Penzo et al. 2017). Rare hereditary dyskeratosis, associated with X-chromosome (X-DC and Hoyeraal–Hreidarsson syndromes), correlates with deficiency of Ψ. The effects of A-to-I editing have been observed primarily in brain tissue or neurons. Therefore, its disruption can cause severe pathologies such as the appearance of brain cancers and neuronal disorders (McCown et al. 2020).

A large number of pathologies of different nature, from cancer to mental disorders, are associated with methylation of RNA nucleoside components (m^1A, m^6A, m^5C), which is regulated by up- or downregulation of specific enzymes' activity (Tong et al. 2018). Methylation of adenosine at N^6 with the formation of m^6A enhances the translation of oncogenes.

$m_2{}^6A$ is another structurally significant modification that participates in the proper functioning of eukaryotic ribosomes. Demethylation of $m_2{}^6A1518$ and $m_2{}^6A1519$ appears lethal for cells and can cause extreme scenarios for the whole organism (McCown et al. 2020).

m^3C is a potential diagnostic marker for cancer as its level in mRNA of hepatocellular carcinoma cells (HCC) is lower than in non-cancerous cells (McCown et al. 2020). This indicates the role of methyl transfer reactions in epigenetic and carcinogenesis processes.

2.4 Isolation and Quantification of Minor RNA Components

Currently, the most sensitive and accurate quantification methods rely on mass spectrometry. Therefore, isotope-labeled minor nucleosides are used as internal standards for the analysis of RNA composition by HPLC–MS methods (Popova and Williamson 2014; Bruckl et al. 2009; Pearson et al. 2011). Mass-spectrometry methods allow the determination and quantification of minor RNA components and comparative analysis of RNA components from different species. In the first step, the whole tRNA is isolated from cells or tissue and then fully digested by nucleases to obtain a mixture of tRNA nucleosides (Bruckl et al. 2009) (Fig. 5, **A**). Then, to quantify certain modifications, precisely known amounts of synthetic isotope-labeled, modified tRNA nucleosides are added. The percentage of tRNA component in a sample is evaluated from the analysis of calibration curves of labeled nucleoside standards (Bruckl et al. 2009). In this sense, nucleosides containing several isotopes (m^6A-D_3, Am-D_3, i^6A-D_2, Arp-D_2, ms^2t^6A-^{15}N, ^{13}C, etc.) (Fig. 5, **B**) possess more abilities for characterization of RNA analytes due to significant difference in retention time with analogous non-labeled nucleosides. The production of heavily labeled compounds can be done by synthetic methods (Földesi et al. 2000) or by metabolic labeling (Popova and Williamson 2014). Using synthetic methods, modified nucleosides containing several isotopes in carbohydrate fragments or heterocyclic bases can be obtained.

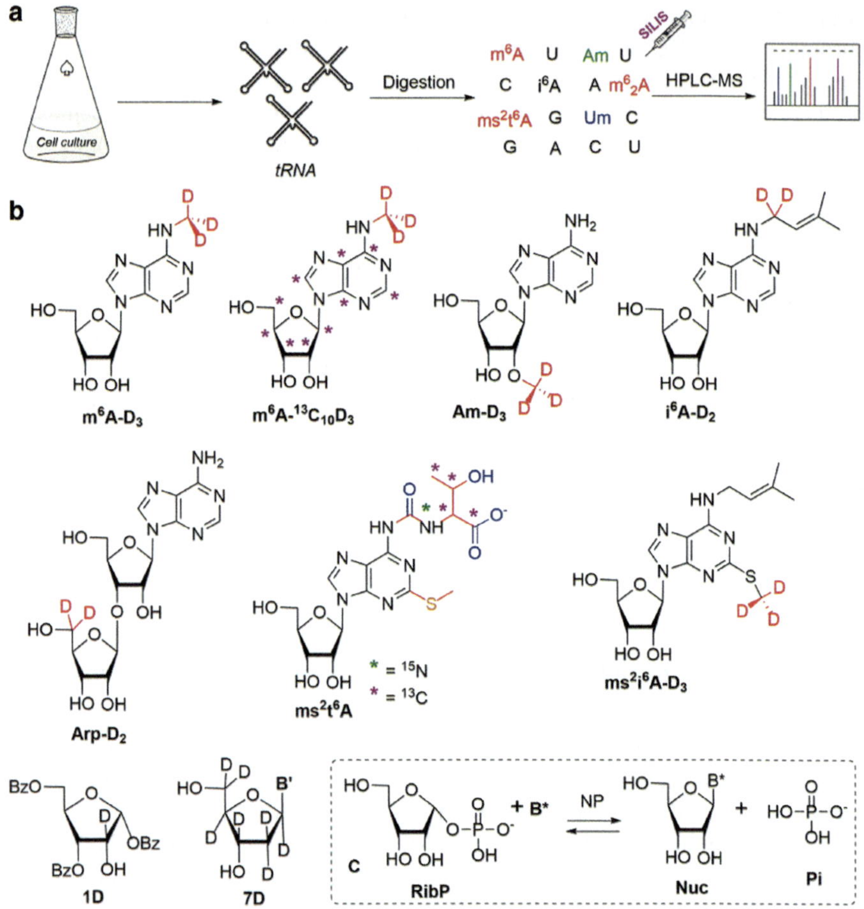

Fig. 5 Stable isotope-labeled internal nucleosides standards (SILIS)

There are two main methods for obtaining nucleoside analogues. The first one is based on the modification of natural compounds. In the second, derivatives of heterocyclic bases or monosaccharide derivatives are obtained at first, and then nucleosides are obtained. In this case, the critical step of synthesis is the formation of an *N*-glycosidic bond. To date, a convenient and effective chemical method based on trimethylsilyl derivatives of nucleic bases and fully acylated ribofuranose in the presence of Lewis acids has been developed by Vorbruggen and co-workers for the production of ribonucleosides (Vorbruggen and Ruh-Pohlenz 2001). Stereoselectivity of this reaction is determined by the 2′-*O*-acyl group involved in forming the acyloxonium ion; the products of these reactions are natural β-nucleosides in which the heterocyclic base is in a *trans*-position concerning the 2′-*O*-acyl group.

Enzymatic methods of formation of the *N*-glycoside bond of synthesis significantly complement the chemical procedures and, in some cases, have undoubted advantages as they allow for reducing the number of steps and replace traditional chemical synthesis with various catalytic methods without the use of toxic organic solvents. Nucleoside phosphorylases (NPs) are used to obtain nucleosides (Fig. 5, **C**). These enzymes catalyze reversible phosphorolysis of nucleosides with the formation of α-D-ribofuranosyl-1-phosphate (RibP) and heterocyclic base (Fig. 5, **C**). The equilibrium of these reactions is shifted toward the formation of nucleosides, and in the case of purine more significantly. The constants for the phosphorolysis of natural pyrimidine nucleosides are approximately 20 times higher than the constants for purine nucleosides (Alexeev et al. 2018). Therefore, based on enzymatic methods, nucleoside derivatives bearing several isotope labels in carbohydrate or heterocyclic base or both structural subunits can be obtained from the corresponding α-D-ribose-1-phosphate (RibP) or PRPP. In many fields, metabolic isotope labeling for the production of SILIS is widespread (Penzo et al. 2017). RNA metabolic transformation in the presence of SILIS-containing nutrients represents an approach for finding out of metabolic pathways of RNA and PAR modification in the presence of various metabolic precursors by their substitution with isotopically labeled analogues and can produce fully D- or ^{15}N, ^{13}C-labeled compounds. As was shown on ^{15}N, ^{13}C-labeled PAR, this method is also representative of three-dimensional structural studies and prediction of complexation with proteins (Schultheisz et al. 2009). The detection efficiency of the HPLC method depends on many, partly non-controllable, factors in mass spectrometry. Therefore, the heavy isotopologues of natural nucleosides containing ^{13}C, ^{15}N, ^{18}O nuclei are characterized by precise retention times of analyte and thus can be used to correct for mass spectrometric detection fluctuations (Penzo et al. 2017). The use of deuterium isotope-labeled internal standards (DILIS) is also possible but is restricted by varying retention times due to their physico-chemical properties. In some cases, this leads to signal fluctuations; thus, a reliable quantification is not possible. An alternative way to quantify minor nucleosides, which can be proposed, is the replacement of SILIS with optical isomers containing pentafuranose moiety in L-configuration as internal standards and further HPLC-analysis on a column with a chiral carrier.

3 Modified Nucleosides and Drug Design

Most emerging and/or neglected viral pathogens have an RNA genome and demonstrate extensive genetic variability and the potential for rapid evolution (Chen and Shakhnovich 2009). The molecular basis of this variability is the rapid rate of virus replication, which makes them sensitive to an increased mutation rate, which ranges from (10^{-4}) to (10^{-5}) mutations per included nucleotide compared to the range of (10^{-8})–(10^{-11}) erroneous inclusions per base pair per DNA replication cycle (Domingo et al. 1996). This fact, in turn, is facilitated by the error-prone nature of RNA-dependent RNA polymerase (RdRp) (Steinhauer et al. 1992). Retroviruses also have a high mutation rate (8.5×10^{-5} mutations per base pair and a replicative cycle) since they use viral DNA to transform their genomes because they use their own reverse transcriptase (RT) to convert their single-stranded genomes into double-stranded DNA, which is then integrated into the host genome and replicated by eukaryotic DNA polymerases (including the proof-reading activity of the host DNA polymerase). Several oncolytic RNA viruses causing cancer belong to retroviral families (hepatitis virus HCV, human T-lymphotropic virus HTLV, Rouse sarcoma virus RSV). The evolution of RNA viruses and cancer cells caused by retroviral action has a lot in common, including the processes of mutation at a high rate and the adaptation of populations to different microenvironments through replication of the most suitable variants. According to the theory of quasi-species prediction, there is an internal limit to the maximum genetic variability of the virus, beyond which any further increase in the number of errors leads to an error catastrophe of the population of the RNA virus becomes unviable (Dapp et al. 2013; Crotty and Andino 2002). Such extinction of a viral population beyond certain threshold values of the mutation rate is called lethal mutagenesis (Perales et al. 2011). It is considered that the lethal mutagenesis threshold is about three mutations *per* genome *per* replication for semi-conservative replication and about six for conservative replication for all genomic sizes (Chen and Shakhnovich 2009). The replication of RNA viruses is very close to the threshold of error catastrophe due to their high mutation rate. The viability of the virus is preserved for a significant percentage of offspring near the error threshold, generating a vast spectrum of mutants, which facilitates the adaptation of RNA viruses, allowing them to be resistant to antiviral therapy (Domingo and Holland 1994). On the other hand, the proximity to the error catastrophe and the RNA viruses' replication rate promotes the use of mutagenic drugs, leading to the former's disappearance. The feasibility of this new antiviral approach was confirmed by experiments on PV, VSV and HIV (Perales et al. 2011). The proposed mechanism of action of mutagenic nucleosides is generally based on their metabolic transformation to active triphosphates (NTP) and their further inclusion into the growing nucleic acid strain with the formation of erroneous complementary pairs, increasing the rate of mutations and the appearance of non-viable strains of the virus (Zenchenko et al. 2021; Bonnac et al. 2013). Ribavirin is a classic example of nucleoside acting by the mechanism of lethal mutagenesis (Hadj Hassine et al. 2022; Zenchenko et al. 2021). Ribavirin mutagenicity is caused by rotation around the C3-carbonyl bond with the formation of s-*cis* and s-*trans* conformers (Fig. 6), which leads to nonspecific base pairing during the replication of RNA viruses. Ribavirin s-*cis*- and s-*trans*-conformers, resembling adenosine and guanosine, respectively, form complementary pairs with uracil and cytosine, dramatically increasing the frequency of mutations leading to the death of the mutated population of RNA viruses.

Another antiviral drug Favipiravir (Fig. 6), which demonstrates activity against SARS-CoV-2, as well as Ribavirin, has a mobile carboxamide group freely rotating around the C–C bond, which makes it a potential "pseudo-base" that mimics both guanine and adenine, forming non-canonical base pairs (Crotty et al. 2000; Hadj Hassine et al. 2022) Janus guanine-cytosine (J-GC) inhibiting HBV is also known, one part of which has one side with a multiplicity of Watson–Crick hydrogen bonds of guanine, and the other side with an array of cytosine bonds.

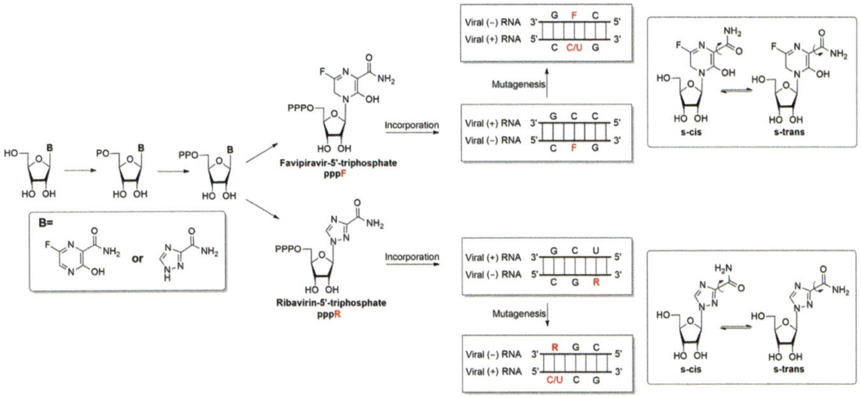

Fig. 6 Structure and mechanism of action of ribavirin (R) and favipiravir (F)

The modified base in J-GC is able to rotate around the glycoside bond, forming two conformers (*syn*/*anti*), which also leads to low-selective base pairing (Yang et al. 2011). Decitabine and its ribosylated analogue 5-azacitidine (5-AZC) cause viral HIV-1 hypermutation through base rearrangement. 5-AZC is metabolically converted to 5-aza-dCTP, which is then embedded in the viral negative DNA chain during reverse transcription. After that, a spontaneous opening of the cytosine ring forms a guanylurea fragment making a base pair with cytosine (Fig. 7).

Thus, cytosine can be included in the positive DNA strand oppose to 5-aza-dC. After integration of the viral DNA into the host cell's genome, 5-aza-dC(Z) insertions are restricted by the host DNA repair mechanism and then replaced by guanosine, forming the generation of viral G → C mutants (Dapp et al. 2009). Keto-tautomeric forms (for G and U) and amino-tautomeric forms (for A and C) of natural nucleotides are the predominant species among natural nucleosides.

Mutagenic effects of several nucleoside drugs are due to tautomeric conversions of modified bases to rare enolic or imino forms (formation of both amino- (analogue

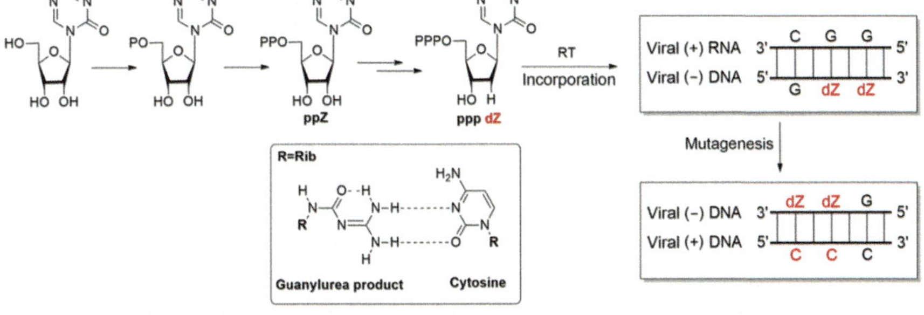

Fig. 7 Structure and mechanism of action of 5-azacytidine

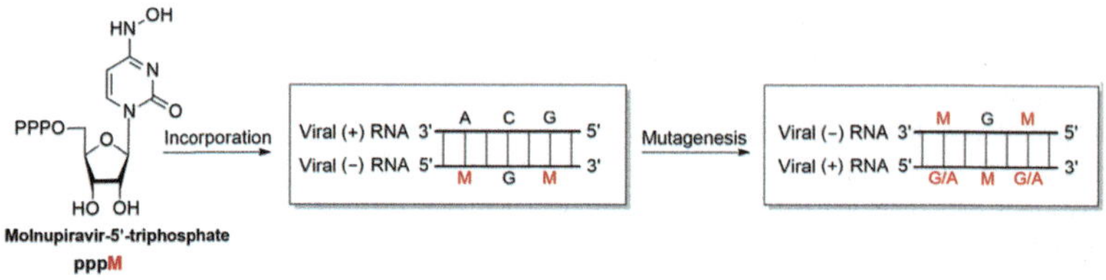

Fig. 8 Mechanism of action of Molnupiravir (M)

C) and imino- (analogue T) tautomers by 5-hydroxycytidine) (Yang et al. 2011; Bonnac et al. 2013). Base ionization has been proposed as a potential source of natural, spontaneous mutations, with 5-bromouracil (BU) and 5-fluorouracil (FU) as typical examples (Bonnac et al. 2013). Lethal mutagenesis is a promising strategy for treating infections caused by RNA viruses.

One of the difficulties in creating drugs against SARS-CoV2 is the development of such a nucleoside that can bypass the viral ExoN proofreading activity, which reduces the antiviral effect of nucleosides embedded in the synthesized RNA-strand (Robson et al. 2020). Molnupiravir was the first mutagenic anticoronaviral drug bypassing ExoN-activity (Zhao et al. 2021). After incorporation of molnupiravir triphosphate (MTP), the obtained negative-chain RNAs containing M nucleotides can subsequently be used as a matrix for obtaining a mutagenized positive chain; respectively, M can form base pairs with guanine (G) or adenine (A), leading to the accumulation of errors in the viral genome and suppression of viral replication (Fig. 8).

The main limitation of molecular targeted therapy, both antiviral and antitumor, is associated with the appearance of resistance. Ribonucleoside analogues, whose mechanism of action is similar to 5-AZC, probably need mutations to be acquired in both RT and RNA polymerase II to appear a high level of resistance. This fact contrasts with deoxynucleoside analogues, such as AZT, which prevent replication by breaking the chain and effectively select any RT mutations that restore viral DNA synthesis (Dapp et al. 2009).

The targeting cellular systems such as cytidine deaminases APOBEC (activation of APOBEC 3) can be considered an alternative approach, reducing the development of viral resistance to nucleoside therapeutics. Members of APOBEC family represent an evolutionarily preserved mechanism of protection against viruses. The primary antiviral activity of these enzymes is the hypermutagenesis of C-U single-stranded viral DNA, which occurs during the synthesis of the negative DNA chain of reverse transcription and leads to hypermutagenesis of G-A in the positive DNA strand. A limitation for developing ribonucleosides as potential antiviral agents are the relatively high concentrations necessary to observe the antiviral effect (competition with natural NTP) and increase the risks of genotoxicity (as proposed for 5-AZCTP). However, it was shown that the genotoxicity was negligible at concentrations necessary to inhibit viral replication, which is likely due to the effectivity of the host DNA repair mechanism. Enhanced mutagenesis is the leading cause of human cancer induction. Therefore, to minimize risks, it is necessary to use mutagenic nucleoside analogues effective at doses that do not cause acute toxic effects since, in addition to lethal mutagenesis, nucleosides and their analogues can act by other mechanisms, inhibiting cellular enzymes and be toxic to cells. There is also a risk of an increase in the frequency of mutations in healthy cells, which can lead to the induction of secondary malignancies. Unlike viral mutagenesis, the harmful effects of lethal mutagenesis of tumors will not manifest themselves for several generations after the inclusion of a mutagenic analogue (Fox and Loeb 2010). Alternatively, to synthetic mutagenic nucleosides, minor tRNA components can be considered as potential drugs. i^6A was tested as an antitumor drug 50 years ago. Despite unsuccessful clinical trials of i^6A (McCown et al. 2020), it was further shown to inhibit human enterovirus 71 type (EV71) reproduction efficiently. A series of compounds with high antienteroviral activity and low cytotoxicity was then elaborated based on i^6A structure (Drenichev et al. 2016).

4 Conclusions

Nucleosides are an important class of natural compounds participating in storing and implementing hereditary information. A library of four basic nucleoside structures is substantially expanded by poly(ADP-ribose) (PAR) and minor RNA components. Disaccharide nucleosides are important participants and regulators of DNA repair, protein biosynthesis and enzymatic activity of PARP and PARG. A non-canonical mono(ADP-ribose) ADPR 5′-cap can be formed in 5′-phosphorylated ends of RNA by enzymatic transfer of ADPR fragment from NAD^+ to RNA and is thought to distinguish viral and host RNA. With the finding of novel plant immunity suppressors based on disaccharide nucleoside molecules, other disaccharide 5′-capping mechanisms can be possible in RNA. Base-modified ribonucleosides, found in tRNA, have diverse structures and functions. Most are methylated compounds, which can biochemically convert to more structurally complex compounds. Minor nucleoside components being involved in altering RNA structure and protein biosynthesis can be markers of human pathologies and metabolic disorders. Nucleoside stable isotope-labeled internal standards (SILIS) based on minor RNA components are widely used to analyze RNA probes and have a high potential for diagnostics of human diseases associated with nucleoside biosynthesis. Synthetic nucleosides which modify RNA are drugs ribavirin, favipiravir, decitabine, molnupiravir and others. In a cell, these compounds convert to active triphosphate forms, which incorporate into viral RNAs changing their properties and causing to inhibition of viral reproduction by lethal mutagenesis mechanisms. N^6-isopentenyladenosine i^6A was shown to have a potent antienteroviral (EV71) effect among minor tRNA components. A series of more selective EV71 inhibitors were then elaborated based on its structure. 2020 Pandemic invoked by new coronaviral infection COVID-19 outlined the need for novel efficient antivirals. In this sense, naturally occurring modified ribonucleosides seem to be a prospective field for research and development of new drugs on their basis.

Acknowledgements This work was funded by Russian Science Foundation, project number 21-14-00346.

References

Alexeev CS, Kulikova IV, Gavryushov S et al (2018) Quantitative prediction of yield in transglycosylation reaction catalyzed by nucleoside phosphorylases. Adv Synth Catal 360:3090–3096
Bednarek P, Winter J, Hamberger B et al (2004) Induction of 3′-O-β-D-ribofuranosyl adenosine during compatible, but not during incompatible, interactions of Arabidopsis thaliana or Lycopersicon esculentum with Pseudomonas syringae pathovar tomato. Planta 218:668–672
Bonnac LF, Mansky LM, Patterson SE (2013) Structure–activity relationships and design of viral mutagens and application to lethal mutagenesis. J Med Chem 56:9403–9414
Bruckl T, Globisch D, Wagner M et al (2009) Parallel isotope-based quantification of modified tRNA nucleosides. Angew Chem Int Ed 48:7932–7934
Chen P, Shakhnovich EI (2009) Lethal mutagenesis in viruses and bacteria. Genetics 183:639–650
Crotty S, Andino R (2002) Implications of high RNA virus mutation rates: lethal mutagenesis and the antiviral drug ribavirin. Microbes Infect 4:1301–1307
Crotty S, Maag D, Arnold JJ et al (2000) The broad-spectrum antiviral ribonucleoside ribavirin is an RNA virus mutagen. Nat Med 6:1375–1379
D'Amours D, Desnoyers S, D'Silva I et al (1999) Poly(ADP-ribosyl)ation reactions in the regulation of nuclear functions. Biochem J 342:249–268
Dapp MJ, Clouser CL, Patterson S et al (2009) 5-Azacytidine can induce lethal mutagenesis in human immunodeficiency virus type 1. J Virol 83:11950–11958
Dapp MJ, Patterson SE, Mansky LM (2013) Back to the future: revisiting HIV-1 lethal mutagenesis. Trend Microbiol 2:56–62
Domingo E, Holland JJ (1994) Mutation rates and rapid evolution of RNA viruses. In: Morse S (ed) The evolutionary biology of viruses, 1st edn. Raven Press, New York
Domingo E, Escarmís C, Sevilla N et al (1996) Basic concepts in RNA virus evolution. FASEB J 10:859–864
Drenichev MS, Mikhailov SN (2015) Poly(ADP-ribose)—a unique natural polymer. Structural features, biological role and approaches to the chemical synthesis. Nucleosides, Nucleotides Nucleic Acids 34:258–276
Drenichev MS, Mikhailov SN (2016) Poly(ADP-ribose): from chemical synthesis to drug design. Bioorg Med Chem Lett 26:3395–3403
Drenichev MS, Oslovsky VE, Mikhailov SN (2016) Cytokinin nucleosides—natural compounds with a unique spectrum of biological activities. Curr Top Med Chem 16:2562–2576
Drenichev MS, Bennett M, Novikov RA et al (2019) A role for 3′-O-β-D-ribofuranosyladenosine in altering plant immunity. Phytochem 157:128–134
Efimtseva EV, Kulikova IV, Mikhailov SN (2007) Disaccharide nucleosides and their incorporation into oligonucleotides. Curr Org Chem 11:337–354
Fahrer J, Kranaster R, Altmeyer M et al (2007) Quantitative analysis of the binding affinity of the poly(ADP-ribose) to specific binding proteins as a function of chain length. Nucleic Acids Res 35:e143
Ferro AM, Oppenheimer NJ (1978) Structure of a poly(adenosine diphosphoribose) monomer 2′-(5′-phosphoribosyl)-5′-adenosine monophosphate. Proc Natl Acad Sci USA 75:809–813
Földesi A, Trifonova A, Kundu MK et al (2000) The synthesis of deuterionucleosides. Nucleosides, Nucleotides Nucleic Acids 19:1615–1656
Fox EJ, Loeb LA (2010) Lethal mutagenesis: targeting the mutator phenotype in cancer. Semin Cancer Biol 20:353–359

Gagne JP, Isabelle M, Lo KS et al (2008) Proteome-wide identification of poly(ADP-ribose) binding proteins and poly(ADP-ribose)-associated protein complexes. Nucleic Acids Res 36:6959–6976
Groslambert J, Prokhorova E, Ahel I (2021) ADP-ribosylation of DNA and RNA. DNA Repair 105:103144
Hadj Hassine I, Ben M'hadheb M, Menéndez-Arias L (2022) Lethal mutagenesis of RNA viruses and approved drugs with antiviral mutagenic activity. Viruses 14:841
Huang S, Jia A, Song W et al (2022) Identification and receptor mechanism of TIR-catalyzed small molecules in plant immunity. Science 377:eabq3297
Jia A, Huang S, Song W et al (2022) TIR-catalyzed ADP-ribosylation reactions produce signaling molecules for plant immunity. BioRxiv. https://doi.org/10.1101/2022.05.02.490369
Jiao X, Doamekpor SK, Bird JG et al (2017) 5′ end nicotinamide adenine dinucleotide cap in human cells promotes RNA decay through DXO-mediated deNADding. Cell 168:1015–1027
Kistemaker HA, Lameijer LN, Meeuwenoord NJ et al (2015a) Synthesis of well-defined adenosine diphosphate ribose oligomers. Angew Chem 54:4915–4918
Kistemaker HA, Overkleeft HS, van der Marel GA et al (2015b) Branching of poly(ADP-ribose): synthesis of the core motif. Org Lett 17:4328–4331
Kolitz SE, Lorsch JR (2010) Eukaryotic initiator tRNA: finely tuned and ready for action. FEBS Lett 584:396–404
Kulikova IV, Muradova DA, Drenichev MS et al (2015) Stereoselective synthesis of 2′-O-α-D-ribofuranosyluridine, a structural fragment of Hellecaucaside A. Chem Nat Comp 51:256–260
Liu Q, Kistemaker HA, Overkleeft HS et al (2017) Synthesis of ribosyl-ribosyl-adenosine-5′,5″,5‴(triphosphate)—the naturally occurring branched fragment of poly(ADP ribose). Chem Commun 53:10255–10258
Manik MK, Shi Y, Li S et al (2022) Cyclic ADP ribose isomers: production, chemical structures, and immune signaling. Science 377:eadc8969
Mathews CK, Van Holde KE (1996) Biochemistry. Benjamin Cummings PC, San Francisco
McCown PJ, Ruszkowska A, Kunkler CN et al (2020) Naturally occurring modified ribonucleosides. WIREs RNA 11:e1595
Messner S, Altmeyer M, Zhao H et al (2010) PARP-1 ADP-ribosylates lysine residues of the core histone tails. Nucleic Acid Res 38:6350–6362
Meyer R, Meyer-Ficca M, Jacobsen E, Jacobsen M (2006) Enzymes in poly(ADP-Ribose) metabolism. In: Burkle A (ed) Poly(ADP-Ribosyl)ation. Molecular biology intelligence unit. Springer Science+Business Media Inc, New York
Minaga T, Kun E (1983) Probable helical conformation of poly(ADP-ribose). J Biol Chem 258:5726–5730
Miwa M, Ishihara M, Takishima S et al (1981) The branching and linear portions of poly(adenosine diphosphate ribose) have the same alpha (1 leads to 2) ribose-ribose linkage. J Biol Chem 256:2916–2921
Moreau C, Ashamu GA, Bailey VC et al (2011) Synthesis of cyclic adenosine 5′-diphosphate ribose analogues: a C2′endo/syn "southern" ribose conformation underlies activity at the sea urchin cADPR receptor. Org Biomol Chem 9:278–290
Munir A, Banerjee A, Shuman S (2018) NAD^+-dependent synthesis of a 5-phospho-ADP-ribosylated RNA/DNA cap by RNA 2-phosphotransferase Tpt1. Nucleic Acids Res 46:9617–9624
Munnur D, Bartlett E, Mikolčević P et al (2019) Reversible ADP-ribosylation of RNA. Nucleic Acid Res 47:5658–5669
Pearson D, Hienzsch A, Wagner M et al (2011) LC-MS based quantification of 2-O-ribosylated nucleosides Ar(p) and Gr(p) in tRNA. Chem Commun 47:5196–5198
Penzo M, Guerrieri AN, Zacchini F et al (2017) RNA pseudouridylation in physiology and medicine: For better and for worse. Genes 8:301
Perales C, Martín V, Domingo E (2011) Lethal mutagenesis of viruses. Curr Opin Virol 1:419–422
Popova AM, Williamson JR (2014) Quantitative analysis of rRNA modifications using stable isotope labeling and mass spectrometry. J Am Chem Soc 136:2058–2069

Robson F, Khan KS, Le TK et al (2020) Coronavirus RNA proofreading: molecular basis and therapeutic targeting. Mol Cell 79:710–727

Schultheisz HL, Szymczyna BR, Williamson JR (2009) Enzymatic synthesis and structural characterization of ^{13}C, ^{15}N-Poly(ADP-ribose). J Am Chem Soc 131:14571–14578

Schultz SK, Kothe U (2021) Partially modified tRNAs for the study of tRNA maturation and function. In: Jackman JE (ed) Methods in enzymology, 1st edn. Academic Press, New York

Steinhauer DA, Domingo E, Holland JJ (1992) Lack of evidence for proofreading mechanisms associated with an RNA virus polymerase. Gene 122:281–288

Sylla B, Gauthier C, Legault J et al (2014) Isolation of a new disaccharide nucleoside from Helleborus caucasicus: structure elucidation and total synthesis of hellecaucaside A and its β-anomer. Carbohydr Res 398:80–89

Tong J, Flavell RA, Li HB (2018) RNA m6A modification and its function in diseases. Front Med 12:481–489

Vorbruggen H, Ruh-Pohlenz C (2001) Handbook of nucleoside synthesis. Wiley, New York

Walters RW, Matheny T, Mizoue LS et al (2017) Identification of NAD+ capped mRNAs in Saccharomyces cerevisiae. PNAS 114:480–485

Watt JM, Thomas MP, Potter BV (2018) Synthetic cADPR analogues may form only one of two possible conformational diastereoisomers. Sci Rep 8:15268

Yang HZ, Pan MY, Jiang DW et al (2011) Synthesis of Janus type nucleoside analogues and their preliminary bioactivity. Org Biomol Chem 9:1516–1522

Zenchenko AA, Drenichev MS, Il'icheva IA et al (2021) Antiviral and antimicrobial nucleoside derivatives: structural features and mechanisms of action. Rus J Mol Biol 55:785–811

Zhao Y, He G, Huang W (2021) A novel model of molnupiravir against SARS-CoV-2 replication: accumulated RNA mutations to induce error catastrophe. Signal Transduct Target Ther 6:1–3

Mikhail S. Drenichev (Engelhardt Institute of Molecular Biology, Russian Academy of Sciences, Moscow, *mdrenichev@mail.ru*) graduated from Moscow State Academy of Fine Chemical Technology (MITHT, Department of Chemistry and Technology of Biologically Active Compounds). He received his master's degree in 2008, studying synthesis and photodynamic properties of bacteriochlorin conjugates as potential anticancer therapeutics. Then he continued his research work at Engelhardt Institute of Molecular Biology, where he received PhD degree in 2013. His PhD work was devoted to an elaboration of human poly(ADP-ribose) polymerase I (PARP-I) inhibitors based on disaccharide nucleosides. Dr. Mikhail S. Drenichev is an author and co-author of more than 30 research papers and 5 scientific inventions. His main research interests are in the field of physical chemistry, chemistry of nucleosides and heterocyclic compounds, oligonucleotide synthesis, DNA repair and biochemical properties of nucleos(t)ide metabolizing enzymes.

Anastasia A. Zenchenko (Engelhardt Institute of Molecular Biology, Russian Academy of Sciences, Moscow, *kolomatchenkoa@yandex.ru*) was graduated from Moscow Technological University (MIREA, Department of Chemistry and Technology of Biologically Active Compounds and Medical Chemistry) in 2017. She received her master's degree, studying antiviral activity of lipophilic purine nucleosides. At present she is a PhD student in the Laboratory of Design and Synthesis of Biologically active compounds, Engelhardt Institute of Molecular Biology. Her research work has primarily focused on the development of chemical synthesis strategy of new biologically active nucleosides as well as study of the antiviral properties of the obtained compounds against a number of epidemiologically significant human RNA viruses and antitumor activity of modified nucleoside derivatives. Her research interests are in the field of bioorganic chemistry, medicinal chemistry and chemistry of nucleosides as well as fine organic synthesis.

Cyril S. Alexeev (Engelhardt Institute of Molecular Biology, Russian Academy of Sciences, Moscow, *cyril.aleks@protonmail.com*) received his master's degree in Moscow State Academy of Fine Chemical Technology (MITHT) from Department of Chemistry and Technology of Biologically Active Compounds in 2005. Then he worked at Engelhardt Institute of Molecular Biology (Moscow), where he received PhD degree in 2012. His PhD research is related to studies of substrate specificity of *Escherichia coli* nucleoside phosphorylases. His professional interests are closely connected to enzymes, natural and modified nucleosides, DNA&RNA components, enzymatic synthesis, development of antitumor and antiviral drugs and biocatalysis. His hobbies are literature, music and travelling. Dr. Cyril S. Alexeev is an author and co-author of more than 20 research papers and 2 scientific inventions.

Multifaceted Functions of RNA m^6A Modification in Modulating Regulated Cell Death

Guankai Zhan, Jinfeng Liu, Jiebo Lin, Jiafeng Chen, Siqi Sun, Yasen Maimaitiyiming, and Chih-Hung Hsu

Contents

Guankai Zhan, Jinfeng Liu, Jiebo Lin contributed equally to this work.

G. Zhan · J. Liu · J. Lin · J. Chen · S. Sun · C.-H. Hsu (✉)
Institute of Genetics, Department of Environmental Medicine, Women's Hospital, Zhejiang University School of Medicine, Hangzhou 310006, China
e-mail: ch_hsu@zju.edu.cn

Y. Maimaitiyiming (✉)
Department of Public Health and Department of Hematology of First Affiliated Hospital, Zhejiang University School of Medicine, Hangzhou 310058, China
e-mail: yasinjan@zju.edu.cn

Department of Neurobiology and Department of Neurology of the First Affiliated Hospital, NHC and CAMS Key Laboratory of Medical Neurobiology, School of Brain Science and Brain Medicine, Zhejiang University School of Medicine, Zhejiang University, Hangzhou 310058, China

J. Barciszewski (ed.), *RNA Structure and Function*, RNA Technologies 14,
https://doi.org/10.1007/978-3-031-36390-0_24

Abstract Cell death is attributed to the unrecoverable disturbance of cells exposed to intracellular or extracellular stimulus. Regulated cell death (RCD), also known as programmed cell death (PCD), involves signaling cascades in which effector molecules participate in and play important roles for maintaining organism homeostasis. As a basic biological phenomenon of cells, RCD displays exceptional influence on the occurrence and development of many pathophysiological processes. N^6-methyladenosine (m^6A) is an important epigenetic modification that regulates RNA fate and subsequent bioprocesses. It occurs extensively on multiple RNA species and is the most abundant internal modification of eukaryotic messenger RNA (mRNA). The abundance of m^6A modification is regulated by its methyltransferases and demethylases, and the diverse biological functions of m^6A are executed by distinct m^6A binding proteins. Accruing evidence shows that m^6A modification participates in eukaryotic cell fate transition and determination. In recent years, the disorder of m^6A modification status has been observed during the initiation and progression of various RCDs, including classical apoptosis, necroptosis, pyroptosis, and the newly described ferroptosis. Due to the importance of m^6A and RCD in cellular and organismal homeostasis, it is worthwhile to recapitulate and organize the current knowledge of the crosstalk between m^6A status and RCDs with distinct biochemical and morphological features, as well as immune consequences. In this chapter, we systematically summarize the regulatory networks of m^6A modification on RCDs to provide the profiling of m^6A-RCD regulatory axis as comprehensively as possible.

Keywords RNA m^6A modification · mRNA decay · Translation regulation · Regulated cell death · Apoptosis · Autophagy · Necroptosis · Pyroptosis · Ferroptosis

1 Introduction

According to recommendations of the Nomenclature Committee on Cell Death in 2018 (Galluzzi et al. 2018), cell death can be divided into accidental cell death (ACD) and regulated cell death (RCD) based on the differences in mechanistic aspects and essential involved factors of the processes. ACD is biologically uncontrolled cell death, which is triggered by overwhelming insults, such as extreme physical, chemical or mechanical factors (Galluzzi et al. 2016b). These detrimental stimuli far exceed the cell's regulatory capacity, leading to immediate cell death (Tang et al. 2019). RCD, also known as programmed cell death (PCD), involves the well-organized signal cascade reaction of effectors, which has unique biochemical and morphological characteristics, as well as immunological consequences (Galluzzi et al. 2018). According to the morphological changes and biological features of dead cells, the classical cell death types are cataloged as follows: 1) Apoptosis (type I cell death), the typical phenotype is the formation of many spherical membrane-encapsulated bodies, containing cytoplasmic organelles and nuclear fragments, called apoptotic bodies (Carneiro and El-Deiry 2020); 2) Autophagy (type II cell death), an evolutionarily conserved intracellular degradation system in eukaryotes, is characterized by the massive accumulation of cytoplasmic double-membrane autophagic vacuoles (autophagosomes) which finally fuse with lysosomes to degrade contents (Mizushima and Komatsu 2011); and 3) Necrosis (type III cell death), also known as ACD due to overwhelming chemical or physical insult as mentioned above, is distinct from apoptosis and autophagy (Chen et al. 2018). Necrotic cells do not form apoptotic bodies or autophagosomes, but typically show increased membrane permeability, organelles deformation or enlargement, damage of plasma membranes, and ultimately cell lysis.

With the development of experimental techniques in biochemistry and molecular biology, previously undiscovered cell death pathways with newly identified mechanisms are beginning to emerge (Tang et al. 2019). More and more studies have found that under specific conditions, certain dead cells showed neither consistent morphological changes nor the corresponding molecular biological changes of the above-mentioned well-cataloged cell death types. Accordingly, several novel cell death types were identified and elucidated in recent decades, such as pyroptosis (Shi et al. 2017b), necroptosis (Shan et al. 2018), parthanatos (Andrabi et al. 2008), ferroptosis (Dixon et al. 2012) and cuproptosis (Tsvetkov et al. 2022).

The regulation of epigenetic modification is one of the main response mechanisms of cells when they are disturbed by internal or external environmental stimuli. Organisms change gene expression in response to environmental stimuli by regulating epigenetic landscape (histone modification, DNA modification, RNA modification, etc.) and ultimately determine the fate of committed cells. m^{6}A is the most abundant internal modification of eukaryotic messenger RNA (mRNA) (Dominissini et al. 2012). It is a dynamic and reversible epitranscriptomic mark subjected to regulation by m^{6}A methyltransferases/m^{6}A Writers (e.g., METTL3/14 methyltransferase complex and METTL16) (Liu et al. 2014; Pendleton et al. 2017), m^{6}A demethylases/m^{6}A Erasers (e.g., FTO and ALKBH5)(Jia et al. 2011; Zheng et al. 2013), and m^{6}A binding proteins/m^{6}A Readers (YTH family proteins and IGF2BP family proteins, etc.) (Hsu et al. 2017; Huang et al. 2018; Shi et al. 2017a; Wang et al. 2014; Xiao et al. 2016). The dynamic and adjustable nature makes m^{6}A a key regulator in RNA fate determination and transition, affecting nearly all aspects of mRNA metabolism, including mRNA splicing, translation, stability, and miRNA maturation (Table 1).

Accumulating evidence suggested that m^{6}A participates in eukaryotic cell fate transition and determination, and the disorder of m^{6}A modification status was observed in the process of various forms of RCDs (Batista et al. 2014; Liu et al. 2022b). For instance, during testosterone synthesis, a process highly correlated with autophagy, m^{6}A modification levels are significantly reduced (Chen et al. 2021), and the negative correlation between m^{6}A modification and autophagy was similarly reported in ischemia/reperfusion (I/R)-treated mouse cardiomyocytes (Song et al. 2019). On the other hand, knockdown of WTAP and/or METTL3 in zebrafish induced a significant reduction of global m^{6}A levels on mRNAs, resulting in embryonic differentiation defects as well as increased apoptosis (Ping et al. 2014). Although R-2-hydroxyglutarate (R-2HG) was reported as an oncometabolite, a study found that

Table 1 Major m^6A modulators and their functions (Shi et al. 2019; Zaccara et al. 2019)

Type	Modulator		Major Function
m^6A Writers	METTL3/14 methyltransferase complex	METTL3	Deposition of m^6A
		METTL14	
		WTAP	
		VIRMA	
		ZC3H13	
		HAKAI	
		RBM15/15B	
	others	METTL16	
m^6A Erasers	m^6A demethylases	FTO	Removal of m^6A
		ALKBH5	
m^6A Readers	YTH family	YTHDF1	Regulating mRNA decay or translation
		YTHDF2	
		YTHDF3	
		YTHDC1	Regulating mRNA nuclear export
		YTHDC2	Regulating mRNA decay and translation
	IGF2BP family	IGF2BP1	Enhancing mRNA stability and translation
		IGF2BP2	
		IGF2BP3	
	hnRNP family	hnRNPA2B1	Regulating pri-miRNA processing
		hnRNPC	Regulating mRNA selective splicing
		hnRNPG	
	others	EIF3	Regulating mRNA translation
		FMR1	Regulating mRNA decay

Abbreviations METTL3, methyltransferase-like 3; METTL14, methyltransferase-like 14; WTAP, Wilms tumor 1-associating protein; VIRMA, Vir like m^6A methyltransferase associated; ZC3H13, zinc finger CCCH-type containing 13; RBM15/15B, RNA binding motif protein 15/15B; METTL16, methyltransferase-like 16; FTO, fat mass and obesity-associated protein; ALKBH5, AlkB homolog 5; YTH family, YTH domain-containing family protein; IGF2BP family, Insulin-like growth factor 2 mRNA binding protein family; hnRNP family, heterogeneous nuclear ribonucleoprotein family; EIF3, Eukaryotic translation initiation factor 3; FMR1, Fragile X messenger ribonucleoprotein 1

R-2HG increased the abundance of global m^6A modification in leukemia cells by inhibiting FTO activity to promote cancer cell apoptosis (Su et al. 2018). Ferroptosis, as a newly identified form of RCD, has been extensively investigated. Ferroptosis inducers have been reported to increase the overall levels of m^6A modification and induce ferroptosis in hepatic stellate cells (Shen et al. 2021).

As a whole, m^{6}A modification plays an important regulatory role in initiation and progression of RCDs. Regarding the diversity of distinct pathways of RCDs and multiple layers of m^{6}A regulation, the relationship (mutual regulation) between RNA m^{6}A modification and RCDs displays a complicated nature. In view of the importance of m^{6}A and RCD in cellular homeostasis, in this chapter, we systematically summarize the regulatory networks of m^{6}A modification on RCDs to provide comprehensive picture of m^{6}A-RCD regulatory axis in distinct cellular contexts.

2 Apoptosis

Apoptosis is the earliest identified form of programmed cell death and can be initiated by intracellular and external damages induced upon a variety of environmental stimuli, such as DNA damage, replication stress, hypoxia, endoplasmic reticulum (ER) stress, or mitotic defects et al. (Lockshin and Zakeri 2001). Mechanistically, apoptosis is mainly activated through mitochondrial pathway and death receptor pathway (Galluzzi et al. 2018). For mitochondrial apoptotic signaling pathways, B-cell lymphoma 2 (BCL-2) family proteins play major roles in regulation of apoptosis by controlling mitochondrial permeability. Upon induction of apoptosis, the Bax (BCL-2-associated X-protein) and/or Bak (BCL2-antagonist/killer) form homodimers or heterodimers and then translocate to the mitochondria, wherein promoting the release of cytochrome c. Then, cytochrome c promotes oligomerization of apoptotic peptidase activating factor 1 (APAF1) leading to apoptosome formation, which recruits and activates the initiator procaspase-9. Further, the active caspase-9 cleaves and activates the executioner caspase-3, -6, and -7, finally leading to apoptosis (Galluzzi et al. 2018). TP53 (p53) also plays a critical role in the mitochondrial apoptosis pathway. On one hand, as a nuclear transcription factor, p53 regulates expression of many apoptosis-related genes. On the other hand, p53 migrates to mitochondria and mediates the mitochondrial apoptosis pathway primarily through direct interactions with BCL-2 and BCL-xL, leading to activation Bax/Bak homodimers (Wei et al. 2021). The XIAP (X-linked inhibitor of apoptosis protein) is an anti-apoptotic protein in mitochondrial apoptotic signaling pathways, which works by inhibiting activation of caspase-3 and -7 (Galluzzi et al. 2016b).

The death receptor pathway mediated apoptosis (extrinsic apoptosis signaling pathway) is generally initiated by activated receptors such as Fas cell surface death receptor (FAS) and tumor necrosis factor (TNF) receptor superfamily. Activated death receptors promote the formation of a dynamic multiprotein complex that responds to different death signals, which is called the death-inducing signaling complex (DISC), such as "complex I" (generally consisting of TRADD, TRAF2/5, c-IAP1/2, RIPK1), and "complex II" (converted from complex I via additional recruitment of FADD and caspase-8). The DISC regulates apoptosis via activating or repressing the functions of caspase-8 (Galluzzi et al. 2016b, 2018).

As the earliest identified and most common form of RCD, apoptosis is recognized to participate in many biological processes, such as tissue regeneration, embryonic development, tumor progression, and so on (Bergmann and Steller 2010). Recently, a notable number of studies revealed the tight correlation between RNA m^{6}A modification and apoptosis. Thus, in this part, we mainly recapitulate the relationship between RNA m^{6}A modification and apoptosis.

2.1 M^{6}A Regulates Intrinsic Signaling Pathways Dependent Apoptosis

In human myeloid leukemia cell lines, m^{6}A modification reportedly promotes MYC, BCL2, and PTEN translation, and METTL3 knockdown induces differentiation and apoptosis of these cells (Vu et al. 2017), suggesting the modification repressed apoptosis in this scenario. Another study also showed that BCL2, an important anti-apoptotic protein in the mitochondrial apoptotic signaling pathway, is regulated by m^{6}A modification via m^{6}A reader IGF2BPs in myeloid leukemia cells through interaction with YBX1, leading to stabilization of m^{6}A-tagged *BCL2* mRNA (Feng et al. 2021). Similarly, knockdown of METTL3 in breast and lung cancer cells can reduce methylation levels of *BCL-2* mRNA and decrease its stability, resulting in apoptosis induction and inhibition of tumor growth (He et al. 2022; Wang et al. 2020a). These studies indicated that m^{6}A modification plays a role in stabilizing *BCL2* mRNA and its protein expression to suppress mitochondrial apoptosis.

Moreover, many studies also show an important role of m^{6}A in regulating p53-mediated mitochondrial apoptosis. The m^{6}A methylome in human and mouse revealed that reducing m^{6}A levels by METTL3 knockdown results in disruption of the p53 signaling pathway and apoptosis (Dominissini et al. 2012). In invariant natural killer T (iNKT) cells, knockdown of METTL14 decreases m^{6}A levels of *XAF1* (XIAP Associated Factor 1) and *TRIM25* (tripartite motif containing 25) mRNAs, resulting in upregulation of XAF1 and downregulation of HMGA1B (High mobility group A1 isoform B) expression, which promotes p53 pathway-induced apoptosis (Cao et al. 2022). Notably, overexpression of METTL14 increased m^{6}A levels and promoted apoptosis in cisplatin-treated proximal tubular cell line HK2 via inducing expression of *p53* mRNA (Zhou et al. 2019). Another in vivo study showed that knockout of m^{6}A demethylase ALKBH5 increases global m^{6}A levels, leading to aberrant spermatogenesis and apoptosis in mouse testes, and the p53 functional interaction network plays a major role in this process (Zheng et al. 2013).

ER stress-induced cell death depends on the core mitochondrial apoptosis pathway. WTAP promotes ER stress-mediated apoptosis under myocardial ischemia/reperfusion (I/R) injury condition by regulating *ATF4* mRNA stability via increasing its m^{6}A levels and upregulating the CHOP (C/EBP homologous protein) expression, which can activate the Bax-involved mitochondrial apoptosis pathway (Wang et al. 2021b). On the other hand, YTHDF2 depletion in triple-negative breast cancer cells initiates apoptosis via stabilizing mRNAs in MAPK pathway, which results in accumulation of unfolded proteins, leading to excessive ER stress (Einstein et al. 2021).

In the other study, FTO KO in osteoblasts increases apoptosis by induction of DNA damage and ER stress (Zhang et al. 2019).

In addition, cytochrome c release is also subjected to regulation by m^{6}A. This is evidenced by the fact that METTL14 depletion reduced m^{6}A modification on gasdermin C (*GSDMC*) transcripts and downregulated GSDMC expression, thereby disrupting mitochondrial membrane potential, and triggering cytochrome c release that activates the pro-apoptotic pathway (Du et al. 2022). Taken together, these studies indicate that the various intrinsic apoptosis regulatory molecules are modulated by RNA m^{6}A modification, which largely enriches the potential "arsenal" to be used to induce or inhibit apoptosis in varying disease conditions.

2.2 M^{6}A Regulates Extrinsic Signaling Pathway-Induced Apoptosis

A previous study has indicated that YTHDF2 deficiency can trigger apoptosis in leukemia stem cells through directly increasing tumor necrosis factor receptor (TNFR2) expression (Paris et al. 2019). Moreover, METTL3 conditional knockout in mice cerebellum resulted in severe developmental defects via upregulation of neuronal proapoptotic genes including *DAPK1*, *FADD*, and *NGFR*, all of which are crucial for extrinsic apoptosis signaling pathway (Wang et al. 2018). Similarly, another study showed that METTL14 deficiency in intestinal epithelial cells promotes TNF-induced cell death by suppressing NF-κB-mediated antiapoptotic pathway (Zhang et al. 2022d). Collectively, these observations suggest that the m^{6}A modification participates in various aspects of extrinsic apoptosis via regulating key genes in these pathways (Fig. 1).

2.3 Clinical Significance of m^{6}A Regulating Apoptosis

Since m^{6}A plays an important role in the regulation of a variety of pathophysiological processes including apoptosis, several inhibitors targeting m^{6}A regulatory machinery (m^{6}A regulator) have been developed to modulate m^{6}A levels. Yankova et al. (2021) identified an inhibitor of METTL3, which is named STM2457. STM2457 showed strong pharmacological inhibition of METTL3's methyltransferase function either in vitro or in vivo (Yankova et al. 2021). Notably, daily treatment with STM2457 significantly increased apoptosis in AML cells and showed curative effect in the AML mouse models. Moreover, it is revealed that R-2HG can inhibit FTO activity and lead to an increase in global m^{6}A level in leukemia cells, thereby destabilizing *MYC/CEBPA* transcripts and promoting AML cells' apoptosis via suppression of relevant pathways (Su et al. 2018). Since RNA m^{6}A modification is demonstrated to play important roles in regulation of apoptosis, m^{6}A modifiers and readers might

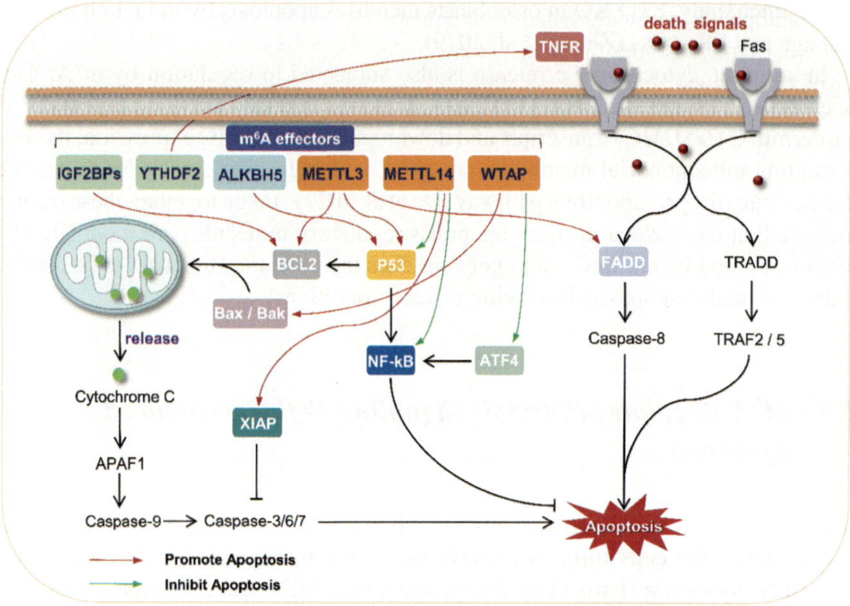

Fig. 1 Schematic diagram of apoptosis regulation by m^{6}A modification. Apoptosis is the programmed cell death induced by intracellular signals (e.g., DNA damage) or external stimuli (e.g., UV light), which lead to activation of the intrinsic or mitochondrial pathway and the extrinsic or death receptor pathway. A variety of m^{6}A-related molecules regulate the fate of key mRNAs in the apoptosis signaling cascade through different biological functions, promoting or inhibiting apoptosis. Abbreviations: TNFR, tumor necrosis factor (TNF) receptor; BCL2, B-cell lymphoma 2; ATF4, activating transcription factor 4; APAF1, apoptotic peptidase activating factor 1; XIAP, X-linked inhibitor of apoptosis protein; FADD, FAS-associated death domain protein; TRADD, TNFR-1-associated death domain; TRAF, TNF receptor-associated factor

have potential to regulate apoptosis in distinct scenarios and could be exploited as novel therapeutic targets in clinic.

3 Autophagy

Autophagy was proposed after the discovery of a "self-eating" phenomenon in cells (Ashford and Porter 1962). It's a self-digestion process by which cells degrade damaged and denatured macromolecules or even whole organelles via lysosomes under the stress of external environmental factors. In humans, abnormal autophagy regulation leads to a variety of pathophysiological conditions, including cancer, neurodegenerative disease, pulmonary disease, and cardiomyopathy (Alharbi et al. 2021). According to different modes of substrate transport to lysosomes, autophagy is divided into at least three forms including macroautophagy, microautophagy, and chaperone-mediated autophagy (Wang et al. 2022a). Macroautophagy is the main autophagy pathway, which is characterized by the transport of cytoplasmic substances to lysosomes through the intermediate bilayer membrane-bound vesicles termed autophagosomes that can fuse with lysosomes to form autophagy lysosomes (Feng et al. 2014). The process of autophagy is mediated by a group of evolutionarily conserved genes (autophagy-related genes, ATG genes) and includes four important stages: (1) Initiation. In this stage, various stimuli lead to the formation of phagocytic vesicles, which involves the ULK1 (UNC51-like kinase-1) and VPS34 (Vacuolar Protein Sorting 34) complexes. (2) Extension. The extension of phagocytic vesicles leads to the formation of autophagosomes with a unique double membrane structure. This step involves two ubiquitin-like coupling pathways, both of which requires ATG5-ATG12 complexes and are catalyzed by ATG7. (3) Fusion. Autophagosomes capture proteins, organelles as well as other substances that need to be degraded or cleared, and combine with lysosomes to form autophagy lysosomes. (4) Degradation. A series of acid hydrolases are involved in the degradation of isolated cytoplasmic substances. The small molecules produced by degradation, especially amino acids, are transported back to the cytoplasm and recycled for protein synthesis to maintain basic cell function under starvation conditions (Liu et al. 2012). Observation of the subcellular structure of autophagosomes by transmission electron microscopy and detection of LC3 and p62 were often used to evaluate autophagy level (Pugsley 2017).

Notably, strictly regulated autophagic responses under non-overwhelming stress conditions most often act as the cellular hub for adaptation to stress, thereby displaying cytoprotective but not cytotoxic effects. Thus, inhibition of autophagy is regarded as a common method to promote the demolishing of cells or to overcome resistance to cell death during cancer treatment (Levy et al. 2017). While, during normal developmental or some pathophysiological conditions, the activation of autophagy indeed contributes to cell death (Anding and Baehrecke 2015; Galluzzi et al. 2016a). Therefore, the term autophagic cell death is generally characterized by persistent and intense activation of autophagy, which is regulated by autophagy process and independent of other forms of cell death (e.g., apoptosis, necroptosis, ferroptosis) (Galluzzi et al. 2018). The epigenetic regulation (modifications) of organisms would be activated in response to environmental changes, and these modifications could directly affect the expression of ATG genes or interfere with the signaling mechanism to regulate autophagy (Lapierre et al. 2015). As a dynamic and reversible epi-marker, RNA m^{6}A modification responds to external stimuli via promptly modulating numerous biological processes, including autophagy.

3.1 M^{6}A Modification Promotes Autophagy

Regulators of the RNA m^{6}A play important roles in the mRNA life cycle. In fact, the m^{6}A binding protein/complex (m^{6}A readers) recognizes mRNA through m^{6}A modification and determines the fate of m^{6}A mRNA, which would further regulate the protein expression of these m^{6}A mRNAs and affect the following biological processes. Many studies have shown that under specific conditions such as diseases, autophagy is closely correlated with m^{6}A levels. For instance, a study showed that after FTO silencing, YTHDF2 can capture m^{6}A containing *eIF4G1* transcripts, resulting in mRNA degradation and decreased expression of eIF4G1 protein, thereby promoting autophagy and reducing tumorigenesis in the oral squamous cell carcinoma (Wang et al. 2021a).

In addition to regulating the initiation of autophagy as described above, m^{6}A levels can regulate signaling pathways and the expression of ATG family proteins, through which affects the extension phase of autophagy. METTL3 positively regulated autophagy and reversed gefitinib resistance by β-element in non-small cell lung cancer (NSCLC) cells by increasing the critical genes of autophagy pathway such as *ATG5* and *ATG7* (Liu et al. 2020). Likewise, HIF1α-activated upregulation of long non-coding RNA *DARS-AS1* (aspartyl-tRNA synthetase (DARS) antisense RNA 1), promoted DARS translation via METTL3- and METTL14-mediated m^{6}A modification, eventually regulating the expression of ATG5 and ATG3 to promote hypoxia-induced cytoprotective autophagy in cervical cancer cells (Shen et al. 2022). On the other hand, knockdown of endogenous YTHDC1 inhibited autophagy in the epidermis and delayed wound healing in the skin tissues of diabetic mice (Liang et al. 2022). Mechanistically, YTHDC1 interacted and cooperated with ELAVL1/HuR complex in modulating the expression of autophagy cargo receptor SQSTM1 (p62) by repressing degradation of m^{6}A modified *SQSTM1* mRNA.

Interestingly, besides participating in regulation of autophagic cell death, through regulating autophagy, m^{6}A also could modulate other types of RCD. For instance, it was found that m^{6}A modification could regulate ferroptosis through autophagy signaling pathway in hepatic stellate cells (HSC) (Shen et al. 2021). Specifically, *BECN1* mRNA could be m^{6}A methylated and subjected to regulation by METTL14 and FTO. The m^{6}A methylated *BECN1* mRNA could be recognized and stabilized by an m^{6}A reader YTHDF1. Enhancement of *BECN1* mRNA stability would consequently activate autophagy to induce HSC ferroptosis. In conclusion, aforementioned studies indicated that via targeting distinct pathways, m^{6}A could regulate autophagy and autophagic cell death.

3.2 *M^{6}A Modification Inhibits Autophagy*

In contrast, in some cases, m^{6}A levels negatively correlate with autophagy as well. The negative regulation of autophagy by m^{6}A also includes multiple mechanisms and pathways of autophagy. Take the initial phase of autophagy as an example, FTO (an m^{6}A demethylase) enhances the stability of *ULK1* transcripts in an m^{6}A-dependent manner through the m^{6}A reader YTHDF2, resulting in upregulation of ULK1 protein to promote the initiation of autophagy (Jin et al. 2018). Transcription factor EB (TFEB) is one of the transcriptional regulators of autophagy, which promotes expression of genes essential for autophagosome formation, and lysosome biogenesis as well as function. In the hypoxia/reoxygenation-treated cardiomyocytes, silencing METTL3 enhanced autophagic flux. Mechanistically, METTL3 methylated *TFEB*, which promoted the association of the m^{6}A reader protein HNRNPD with *TFEB* pre-mRNA and subsequently decreased the expression levels of TFEB, while TFEB-mediated lysosomes are required for the completion of autophagy (Song et al. 2019). Another study found that in drosophila and human cells, mTORC1 activates CCT (the cytosolic chaperonin-containing t-complex polypeptide 1) to stabilize m^{6}A methyltransferase complex, thereby increasing m^{6}A levels on the mRNAs encoding autophagy-related genes, leading to their degradation and suppression of autophagy (Tang et al. 2021). Additionally, Li et al. (2020a, b) found that in the nucleus pulposus cells, the m^{6}A level was decreased via ALKBH5, and lower m^{6}A levels on *FIP200* (focal adhesion kinase family interacting protein of 200 kD) transcripts led to a reduction of YTHDF2-mediated *FIP200* mRNA degradation, thereby resulting in increased autophagic flux (Li et al. 2020a).

Influencing the extension of autophagy by regulating ATG family proteins via m^{6}A-dependent manners has also been reported. During lipogenesis process, *ATG5* and *ATG7* transcripts with higher m^{6}A levels were captured by YTHDF2 after FTO silencing, which resulted in mRNA degradation and reduction of ATG5 and ATG7 protein expression, thereby suppressing autophagy and adipogenesis (Wang et al. 2020b). In colorectal cancer, glutaminolysis inhibition upregulated FTO to abrogate the m^{6}A modification of *ATF4* mRNA, thereby compromising YTHDF2-mediated RNA decay, inducing autophagy and enhancing the antitumor efficacy, which could be reversed by increasing m^{6}A modification (Han et al. 2021). It is reported that in liver cancer, depletion of METTL3 under hypoxia promotes sorafenib resistance, which associates with the expression of angiogenesis genes and autophagy-related pathways (Lin et al. 2020). Mechanistically, depletion of METTL3 lowered the m^{6}A modification level on *FOXO3* transcripts and decreased *FOXO3* mRNA stability by abolishing YTHDF1-mediated stabilization, resulting in activation of angiogenesis genes and autophagy-associated pathways. Accordingly, upregulation of METTL3 or overexpression of FOXO3 restored m^{6}A-dependent sorafenib sensitivity.

The discovery of these new m^{6}A-dependent autophagy regulatory mechanisms also provides the potential diagnosis target and therapeutic strategy for autophagy-related diseases. Besides abovementioned studies, in proteinuric kidney disease, METTL14 knockdown reduced m^{6}A level on *Sirt1* mRNA, thereby increased Sirt1 (a well-known protective deacetylase in proteinuric kidney diseases) expression levels in injured podocytes and resulted in the activation of autophagy, indicating METTL14 as a potential target for the diagnosis and therapy of proteinuric kidney disease (Lu et al. 2021). In addition, platelet-derived growth factor (PDGF) signaling upregulates global m^{6}A levels in glioblastoma cells through the transcriptional control of METTL3. PDGF-METTL3 axis represses tumor suppressor Optineurin (OPTN, as a conserved autophagy receptor) by m^{6}A-medaited *OPTN* mRNA degradation to inhibit glioblastoma stem cell (GSC) mitophagy, making the PDGF-METTL3-*OPTN* signaling as a potential glioblastoma therapeutic target (Lv et al. 2022). In

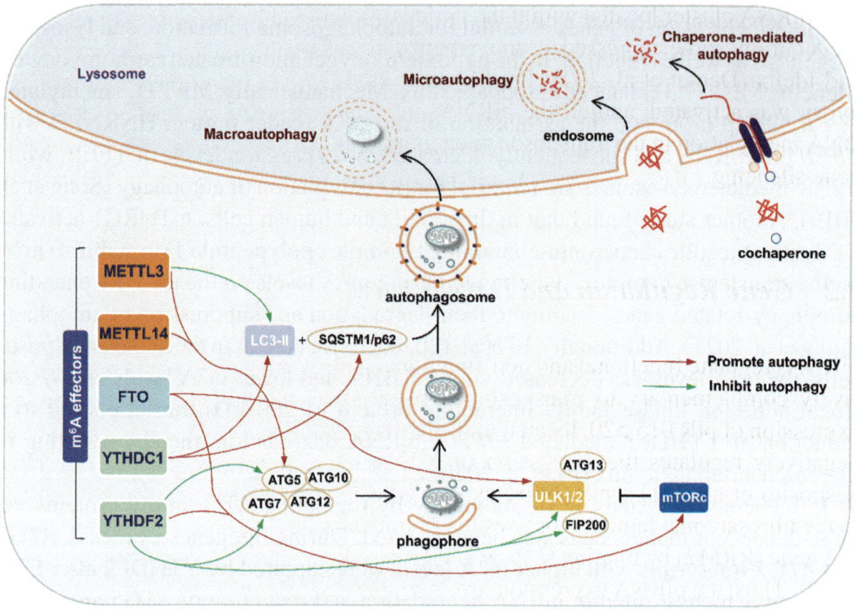

Fig. 2 Schematic diagram of autophagy regulation by m^6 A modification. Autophagy is mainly divided into three types: microautophagy, macroautophagy, and chaperone-mediated autophagy. The occurrence of autophagy mainly includes four processes: (1) The formation of phagocytic vesicles; (2) The extension of phagocytic vesicles; (3) Autophagy lysosomes formation; (4) Degradation of autophagic contents. m^6A-related molecules regulate the fate of key molecules in autophagy pathway through different biological functions, and promote or inhibit autophagy (the model diagram takes regulating microautophagy pathway as an example). Abbreviations: LC3-II, Microtubule-associated protein 1 light chain 3-II; SQSTM1, Sequestosome 1; ULK1/2, Unc-51 like autophagy activating kinase 1/2; FIP200, Focal adhesion kinase family interacting protein of 200 kD; mTORc, mammalian target of rapamycin complex

conclusion, m^6A has multifaceted roles in the regulation of autophagy amid various physiological and pathological conditions (Fig. 2).

3.3 Autophagy Reversely Regulates m^6A Related Molecules

In addition to what has been mentioned above, autophagy also plays a role in regulation of m^6A modifications. Inhibition of FTO suppresses melanoma tumorigenicity and the expression of melanoma cell-intrinsic genes *PD-1*, *CXCR4*, and *SOX10* through an m^6A/YTHDF2-dependent manner. And knockdown of the autophagy essential genes *ATG5* or *ATG7* reduced starvation-induced FTO and PD-1 expression, suggesting that advanced melanoma tumorigenesis by upregulating FTO could be regulated by autophagy-related pathways (Yang et al. 2019c). In another study, UVB irradiation downregulates METTL14 protein through NBR1- (autophagy receptor cargo) dependent selective autophagy, and decreases m^{6}A methylation and translation of the *DDB2* transcripts in a YTHDF1-dependent manner, resulting in the decrease of global genome repair and the increase of UVB-induced skin tumorigenesis (Yang et al. 2021). Likewise, arsenic promotes FTO protein stability by impairing selective autophagy and thus induces malignant transformation and tumorigenesis (Cui et al. 2021). These findings highlight the crosstalk between autophagy and m^{6}A modulators.

4 Necroptosis

Necroptosis is a form of RCD that depends on the sequential phosphoactivation of receptor-interacting serine/threonine kinase 3 (RIPK3) and mixed lineage kinase domain-like pseudokinase (MLKL), which could be resulted from activation of specific death receptors (e.g., FAS, TNFR1), or pathogen recognition receptors (e.g., Toll-like receptor 3 (TLR3), TLR4), and Z-DNA binding protein 1 (ZBP1, also known as DNA-dependent activator of IFN-regulatory factors, DAI) by corresponding cellular or environmental factors (Bertheloot et al. 2021). Molecularly, phosphorylation and subsequent activation of MLKL results in the formation of MLKL oligomers (mostly trimers or tetramers) that relocate to the plasma membrane and trigger plasma membrane permeabilization via binding specific phosphatidylinositol phosphate species (Hildebrand et al. 2014). Following membrane rupture, necroptotic cells release damage-associated molecular patters (DAMPs) to induce immune responses, which is regarded as an important viral defense mechanism (Kaczmarek et al. 2013).

Notably, when caspase-8 is inactive or inhibited, activation of TNFR1 generally results in necroptosis, wherein the activated RIPK1 in the DISC recruits and activates RIPK3, which involves physical interaction via their respective RIP homotypic interaction motif (RHIM) domains and the catalytic activity of RIPK1 (Li et al. 2012). Therefore, the RIPK1 inhibitor necrostatin-1 (nec-1) could robustly inhibit TNFR1-induced necroptosis. The cellular inhibitor of apoptosis proteins (c-IAPs) also inhibits necroptosis via ubiquitinating and degrading RIPK1 (Vanlangenakker et al. 2011). While, in case of TLR3 activation by double-stranded RNA (dsRNA), TLR4 activation by lipopolysaccharides (LPS), or ZBP1 activation by cytosolic DNA, RIPK3 can be directly activated via RHIM-dependent interaction with the toll-like receptor adaptor molecule 1 (TICAM1, also known as TRIF) or the ZBP1, resulting in activation of MLKL and subsequent necroptosis (Bertheloot et al. 2021; Maelfait et al. 2017). Thus, RIPK3 and MLKL inhibitors but not RIPK1 inhibitors are effective in inhibiting necroptosis in these contexts. Therefore, the TNFR1 extrinsic signaling pathway-induced apoptosis and necroptosis should be carefully distinguished according to the activation status of RIPK3 and MLKL but not merely based on the phosphorylation state of RIPK1.

Since necroptosis can function as an alternative RCD of apoptosis when caspase-8 is inhibited, it represents an effective form of RCD to deal with apoptosis-resistant neoplasms (Gong et al. 2019). Recently, accumulating evidence suggests necroptosis as an important cellular response in the modulation of cancer initiation, progression, and metastasis (Zhu et al. 2019). Here, we describe the current understanding of m^6A RNA-mediated regulation of necroptosis in colorectal cancer (CRC). Oxaliplatin (OX) is a widely used drug in the treatment of colorectal cancer (CRC), but the treated patients might develop resistance to OX, which is a major challenge for the treatment of advanced CRC (Greenlee et al. 2021). It has been suggested that tumor-associated macrophages (TAMs) play an important role in the tumor microenvironment (TME) and contributes to OX resistance of CRC, but the underlying mechanism remains largely unknown (DeBerardinis 2020). A study found that the density of TAMs infiltrated into the CRC tissues was significantly higher in OX-resistant patients as compared to the OX-sensitive patients (Lan et al. 2021). Interestingly, both the total m^6A levels and the expression of METTL3 were significantly increased in the CRC tissues from OX-resistant patients, and the M2-polarized TAMs promoted the OX resistance through upregulating METTL3-mediated m^6A modification on TNF receptor-associated factor 5 (*TRAF5*) gene and decreasing its protein expression, leading to resistance of CRC cells to necroptosis. Although current evidence shows that the m^6A RNA modification participated in regulation of necroptosis, future studies are needed to have a more complicative picture of m^6A-medaited regulation of necroptosis.

5 Pyroptosis

Pyroptosis is a form of RCD driven by the pore-forming fragment of (active) gasdermin (GSDM) protein family members (among which gasdermin D (GSDMD) is the most-studied and well-documented), which is proteolytically cleaved by inflammatory caspases (e.g., caspase-1 or caspase-4/-5) in response to activation of the inflammasome (a multiprotein complex that detects infectious as well as sterile insults and responds to them via releasing interleukin (IL) 1 family members including IL-1β and IL-18) (Galluzzi et al. 2018; Su et al. 2022; Tang et al. 2019). Thus, pyroptosis plays an important context-dependent role in pathophysiological conditions including cancer (Wei et al. 2022). On one hand, as a type of proinflammatory cell death, pyroptosis favors formation of a suitable inflammatory microenvironment for tumorigenesis and tumor cell growth (Xia et al. 2019). On the other hand, pyroptosis can inhibit the occurrence and development of tumors via timely killing of malignant cells in their infancy, and therefore targeted induction of pyroptosis in tumor cells via proper biochemical approaches is considered a potential cancer treatment strategy (Yang et al. 2022a). Hence, to understand these seemingly contradictory views of pyroptosis, examining the different layers of pyroptosis regulatory mechanisms would be necessary.

Each distinct inflammasome is composed of a unique sensor (i.e., one of NLR family pyrin domain-containing proteins (NLRPs) such as NLRP1, NLRP3, NLRP6 among others), an adaptor (apoptosis-associated speck-like protein containing a

CARD, ASC), and a proinflammatory caspase (e.g., caspase-1 or caspase-4/-5), which interact through homotypic bindings of caspase recruitment domains (CARDs) or PYRIN domains (PYDs) (Zheng et al. 2020). Notably, human caspase-4 and caspase-5 are functional homologs of mouse caspase-11 (Knodler et al. 2014). Generally, the damage-associated molecular patterns (DAMPs) generated by other RCD types could be recognized by the sensors and activates inflammasomes and initiates pyroptosis (Gao et al. 2018; Kaczmarek et al. 2013); thus, there are interconnection between pyroptosis and other types of RCDs (Wu et al. 2020). Recently, quite a number of studies revealed the association between RNA m^{6}A modification and pyroptosis in various physiological and pathological conditions. In this part, we mainly recapitulate the current understanding of the relationship between RNA m^{6}A modification and pyroptosis.

5.1 M^{6}A Modification Promotes Pyroptosis

Liver fibrosis is manifested as the excessive accumulation of extracellular matrix proteins including collagen that occurs in chronic liver injury (Mederacke et al. 2022). It has been reported that pro-inflammatory M1 macrophages contribute to liver fibrosis via activating hepatic stellate cells, but the underlying mechanism is unclear (Krenkel and Tacke 2017). In Kupffer cells (KCs) isolated from liver fibrosis animal model or in vitro M1-polarized macrophages, METTL3 was upregulated, which increased the long non-coding RNA *MALAT1* (metastasis-associated lung adenocarcinoma transcript 1) levels via m^{6}A modification, thereby promoted *USP8* (ubiquitin-specific peptidase 8) mRNA degradation through the interaction with PTBP1 (polypyrimidine tract-binding protein 1). Subsequently, reduction of USP8 expression regulated the ubiquitination and protein stability of TAK1 (transforming growth factor β-activated kinase 1, an essential regulator for pyroptosis in macrophages), which promoted pyroptosis and inflammation of macrophages (Shu et al. 2021). Hence, targeting individual components of METTL3/MALAT1/PTBP1/USP8/TAK1 axis may benefit the treatment of liver fibrosis which merits further clinical investigation.

Cardiomyocyte pyroptosis is one of the major causes behind myocardial ischemia/reperfusion (MI/R) injury. A recent study found that METTL3 expression was elevated in MI/R rats and oxygen–glucose deprivation/reoxygenation (OGD/R) treated cardiomyocytes, which increased the total m^{6}A levels, promoted *DGCR8* (DiGeorge syndrome critical region gene 8) binding to pre-miR-143-3p, and enhanced miR-143-3p expression to suppress *PRKCE* (protein kinase C epsilon) transcription, further aggravating cardiomyocyte pyroptosis and MI/R injury (Wang et al. 2022b). Another study found that high glucose-induced upregulation of WTAP by histone acetyltransferase p300 promotes the m^{6}A methylation of *NLRP3* mRNA to upregulate NLRP3 inflammasome activation in diabetic nephropathy (DN), thereby inducing pyroptosis and inflammation (Lan et al. 2022).

Accumulating evidence suggests a correlation among macrophage pyroptosis, circRNA regulation, and atherosclerosis (AS). IFN regulatory factor-1 (IRF-1) expression reportedly promotes macrophage pyroptosis in patients with acute coronary syndrome (ACS), but the underlying mechanisms remained elusive (Guo et al. 2019; Yang et al. 2019a). One study found that the expression levels of hsa_circ_0029589 were decreased in macrophages taken from patients with ACS, while the levels of m^6A modification on hsa_circ_0029589 were significantly elevated. Further analysis revealed that overexpression of IRF-1 suppresses hsa_circ_0029589 expression via promoting its m^6A modification by upregulating METTL3 in macrophages, leading to pyroptosis. Accordingly, either overexpression of hsa_circ_0029589 or inhibition of METTL3 significantly increased the expression of hsa_circ_0029589 and attenuated macrophage pyroptosis (Guo et al. 2020). Thus, increasing hsa_circ_0029589 expression via modulating m^6A modification could be a potential approach to alleviate macrophage pyroptosis in ACS patients. These observations suggest a novel mechanism by which IRF-1 facilitates macrophage pyroptosis in ACS and AS by inhibiting circ_0029589 through promoting its m^6A modification.

5.2 *M^6A Modification Inhibits Pyroptosis*

Currently, the underlying regulatory mechanism of m^6A in diabetic cardiomyopathy (DCM) is very limited. A recent study found that METTL14 downregulates lncRNA *TINCR* (terminal differentiation-induced non-coding RNA) in an m^6A-dependent manner, which further decreases the expression of key pyroptosis-related protein NLRP3 and suppresses pyroptosis in cardiomyocyte to alleviate DCM (Meng et al. 2022). Mechanistically, METTL14 increased the m^6A methylation levels of lncRNA *TINCR*, resulting in downregulation of *TINCR* via the m^6A reader protein YTHDF2. Conversely, downregulation of METTL14 in cardiomyocytes and heart tissues promoted pyroptosis and DCM progression. These findings reveal the novel role of METTL14-mediated m^6A methylation in lncRNA regulation and pyroptosis to influence DCM, which extends the understanding regarding involvement of the pyroptosis epigenetic regulation in DCM progression. Thus, key m^6A modifiers can protect cells from pyroptosis via blocking inflammasome activation.

Sepsis is a potentially life-threatening illness that initiated by the body's extreme response (damaging its own tissues) to an infection (Singer et al. 2016), and pyroptosis plays a crucial role in sepsis (Gao et al. 2018). A recent study found that mRNA levels of *WWP1* (WW domain containing E3 ubiquitin protein ligase 1) and *YTHDF1* were downregulated in clinical sepsis samples, as well as cell line and mice models, which augmented pyroptosis-dependent sepsis. Mechanistically, YTHDF1 downregulation resulted in a reduction of WWP1 translation in an m^6A-dependent manner, thereby inhibited ubiquitination of NLRP3, through which increased NLRP3 protein levels and promoted pyroptosis (Zhang et al. 2022c).

Taken together, upon various pathological conditions, alteration in expression of m^6A modifiers resulted in changes in RNA m^6A modification levels of key

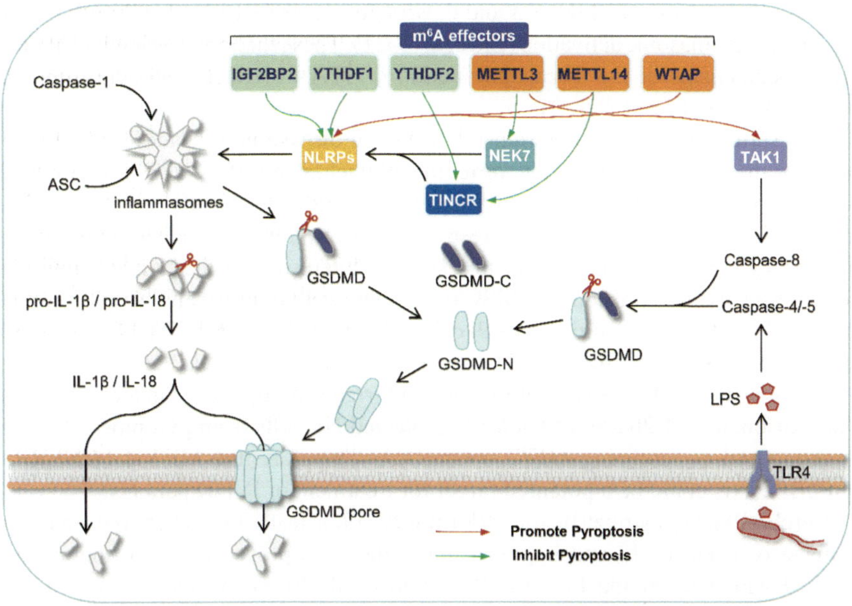

Fig. 3 Schematic diagram showing roles of m^{6}A modification in regulating pyroptosis. Pyroptosis is driven by the pore-forming fragment of (active) gasdermin (GSDM) protein family members, which is proteolytically cleaved by inflammatory caspases in response to activation of the inflammasome. Various m^{6}A modification-related molecules regulate the fate of key molecules in pyroptosis pathway through different biological functions, and promote or inhibit pyroptosis. Abbreviations: NLRPs, NOD-like receptor thermal protein domain associated proteins; NEK7, NIMA (never in mitosis gene A)-related kinase 7; TINCR, terminal differentiation-induced cornification regulator; TAK1, transforming growth factor-beta (TGF-beta)-activated kinase 1; ASC, apoptosis-associated speck-like protein containing a CARD; GSDMD, Gasdermin D; TLR4, Toll-like receptor 4

pyroptosis-related genes to regulate pyroptosis, which would greatly affect the disease progression. Thus, targeting cellular m^{6}A machinery represent a feasible new strategy for coping with such m^{6}A-mediated pyroptosis-related diseases (Fig. 3).

5.3 Treatment of Pyroptosis-Mediated Adverse Effects by Regulating m^{6}A Modification

NIMA (never in mitosis gene A)-related kinase 7 (NEK7), a protein that binds to the leucine-rich repeat domain of NLRP3 in a kinase-independent manner, is reportedly involved in activation of the NLRP3 inflammasome (He et al. 2016). A study indicated that the total flavones of abelmoschus manihot (TFA) could ameliorate pyroptosis and injury in podocytes under high glucose conditions by upregulating METTL3-dependent m^{6}A modification and stabilization of *PTEN* mRNA, which leads to deactivation of PI3K-Akt and downregulation of NEK7, thereby inhibiting NLRP3 inflammasome activation (Liu et al. 2021). These findings revealed the underlying mechanisms of the therapeutic action of TFA to reduce podocyte injury in diabetic kidney disease (DKD).

Intervertebral disc degeneration (IVDD) is a disease manifested as the breakdown of the discs supporting the vertebrae (Ravichandran et al. 2022). It is one of the most frequent causes of severe back pain worldwide, but the current clinical interventions for IVDD are mainly restricted to symptom release. Therefore, novel therapeutic options are needed in this setting. Interestingly, the human umbilical cord mesenchymal stem cell (hucMSC) exosomes were found to effectively improve the viability of nucleus pulposus (NP) cells. Mechanistically, METTL14 was highly expressed in NP cells from IVDD patients, which stabilize *NLRP3* mRNA in an m^6A reader IGF2BP2-dependent manner and trigger pyroptosis (Yuan et al. 2021). The exosomal miR-26a-5p of hucMSC protected NP cells from pyroptosis through targeting and destabilizing *METTL14* mRNA. Therefore, the METTL14/NLRP3/IGF2BP2 axis could be a potential target for treatment of IVDD patients.

Cerebral ischemia/reperfusion (I/R) injury often leads to irreversible neuronal damage or even death, and hypothermia is the only proven effective therapeutic method with unclear mechanisms (Donnino et al. 2015). A study observed that hypothermia remarkably increased cellular viability, reduced the expression of pyroptosis-related proteins including NLRP3, ASC, cleaved caspase-1, and GSDMD p30, along with decreased secretion of the pro-inflammatory cytokines, IL-1β and IL-18 in primary hippocampal neurons. Mechanistically, hypothermia resulted in increased m^6A modification on *PTEN* mRNA and promoted its protein expression, which inhibited PI3K-AKT-GSK3β signaling pathway to inhibit NLRP3 inflammasome activation (Diao et al. 2020). Sepsis brain injury (SBI) is one of the major causes of death in critically ill patients (Wang et al. 2019), and emodin shows a therapeutic effect on SBI with unknown mechanism. Recently, it was reported that emodin inhibits the inflammation and pyroptosis of LPS-treated 1321N1 cells by increasing METTL3 expression to reduce *NLRP3* mRNA level in an m^6A-dependent manner (Wang et al. 2022b).

Collectively, in most non-cancer disease types, activation of pyroptosis and inflammation promotes disease progression. Several drugs or physical interventions inhibit inflammasome activation via altering m^6A modification to reduce pyroptosis and exert therapeutic efficacy. This new layer of pyroptosis regulation by affecting m^6A modification via drug/physical treatments merits further investigation.

6 Ferroptosis

Erastin was first discovered in 2003 (Dolma et al. 2003); like camptothecin (CPT), erastin can selectively kill tumor cells expressing RasV12 protein, but their lethal mechanisms are different: CPT induced cell death is manifested by DNA fragmentation and caspase protein family activation, which can be repressed by caspase inhibitors; however, there is no typical apoptotic phenomenon in erastin-induced cell death, and cannot be reversed by caspase inhibitors, suggesting that this is a novel non-apoptotic form of cell death (Yagoda et al. 2007). In 2008, two new compounds RSL3 and RSL5 were found to induce iron-dependent non-apoptotic cell death, similar to erastin, in tumor cells containing constitutively active mutant RAS family of small GTPases (HRAS, NRAS, and KRAS), and this kind of cell death could be inhibited by iron chelators (Yang and Stockwell 2008). Based on the unique phenotype of death manifestation, this type of cell death was officially termed ferroptosis in 2012 (Dixon et al. 2012). Morphologically, ferroptosis is characterized by reduction of mitochondrial volume, reduction or disappearance of mitochondrial cristae, and mitochondrial membrane rupture, but no significant changes in chromatin. Biochemically, ferroptosis is characterized by iron-dependent accumulation of lipid peroxides, leading to disruption of the lipid bilayer and ultimately cell death (Stockwell 2022). The free iron ions released into the cytoplasm by the divalent metal transporter protein (DMT1) interact with hydrogen peroxide and other peroxides through Fenton reaction, releasing a large quantity of reactive oxygen species (ROS). This leads to extensive peroxidation of polyunsaturated fatty acids (PUFAs) that make up the biofilm, resulting in the destruction of the biofilm, such as the rupture of the outer mitochondrial membrane, and eventually cell death.

Population-based correlation analyses showed that m^{6}A-related regulatory genes were highly associated with the expression of ferroptosis-associated genes in various disease models such as cancers and acute kidney injury (Ni et al. 2022). On the other hand, disordered levels of m^{6}A modification were observed in ferroptotic cells, suggesting a potential regulatory relationship between m^{6}A modification and ferroptosis. Therefore, sorting out and summarizing the regulatory network of m^{6}A modification and ferroptosis is necessary for a deeper understanding of the related fields.

6.1 M^{6}A Modification Promotes Ferroptosis

RNA m^{6}A modification is dynamic and reversible, and different abundance of m^{6}A modifications on a specific transcript would induce distinct biological outcomes of m^{6}A modified RNA. Growing number of evidences reveal a correlation between increased levels of m^{6}A modification and induction of ferroptosis in several scenarios. In an in vitro model of doxorubicin (DOX) induced cardiomyocyte ferroptosis, the overall levels of m^{6}A modification and the expression of METTL14 were significantly upregulated compared to the control group (Zhuang et al. 2021). Another study reported that in HER2-positive breast cancer cells, the overall m^{6}A levels in trastuzumab-resistant cells were significantly decreased compared to that in trastuzumab-sensitive cells (Zou et al. 2022). Mechanistically, m^{6}A reader YTHDC2 downregulates *FGFR4* (fibroblast growth factor receptor 4) mRNA stability in an m^{6}A-dependent manner, which in turn causes ferroptosis by inhibiting cystine and iron (Fe^{2+}) uptake through the β-catenin/TCF4-SLC7A11/FPN1 axis.

The balance of intracellular lipid ROS production and iron homeostasis is essential for cell survival, and an imbalance of either may induce ferroptosis. Coenzyme Q (CoQ) oxidoreductase FSP1 was found in 2019 to inhibit ferroptosis independently of the previously described glutathione peroxidase (GPX4) pathway, which prevents accumulation of membrane phospholipid hydroperoxides using glutathione (GSH) as a cofactor (Bersuker et al. 2019; Doll et al. 2019). Song et al. (2021) reported that inhibition of miR-4443 downregulates FSP1 protein levels in an m^{6}A-dependent manner through METTL3, and induces the accumulation of lipid peroxides in non-small cell lung cancer cells, leading to ferroptosis. As mentioned above, iron homeostasis imbalance is another important factor inducing ferroptosis (Song et al. 2021). Shen et al. (2021) found that YTHDF1 promotes *BECN1* mRNA stability in an m^{6}A-dependent manner, which enhances autophagic ferritin degradation (ferritinophagy), induces excess iron release, and ultimately leads to ferroptosis in hepatic stellate cells. Thus, induction of m^{6}A modification-dependent ferroptosis is a potential treatment strategy for the management of liver fibrosis (Shen et al. 2021).

6.2 M^{6}A Modification Inhibits Ferroptosis

The biological function of m^{6}A modification is executed by different m^{6}A-binding proteins (m^{6}A readers), which confer m^{6}A modification varying functions. Therefore, through context-dependent manners, m^{6}A modification multifariously regulates ferroptosis. Scilicet, m^{6}A modification could both promote and suppress ferroptosis dependent on distinct physiological and pathological conditions. Cysteine is an important amino acid with multiple roles in protein synthesis and redox maintenance (Combs and DeNicola 2019). Most cancer cells rely on the cystine transporter system X_c^- (consisting of the catalytic subunit SLC7A11 and the chaperone subunit SLC3A2) to import cystine from the extracellular environment, which is then converted to cysteine (the rate-limiting substrate for glutathione (GSH) synthesis) by a reduction reaction that consumes NADPH (Stipanuk et al. 2006). Therefore, SLC7A11-mediated cystine uptake plays a critical role in suppressing oxidative responses and maintaining cell survival under oxidative stress conditions. Increasing evidence has found that ferroptosis can be induced by downregulating the expression of SLC7A11 (Koppula et al. 2021). More importantly, negative or positive regulation of *SLC7A11* mRNA by m^{6}A modification was found in various diseases. For instance, METTL3/IGF2BP1 increases ferroptosis resistance in hepatoblastoma cells by inhibiting *SLC7A11* mRNA deadenylation and promoting its mRNA stability in an m^{6}A-dependent manner (Liu et al. 2022a). In contrast, m^{6}A demethylase FTO promotes ferroptosis in papillary thyroid carcinoma cells by downregulating *SLC7A11* mRNA in an m^{6}A-dependent manner (Ji et al. 2022). Furthermore, NKAP (NF-κB activating protein), a recently identified potential m^{6}A binding protein (Zhang et al. 2019), was found to promote *SLC7A11* mRNA splicing and maturation, through which inhibited glioblastoma cells' ferroptosis (Sun et al. 2022).

In addition to regulating ferroptosis-inhibiting pathways such as SLC7A11/GPX4, m^{6}A modification can also cause ferroptosis resistance by negatively regulating ferroptosis-promoting pathways. ACSL4 (Acyl-CoA synthetase long-chain family member 4) synergistically triggers ferroptosis by synthesizing easily oxidizable membrane phospholipids using arachidonic acid (AA) as a substrate (Yuan et al. 2016). YTHDF2 indirectly promotes the ubiquitination and degradation of ACSL4 in an m^{6}A-dependent manner, enhancing ferroptosis resistance in gastric cancer (Yang et al. 2022b). Collectively, m^{6}A modification could modulate both ferroptosis-promoting and inhibiting pathways to regulate this type of RCD (Fig. 4).

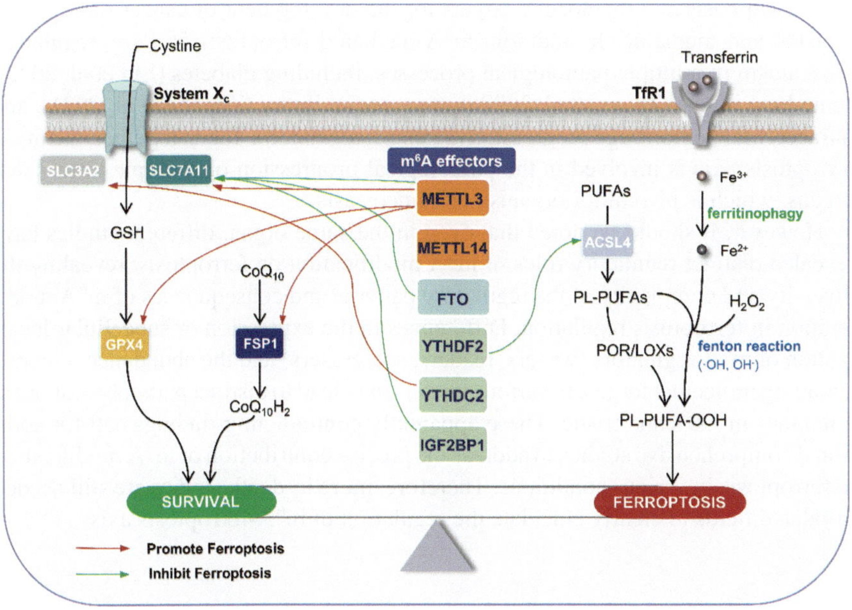

Fig. 4 Schematic diagram showing roles of m^{6}A modification in regulating ferroptosis. The core molecular mechanism of ferroptosis is the regulation of the balance between oxidative damage and antioxidant defense. The free iron ions released into the cytoplasm by transferrin and divalent metal transporter catalyze PUFAs to generate excessive lipid peroxides through Fenton reaction, inducing ferroptosis. On the other hand, three major systems resisting iron-dependent cell death include cysteine/GSH/GPX4 axis as well as GPX4 independent pathways such as NAD(P)H/FSP1/CoQ_{10} axis. Various m^{6}A modification-related molecules regulate the fate of key participants in ferroptosis pathways through different m^{6}A-dependent biological functions, and promote or inhibit ferroptosis. Abbreviations: System X_c^-, cystine/glutamate antiporter system; TfR1, Transferrin receptor 1; PUFAs, polyunsaturated fatty acids; PL-PUFAs, phospholipid-bound polyunsaturated fatty acids; SLC3A2, solute carrier family 3 member 2; SLC3A11, solute carrier family 3 member 11; ACSL4, acyl-CoA synthetase long-chain family member 4; POR, cytochrome P450 oxidoreductase; LOX, lipoxygenases; PL-PUFA-OOH, phospholipid hydroperoxide; GPX4, glutathione peroxidase 4; CoQ_{10}, Coenzyme Q, also known as ubiquinone; FSP1, ferroptosis suppressor protein 1

6.3 M^{6}A Modification Regulates Ferroptosis in Various Diseases

As a new type of RCD, ferroptosis has just been extensively studied in recent years, and the mechanisms of ferroptosis regulation have been continuously elucidated. Its important regulatory role in various diseases has been paid more and more attention. In particular, m^{6}A modification plays an important role in regulating the occurrence of ferroptosis in a variety of tumors and cancers, such as breast cancer (Zou et al. 2022), lung cancer (Song et al. 2021), gastric cancer (Yang et al. 2022b), leukemia (Du et al. 2021), and thyroid cancer (Ji et al. 2022). These studies provide a potential theoretical basis and therapeutic targets for the development of cancer biology and translational medicine. In addition, m^{6}A-mediated ferroptosis is a key regulatory mechanism in multiple pathological processes, including diabetes (Lin et al. 2022), acute lung injury (Zhang et al. 2022a), myocardial injury (Zhuang et al. 2021), and intracerebral hemorrhage progression (Zhang et al. 2022b). Taken together, the m^{6}A/ferroptosis axis is involved in the pathological progression of multiple tissues and organs, which is pivotal for organismal homeostasis.

However, it should be noted that even in the same organ, different studies have revealed distinct regulatory roles of m^{6}A modification on ferroptosis, revealing the diversity and complexity of the regulatory patterns and consequences of m^{6}A modification in ferroptosis regulation. Differences in the expression or subcellular localization of m^{6}A regulators (writers, readers, and erasers) and the abundance of corresponding molecular targets in various models may lead to distinct pathophysiological outcomes in the same tissue. These apparently contradicting findings call for additional comprehensive studies to address the precise contribution of m^{6}A modification to ferroptosis in various conditions. Therefore, more in-depth studies are still needed in related fields to clearly elucidate the regulation of m^{6}A-ferroptosis axis.

7 Concluding Remarks

Regulated cell death occurs as a planned regulatory response of cells to stimuli in order to maintain homeostasis of tissue morphology and function (Galluzzi et al. 2018). To date, more than 150 forms of RNA modifications on endogenous RNAs (including mRNAs, rRNAs, tRNAs, snRNAs, etc.) have been identified within cells (Li et al. 2020b). Characterized by its dynamically reversible nature, m^{6}A modification finely tunes many of the links in the cell death cascade to modulate different types of regulated cell death. In addition to m^{6}A modification, other RNA modifications, such as m^{1}A modification, are also able to influence RCD by regulating the fate of committed RNA species (Chen et al. 2019). Given the high abundance of m^{6}A modification in eukaryotic organisms, this chapter provides a paradigm and reference for the m^{6}A epi-marker regulation of RCDs. The key factors regulating RCD by m^{6}A first depend on the abundance of the m^{6}A modification. The expression levels and enzymatic activities of m^{6}A methyltransferases or demethylases can regulate the overall abundance of m^{6}A to affect the metabolism of multiple key molecules in the regulatory pathways of various RCD subroutines (Table 2).

Besides the abovementioned RCD types, there are many other RCD subroutines such as lysosomal cell death (Aits and Jäättelä 2013), entosis (Overholtzer et al. 2007), parthanatos (Andrabi et al. 2008), autosis (Liu and Levine 2015), oxeiptosis (Holze et al. 2018), alkaliptosis (Song et al. 2018) and the recently identified cuproptosis (Tsvetkov et al. 2022). However, currently, there are no literatures available regarding the interplay between m^{6}A modification and these RCD types. Hence, future studies are recommended to fill in this gap to better understand whether and how m^{6}A-mediated mechanisms participate in regulation of these cell death types.

Notably, in some stress conditions, RCD can also be regulated by specific position of m^{6}A modification on particular mRNA molecules. Thus, simply modulating total m^{6}A levels by affecting methyltransferases or demethylases may not be suitable for selectively intervening a particular m^{6}A-modified mRNA. To address this concern, modern techniques at single methylation site resolution such as dCas13-based strategies to specifically alter the levels of m^{6}A modification at specific positions on target molecules are developed and optimized (Su et al. 2021; Yang et al. 2019b). Intervening m^{6}A modification of key RCD regulatory molecules by tools containing dCas13 fused with m^{6}A methyltransferases (i.e., METTL3, METTL14) or demethylases (i.e., ALKBH5, FTO) is a promising approach to modulate the progress of RCD which merits further investigation.

On the other hand, the expression level and localization of m^{6}A binding proteins is another catalog of key factors in regulating RCD. The fate of m^{6}A-modified RNA basically depends on its m^{6}A binding protein. Different binding proteins have different biological functions, and the same binding protein can also perform distinct functions under varying contexts. In addition, differences in the cellular microenvironment can affect the subcellular localization of m^{6}A-binding proteins. For instance, heat shock-induced nuclear translocation of YTHDF2 preserves 5'UTR methylation of stress-inducible transcripts by inhibiting demethylation by FTO (Zhou et al. 2015). Furthermore, the nuclear translocation of YTHDF2 under heat shock conditions results in decreased *Notch1* mRNA decay and inhibition of apoptosis (Lee et al. 2021).

In conclusion, the regulatory modes of RNA m^{6}A modification appear diverse due to its dynamic and reversible properties. In most cases, researchers only focus on the regulatory mechanism of a specific m^{6}A modifying protein on a single mRNA molecule of a specific RCD type. However, it is worth noting that different cell death modalities are also deeply interconnected (Bedoui et al. 2020). As the most extensive mRNA modification, m^{6}A determines the fate of numerous coding and non-coding RNAs, so it is likely that a single m^{6}A regulator affects more than one RCD via multiple bioprocesses in the same organism. Last but not least, extrapolation of experimental conclusions based on cell lines for cell death studies also requires extra caution due to the fact that cell line-based experiments are deviated from the corresponding microenvironment in vivo. Thus, the research in the field of m^{6}A

Table 2 Molecular mechanisms of RCD regulation by RNA m^6A modification

Types of RCD	Regulation	m^6A modulators	Regulatory mechanisms	Disease models	Refs.
Apoptosis	Inducing apoptosis	FTO/ YTHDF2	Inhibition of FTO activity increased m^6A levels, through which improved the stability of *MYC /CEBPA* transcripts, leading to apoptosis induction	Pan-cancer	(Su et al. 2018)
		ALKBH5	Deficiency of ALKBH5 increased m^6A modification, leading to aberrant spermatogenesis and apoptosis in mouse testis	Mouse infertility	(Zheng et al. 2013)
		METTL3	METTL3 promoted m^6A modification of *TFEB* mRNA and decreased TFEB level via m^6A reader HNRNPD, thereby inhibiting autophagy and enhancing apoptosis	H/R-treated cardiomyocytes	(Song et al. 2019)
		ALKBH5	m^6A modification on *TRAF1* transcripts decreased its mRNA stability to promote apoptosis	Multiple myeloma	(Qu et al. 2022)
	Inhibiting apoptosis	METTL3 /YTHDF2	METTL3 promoted m^6A modification of *PTEN* mRNA and decreased its stability via YTHDF2, thereby inhibited apoptosis	Prostatic hyperplasia	(Li et al. 2022a)
		METTL14	m^6A modification regulated key genes in the p53-mediated apoptosis signaling pathway, leading to resistance to apoptosis	Invariant natural killer T cells	(Cao et al. 2022)
		WTAP	Circ0008399/WTAP complex increased expression of TNFAIP3 by increasing its mRNA stability in an m^6A-dependent manner, leading to inhibition of bladder cancer cell apoptosis	Bladder cancer	(Wei et al. 2021)

(continued)

Table 2 (continued)

Types of RCD	Regulation	m^{6}A modulators	Regulatory mechanisms	Disease models	Refs.
		ALKBH5	m^{6}A methylation of *KCNK15-AS1* decreased its stability, thereby promoted pancreatic cancer progression and apoptosis resistance via regulating KCNK15 and PTEN/AKT signaling	Pancreatic cancer	(He et al. 2021)
		METTL3 /WTAP	Knockdown of WTAP/METTL3 resulted in reduction of global m^{6}A modification, and induced apoptosis along with defective tissue differentiation in zebrafish embryos	Tissue differentiation	(Ping et al. 2014)
		YTHDF2	m^{6}A modification decreased *TNFRSF1B* mRNA stability via YTHDF2 to inhibit TNF induced apoptosis	AML cells	(Paris et al. 2019)
		YTHDF2	Inhibition of YTHDF2 triggered apoptosis via m^{6}A dependent manner in MYC-driven breast cancer	Triple-negative breast cancer	(Einstein et al. 2021)
		METTL14	m^{6}A modification regulated the stability of *NFKBIA* mRNA and modulated the NF-κB pathway to impede colonic cell death	Inflammatory bowel disease	(Zhang et al. 2022d)
Autophagy	Inducing autophagy	FTO/ YTHDF2	YTHDF2 captured *eIF4G1* transcripts containing m^{6}A, resulting in mRNA degradation, thereby promoting autophagy	Oral squamous cell carcinoma	(Wang et al. 2021a)
		YTHDC1	m^{6}A reader YTHDC1 promotes autophagy by positively regulating *SQSTM1* mRNA stability in an m^{6}A dependent manner	Diabetic skin	(Liang et al. 2022)
		METTL4/ FTO/ YTHDF1	YTHDF1 promotes *BECN1* mRNA stability and autophagy activation via recognizing the m^{6}A binding site within *BECN1* coding regions	Liver fibrosis	(Shen et al. 2021)

(continued)

Table 2 (continued)

Types of RCD	Regulation	m^6A modulators	Regulatory mechanisms	Disease models	Refs.
	Inhibiting autophagy	ALKBH5/ YTHDF2	m^6A modification on *FIP200* led to YTHDF2-mediated mRNA decay, thereby inhibiting autophagic flux	Intervertebral disc degeneration	(Li et al. 2020a)
		FTO/ YTHDF2	m^6A modification destabilized *ATG5* and *ATG7* transcripts via YTHDF2, thereby inhibited autophagy	Adipogenesis	(Wang et al. 2020b)
		FTO/ YTHDF2	m^6A modification on *ATF4* mRNA led to YTHDF2-mediated RNA decay, thereby inhibiting ATF4 dependent autophagy	Colorectal cancer	(Han et al. 2021)
		METTL14	m^6A modification on *SIRT1* mRNA increased its protein levels and resulted in the activation of autophagy	Proteinuric kidney disease	(Lu et al. 2021)
		METTL14/ ALKBH5/ YTHDF1	m^6A modification promoted translation of *PPM1A* but decreased RNA stability of *CAMKK2*, resulted in reduced AMPK activity and subsequent autophagy inhibition	Testosterone synthesis in Leydig cells	(Chen et al. 2021)
		METTL3	PDGF ligands stimulated *EGR1* transcription to upregulate METTL3, increasing *OPTN* mRNA decay to inhibit mitophagy	Glioblastoma stem cells	(Lv et al. 2022)
Necroptosis	Inhibiting necroptosis	METTL3	METTL3 mediated m^6A modification on *TRAF5* gene and decreased its protein expression, leading to resistance of colorectal cancer cells to necroptosis	Colorectal cancer	(Lan et al. 2021)
Pyroptosis	Inducing pyroptosis	METTL3	METTL3 increased lncRNA *MALAT1* level via m^6A, promoted *USP8* mRNA decay, and decreased its protein expression, thereby stabilizing TAK1 protein to promote pyroptosis	Liver fibrosis	(Shu et al. 2021)

(continued)

Table 2 (continued)

Types of RCD	Regulation	m^{6}A modulators	Regulatory mechanisms	Disease models	Refs.
		METTL3	IRF-1 suppressed hsa_circ_0029589 expression via promoting its m^{6}A modification by upregulating METTL3 in macrophages, leading to pyroptosis	Atherosclerosis and acute coronary syndrome	(Guo et al. 2020)
		METTL3	METTL3 promoted DGCR8 binding to pre-miR-143-3p via m^{6}A modification, and enhanced miR-143-3p expression to suppress *PRKCE* transcription, aggravating cardiomyocyte pyroptosis	Myocardial ischemia/reperfusion (MI/R) injury	(Wang et al. 2022b)
	Inhibiting pyroptosis	YTHDF1	m^{6}A modification promotes *WWP1* translation in an YTHDF1 dependent manner, thereby increasing ubiquitination of NLRP3 and inhibiting pyroptosis	Sepsis	(Zhang et al. 2022c)
		METTL14/YTHDF2	m^{6}A modification decreased lncRNA *TINCR* level via YTHDF2, which further decreased the expression of NLRP3	Diabetic cardiomyopathy	(Meng et al. 2022)
Ferroptosis	Inducing ferroptosis	YTHDF1	YTHDF1 promoted *BECN1* mRNA stability in an m^{6}A-dependent manner and activated ferritinophagy	Liver fibrosis	(Shen et al. 2021)
		METTL14/IGF2BP1	DOX upregulated METTL14 and increased the stability of lncRNA *KCNQ1OT1* through IGF2BP1, sponging miR-7-5p to increase transferrin level	Cardiomyopathy	(Zhuang et al. 2021)
		METTL3	HGHF increased m^{6}A methylation on *ASK1*, triggering ferroptosis-related osteoblastic cell death through ASK1-p38 pathway	Diabetic bone loss	(Lin et al. 2022)
		METTL3/YTHDF2	NETs treatment resulted in METTL3 upregulation, leading to YTHDF2-mediated *GPX4* mRNA decay and ferroptosis	Acute lung injury	(Zhang et al. 2022a)

(continued)

Table 2 (continued)

Types of RCD	Regulation	m^6A modulators	Regulatory mechanisms	Disease models	Refs.
		METTL3	METTL3 overexpression decreased *SLC7A11* and *FSP1* mRNAs' stability to promote ferroptosis	Aortic dissection	(Li et al. 2022b)
		METTL14/ YTHDC2	Extensive m^6A modification induces ferroptosis through YTHDC2-mediated downregulation of *FGFR4* stability	Breast cancer	(Zou et al. 2022)
	Inhibiting ferroptosis	YTHDF2	lncRNA-*CBSLR* recruited YTHDF2 protein, decreasing *CBS* mRNA stability in an m^6A-dependent manner to block ferroptosis	Gastric cancer	(Yang et al. 2022b)
		METTL3/ YTHDF1	YTHDF1 stabilized *SLC7A11* mRNA and promoted its translation, thus increasing LUAD cell proliferation and inhibiting ferroptosis	Lung adenocarcinoma	(Xu et al. 2022)
		METTL3/ IGF2BP1	IGF2BP1 promoted *SLC7A11* mRNA stability and increased its expression, enhancing hepatoblastoma ferroptosis resistance	Hepatoblastoma	(Liu et al. 2022a)
		NKAP	RNA binding protein NKAP protects glioblastoma cells from ferroptosis by promoting *SLC7A11* mRNA splicing in an m^6A-dependent manner	Glioblastoma	(Sun et al. 2022)

Abbreviations: TFEB, the transcription factor EB; HNRNPD, heterogeneous nuclear ribonucleoprotein D; TRAF1, TNF receptor associated factor 1; PTEN, phosphatase and tensin homolog; TNFAIP3, TNF alpha induced protein 3; KCNK15-AS1, lncRNA KCNK15 and WISP2 antisense RNA 1; TNFRSF1B, TNF receptor superfamily member 1B; NFKBIA, NFKB Inhibitor Alpha; eIF4G1, eukaryotic translation initiation factor 4 gamma 1; SQSTM1, sequestosome 1, also known as p62; BECN1, beclin 1; FIP200, focal adhesion kinase family interacting protein of 200 kDa; SIRT1, sirtuin 1; PPM1A, magnesium-dependent phosphatase 1 A; CAMKK2, calcium/calmodulin-dependent protein kinase 2; PDGF, platelet-derived growth factors; MALAT1, metastasis associated lung adenocarcinoma transcript 1; USP8, ubiquitin specific peptidase 8; TAK1, transforming growth factor β-activated kinase 1; IRF-1, Interferon regulatory factor 1; DGCR8, DiGeorge syndrome critical region gene 8; PRKCE, protein kinase C epsilon type; WWP1, WW domain-containing E3 ubiquitin protein ligase 1; TINCR, terminal differentiation-induced non-coding RNA; KCNQ1OT1, potassium voltage-gated channel (KCNQ1) opposite strand/antisense transcript 1; HGHF, high glucose and high fat; NETSs, neutrophil extracellular traps; ASK1, apoptosis signal-regulating kinase 1; GPX4, glutathione peroxidase 4; FSP1, ferroptosis suppressor protein 1; SLC7A11, Solute carrier family 7 member 11; FGFR4, fibroblast growth factor receptor 4; CBS, Cystathionine-beta-Synthase; LUAD, lung adenocarcinoma; NKAP, NFKB activating protein

-RCD is still in the early stage. Whether there is regulatory rule/law of m^{6}A modification on RCD and how other RNA modifications participate in RCD regulation still needs further exploration. In the past few decades, corresponding clinical treatment strategies for each key target of RCD have been continuously launched (Zeng et al. 2022), and small molecule inhibitors targeting key m^{6}A regulators have also been gradually developed (Gu et al. 2020). Teasing out the network and regulatory modes of m^{6}A modification in regulating RCD not only broadens our understanding of RNA modification but also reveals a new layer of intervening approach regarding regulated cell death. More importantly, the m^{6}A-RCD regulatory network provides us with new inspiration in developing disease prevention and treatment strategies; namely, it provides more novel treatment options for diseases such as acute organ injury that need to prevent excessive RCD, and for fields such as cancer therapy that need to promote RCD.

References

Aits S, Jäättelä M (2013) Lysosomal cell death at a glance. J Cell Sci 126:1905–1912

Alharbi YM, Bima AI, Elsamanoudy AZ (2021) An Overview of the Perspective of Cellular Autophagy: Mechanism, Regulation, and the Role of Autophagy Dysregulation in the Pathogenesis of Diseases. J Microsc Ultrastruct 9:47–54

Anding AL, Baehrecke EH (2015) Autophagy in Cell Life and Cell Death. Curr Top Dev Biol 114:67–91

Andrabi SA, Dawson TM, Dawson VL (2008) Mitochondrial and nuclear cross talk in cell death: parthanatos. Ann NY Acad Sci 1147:233–241

Ashford TP, Porter KR (1962) Cytoplasmic components in hepatic cell lysosomes. J Cell Biol 12:198–202

Batista PJ, Molinie B, Wang J et al (2014) m(6)A RNA modification controls cell fate transition in mammalian embryonic stem cells. Cell Stem Cell 15:707–719

Bedoui S, Herold MJ, Strasser A (2020) Emerging connectivity of programmed cell death pathways and its physiological implications. Nat Rev Mol Cell Biol 21:678–695

Bergmann A, Steller H (2010) Apoptosis, stem cells, and tissue regeneration. Sci Signal 3:re8

Bertheloot D, Latz E, Franklin BS (2021) Necroptosis, pyroptosis and apoptosis: an intricate game of cell death. Cell Mol Immunol 18:1106–1121

Cao L, Morgun E, Genardi S et al (2022) METTL14-dependent m(6)A modification controls iNKT cell development and function. Cell Rep 40:111156

Carneiro BA, El-Deiry WS (2020) Targeting apoptosis in cancer therapy. Nat Rev Clin Oncol 17:395–417

Chen Q, Kang J, Fu C (2018) The independence of and associations among apoptosis, autophagy, and necrosis. Signal Transduct Target Ther 3:18

Chen Y, Wang J, Xu D et al (2021) m(6)A mRNA methylation regulates testosterone synthesis through modulating autophagy in Leydig cells. Autophagy 17:457–475

Chen Z, Qi M, Shen B et al (2019) Transfer RNA demethylase ALKBH3 promotes cancer progression via induction of tRNA-derived small RNAs. Nucleic Acids Res 47:2533–2545

Combs J A, DeNicola G M (2019) The Non-Essential Amino Acid Cysteine Becomes Essential for Tumor Proliferation and Survival. Cancers (Basel) 11:

Cui YH, Yang S, Wei J et al (2021) Autophagy of the m(6)A mRNA demethylase FTO is impaired by low-level arsenic exposure to promote tumorigenesis. Nat Commun 12:2183

DeBerardinis RJ (2020) Tumor Microenvironment, Metabolism, and Immunotherapy. N Engl J Med 382:869–871

Diao MY, Zhu Y, Yang J et al (2020) Hypothermia protects neurons against ischemia/reperfusion-induced pyroptosis via m6A-mediated activation of PTEN and the PI3K/Akt/GSK-3β signaling pathway. Brain Res Bull 159:25–31

Dixon SJ, Lemberg KM, Lamprecht MR et al (2012) Ferroptosis: an iron-dependent form of nonapoptotic cell death. Cell 149:1060–1072

Dolma S, Lessnick SL, Hahn WC et al (2003) Identification of genotype-selective antitumor agents using synthetic lethal chemical screening in engineered human tumor cells. Cancer Cell 3:285–296

Dominissini D, Moshitch-Moshkovitz S, Schwartz S et al (2012) Topology of the human and mouse m6A RNA methylomes revealed by m6A-seq. Nature 485:201–206

Donnino MW, Andersen LW, Berg KM et al (2015) Temperature Management After Cardiac Arrest: An Advisory Statement by the Advanced Life Support Task Force of the International Liaison Committee on Resuscitation and the American Heart Association Emergency Cardiovascular Care Committee and the Council on Cardiopulmonary, Critical Care, Perioperative and Resuscitation. Circulation 132:2448–2456

Du J, Sarkar R, Li Y et al (2022) N(6)-adenomethylation of GsdmC is essential for Lgr5(+) stem cell survival to maintain normal colonic epithelial morphogenesis. Dev Cell 57:1976-1994.e1978

Du Y, Han M, Cao K et al (2021) Gold Nanorods Exhibit Intrinsic Therapeutic Activity via Controlling N6-Methyladenosine-Based Epitranscriptomics in Acute Myeloid Leukemia. ACS Nano 15:17689–17704

Einstein JM, Perelis M, Chaim IA et al (2021) Inhibition of YTHDF2 triggers proteotoxic cell death in MYC-driven breast cancer. Mol Cell 81:3048-3064.e3049

Feng M, Xie X, Han G et al (2021) YBX1 is required for maintaining myeloid leukemia cell survival by regulating BCL2 stability in an m6A-dependent manner. Blood 138:71–85

Feng Y, He D, Yao Z et al (2014) The machinery of macroautophagy. Cell Res 24:24–41

Galluzzi L, Bravo-San Pedro JM, Blomgren K et al (2016a) Autophagy in acute brain injury. Nat Rev Neurosci 17:467–484

Galluzzi L, Bravo-San Pedro JM, Kepp O et al (2016b) Regulated cell death and adaptive stress responses. Cell Mol Life Sci 73:2405–2410

Galluzzi L, Vitale I, Aaronson SA et al (2018) Molecular mechanisms of cell death: recommendations of the Nomenclature Committee on Cell Death 2018. Cell Death Differ 25:486–541

Gao YL, Zhai JH, Chai YF (2018) Recent Advances in the Molecular Mechanisms Underlying Pyroptosis in Sepsis. Mediators Inflamm 2018:5823823

Gong Y, Fan Z, Luo G et al (2019) The role of necroptosis in cancer biology and therapy. Mol Cancer 18:100

Greenlee JD, Lopez-Cavestany M, Ortiz-Otero N et al (2021) Oxaliplatin resistance in colorectal cancer enhances TRAIL sensitivity via death receptor 4 upregulation and lipid raft localization. Elife 10:

Gu J, Xu J, You Q et al (2020) Recent developments of small molecules targeting RNA m(6)A modulators. Eur J Med Chem 196:112325

Guo M, Yan R, Ji Q et al (2020) IFN regulatory Factor-1 induced macrophage pyroptosis by modulating m6A modification of circ_0029589 in patients with acute coronary syndrome. Int Immunopharmacol 86:106800

Guo M, Yan R, Yao H et al (2019) IFN Regulatory Factor 1 Mediates Macrophage Pyroptosis Induced by Oxidized Low-Density Lipoprotein in Patients with Acute Coronary Syndrome. Mediators Inflamm 2019:2917128

Han S, Zhu L, Zhu Y et al (2021) Targeting ATF4-dependent pro-survival autophagy to synergize glutaminolysis inhibition. Theranostics 11:8464–8479

He Y, Wang W, Xu X et al (2022) Mettl3 inhibits the apoptosis and autophagy of chondrocytes in inflammation through mediating Bcl2 stability via Ythdf1-mediated m(6)A modification. Bone 154:116182

He Y, Yue H, Cheng Y et al (2021) ALKBH5-mediated m(6)A demethylation of KCNK15-AS1 inhibits pancreatic cancer progression via regulating KCNK15 and PTEN/AKT signaling. Cell Death Dis 12:1121
He Y, Zeng MY, Yang D et al (2016) NEK7 is an essential mediator of NLRP3 activation downstream of potassium efflux. Nature 530:354–357
Hildebrand JM, Tanzer MC, Lucet IS et al (2014) Activation of the pseudokinase MLKL unleashes the four-helix bundle domain to induce membrane localization and necroptotic cell death. Proc Natl Acad Sci USA 111:15072–15077
Holze C, Michaudel C, Mackowiak C et al (2018) Oxeiptosis, a ROS-induced caspase-independent apoptosis-like cell-death pathway. Nat Immunol 19:130–140
Hsu PJ, Zhu Y, Ma H et al (2017) Ythdc2 is an N(6)-methyladenosine binding protein that regulates mammalian spermatogenesis. Cell Res 27:1115–1127
Huang H, Weng H, Sun W et al (2018) Recognition of RNA N(6)-methyladenosine by IGF2BP proteins enhances mRNA stability and translation. Nat Cell Biol 20:285–295
Ji FH, Fu XH, Li GQ et al (2022) FTO Prevents Thyroid Cancer Progression by SLC7A11 m6A Methylation in a Ferroptosis-Dependent Manner. Front Endocrinol (Lausanne) 13:857765
Jia G, Fu Y, Zhao X et al (2011) N6-methyladenosine in nuclear RNA is a major substrate of the obesity-associated FTO. Nat Chem Biol 7:885–887
Jin S, Zhang X, Miao Y et al (2018) m(6)A RNA modification controls autophagy through upregulating ULK1 protein abundance. Cell Res 28:955–957
Kaczmarek A, Vandenabeele P, Krysko DV (2013) Necroptosis: the release of damage-associated molecular patterns and its physiological relevance. Immunity 38:209–223
Knodler LA, Crowley SM, Sham HP et al (2014) Noncanonical inflammasome activation of caspase-4/caspase-11 mediates epithelial defenses against enteric bacterial pathogens. Cell Host Microbe 16:249–256
Koppula P, Zhuang L, Gan B (2021) Cystine transporter SLC7A11/xCT in cancer: ferroptosis, nutrient dependency, and cancer therapy. Protein Cell 12:599–620
Krenkel O, Tacke F (2017) Liver macrophages in tissue homeostasis and disease. Nat Rev Immunol 17:306–321
Lan H, Liu Y, Liu J et al (2021) Tumor-Associated Macrophages Promote Oxaliplatin Resistance via METTL3-Mediated m(6)A of TRAF5 and Necroptosis in Colorectal Cancer. Mol Pharm 18:1026–1037
Lan J, Xu B, Shi X et al (2022) WTAP-mediated N(6)-methyladenosine modification of NLRP3 mRNA in kidney injury of diabetic nephropathy. Cell Mol Biol Lett 27:51
Lapierre LR, Kumsta C, Sandri M et al (2015) Transcriptional and epigenetic regulation of autophagy in aging. Autophagy 11:867–880
Lee B, Lee S, Shim J (2021) YTHDF2 Suppresses Notch Signaling through Post-transcriptional Regulation on Notch1. Int J Biol Sci 17:3776–3785
Levy JMM, Towers CG, Thorburn A (2017) Targeting autophagy in cancer. Nat Rev Cancer 17:528–542
Li G, Song Y, Liao Z et al (2020a) Bone-derived mesenchymal stem cells alleviate compression-induced apoptosis of nucleus pulposus cells by N6 methyladenosine of autophagy. Cell Death Dis 11:103
Li J, McQuade T, Siemer AB et al (2012) The RIP1/RIP3 necrosome forms a functional amyloid signaling complex required for programmed necrosis. Cell 150:339–350
Li J, Yao H, Huang J et al (2022a) METTL3 promotes prostatic hyperplasia by regulating PTEN expression in an m(6)A-YTHDF2-dependent manner. Cell Death Dis 13:723
Li N, Yi X, He Y et al (2022b) Targeting Ferroptosis as a Novel Approach to Alleviate Aortic Dissection. Int J Biol Sci 18:4118–4134
Li Q, He W, Wan G (2020b) Methyladenosine Modification in RNAs: Classification and Roles in Gastrointestinal Cancers. Front Oncol 10:586789
Liang D, Lin WJ, Ren M et al (2022) m(6)A reader YTHDC1 modulates autophagy by targeting SQSTM1 in diabetic skin. Autophagy 18:1318–1337

Lin Y, Shen X, Ke Y et al (2022) Activation of osteoblast ferroptosis via the METTL3/ASK1-p38 signaling pathway in high glucose and high fat (HGHF)-induced diabetic bone loss. Faseb j 36:e22147
Lin Z, Niu Y, Wan A et al (2020) RNA m(6) A methylation regulates sorafenib resistance in liver cancer through FOXO3-mediated autophagy. Embo j 39:e103181
Liu BH, Tu Y, Ni GX et al (2021) Total Flavones of Abelmoschus manihot Ameliorates Podocyte Pyroptosis and Injury in High Glucose Conditions by Targeting METTL3-Dependent m(6)A Modification-Mediated NLRP3-Inflammasome Activation and PTEN/PI3K/Akt Signaling. Front Pharmacol 12:667644
Liu J, Yue Y, Han D et al (2014) A METTL3-METTL14 complex mediates mammalian nuclear RNA N6-adenosine methylation. Nat Chem Biol 10:93–95
Liu L, He J, Sun G et al (2022a) The N6-methyladenosine modification enhances ferroptosis resistance through inhibiting SLC7A11 mRNA deadenylation in hepatoblastoma. Clin Transl Med 12:e778
Liu L, Li H, Hu D et al (2022b) Insights into N6-methyladenosine and programmed cell death in cancer. Mol Cancer 21:32
Liu S, Li Q, Li G et al (2020) The mechanism of m(6)A methyltransferase METTL3-mediated autophagy in reversing gefitinib resistance in NSCLC cells by β-elemene. Cell Death Dis 11:969
Liu Y, Levine B (2015) Autosis and autophagic cell death: the dark side of autophagy. Cell Death Differ 22:367–376
Liu ZX, Chen J, Zhang DY (2012) The Mechanism of Autophagy Regulation and The Role of Autophagy in Alzheimer′s Disease*. Progress in Biochemistry and Biophysics 39:726–733
Lockshin RA, Zakeri Z (2001) Programmed cell death and apoptosis: origins of the theory. Nat Rev Mol Cell Biol 2:545–550
Lu Z, Liu H, Song N et al (2021) METTL14 aggravates podocyte injury and glomerulopathy progression through N(6)-methyladenosine-dependent downregulating of Sirt1. Cell Death Dis 12:881
Lv D, Gimple RC, Zhong C et al (2022) PDGF signaling inhibits mitophagy in glioblastoma stem cells through N(6)-methyladenosine. Dev Cell 57:1466-1481.e1466
Maelfait J, Liverpool L, Bridgeman A et al (2017) Sensing of viral and endogenous RNA by ZBP1/DAI induces necroptosis. Embo j 36:2529–2543
Mederacke I, Filliol A, Affo S et al (2022) The purinergic P2Y14 receptor links hepatocyte death to hepatic stellate cell activation and fibrogenesis in the liver. Sci Transl Med 14:eabe5795
Meng L, Lin H, Huang X et al (2022) METTL14 suppresses pyroptosis and diabetic cardiomyopathy by downregulating TINCR lncRNA. Cell Death Dis 13:38
Mizushima N, Komatsu M (2011) Autophagy: renovation of cells and tissues. Cell 147:728–741
Ni L, Bai R, Zhou Q et al (2022) The Correlation between Ferroptosis and m6A Methylation in Patients with Acute Kidney Injury. Kidney Blood Press Res 47:523–533
Overholtzer M, Mailleux AA, Mouneimne G et al (2007) A nonapoptotic cell death process, entosis, that occurs by cell-in-cell invasion. Cell 131:966–979
Paris J, Morgan M, Campos J et al (2019) Targeting the RNA m(6)A Reader YTHDF2 Selectively Compromises Cancer Stem Cells in Acute Myeloid Leukemia. Cell Stem Cell 25:137-148.e136
Pendleton KE, Chen B, Liu K et al (2017) The U6 snRNA m(6)A Methyltransferase METTL16 Regulates SAM Synthetase Intron Retention. Cell 169:824-835.e814
Ping XL, Sun BF, Wang L et al (2014) Mammalian WTAP is a regulatory subunit of the RNA N6-methyladenosine methyltransferase. Cell Res 24:177–189
Pugsley HR (2017) Assessing Autophagic Flux by Measuring LC3, p62, and LAMP1 Co-localization Using Multispectral Imaging Flow Cytometry. J Vis Exp
Qu J, Hou Y, Chen Q et al (2022) RNA demethylase ALKBH5 promotes tumorigenesis in multiple myeloma via TRAF1-mediated activation of NF-κB and MAPK signaling pathways. Oncogene 41:400–413
Ravichandran D, Pillai J, Krishnamurthy K (2022) Genetics of intervertebral disc disease: A review. Clin Anat 35:116–120

Shan B, Pan H, Najafov A et al (2018) Necroptosis in development and diseases. Genes Dev 32:327–340
Shen M, Li Y, Wang Y et al (2021) N(6)-methyladenosine modification regulates ferroptosis through autophagy signaling pathway in hepatic stellate cells. Redox Biol 47:102151
Shen W, Zhu M, Wang Q et al (2022) DARS-AS1 recruits METTL3/METTL14 to bind and enhance DARS mRNA m(6)A modification and translation for cytoprotective autophagy in cervical cancer. RNA Biol 19:751–763
Shi H, Wang X, Lu Z et al (2017a) YTHDF3 facilitates translation and decay of N(6)-methyladenosine-modified RNA. Cell Res 27:315–328
Shi H, Wei J, He C (2019) Where, When, and How: Context-Dependent Functions of RNA Methylation Writers, Readers, and Erasers. Mol Cell 74:640–650
Shi J, Gao W, Shao F (2017b) Pyroptosis: Gasdermin-Mediated Programmed Necrotic Cell Death. Trends Biochem Sci 42:245–254
Shu B, Zhou YX, Li H et al (2021) The METTL3/MALAT1/PTBP1/USP8/TAK1 axis promotes pyroptosis and M1 polarization of macrophages and contributes to liver fibrosis. Cell Death Discov 7:368
Singer M, Deutschman CS, Seymour CW et al (2016) The Third International Consensus Definitions for Sepsis and Septic Shock (Sepsis-3). JAMA 315:801–810
Song H, Feng X, Zhang H et al (2019) METTL3 and ALKBH5 oppositely regulate m(6)A modification of TFEB mRNA, which dictates the fate of hypoxia/reoxygenation-treated cardiomyocytes. Autophagy 15:1419–1437
Song X, Zhu S, Xie Y et al (2018) JTC801 Induces pH-dependent Death Specifically in Cancer Cells and Slows Growth of Tumors in Mice. Gastroenterology 154:1480–1493
Song Z, Jia G, Ma P et al (2021) Exosomal miR-4443 promotes cisplatin resistance in non-small cell lung carcinoma by regulating FSP1 m6A modification-mediated ferroptosis. Life Sci 276:119399
Stipanuk MH, Dominy JE Jr, Lee JI et al (2006) Mammalian cysteine metabolism: new insights into regulation of cysteine metabolism. J Nutr 136:1652s–1659s
Stockwell BR (2022) Ferroptosis turns 10: Emerging mechanisms, physiological functions, and therapeutic applications. Cell 185:2401–2421
Su M, Chen C, Li S et al (2022) Gasdermin D-dependent platelet pyroptosis exacerbates NET formation and inflammation in severe sepsis. Nat Cardiovasc Res 1:732–747
Su R, Dong L, Li C et al (2018) R-2HG Exhibits Anti-tumor Activity by Targeting FTO/m(6)A/MYC/CEBPA Signaling. Cell 172:90-105.e123
Su Y, Maimaitiyiming Y, Wang L et al (2021) Modulation of Phase Separation by RNA: A Glimpse on N(6)-Methyladenosine Modification. Front Cell Dev Biol 9:786454
Sun S, Gao T, Pang B et al (2022) RNA binding protein NKAP protects glioblastoma cells from ferroptosis by promoting SLC7A11 mRNA splicing in an m(6)A-dependent manner. Cell Death Dis 13:73
Tang D, Kang R, Berghe TV et al (2019) The molecular machinery of regulated cell death. Cell Res 29:347–364
Tang HW, Weng JH, Lee WX et al (2021) mTORC1-chaperonin CCT signaling regulates m(6)A RNA methylation to suppress autophagy. Proc Natl Acad Sci USA 118:e2021945118
Tsvetkov P, Coy S, Petrova B et al (2022) Copper induces cell death by targeting lipoylated TCA cycle proteins. Science 375:1254–1261
Vanlangenakker N, Vanden Berghe T, Bogaert P et al (2011) cIAP1 and TAK1 protect cells from TNF-induced necrosis by preventing RIP1/RIP3-dependent reactive oxygen species production. Cell Death Differ 18:656–665
Vu LP, Pickering F, Cheng Y et al (2017) The N(6)-methyladenosine (m(6)A)-forming enzyme METTL3 controls myeloid differentiation of normal hematopoietic and leukemia cells. Nat Med 23:1369–1376
Wang CX, Cui GS, Liu X et al (2018) METTL3-mediated m6A modification is required for cerebellar development. PLoS Biol 16:e2004880

Wang F, Liao Y, Zhang M et al (2021a) N6-methyladenosine demethyltransferase FTO-mediated autophagy in malignant development of oral squamous cell carcinoma. Oncogene 40:3885–3898
Wang H, Xu B, Shi J (2020a) N6-methyladenosine METTL3 promotes the breast cancer progression via targeting Bcl-2. Gene 722:144076
Wang J, Zhang J, Ma Y et al (2021b) WTAP promotes myocardial ischemia/reperfusion injury by increasing endoplasmic reticulum stress via regulating m(6)A modification of ATF4 mRNA. Aging (Albany NY) 13:11135–11149
Wang L, Klionsky D J, Shen H M (2022a) The emerging mechanisms and functions of microautophagy. Nat Rev Mol Cell Biol
Wang L, Liang Q, Lin A et al (2019) Borneol alleviates brain injury in sepsis mice by blocking neuronal effect of endotoxin. Life Sci 232:116647
Wang X, Li Y, Li J et al (2022b) Mechanism of METTL3-Mediated m(6)A Modification in Cardiomyocyte Pyroptosis and Myocardial Ischemia-Reperfusion Injury. Cardiovasc Drugs Ther
Wang X, Lu Z, Gomez A et al (2014) N6-methyladenosine-dependent regulation of messenger RNA stability. Nature 505:117–120
Wang X, Wu R, Liu Y et al (2020b) m(6)A mRNA methylation controls autophagy and adipogenesis by targeting Atg5 and Atg7. Autophagy 16:1221–1235
Wei H, Qu L, Dai S et al (2021) Structural insight into the molecular mechanism of p53-mediated mitochondrial apoptosis. Nat Commun 12:2280
Wei X, Xie F, Zhou X et al (2022) Role of pyroptosis in inflammation and cancer. Cell Mol Immunol 19:971–992
Wu C, Zhou L, Yuan H et al (2020) Interconnections among major forms of regulated cell death. Apoptosis 25:616–624
Xia X, Wang X, Cheng Z et al (2019) The role of pyroptosis in cancer: pro-cancer or pro-"host"? Cell Death Dis 10:650
Xiao W, Adhikari S, Dahal U et al (2016) Nuclear m(6)A Reader YTHDC1 Regulates mRNA Splicing. Mol Cell 61:507–519
Xu Y, Lv D, Yan C et al (2022) METTL3 promotes lung adenocarcinoma tumor growth and inhibits ferroptosis by stabilizing SLC7A11 m(6)A modification. Cancer Cell Int 22:11
Yagoda N, von Rechenberg M, Zaganjor E et al (2007) RAS-RAF-MEK-dependent oxidative cell death involving voltage-dependent anion channels. Nature 447:864–868
Yang F, Bettadapura SN, Smeltzer MS et al (2022a) Pyroptosis and pyroptosis-inducing cancer drugs. Acta Pharmacol Sin 43:2462–2473
Yang H, Hu Y, Weng M et al (2022b) Hypoxia inducible lncRNA-CBSLR modulates ferroptosis through m6A-YTHDF2-dependent modulation of CBS in gastric cancer. J Adv Res 37:91–106
Yang L, Yang F, Zhao H et al (2019a) Circular RNA circCHFR Facilitates the Proliferation and Migration of Vascular Smooth Muscle via miR-370/FOXO1/Cyclin D1 Pathway. Mol Ther Nucleic Acids 16:434–441
Yang LZ, Wang Y, Li SQ et al (2019b) Dynamic Imaging of RNA in Living Cells by CRISPR-Cas13 Systems. Mol Cell 76:981-997.e987
Yang S, Wei J, Cui YH et al (2019c) m(6)A mRNA demethylase FTO regulates melanoma tumorigenicity and response to anti-PD-1 blockade. Nat Commun 10:2782
Yang WS, Stockwell BR (2008) Synthetic lethal screening identifies compounds activating iron-dependent, nonapoptotic cell death in oncogenic-RAS-harboring cancer cells. Chem Biol 15:234–245
Yang Z, Yang S, Cui YH et al (2021) METTL14 facilitates global genome repair and suppresses skin tumorigenesis. Proc Natl Acad Sci USA 118:e2025948118
Yankova E, Blackaby W, Albertella M et al (2021) Small-molecule inhibition of METTL3 as a strategy against myeloid leukaemia. Nature 593:597–601
Yuan H, Li X, Zhang X et al (2016) Identification of ACSL4 as a biomarker and contributor of ferroptosis. Biochem Biophys Res Commun 478:1338–1343

Yuan X, Li T, Shi L et al (2021) Human umbilical cord mesenchymal stem cells deliver exogenous miR-26a-5p via exosomes to inhibit nucleus pulposus cell pyroptosis through METTL14/NLRP3. Mol Med 27:91
Zaccara S, Ries RJ, Jaffrey SR (2019) Reading, writing and erasing mRNA methylation. Nat Rev Mol Cell Biol 20:608–624
Zeng Q, Ma X, Song Y et al (2022) Targeting regulated cell death in tumor nanomedicines. Theranostics 12:817–841
Zhang H, Liu J, Zhou Y et al (2022a) Neutrophil extracellular traps mediate m(6)A modification and regulates sepsis-associated acute lung injury by activating ferroptosis in alveolar epithelial cells. Int J Biol Sci 18:3337–3357
Zhang L, Wang X, Che W et al (2022b) Methyltransferase-like 3 silenced inhibited the ferroptosis development via regulating the glutathione peroxidase 4 levels in the intracerebral hemorrhage progression. Bioengineered 13:14215–14226
Zhang Q, Riddle RC, Yang Q et al (2019) The RNA demethylase FTO is required for maintenance of bone mass and functions to protect osteoblasts from genotoxic damage. Proc Natl Acad Sci USA 116:17980–17989
Zhang S, Guan X, Liu W et al (2022c) YTHDF1 alleviates sepsis by upregulating WWP1 to induce NLRP3 ubiquitination and inhibit caspase-1-dependent pyroptosis. Cell Death Discov 8:244
Zhang T, Ding C, Chen H et al (2022d) m(6)A mRNA modification maintains colonic epithelial cell homeostasis via NF-κB-mediated antiapoptotic pathway. Sci Adv 8:eabl5723
Zheng D, Liwinski T, Elinav E (2020) Inflammasome activation and regulation: toward a better understanding of complex mechanisms. Cell Discov 6:36
Zheng G, Dahl JA, Niu Y et al (2013) ALKBH5 is a mammalian RNA demethylase that impacts RNA metabolism and mouse fertility. Mol Cell 49:18–29
Zhou J, Wan J, Gao X et al (2015) Dynamic m(6)A mRNA methylation directs translational control of heat shock response. Nature 526:591–594
Zhou P, Wu M, Ye C et al (2019) Meclofenamic acid promotes cisplatin-induced acute kidney injury by inhibiting fat mass and obesity-associated protein-mediated m(6)A abrogation in RNA. J Biol Chem 294:16908–16917
Zhu F, Zhang W, Yang T et al (2019) Complex roles of necroptosis in cancer. J Zhejiang Univ Sci B 20:399–413
Zhuang S, Ma Y, Zeng Y et al (2021) METTL14 promotes doxorubicin-induced cardiomyocyte ferroptosis by regulating the KCNQ1OT1-miR-7–5p-TFRC axis. Cell Biol Toxicol
Zou Y, Zheng S, Xie X et al (2022) N6-methyladenosine regulated FGFR4 attenuates ferroptotic cell death in recalcitrant HER2-positive breast cancer. Nat Commun 13:2672

Incorporation of Pseudouridine into RNA for Biochemical and Biophysical Studies

Tristan Sanford, Andrew Riley, Melanie Clawson, Kylie Raasch, Ridwan Oyebamiji, and Minako Sumita

Contents

Abstract Modification of RNA molecules has a significant effect on their structure and function. Many different examples of RNA modifications have been observed and each contributes in various ways to ensure proper biological activity. One of the most common modifications is pseudouridine which occurs in key dynamic locations in many RNAs. Despite its prevalence in natural RNA sequences, organic synthesis of pseudouridine has been challenging because of the stereochemistry requirement and the sensitivity of reaction steps to moisture. Herein, we describe the

T. Sanford · A. Riley · K. Raasch · M. Sumita (✉)
Department of Chemistry, Southern Illinois University Edwardsville, Edwardsville, IL 62026, USA
e-mail: msumita@siue.edu

M. Clawson · R. Oyebamiji · M. Sumita
Department of Biological Science, Southern Illinois University Edwardsville, Edwardsville, IL 62026, USA

T. Sanford
Department of Microbiology and Immunology, University of Oklahoma Health Sciences Center, Oklahoma City, OK 73104, USA

A. Riley
Department of Cell and Developmental Biology, University of Illinois Urbana-Champaign, Urbana, IL 61801, USA

J. Barciszewski (ed.), *RNA Structure and Function*, RNA Technologies 14,
https://doi.org/10.1007/978-3-031-36390-0_25

semi-enzymatic synthetic route for the synthesis of pseudouridine using adenosine-5'-monophosphate and uracil as the starting materials and a reverse reaction catalyzed by a pseudouridine monophosphate glycosidase. Moreover, we describe the conversion from nucleoside (pseudouridine) to nucleotide triphosphate (pseudouridine-5'-triphosphate) to incorporate into RNA via in vitro transcription for biochemical and biophysical studies.

Keywords RNA modification · Synthesis of modified nucleoside · Pseudouridine · Pseudouridine monophosphate glycosidase

List of Abbreviations

ACE	Bis(2-acetoxyethoxy)methyl
AMP	Adenosine-5′-Monophosphate
DMT	Dimethoxytrityl
Et_3NHHCO_3	Triethylammonium Bicarbonate
$(HNBu_3)_2H_2P_2O_7$	*bis*-(Tri-*n*-butylammonium)pyrophosphate
NTP	Nucleoside Triphosphate
$POCl_3$	Phosphorous Oxychloride
$PSCl_3$	Thiophosphoryl Chloride
R5P	Ribose-5′-Phosphate
Ψ	Pseudouridine
ΨMP	Pseudouridine-5′-Monophosphate
ΨMP Glycosidase	Pseudouridine Monophosphate Glycosidase
Ψ Synthase	Pseudouridine Synthase
ΨTP	Pseudouridine-5′-Triphosphate
snoRNA	Small Nucleolar RNA
snoRNP	Small Nucleoprotein Particle
TBDMS	*tert*-Butyldimethylsilyl
TFAA	Trifluoroacetic Anhydride
TLR7	Toll-like Receptor 7
TOM	2′-O-triisopropylsilyloxymethyl

1 Introduction

Studies of RNA modification have shown that a large number of diseases are linked to mutations in the structure of RNA and RNA modification enzymes (Jonkhout et al. 2017). Therefore, the research of RNA modification in nature will elucidate the emergence of certain diseases such as cancer, birth defects, and neurological disorders. Furthermore, RNA modification studies will explain the mechanisms associated with the formation of these diseases and contribute to the development of new and effective treatments.

More than 100 RNA modifications have been discovered in nature and each contributes a wide range of biological functions. Modified bases also exist naturally in DNA, but RNA modifications are far more diverse. Because RNA has more complex structures and functions, RNA may need modifications to adjust its structures and functions in the process of evolution. RNA modifications are known for the stabilization of RNA structures, RNA-protein interactions, and regulation of RNA processing and translation (Schaefer et al. 2017). These modifications are developed by several different enzymes. If the enzymes are mutated or not folded properly, diseases such as cancer, cardiovascular disease, metabolic disease, and neurological disorders can arise (Jonkhout et al. 2017). Investigation of these modifications can provide further insight into preventing diseases and thus lead to cures and treatments. RNA modifications occur post-transcriptionally to change the RNA structure and function. Each RNA modification has important cellular functions such as stabilizing RNA strands and increasing functional effectiveness. For example, methylation at the 5′-end guanosine base of mRNA serves as a 5′-cap that protects from degradation by nucleases and gives mRNAs a longer lifetime in eukaryotes (Ramanathan et al. 2016). Some RNA modifications feature unique structural alterations. For example, inosine is formed by the deamination of adenosine at the C6 position and is primarily found in the anticodon region of tRNA. Inosine is unique as it can base-pair with relative stability to any nucleic acid with similar affinity making it a universal anticodon recognizer (Rafels-Ybern et al. 2018).

Many natural and unnatural modified nucleotides are commercially available. However, synthesis of modified nucleotides is necessary due to commercial availability limits and isotope labeling requirements. Modified nucleotides can be synthesized chemically or enzymatically. Enzymatic synthesis is specific to substrates, has no side reactions, and results in high yields of the desired product. However, we can design artificial nucleotides and synthesize them through chemical synthesis. When nucleotides are synthesized, they need to be incorporated into RNA sequences. There are two methods to incorporate a modified nucleotide into RNAs—in vitro transcription and solid-phase synthesis—each with advantages and disadvantages. In vitro transcription is more cost-effective; however, modified nucleotides cannot be incorporated into a specific position within RNA. Also, the yield of RNA production will be lower with modified nucleotides compared to the canonical four nucleotides. On the other hand, solid-phase synthesis can incorporate modified nucleotides into a specific position, but it is not suitable for more than 100 nucleotide-long RNA sequences. Herein, we focus on organic and enzymatic syntheses of pseudouridine and discussion of the semi-enzymatic synthetic route to incorporate pseudouridine into RNA for biochemical and biophysical chemistry research.

2 Pseudouridine

The first discovered among the most common modifications is pseudouridine (Cohn 1960; Ofengand et al. 1995; Charette and Gray 2000). Pseudouridine exists in all three biological domains and about 0.2~0.7% of the uridine nucleotides are isomerized to pseudouridine in mammalian cell lines and tissues (Roundtree et al. 2017). The modified nucleotide is found in many RNA classes, such as tRNA, rRNA, and even mRNA. In tRNA, the nearly universally conserved pseudouridine at residue 55 (Ψ55) is located in the TΨC stem-loop and stabilizes the tRNA structural motifs (Davis and Poulter 1991). Interestingly, higher organisms have more pseudouridine residues in their large subunit rRNAs. Eukaryotic large subunit rRNAs contain 0.9–1.4% pseudouridines compared to 0.03–0.4% in bacteria and the pseudouridine residues tend to be clustered in the region of the peptidyl-transferase center (Maden and Wakeman 1988; Brimacombe et al. 1993; Charette and Gray 2000). Pseudouridine has its glycosidic bond between a base and the sugar that is isomerized from a standard C1′-N1 found in uridine to a C1′-C5 bonding scheme (Fig. 1). The C–C glycosidic bond increases the rotational freedom between the base and sugar compared to the N–C glycosidic bonding; therefore, the conformation of pseudouridine is likely more flexible than uridine (Charette and Gray 2000). Conformational studies suggest that the *syn*-conformation of pseudouridine is slightly preferred over the *anti*-conformation (Hurd and Reid 1977; Neumann et al. 1980). The hydrogen-bonding pattern between pseudouridine and adenosine residues can be the same between *syn*- and *anti*-conformations. When pseudouridine is in an *anti*-conformation, the N1 imino proton can form water-mediated hydrogen bonding with the 5′-side of the phosphodiester bond (Davis 1995; Charette and Gray 2000; Newby and Greenbaum 2002). Pseudouridine increases stacking with neighboring bases as compared to uridine (Davis 1995; Yarian et al. 1999). On the other hand, not all pseudouridine residues involve stabilizing RNA structures. For example, position 1915 of the *E. coli* 23S rRNA is known to destabilize the helix 69 loop structure (Meroueh et al. 2000). In addition, many studies have been performed by knocking out pseudouridine modification enzymes and observing the effects. While this knockout is not always lethal, studies show a significant loss of activity for RNA systems when pseudouridine cannot be formed (Spenkuch et al. 2014).

Recently, due to the COVID-19 pandemic, a derivative of pseudouridine (N1-methylpseudouridine) has been used for the development of mRNA vaccines against SARS-CoV-2. The N1-methylpseudouridine derivative was used in the vaccines as the modified nucleotide enhances antigen expression and reduces immunogenicity due to the inhibition of signal transduction (Anderson et al. 2010; Andries et al. 2015; Pardi et al. 2018). The mRNA with N1-methylpseudouridine sequences perturbs the interaction with Toll-like receptor 7 (TLR7), ultimately suppressing the immune response (Nance and Meier 2021). Thus, pseudouridine has been studied for its structural effects on RNA and functional effects in the cell. However, there are many

Fig. 1 The structure of pseudouridine (Ψ)

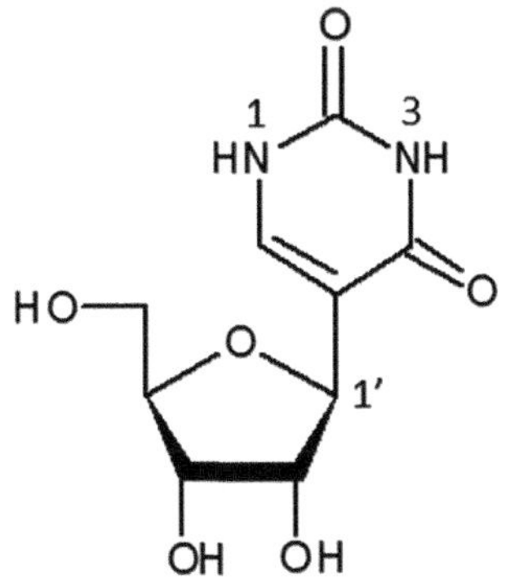

unknowns regarding the function of pseudouridine within RNAs and it is difficult to predict the structural and cellular functions of individual pseudouridine residues (Borchardt et al. 2020).

3 Synthesis of Pseudouridine

3.1 Organic Synthesis of Pseudouridine

Organic synthesis has developed many reaction schemes for natural and unnatural products. The main advantage is to cost-effectively synthesize large-scale products. Organic synthesis can also yield isotope-labeled products for NMR characterization. In addition, we can design and synthesize the derivatives to improve the biological function and expose new properties. Therefore, organic chemists have been attempting to develop a synthetic method to obtain large quantities of modified nucleosides. The challenges of organic synthesis of pseudouridine are the protection of functional groups, the stereoselectivity requirement, and the moisture-sensitive reaction steps.

Both uracil and ribose contain several functional groups each with varying reactivities that can interfere with the reaction and produce undesired side products. To overcome this challenge, protection and deprotection steps were added to obtain higher yields. The coupling reaction between base and ribose could produce both α and β anomers; however, the β anomers more commonly exist in nature. Therefore, stereoselectivity is crucial for the organic synthesis of pseudouridine. Unfortunately, many organic reagents are moisture sensitive, so controlling moisture is very essential to obtain high-yield products.

The first synthesis of pseudouridine was reported in 1961 (Shapiro 1961). Shapiro et al., successfully coupled 2,3,5-tri-*O*-benzyl-D-ribofuranosyl chloride and 2,4-dimethoxy-pyrimidine-5-lithium to synthesize the nucleoside in a total of 5 reaction steps; however, the yield was only 2% and the product contains both α and β anomers (Shapiro 1961). The significance of this work is that the lithiated pyrimidine was used to form the C–C anomeric bond. A decade later, a new synthetic route was developed in which 2,4-di-*tert*-butoxypyrimidine-5-lithium and 2,4:3,5-di-*O*-benzylidene-aldehydo-D-ribose were coupled to form the C–C anomeric bond. Similar to Shapiro et al., the crude product also contained both α and β anomers, but the desired β anomer was separated by chromatography. The yield increased to 18%; however, the number of reaction steps increased to more than 10 steps due to the synthesis of the starting materials (Lerch et al. 1971).

In 1999, the organic synthesis of pseudouridine using ribonolactone and 2,4-dimethoxy-pyrimidine-5-lithium successfully produced only the β anomer and increased the yield to 20% while reducing the number of reaction steps to 7 (Grohar and Chow 1999). A few years later, the original synthesis was improved by coupling protected ribonolactone and 5-iodo-2,4-dimethoxypyrimidine. To overcome the stereoselectivity difficulty, the authors used a ribose ring-opening and ring-closing reaction with Zn^{2+} chelation after the coupling reaction between base and ribose (Fig. 2). The yield was significantly improved to 46% of the β anomer (Hanessian and Machaalani 2003). Recently, the Zn^{2+} chelation technique has been used to synthesize pseudouridine derivatives (Chang et al. 2008). Overall, organic synthesis of pseudouridine is time-consuming, requires multiple reaction steps, and has low overall yields.

3.2 Enzymatic Synthesis of Pseudouridine

In nature, pseudouridine is formed via a pseudouridine synthase (Ψ synthase). Pseudouridine synthases isomerize specific uridine residues in an RNA sequence and do not require any energy to perform their functions (Ofengand 2002). There are six classified Ψ synthase families including TruD, TruA, TruB, RsuA, RluA, and PUS10. RluA is in both bacteria and eukaryotes. The four families, TruD, TruA, TruB, and RsuA, only exist in bacteria, whereas PUS10 only exists in eukaryotes (Penzo et al. 2017; Stockert et al. 2021). All Ψ synthases work by cleaving the N1-C1′ glycosidic bond of uridine and implementing a 180° rotation of the nitrogenous base along the N3-C6 axis. This forms a new C5-C1′ glycosidic bond within RNA sequences and not free nucleotides (Yu and Meier 2014). Bacteria and eukaryotes have different mechanisms for modifying their RNA by Ψ synthases. Eukaryotic Ψ synthases catalyze the isomerization independently or with small nucleolar RNAs (snoRNAs), specifically, H/ACA box snoRNAs. These guide snoRNAs, with assistance from four small nucleolar ribonucleoprotein particles (snoRNPs), aid in site recognition by Ψ synthases (Bousquet-Antonelli et al. 1997; Lafontaine and Tollervey 1998; Kiss 2001). Prokaryotic Ψ synthases and PUS10 in eukaryotes are guide-RNA-independent enzymes. The enzymes rely on substrate recognition sites located within the enzyme for their binding specificity (Rintala-Dempsey and Kothe 2017; Zhao et al. 2018). However, these enzymes still require specific RNA sequences to convert uridine to pseudouridine. It is possible to extract pseudouridine-5’-monophosphate from RNAs in the

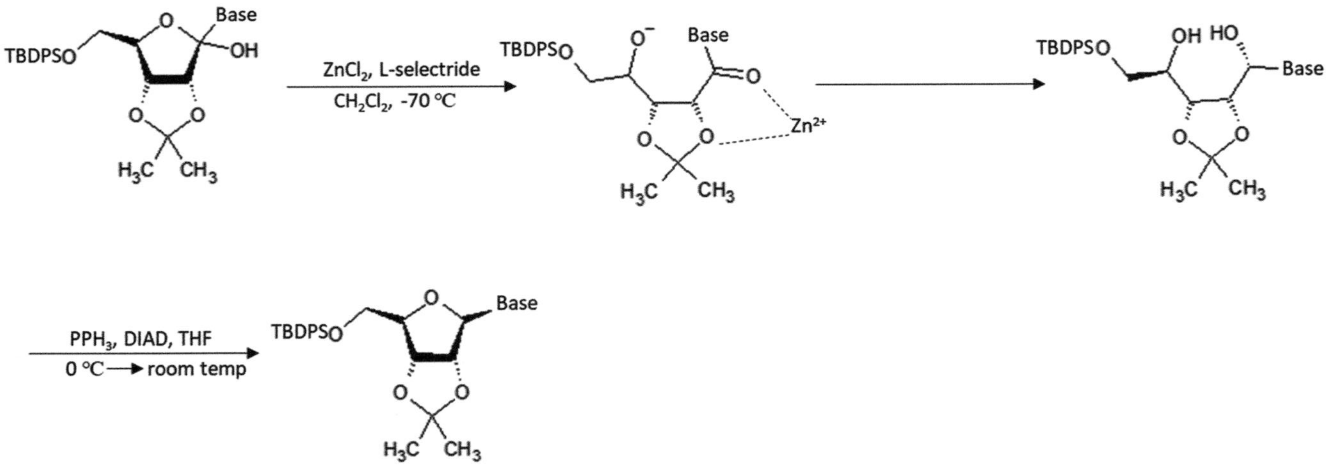

Fig. 2 The ring opening and closing steps of pseudouridine synthesis. The zinc ion is added to chelate two oxygens at 1′ and 2′ of the opened ribose ring (Hanessian and Machaalani 2003)

cells by using the enzymes. However, the quantities of pseudouridine are far less abundant than the other four canonical nucleosides (Roundtree et al. 2017). The mechanism of pseudouridine formation involves an aspartic acid residue within the active site of Ψ synthases (Fig. 3). In TruA, the carboxylate group of Asp-60 attacks the C6 position of uridine, which leads to cleavage of the N1-C1′ glycosidic bond. The base attached to Asp-60 is rotated 180° and a new C1′-C5 bond forms between the sugar and the base followed by the release of Asp-60 (Foster et al. 2000).

3.3 *Semi-Enzymatic Synthesis of Pseudouridine*

Using an enzymatic reaction step in organic synthesis has several advantages. In organic synthesis, a strong acid or base, high temperature, and high pressure can be used to facilitate the reaction. However, biomolecules are sensitive to these "harsh" conditions. Enzymatic reactions can speed up syntheses under mild conditions such as low temperature (usually less than 40 °C), normal atmospheric pressure, and mild buffers with neutral pH. Because enzymatic reactions are specific, protection and deprotection are not required. Therefore, the overall number of reaction steps will be less than the organic synthesis scheme. Additionally, the solvent of enzymatic reactions is usually a buffer, so the challenge of using moisture-sensitive reagents in the reaction is avoided. However, the enzymatic reaction yield depends on the activity of the enzyme. If the desired enzyme is not commercially available, extraction from cell culture is required. When extracting protein from a cell culture, the amount of enzyme can vary between batches of bacterial cultures because of misfolded or denatured enzymes. Therefore, cloning and overexpression are used to produce the desired enzyme more efficiently. Thus, enzymatic reactions have many advantages but also have drawbacks. It is important to compensate for the disadvantages of enzymatic reactions to design a semi-enzymatic synthetic scheme. Our semi-enzymatic synthetic route of pseudouridine (Fig. 4) uses adenosine-5′-monophosphate (AMP) and uracil as the starting materials and a reverse reaction catalyzed by recombinant ΨMP glycosidase (Riley et al. 2021).

3.3.1 Pseudouridine Monophosphate Glycosidase

Pseudouridine monophosphate glycosidase (ΨMP glycosidase) is a metabolic enzyme used in the nucleotide catabolic pathway in bacteria (Thapa et al. 2014). Some eukaryotes have a similar enzyme, but it functions as both a kinase and ΨMP glycosidase (Huang et al. 2012). However, higher organisms, including humans, do not have ΨMP glycosidase or similar enzymes (Feng et al. 2005). In nature, the function of ΨMP glycosidase is decoupling the anomeric bond of pseudouridine-5′-monophosphate (Fig. 4 Step 2 reverse reaction) to uracil and ribose-5′-monophosphate in the nucleotide degradation pathway. Higher organisms do not degrade pseudouridine because of the lack of ΨMP glycosidase or similar enzymes,

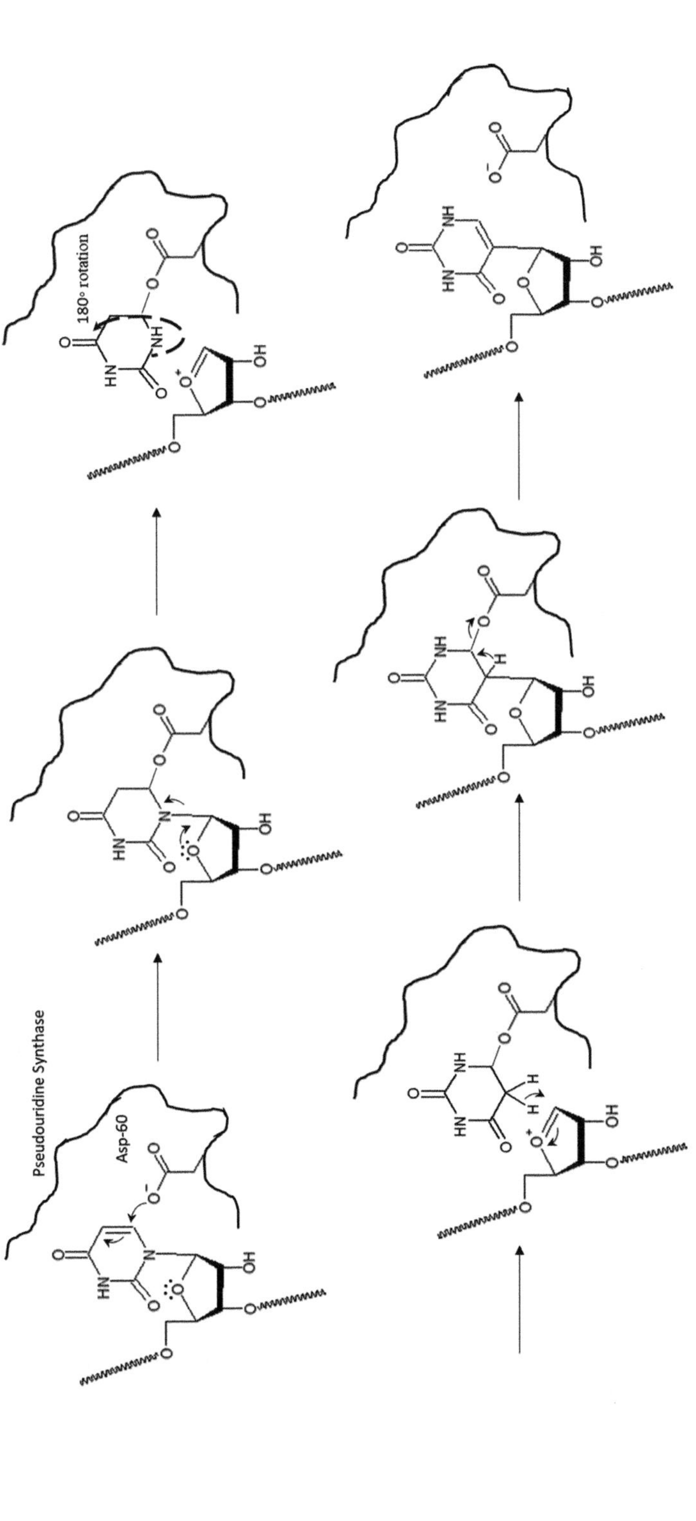

Fig. 3 Mechanism of pseudouridine synthase. An aspartic acid residue located within the active site of the pseudouridine synthase attacks C6, thus breaking the N1-C1′ glycosidic bond. The nitrogenous base rotates and the lone pair electrons delocalize resulting in the new C5-C1′ glycosidic bond releasing the nucleotide from the enzyme (Gu et al. 1999)

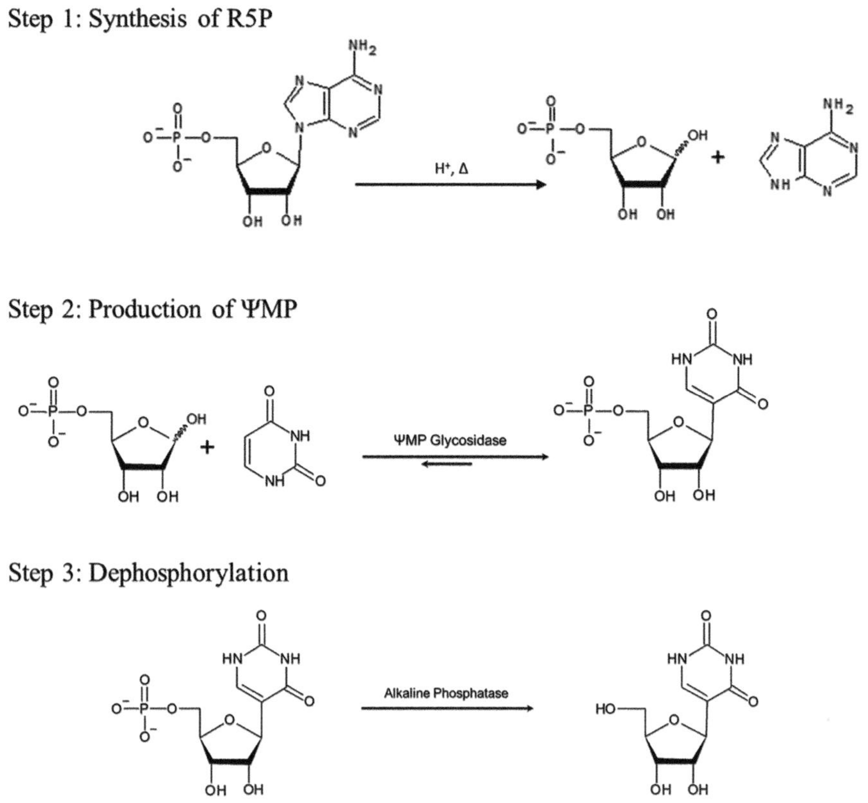

Fig. 4 Three steps of the semi-enzymatic synthesis of pseudouridine (Riley et al. 2021)

but eliminate pseudouridine in urine (Drahovsky et al. 1964). The concentration of urinary pseudouridine is 23.8 ± 4.9 nmol per 1 μmol of creatinine, but is found to be significantly higher in cancer patients (Tamura et al. 1988). Therefore, pseudouridine can be a cancer marker for early detection and progression of the disease.

The structure of ΨMP glycosidase is a homotrimer with a molecular weight of 106.2 kDa (Fig. 5a). Each subunit (αβα motif) is composed of 13 α-helices, 11 β-sheets, one manganese cofactor-binding site, and 312 amino acids. The trimer complex is facilitated by two sets of α-helices: α5 and α6 from one promoter and α9 and α10 from the adjacent promoter (Huang et al. 2012). The active site of the enzyme is a cleft formed between α-helices 12 and 13 and β-sheets 2, 7, and 8 (Huang et al. 2012). The substrate, however, does not directly interact with most of the residues found within the active site (Fig. 5b and c). Instead, most of the interactions between the active site and the substrate are through water-mediated hydrogen bonding and the salt bridge formed by the manganese cofactor. The cofactor (Mn^{2+}) binding site is located within the active site where the cofactor aids in the stabilization of the phosphate group and proper orientation of the modified nucleotide. The salt bridge

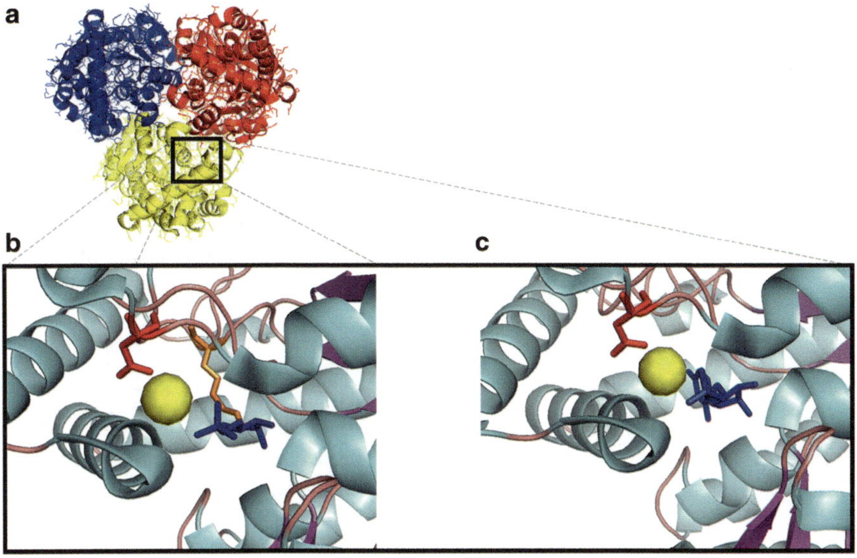

Fig. 5 Crystal structure of pseudouridine monophosphate glycosidase (ΨMP glycosidase). **a** ΨMP glycosidase overall structure. The structure is a homotrimer (PDB 4GIK). **b** The active site of ΨMP glycosidase with ribose-5′-monophosphate (blue) (PDB 4GIK). Two important amino acid residues, Asp-145 (red) and Lys-166 (orange) are shown. The yellow sphere indicates the cofactor, Mn^{2+}. **c** The active site of ΨMP glycosidase with pseudouridine-5′-monophosphate (blue) (PDB 4GIM). Asp-145 is shown in red and Mn^{2+} is the yellow sphere. Lys-166 is not shown due to mutation (Huang et al. 2012)

is formed in an octahedral conformation between water-mediated hydrogen bonding and Asp-145.

Currently, the proposed mechanism suggests that the lone pair electrons associated with N1 are delocalized and thus result in the formation of a double-bonded glycosidic linkage and opening of the ribose ring (Fig. 6). The ring-opening step is evident in multiple snapshots of the crystal structure (Huang et al. 2012). The presence of Lys-166 then attacks the anomeric carbon, creating the key intermediate of the two degradation products as uracil has been completely removed from the ribose. Hydrolysis then releases the ribose where the ring structure is restored thereby producing the two final degradation products, ribose-5′-phosphate and uracil.

Despite the glycosidic properties of the enzyme, the reverse reaction can occur as well. While ΨMP glycosidase functions in favor of degradation, the equilibrium strongly favors ΨMP synthesis with an equilibrium constant of 2.3×10^{-4} M for the degradation reaction (Preumont et al. 2008). The enzymatic glycosylation reaction of ΨMP is favored in nature and may be due to the further breakdown of products (uracil and R5P) by the catabolic pathway.

Due to ΨMP glycosidase not being commercially available, recombinant protein technology is necessary to express a large quantity of enzyme (Huang et al. 2012). Simply, the encoded gene (psuG, yeiN is a synonym) is amplified by polymerase

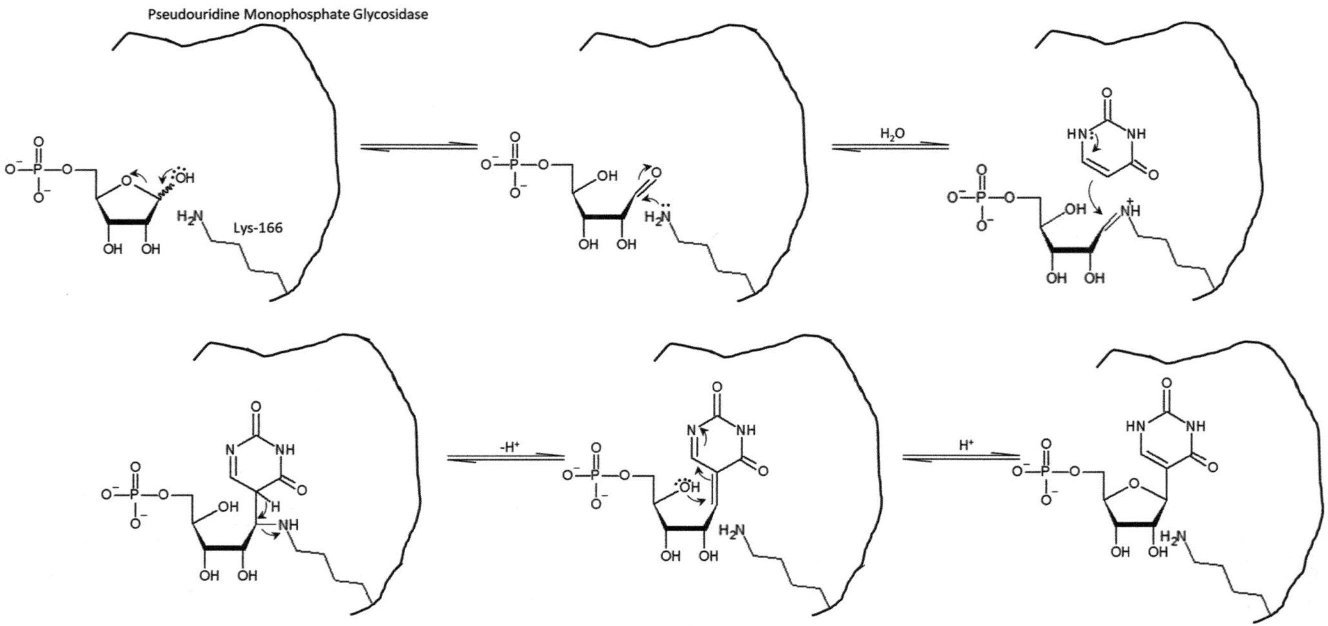

Fig. 6 Proposed synthetic mechanism of pseudouridine monophosphate glycosidase. The lone pair electrons of OH at C1' delocalize and disrupt the ribose ring structure forming a double bond. The C5 of uracil and C1' forms a bond after lysine-166 attacks the anomeric carbon (Huang et al. 2012)

chain reaction (PCR). Once amplified, the gene can be inserted into a DNA vector to form a recombinant plasmid. The plasmid is transformed into *E. coli* cell lines for both plasmid propagation and protein expression. By using this cloning and overexpression method, ΨMP glycosidase is collected in high yields.

Synthetic Strategies

The starting materials of our developed semi-enzymatic synthesis was uracil and ribose-5′-monophosphate, both of which were commercially available. Enzymatically, R5P could be synthesized from ribose and ATP with ribokinase, which is also commercially available. However, the organic synthesis of R5P is relatively easy, quick, and economical. Therefore, we selected to synthesize R5P for our first reaction step by AMP depurination with acid and heat (Fig. 4 Step 1). After purification by ion-exchange chromatography, the yield was 95% (Riley et al. 2021). The product, R5P, showed both α and β anomers by NMR spectroscopy. The β anomer is favored three times more as compared to the α anomer. Commercially available R5P is also a mixture of α and β anomers. For ribofuranose, the ratio of α and β anomers is 1:3 (Bhutani 2020). However, the stereochemistry of R5P does not affect the production of the desired β anomer because of the enzymatic reaction mechanism by Lys-166 as previously described (Fig. 6).

Step 2 is the coupling reaction between uracil and R5P. In this step, the enzymatic reaction has several advantages over organic synthesis. Because the enzymatic reaction is specific, none of the protection reactions are required. Enzymatic reaction with pseudouridine monophosphate glycosidase produces the desired β-pseudouridine. Our NMR analysis confirmed the production of the β anomer. Thus, all of the challenges in organic synthesis could be overcome with the use of an enzyme. The only disadvantage was that ΨMP glycosidase is not commercially available, so the enzyme needed to be obtained by cloning and overexpression from *E. coli* (Huang et al. 2012). The starting materials, uracil and R5P, were mixed in 25 mM HEPES, pH 7.1 buffer with ΨMP glycosidase and Mn^{2+} as a cofactor. After about 24 h, the reaction was terminated and the crude product was purified by column chromatography. The yield was 75% (Riley et al. 2021). We selected an enzymatic reaction by alkali phosphatase (commercially available) for the final dephosphorylation reaction (Fig. 4 Step 3), which resulted in a yield of 97% (Riley et al. 2021).

4 Incorporation of a Modified Nucleotide into RNA Sequence

4.1 *In Vitro Transcription*

The direction of transcription is from the 5′ end to the 3′ end. T7 RNA polymerase binds to double-stranded DNA at both the template and promoter. The mechanism of the elongation reaction is a nucleophilic displacement of pyrophosphate from NTP by the 3'-hydroxyl group of RNA. In vitro transcription is relatively cost-effective compared to chemical synthesis. However, the desired RNA sequence should generally start with one or more guanine residues at the 5' end. In addition, RNA sequences shorter than about 10 nucleotides are difficult to synthesize via in vitro transcription (Milligan and Uhlenbeck 1989). When uridine-5'-triphosphate can be replaced by pseudouridine-5'-triphosphate, in vitro transcription has advantages although the yield is lower than using canonical bases. In this case, all uridine residues in the RNA sequence are replaced by pseudouridine.

Nucleotide-5'-triphosphates are required for in vitro transcription; however, synthesis and analysis of nucleotide-5'-triphosphates presents many challenges (Burgess and Cook 2000). Despite the obstacles, there are two efficient ways to synthesize nucleoside triphosphates through either enzymatic or chemical synthesis. Although several methods for chemical synthesis exist, the most widely used method is the "one-pot, three-step" procedure (Fig. 7) (Burgess and Cook 2000; Mitsui et al. 2003). A nucleoside is first converted to a nucleotide-5'-monophosphate with phosphorous oxychloride ($POCl_3$) and trimethyl- or triethylphosphate as the solvent. The solvent, trimethyl- or triethylphosphate, accelerates nucleoside phosphorylation (Yoshikawa et al. 1967). The product of this procedure is a nucleotide chlorophosphite, which is used directly in the second step of the reaction with pyrophosphate. One advantage of this method is that protection of the 2'- and 3'-hydroxyl groups is not required because the major product is a 5'-monophosphate (about 90%). In the second step, *bis*-(tri-*n*-butylammonium)pyrophosphate ($(HNBu_3)_2H_2P_2O_7$) is added to the reaction mixture. The reaction is quenched with triethylammonium bicarbonate (Et_3NHHCO_3) after a short period of time (within 30 min) (Ludwig and Eckstein 1989; Mitsui et al. 2003). Nucleoside-5'-thiotriphosphates can also be synthesized by using thiophosphoryl chloride ($PSCl_3$) instead of $POCl_3$ in the first step (Ludwig and Eckstein 1989) when the starting material is the nucleoside (the product of Fig. 4 Step 3). The advantage of the previously described semi-enzymatic synthesis of pseudouridine is that the product of Step 2 yields a monophosphate product. Therefore, the first phosphorylation is already complete. In this case, the phosphate group is activated by 1-methylimidazole and trifluoroacetic anhydride (TFAA), then coupled with pyrophosphate to produce a triphosphate (Mohamady and Jakeman 2005).

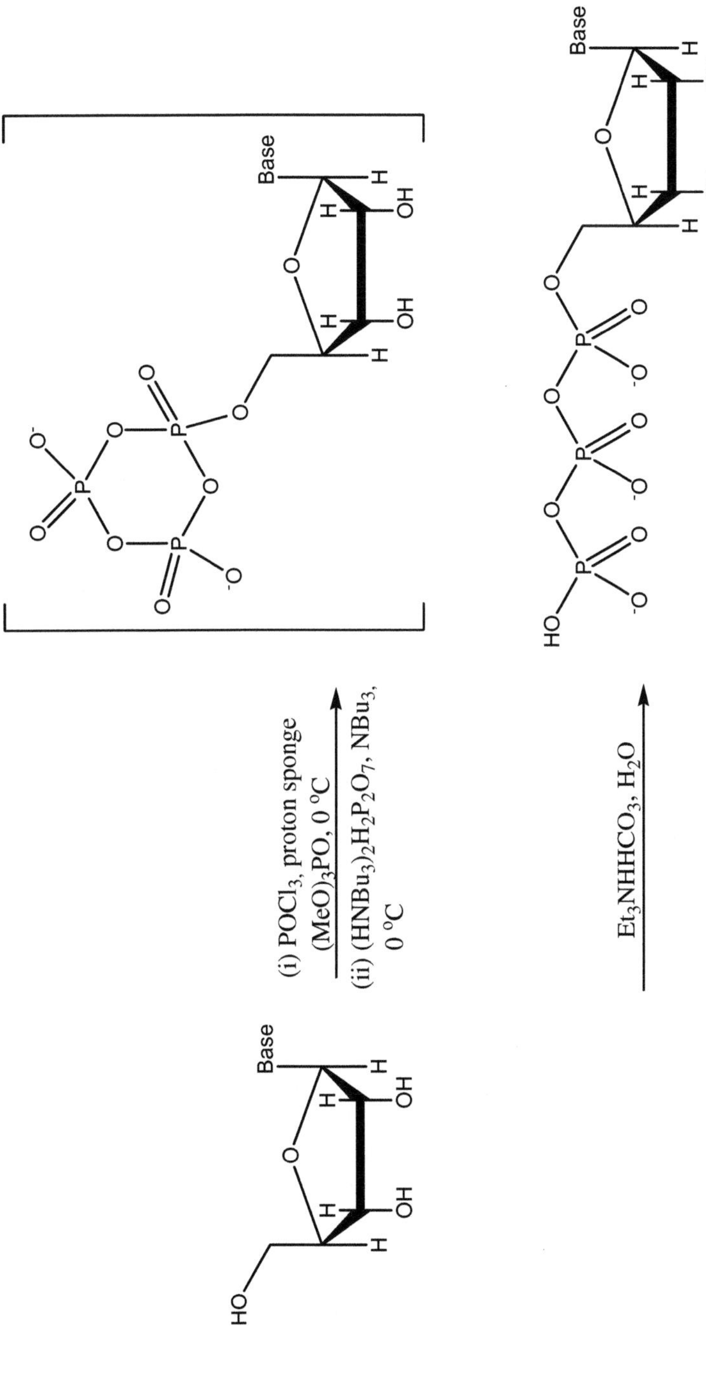

Fig. 7 The reaction scheme for nucleoside-5'-triphosphate synthesis by the "one-pot, three-step" method (Burgess and Cook 2000)

4.2 *Solid-Phase Synthesis*

The building blocks of solid-phase synthesis of RNA are phosphoramidites. To obtain high-yield nucleotide synthesis, each phosphoramidite is protected by an acid-labile dimethoxytrityl (DMT) group at the 5'-OH. The phosphoramidite at the 3'-OH is also protected by the base-labile cyanoethyl group. For RNA, the 2'-OH needs to be protected by various groups, such as *tert*-butyldimethylsilyl (TBDMS), 2'-O-triisopropylsilyloxymethyl (TOM), and bis(2-acetoxyethoxy)methyl (ACE) orthoester groups (Fig. 8).

Solid-phase synthesis of RNA procedes in the 3'to 5' direction. The 3'-OH is attached to a solid support and phosphoramidites are used to synthesize RNA sequences with modified bases such as pseudouridine on a DNA/RNA synthesizer (Meroueh et al. 2000; Scaringe 2000). The 3'-OH group of the first nucleotide is attached to a solid support and the 5'-OH group is protected. The DMT protecting group is removed by the addition of a fluoride source. Then, the phosphoramidite, which has the 2'-OH group protected, is introduced and couples with the first nucleotide. To prevent further coupling with the failed sequences, a capping reaction at the 5' end is carried out. Upon removal of the DMT-protecting group, the clear solution changes to orange allowing for the coupling efficiency of each phosphoramidite to be monitored by UV spectroscopy (Francis and Resendiz 2017). These procedures are repeated until the desired RNA sequence and length is obtained. Finally, the RNA product is cleaved from the solid support. At this point, the 2'-protecting groups are still attached which serves as an advantage for the newly synthesized oligonucleotide as it increases the RNA stability. When RNA is ready for use, the 2'-protection will be removed. The major advantages of solid-phase synthesis are the incorporation of modified bases at specific positions and its ability to start from any base. However, solid-phase synthesis has a length limit (usually less than 100 nucleotides) because of the solid support.

5 Conclusions

Modified nucleotides in RNA, including pseudouridine, are still mysterious and interesting RNA residues to investigate biochemically and biophysically. RNA synthesis with modified nucleotides is necessary to investigate its structural and functional effects in RNA. However, many modified nucleotides are not commercially available, and it is necessary to synthesize and incorporate them into RNA sequences. The introduction of modified nucleotides into various RNA sequences may also lead to RNA designs with novel functions that can be used in biotechnology and pharmaceutical engineering. Considering both advantages and disadvantages of chemical and enzymatic syntheses, we introduced the semi-enzymatic synthesis of pseudouridine using the reverse reaction catalyzed by pseudouridine monophosphate glycosidase.

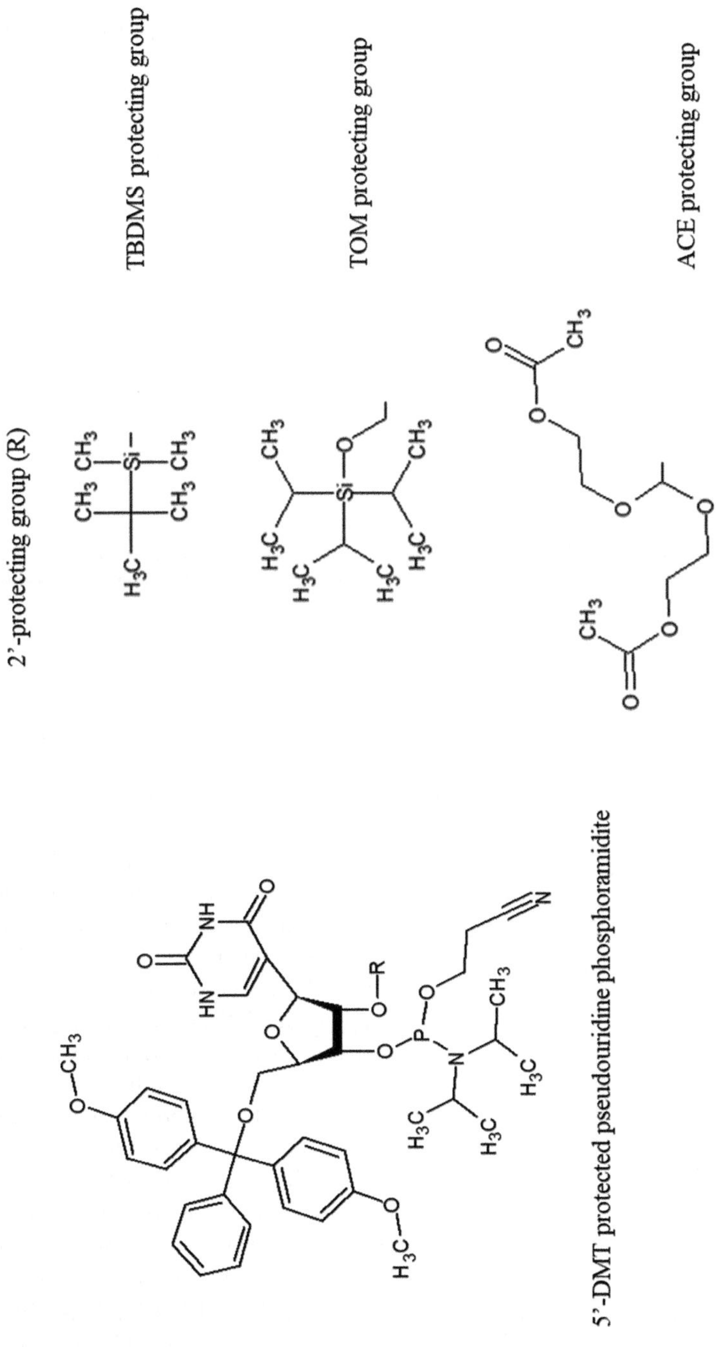

Fig. 8 The structure of pseudouridine phosphoramidite (left) and common 2'-protecting groups (right)

The synthetic route overcomes the challenges of stereoselectivity and moisture sensitivity of the coupling step between uracil and sugar by using the enzymatic reaction. The reaction has only 3 steps and the overall yield of the desired β-pseudouridine is 68% (Fig. 4). The product of the second step is pseudouridine-5'-monophosphate that can be converted to a triphosphate to use in in vitro transcription. The product of the final step is pseudouridine nucleoside which can be converted to a phosphoramidite to be used in solid-phase synthesis.

Acknowledgements This work was supported by faculty startup funds and graduate student research grants from Southern Illinois University Edwardsville.

References

Anderson BR, Muramatsu H, Nallagatla SR et al (2010) Incorporation of pseudouridine into mRNA enhances translation by diminishing PKR activation. Nucleic Acids Res 38:5884–5892

Andries O, Mc Cafferty S, De Smedt SC et al (2015) N1-methylpseudouridine-incorporated mRNA outperforms pseudouridine-incorporated mRNA by providing enhanced protein expression and reduced immunogenicity in mammalian cell lines and mice. J Controlled Release 217:337–344

Bhutani SP (2020) Chemistry of Biomolecules. CRC Press, Boca Raton, FL

Bousquet-Antonelli C, Henry Y, G'Elugne JP et al (1997) A small nucleolar RNP protein is required for pseudouridylation of eukaryotic ribosomal RNAs. EMBO J 16:4770–4776

Brimacombe R, Mitchell P, Osswald M et al (1993) Clustering of modified nucleotides at the functional center of bacterial ribosomal RNA. FASEB J 7:161–167

Burgess K, Cook D (2000) Syntheses of nucleoside triphosphates. Chem Rev 100:2047–2060

Chang YC, Herath J, Wang TH et al (2008) Synthesis and solution conformation studies of 3-substituted uridine and pseudouridine derivatives. Bioorg Med Chem 16:2676–2686

Charette M, Gray MW (2000) Pseudouridine in RNA: what, where, how, and why. IUBMB Life 49:341–351

Cohn WE (1960) Pseudouridine, a carbon-carbon linked ribonucleoside in ribonucleic acids: isolation, structure, and chemical characteristics. J Biol Chem 235:1488–1498

Davis DR (1995) Stabilization of RNA stacking by pseudouridine. Nucleic Acids Res 23:5020–5026

Davis DR, Poulter CD (1991) ^{1}H–^{15}N NMR studies of Escherichia coli $tRNA^{Phe}$ from hisT mutants: a structural role for pseudouridine. Biochemistry 30:4223–4231

Drahovsky D, Winkler A, Skoda J (1964) Increased Urinary Pseudouridine Excretion in Rats Following Irradiation. Nature 201:411–412

Feng B, Zheng MH, Zheng YF et al (2005) Normal and modified urinary nucleosides represent novel biomarkers for colorectal cancer diagnosis and surgery monitoring. J Gastroenterol Hepatol 20:1913–1919

Foster PG, Huang L, Santi DV et al (2000) The structural basis for tRNA recognition and pseudouridine formation by pseudouridine synthase I. Nat Struct Biol 7:23–27

Francis AJ, Resendiz MJE (2017) Protocol for the Solid-phase Synthesis of Oligomers of RNA Containing a 2'-O-thiophenylmethyl Modification and Characterization via Circular Dichroism. J vis Exp 28:e56189

Grohar PJ, Chow CS (1999) A Practical Synthesis of the Modified RNA Nucleoside Pseudouridine. Tet Lett 40:2049–2052

Gu X, Liu Y, Santi DV (1999) The mechanism of pseudouridine synthase I as deduced from its interaction with 5-fluorouracil-tRNA. Proc Natl Acad Sci USA 96:14270–14275

Hanessian S, Machaalani R (2003) A highly stereocontrolled and efficient synthesis of α- and β-pseudouridines. Tetrahedron Lett 44:321–8323
Huang S, Mahanta N, Begley TP et al (2012) Pseudouridine monophosphate glycosidase: a new glycosidase mechanism. Biochemistry 51:9245–9255
Hurd RE, Reid BR (1977) NMR spectroscopy of the ring nitrogen protons of uracil and substituted uracils; relevance to AΨ base pairing in the solution structure of transfer RNA. Nucleic Acids Res 4:2747–2755
Jonkhout N, Tran J, Smith MA et al (2017) The RNA modification landscape in human disease. RNA 23:1754–1769
Kiss T (2001) Small nucleolar RNA-guided post-transcriptional modification of cellular RNAs. EMBO J 20:3617–3622
Lafontaine DL, Tollervey D (1998) Birth of the snoRNPs: the evolution of the modification-guide snoRNAs. Trends Biochem Sci 23:383–388
Lerch U, Burdon MG, Moffatt JG (1971) C-glycosyl nucleosides. I. Studies on the synthesis of pseudouridine and related compounds. J Org Chem 36:1507–1513
Ludwig J, Eckstein F (1989) Rapid and efficient synthesis of nucleoside 5'-o-(1-thiotriphosphates), 5'-triphosphates and 2',3'-cyclophosphorothioates using 2-chloro-4H-1,3,2-benzodioxaphosphorin-4-one. J Org Chem 54:631–635
Maden EH, Wakeman JA (1988) Pseudouridine distribution in mammalian 18S ribosomal RNA. A major cluster in the central region of the molecule. Biochem J 249:459–464
Meroueh M, Grohar PJ, Qiu J et al (2000) Unique structural and stabilizing roles for the individual pseudouridine residues in the 1920 region of *Escherichia coli* 23S rRNA. Nucleic Acids Res 28:2075–2083
Milligan JF, Uhlenbeck OC (1989) Synthesis of small RNAs using T7 RNA polymerase. Methods Enzymol 180:51–62
Mitsui T, Kitamura A, Kimoto M et al (2003) An unnatural hydrophobic base pair with shape complementarity between pyrrole-2-carbaldehyde and 9-methylimidazo[(4,5)-b]pyridine. J Am Chem Soc 125:5298–5307
Mohamady S, Jakeman DL (2005) An improved method for the synthesis of nucleoside triphosphate analogues. J Org Chem 70:10588–10591
Nance KD, Meier, JL (2021) Modifications in an emergency: the role of N1-Methylpseudouridine in COVID-19 vaccines. ACS Cent Sci 7:748–756
Neumann JM, Bernassau JM, Gueron M et al (1980) Comparative conformations of uridine and pseudouridine and their derivatives. Euro J Biochem 108:457–463
Newby MI, Greenbaum NL (2002) Investigation of Overhauser effects between pseudouridine and water protons in RNA helices. Proc Natl Acad Sci USA 99:12697–12702
Ofengand J (2002) Ribosomal RNA pseudouridines and pseudouridine synthases. FEBS Lett 514:17–25
Ofengand J, Bakin A, Wrzesinski J et al (1995) The pseudouridine residues of ribosomal RNA. Biochem Cell Biol 73:915–924
Pardi N, Hogan MJ, Naradikian MS et al (2018) Nucleoside-modified mRNA vaccines induce potent T follicular helper and germinal center B cell responses. J Exp Med 215:1571–1588
Penzo M, Guerrieri AN, Zacchini F et al (2017) RNA Pseudouridylation in Physiology and Medicine: For Better and for Worse. Genes (basel) 8:301
Preumont A, Snoussi K, Stroobant V et al (2008) Molecular identification of pseudouridine-metabolizing enzymes. J Biol Chem 283:25238–25246
Rafels-Ybern A, Torres AG, Grau-Bove X et al (2018) Codon adaptation to tRNAs with Inosine modification at position 34 is widespread among Eukaryotes and present in two Bacterial phyla. RNA Biol 15:500–507
Ramanathan A, Robb GB, Chan SH (2016) mRNA capping: biological functions and applications. Nucleic Acids Res 44:7511–7526
Riley AT, Sanford TC, Woodard AM et al (2021) Semi-enzymatic synthesis of pseudouridine. Bioorg Med Chem Lett 44:128105

Rintala-Dempsey AC, Kothe U (2017) Eukaryotic stand-alone pseudouridine synthases - RNA modifying enzymes and emerging regulators of gene expression? RNA Biol 14:1185–1196

Roundtree IA, Evans ME, Pan T et al (2017) Dynamic RNA Modifications in Gene Expression Regulation. Cell 169:1187–1200

Scaringe SA (2000) Advanced 5'-silyl-2'-orthoester approach to RNA oligonucleotide synthesis. Methods Enzymol 317:3–18

Schaefer M, Kapoor U, Jantsch MF (2017) Understanding RNA modifications: the promises and technological bottlenecks of the 'epitranscriptome'. Open Biol 7

Shapiro R, Chambers RW (1961) Synthesis of Pseudouridine. J Am Chem Soc 83:3920–3921

Spenkuch F, Motorin Y, Helm M (2014) Pseudouridine: still mysterious, but never a fake (uridine)! RNA Biol 11:1540–1554

Stockert JA, Weil R, Yadav KK et al (2021) Pseudouridine as a novel biomarker in prostate cancer. Urol Oncol 39:63–71

Tamura S, Fujioka H, Nakano T et al (1988) Urinary pseudouridine as a biochemical marker in the diagnosis and monitoring of primary hepatocellular carcinoma. Am J Gastroenterol 83:841–845

Thapa K, Oja T, Metsa-Ketela M (2014) Molecular evolution of the bacterial pseudouridine-5'-phosphate glycosidase protein family. FEBS J 281:4439–4449

Yarian CS, Basti MM, Cain RJ et al (1999) Structural and functional roles of the N1- and N3-protons of Ψ at tRNA's position 39. Nucleic Acids Res 27:3543–3549

Yoshikawa M, Kato T, Takenishi T (1967) A novel method for phosphorylation of nucleosides to 5'-nucleotides. Tet Lett 8:5065–5068

Yu YT, Meier UT (2014) RNA-guided isomerization of uridine to pseudouridine–pseudouridylation. RNA Biol 11:1483–1494

Zhao Y, Dunker W, Yu YT et al (2018) The Role of Noncoding RNA Pseudouridylation in Nuclear Gene Expression Events. Front Bioeng Biotechnol 6:8

Molecular Dynamics Simulations of Chemically Modified Ribonucleotides

Valerio Piomponi, Mattia Bernetti, and Giovanni Bussi

Contents

Abstract Post-transcriptional modifications are crucial for RNA function, with roles ranging from the stabilization of functional RNA structures to modulation of RNA–protein interactions. Additionally, artificially modified RNAs have been suggested as optimal oligonucleotides for therapeutic purposes. The impact of chemical modifications on secondary structure has been rationalized for some of the most common modifications. However, the characterization of how the modifications affect the three-dimensional RNA structure and dynamics and its capability to bind proteins is still highly challenging. Molecular dynamics simulations, coupled with enhanced sampling methods and integration of experimental data, provide direct access to RNA structural dynamics. In the context of RNA chemical modifications, alchemical simulations where a wild type nucleotide is converted to a modified one are particularly common. In this Chapter, we review recent molecular dynamics studies of modified

V. Piomponi · G. Bussi (✉)
Scuola Internazionale Superiore di Studi Avanzati-SISSA, via Bonomea 265, 34136 Trieste, Italy
e-mail: bussi@sissa.it

M. Bernetti
Computational and Chemical Biology, Italian Institute of Technology, 16152 Genova, Italy

Department of Pharmacy and Biotechnology, Alma Mater Studiorum-University of Bologna, 40126 Bologna, Italy

J. Barciszewski (ed.), *RNA Structure and Function*, RNA Technologies 14,
https://doi.org/10.1007/978-3-031-36390-0_26

ribonucleotides. We discuss the technical aspects of the reviewed works, including the employed force fields, enhanced sampling methods, and alchemical methods, in a way that is accessible to experimentalists. Finally, we provide our perspective on this quickly growing field of research. The goal of this Chapter is to provide a guide for experimentalists to understand molecular dynamics works and, at the same time, give molecular dynamics experts a solid review of published articles that will be a useful starting point for new research.

Keywords RNA · Chemical modification · Molecular dynamics simulations

1 Introduction

RNA molecules are sequences of four common nucleotides: adenosine (A), uridine (U), cytidine (C), and guanosine (G). However, a number of naturally occurring or artificially synthesized nucleotides can be incorporated as well (Fig. 1). Naturally occurring modifications are often chemical marks on cellular RNA, are regulated and recognized by proteins known as *writers* and *readers*, respectively, and are usually referred to as *post-transcriptional* modifications. More than a hundred of them have been identified to date. Transfer RNAs (tRNAs) are known to be heavily modified with a variety of modifications found both in the anticodon region and in the tRNA-body region (Suzuki 2021). Ribosomal RNA (rRNA) is also extensively edited after transcription (Jiang et al. 2016). Recent technical advances revealed widespread modifications also on messenger RNAs (mRNAs) (Gilbert et al. 2016). In general, post-transcriptional modifications have two roles: (i) they allow correct folding of noncoding RNAs (e.g., tRNA and rRNA) into their functional structure and (ii) they affect the target specificity of RNA–RNA and RNA–protein interactions. In addition to naturally occurring modifications, a number of artificially modified nucleotides have been studied, mostly in the context of oligonucleotide design (Wan and Seth 2016). Synthetic oligonucleotides hold great potential as innovative therapeutic strategies, including small interfering RNAs (siRNAs), antisense oligonucleotides, microRNAs, and aptamers. However, intrinsic limitations in terms of instability, immunogenicity, and poor pharmacokinetic properties hamper their use. Therefore, artificial modifications of RNA nucleotides were explored to overcome these limitations and optimize the oligonucleotide properties. A remarkable example in this respect, though through the artificial insertion of a natural modification, is pseudouridine (Ψ) in mRNA-vaccine technology (Karikó et al. 2008).

Although the research on RNA modifications has been exponentially increasing in the past years, computational studies on modified RNAs are still limited, even in the relatively simpler context of secondary structure prediction (Tanzer et al. 2019). Additional complexity is present in the study of the impact of modifications on tertiary structure. Models for tertiary structure predictions are typically trained on available structural datasets (Parisien and Major 2008; Townshend et al. 2021), and the statistics available on modified nucleotides is scarce. Furthermore, methods trained on static structures give limited access to structural dynamics. In this respect, molecular dynamics (MD) simulations (Šponer et al. 2018) are a promising tool since (i) they give direct access to dynamics and (ii) they are grounded in physics-based models, which could possibly be capable to describe systems for which the amount of reference experimental structures is limited.

The aim of this Chapter is to review recent applications of MD simulations to the study of chemically modified ribonucleotides. In particular, we will first describe the basic principles of MD simulations, including advanced methods to enhance sampling and compute mutation free energies. Then, we will review selected applications of MD to the study of the effect of modifications on RNA structural dynamics and RNA recognition. Finally, we will provide our perspective on the field. A recent related review complements ours by providing a different perspective (D'Esposito et al. 2022).

2 Molecular Dynamics Simulations

2.1 Standard Molecular Dynamics Simulations

Molecular dynamics simulations are a natural tool to characterize RNA structural dynamics (Šponer et al. 2018). A key ingredient of any molecular dynamics (MD) simulations is the employed force field, which is a function that computes the forces on the atoms given their positions. Since evaluating the force field is the computational bottleneck of MD simulation, its functional form has to be chosen with compromises, so as to be accurate enough to describe the relevant chemistry but not too expensive. The functional form of the commonly used AMBER force field is the following one:

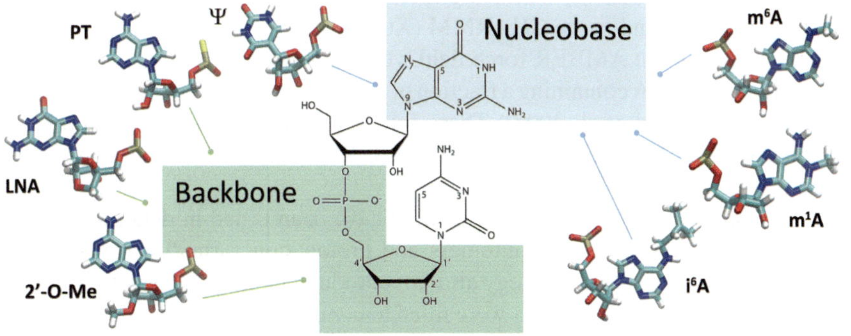

Fig. 1 Schematic representing some of the nucleotide modifications that are discussed in this Chapter

$$
\begin{aligned}
E = & \sum_{\text{bonds}} \frac{1}{2} k_b (r - r_0)^2 + \sum_{\text{angles}} \frac{1}{2} k_a (a - a_0)^2 \\
& + \sum_{\text{torsions}} \sum_{n} \frac{V_n}{2} (1 + \cos(n\phi - \delta)) \\
& + \sum_{LJ} 4\epsilon_{ij} \left(\left(\frac{\sigma_{ij}}{r_{ij}} \right)^{12} - \left(\frac{\sigma_{ij}}{r_{ij}} \right)^{6} \right) \\
& + \sum_{\text{electrostatics}} \frac{q_i q_j}{r_{ij}}
\end{aligned}
\tag{1}
$$

Here, k_b, k_a, and V_n control the so-called bonded interactions (bonds, angles, and torsional angles, respectively), σ and ϵ control Lennard-Jones potentials, and charges q control electrostatic interactions.

The parameters of the force field are heavily system dependent and are derived using a mixture of accurate quantum chemical calculations and of experimental data (Fröhlking et al. 2020). The two main families of force fields used for nucleic acids are AMBER and CHARMM, both of which have evolved in multiple revised versions during the past decades.

2.2 Force Fields for Chemically Modified Nucleotides

Specific force-field parameters should be derived for each type of modification. The AMBER family of force fields offers a well-defined recipe for arbitrary molecules. In particular, charges are obtained by fitting the electrostatic potential, and torsional parameters by fitting the energy profiles associated with bond rotations. For the CHARMM force field, the procedure is more complex and targets both quantum mechanical data on nucleosides and experimental data on nucleosides or oligonucleosides. Luckily, parameters obtained using these procedures have been published for approximately 100 naturally occurring modified nucleotides, both in the AMBER (Aduri et al. 2007) and in the CHARMM (Xu et al. 2016b) frameworks.

Parameters for the AMBER force field were validated performing standard MD simulations of a tRNA containing a fraction of the modifications for which parameters were reported (Aduri et al. 2007). These force-field parameters have been used in several later MD simulations using the AMBER force field, and are also used in modeling tools (Stasiewicz et al. 2019).

Parameters for the CHARMM force field have been tested in detail for thirteen modifications. Parameters for tautomers and protonation variants have also been included. The charge-fitting strategy aims at reproducing interactions with water and correct dipole moments. Torsions were fitted to computing potential energy surfaces with quantum mechanical methods. Simulations of nucleosides and trinucleotides were compared with nuclear magnetic resonance (NMR) data, when available. These force-field parameters have been used in several later MD simulations using the CHARMM force field.

In addition to these two works, it is relevant to mention that for many of the applications discussed below the authors developed and tested new sets of force-field parameters specific for a single or a few modifications.

2.3 Enhanced Sampling Methods

RNA molecules are often characterized by conformational ensembles composed of multiple partly heterogeneous structures or substates that are relevant for function (Ganser et al. 2019). MD simulations can access at most the multi-microsecond timescale with current resources. Thus, changes in tertiary interactions and base-pairing patterns cannot be directly simulated with MD. To circumvent this problem, enhanced sampling methods can be used.

Enhanced sampling methods (Mlýnský and Bussi 2018; Hénin et al. 2022) are roughly classified into two categories. One category includes methods based on heating the system so as to accelerate the exploration of the conformational space. A representative of these methods is parallel tempering, also known as temperature replica exchange (Sugita and Okamoto 1999). These methods are typically very expensive and can thus be fruitfully applied only for sampling small oligomers. The other category includes methods based on adding biasing forces on specifically chosen degrees of freedom, or collective variables. A representative of these methods is umbrella sampling, usually performed by combining multiple windows (Kumar et al. 1992) so as to progressively convert the system from an initial conformation to a final one. These methods can be used to accelerate relevant events if sufficient prior information about the process is given. More recent variants of the mentioned methods exist, but have not been used to simulate modified ribonucleotides yet.

2.4 Alchemical Methods

Alchemical methods allow simulating trajectories where molecular species are mutated to different ones and the free energy associated with the transformation can be computed (Mey et al. 2020). Most MD codes support these methods, but setting up the simulations is usually more complex than for standard MD. The intermediate states might be simulated independently of each other or with a more robust replica-exchange procedure (Meng et al. 2011). Simulations should be then repeated in different structural contexts. For instance, the conversion between (unmodified) A and (modified) m^6A can be performed in a single strand and in a duplex. The difference between the free-energy changes computed in the two contexts corresponds to the stabilization of the duplex resulting from the additional methylation (Fig. 2). Similarly, the impact of the modification on the affinity between the studied RNA and a protein can be estimated.

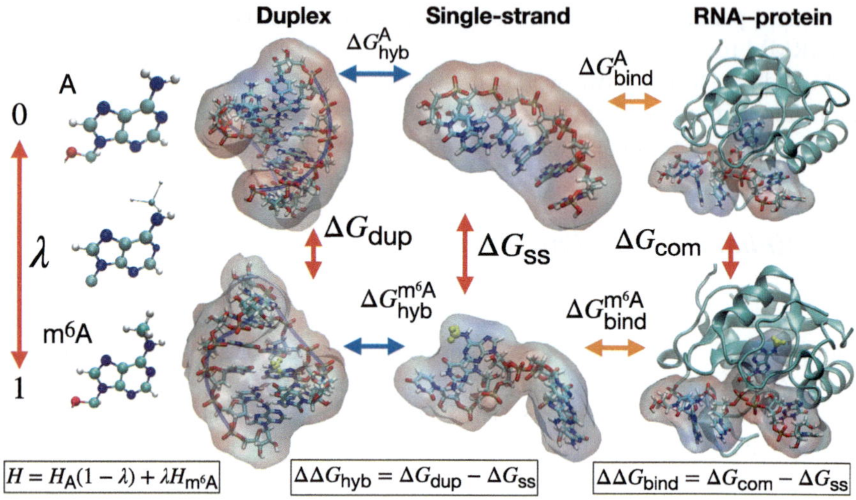

Fig. 2 Schematic representation of alchemical simulation protocols. Transformation A↔m^{6}A (vertical red arrows) can be used to compute the free-energy change in different structural contexts (duplex, ΔG_{dup}, single strand, ΔG_{ss}, and RNA–protein complex, ΔG_{com}). The Hamiltonian function (H) is interpolated between the two physical systems. Experiments (horizontal blue and orange arrows) report hybridization free energies (ΔG_{hyb}) and free energies of binding (ΔG_{bind}). $\Delta\Delta G$s can be directly compared between simulation and experiment. E.g., $\Delta G_{dup} - \Delta G_{ss} = \Delta G_{hyb,m^6A} - \Delta G_{hyb,A}$

The results of alchemical simulations should be judged with care. Specifically, if there are slow degrees of freedom coupled with the alchemical change, the result might be affected by artifacts. A typically difficult situation arises when one or both the alchemical states correspond to flexible conformations whose extensive sampling is difficult. A possible way to alleviate this problem consists of combining alchemical simulations with enhanced sampling methods, as done, for instance, in alchemical metadynamics (Hsu et al. 2023).

3 Applications

3.1 Validation and Fitting of Force Fields Against Experimental Data

Quantitative validations are crucial to assess the capability of force-field parameters to generate structural ensembles compatible with experiments.

It has been shown that the Aduri parametrization is not able to reproduce substate populations for a number of modified uridines (Deb et al. 2014). A reparametrization for Ψ, s^{2}U and s^{4}U has been provided (Deb et al. 2016), and it has been discussed how changes in the Lennard-Jones parameters could affect the relative population of different sugar conformations. These parameters can be transferred to other modified uridines (Dutta et al. 2020). More recently, charges for Ψ and three different methylated versions of Ψ were reparametrized and validated simulating single-stranded oligonucletides and obtaining conformational and hydration properties in agreement with experimental data (Dutta et al. 2022).

Aduri parameters for m^6A were validated using alchemical simulations, confirming their capability to reproduce melting experiments (Hurst and Chen 2021). However, a later work from our group using a similar protocol on a larger validation set showed that modifications to the charges are necessary to simultaneously reproduce thermodynamic data and *syn/anti* balance for the methyl group (Piomponi et al. 2022).

3.2 *Effects of Post-transcriptional Modifications on RNA Structure and Dynamics*

Here we review MD simulation studies aimed at elucidating the impact of post-transcriptional modifications on the structure and dynamics of tRNAs, rRNAs, and other species.

3.2.1 Molecular Simulations of Anticodon Stem-Loops and of Entire tRNA Molecules

A number of works used MD simulations to investigate the structural role of modifications on entire tRNAs, suggesting them to be crucial for the stabilization of the functional structure. An extensive study on three tRNAs compared all-modified with non-modified tRNAs (Zhang et al. 2014). Modifications were shown to increase the rigidity of the anticodon stem-loop, presumably facilitating pairing with mRNA during translation. Overall, the effect of the modifications was suggested to be non-trivial, making tRNA more rigid in some regions and more flexible in other regions. An independent work also reported that the presence of modifications and of Mg^{2+} ions makes the functional tRNA structure more stable (Xu et al. 2016a). The effect of modifications in position 37 in the anticodon loop (including G to A, N2-methylguanosine, m^2G, N2-dimethylguanosine, m^2_2G, N6-isopentenyladenosine, i^6A, and others) was also characterized, showing that a single modified position could crucially stabilize the functional structure of the anticodon loop (see Fig. 3) (Sonawane et al. 2016; Prabhakar et al. 2021).

The effect of modifications on the anticodon loop was addressed in several other works, where simulations were performed on smaller fragments. Notably, $tRNA_3^{Lys}$ is a primer for reverse transcription in the life cycle of human immunodeficiency virus (HIV) and displays modifications at positions 34, 37, and 39. These modifications

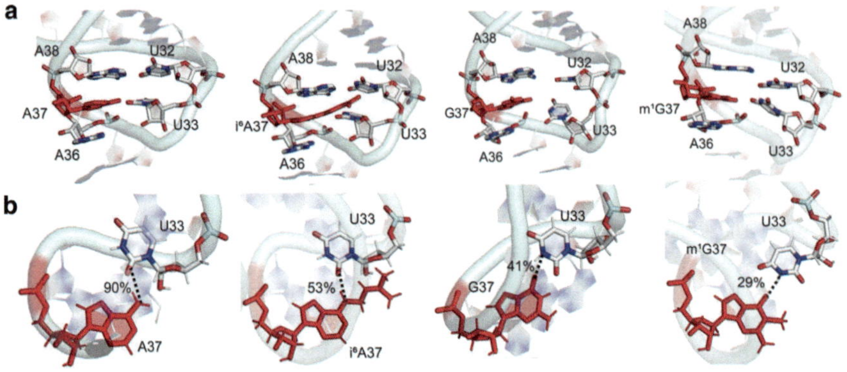

Fig. 3 **a** Stacking interactions between unmodified A37 and modified i^{6}A37/G37/m^{1}G37 and neighboring nucleobases in the anticodon loop. **b** Hydrogen-bonding interactions (with occupancies) between base A37/i^{6}A37/G37/m^{1}G37 and Watson-Crick base U33 as obtained with MD simulations. Adapted with permission from Prabhakar et al. (2021)

are known to be important to stabilize interaction with HIV A-loop. MD simulations of the anticodon loop in complex with HIV A-loop were used to dissect the contribution of these modifications, showing that modifications at positions 34 and 39 are required for stabilizing the loop structure, whereas modifications at positions 34 and 37 are specifically stabilizing the complex with the A-loop (Galindo-Murillo et al. 2016). Given the relevance of this complex, another study used MD simulations to characterize the interaction of these modified nucleotides with a computationally designed peptide (Xiao et al. 2016). The interactions between pairs of amino acids and nucleotides were separately analyzed to show how the peptide can specifically recognize the modified tRNA compared to decoy sequences. Interestingly, the modification at position 37 is also relevant for binding one of the three synonymous codons for Lysine (AAG). MD has been used to dissect the effect of this modification in binding with a model mRNA sequence, specifically monitoring how the modification tuned the Van der Waals interactions and the hydration of the stem-loop (Vangaveti et al. 2022). Other modifications of the anticodon loop were shown to be important for codon recognition. For instance, modifications at positions 32 and 37 were shown to be crucial for the overall loop dynamics to accommodate the pairing of inosine, another modified nucleotide, at wobble position (Vangaveti et al. 2020). Finally, a synthesis protocol has been introduced for geranylated nucleotides, recently identified in bacteria (Wang et al. 2016). This work also includes MD simulations of geranylated nucleotides in RNA and DNA duplexes and in anticodon–codon pairs, including the ribosomal subunit with all its associated proteins, showing how geranylation of udirine affects its capability to bind nucleobases at the third position of the codon, specifically forming a weaker hydrogen bonding with A, while maintaining pairing with G.

3.2.2 Molecular Simulations of Entire Ribosomes

Ribosomal RNAs contain a large number of post-transcriptional modifications. Therefore, MD simulations of the ribosome usually take these modifications into account. However, in many cases, the modifications are not directly involved in the process of interest. Simulations of the *E. coli* ribosome in multiple variants were used to characterize the relationship between mutations and modifications and resistance to macrolides, a widely prescribed class of antibiotics (Pavlova et al. 2017). Several modifications of an adenine in the exit tunnel were simulated (G, m^6A and m^6_6A) in complex with a number of macrolides. The modifications were found to weaken the interaction of the ribosome with the macrolides, thus providing a microscopic explanation for the observed antibiotic resistance.

3.2.3 Molecular Simulations of Model Molecules

A number of papers focused on the effect of post-transcriptional modifications on the energetics and structure of model molecules, ranging from individual base pairs to duplexes.

Some works addressed the dynamics of individual or paired nucleotides. Short MD simulations of modified nucleotides (m^2G and m^2_2G) were reported and compared with crystal structures (Bavi et al. 2013). When a single methyl group is present (m^2G), it can rotate thus allowing either canonical or non-canonical pairing. Dimethylation of G, instead, was reported to inhibit pairing with C, so that m^2_2G is suggested to only pair with A or U. Umbrella-sampling simulations were used to characterize the effect of modifications on base-pairing free energy and geometry (Vendeix et al. 2009). In particular, computations showed how the 5-oxyacetic acid uridine is able to form mismatch base pairs with U and C, thanks to the coordination with a water molecule, with a geometry compatible with an A-form helix, as expected from experiments. The thermodynamic stability of multiple modified nucleotides was studied using a combination of MD simulations and quantum mechanical calculations (Hopfinger et al. 2020). Here, a protocol for accurate prediction of dash nearest-neighbor parameters for modified nucleotides was proposed. The method could quantify both the stacking and hydrogen bonding contribution to the energy difference between natural and modified nucleotides.

Several works specifically studied pseudouridine (Ψ). A few of them addressed the impact of U to Ψ modification in the structure and stability of duplexes containing CUG or CΨG repeats. Enhanced sampling calculations were used to estimate the free-energy change associated with the opening of the central base pair, which was found to be larger in the presence of Ψs in one or in both the strands, relating the result to a change in the water coordination of the modified nucleotide (deLorimier et al. 2014). A later work from the same group showed that the rigidity of Ψ in the context of CUG and CCUG repeats resulted in lower affinity with a RNA–binding protein (deLorimier et al. 2017). Similar results were obtained using a different force field and thereby confirmed by NMR experiments (Deb et al. 2019). Another modification

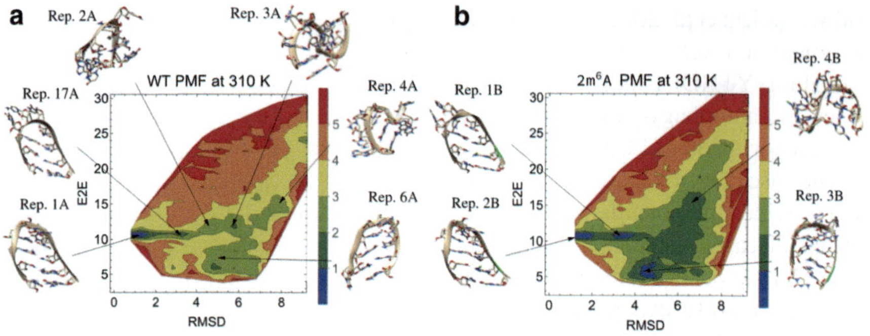

Fig. 4 Free-energy landscapes associated with the folding of a hairpin loop, either unmodified (**a**) or N6-methylated at position 2 (**b**). The free energy is reported in kcal/mol and shown as a function of two analyzed variables, end-to-end distance (E2E) and root-mean-square deviation (RMSD) from the native structure. Selected representative structures are reported. Whereas the native structure is long-lived in the modified case (structure 2B), additional metastable conformations appear (1B and 3B). Adapted with permission from Hurst and Chen (2021)

of uridine, s^2U, has been studied, showing that intercalation of the sulfur atom can induce order in single-stranded RNAs with consecutive uridines (Sarkar et al. 2020).

Another commonly found modification is m^6A. The free-energy landscape of a hairpin composed of a tetraloop and three base pairs with or without modifications, including m^6A, was simulated with the Aduri parameters (Hurst and Chen 2021). Simulations were accelerated using temperature replica exchange MD. Remarkably, methylations at different positions were shown to stabilize or destabilize the hairpin structure in specific contexts (see Fig. 4), which may provide a tool for RNA nanotherapeutic design.

The consequences of N1 adenosine and guanosine methylation (m^1A and m^1G) on duplex dynamics was studied with MD (Zhou et al. 2016). In particular, this work shows that this modification changes the energetic balance between Watson–Crick and Hoogsteen pairing, with the result of being not tolerated in RNA, where it can induce duplex melting. We note that, compared with other common nucleobase methylations, m^1A shifts the nucleotide charge by +1. Interestingly, the N1-methylation of A9 in a tRNA has been shown to be necessary to suppress a non-functional hairpin structure, thus restoring the functional cloverleaf structure (Voigts-Hoffmann et al. 2007). This happens in fact by destabilizing the duplex formed by an extended acceptor stem.

A few papers addressed inosine (I). Given the similarity in the hydrogen-bond pattern formed by I and G, I can pair with both U and C, though the I·C pair only forms two hydrogen bonds. The thermodynamics of I·C pairs was studied using MD simulations of RNA and DNA duplexes subjected to alchemical calculations (Krepl et al. 2013). The thermodynamic cycle was performed by converting a G·C pair to a I·C pair and computing the corresponding destabilization of the duplex, and results were correlated with experimental thermodynamic data. A more systematic study

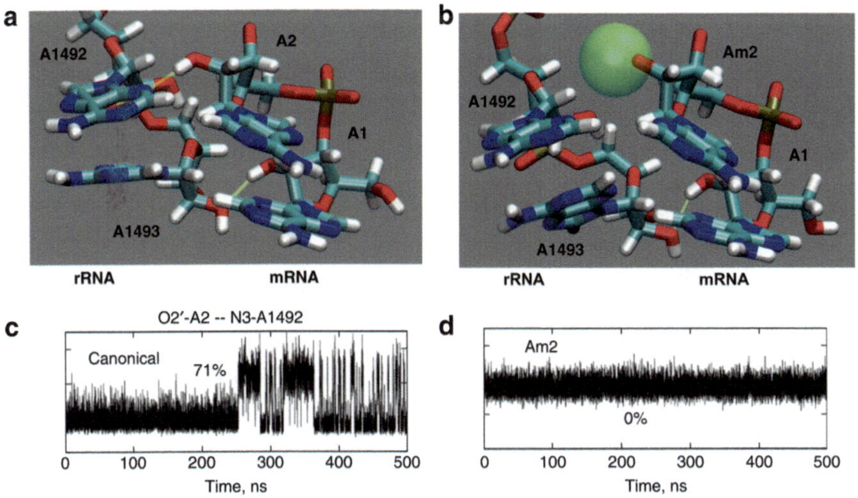

Fig. 5 Molecular dynamics modeling of 2′-OH methylations of mRNA at the ribosomal A-site. Representative configurations of the canonical (**a**) and methylated (**b**) structures, where the methyl group is indicated as a yellow-green sphere. Time series of the donor–acceptor distances (**c**) in the unmodified case and of the A1492–methyl-group distance in the methylated case (**d**). Adapted with permission from Elliott et al. (2019)

including more sequences as well as new experimental data confirmed that MD can correctly predict differential stability of G·C and I·C pairs (Sakuraba et al. 2020). I·U pairs are less stable than I·C pairs. MD simulations of I·U pairs showed that dynamics of tandem I·U pairs depend on the neighboring base, with UII being the most rigid sequence, with a limited impact on structure with respect to unmodified nucleotides (Špačková and Réblová 2018).

Finally, not only tRNA and rRNA can be modified, as discussed in previous Sections, but also mRNA. The effects of 2′-O-methylation in mRNA on gene expression were investigated using a number of experimental essays (Elliott et al. 2019). Experimental results were supported by MD simulations of tRNA in complex with portions of rRNA and mRNA. Individual 2′-O-methylations in a model codon were sufficient to destabilize binding between mRNA and tRNA. Whereas these results are only qualitative, they suggest that 2′-O-methylations could disrupt a wide range of other RNA interactions that are dependent on the 2′-OH group (see Fig. 5).

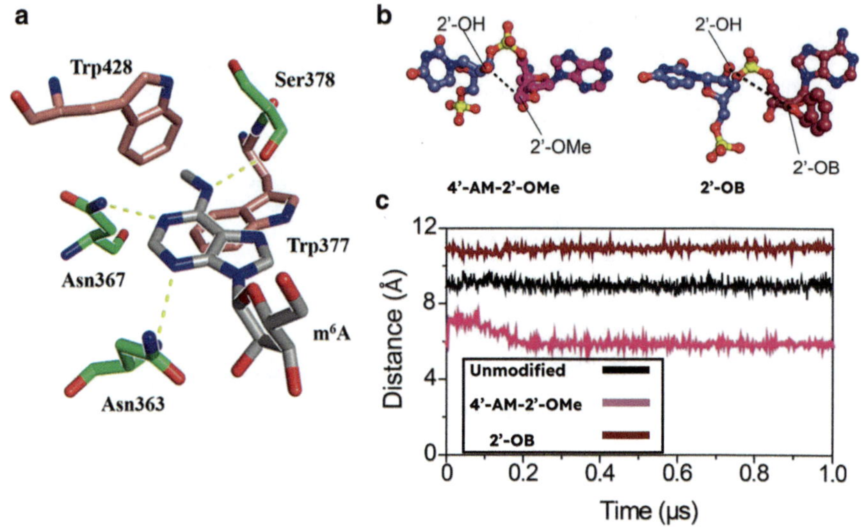

Fig. 6 **a** Detailed interactions between m^{6}A and pocket of YTHDC1. Adapted with permission from (Li et al., 2019). The full structure of this complex can be seen in Fig. 2. **b** The conformation of dinucleotide junction for 4′-AM-2′-OMe (magenta) and 2′-OB (ruby) modified siRNA2-hAGO2 complexes. **c** Time dependent variations of distance between 2′-oxygen atoms of the unmodified (black), 4′-AM-2′-OMe (magenta), and 2′-OB modified (brown) siRNA2-hAGO2 complexes. Adapted with permission from Harikrishna and Pradeepkumar (2017)

3.3 Effects of Post-transcriptional Modifications on Interaction with Proteins

3.3.1 RNA Recognition by Reader Domain YTHDC1

The YTHDC1 is one of the most studied m^{6}A readers. The RNA–protein complex has been solved for different oligonucleotide sequences, invariably showing that m^{6}A is captured in an aromatic cage, with the flanking nucleotides laying at the RNA–protein surface. Several MD studies have characterized the hydrogen-bond networks formed in the complex (Li et al. 2019, 2021, 2022; Krepl et al. 2021; Zhou et al. 2022).

Simulations with the CHARMM force field were reported, using a 5-nucleotides RNA single strand (5′-GG(m^{6}A)CU-3′), highlighting the role of two tryptophan residues in the pocket, respectively stacking on m^{6}A and stabilizing its methyl group (Fig. 6a) (Li et al. 2019). The role of the flanking nucleotides was studied by both analyzing their fluctuations in the simulations and performing experiments with shorter variants. This work was extended by performing alchemical calculations to estimate the complex stabilization induced by m^{6}A, using as a reference an isolated nucleoside (Li et al. 2021). Stabilization was slightly overestimated with respect to reference experimental data. A crucial water molecule was identified in the binding site and also studied with alchemical methods.

A related study with a similar RNA sequence (5′- CG(m^6A)CAC-3′) used AMBER parameters and specifically developed charges (Krepl et al. 2021). The authors performed alchemical calculations, identifying a water molecule entering the binding site in a position occupied by the methyl group. Also, in this case, the stabilization of the complex was overestimated when compared to experiment. Results were shown to be dependent on the protocol used to initialize the alchemical calculation.

3.3.2 Other RNA–Protein Interactions

The interaction between RNA strands with a single oxidized G (8-oxoG) and a polynucleotide phosphorylase implicated in RNA turnover was characterized using MD simulations (Gonzalez-Rivera et al. 2020). Interestingly, a protocol was developed based on relatively short simulations to assess the affinity of mutated protein sequences for both the modified and unmodified RNA, validating the optimized sequences by affinity measurements. The protocol was based on implicit solvent calculations of the RNA–protein interaction, thus with more approximations than the explicit solvent protocols mentioned in the previous Section; nonetheless, it was capable of identifying two binding regions within the polynucleotide phosphorylase channel and to reasonably predict the affinity of many mutant sequences.

3.4 Molecular Dynamics Simulations of Synthetic Nucleic Acids

Synthetic nucleic acids often display backbone modifications that are introduced to facilitate their therapeutic use, often aimed at modulating degradation and/or affinity with the target sequences. MD simulations have been used to understand how the modifications impact backbone flexibility and hybridization energies. siRNAs including a 4′-C-aminomethyl-2′-O-methyl modification were syntesized and analyzed using MD simulations (Gore et al. 2012). The simulations showed that the modification induces changes in the sugar puckering, flexibility in groove dimension, and disturbances in hydrogen bonding and base stacking, resulting in lower duplex stability as confirmed experimentally. In a later work, plain MD simulations of siRNA in complex with Argonaute 2 were used to elucidate how a number of synthetic modifications affect protein–RNA interactions (Fig. 6b) (Harikrishna and Pradeepkumar 2017). Simulations were performed on the microsecond timescale, showing that the minimum timescale to see this conformational variability is of the order of 300 ns (Fig. 6c). A number of 2′-O-modified nucleotides were characterized with MD simulations (Masaki et al. 2010). These modifications affect the sugar flexibility and thus the fluctuations of duplex helical parameters. The authors identified linear relationships between the predicted fluctuations and the duplex thermal stability, suggesting the possibility of predicting the thermal stability of 2′-O-modified duplexes at the computer-aided molecular design stage. This idea was then generalized to other modifications, also covering the nucleobase (Masaki et al. 2012).

A modified nucleotide that can be used at the 5′ termini of oligonucleotides so as to distinguish complementary polynucleotides depending on their length, for instance, to distinguish microRNAs and pre-microRNAs, was developed (Seio et al. 2012). In this work, MD simulations were used to construct structural models of the resulting double helix and to characterize the interaction of the modified nucleotide with the terminal phosphate of the recognized RNA.

Phosphorothioate (PT) modification of RNA backbone is generally considered an artificial modification, though it has been recently reported to occur naturally. The effect of PT in a riboswitch in which the modification was artificially included was studied and results were compared with NMR data (Zhang et al. 2021). The authors observed that existing force-field parameters cannot recapitulate the interactions seen in the experiment, and suggested that polarizable force fields might be necessary to describe this modification. Interestingly, an extension of the AMOEBA polarizable force field to a number of modified nucleotides, including PT, has been proposed (Jing et al. 2019). Alchemical calculations were used to compute changes in duplex hybridization free energy induced by the modification, reporting results in general agreement with experimental data.

Other interesting synthetic backbone modifications are locked nucleic acids (LNAs). Simulations with these modifications have been mostly conducted on LNAs within DNA or fully modified sequences and are thus not covered here.

4 Discussion and Challenges Ahead

A number of issues can be highlighted in our survey.

MD simulations of nonmodified RNAs are routinely performed by many groups, thus providing a significant mass of reference work. On the contrary, simulations of modified nucleotides are sparse, with most modifications simulated in a handful of papers. Hence, force fields have not been validated to the same extent. Researchers approaching the field should carefully validate parameters and be ready to develop new ones. Integration of experimental data (Bernetti and Bussi 2022) could provide a significant step forward.

Advanced sampling techniques are used rarely in this field. However, flexible and structurally heterogeneous RNA molecules require enhanced sampling methods to be faithfully characterized. Possibly, this shortfall is a consequence of the fact that being able to carefully design studies on systems with complex modified nucleotides require a deep biochemistry knowledge that is not commonly available in the community using enhanced sampling methods, and vice versa. Collaborative works could open the way to the application of state-of-the-art simulation methods in this field.

Nevertheless, the presented applications show that MD simulations are at the level of being useful in the interpretation and design of experiments. Given the growing biological relevance of RNA modifications, we are confident that synergy between simulation and experiment will become even stronger in the coming future.

Acknowledgements Zhengyue Zhang and Stefano Bosio are acknowledged for reading a draft of this Chapter and providing useful suggestions.

References

Aduri R, Psciuk BT, Saro P et al (2007) AMBER force field parameters for the naturally occurring modified nucleosides in RNA. J Chem Theory Comput 3:1464–1475

Bavi RS, Sambhare SB, Sonawane KD (2013) MD simulation studies to investigate ISO-energetic conformational behaviour of modified nucleosides m^2G and $M^2{}_2G$ present in tRNA. Comput Struct Biotechnol J 5:e201302015

Bernetti M, Bussi G (2022) Integrating experimental data with molecular simulations to investigate RNA structural dynamics. Curr Opin Struct Biol 78:102503

Deb I, Sarzynska J, Nilsson L, Lahiri A (2014) Conformational preferences of modified uridines: comparison of AMBER derived force fields. J Chem Inf Model 54:1129–1142

Deb I, Pal R, Sarzynska J, Lahiri A (2016) Reparameterizations of the χ torsion and Lennard-Jones σ parameters improve the conformational characteristics of modified uridines. J Comput Chem 37:1576–1588

Deb I, Popenda Ł, Sarzyńska J et al (2019) Computational and NMR studies of RNA duplexes with an internal pseudouridine-adenosine base pair. Sci Rep 9:16278

deLorimier E, Coonrod LA, Copperman J et al (2014) Modifications to toxic CUG RNAs induce structural stability, rescue mis-splicing in a myotonic dystrophy cell model and reduce toxicity in a myotonic dystrophy zebrafish model. Nucleic Acids Res 42:12768–12778

deLorimier E, Hinman MN, Copperman J et al (2017) Pseudouridine modification inhibits muscleblind-like 1 (mbnl1) binding to CCUG repeats and minimally structured RNA through reduced RNA flexibility. J Biol Chem 292:4350–4357

Dutta N, Sarzynska J, Lahiri A (2020) Molecular dynamics simulation of the conformational preferences of pseudouridine derivatives: improving the distribution in the glycosidic torsion space. J Chem Inf Model 60:4995–5002

Dutta N, Deb I, Sarzynska J, Lahiri A (2022) Data-informed reparameterization of modified RNA and the effect of explicit water models: application to pseudouridine and derivatives. J Comput Aided Mol Des 36:205–224

D'Esposito RJ, Myers CA, Chen AA, Vangaveti S (2022) Challenges with simulating modified RNA: insights into role and reciprocity of experimental and computational approaches. Genes 13:540

Elliott BA, Ho HT, Ranganathan SV et al (2019) Modification of messenger RNA by 2′-O-methylation regulates gene expression in vivo. Nat Commun 10:1–9

Fröhlking T, Bernetti M, Calonaci N, Bussi G (2020) Toward empirical force fields that match experimental observables. J Chem Phys 152:230902

Galindo-Murillo R, Davis DR, Cheatham TE III (2016) Probing the influence of hypermodified residues within the $tRNA^{Lys}$ anticodon stem loop interacting with the A-loop primer sequence from HIV-1. Biochim Biophys Acta 1860:607–617

Ganser LR, Kelly ML, Herschlag D, Al-Hashimi HM (2019) The roles of structural dynamics in the cellular functions of RNAs. Nat Rev Mol Cell Biol 20:474–489

Gilbert WV, Bell TA, Schaening C (2016) Messenger RNA modifications: form, distribution, and function. Science 352:1408–1412

Gonzalez-Rivera JC, Orr AA, Engels SM et al (2020) Computational evolution of an RNA-binding protein towards enhanced oxidized-RNA binding. Comput Struct Biotechnol J 18:137–152

Gore KR, Nawale GN, Harikrishna S et al (2012) Synthesis, gene silencing, and molecular modeling studies of 4'-c-aminomethyl-2′-o-methyl modified small interfering RNAs. J Org Chem 77:3233–3245

Harikrishna S, Pradeepkumar P (2017) Probing the binding interactions between chemically modified siRNAs and human argonaute 2 using microsecond molecular dynamics simulations. J Chem Inf Model 57:883–896

Hénin J, Lelièvre T, Shirts MR et al (2022) Enhanced sampling methods for molecular dynamics simulations [article v1.0]. Living J Comp Mol Sci 4:1583

Hopfinger MC, Kirkpatrick CC, Znosko BM (2020) Predictions and analyses of RNA nearest neighbor parameters for modified nucleotides. Nucleic Acids Res 48:8901–8913

Hsu WT, Piomponi V, Merz PT, Bussi G, Shirts MR (2023) Adding alchemical variables to metadynamics to enhance sampling in free energy calculations. J Chem Theory Comput 19:1805–1817. arXiv:2206.01329

Hurst T, Chen SJ (2021) Deciphering nucleotide modification-induced structure and stability changes. RNA Biol 18:1920–1930

Jiang J, Seo H, Chow CS (2016) Post-transcriptional modifications modulate rRNA structure and ligand interactions. Acc Chem Res 49:893–901

Jing Z, Qi R, Thibonnier M, Ren P (2019) Molecular dynamics study of the hybridization between RNA and modified oligonucleotides. J Chem Theory Comput 15:6422–6432

Karikó K, Muramatsu H, Welsh FA et al (2008) Incorporation of pseudouridine into mRNA yields superior nonimmunogenic vector with increased translational capacity and biological stability. Mol Ther 16:1833–1840

Krepl M, Otyepka M, Banáš P, Šponer J (2013) Effect of guanine to inosine substitution on stability of canonical DNA and RNA duplexes: molecular dynamics thermodynamics integration study. J Phys Chem B 117:1872–1879

Krepl M, Damberger FF, von Schroetter C et al (2021) Recognition of N6-methyladenosine by the YTHDC1 YTH domain studied by molecular dynamics and NMR spectroscopy: The role of hydration. J Phys Chem B 125:7691–7705

Kumar S, Rosenberg JM, Bouzida D et al (1992) The weighted histogram analysis method for free-energy calculations on biomolecules. I. the method. J Comput Chem 13:1011–1021

Li Y, Bedi RK, Wiedmer L et al (2019) Flexible binding of m^6A reader protein YTHDC1 to its preferred RNA motif. J Chem Theory Comput 15:7004–7014

Li Y, Bedi RK, Wiedmer L et al (2021) Atomistic and thermodynamic analysis of N6-methyladenosine (m^6A) recognition by the reader domain of YTHDC1. J Chem Theory Comput 17:1240–1249

Li Y, Bedi RK, Nai F et al (2022) Structure-based design of ligands of the m^6A-RNA reader YTHDC1. Eur J Med Chem 5:100057

Masaki Y, Miyasaka R, Ohkubo A et al (2010) Linear relationship between deformability and thermal stability of 2′-O-modified RNA hetero duplexes. J Phys Chem B 114:2517–2524

Masaki Y, Miyasaka R, Hirai K et al (2012) Prediction of the stability of modified RNA duplexes based on deformability analysis: oligoribonucleotide derivatives modified with 2′-O-cyanoethyl-5-propynyl-2-thiouridine as a promising component. Chem Commun 48:7313–7315

Meng Y, Sabri Dashti D, Roitberg AE (2011) Computing alchemical free energy differences with hamiltonian replica exchange molecular dynamics (H-REMD) simulations. J Chem Theory Comput 7:2721–2727

Mey ASJS, Allen BK, Bruce McDonald HE et al (2020) Best practices for alchemical free energy calculations [article v1.0]. Living J Comp Mol Sci 2:18378

Mlýnský V, Bussi G (2018) Exploring RNA structure and dynamics through enhanced sampling simulations. Curr Opin Struct Biol 49:63–71

Parisien M, Major F (2008) The MC-Fold and MC-Sym pipeline infers RNA structure from sequence data. Nature 452:51–55

Pavlova A, Parks JM, Oyelere AK, Gumbart JC (2017) Toward the rational design of macrolide antibiotics to combat resistance. Chem Biol Drug Des 90:641–652

Piomponi V, Fröhlking T, Bernetti M, Bussi G (2022) Molecular simulations matching denaturation experiments for N6-methyladenosine. ACS Cent Sci 8:1218–1228

Prabhakar PS, Takyi NA, Wetmore SD (2021) Posttranscriptional modifications at the 37th position in the anticodon stem-loop of tRNA: structural insights from MD simulations. RNA 27:202–220

Sakuraba S, Iwakiri J, Hamada M et al (2020) Free-energy calculation of ribonucleic inosines and its application to nearest-neighbor parameters. J Chem Theory Comput 16:5923–5935

Sarkar AK, Sarzynska J, Lahiri A (2020) Ensemble allosteric model for the modified wobble hypothesis. J Phys Chem Lett 11:6337–6343

Seio K, Kurohagi S, Kodama E et al (2012) Short-RNA selective binding of oligonucleotides modified using adenosine and guanosine derivatives that possess cyclohexyl phosphates as substituents. Org Biomol Chem 10:994–1006

Sonawane KD, Bavi RS, Sambhare SB, Fandilolu PM (2016) Comparative structural dynamics of $tRNA^{Phe}$ with respect to hinge region methylated guanosine: a computational approach. Cell Biochem Biophys 74:157–173

Špačková N, Réblová K (2018) Role of inosine-uracil base pairs in the canonical RNA duplexes. Genes 9:324

Šponer J, Bussi G, Krepl M et al (2018) RNA structural dynamics as captured by molecular simulations: a comprehensive overview. Chem Rev 118:4177–4338

Stasiewicz J, Mukherjee S, Nithin C, Bujnicki JM (2019) QRNAS: software tool for refinement of nucleic acid structures. BMC Struct Biol 19:1–11

Sugita Y, Okamoto Y (1999) Replica-exchange molecular dynamics method for protein folding. Chem Phys Lett 314:141–151

Suzuki T (2021) The expanding world of tRNA modifications and their disease relevance. Nat Rev Mol Cell Biol 22:375–392

Tanzer A, Hofacker IL, Lorenz R (2019) RNA modifications in structure prediction-status quo and future challenges. Methods 156:32–39

Townshend RJ, Eismann S, Watkins AM et al (2021) Geometric deep learning of RNA structure. Science 373:1047–1051

Vangaveti S, Cantara WA, Spears JL et al (2020) A structural basis for restricted codon recognition mediated by 2-thiocytidine in tRNA containing a wobble position inosine. J Mol Biol 432:913–929

Vangaveti S, Ranganathan SV, Agris PF (2022) Physical chemistry of a single tRNA-modified nucleoside regulates decoding of the synonymous lysine wobble codon and affects type 2 diabetes. J Phys Chem B 126:1168–1177

Vendeix FA, Munoz AM, Agris PF (2009) Free energy calculation of modified base-pair formation in explicit solvent: a predictive model. RNA 15:2278–2287

Voigts-Hoffmann F, Hengesbach M, Kobitski AY et al (2007) A methyl group controls conformational equilibrium in human mitochondrial $tRNA^{Lys}$. J Am Chem Soc 129:13382–13383

Wan WB, Seth PP (2016) The medicinal chemistry of therapeutic oligonucleotides. J Med Chem 59:9645–9667

Wang R, Vangaveti S, Ranganathan SV et al (2016) Synthesis, base pairing and structure studies of geranylated RNA. Nucleic Acids Res 44:6036–6045

Xiao X, Zhao B, Agris PF, Hall CK (2016) Simulation study of the ability of a computationally-designed peptide to recognize target $tRNA^{Lys3}$ and other decoy tRNAs. Protein Sci 25:2243–2255

Xu Y, MacKerell AD Jr, Nilsson L (2016) Structural effects of modified ribonucleotides and magnesium in transfer RNAs. Bioorg Med Chem 24:4826–4834

Xu Y, Vanommeslaeghe K, Aleksandrov A et al (2016) Additive CHARMM force field for naturally occurring modified ribonucleotides. J Comput Chem 37:896–912

Zhang X, Walker RC, Phizicky EM, Mathews DH (2014) Influence of sequence and covalent modifications on yeast tRNA dynamics. J Chem Theory Comput 10:3473–3483

Zhang Z, Vögele J, Mráziková K et al (2021) Phosphorothioate substitutions in RNA structure studied by molecular dynamics simulations, QM/MM calculations, and NMR experiments. J Phys Chem B 125:825–840
Zhou H, Kimsey IJ, Nikolova EN et al (2016) m1A and m1G disrupt A-RNA structure through the intrinsic instability of Hoogsteen base pairs. Nat Struct Mol Biol 23:803–810
Zhou W, Han Z, Wu Z et al. (2022) Specific recognition between YTHDF3 and m^{6}A-modified RNA: An all-atom molecular dynamics simulation study. Proteins: Struct Funct Bioinf 90:1965–1972

Ribonucleases for Sequencing and Characterization of RNA by LC–MS

Ivan R. Corrêa Jr., Eric J. Wolf, Erbay Yigit, and S. Hong Chan

Contents

Abstract The global deployment of mRNA vaccines against SARS-CoV-2 and the projected expansion of therapeutic applications of synthetic mRNA call for robust and high-precision analytical methods to evaluate attributes that are crucial to the safety and efficacy of the mRNA drug substances. Liquid chromatography–mass spectrometry (LC–MS) is one of the few techniques that can provide a direct and high-confidence readout of the identity and incorporation efficiency of the 5′ cap, length of the poly(A) tail, nucleotide sequence, and modification profile of synthetic mRNA molecules. Prior to LC–MS analysis, the RNA molecules are partially digested by specific endoribonucleases into oligonucleotides that are suitable for charge state-dependent fragmentation and mass deconvolution. The most commonly used endoribonuclease for RNA sequence mapping is the guanosine-specific RNase T1. RNase T1 has been employed for analysis of mRNA, rRNA, and tRNA as well as for mRNA poly(A) tail length verification. For mRNA 5′ cap analysis, selective excisions using probe-restrained RNase H or (deoxy)ribozymes are typically required. In this chapter, we will review the application of endoribonucleases for mRNA analysis, with emphasis on a recently characterized endoribonuclease derived from human RNase 4. We will also discuss the latest methods to assess 5′ cap and poly(A) tail incorporation in synthetic mRNA. Finally, we will highlight why more enzymatic

I. R. Corrêa Jr. · E. J. Wolf · E. Yigit · S. H. Chan (✉)
New England Biolabs, Inc, 240 County Road, Ipswich, MA 01938, USA
e-mail: chan@neb.com

J. Barciszewski (ed.), *RNA Structure and Function*, RNA Technologies 14,
https://doi.org/10.1007/978-3-031-36390-0_27

tools are needed and how they can contribute to improving the quality of synthetic RNA analysis, and to help understand the biology of RNA modifications in the cell.

Keywords RNA sequence mapping · RNA modification mapping · mRNA analysis · Cap analysis · Poly(A) tail analysis · LC–MS/MS

1 Introduction

The SARS-CoV-2 pandemic had greatly accelerated the scientific and technological advances of synthetic mRNA and mRNA vaccines (Dolgin 2021). Decades of basic research on the biology of mRNA, development of mRNA manufacturing, and delivery methods have laid the foundation for the rapid worldwide deployment of mRNA vaccines in response to the pandemic. Synthetic mRNA is produced in cell-free enzymatic reactions, where a phage RNA polymerase, such as T7 RNA polymerase, is used to transcribe RNA molecules from a DNA template using nucleoside triphosphates as building blocks. Synthetic mRNA molecules recapitulate the functional elements of cellular mRNA—the 5′ cap structure, the 3′ end poly(A) tail, the coding region that contains the gene to be expressed in the cell, and the untranslated regions that flank the coding region (5′ UTR and 3′ UTR). Major technological breakthroughs such as the reduced immunogenicity by pseudouridine substitutions (Karikó et al. 2008) and lipid nanoparticle for delivery helped realize the public health impact of synthetic RNA in response to the SARS-CoV-2 pandemic.

As the scientific and biopharmaceutical communities are rapidly expanding research, development, and manufacturing capacity to meet the global demand for infectious diseases, future applications such as cancer vaccines, protein replacement, and systemic mRNA therapeutics, analytical methodologies are being developed and formalized to assess the quality of the mRNA drug substances and their formulations with lipid nanoparticles (LNP). These analytical assays are not only important for product quality control but also essential in guiding development processes. While the characterization of the RNA primary sequence and the presence of modified nucleotides is critical for all RNA modalities, including antisense oligonucleotide (ASO), silencing RNA (siRNA) as well as single-guide RNA (sgRNA) of CRISPR genome editing systems (Hu et al. 2020; Uddin et al. 2020), the identity of the 5′ cap, the extent of cap incorporation, and the length of the poly(A) tails are crucial for the safety and efficacy of mRNA drug substances. Here, we will review some important aspects and recent advances in the analytical strategies for characterization of biologically relevant synthetic RNA drugs, with focus on enzymes that facilitate the analyses using mass spectrometry (MS) technologies.

2 Analytical Strategies for Sequencing and Modification Analysis of Synthetic RNA

The identity and modification profile of the RNA primary sequence, the extent of 5′ cap incorporation, and the length of the poly(A) tail are important attributes of the quality, safety, and efficacy of synthetic mRNA drug substances. The availability of analytical tools for direct and accurate characterization of these attributes is essential to advance the research, development, and manufacturing of mRNA vaccines and therapeutics. Modification mapping and direct sequence analysis of full-length synthetic mRNA molecules is challenging because of their size and complexity. Next-Generation Sequencing (NGS) techniques, which are often employed for gene expression analysis of cell populations or single cells, involve cDNA synthesis using a reverse transcriptase followed by PCR amplification. NGS modification detection methods are often based on modification-specific antibodies or modification-associated reverse transcription errors or stops that may lack the base resolution, sequence coverage, or quantitative accuracy required for quality control purposes. Direct RNA sequencing using nanopores is a promising platform for epitranscriptomics and full-length RNA analysis. However, extensive development of robust data deconvolution algorithms and perhaps the nanopores themselves may be needed to facilitate unambiguous identification of RNA modifications.

Mass spectrometry (MS)-based sequencing and modification mapping is based upon collision-, photon-, or electron-induced fragmentation of oligonucleotides to produce a series of fragment ions whose masses allow inference of their nucleotide composition and subsequent RNA sequence reconstruction (Yolųc et al. 2021). Furthermore, by comparing the mass data against a reference sequence, it is possible to precisely locate mass-altering RNA modifications (Fig. 1). As such, MS is an invaluable technique for direct and comprehensive characterization of nucleic acids and their chemical modifications. However, several technical challenges remain, both in terms of instrument resolution and fast-scanning capabilities and software packages for data processing. In addition, the analysis is further complicated by the large number of possible negative charge states that oligonucleotides can occupy. In fact, the application of MS as a sequencing tool to date is limited to oligonucleotides that are shorter than 50–60 nt.

Tandem liquid chromatography–mass spectrometry (LC–MS/MS) has long been the method of choice for identity and impurity analysis of small synthetic oligonucleotides, including antisense oligonucleotides (ASOs), silencing RNA (siRNA), and micro-RNA (miRNA), all of which range from 18 to 30 nt in length. LC–MS/MS has also been commonly deployed for modification mapping in longer sequences such as tRNA molecules from biological samples (76–90 nt) aided by endoribonucleases that partially digest the tRNA molecules into signature oligonucleotide cleavage products (Hossain and Limbach 2007). More recently, this approach has been extended to even larger molecules, such as synthetic mRNAs comprising a few thousand nucleotides (Jiang et al. 2019).

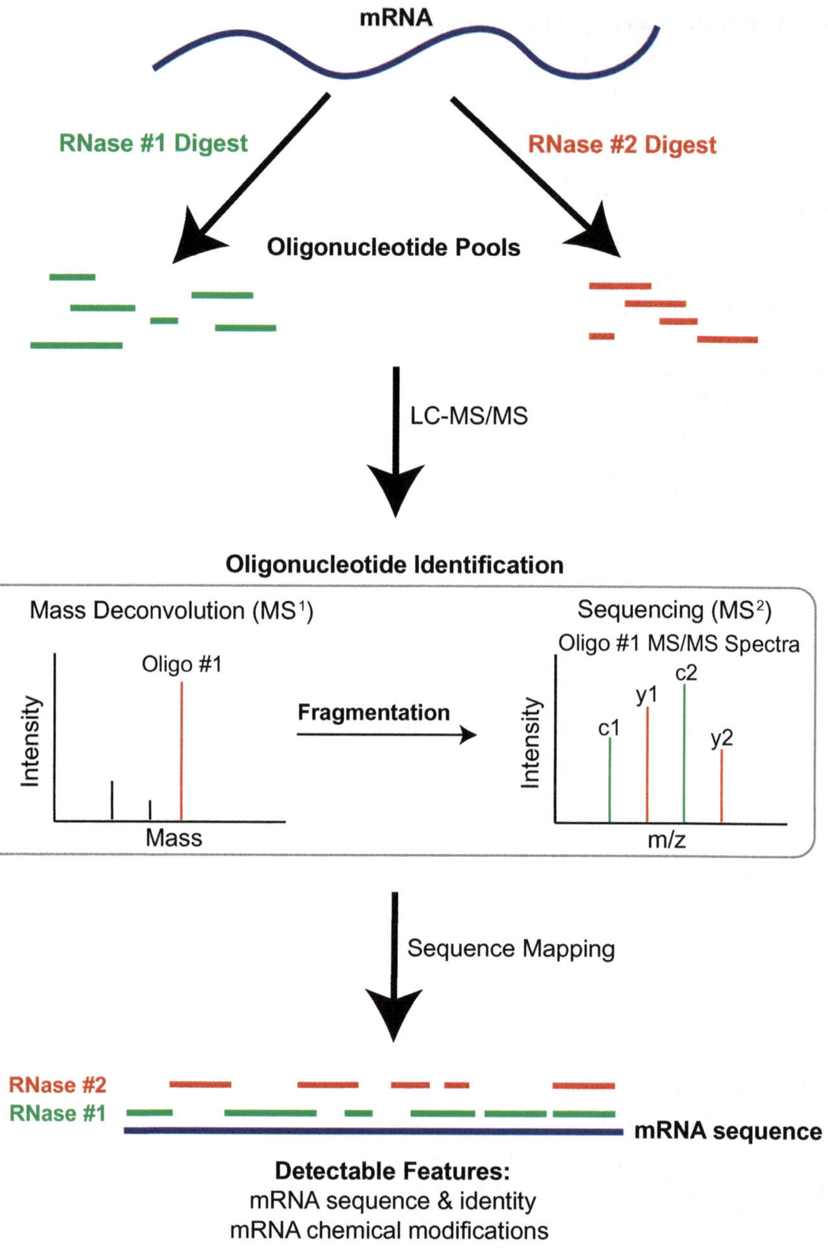

Fig. 1 A general workflow for RNA sequence mapping by LC–MS/MS. An mRNA sample is first digested using one or more endoribonucleases in parallel to generate oligonucleotide cleavage products suitable for LC–MS/MS analysis. The oligonucleotides are subjected to chromatographic separation followed by mass spectrometric fragmentation and detection. After mass matching and sequence assignment, the oligonucleotide sequences are mapped to the reference sequence and the rate of coverage is determined

Some of the analytical methods used in the quality assessment of RNA therapeutics, while originally designed in academic research labs, have now been adapted to the regulated pharmaceutical space, where accuracy, reproducibility, and robustness are imperative. LC–MS/MS addresses all of these attributes with the added advantage of acquiring data directly from the RNA molecules. In the following sections, we will discuss how endoribonucleases facilitate the use of LC–MS/MS in synthetic mRNA analysis as well as new approaches that aim at further improving the throughput and mapping coverage of existing workflows.

3 Application of Endoribonucleases in RNA Analysis

MS-based approaches for RNA sequencing and modification mapping require the conversion of full-length RNA into short oligonucleotides, whose size range (typically 4–40 nt) is ideally suitable for MS/MS spectral deconvolution. The controlled conversion of an RNA-of-interest into an oligonucleotide pool with an appropriate size distribution is carried out using nucleotide- or sequence-specific endoribonucleases. The workflow is similar to that of bottom-up proteomics: the full-length RNA is first digested with one or more endoribonucleases in parallel reactions, optionally in denaturing conditions, followed by separation of the resulting cleavage oligonucleotides through LC, and subsequent analysis by multistage MS. In contrast to proteomics, however, nucleic acid mass spectrometry is a developing technology. Recent advancements in instrumental hardware, including new fragmentation modalities and ion-mobility separations techniques, have improved the resolution, speed, and upper limit of mass detection. In addition, more open-source software tools have become available to support RNA sequence annotation (Wein et al. 2020; D'Ascenzo et al. 2022). Nonetheless, there is a need of robust enzymatic tools for processing RNA into oligonucleotides suitable for LC–MS/MS analysis.

RNase T1 (G↓N), RNase A (U/C↓N), and RNase U2 (A/G↓N) are some of the commonly used endoribonucleases in RNA mapping applications (Hossain and Limbach 2007; Goyon et al. 2021). Nucleotide-specific endoribonucleases such as those described in Table 1 can cleave RNA up to once every 2–4 nucleotides. If allowed to go to completion, these enzymes will generate a large fraction of cleavage products that are too short to render meaningful MS sequencing data. Not only may short oligonucleotides lead to ambiguity in the annotation of isomeric (isobaric) sequences but also may be mapped to multiple locations across a given sequence, thus significantly limiting the mapping coverage and accuracy. Subjecting an RNA sample to a dilution series of a frequent cutting endoribonuclease can help mitigate the issue by allowing the formation of oligonucleotide products with one or more missed cleavage sites (Vanhinsbergh et al. 2022). Although this approach is effective at generating longer oligonucleotides, in practice, the method is difficult to implement, and the consistency in the results will be highly dependent on the enzyme preparation and the RNA sequence itself.

Table 1 Endoribonucleases used for RNA modification mapping and sequencing by LC–MS/MS. The sequence specificities of the endoribonucleases are indicated as motifs where Y refers to pyrimidines; R refers to purines; H refers to A, U, or C. Down arrows represent cleavage sites. Nu refers to nucleotides. The 5′ group and the 3′ group of the cleavage fragments are indicated under the respective categories, where >p refers to a 2′,3′-cyclic phosphate, p refers to a linear 3′-phosphate, and OH refers to a hydroxyl group (note that endoribonuclease digestion occurs via a 2′,3′-cyclic phosphate intermediate, and this intermediate may be further converted to a linear 3′-phosphate depending on the enzyme concentration and/or reaction time)

	5′ FRAG		3′ FRAG			
Enzyme	Nu	3′ group	5′ group	Nu	Motif	Notes
RNase T1	G	>p, p	OH	N	G↓N	
RNase A	U, C	>p, p	OH	N	Y↓N	
RNase U2	A, G	>p, p	OH	N	R↓N	Five-fold preference for A over G
Colicin E5	G	>p, p	OH	U	G↓U	Physiological target sequence is UQU of tRNAs
MazF	A	>p, p	OH	C	N↓ACA	Does not cleave m^{6}ACA
MC1	A, U, C	>p, p	OH	U	H↓U	Cleaves prior to Ψ, m^{5}U, and Um. Cleaves prior to s^{4}U and m^{5}Um with lower efficiency. Does not cleave prior to m$^1\Psi$
Cusativin	C*	>p, p	OH	U*	–	Preferentially cleaves after cytidine (except CpC and CpA) and prior to uridine (except GpU) sites
Human RNase 4	U	>p, p	OH	A,G	U↓R	Cleaves after Ψ, m$^1\Psi$, mo^{5}U, and D. Cleaves s^{4}U and m^{5}U with lower efficiency. Does not cleave prior to Um

*Most preferred substrates for Cusativin

On the other hand, the use of endoribonucleases that inherently cut less frequently, such as Colicin E5 (G↓U) and MazF (N↓ACA) (Table 1), can generate oligonucleotide products that are longer in length on average. These two- and three-base cutters that can cleave RNA up to once every 16–64 nucleotides have been used in parallel with more frequent cutting endoribonucleases to improve RNA sequence mapping coverage (Jiang et al. 2019). In fact, parallel endoribonuclease digestion has been shown to overcome the limited sequence coverage of individual endoribonucleases for long, complex RNA species. In this approach, an RNA-of-interest is separately digested with enzymes of complementary nucleotide cleavage specificities. The resulting sequencing data is then combined to produce a fractional coverage map of the RNA. While this approach has been used to improve the mapping coverage of tRNA, Jiang et al. were the first to report the application of parallel endoribonuclease digestion for sequencing of therapeutic mRNAs by LC–MS/MS (Jiang et al. 2019). By combining data from digestions with RNase T1, Colicin E5, and MazF, the authors achieved mapping coverages of 73–87% for three synthetic mRNA sequences (encoding for erythropoietin, firefly luciferase, and α-catenin) ranging from 745 to 2884 nt. They also reported the ability to detect single nucleotide polymorphisms (SNPs) with abundance as low as 1%.

Parallel endoribonuclease digestion has also been successfully applied to high-resolution mapping of double-stranded RNA (dsRNA) and sgRNA. Nwokeoji et al. developed a rapid and highly reproducible testing for identity and in-process analysis of dsRNA sequences by parallel digestions with RNase T1 and RNase A (Nwokeoji et al. 2019). Their approach permitted batch variability analysis of two biomanufactured dsRNAs, 561 bp (base pair) and a 686 bp long, sharing 72% sequence homology. Similarly, Goyon et al. utilized parallel digestions with RNases T1, A, and U2 to achieve complete sequencing of a 100 nt long sgRNA (Goyon et al. 2021). An interesting feature of their approach was the use of hydrophilic interaction LC (HILIC)–MS to separate and identify the digestion products, obviating the need for ion-pairing reagents in the mobile phase.

4 Novel Endoribonuclease Strategies to Improve Sequence Mapping

A recurring limitation with many frequent cutting endoribonucleases, including RNase T1 and RNase A, is the inability to stop the digestion reaction completely using common RNase inhibitors or denaturants. Vanhinsbergh et al. recently reported the use of RNase T1 immobilized on magnetic beads as a strategy to produce longer cleavage products and prevent contamination of chromatographic columns with active endoribonucleases (Vanhinsbergh et al. 2022). In their approach, controlled partial digestion was achieved by physically removing the endoribonuclease from the reaction vessel, resulting in a remarkable improvement in sequence coverage (>80%) of large mRNAs. The authors chose the mRNA encoding for the SARS-CoV-2 spike protein and demonstrated that partial digestion with immobilized RNase T1 led to greater than 86% sequence coverage compared to only 10–25% with in-solution complete digestion. Although the method is relatively simple and amenable to automation, the reliance on partial digestion still calls for careful titration of the endoribonuclease-specific activity prior to analysis of any given RNA sequence.

Another strategy recently employed to increase the number of uniquely mappable cleavage products and thus reduce gaps in sequence coverage is the use of endoribonucleases with cutting frequency that lies between once every nucleotide to once every two nucleotides. These so-called 1.5-base cutters can cleave RNA on average once every eight nucleotides. Interestingly, this cutting frequency was predicted to yield the highest sequence coverage in a complete digestion of 1000 randomly selected human mRNA transcript sequences (<5kB) (Wolf et al. 2022). An example of such endoribonuclease is the uridine-specific human RNase 4 (hRNase 4). The hRNase 4 cuts at U↓R sites (wherein R is adenosine or guanosine) and generates oligonucleotide products that are twice as long as those of RNase T1. As a result, hRNase 4 alone allows for 60–80% sequence coverage of synthetic RNA transcripts ranging from 800 to 4000 nt (Fig. 2). Furthermore, hRNase 4 tolerates well many uridine nucleobase modifications. On this basis, hRNase 4 was deployed for the analysis of mRNAs fully substituted with 1-methylpseudouridine ($m^1\Psi$) or 5-methoxyuridine (mo^5U), as well as mRNAs selectively depleted of uridine, which are two common schemes used to reduce immunogenicity of synthetic mRNA (Wolf et al. 2022). Other endoribonucleases with potential for application in RNA sequence mapping include the cytidine-specific ribonuclease Cusativin and the uridine-specific ribonuclease MC1 (Table 1) (Grünberg et al. 2022).

5 Synthetic mRNA 5′ Cap Analysis

The mRNA 5′ cap is an essential modification for the efficacy and stability of mRNA therapeutics and vaccines (Pardi et al. 2018; VanBlargan et al. 2018). The 5′ cap structures are required for efficient mRNA translation (Gonatopoulos-Pournatzis and Cowling 2014; Ramaswamy et al. 2017), evasion of innate immune surveillance (Hyde and Diamond 2015), and regulation of 5′-mediated decay (Mugridge et al. 2018). Hence, the identity of the 5′ cap and the extent of its incorporation into synthetic mRNA are considered critical quality attributes in synthetic mRNA.

The 5′ cap consists of an *N*7-methylated guanosine linked to the 5′ end of RNA through a 5′–5′ triphosphate group. In metazoans, the m^7GpppN cap (Cap-0) is further modified by a cap-specific 2′-*O*-methyltransferase (MTase) to a m^7GpppNm structure (Cap-1). The 5′ cap structure can be incorporated into synthetic RNA by well-established methods that are readily adaptable and scalable: (i) co-transcriptional incorporation of a chemically synthesized cap structure, such as the anti-reverse cap analog (ARCA) or capped trinucleotides (Henderson et al. 2021), or (ii) post-transcriptional capping using a mRNA capping enzyme (Fuchs et al. 2016).

The 5′ cap structure only accounts for an approximately 600 Da mass change over hundreds of kilodaltons of a synthetic mRNA molecule. Direct detection of such a relatively small mass change is outside of the resolution range of most analytical methods. To overcome this problem, a recurring approach is to generate pre-defined fragments from the 5′ end of a synthetic mRNA to sizes that are amenable to polyacrylamide gel electrophoresis (PAGE) for a quick estimation of the capping efficiency, or LC–MS for a more quantitative determination of the presence of Cap-0, Cap-1, or any intermediate products resulting from the chosen capping strategy (Beverly et al. 2016; Chan et al. 2022). Among the available methods for generating pre-defined mRNA 5′ fragments are deoxyribozymes (Cairns et al. 2003), ribozymes (Vlatkovic

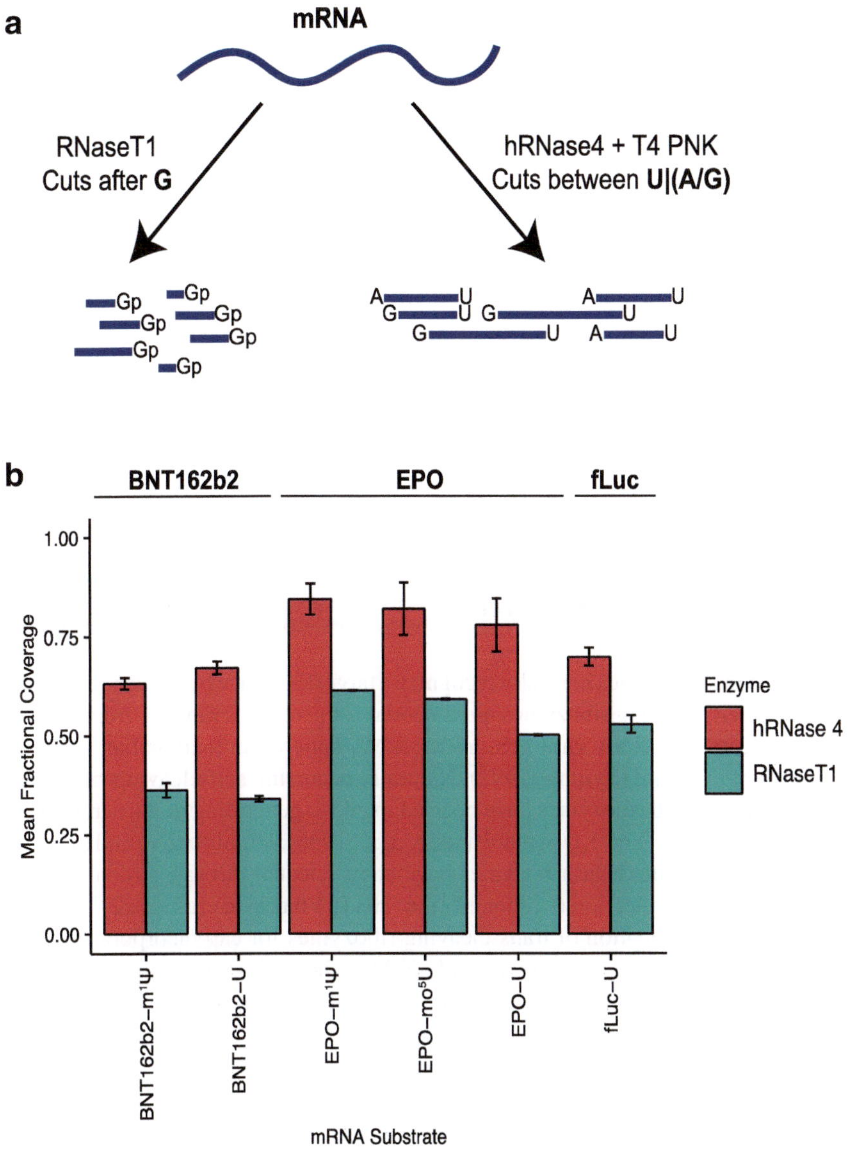

Fig. 2 A comparison of mRNA sequence coverage upon LC–MS/MS analysis of hRNase 4 and RNase T1 digests. (A) A schematic illustrating the primary cleavage specificity of RNase T1 and hRNase 4. T4 PNK is optionally used along with hRNase 4 to resolve mixtures of 2′,3′-cyclic phosphorylated and 3′-phosphorylated cleavage products into homogeneous 3′-hydroxylated products, thereby reducing the complexity of the sample. (B) Fractional sequence coverage of $m^1\Psi$- or U-modified BNT162b2 COVID-19 vaccine mRNA (4,187 nt), $m^1\Psi$-, mo^5U-, or U-modified erythropoietin (EPO) mRNA (800 nt), and firefly luciferase (Fluc) mRNA (1,766 nt) upon digestion with either hRNase 4 or RNase T1. Adapted from Wolf et al. (2022)

et al. 2022), and DNA–RNA chimera-guided RNase H cleavage (Beverly et al. 2016; Chan et al. 2022).

5.1 *Deoxyribozymes*

Trans-acting deoxyribozymes (also called DNAzymes) are single-stranded DNA enzymes artificially evolved and selected in vitro to perform chemical catalysis. DNAzymes have been selected to cleave RNA in a sequence-dependent manner through sequence complementarity at high efficiency (Breaker and Joyce 1994; Cairns et al. 2003). The 10–23 deoxyribozyme is one of the scaffolds commonly used to generate site-specific cleavage of synthetic mRNA for cap analysis (Cairns et al. 2003). DNAzyme-based cleavage has also been utilized to detect and quantify NAD^+- and NADH-capped mitochondrial RNA in vivo and in synthetic standards (Bird et al. 2018).

5.2 *Trans-Cleaving Ribozymes*

Trans-cleaving ribozymes normally recognize a target sequence by sequence complementarity. Self-cleaving ribozymes are essential for processes ranging from rolling-circle replication of RNA viral genomes to RNA splicing in mammalian cells and are abundant in nature (Symons 1997). Naturally occurring self-cleaving ribozymes such as hammerhead ribozymes have been adapted for generating in vitro transcripts with defined 3′ ends (Ferré-D'Amaré and Doudna 1996). Highly active and selective trans-cleaving hammerhead ribozymes have been selected through structure-based design and in vitro selection (Carbonell et al. 2011; Vlatkovic et al. 2022). A recent example of the application of trans-cleaving ribozymes for cap incorporation analysis has been published (Vlatkovic et al. 2022). The authors designed a series of ribozymes (35–47 nt long) that cleaved off 10–30 nt long oligonucleotide fragments from the 5′ end of the target mRNA. The approach was initially deployed for enzymatic or co-transcriptional capping quantification of U- and $m^1\Psi$-containing EPO mRNA by PAGE and LC–MS analysis, and then extended to other synthetic mRNA transcripts ranging from 1.1 to 9.4 kb in length.

5.3 *RNase H*

RNase H (RNase H1) is an endonuclease that degrades RNA primers from the Okazaki fragments of replicating DNA and processes R-Loops in the cell (Broccoli et al. 2004; Parajuli et al. 2017). Early in vitro studies indicated that cleavage by RNase H from *E. coli* was restrained to specific sites on a ssRNA by the presence of a DNA–RNA chimera with sequence complementarity to the target sequence (Lapham and Crothers 1996; Yu et al. 1997). These observations have been leveraged to generate pre-defined 5′ fragments for synthetic mRNA 5′ cap analysis using RNase H from both *E. coli* and *Thermus thermophilis* (Beverly et al. 2016; Chan et al. 2022). RNase H, however, tends to cleave one or more nucleotides away from the 5′ or 3′ of the chimeric probe target site, thereby complicating quantitative analysis of cap incorporation (Lapham and Crothers 1996; Yu et al. 1997; Vlatkovic et al. 2022). A simple RNase H probe selection scheme has recently been shown to improve the specificity of RNase H cleavage (Chan et al. 2022). After enrichment using streptavidin magnetic beads, the resulting oligonucleotide cleavage can be subjected to Cap-1 analysis using LC–MS. Alternatively, the RNase H cleavage products can be filled in with a fluorescently labeled deoxynucleotide using Klenow DNA Polymerase I for Cap-0 analysis using capillary electrophoresis (Chan et al. 2022) (Fig. 3).

6 Poly(A) Tail Analysis in Synthetic mRNA

The poly(A) tail is another critical structural element of the mRNA molecules that contributes to their stability and efficiency of translation. The length of the poly(A) tail has been demonstrated through extensive experimentation to impact mRNA function (Jalkanen et al. 2014). The poly(A) tail can be added to a synthetic mRNA (i) co-transcriptionally from a plasmid encoding a 3′ end polyadenosine tract or (ii) post-transcriptionally using a polyadenosine polymerase.

The endoribonuclease RNase T1 is the enzyme most commonly used in LC–MS-based poly(A) tail analyses. With high specificity towards guanosine, RNase T1 cleaves synthetic mRNA molecules into small fragments leaving the long poly(A) tail intact. The cleaved poly(A) tail is typically enriched using immobilized poly(dT) oligonucleotides (e.g., oligo d(T)25 magnetic beads) prior to LC–MS analysis (Beverly et al. 2018). Recently, Strezsak et al. developed a new method that does not require oligo dT-based purification, potentially reducing hybridization bias previously reported for longer poly(A) tails (Strezsak et al. 2022). Moreover, their method provided improved chromatographic buffer conditions that enabled a better control of oligonucleotide charge stages and metal adducts, resulting in a more reliable size distribution validation of poly(A) tails from 50 to 150 nt. Direct and indirect deep sequencing methods for poly(A) tail analysis have been reviewed elsewhere (Chang et al. 2014; Liu et al. 2019; Brouze et al. 2022).

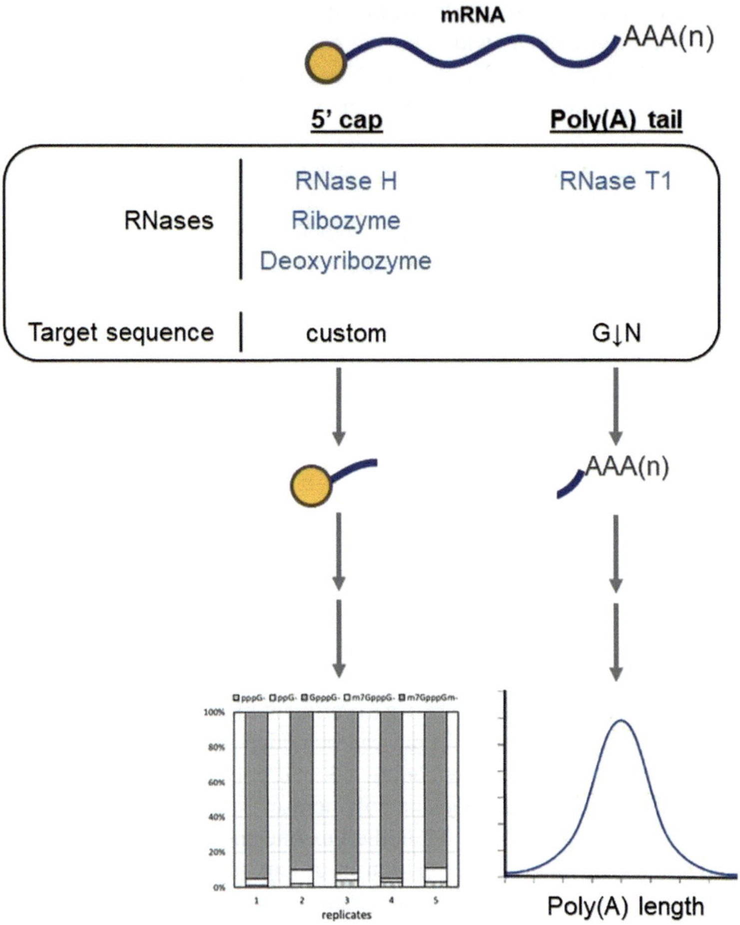

Fig. 3 mRNA 5′ cap and tail analyses. To assess the level of 5′ cap incorporation, a pre-defined fragment is cleaved near the 5′ end of the synthetic transcript using RNase H, a hammerhead ribozyme, or a deoxyribozyme. For poly(A) tail analysis, a nucleotide-specific endoribonuclease, for instance, RNase T1, is used to cut near to and release the poly(A) tail. After enrichment, the ratios and identities of the 5′ ends (e.g., Cap-0, Cap-1, ppp, pp, etc.) or the length of the poly(A) tail is profiled by LC–MS/MS

7 Conclusions and Outlook

Mass spectrometry (MS) offers great advantages for the quality assessment of therapeutic mRNA molecules: direct and accurate mass measurements allow for unambiguous identification of the target RNA molecules and modifications thereof. To date, most MS workflows for mRNA sequencing analysis are based on shotgun approaches and require the generation of short oligonucleotide fragments (4–40 nt) through controlled digestion with endoribonucleases. Table 1 summarizes some of the nucleotide- and sequence-specific endoribonucleases that have been used (RNases T1, A, U2, Colicin E5, MazF, and hRNase 4) or have the potential to be used (MC1 and Cusativin) in mRNA sequence mapping.

Nevertheless, additional endoribonucleases are needed to create a more robust toolset for RNA sequence and modification mapping applications. The current generation of mRNA vaccines and therapeutic mRNA is fully or partially substituted with pseudouridine and other derivatives. Recombinant MC1 activity (A/U/C↓U) is impeded by the presence of $m^1\Psi$ and to a lesser extent s^4U and $m^{5,2}Um$ (Wolf et al., 2022). hRNase 4 (U↓A/G) is not sensitive to Ψ, dihydrouridine (D), or $m^1\Psi$, but its activity is blocked by 2′-*O*-Methyluridine (Um) at the target site (Wolf et al. 2022). In addition, enzymatic digestion can be affected by local RNA structures. As an example, sequence mapping of an EPO synthetic transcript using RNase T1 (G↓N) results in coverage gaps as large as 100 nt; hRNase 4, likewise, leaves a gap of about 50 nt (Wolf et al. 2022). Modified nucleotides can further impact endoribonuclease activities directly by changing enzyme binding affinity or indirectly through altering local RNA structures. These problems can be exacerbated by the fact that sequence or nucleotide specificity of some endoribonucleases may be affected by changes in the enzyme and/or RNA concentration and reaction conditions (buffer, temperature, incubation time, additives, etc.).

MS has been frequently used to identify and map the position of modified nucleotides within tRNA and rRNA sequences. These molecules not only are highly structured but also heavily modified, both of which can impede the activity of most endoribonucleases. This is particularly problematic when multiple modifications are clustered at a particular site. Such inhibition may result in oligonucleotide fragments that do not contain a given target modification or long fragments containing multiple modifications that make annotation of modified sites challenging. Hence, the availability of more endoribonucleases that (i) have complementary nucleotide specificities, (ii) are non-discriminatory against modified nucleotides, and (iii) can cleave highly structured sequences with or without the help from denaturing agents is highly desirable to improve coverage in RNA sequence mapping.

Although it is possible to computationally predict endoribonuclease homologues possessing similar nucleotide or sequence specificity, biochemical properties such as sensitivity to ribonucleoside modifications and secondary structures (such as double-stranded regions) and resistance to denaturants must be determined empirically. Therefore, massive screening campaigns with deep sequencing-based readout could help speed up the discovery and characterization of novel enzymes (Miyamoto et al. 2016a, b). Thoughtful screening design can also help select novel endoribonucleases that are insensitive to specific RNA modifications or tolerant to heat or chemical denaturation. Helper proteins such as RNA helicases or single-stranded RNA binding proteins may be considered as a means to facilitate access to structured regions of RNA.

Furthermore, the endoribonuclease's stability and physical purity have a big impact on the quality and reproducibility of RNA analysis. Because recombinant endoribonucleases are often toxic to the expression host, these enzymes are particularly challenging to produce at high yields and high-purity levels. Some may further require posttranslational modifications, such as disulfide bridges or glycosylation, for proper folding and activity. In some cases, enzyme preparations contaminated with nuclease activities originating from the expression host can lead to misidentification of sequence specificity. Bacterial periplasmic expression or yeast excretory expression can be a generalizable method to express and purify endoribonucleases with high quality by minimizing the toxic effect of the endoribonucleases to the host cells and helping disulfide bond formation (Grünberg et al. 2022).

In addition to more robust enzymatic tools, technological progress in MS instrumentation and software are critical to further advance sequence mapping of mRNA, in particular of oligonucleotide fragments that are much longer than ~55 nt. Such analyses will certainly necessitate the use of a new generation of endoribonucleases that can digest full-length RNA molecules into longer cleavage fragments in a controllable manner.

The availability of an array of DNA restriction enzymes in the 1970s enabled the mapping of genes and set the stage for the age of molecular biology (Roberts 2005). It is not difficult to envision that the characterization of endoribonucleases with unique sequence specificities, and robust biochemical and biophysical properties, will help open new avenues for RNA technological development and scientific discovery.

References

Beverly M, Dell A, Parmar P et al (2016) Label-free analysis of mRNA capping efficiency using RNase H probes and LC–MS. Anal Bioanal Chem 408:5021–5030

Beverly M, Hagen C, Slack O (2018) Poly A tail length analysis of in vitro transcribed mRNA by LC–MS. Anal Bioanal Chem 410:1667–1677

Bird JG, Basu U, Kuster D et al (2018) Highly efficient 5′ capping of mitochondrial RNA with NAD+ and NADH by yeast and human mitochondrial RNA polymerase. Elife 7:e42179

Breaker RR, Joyce GF (1994) A DNA enzyme that cleaves RNA. Chem Biol 1:223–229

Broccoli S, Rallu F, Sanscartier P et al (2004) Effects of RNA polymerase modifications on transcription-induced negative supercoiling and associated R-loop formation. Mol Microbiol 52:1769–1779

Brouze A, Krawczyk PS, Dziembowski A et al (2022) Measuring the tail: Methods for poly(A) tail profiling. Wiley Interdiscip Rev RNA e1737

Cairns MJ, King A, Sun L (2003) Optimisation of the 10–23 DNAzyme–substrate pairing interactions enhanced RNA cleavage activity at purine–cytosine target sites. Nucleic Acids Res 31:2883–2889

Carbonell A, Flores R, Gago S (2011) Trans -cleaving hammerhead ribozymes with tertiary stabilizing motifs: In vitro and in vivo activity against a structured viroid RNA. Nucleic Acids Res 39:2432–2444

Chan SH, Whipple JM, Dai N et al (2022) RNase H-based analysis of synthetic mRNA 5′ cap incorporation. RNA (new York, NY) 28:1144–1155

Chang H, Lim J, Ha M, Kim VN (2014) TAIL-seq: Genome-wide determination of poly(A) tail length and 3′ end modifications. Mol Cell 53:1044–1052

D'Ascenzo L, Popova AM, Abernathy S et al (2022) Pytheas: A software package for the automated analysis of RNA sequences and modifications via tandem mass spectrometry. Nat Commun 13:2424

Dolgin E (2021) The tangled history of mRNA vaccines. Nature 597:318–324

Ferré-D'Amaré AR, Doudna JA (1996) Use of Cis- and trans-ribozymes to remove 5′ and 3′ heterogeneities from milligrams of in vitro transcribed RNA. Nucleic Acids Res 24:977–978

Fuchs A-L, Neu A, Sprangers R (2016) A general method for rapid and cost-efficient large-scale production of 5′ capped RNA. RNA 22:1454–1466

Gonatopoulos-Pournatzis T, Cowling VH (2014) Cap-binding complex (CBC). Biochem J 457:231–242

Goyon A, Scott B, Kurita K et al (2021) Full sequencing of CRISPR/Cas9 single guide RNA (sgRNA) via parallel ribonuclease digestions and hydrophilic interaction liquid chromatography–high-resolution mass spectrometry analysis. Anal Chem 93:14792–14801

Grünberg S, Wolf EJ, Jin J et al (2022) Enhanced expression and purification of nucleotide-specific ribonucleases MC1 and Cusativin. Protein Expres Purif 190:105987

Henderson JM, Ujita A, Hill E et al (2021) Cap 1 messenger RNA synthesis with co-transcriptional CleanCap® analog by in vitro transcription. Curr Protoc 1:e39

Hossain M, Limbach PA (2007) Mass spectrometry-based detection of transfer RNAs by their signature endonuclease digestion products. RNA 13:295–303

Hu B, Zhong L, Weng Y et al (2020) Therapeutic siRNA: State of the art. Signal Transduct Target Ther 5:101

Hyde JL, Diamond MS (2015) Innate immune restriction and antagonism of viral RNA lacking '2-O methylation. Virology 479:66–74

Jalkanen AL, Coleman SJ, Wilusz J (2014) Determinants and implications of mRNA poly(A) tail size—Does this protein make my tail look big? Semin Cell Dev Biol 34:24–32

Jiang T, Yu N, Kim J et al (2019) Oligonucleotide sequence mapping of large therapeutic mRNAs via parallel ribonuclease digestions and LC–MS/MS. Anal Chem 91:8500–8506

Karikó K, Muramatsu H, Welsh FA et al (2008) Incorporation of pseudouridine into mRNA yields superior nonimmunogenic vector with increased translational capacity and biological stability. Mol Ther 16:1833–1840

Lapham J, Crothers DM (1996) RNase H cleavage for processing of in vitro transcribed RNA for NMR studies and RNA ligation. RNA New York NY 2:289–296

Liu Y, Nie H, Liu H et al (2019) Poly(A) inclusive RNA isoform sequencing (PAIso−seq) reveals wide-spread non-adenosine residues within RNA poly(A) tails. Nat Commun 10:5292

Miyamoto T, Kato Y, Sekiguchi Y et al (2016a) Characterization of MazF-mediated sequence-specific RNA cleavage in pseudomonas putida using massive parallel sequencing. PLoS ONE 11:e0149494

Miyamoto T, Yokota A, Tsuneda S et al (2016b) AAU-specific RNA cleavage mediated by MazF toxin endoribonuclease conserved in nitrosomonas Europaea. Toxins 8:174

Mugridge JS, Tibble RW, Ziemniak M et al (2018) Structure of the activated Edc1-Dcp1-Dcp2-Edc3 mRNA decapping complex with substrate analog poised for catalysis. Nat Commun 9:1152

Nwokeoji AO, Earll ME, Kilby PM et al (2019) High resolution fingerprinting of single and double-stranded RNA using ion-pair reverse-phase chromatography. J Chromatogr B 1104:212–219

Parajuli S, Teasley DC, Murali B et al (2017) Human ribonuclease H1 resolves R-loops and thereby enables progression of the DNA replication fork. J Biol Chem 292:15216–15224

Pardi N, Hogan MJ, Porter FW et al (2018) mRNA vaccines—a new era in vaccinology. Nat Rev Drug Discov 17:261–279

Ramaswamy S, Tonnu N, Tachikawa K et al (2017) Systemic delivery of factor IX messenger RNA for protein replacement therapy. Proc National Acad Sci USA 114:E1941–E1950

Roberts RJ (2005) How restriction enzymes became the workhorses of molecular biology. Proc National Acad Sci USA 102:5905–5908

Strezsak SR, Pimentel AJ, Hill IT et al (2022) Novel mobile phase to control charge states and metal adducts in the LC/MS for mRNA characterization assays. ACS Omega 7:22181–22191

Symons RH (1997) Plant pathogenic RNAs and RNA catalysis. Nucleic Acids Res 25:2683–2689

Uddin F, Rudin CM, Sen T (2020) CRISPR gene therapy: Applications, limitations, and implications for the future. Frontiers Oncol 10:1387

VanBlargan LA, Himansu S, Foreman BM et al (2018) An mRNA vaccine protects mice against multiple tick-transmitted flavivirus infections. Cell Rep 25:3382-3392.e3

Vanhinsbergh CJ, Criscuolo A, Sutton JN et al (2022) Characterization and sequence mapping of large RNA and mRNA therapeutics using mass spectrometry. Anal Chem 94:7339–7349

Vlatkovic I, Ludwig J, Boros G et al (2022) Ribozyme assays to quantify the capping efficiency of in vitro-transcribed mRNA. Pharm 14:328

Wein S, Andrews B, Sachsenberg T et al (2020) A computational platform for high-throughput analysis of RNA sequences and modifications by mass spectrometry. Nat Commun 11:926

Wolf EJ, Grünberg S, Dai N et al (2022) Human RNase 4 improves mRNA sequence characterization by LC–MS/MS. Nucleic Acids Res. https://doi.org/10.1093/nar/gkac632

Yoluç Y, Ammann G, Barraud P, et al (2021) Instrumental analysis of RNA modifications enhanced reader. Crit Rev Biochem Mol Biol 56:178–204

Yu YT, Shu MD, Steitz JA (1997) A new method for detecting sites of 2′-O-methylation in RNA molecules. RNA (new York, NY) 3:324–331

RNA-Processing DNAzymes

Ingrid Span and Manuel Etzkorn

Contents

Abstract The great potential of nucleic acids as therapeutics has been recognized for a while but has experienced a tremendous attention with the recent development of RNA vaccines. In contrast to protein-targeting strategies, nucleic acid-based approaches often have the advantage that the required target selectivity is not realized via matching a specific structure, but plainly via the primary sequence of the applied RNA or DNA construct. This sequence is then either directly processed or comprises an additional unit capable of processing a target molecule. The latter is true for a number of DNA sequences, called DNAzymes, that are capable of both binding and processing a target with a high selectivity. While the mRNA technology has the inherent strength of bringing something into the system, RNA-processing DNA catalysts such as RNA-cleaving DNAzymes have the inherent strength of taking

I. Span (✉)
Department of Chemistry and Pharmacy, Bioinorganic Chemistry, Friedrich Alexander University Erlangen-Nürnberg, Erlangen, Germany
e-mail: ingrid.span@fau.de

M. Etzkorn (✉)
Institute of Physical Biology, Heinrich Heine University Düsseldorf, Düsseldorf, Germany
e-mail: manuel.etzkorn@hhu.de

Institute of Biological Information Processing (IBI-7: Structural Biochemistry), Forschungszentrum Jülich, Jülich, Germany

Jülich Center for Structural Biology (JuStruct), Forschungszentrum Jülich, Jülich, Germany

J. Barciszewski (ed.), *RNA Structure and Function*, RNA Technologies 14,
https://doi.org/10.1007/978-3-031-36390-0_28

something out of the system. Consequently, the DNAzyme technology has the potential to emerge as counterpart to the mRNA technology. However, and in line with the endeavors that were required for the success of the mRNA technology, specific improvements need to be realized to unravel the full potential of RNA-processing DNAzymes. This review provides an overview of recent findings and remaining limitations.

Keywords DNAzymes · DNA catalysts · RNA-cleaving · DNA structure · DNA–RNA complexes

1 Introduction

1.1 Background and Applications

Nucleic acids are responsible for the storage and transport of genetic information inside cells. DNA carries the genetic instructions for the development, functioning, growth, and reproduction of all known organisms and many viruses. As such DNA is also designed to be stable in a cellular environment. RNA plays a more dynamic and transient role. In the form of a messenger molecule, mRNA passes genetic information on to the protein biosynthesis machinery. Therefore, mRNA is an indispensable link between genes and proteins as the main gene products. Moreover, non-coding RNAs including transfer RNAs (tRNAs) (Hoagland et al. 1958) and ribosomal RNAs (rRNAs) (Littlefield et al. 1955; Roberts 1958; McQuillen et al. 1959) play important roles in protein synthesis. The primary function of small RNAs, such as microRNAs (miRNAs) (Lee et al. 1993; Ruvkun 2001) and small-interfering RNAs (siRNAs) (Hamilton and Baulcombe 1999), is RNA degradation and post-transcriptional gene silencing, while small nuclear RNAs (snRNAs) (Hadjiolov et al. 1966) play a role in the processing of pre-messenger RNA (hnRNA) in the nucleus. For a long time, it was assumed that the biological role of nucleic acids is limited to encoding and translating the hereditary information, while proteins and protein assemblies are the biopolymers that perform catalysis inside the cell. The discovery of catalytically active RNA enzymes (ribozymes) has shown that nucleic acids are also capable of accelerating chemical transformations. This raised the question if Nature also evolved DNA enzymes (DNAzymes) as biocatalysts (Kruger et al. 1982)?

The chemical structures of DNA and RNA are very similar, differing only by the absence of the 2′-hydroxyl group on the ribose of each nucleotide in DNA and the presence of T or U nucleobases. These seemingly small differences have a dramatic impact on the structure and the function of the biopolymers. In its role as the long-term storage of genetic information, DNA is double-stranded most of the time. The DNA duplex is held together by Watson–Crick base pairing and displays a regular structure with the surface consisting of the backbone phosphates. The high stability and chemically inert nature of DNA derives from the formation of the duplex conformation

(Fig. 1a), which is favorable for its biological function as storage molecule, but detrimental for a catalyst. RNA—as the transient genetic messenger with various other biological roles—is present as single-stranded polymer most of the time, leading to a lower stability and lifetime inside cells. RNA molecules can adopt defined three-dimensional structures through intra- or intermolecular Watson–Crick base pairing leading to a large variety of molecular architectures. It is therefore not surprising that RNA strands can perform enzymatic reactions, just like protein enzymes, as both molecules consist of a defined sequence of monomers (nucleotides or amino acids) and adopt three-dimensional conformations (Fig. 1b) (Silverman 2016).

DNA duplexes have unfavorable properties for catalysis, however, DNA strands in its single-stranded form are capable of adopting similar conformations as observed in ribozymes and accelerate chemical transformations. DNAzymes have not been discovered in Nature so far. They have been identified in the laboratory by in vitro selection methods, a process in which random sequences are tested in parallel for a desired function. The first DNAzyme was reported by Breaker and Joyce in (1994). This DNAzyme catalyzes the Pb^{2+}-dependent cleavage of an RNA phosphodiester linkage within its oligonucleotide sequence, similar to the activity of a ribonuclease (Cuchillo et al. 2011). Since this exciting discovery, many DNAzymes with new functionalities have been identified (for reviews see (Breaker 1997; Silverman 2009, 2015, 2016; Hollenstein 2015; Morrison et al. 2018; Ma and Liu 2020; Micura and Höbartner 2020)). Despite the great interest of the scientific community in DNAzymes, structure elucidation has been challenging and the promising results in vitro are often not transferable to in vivo systems (Silverman 2016; Victor et al. 2018). In this review, we focus on reporting the recent progress that has been made in understanding the structure and function of DNAzymes and improving the activity of DNAzymes inside cells.

1.2 DNAzyme Diversity

DNAzymes have a great potential for catalyzing a broad variety of chemical reactions, which is mostly limited by the ability to set up a suitable in vitro selection process. Most of the transformations mediated by DNAzymes involve oligonucleotide substrates. RNA-cleaving enzymes are arguably the best-characterized class of DNAzymes. They can be used to manipulate large RNAs (Pyle et al. 2000), treat diseases related to high gene expression (Baum and Silverman 2008), and in biosensors for detecting metal ions (Ma and Liu 2020).

In addition to RNA-cleaving, DNAzymes with oligonucleotide substrates can perform the ligation (Cuenoud and Szostak 1995) or cleavage (Chandra et al. 2009; Gu et al. 2013) of DNA, the ligation of RNA strands (Flynn-Charlebois et al. 2003), and the photoreversion of thymine dimers (Chinnapen and Sen 2004). DNA-mediated catalysis can also involve non-oligonucleotide substrates including peroxide reduction mimicking peroxidase activity (Travascio et al. 1998) and peptide side chain as

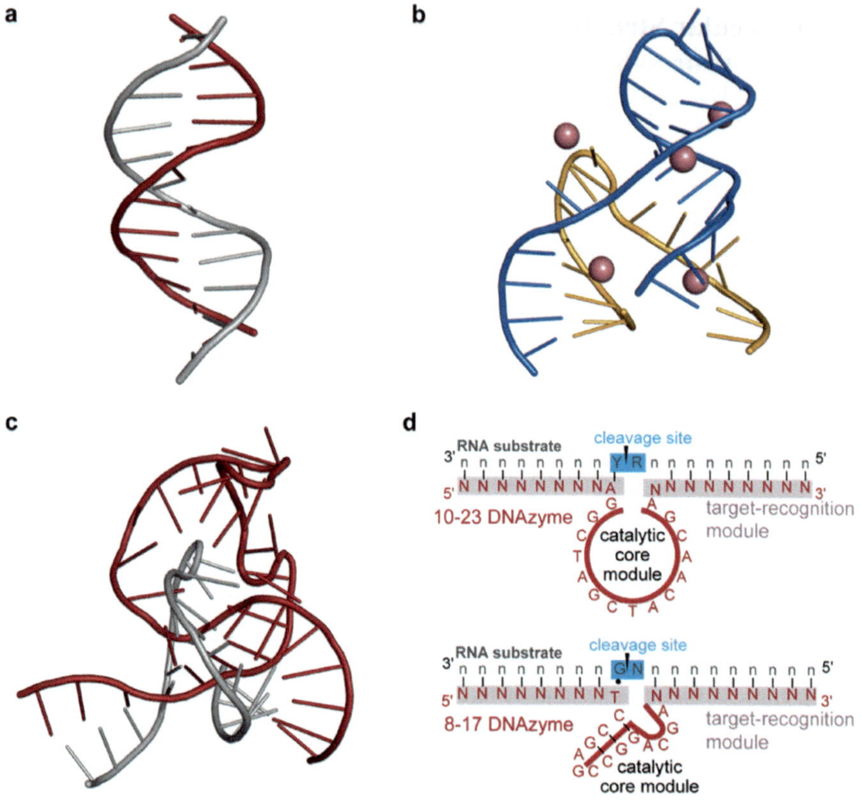

Fig. 1 Architectures of different nucleic acid systems. **a** DNA double helix (Fratini et al. 1982) in which the functional groups of the nucleobase regions are located in the core of the helix, generating a very stable but catalytically unfavorable structure. **b** A hammerhead ribozyme (Dunham et al. 2003) adopting a three-dimensional structure that considerably differs from the double helix arrangement. A single-stranded RNA sequence (blue) will generate a catalytically active structure upon complex formation with the nucleotide sequence of the RNA substrate (yellow). The structure is stabilized via a combination of base-pairing in different stem loops and interactions with metal ions (spheres). **c** RNA-cleaving 10–23 DNAzyme (red) in pre-catalytic complex with its RNA substrate (grey) (Borggräfe et al. 2022b). The DNAzyme architecture shows some similarities but also clear differences to the ribozyme. **d** Schematic representation of two different types of RNA-cleaving DNAzymes, i.e., the 10–23 and the 8–17 DNAzymes that either do not or do form stable base-pairing in their catalytic core, respectively. In addition to their core module, the requirements for the RNA target cleavage site (blue) also differ (Santoro and Joyce 1997a; Silverman 2016)

well as backbone modifications (Pradeepkumar et al. 2008; Wong et al. 2011; Chandrasekar and Silverman 2013; Walsh et al. 2013, 2015; Chu et al. 2014; Chandrasekar et al. 2015; Silverman 2015).

2 Molecular Structure and Requirements Underlying RNA-Processing

DNAzymes are single-stranded DNA molecules that were identified by in vitro selection processes (Breaker and Joyce 1994; Cuenoud and Szostak 1995; Santoro and Joyce 1997b; Silverman 2016). Their catalytic function depends on their three-dimensional structure, which single-stranded DNA adapts in a similar way to ribozymes.

High-resolution structural information on DNAzymes is scarce. Currently, only four structures have been reported including (i) the 10–23 DNAzyme in a catalytically irrelevant crystallographic artifact formed due to a palindromic element in the loop (Nowakowski et al. 1999), (ii) the crystal structure of the RNA-ligating 9DB1 DNAzyme in post-catalytic complex with its RNA substrate (Ponce-Salvatierra et al. 2016), (iii) the crystal structure of the pre-catalytic complex of the 8–17 DNAzyme in complex with a biologically inactive DNA substrate (Liu et al. 2017), and (iv) a NMR-derived ensemble of the pre-catalytic complex of the 10–23 DNAzyme with its RNA substrate (Borggräfe et al. 2022b). In the following general structural elements of RNA-processing, DNAzymes as well as key features of the already obtained structural insights and their consequences for RNA-processing will be discussed. To this end, the DNAzymes will be dissected into their two basic modules, i.e., an RNA-recognition module also referred to as binding arm(s) and a catalytic core module also referred to as catalytic loop (Fig. 1d).

2.1 The Catalytic Core Module

Different classes of DNAzymes characterized by different catalytic core modules have been identified (Silverman 2016). One prominent difference in the catalytic loop is, e.g., whether or not the nucleotide sequence will promote stable base pairing. The two well-known DNAzymes 10–23 and 8–17, which both have been named according to their identifier in the respective in vitro selection assay, particularly differ in this respect (Fig. 1d).

From a structural point of view, the catalytic core module of DNA catalysts needs to generate distinct structural elements to enable RNA processing. First, the RNA-substrate needs to be aligned, i.e., the DNA needs to fixate the RNA and expose the relevant moieties of the processed RNA site (e.g., the cleavage or ligation site) towards the DNA's 'active center'. While substrate binding will be realized by the DNAzyme's recognition module, substrate alignment will be generally different from standard RNA:DNA duplex formation and therefore the catalytic core module will be its driving force. Second, a catalytically active conformation needs to be induced/overpopulated at the processed RNA site, e.g., the 'in-line attack conformation' for RNA-cleaving DNAzymes. Third, an active environment needs to be generated, i.e., the active groups need to be positioned around the processed site. The latter could involve functional groups of the DNA and/or metal ion cofactors.

The inherent differences in the systems and applied conditions used to elucidate structural details of DNA catalysts so far make it difficult to identify central features responsible for the above-discussed structural elements. Still, the insights into the RNA-cleaving 10–23 (Borggräfe et al. 2022b) and 8–17 DNAzyme (Liu et al. 2017), indicated similarities in their RNA-cleavage mechanism. In this respect, the molecular arrangements of the two different DNAzymes promote the positioning of similar functional groups in proximity to the cleavage site. In particular, a catalytic guanine acting as potential general base to deprotonate the 2′-OH of the processed RNA nucleotide was identified as common feature (Borggräfe et al. 2022a). In addition, comparable stacking interactions between catalytic loop nucleotides and the nucleotides of the processed site are found in MD simulations of the 10–23 DNAzyme (i.e. between G6 and rG0) (Borggräfe et al. 2022b) and also in the crystal structure of the RNA-ligating 9DB1 DNAzyme (i.e. between dG27 and rG1) (Ponce-Salvatierra et al. 2016). These interactions appear to support the stabilization of a processable RNA conformation in both cases.

Unlike to the crystallographic DNAzyme structures, the NMR-derived ensemble of the 10–23 DNAzyme in pre-catalytic complex with its RNA substrate does not capture a distinct conformation with fixed inter-nucleotide distances in the catalytic loop. Instead, a characteristic fold with increased structural plasticity in key regions of the catalytic loop was identified. The observed structural plasticity is likely a direct consequence of the absence of stable base pairing interactions in the catalytic loop module, which is a characteristic feature of the 10–23 DNAzyme. Furthermore, NMR and MD simulations show that the respective DNA:RNA ensemble is further dynamically modulated by metal-ions (Borggräfe et al. 2022a, b). While this complicates the extraction of specific structural details, the data may bring a new angle to our perception of the molecular features of DNAzymes, which may not be seen as static switches but dynamic ensembles that sample a larger structural space including several active and inactive conformations. In this picture, the stabilization of active conformations or the elimination of inactive conformations may be promising routes to develop new generation of DNAzymes with increased activity.

NMR insights of the 10–23 DNAzyme in the absence of its RNA substrate indicate a higher degree of sequential π–π stacking in the catalytic core module as compared to the recognition module of the single-stranded DNA. It is currently not clear whether this is a general feature or whether the low degree of stacking interactions is a consequence of the specific binding arm sequences. Interestingly, the DNAzyme in this study shows a comparatively high activity (i.e., full substrate cleavage after about 30 min (Rosenbach et al. 2020a), whereas e.g., another reported 10–23 DNAzyme variant with different binding arms but identical catalytic core module cleaves less than 5% of its substrate under similar conditions (Taylor et al. 2022)). This could indicate one of possibly many not well-characterized features comprising the RNA-recognition module, which may impact the DNAzyme's in vitro and in vivo activity.

2.2 The RNA-Recognition Module

In most cases, the RNA-recognition module contains two substrate-specific binding arms that flank the catalytic loop and can be adjusted to bind the target RNA via Watson–Crick base pairing with high selectivity. This arm-loop-arm arrangement further promotes a catalytic turnover due to the resulting difference in the respective dissociation temperatures for educt and product RNA. In this respect, the precatalytic DNA–RNA educt complex is generally assumed to display a dissociation temperature that is determined by the combined sequences of both binding arms. In contrast, the two product RNA sequences are assumed to show considerably reduced dissociation temperatures that are roughly determined by their individual sequences. Hence, when designing new DNAzymes, the length and composition of their RNA-recognition sequences should be considered to allow formation of thermodynamically stable DNAzyme:RNA-educt complexes and to allow subsequent product release at their working temperature (e.g., typically at 37 °C for therapeutic applications).

In general, DNAzymes have specific requirements for the target's cleavage site, e.g., the need of a purine–pyrimidine dinucleotide pair in the case of the 10–23 DNAzyme (Santoro and Joyce 1997a) (Fig. 1d, upper panel). Due to the high abundance of these features, it can be assumed that nearly all naturally occurring RNAs and in particular mRNAs can be targeted by DNAzymes. As outlined above, substrate binding and product release are thermodynamic processes that will depend on the length and composition of the substrate recognition module. When targeting longer RNA sequences with structured regions, the accessibility of the RNA target region will also affect catalytic activity (Cairns and Sun 2004). Surprisingly, this is often neglected, e.g., even in recent DNAzyme builder web application (Mohammadi-Arani et al. 2022). In general, DNAzymes will always thermodynamically compete with e.g., secondary structure formation of the targeted RNA sequence. Therefore, neglecting the RNA secondary structure may still result in partly active variants. However, since the sequence requirement for the cleavage site is generally very moderate, a standard mRNA contains a large number of possible target regions. Therefore, it appears unnecessarily ineffective to not apply suitable selection algorithms to identify the thermodynamically favored regions within the large pool of possible targets. This selection can be made simply by predicting the globular secondary structure of the full-length target RNA (in equilibrium) and/or by considering the local secondary structure formation following RNA unwinding during translation (Workman and Krogh 1999; Rivas and Eddy 2000; Steger and Victor 2022). While, so far, we preferred the latter (see e.g., (Steger and Victor 2022) for more detailed discussion), it is still not well-studied, which target-region selection algorithm is most effective and how much a thermodynamically well-selected target region will indeed improve in vivo activity. In cases where the target region is restricted, e.g., when selectively targeting an allele-specific point mutation, longer binding arms may be required to be able to compete with the secondary structure formation at this particular RNA region (Taylor et al. 2022).

In general, assessment of the in vitro activity of RNA-processing DNAzymes will be strongly dependent on the applied experimental conditions. In this respect temperature, pH, and the often-required concentration of divalent metal ions are arguably the most important experimental parameters. In addition, one should, however, also keep in mind that high concentrations of RNA/DNA will noticeably increase the respective RNA–DNA, DNA–DNA, or DNA–RNA dissociation temperatures (Steger and Victor 2022; Borggräfe et al. 2022b). Furthermore, the ionic strength of the buffer system will directly impact complex formation and activity (Record et al. 1976; Rosenbach et al. 2020a, b; Steffen et al. 2020; Borggräfe et al. 2022b).

Experimental evidence also revealed that the catalytic core module of the 10–23 DNAzyme has a substantial impact on the metal-ion binding properties of the substrate recognition module (Borggräfe et al. 2022b). This highlights a possible structural link between both modules, which may also be affected by the respective composition of the binding arm sequences. Overall and in addition to the explicit restrictions in the cleavage site, the design of the RNA-recognition module appears to have a considerable impact on activity. However, a better description of the underlying features and their implication in selecting the best RNA target regions is urgently needed.

3 Activity in Cells

DNAzymes have a great potential as gene-silencing agents in therapeutic applications and for the treatment of viral infections (Pradeepkumar and Höbartner 2012; Zhang 2018). Previous clinical trials for basal cell carcinoma and asthma have already shown that DNAzymes exhibit a high safety profile and tolerability in human patients (Cho et al. 2013; Homburg et al. 2013; Cao et al. 2014). Furthermore, other preclinical investigations have revealed that DNAzymes are non-toxic to the cell and do not elicit an innate immune response (Cai et al. 2012; Dicke et al. 2012).

DNA oligonucleotides generally have a higher biostability as compared to RNA molecules and can be produced considerably more cost-effective. Furthermore, DNAzymes are capable of achieving a high catalytic activity independent of accessory proteins. Despite these favorable characteristics and the promising results in biochemical studies, the transition from laboratory tools to clinical reagents has been hampered due to their reduced activity in a cellular environment.

Numerous efforts have been made to understand why the high catalytic activity of RNA-cleaving DNAzymes in vitro does not correlate with the activity in vivo. It has been suggested that the limiting factor is the low concentration of free magnesium cations (Mg^{2+}) inside the cell (Victor et al. 2018). Mg^{2+} is an essential cofactor in various physiological processes, due to its sufficient natural abundance and its chemical properties. The ionic radius for the Mg^{2+} cation is small (0.65 Å), which results in a high charge density and, therefore, in a high polarizing power. Despite the relatively high abundance of Mg^{2+} inside cells, the strong bonding and the presence of many binding partners lead to a low concentration of free Mg^{2+} ions. Amongst these interaction partners, organic phosphates such as adenosine triphosphate (ATP) capture high levels of the cellular Mg^{2+}. This study demonstrated that Mg^{2+} binding to free ATP outcompetes the DNA catalyst and that stoichiometric amounts of ATP to Mg^{2+} inhibited the activity of the 10–23 DNAzyme.

A more detailed investigation of the effects of metal ions on the catalytic activity of the 10–23 DNAzyme revealed that monovalent cations inside the cell also modulate catalysis (Rosenbach et al. 2020a). Potassium (K^+) is the major intracellular cation, with 98% of the total pool being located in the cells at a concentration of 140–150 mM (Zacchia et al. 2016). While sodium (Na^+) and K^+ ions are involved in unspecific charge neutralization (Pechlaner and Sigel 2012), they also have a strong impact on the interaction of divalent cations with nucleic acids (Record et al. 1976). In this study, FRET-based assays showed that the presence of Na^+ slows down the Mg^{2+}-mediated catalytic reaction. At conditions representing a physiological relevant ionic strength, the "observed rate constant" (k_{obs}) is reduced by 40% compared to the reactions carried out in the absence of monovalent ions. Interestingly, the reaction rate of the DNAzyme in the presence of manganese cations (Mn^{2+}) is increasing in the presence of Na^+, suggesting different modes of action for the divalent cations. Taken together, this work shed light on the complex interplay of mono- and divalent metal ions and its implications on the structure and function of the 10–23 DNAzyme.

To overcome the persisting limitations, great efforts have been reported to modify and improve the performance of DNAzymes in a physiological environment. In vitro selection has led to numerous promising catalytically active DNA sequences for specific chemical reactions (Silverman 2016). Variation of the RNA recognition module in terms of length can improve the reaction rate. However, the range of parameters for boosting the efficiency of the DNA catalyst when using only the four naturally occurring standard nucleotide building blocks is limited. Thus, non-standard (naturally occurring or non-natural) building blocks have emerged as promising strategies for tuning the in vivo performance of DNAzymes. Three different classes of modifications have been introduced in DNAzymes: Alterations of the nucleotide linkage, the ribose moiety, or the nucleobase's functional groups. In naturally occurring DNA strands, the building blocks are linked by phosphate groups, in which four oxygen atoms coordinate the phosphate in a tetrahedral coordination environment. The free oxygen atoms can be replaced by sulfur atoms, resulting in phosphorothioate linkages, while the bridging oxygen atoms can be exchanged by amine groups, leading to phosphoroamidite linkages (Fig. 2). DNAzymes with phosphorothioate linkages (Fig. 2) have been synthesized with the aim to increase their stability against nucleolytic attack. The exchange of an oxygen atom against its heavier homolog sulfur has indeed led to more nuclease-resistant DNAzymes (Schubert et al. 2003; Appaiahgari and Vrati 2007). The replacement of the oxygen by sulfur, which according to the Pearson concept of hard and soft acids and bases is classified as a soft ligand, has led to a higher affinity of the DNAzyme for soft, thiophilic metals, such as cadmium (Cd^{2+}) (Huang and Liu 2015). Furthermore, replacement of one or two nucleotides at both the 3′- and 5′-termini of the DNAzyme by N3′-P5′ phosphoramidate nucleotides also enhanced its nuclease resistance.

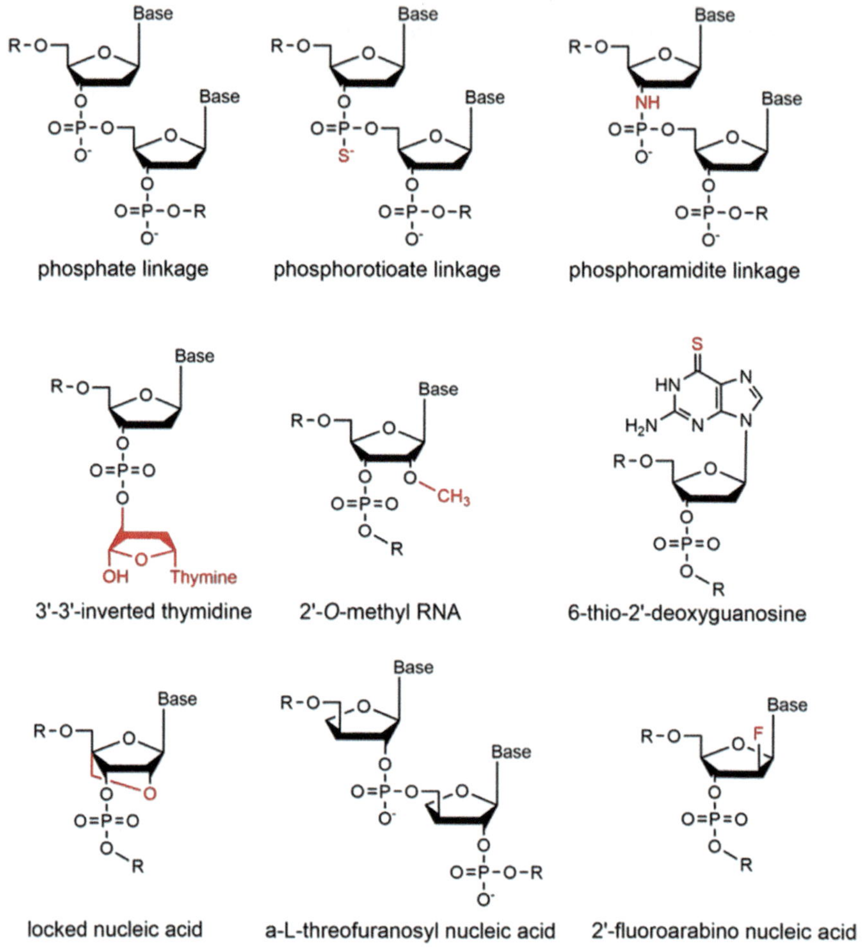

Fig. 2 Selected chemical modifications of nucleotides that have been introduced into DNAzymes

In addition to the phophorothioate linkages, incorporation of 3′-3′-inverted thymidine nucleotides, 2′-O-methyl RNA, 6-thio-dG, or locked nucleic acids (LNAs) into DNAzymes have resulted in enhanced cleavage activity and/or nuclease stability (Vester et al. 2002; Schubert et al. 2003; Borggräfe et al. 2022b).

Most of these non-standard nucleotides or linkages were introduced at the RNA-recognition module with the goal to increase the affinity of the DNA fragments to the RNA substrate. The increased affinity is beneficial for the association step and the formation of the DNAzyme:RNA complex. However, the rate of the catalytic reaction also depends on product release. The modest performance of some of the mentioned modified DNAzymes in cells is proposed to derive from product inhibition with the DNAzyme:product complex failing to dissociate from the postcatalytic state (Wang et al. 2021).

To overcome these limitations, new DNAzymes with non-natural modifications so-called xeno-nucleic acids (XNAs) have attracted a considerable amount of attention. This new molecular chemotype is believed to display physicochemical properties that are capable of achieving an enhanced biological stability without sacrificing the catalytic activity under multiple-turnover conditions (Wang et al. 2021). The XNAs contain three different non-natural building blocks: LNAs, α-L-threofuranosyl nucleic acid (TNAs) and 2′-fluoroarabino nucleic acids (FANA).

LNAs have gained attention in several applications due to their attractive features. These non-natural nucleotides have a restricted conformation, which leads to an increased helical thermostability, excellent mismatch discrimination when hybridized with RNA (or DNA), and resistance towards exonucleolytic degradation (Vester et al. 2002). DNAzymes containing LNA building blocks, referred to as LNAzymes, have been shown to increase cleaving activity compared to that of a reference DNAzyme (Vester et al. 2002). However, the LNAzymes were also found to exhibit significant cell toxicity (Jakobsen et al. 2007).

FANA has shown promising characteristics as building block for DNAzymes. FANA contains a fluorine atom at the 2′ position of a 2′-deoxyarabinose sugar. FANA–RNA duplexes are typically ~1.2 °C more stable per base pair than the equivalent DNA–RNA duplexes, which results in increased association kinetics for DNAzymes (Wang et al. 2021). As a result, FANA modifications will modulate catalytic turnover rates at a given temperature as compared to their DNA analogs. Consequently, when using an assay temperature that is favored by the FANA construct, the respective DNA-based reference may perform well below its optimal activity. Therefore, a temperature series and/or an adjusted DNA length should ideally be used to assess whether increased in vitro turnover activity is relevant for in vivo applications.

α-L-threofuranosyl nucleic acid (TNA) contain vicinally connected (3′ → 2′) phosphodiester bridges and are capable of cross-pairing with RNA and DNA (Schöning et al. 2000). Incorporation of TNAs into DNAzymes leads to an enhanced stabilization of the backbone structure against exonucleolytic degradation and, thus, to an increased biological stability (Culbertson et al. 2016). A recent study has screened XNAs that comprise the three different nucleic acid building blocks: DNA, FANA, and TNA (Wang et al. 2021). It was reported that this X10–23 variant achieves a ~50-fold increase in multiple-turnover activity (compared to a not length or temperature-adjusted DNA-based reference) under simulated physiological conditions and enhances the biological stability of the backbone structure >100-fold under stringent nuclease conditions. The authors also report promising results in human cancer cell lines (Nguyen et al. 2021) and engaged in an important scientific discussion regarding the role of RNase H as alternative mechanism for RNA degradation (Spitale and Chaput 2022; Taylor and Holliger 2022).

In parallel new selection and production routines for conventional DNAzymes (Sednev et al. 2022), FANA (Gerber et al. 2022; Taylor et al. 2022), LNA and/ or all 2′OMe-RNA (Freund et al. 2022; Hervey et al. 2022) based catalysts have been develop including very promising improvements under simulated physiological conditions.

4 Conclusions and Outlook

With a growing number of high-resolution structural insights and an increased toolset for the incorporation of non-standard DNA building blocks, the field of RNA-processing DNAzymes currently undergoes an exciting revival. In this respect, structure-guided rational design strategies, which exploit the progress in the mechanistic understanding of the systems and combine it with rational positioning of innovative synthetic modifications, offer new routes to overcome persisting limitations of the DNAzyme technology. The development of such next-generation DNAzymes will likely benefit from modifications in the catalytic-core module as well as in the substrate-recognition module. Still, both aspects will further benefit from dedicated studies to answer the remaining questions, e.g., regarding the role of the target sequence composition and/or accessibility. Based on the recent successes and the rather well-defined open questions, it can, however, be anticipated that the new routes will soon also promote the translation of new DNAzyme variants from basic research to therapeutic applications.

References

Appaiahgari MB, Vrati S (2007) DNAzyme-mediated Inhibition of Japanese encephalitis virus replication in mouse brain. Mol Ther 15:1593–1599

Baum DA, Silverman SK (2008) Deoxyribozymes: useful DNA catalysts in vitro and in vivo. Cell Mol Life Sci 65:2156–2174

Borggräfe J, Gertzen CGW, Viegas A et al (2022a) The architecture of the 10–23 DNAzyme and its implications for DNA-mediated catalysis. FEBS J in Press: https://doi.org/10.1111/febs.16698

Borggräfe J, Victor J, Rosenbach H et al (2022b) Time-resolved structural analysis of an RNA-cleaving DNA catalyst. Nature 601:144–149

Breaker RR (1997) DNA enzymes. Nat Biotechnol 15:427–431

Breaker RR, Joyce GF (1994) A DNA enzyme that cleaves RNA. Chem Biol 1:223–229

Cai H, Santiago FS, Prado-Lourenco L et al. (2012) DNAzyme targeting c-jun suppresses skin cancer growth. Sci Transl Med 4:139ra82

Cairns MJ, Sun L-Q (2004) Target-site selection for the 10–23 DNAzyme. Methods Mol Biol 252:267–277

Cao Y, Yang L, Jiang W et al (2014) Therapeutic evaluation of Epstein-barr virus-encoded latent membrane protein-1 targeted DNAzyme for treating of nasopharyngeal carcinomas. Mol Ther 22:371–377

Chandra M, Sachdeva A, Silverman SK (2009) DNA-catalyzed sequence-specific hydrolysis of DNA. Nat Chem Biol 5:718–720

Chandrasekar J, Silverman SK (2013) Catalytic DNA with phosphatase activity. Proc Natl Acad Sci USA 110:5315–5320

Chandrasekar J, Wylder AC, Silverman SK (2015) Phosphoserine lyase deoxyribozymes: DNA-catalyzed formation of dehydroalanine residues in peptides. J Am Chem Soc 137:9575–9578

Chinnapen DJF, Sen D (2004) A deoxyribozyme that harnesses light to repair thymine dimers in DNA. Proc Natl Acad Sci USA 101:65–69

Cho EA, Moloney FJ, Cai H et al (2013) Safety and tolerability of an intratumorally injected DNAzyme, Dz13, in patients with nodular basal-cell carcinoma: A phase 1 first-in-human trial (DISCOVER). Lancet 381:1835–1843

Chu C-C, Wong OY, Silverman SK (2014) A generalizable DNA-catalyzed approach to peptide-nucleic acid conjugation. ChemBioChem 15:1905–1910
Cuchillo CM, Nogués MV, Raines RT (2011) Bovine pancreatic ribonuclease: Fifty years of the first enzymatic reaction mechanism. Biochemistry 50:7835–7841
Cuenoud B, Szostak JW (1995) A DNA metalloenzyme with DNA ligase activity. Nature 375:611–614
Culbertson MC, Temburnikar KW, Sau SP et al (2016) Evaluating TNA stability under simulated physiological conditions. Bioorg Med Chem Lett 26:2418–2421
Dicke T, Pali-Schöll I, Kaufmann A et al (2012) Absence of unspecific innate immune cell activation by GATA-3-specific DNAzymes. Nucleic Acid Ther 22:117–126
Dunham CM, Murray JB, Scott WG (2003) A helical twist-induced conformational switch activates cleavage in the hammerhead ribozyme. J Mol Biol 332:327–336
Flynn-Charlebois A, Wang Y, Prior TK et al (2003) Deoxyribozymes with 2′-5′ RNA ligase activity. J Am Chem Soc 125:2444–2454
Fratini AV, Kopka ML, Drew HR et al (1982) Reversible bending and helix geometry in a B-DNA dodecamer: CGCGAATTBrCGCG. J Biol Chem 257:14686–14707
Freund N, Taylor AI, Arangundy-Franklin S et al (2022) A two-residue nascent-strand steric gate controls synthesis of 2′-O-methyl- and 2′-O-(2-methoxyethyl)-RNA. Nat Chem in Press. https://doi.org/10.1038/s41557-022-01050-8
Gerber PP, Donde MJ, Matheson NJ et al (2022) XNAzymes targeting the SARS-CoV-2 genome inhibit viral infection. Nat Commun 13:6716
Gu H, Furukawa K, Weinberg Z et al (2013) Small, highly active DNAs that hydrolyze DNA. J Am Chem Soc 135:9121–9129
Hadjiolov AA, Venkov PV, Tsanev RG (1966) Ribonucleic acids fractionation by density-gradient centrifugation and by agar gel electrophoresis: A comparison. Anal Biochem 17:263–267
Hamilton AJ, Baulcombe DC (1999) A species of small antisense RNA in posttranscriptional gene silencing in plants. Science 286:950–952
Hervey JRD, Freund N, Houlihan G et al (2022) Efficient synthesis and replication of diverse sequence libraries composed of biostable nucleic acid analogues. RSC Chem Biol 3:1209–1215
Hoagland MB, Stephenson ML, Scott JF et al (1958) A soluble ribonucleic acid intermediate in protein synthesis. J Biol Chem 231:241–257
Hollenstein M (2015) DNA catalysis: The chemical repertoire of DNAzymes. Molecules 20:20777–20804
Homburg U, Turowska A, Kuhlmann J et al (2013) Safety profile and pharmacokinetics of SB010, an inhaled GATA-3-specific DNAzyme, in phase I clinical trials in healthy and asthmatic subjects. Eur Respir J 42:4858
Huang P-JJ, Liu J (2015) Rational evolution of Cd2+-specific DNAzymes with phosphorothioate modified cleavage junction and Cd2+ sensing. Nucleic Acids Res 43:6125–6133
Jakobsen MR, Haasnoot J, Wengel J et al (2007) Efficient inhibition of HIV-1 expression by LNA modified antisense oligonucleotides and DNAzymes targeted to functionally selected binding sites. Retrovirology 4:29
Kruger K, Grabowski PJ, Zaug AJ et al (1982) Self-splicing RNA: Autoexcision and autocyclization of the ribosomal RNA intervening sequence of tetrahymena. Cell 31:147–157
Lee RC, Feinbaum RL, Ambros V (1993) The C. elegans heterochronic gene lin-4 encodes small RNAs with antisense complementarity to lin-14. Cell 75:843–854
Littlefield JW, Keller EB, Gross J et al (1955) Studies of cytoplamsic ribonucleoprotein particles from the liver of the rat. J Biol Chem 217:111–124
Liu H, Yu X, Chen Y et al (2017) Crystal structure of an RNA-cleaving DNAzyme. Nat Commun 8:2006
Ma L, Liu J (2020) Catalytic nucleic acids: biochemistry, chemical biology, biosensors, and nanotechnology. iScience 23:100815
McQuillen K, Roberts RB, Britten RJ (1959) Synthesis of nascent protein by ribosomes in *Escherichia Coli*. Proc Natl Acad Sci USA 45:1437–1447

Micura R, Höbartner C (2020) Fundamental studies of functional nucleic acids: Aptamers, riboswitches, ribozymes and DNAzymes. Chem Soc Rev 49:7331–7353

Mohammadi-Arani R, Javadi-Zarnaghi F, Boccaletto P et al (2022) DNAzymeBuilder, a web application for in situ generation of RNA/DNA-cleaving deoxyribozymes. Nucleic Acids Res 50:W261–W265

Morrison D, Rothenbroker M, Li Y (2018) DNAzymes: Selected for applications. Small Methods 2:1700319

Nguyen K, Wang Y, England WE et al (2021) Allele-specific RNA knockdown with a biologically stable and catalytically efficient XNAzyme. J Am Chem Soc 143:4519–4523

Nowakowski J, Shim PJ, Prasad GS, a. (1999) Crystal structure of an 82-nucleotide RNA-DNA complex formed by the 10–23 DNA enzyme. Nat Struct Biol 6:151–156

Pechlaner M, Sigel RKO (2012) Characterization of metal ion-nucleic acid interactions in solution. Metal Ions in Life Sciences, pp 1–42

Ponce-Salvatierra A, Wawrzyniak-Turek K, Steuerwald U et al (2016) Crystal structure of a DNA catalyst. Nature 529:231–234

Pradeepkumar PI, Höbartner C, Baum DA et al (2008) DNA-catalyzed formation of nucleopeptide linkages. Angew Chemie - Int Ed 47:1753–1757

Pradeepkumar PI, Höbartner C (2012) RNA-cleaving DNA enzymes and their potential therapeutic applications as antibacterial and antiviral agents. From Nucleic Acids Sequences to Molecular Medicine Berlin, Heidelberg: Springer Berlin Heidelberg. pp 371–410

Pyle AM, Chu VT, Jankowsky E et al (2000) Using DNAzylnes to cut, process, and map RNA molecules for structural studies or modification. Methods Enzymol 317:140–146

Record MT, Lohman TM, de Haseth P (1976) Ion effects on ligand–nucleic acid interactions. J Mol Biol 107:145–158

Rivas E, Eddy SR (2000) Secondary structure alone is generally not statistically significant for the detection of noncoding RNAs. Bioinformatics 16:583–605

Roberts RB (1958) Microsomal particles and protein synthesis; papers presented at the First Symposium of the Biophysical Society, at the Massachusetts Institute of Technology, Cambridge, February 5, 6, and 8, 1958. New York: Published on behalf of the Washington Academy of Sciences, Washington, D.C., by Pergamon Press

Rosenbach H, Borggräfe J, Victor J et al (2020a) Influence of monovalent metal ions on metal binding and catalytic activity of the 10–23 DNAzyme. Biol Chem 402:99–111

Rosenbach H, Victor J, Etzkorn M et al (2020b) Molecular features and metal ions that influence 10–23 DNAzyme activity. Molecules 25:3100

Ruvkun G (2001) Glimpses of a tiny RNA world. Science 294:797–799

Santoro SW, Joyce GF (1997a) A general purpose RNA-cleaving DNA enzyme. Proc Natl Acad Sci USA 94:4262–4266

Santoro SW, Joyce GF (1997b) A general purpose RNA-cleaving DNA enzyme. Proc Natl Acad Sci USA 94:4262–4266

Schöning K-U, Scholz P, Guntha S et al (2000) Chemical etiology of nucleic acid structure: The α-threofuranosyl-(3′→2′) oligonucleotide system. Science 290:1347–1351

Schubert S, Gül DC, Grunert H-P et al (2003) RNA cleaving "10–23" DNAzymes with enhanced stability and activity. Nucleic Acids Res 31:5982–5992

Sednev MV, Liaqat A, Höbartner C (2022) High-throughput activity profiling of RNA-cleaving DNA catalysts by deoxyribozyme sequencing (DZ-seq). J Am Chem Soc 144:2090–2094

Silverman SK (2009) Deoxyribozymes: Selection design and serendipity in the development of DNA catalysts. Acc Chem Res 42:1521–1531

Silverman SK (2015) Pursuing DNA catalysts for protein modification. Acc Chem Res 48:1369–1379

Silverman SK (2016) Catalytic DNA: Scope, applications, and biochemistry of deoxyribozymes. Trends Biochem Sci 41:595–609

Spitale RC, Chaput JC (2022) Reply to: On gene silencing by the X10–23 DNAzyme. Nat Chem 14:859–861

Steffen FD, Khier M, Kowerko D et al (2020) Metal ions and sugar puckering balance single-molecule kinetic heterogeneity in RNA and DNA tertiary contacts. Nat Commun 11:104

Steger G, Victor J (2022) Design of a DNAzyme. Methods Mol Biol, DNAzymes: 47–63

Taylor AI, Holliger P (2022) On gene silencing by the X10–23 DNAzyme. Nat Chem 14:855–858

Taylor AI, Wan CJK, Donde MJ et al (2022) A modular XNAzyme cleaves long, structured RNAs under physiological conditions and enables allele-specific gene silencing. Nat Chem 14:1295–1305

Travascio P, Li Y, Sen D (1998) DNA-enhanced peroxidase activity of a DNA aptamer–hemin complex. Chem Biol 5:505–517

Vester B, Lundberg LB, Sørensen MD et al (2002) LNAzymes: Incorporation of LNA-type monomers into DNAzymes markedly increases RNA cleavage. J Am Chem Soc 124:13682–13683

Victor J, Steger G, Riesner D (2018) Inability of DNAzymes to cleave RNA in vivo is due to limited Mg2+ concentration in cells. Eur Biophys J 47:333–343

Walsh SM, Sachdeva A, Silverman SK (2013) DNA catalysts with tyrosine kinase activity. J Am Chem Soc 135:14928–14931

Walsh SM, Konecki SN, Silverman SK (2015) Identification of sequence-selective tyrosine kinase deoxyribozymes. J Mol Evol 81:218–224

Wang Y, Nguyen K, Spitale RC et al (2021) A biologically stable DNAzyme that efficiently silences gene expression in cells. Nat Chem 13:319–326

Wong OY, Pradeepkumar PI, Silverman SK (2011) DNA-catalyzed covalent modification of amino acid side chains in tethered and free peptide substrates. Biochemistry 50:4741–4749

Workman C, Krogh A (1999) No evidence that mRNAs have lower folding free energies than random sequences with the same dinucleotide distribution. Nucleic Acids Res 27:4816–4822

Zacchia M, Abategiovanni ML, Stratigis S et al (2016) Potassium: From physiology to clinical implications. Kidney Dis 2:72–79

Zhang J (2018) RNA-cleaving DNAzymes: Old catalysts with new tricks for intracellular and in vivo applications. Catalysts 8:550

RNA Nanotechnology for Chemotherapy and Immunotherapy

Cristian Guzman, Daniel W. Binzel, Dan Shu, Richard Nho, and Peixuan Guo

Contents

Abstract RNA nanotechnology is the bottom-up self-assembly of nanoscale RNA structures, the main framework of which is mainly composed of RNA. Scaffolds, ligands, therapeutics, and modulators can all be composed of RNA. Classical RNA research has focused on interactions and 2D/3D structures within RNA. RNA nanotechnology can elucidate and exploit RNA-RNA interactions and quaternary (4D) structures. RNA technology is like "Lego bricks". Ideal materials for building blocks should have the following attributes: 1. Diversity. 2. Capable of constructing structures with various shapes, sizes, and modifiable stoichiometry. 3. Mix together

C. Guzman · D. W. Binzel · D. Shu · P. Guo (✉)
College of Pharmacy, The Ohio State University, Columbus, OH 43210, USA
e-mail: guo.1091@osu.edu

Center for RNA Nanobiotechnology and Nanomedicine, The Ohio State University, Columbus, OH 43210, USA

D. W. Binzel · D. Shu · P. Guo
James Comprehensive Cancer Center, The Ohio State University, Columbus, OH 43210, USA

R. Nho · P. Guo
Dorothy M. Davis Heart and Lung Research Institute, The Ohio State University, Columbus, OH 43210, USA

College of Medicine, The Ohio State University, Columbus, OH 43210, USA

J. Barciszewski (ed.), *RNA Structure and Function*, RNA Technologies 14,
https://doi.org/10.1007/978-3-031-36390-0_29

to self-assemble. 4. Thermodynamically, chemically, and enzymatically stable, with long shelf life. The potential of RNA NPs (nanoparticles) in various drug delivery applications including anticancer drug delivery is enormous. Therefore, RNA NPs can be developed in cancer research for chemotherapy and immunotherapy. The next decade is expected to witness many clinical trials in this field. RNA instability is no longer an issue and can be overcome through various chemical modifications. If we look at the big picture, the therapeutic potential of RNA, well-defined RNA NPs, and recent successes in stabilizing them not only support that RNA nanotechnology at the beginning of a new era but also suggest that future therapeutic prospects may predominate. Only 1.5% of the human genome encodes proteins. A significant portion of the remaining 98.5% of so-called "junk DNA" codes for important small or long non-coding RNAs. All substances that are toxic to cells are known to stop cancer growth. Currently, most cancer chemotherapies are highly toxic to normal cells and viral organoids. RNA nanoparticles have been shown to be a motile, dynamic, and deformable material that spontaneously runs enriched with cancer vasculature and is rapidly excreted into the urine via the glomerulus with little accumulation in vital organs. Therefore, RNA nanoparticles can efficiently and specifically target tumor tissues while being rapidly cleared from the body. Thus, RNA technology can convert toxic drugs in chemotherapy into non-toxic drugs. The promise of RNA as the third milestone in drug development has come true. Importantly, RNA itself can be used as a drug, such as siRNA, miRNA, aptamer, ribozyme, anti-miRNA, etc. Another area of interest is the use of small chemical drugs as ligands targeting non-coding RNA or mRNA. This field has been attractive and emerging but progress has been slow. This segment is expected to witness significant growth in the coming years.

Keywords RNA nanotechnology · Immunology · RNA · Drug delivery · Cancer therapy

1 Introduction to RNA Nanotechnology

RNA nanotechnology is the study of nanometer-sized RNA designed for the purpose of furthering medicinal research—cancer therapeutics and mRNA vaccines (Afonin et al. 2010; Guo 2010; Guo et al. 1998; Jaeger and Leontis 2000; Shapiro et al. 2008). DNA nanotechnology has been researched extensively for the past 30 years and has led toward modern understandings of RNA nanotechnology (Lee and Mao 2004; Rothemund 2006; Seeman 2005). DNA, in contrast to RNA, forms base pairs with Adenine to Thymine, and Cytosine to Guanine. RNA, however, primarily base pairs with Adenine to Uracil, and Cytosine to Guanine. DNA's complex formation of the DNA helix through Watson Crick base pairing makes it less flexible when compared to RNA (Binzel et al. 2021; Li et al. 2022). RNA exists in either one strand—the most common—and in other double-stranded conformations. RNA has been proven to have more extensive in vivo applications than DNA through its versatility, durability, and stereospecificity (Guo 2010).

RNA nanotechnology was first discovered in 1998 and was contrasted to DNA by its ability to fold into many different chemical structures—RNA loops and motifs (Guo et al. 1998). RNA follows canonical and noncanonical base pairing and has in vivo attributes that allow for structural versatility of the RNA molecule (Afonin et al. 2010; Guo 2010; Guo et al. 1998; Jaeger and Leontis 2000; Shapiro et al. 2008). RNA is different than DNA via methods of designing and manipulating its structure amongst different base pairs. When comparing the many heteroduplexes and homoduplexes—DNA-DNA, RNA-DNA, RNA-RNA—of RNA and DNA, it has been found that the RNA-RNA homoduplex is amongst the most stable of the helix's conformations (Jasinski et al. 2017; Piao et al. 2018). This fact has led scientists to pursue applications in the ever-growing field of RNA nanotechnology.

RNA nanotechnology's main therapeutic advantage, amongst many others, is its ability to form multivalent nanostructures (Shu et al. 2011; Wang et al. 2020; Westhof et al. 1996). It can be manipulated to produce nanostructures that enhance the chemical and physical properties of specific compounds (Guo 2010). For example, folate conjugated onto RNA nanoparticles (NPs) allows for to the targeting of specific cancer cells (Lee et al. 2017; Rychahou et al. 2015; Shu et al. 2011). It not only supplies a means of direction for cancer cells but also for its use in siRNA delivery and cancer cell regression (Jasinski et al. 2017). Another advantage is its ability to keep chemicals stable under low pHs—enabling survivability in the low acidity of the stomach and endosomes of intended cells (Ghimire et al. 2020; Piao et al. 2018). Furthermore, when RNA nanoparticles are applied to exosomes to display cancer-targeting ligands such as folate, they prevent endosome trapping of loaded cargos, which is common with forms of DNA nanotechnology (Pi et al. 2018; Zheng et al. 2019). RNA therapeutics has improved upon many previous concerns of DNA nanotechnology (Binzel et al. 2021). RNA nanotechnology, therefore, can pave the way toward making hydrophobic drugs more soluble and toxic drugs safer through the conjugation of fluorescent dyes, molecular probes, and cancer therapeutics (Guo et al. 2020; Jasinski et al. 2018; Piao et al. 2019).

2 RNA Conjugation to Make Hydrophobic Drugs More Soluble

The problems anticancer agents face today are the hydrophobic drug insolubility problem in aqueous solutions as well as nonspecific delivery to the cancer cells—causing toxicity. Insoluble hydrophobic drugs, in this case, are commonly shown to be highly efficacious in treating cancer, but their poor bioavailability limits their in vivo delivery. There, however, has been an attempt to move towards completely water-soluble nanocarriers for these hydrophobic anticancer drugs. 2′-Fluoro pyrimidine has been introduced into the 3WJ RNA—scaffold used to produce RNA nanoparticles (Piao et al. 2019). This chemical conjugate not only enhances the chemical and enzymatic stability of RNA nanoparticles but also allows for safe delivery of the conjugated chemotherapeutics such as camptothecin or paclitaxel (Guo et al. 2020; Piao et al. 2019).

The use of RNA nanotechnology has been shown to enhance the water solubility of many anticancer drugs such as paclitaxel (PTX) and camptothecin (CPT) (Guo et al. 2020; Piao et al. 2019). Camptothecin, specifically, has been conjugated to 3WJ-RNA as CPT is a rather potent anticancer drug. The conjugated camptothecin to RNA nanoparticles (CPT-RNA) has shown considerable tumor suppression in cancer cells when tagged with folate. However, there are two major problems with CPT that prevent it from being used widely in cancer treatment and that is its poor solubility and limited loading capacity. Similarly, this problem was countered with conjugation of PTX to substituents such as Vitamin E (Fu et al. 2015). This conjugation provided favorable hydrophobic interactions and longer circulation times within the body (Fig. 1). The next step, RNA therapeutics needed was not only a way to control drug release but also to increase solubility and targeting of specific cancer cells (Piao et al. 2019). Following this study, studies were performed with the conjugation of CPT to RNA, to address the substantial water-solubility properties of RNA. The 3WJ-RNA that camptothecin is conjugated to increases the loading capacity of traditional RNA-CPT. The results of this conjugation were recorded via native polyacrylamide gel electrophoresis analysis (Piao et al. 2019). The assembly of 7 CPT prodrugs—7CPT-3WJ—nanoparticles showed little aggregation in the gel assay compared to normal RNA-CPT, thus demonstrating the advantage of the 3WJ-RNA nanoparticle in increasing loading capacity. The high-loaded 7 CPT prodrugs triggered cancer cell death and tumor regression as expected.

Another complex of interest is the RNA-PTX conformation, which is stable and rigid in nature. In contrast to camptothecin, intravenous injections of the RNA-PTX nanoparticle inhibit breast and liver cancer growth and show no signs of toxicity and immune responses in mice, specifically (Guo et al. 2020; Wang et al. 2020). Alternatively, the 3WJ had to be redesigned due to reduced thermostability and limited loading capacity. This led to the usage of the 4WJ-RNA—a more stable conformation of four RNA strands (Guo et al. 2020). The 4WJ-RNA allowed for

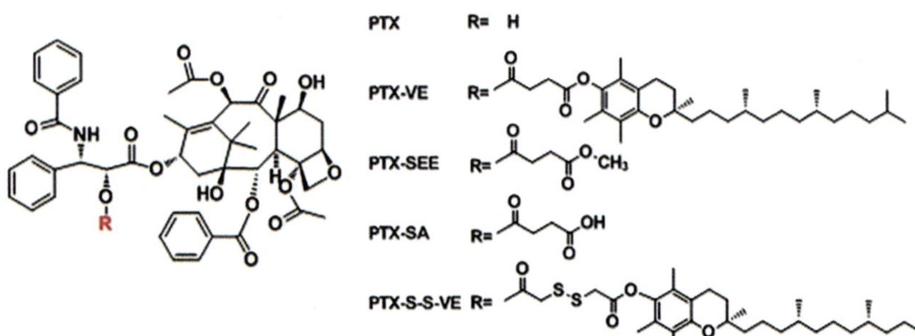

Fig. 1 Various modifications have been made to PTX which enhance its pharmacokinetic properties: solubility, cell targeting, etc. Reproduced with permission under Creative commons Attribution 4.0 International License from (Fu et al. 2015) Copyright © 2015, Fu et al.

loads up to 24 PTX molecules. Like, camptothecin, PTX conjugated with RNA has been shown to greatly reduce relative toxicity and overcome water insolubility. These specific nanoparticles showed enhanced therapeutic efficiencies as well as fewer to no side effects (Binzel et al. 2021).

Toxicity is shown within cells that are exposed to the cancer therapeutic drugs of PTX and camptothecin. The targeting ability and low water solubility of these drugs have proven problematic within biodistribution and pharmacokinetic studies. With this toxicity comes the problem of adverse side effects in the human/mice models. The RNA nanoparticles conjugated with these anticancer agents have led to decreased toxicity and increased water solubility of these cancer therapeutics (Guo et al. 2020).

3 The Dynamic and Motile Nature of RNA Make Them Efficient in Specific Cancer Targeting

It is imperative to continue discussing RNA NP's role in reducing drug toxicity, but first analysis of RNA's structure will lead to a more defined understanding. RNA exists as an elastomer—due to its enhanced elasticity and viscosity—that can build into rubbery RNA antibody like structures to provide therapeutic advantages such as a compelling EPR—enhanced permeability and retention—effect and enhanced tumor targeting capabilities (Fig 2) (Ghimire et al. 2020). The rubbery attribute of RNA has been tested with optical tweezers and in vivo tumor targeting. For example, within a 20 nm RNA nano-square was stretched out to 28 nm—a 1 to fourfold increase—and was also able to revert to its original size upon discontinued stretching. This rubbery nano-square is stretchable and shrinkable with multi-sequence relaxation and extension repeats (Ghimire et al. 2020). In vitro testing has allowed the RNA nanoparticles: enhanced tumor targeting retention effects, enhanced EPR effects, and less accumulation in tested organs. The comparison of gold, RNA, and DNA nanoparticles altogether displays these mechanisms in vivo. Lastly, the deformative property of RNA allows nanoparticles to squeeze through the tumor vasculature to improve the EPR effect (Ghimire et al. 2020). This, however, only scratches the surface when it comes to RNA.

Other important attributes of RNA nanoparticles involve its favorable biodistribution, pharmacokinetic and pharmacodynamic profiles when compared to gold and DNA nanoparticles. For starters, the negative charge of RNA prevents accumulation to negatively charged cell membranes thus preventing nonspecific interactions with non-targeted cells, while allowing binding to only targeted cell receptors (Binzel et al. 2021; Khisamutdinov et al. 2014a; Li et al. 2015). With this, the RNA nanoparticle easily penetrates the targeted cells while also undergoing rapid renal excretion by slipping through glomerulus filtration (Ghimire et al. 2020; Li et al. 2022). Another attribute of RNA nanoparticles is their defined size and stoichiometrical arrangements

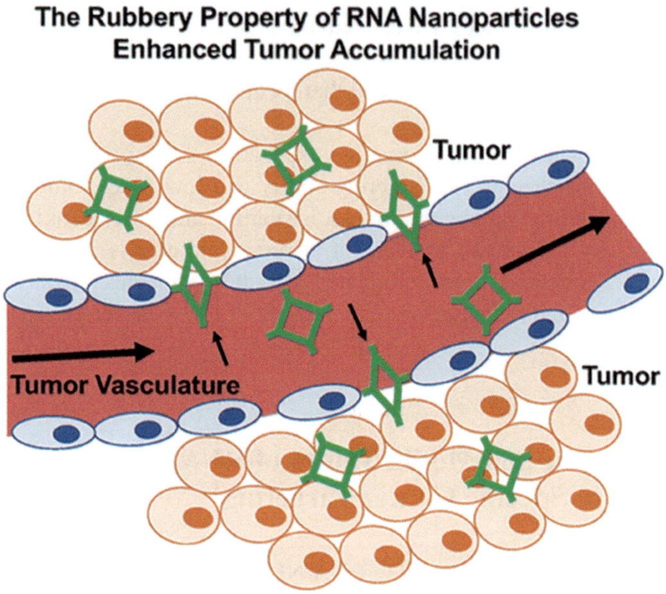

Fig. 2 Demonstration of elastic RNA NP's ability to penetrate tumor vasculature and adhere to the tumor surface. Reproduced with permission from (Ghimire et al. 2020). Copyright © 2020, American Chemical Society

(Khisamutdinov et al. 2014a; Li et al. 2015). The branched shape of RNA nanoparticles creates a one-way trafficking mechanism through tumor penetration and the EPR effect. Large-scale production of RNA can thus be made possible through solid-phase synthesis—self-assembly of RNA nanoparticles from a chemical precursor (Kempe et al. 1985; Sakatsume et al. 1991). In reference to immunological responses, they are wholly dependent on the specific sequence, shape, and size of the RNA nanoparticles (Guo et al. 2017). One important aspect in immunotherapy is the RNA nanoparticles' ability to be synthesized with no inherent immune response thus reducing cytokine interference. The versatility and multivalency of RNA nanoparticles allow for simultaneous targeting and delivery of attached ligands for synergistic effects. Targeted delivery is possible with cancer-specific ligands such as folate and EpCam (Lee et al. 2017; Pang et al. 2020; Rychahou et al. 2015; Xu et al. 2019; Zheng et al. 2019). The enhanced solubility of RNA nanoparticles helps in bypassing cellular aggregation, thus demonstrating lower to nonexistent toxicology profiles. Lastly, the thermostability of RNA allows these nanoparticles to be existent at low extracellular concentrations within the body (Khisamutdinov et al. 2014a; Piao et al. 2018; Shu et al. 2011). These many attributes lead us towards expanding upon the field of RNA nanotechnology's role in immunotherapy.

4 The Dynamics and Motile Nature of RNA Make Them Quick in Renal Excretion

As mentioned previously, the main obstacle that makes anticancer drugs—PTX or camptothecin—problematic is their relative toxicities within the human body. These anticancer therapeutics pose a problem due to their limited bioavailability, high dose requirements, adverse side effects, and non-specific targeting (Senapati et al. 2018). The first focus must be on specific drug targeting via RNA conjugation and then reducing adverse side effects adjacently. With size kept constant with respect to gold and iron nanoparticles, RNA nanoparticles have exhibited enhanced cancer targeting and less accumulation within the body comparatively (Ghimire et al. 2020).

Renal excretion is the most effective parameter for preventing the toxicity from these nanoparticles. The reason for the focus on renal excretion is due to water-soluble drugs being eliminated solely from the liver. Thus, to test whether RNA nanoparticles are toxic, researchers compared their elimination to the upper limit of renal excretion, 5.5 nm (Ghimire et al. 2020). RNA nanoparticles passed the renal filtration test—staying below the upper limit—and maintained their structure in the urine. The non-toxicity of RNA nanoparticles has been demonstrated due to their rapid clearance from the body and little to no accumulation in the body. In Fig 3, we can get a more accurate estimate of how size and charge of the NP plays a role in renal excretion.

Another key factor in the non-toxicity of RNA nanoparticles is their ability to specifically accumulate in carcinogenic cells through enhanced permeability and retention—EPR—effect (Senapati et al. 2018). The factors that influence the EPR effect include but are not limited to molecular size and surface charge. Molecular sizes greater than 40 kDa are preferred for the EPR effect to occur (Fang et al. 2011; Maeda 2001; Matsumura and Maeda 1986). In addition to this, RNA nanoparticles before conjugation range in size from 10 to 40 kDa. Surface charge, however, is the principal factor concerning the targeted delivery of RNA nanoparticles. High positive charges have displayed non-selectivity as most blood vessels have a high negative charge, thus clearing non-specific, positively charged compounds rapidly from blood circulation (Maeda 2001). Conjugating chemical carriers to these nanoparticles promotes specificity—folate to folate receptors. The binding of chemical conjugates to RNA nanoparticles, therefore leads to fewer adverse side effects due to their ability to attach to the chemically marked receptors/cells.

An introduction into unique RNA aptamers—oligonucleotides, like protein Abs in their ability to recognize ligands through specific binding pockets—is apparent when addressing RNA nanoparticle body clearance (Guo 2005; Kang and Lee 2013; Thiviyanathan and Gorenstein 2012). RNA has demonstrated the ability to form many different aptamers with variable functions due to its diverse nature in structuring. As an example of RNA's structuring, the phi29 pRNA strand can naturally form into

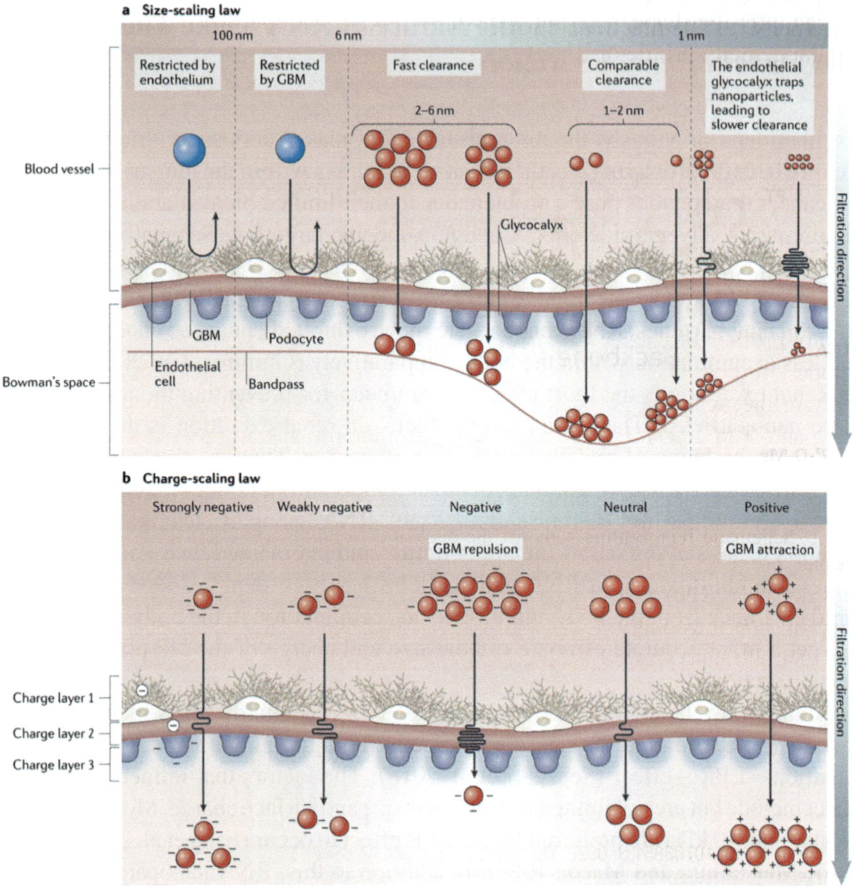

Fig. 3 Representation of the glomerulus in the kidney and how size determines a NPs ability to penetrate the membrane affecting clearance via renal excretion (**a**). The endothelial glycocalyx is negatively charged making positively charged NPs to clear through the kidney quicker. Charge and size have a major impact on the excretion of NPs (**b**). Reproduced with permission from (Du et al. 2018). Copyright © 2018, Springer Nature Limited

dimers and hexamers (Guo et al. 1998). From this, the pRNA 3WJ motif was discovered which is thermodynamically stable under high acid concentrations and temperature conditions (Shu et al. 2011). Using phi29 as a template, various nanoparticles—with each an independent functionality and folding pattern—have been synthesized for intended purposes in efficient cancer targeting and gene silencing (Haque et al. 2017; Khisamutdinov et al. 2015, 2014a, 2016; Li et al. 2016). RNA nanoparticles have bypassed renal depletion cutoff sizes and have minimized clearance through macrophage endocytosis by the production of 10–20 nm in diameter nanoparticles (Ghimire et al. 2020). These size characteristics allow for tumor penetration via the EPR effect and attached ligands to bypass receptor-mediated endocytosis. Thus, one can infer the shape, size, and structures of RNA nanoparticles greatly affect the immunogenicity profile of RNA.

5 Methods for Conjugating Chemical Drugs to RNA

Conjugation is the process of combining sequences of protein, lipids, or nucleic acid conjugates to the RNA duplex for the purpose of altering RNA's function (Paredes et al. 2011). This process has been used to alter the hydrophobicity and toxicity profiles of cancer therapeutics. Specifically, RNA conjugation with simple isotopes, fluorescent markers, and hydrophobic drug molecules, has been used extensively in the field of RNA nanotechnology (Jasinski et al. 2018; Piao et al. 2019; Wang and Guo 2021). Conjugations can be either site-specific on RNA—5' and 3' ends—or happen via direct nucleic acid conjugations—altering bases, sugars, and phosphate backbones. Site-specific conjugations can occur on RNA through known methods of synthetic chemistry, post-synthetic conjugative chemistries, and enzymatic approaches (Guo et al. 2012b). There are, however, several ways that conjugations occur on RNA, and these include click chemistry modifications and reactions involving the introduction of fluorescent dyes and molecular probes (Binzel et al. 2021; Paredes and Das 2011).

The most common and useful method for RNA conjugation is that of Solid Phase Synthesis—introducing a reactive functional group during the synthesis of an RNA strand (Binzel et al. 2021). Examples of Solid Phase Synthesis include the addition of fluorescent markers and molecular probes—thiols, amides, alkynes, and aldehydes (Binzel et al. 2021). An example of this is the direct modification of 2' hydroxy of RNA—protecting group—on the RNA molecule (Beaucage and Reese 2009).

Click Chemistry is defined as high-yielding reactions that are wide in scope, stereospecific, easy to perform, and can be performed in easily removable solvents (El-Sagheer and Brown 2012; Paredes and Das 2011). Click chemistry has been used over the years for the conjugation of specific molecules to RNA. The attribute that must be considered is the high specificity concerning these reactions. The most common click chemistry reactions involve copper-catalyzed azide-alkyne cycloaddition, strain-promoted click chemistry, and the inverse Diels Alder reaction (Piao et al. 2019).

The copper-catalyzed azide-alkyne cycloaddition (CuAAC) involves an orthogonal ligation reaction where the reactivity of the azide as a 1,3—dipole allows for the cycloaddition of a 1,2,3—triazole linkage (Fig. 4) (Breve et al. 2020; Presolski et al. 2011). The importance of this reaction is the use of Cu(l) as a chemical catalyst. The introduced copper acts as a catalyst for the reaction thus taking place seven times faster than the traditional cycloaddition (Meldal and Tornoe 2008). This reaction is one of the most used click chemistry techniques due to its overall efficiency and the ability of the copper atom to associate with other alkyne-based ligands. The second reaction to be discussed is that of strain-promoted click chemistry.

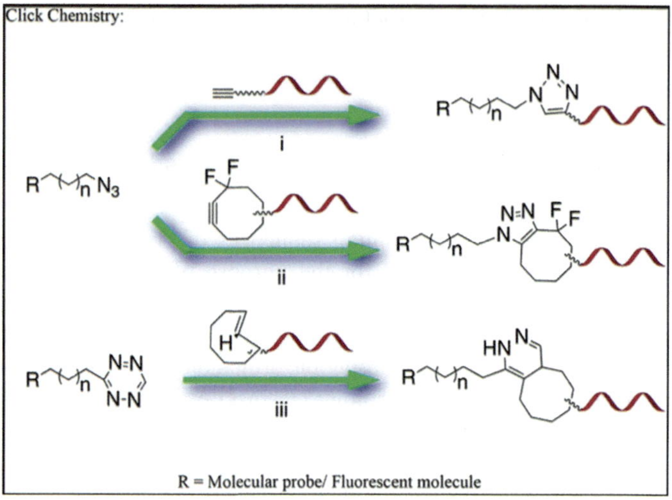

Fig. 4 Examples of common click chemistry modifications made to an azide with varying differences in stoichiometry to conjugate chemical drugs to RNA. Reproduced with permission from (Binzel et al. 2021). Copyright © 2021, Binzel et al. Published by American Chemical Society

Strain promoted click chemistry, or the Copper-free click reaction involves introducing an alkyne into a strained cyclooctyne ring (Agard et al. 2004; Chio and Bane 2020). This reaction, although occurring at a slower rate than the CuAAC reaction, improves the reaction kinetics of the copper-catalyzed azide-alkyne cycloaddition by introducing electron-withdrawing fluorine. Thus, creating a DIFO group or a difluorinated cyclooctyne ring. Both the above reactions are used in the labeling of specific biomolecules.

The last reaction to discuss is that of an inverse Diels Alder reaction (Gregoritza and Brandl 2015; Oluwasanmi and Hoskins 2021). This reaction involves a diene (electron-rich group) reacting with a dienophile (electron-poor group). Specifically, the inverse Diels–Alder reaction involves a mechanical overlap between the highest occupied molecular orbital of the dienophile with the unoccupied molecular orbital of the diene (Senapati et al. 2018). The most used and common reaction is the inverse Diels–Alder reaction involves the use of tetracycline cycloaddition—which enhances the kinetic parameters of the reaction while also being catalyst-free, unlike the other reactions.

Chemical conjugation can also be performed via reactions with thiol, NHS (N-hydroxysuccinimide ester), and EDC (N-(3-(dimethylamino)propyl)-N′-ethyl carbodiimide) (Binzel et al. 2021). These are post-synthetic reactions that occur after the synthesis of the RNA strand. These reactions are used specifically for the conjugation of fluorescent markers and molecular probes to the oligonucleotide of interest. The reactions with thiol, NHS, and EDC have high reaction yields and are easy to perform, thus making them important for RNA conjugation.

Conjugation can also be performed with different reaction parameters by introducing fluorescent dyes, specific targeting ligands, and drugs. The addition of fluorescent dyes is a common approach to RNA conjugation that supplies the opportunity for post-translational modifications to RNA. These fluorescent dyes are coupled to the RNA, thus heeding the production of single-labeled nanoparticles (Binzel et al. 2021; Jasinski et al. 2018). The fluorescent dyes used allow for the detection and visualization of RNA conjugates. When testing different samples—for example, Cy3 and Alexa-647 RNA strands—the fluorescent dye that is used depends upon the desired application.

Another strategy that is employed often is the use of targeting ligands. This type of reaction is most successfully seen in cancer-targeting probes. These targeting ligands are conjugated to the RNA nanoparticle and can target specific cancer receptors. For example, folate-tagged nanoparticles attach to specific folic acid receptors in the body (Zheng et al. 2019). Tumor-targeting ligands are essential to cancer chemotherapeutics and prevent toxic adverse effects from cytotoxic drugs such as Paclitaxel when conjugated to RNA (Guo et al. 2020; Wang et al. 2020). The conjugation of chemical drugs to RNA has been tested in mouse models (in vivo testing), and its pharmacokinetics and therapeutic drug efficiencies have been recorded (Guo et al. 2020). An example of conjugating chemical drugs to RNA takes place when Paclitaxel (PTX) was conjugated to 4WJ nanoparticles. This allowed for enhanced solubility of the hydrophobic drug while also increasing the therapeutic efficiency of the PTX. With experimentation in mouse models, it was shown that the PTX conjugation enhanced water solubility up to 32,000 times (Guo et al. 2020). RNA, one can infer, can make these chemical drugs more soluble due to its charge separation, thus solving the water insolubility dilemma.

Spherical nucleic acids (SNA)—nanoparticle cores with oligonucleotide sequences—can be conjugated to RNA to prevent nuclease degradation and internalized within the target cell thereof (Barnaby et al. 2016; Jensen et al. 2013; Tommasini-Ghelfi et al. 2019). SNAs (Spherical Nucleic Acids) specifically target two distinct RNA targets in a live cell, while also allowing for RNA nanoparticle diffusion through the blood–brain barrier (Jensen et al. 2013; Tommasini-Ghelfi et al. 2019). For example, exosome-encapsulated SNAs can deliver cancer therapeutics into the targeted cancer cell thus supplying vehicles to be delivered directly to the cancer cells (Alhasan et al. 2014).

6 Bispecific RNA Nanoparticles for Cancer Treatment

The introduction of cancer vaccines, in the late twentieth century, led to a somewhat promising way to approach tumor regression via the adaptive immune response (Rausch et al. 2014; Silva et al. 2015). Some common approaches to these cancer vaccines are delivering a large amount of antigen to dendritic cells, and the induction of T-lymphocyte responses at the site of the tumor. With this came the immunology approach of introducing tumor-reactive T-cells into the body (Saxena et al. 2021).

These cancer vaccines, however, failed to recruit large amounts of T-cells to produce lasting immunity, while also ignoring many immune system mechanisms such as immune suppression and escape. This problem, here, demonstrates the need for a new approach towards cancer vaccines.

An approach that seems promising due to high selectivity for tumor targets, reduced toxicity within in vitro studies, and high solubilizing potential of the drugs, is the use of bispecific and trispecific antibodies (Seung et al. 2022; Shim 2020; Thakur et al. 2018; Wu et al. 2020; Wu and Cheung 2018; Zhao et al. 2022). The main commercial approach to this is in the form of protein Abs which is an engineered complex that contains two variable regions which attach to different molecular targets. These protein complexes can attach to molecular targets causing apoptosis through the recruitment of T-cells to the site of action (Seung et al. 2022; Wu et al. 2020). The advantages of this protein complex include but are not limited to low immunogenicity profiles, little to no aggregation, and improved vascular diffusion and clearance. Protein Abs have been researched extensively in the past decade and has unfortunately presented problems in the form of mass production, due to the bulkiness of the complex, and many unsuccessful clinical trials thereof (Shim 2020; Thakur et al. 2018).

Therefore, an innovative approach to introducing RNA antibodies has been established. It was proposed to introduce a thermodynamically stable RNA conformation in the form of pRNA 3WJ with conjugated immunolical ligands (Fig. 5) (Guo 2010; Shu et al. 2011). This improves upon the protein Abs by adding multivalent RNA antibodies with three distinct arms for holding functional RNA modules (Shu et al. 2021). The addition of RNA to the established protein Ab field complex helps in reducing toxicity and immunogenicity. There are multiple advantages with the RNA protein complex including specific targeting with the negative charge of RNA—will not attach randomly to other cells—inherent stability and stereospecificity of RNA nanoparticles thus reducing side effects, and large-scale production capabilities in a cell-free environment—protein Abs requires a cellular environment to be produced.

The EPR effect of RNA nanoparticles strongly facilitates our RNA antibody to reach and accumulate in cancer vascular where T-cells with checkpoints reside (Li et al. 2022). The EPR effect has been demonstrated as the major pathway for nanoscale particles to tumor sites (Fang et al. 2011; Maeda 2001). Many of the chemicals excreted by tumor sites, such as nitric oxide, aid in the ability of nanoparticles to penetrate inflamed cancer tissue. The targeting of these bispecific nanoparticles, however, is mostly dependent on particle size. With this is mind, it is important to keep these structures confined to the 5–100 nm range; thus, allowing the nanoparticles to bypass the tumor depletion cutoff sizes. The small size and elasticity of RNA nanoparticles allow for a strong or enhanced EPR effect with the avoidance of accumulation in the liver or spleen (Ghimire et al. 2020). Nanoparticles in the 10–50 nm range are preferred due to their fast renal clearance (Ghimire et al. 2020). It is important to remember 50 nm as a cutoff in size due to entrapment by Kupffer cells and macrophages (Zhang et al. 2016). Keeping these constructed nanoparticles in the 10–20 nm is ideal as it easily penetrates tumor vasculature while conferring the

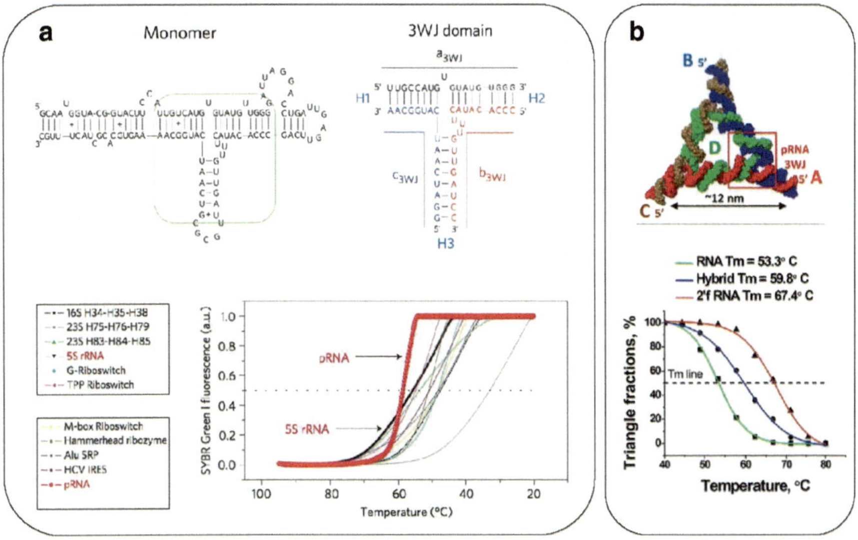

Fig. 5 Discovery of the thermodynamically stable three-way junction (3WJ) of phi29 pRNA that leads to a way in constructing stable RNA nanoparticles for in vivo *applications* (Guo et al. 1987). Reproduced with permission from (Shu et al. 2011). Copyright © 2011, Nature Publishing Group. Reproduced with permission under Creative Commons CC-BY from (Khisamutdinov et al. 2014a). Copyright © 2014, American Chemical Society

advantage of tumor selectivity over protein Abs. As in the mRNA vaccines, systemically injected RNA nanoparticles can bypass various organs to prevent damage to vital bodily organs.

The novelty of this work was the transformation of "protein" bispecific and trispecific Ab platforms into "RNA" bispecific and trispecific Abs composed of two or three different binding modules fused into a single RNA nanoparticle. RNA antibodies have several advantages: low immunogenicity, modular design, no aggregation, chemical synthesis at an industrial scale, adjustable arm size, and nanoscale size to facilitate rapid vascular accumulation while fast clearance thus with low toxicity and side-effect (Binzel et al. 2021; Jasinski et al. 2017; Xu et al. 2018). The multi-functional 3WJ nanoparticles were designed to harbor sequences that allow binding to (i) specific checkpoint on T-cells to activate T cell or specific T cell surface antigen to trigger release of cytokines, (ii) tumor-specific surface marker to bridge the cancer cells and cytotoxic T cells (Shu et al. 2021). The idea and technology are highly innovative with great potential to make transformative advances in cancer therapy.

7 RNA Aptamers as Antibody to Regulate Immune Check Point in Cells Such as Treg Cells

Immunotherapy has become a revolutionary subject in cancer therapy over the past few years (Saxena et al. 2021; Wu and Cheung 2018). For the immunotherapy to be more effective, the physiological role of various immune cells needs to be considered. One of the recent approaches is to target the negative regulator of cancer immunotherapy. Several tumor-associated immune cells such as myeloid-derived suppressor cells (MDSCs), tumor-associated macrophages (TAMs), regulatory T cells (Tregs) inhibit cancer cell destruction (Bockamp et al. 2020). Thus, targeting or inactivation of these cells has been proposed as an effective anti-cancer immunotherapy. Among them, the function of Tregs must be considered for the development of an effective anti-cancer immunotherapy. Although Tregs prevent autoimmune disease, they contribute to tumor-specific T cell tolerance, and the increased presence of Tregs has been linked to poor clinical prognosis in various types of cancer (Manrique-Rincon et al. 2021). Tregs is known to destroy immune effector cells by utilizing granzyme and perforin. Tregs also suppresses the function of effector T cells. Thus, targeting of Tregs, specifically located in the tumor microenvironment by NPs has been proposed to improve the outcomes of immunotherapy. A recent study showed that the administration of 4-1BB-TS6 aptamer to tumor-challenged mice improves anti-tumor responses mediated GVAX immunotherapy (Manrique-Rincon et al. 2021), suggesting that Treg targeting by aptamer can be a promising anticancer approach.

Recently, immunotherapy has also become a popular point of research in the field of cancer treatment due to the discovery that monoclonal antibodies (mAbs) targeting co-inhibitory immune checkpoint proteins can improve the body's immune response against cancer cells (Buss et al. 2012; Kimiz-Gebologlu et al. 2018). Such checkpoint inhibitors restore T cell proliferation and stimulate effector functions, such as the release of effector cytokines and cytotoxic granules. Co-stimulatory checkpoint molecules, such as 4-1BB and CD28, are responsible for the proper activation of T lymphocytes (Cappell and Kochenderfer 2021; Chester et al. 2018; Roselli et al. 2021; Seung et al. 2022). As of April 2018, about 25 agonist antibodies targeting immune co-stimulatory molecules have been clinically tested for use in cancer therapy (Kaviani et al. 2022).

Although immune checkpoint-targeting antibodies (Abs) could boost anticancer immune responses. Certain protein-based immunotherapies revealed side effects and unfavorable biodistribution, so effective non-protein options with lower side effects are highly sought after (Thakur et al. 2018). RNA's ability to form various three-dimensional configurations allows for the creation of a variety of ligands to bind different cell receptors. The rubber-like properties of RNA NPs allow for swift lodging to cancer vasculature with little accumulation in vital organs, resulting in favorable PK/PD profile and safe pharmacological parameters (Ghimire et al. 2020). Multi-specific drugs are expected to be the fourth wave of biopharmaceutical innovation. Here, we report the development of multi-specific Ab-like RNA NPs carrying multiple ligands for immunotherapy (Shu et al. 2021). The stoichiometries and stereo conformations of the checkpoint-activating RNA NPs were optimized for T cell activation. Comparing to mono- and bi-specific RNA NPs, the tri-specific Ab-like RNA NPs bound to the trimeric T cell receptor with the highest efficiency, showed the optimal T cells activation, and promoted the strongest anti-tumor function of immune cells. Animal trails demonstrated that the tri-specific RNA NPs inhibited cancer growth (Shu et al. 2021). This Ab-like RNA NPs platform represents an alternative to protein Abs for tumor immunotherapy.

Thus, evidently, the development of cancer immunotherapy was a major advancement in both the fields of immunology and oncology. However, some drawbacks limit the progress of the usual protein-based options in clinical trials. For example, Fc-receptors of IgG-like bi-specific antibodies may be immunogenic, and non-specific interactions between bi-specific Abs with white blood cells may possibly change their bio-distribution (Piao et al. 2019). Additionally, toxicological effects like irAEs pose a major challenge for immunomodulation attempts.

The amphoteric ionization property and intrinsic hydropathic nature of proteins make them relatively bulky (Shu et al. 2021). This leads them to often form aggregates in vital organs, resulting in nonspecific binding to vital organs. Such adverse effects can lead to systemic activation of the immune system, possibly causing fulminant and even fatal toxicological effects to occur with protein Abs. The known drawbacks of protein Abs can hamper their otherwise great therapeutic potential. Therefore, effective non-protein cancer-specific immunotherapy drugs are desirable as they would not produce such adverse effects.

Recent developments in protein Abs research and peptide pharmaceutics have made it possible to develop multi-specific protein Ab platforms to address the toxicity of immunotherapy and increase antibody efficacy. Multi-specific drugs have been predicted to be the fourth wave of biopharmaceutical innovation. Multi-specific Abs are re-engineered protein-based reagents with two or more variable regions that can bind both immune cells and cancer cells more precisely and effectively (Chiu and Gilliland 2016; Elgundi et al. 2017; Jachimowicz et al. 2014). As a result, they have achieved great success, such as in the cases of B cell acute lymphoblastic leukemia (Shi et al. 2022; Wudhikarn et al. 2022).

The emerging protein-free aptamers represent a promising platform for targeted immunotherapy (Chen et al. 2022; Zhang et al. 2021; Zhu and Chen 2018). Compared to protein-based immunotherapy mAbs, chemically synthesized aptamers possess many advantages as targeting reagents: low cost, faster SELEX selection, low immunogenicity, rapid tissue penetration, and long-term stability (Zhou and Rossi 2017). All well-known checkpoint molecules were developed as blocking aptamers and showed comparable effects to those of Abs in mouse models (Shu et al. 2021; Zhang et al. 2021). However, one of the biggest obstacles in the development of such an aptamer delivery system is its various and complex scaffold selection system. Thus, the creation of a stable and flexible scaffold that can effectively deliver different aptamers for cancer immunotherapy is essential.

The concept of RNA nanotechnology was proposed in 1998 and has now developed into a mature field with high potential as a novel therapeutic platform (Binzel et al. 2021; Guo et al. 1998). RNA nanotechnology is the bottom-up self-assembly of nanometer-scale structures, which can include scaffolds, ligands, therapeutics, and regulators, composed mainly of RNA (Binzel et al. 2021). By utilizing RNA nanotechnology, we have constructed a variety of RNA NPs and demonstrated that RNA NPs harboring different functional modules retain their folding capabilities and independent functionalities (Binzel et al. 2021; Guo 2005, 2010; Guo et al. 2012a, 2012b; Shu et al. 2014; Xu et al. 2018). The rubber-like properties of RNA nanoparticles (NPs) allow for swift lodging to cancer vasculature with little accumulation in vital organs, resulting in favorable PK/PD profile and safe pharmacological parameters (Ghimire et al. 2020). The immunogenicity of RNA NPs is dependent on their shape, size, and sequence (Guo et al. 2017). Thus, they can be manipulated to produce little or no immune response or toxicological effects (Khisamutdinov et al. 2014b). RNA-based nanostructures have diverse functions, including gene expression, gene regulation, and the specific binding to different molecules. Also, due to the promising advancements, RNA nanotechnology was used for immunotherapy and immunomodulation studies (Shu et al. 2021). Using the 3WJ as a scaffold from the pRNA of the Phi29 bacteriophage, we have constructed robust RNA NPs with various tumor-targeting aptamers. We have demonstrated that our 3WJ scaffold with EGFR, HER2, PSMA, Annexin A2, CD133, and folate aptamers are able to bind to and enter tumor cells specifically and target solid tumors in animals with little or no accumulation in vital organs (Binzel et al. 2016; Guo et al. 2020; Pi et al. 2017; Shu et al. 2011, 2015; Yin et al. 2019; Zhang et al. 2017).

Ab-like RNA NPs can be designed, assembled, analyzed, and characterized to carry specific checkpoint activators using ultra-stable pRNA 3WJ as a nano-scaffold (Fig. 6) (Shu et al. 2021). The usage of pRNA 3WJ as a nano-scaffold is a promising platform for aptamer-based checkpoint immunotherapy applications. We investigated the binding of multi-specific Ab-like RNA NPs to T cells and tumor cells. Optimization of the stoichiometries and stereo conformations of the checkpoint activators in the 3WJ scaffold was placed under scrutiny. It was found that Ab-like RNA NPs could effectively stimulate T cell activation and proliferation in vitro and inhibit cancer growth in animal trials (Shu et al. 2021). The results suggested that the ultra-stable Y-shaped pRNA 3WJ nano-scaffold would be a promising vehicle for Ab-like RNA NPs construction for the purpose of harboring RNA aptamers for immune checkpoint binding in cancer immunotherapy.

8 New Technology to Display T Cell Ligands on Cancer Cells

The orientation of arrow-shaped RNA has been altered to control ligand display on extracellular vesicle membranes for specific cell targeting (Pi et al. 2018). This surface-displaying technology can be applied to the display of cancer cells, and immune cells such as T cells and B cells. Placing membrane-anchoring cholesterol

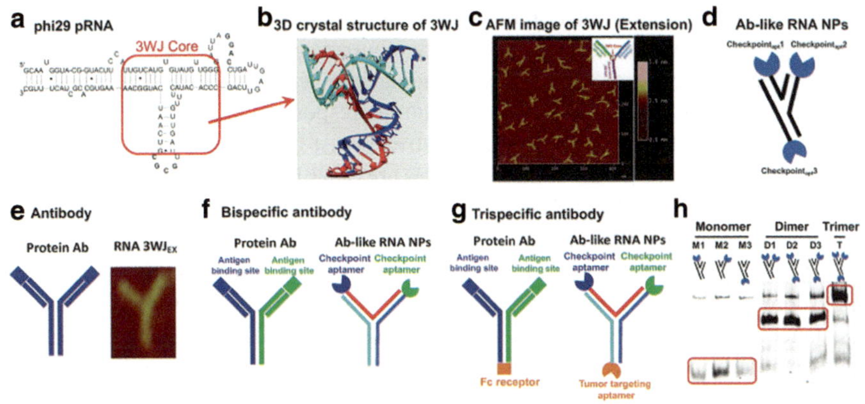

Fig. 6 The construction of bispecific and trispecific antibodies using Protein Ab as a template. Reproduced with permission under Creative commons Attribution 4.0 International License from (Shu et al. 2021). Copyright © 2021, Shu et al.

at the tail of the arrow results in display of RNA aptamer or folate on the outer surface of the extracellular vesicle (Pi et al. 2018). Taking advantage of the RNA ligand for specific targeting and extracellular vesicles for efficient membrane fusion, the resulting ligand-displaying extracellular vesicles can be used to display the ligand in cancer cells for T-cell recognition. We have demonstrated that extracellular vesicles displaying tumor targeting aptamers bind to prostate-specific membrane antigen, and loaded with survivin siRNA, inhibited prostate cancer xenograft (Pi et al. 2018). The same extracellular vesicle instead displaying epidermal growth-factor receptor aptamer inhibited orthotopic breast and lung cancer models (Li et al. 2021; Pi et al. 2018). Likewise, surviving siRNA-loaded and folate-displaying extracellular vesicles inhibited patient-derived colorectal cancer xenograft (Pi et al. 2018). Application of this surface-displaying technology will expand the field in immunotherapy.

9 Summary and Perspectives

Cancer therapeutics have long suffered from poor bioavailability, poor solubility, high toxicity, and high immune responses all leading to poor efficacy within the human body (Friberg and Nystrom 2016; Guo et al. 2012a; Shi et al. 2017; Venditto and Szoka 2013). The field of drug delivery's goal is to safely deliver these potent therapeutics to disease sites while significantly reducing healthy organ affects. Nanotechnology, especially RNA nanotechnology, demonstrates promise of achieving this goal by easily conjugating cancer drugs and immunotherapies and working towards providing specific caner targeting. Here, we highlight the recent advances in RNA nanotechnology develop unique nanostructures that can conjugate and incorporate chemical anti-cancer drugs and immunotherapy components. RNA nanoparticles have demonstrated high cancer targeting efficiency through spontaneous and active targeting that allows for the delivery of high payloads of therapeutics. The conjugation of paclitaxel to thermo-stable RNA nanoparticles led to a 32,000-fold increase in paclitaxel solubility for greatly reduced immune responses and toxicities to healthy organs, while greatly affecting triple negative breast cancer tumors in mice (Guo et al. 2020). Additionally, RNA nanoparticles are tunable to elicit or have no immune responses (Guo et al. 2017). Construction of immunoadjuvant RNA nanoparticles increased cytokine responses that can complement cancer therapies (Khisamutdinov et al. 2014b). Furthermore, the direct conjugation of immunotherapeutic components led to control of cancer tumors by linking T-cells to tumor cells to overcome immune resistance mechanisms (Chandler et al. 2020; Shu et al. 2021). Overall, the field of RNA nanotechnology has recently shown great promise in achieving the dream of drug delivery and significantly improving the treatment of cancers and safety profiles.

Acknowledgements The work was partially supported by NIH grant U01CA207946 to P.G and R01CA257961 to D. S. and D.W.B. The content is solely the responsibility of the authors and does not necessarily represent the official views of NIH. P.G.'s Sylvan G. Frank Endowed Chair position in Pharmaceutics and Drug Delivery is funded by the CM Chen Foundation. P.G. is the consultant, licensor, and grantee of Oxford Nanopore Technologies; as well as the cofounder, member of the board of directors of the ExonanoRNA, LLC.

References

Afonin KA, Bindewald E, Yaghoubian AJ et al (2010) In vitro Assembly of Cubic RNA-based Scaffolds Designed in silico. Nat Nanotechnol 5:676–682

Agard NJ, Prescher JA, Bertozzi CR (2004) A strain-promoted [3 + 2] azide-alkyne cycloaddition for covalent modification of biomolecules in living systems. J Am Chem Soc 126:15046–15047

Alhasan AH, Patel PC, Choi CH et al (2014) Exosome encased spherical nucleic acid gold nanoparticle conjugates as potent microRNA regulation agents. Small 10:186–192

Barnaby SN, Perelman GA, Kohlstedt KL et al (2016) Design considerations for RNA spherical nucleic acids (SNAs). Bioconjug Chem 27:2124–2131

Beaucage SL, Reese CB (2009) Recent advances in the chemical synthesis of RNA. Curr Protoc Nucleic Acid Chem Chapter 2:Unit 2 16:1–31

Binzel D, Shu Y, Li H et al (2016) Specific delivery of mirna for high efficient inhibition of prostate cancer by RNA nanotechnology. Mol Ther 24:1267–1277

Binzel DW, Li X, Burns N et al (2021) Thermostability, tunability, and tenacity of RNA as rubbery anionic polymeric materials in nanotechnology and nanomedicine-specific cancer targeting with undetectable toxicity. Chem Rev 121:7398–7467

Bockamp E, Rosigkeit S, Siegl D et al (2020) Nano-Enhanced Cancer Immunotherapy: Immunology Encounters Nanotechnology. Cells 9:2102

Breve TG, Filius M, Araman C et al (2020) Conditional Copper-catalyzed azide-alkyne cycloaddition by catalyst encapsulation. Angew Chem Int Ed Engl 59:9340–9344

Buss NA, Henderson SJ, McFarlane M et al (2012) Monoclonal antibody therapeutics: history and future. Curr Opin Pharmacol 12:615–622

Cappell KM, Kochenderfer JN (2021) A comparison of chimeric antigen receptors containing CD28 versus 4–1BB costimulatory domains. Nat Rev Clin Oncol 18:715–727

Chandler M, Johnson MB, Panigaj M et al (2020) Innate immune responses triggered by nucleic acids inspire the design of immunomodulatory nucleic acid nanoparticles (NANPs). Curr Opin Biotechnol 63:8–15
Chen Z, Zeng Z, Wan Q et al (2022) Targeted immunotherapy of triple-negative breast cancer by aptamer-engineered NK cells. Biomaterials 280:121259
Chester C, Sanmamed MF, Wang J et al (2018) Immunotherapy targeting 4–1BB: mechanistic rationale, clinical results, and future strategies. Blood 131:49–57
Chio TI, Bane SL (2020) Click Chemistry Conjugations. Methods Mol Biol 2078:83–97
Chiu ML, Gilliland GL (2016) Engineering antibody therapeutics. Curr Opin Struct Biol 38:163–173
Du B, Yu M, Zheng J (2018) Transport and interactions of nanoparticles in the kidneys. Nat Review Mat 3:358–374
Elgundi Z, Reslan M, Cruz E et al (2017) The state-of-play and future of antibody therapeutics. Adv Drug Deliv Rev 122:2–19
El-Sagheer AH, Brown T (2012) Click nucleic acid ligation: applications in biology and nanotechnology. Acc Chem Res 45:1258–1267
Fang J, Nakamura H, Maeda H (2011) The EPR effect: unique features of tumor blood vessels for drug delivery, factors involved, and limitations and augmentation of the effect. Adv Drug Deliv Rev 63:36–151
Friberg S, Nystrom AM (2016) NANOMEDICINE: will it offer possibilities to overcome multiple drug resistance in cancer? J Nanobiotechnology 14:17
Fu Q, Wang Y, Ma Y et al (2015) Programmed hydrolysis in designing paclitaxel prodrug for nanocarrier assembly. Sci Rep 5:12023
Ghimire C, Wang H, Li H et al (2020) RNA nanoparticles as rubber for compelling vessel extravasation to enhance tumor targeting and for fast renal excretion to reduce toxicity. ACS Nano 14:13180–13191
Gregoritza M, Brandl FP (2015) The Diels-Alder reaction: A powerful tool for the design of drug delivery systems and biomaterials. Eur J Pharm Biopharm 97:438–453
Guo P (2005) RNA nanotechnology: engineering, assembly and applications in detection, gene delivery and therapy. J Nanosci Nanotechnol 5:1964–1982
Guo P (2010) The emerging field of RNA nanotechnology. Nat Nanotechnol 5:833–842
Guo PX, Erickson S, Anderson D (1987) A small viral RNA is required for in vitro packaging of bacteriophage Phi29 DNA. Science 236:690–694
Guo P, Zhang C, Chen C et al (1998) Inter-RNA interaction of phage Phi29 pRNA to form a hexameric complex for viral DNA transportation. Mol Cell 2:149–155
Guo P, Haque F, Hallahan B et al (2012a) Uniqueness, advantages, challenges, solutions, and perspectives in therapeutics applying RNA nanotechnology. Nucleic Acid Ther 22:226–245
Guo P, Shu Y, Binzel D et al (2012b) Synthesis, conjugation, and labeling of multifunctional pRNA nanoparticles for specific delivery of siRNA, drugs, and other therapeutics to target cells. Methods Mol Biol 928:197–219
Guo S, Li H, Ma M et al (2017) Size, shape, and sequence-dependent immunogenicity of RNA nanoparticles. Mol Ther Nucleic Acids 9:399–408
Guo S, Vieweger M, Zhang K et al (2020) Ultra-thermostable RNA nanoparticles for solubilizing and high-yield loading of paclitaxel for breast cancer therapy. Nat Commun 11:972–982
Haque F, Xu C, Jasinski DL et al (2017) Using planar Phi29 pRNA three-way junction to control size and shape of RNA nanoparticles for biodistribution profiling in mice. Methods Mol Biol 1632:359–380
Jachimowicz RD, Borchmann S, Rothe A (2014) Multi-specific antibodies for cancer immunotherapy. BioDrugs 28:331–343
Jaeger L, Leontis NB (2000) Tecto-RNA: one-dimensional self-assembly through tertiary interactions. Angew Chem Int Ed Engl 39:2521–2524
Jasinski D, Haque F, Binzel DW et al (2017) Advancement of the emerging field of RNA nanotechnology. ACS Nano 11:1142–1164

Jasinski DL, Yin H, Li Z et al (2018) Hydrophobic effect from conjugated chemicals or drugs on in vivo biodistribution of RNA nanoparticles. Hum Gene Ther 29:77–86

Jensen SA, Day ES, Ko CH et al (2013) Spherical nucleic acid nanoparticle conjugates as an RNAi-based therapy for glioblastoma. Sci Transl Med 5:209ra152

Kang KN, Lee YS (2013) RNA aptamers: a review of recent trends and applications. Adv Biochem Eng Biotechnol 131:153–169

Kaviani E, Hosseini A, Mahmoudi Maymand E et al (2022) Triggering of lymphocytes by CD28, 4–1BB, and PD-1 checkpoints to enhance the immune response capacities. PLoS ONE 17:e0275777

Kempe T, Sundquist WI, Chow F et al (1985) Chemical and enzymatic biotin-labeling of oligodeoxyribonucleotides. Nucleic Acids Res 13:45–57

Khisamutdinov EF, Jasinski DL, Guo P (2014a) RNA as a boiling-resistant anionic polymer material to build robust structures with defined shape and stoichiometry. ACS Nano 8:4771–4781

Khisamutdinov EF, Li H, Jasinski DL et al (2014b) Enhancing Immunomodulation on Innate Immunity by Shape Transition Among RNA Triangle, Square and Pentagon Nanovehicles. Nucleic Acids Res 42:9996–10004

Khisamutdinov EF, Bui MN, Jasinski D et al (2015) Simple method for constructing RNA triangle, square, pentagon by tuning interior RNA 3WJ angle from 60 degrees to 90 degrees or 108 degrees. Methods Mol Biol 1316:181–193

Khisamutdinov EF, Jasinski DL, Li H et al (2016) Fabrication of RNA 3D nanoprisms for loading and protection of small RNAs and model drugs. Adv Mater 28:10079–10087

Kimiz-Gebologlu I, Gulce-Iz S, Biray-Avci C (2018) Monoclonal antibodies in cancer immunotherapy. Mol Biol Rep 45:2935–2940

Lee SH, Mao C (2004) DNA nanotechnology. Biotechniques 37:517–519

Lee TJ, Yoo JY, Shu D et al (2017) RNA nanoparticle-based targeted therapy for glioblastoma through inhibition of oncogenic miR-21. Mol Ther 25:1544–1555

Li H, Lee T, Dziubla T et al (2015) RNA as a stable polymer to build controllable and defined nanostructures for material and biomedical applications. Nano Today 10:631–655

Li H, Zhang K, Pi F et al (2016) Controllable self-assembly of RNA tetrahedrons with precise shape and size for cancer targeting. Adv Mater 28:7501–7507

Li Z, Yang L, Wang H et al (2021) Non-small-cell lung cancer regression by siRNA delivered through exosomes that display EGFR RNA aptamer. Nucleic Acid Ther 31:364–374

Li X, Bhullar AS, Binzel DW et al (2022) The dynamic, motile and deformative properties of RNA nanoparticles facilitate the third milestone of drug development. Adv Drug Deliv Rev 186:114316

Maeda H (2001) The enhanced permeability and retention (EPR) effect in tumor vasculature: the key role of tumor-selective macromolecular drug targeting. Adv Enzyme Regul 41:189–207

Manrique-Rincon AJ, Ruas LP, Fogagnolo CT et al (2021) Aptamer-mediated transcriptional gene silencing of Fox p 3 inhibits regulatory T cells and potentiates antitumor response. Mol Ther Nucleic Acids 25:143–151

Matsumura Y, Maeda H (1986) A new concept for macromolecular therapeutics in cancer chemotherapy: mechanism of tumoritropic accumulation of proteins and the antitumor agent smancs. Cancer Res 46:6387–6392

Meldal M, Tornoe CW (2008) Cu-catalyzed azide-alkyne cycloaddition. Chem Rev 108:2952–3015

Oluwasanmi A, Hoskins C (2021) Potential use of the Diels-Alder reaction in biomedical and nanomedicine applications. Int J Pharm 604:120727

Pang L, Shah H, Wang H et al (2020) EpCAM-targeted 3WJ RNA nanoparticle harboring delta-5-desaturase siRNA inhibited lung tumor formation via DGLA peroxidation. Mol Ther Nucleic Acids 22:222–235

Paredes E, Das SR (2011) Click chemistry for rapid labeling and ligation of RNA. ChemBioChem 12:125–131

Paredes E, Evans M, Das SR (2011) RNA labeling, conjugation and ligation. Methods 54:251–259

Pi F, Zhang H, Li H et al (2017) RNA nanoparticles harboring Annexin A2 aptamer can target ovarian cancer for tumor-specific doxorubicin delivery. Nanomedicine 13:1183–1193
Pi F, Binzel DW, Lee TJ et al (2018) Nanoparticle orientation to control RNA loading and ligand display on extracellular vesicles for cancer regression. Nat Nanotechnol 13:82–89
Piao X, Wang H, Binzel DW et al (2018) Assessment and comparison of thermal stability of phosphorothioate-DNA, DNA, RNA, 2'-F RNA, and LNA in the context of Phi29 pRNA 3WJ. RNA 24:67–76
Piao X, Yin H, Guo S et al (2019) RNA nanotechnology to solubilize hydrophobic antitumor drug for targeted delivery. Adv Sci 6:1900951
Presolski SI, Hong VP, Finn MG (2011) Copper-catalyzed azide-alkyne click chemistry for bioconjugation. Curr Protoc Chem Biol 3:153–162
Rausch S, Schwentner C, Stenzl A et al (2014) mRNA vaccine CV9103 and CV9104 for the treatment of prostate cancer. Hum Vaccin Immunother 10:3146–3152
Roselli E, Boucher JC, Li G et al (2021) 4–1BB and optimized CD28 co-stimulation enhances function of human mono-specific and bi-specific third-generation CAR T cells. J Immunother Cancer 9:e003354
Rothemund PW (2006) Folding DNA to create nanoscale shapes and patterns. Nature 440:297–302
Rychahou P, Haque F, Shu Y et al (2015) Delivery of RNA nanoparticles into colorectal cancer metastases following systemic administration. ACS Nano 9:1108–1116
Sakatsume O, Yamaguchi T, Takaku H (1991) Solid phase synthesis of mRNA by the phosphoramidite approach using 2'-O-1-(2-chloroethoxy)ethyl protection and its stability in *E. Coli* system. Nucleic Acids Symp Ser 33–36
Saxena M, van der Burg SH, Melief CJM et al (2021) Therapeutic cancer vaccines. Nat Rev Cancer 21:360–378
Seeman NC (2005) Structural DNA nanotechnology: an overview. Methods Mol Biol 303:143–166
Senapati S, Mahanta AK, Kumar S et al (2018) Controlled drug delivery vehicles for cancer treatment and their performance. Signal Transduct Target Ther 3:7
Seung E, Xing Z, Wu L et al (2022) A trispecific antibody targeting HER2 and T cells inhibits breast cancer growth via CD4 cells. Nature 603:328–334
Shapiro BA, Bindewald E, Kasprzak W et al (2008) Protocols for the in silico Sesign of RNA Nanostructures. Methods Mol Biol 474:93–115
Shi J, Kantoff PW, Wooster R et al (2017) Cancer Nanomedicine: Progress, Challenges and Opportunities. Nat Rev Cancer 17:20–37
Shi Z, Zhu Y, Zhang J et al (2022) Monoclonal antibodies: new chance in the management of B-cell acute lymphoblastic leukemia. Hematology 27:642–652
Shim H (2020) Bispecific Antibodies and Antibody-Drug Conjugates for Cancer Therapy: Technological Considerations. Biomolecules 10:360
Shu D, Shu Y, Haque F et al (2011) thermodynamically stable RNA three-way junction for constructing multifunctional nanoparticles for delivery of therapeutics. Nat Nanotechnol 6:658–667
Shu Y, Pi F, Sharma A et al (2014) Stable RNA nanoparticles as potential new generation drugs for cancer therapy. Adv Drug Deliv Rev 66:74–89
Shu D, Li H, Shu Y et al (2015) Systemic Delivery of Anti-miRNA for Suppression of Triple Negative Breast Cancer Utilizing RNA Nanotechnology. ACS Nano 9:9731–9740
Shu D, Zhang L, Bai X et al (2021) Stoichiometry of multi-specific immune checkpoint RNA Abs for T cell activation and tumor inhibition using ultra-stable RNA nanoparticles. Mol Ther Nucleic Acids 24:426–435
Silva JM, Zupancic E, Vandermeulen G et al (2015) In vivo delivery of peptides and toll-like receptor ligands by mannose-functionalized polymeric nanoparticles induces prophylactic and therapeutic anti-tumor immune responses in a melanoma model. J Control Release 198:91–103
Thakur A, Huang M, Lum LG (2018) Bispecific antibody based therapeutics: strengths and challenges. Blood Rev 32:339–347

Thiviyanathan V, Gorenstein DG (2012) Aptamers and the next generation of diagnostic reagents. Proteomics Clin Appl 6:563–573

Tommasini-Ghelfi S, Lee A, Mirkin CA et al (2019) Synthesis, physicochemical, and biological evaluation of spherical nucleic acids for RNAi-based therapy in glioblastoma. Methods Mol Biol 1974:371–391

Venditto VJ, Szoka FC Jr (2013) Cancer nanomedicines: so many papers and so few drugs! Adv Drug Deliv Rev 65:80–88

Wang H, Guo P (2021) Radiolabeled RNA nanoparticles for highly specific targeting and efficient tumor accumulation with favorable in vivo biodistribution. Mol Pharm 18:2924–2934

Wang H, Ellipilli S, Lee WJ et al (2020) Multivalent rubber-like rna nanoparticles for targeted co-delivery of paclitaxel and MiRNA to silence the drug efflux transporter and liver cancer drug resistance. J Control Release 330:173–184

Westhof E, Masquida B, Jaeger L (1996) RNA tectonics: towards RNA design. Fold Des 1:R78-88

Wu Z, Cheung NV (2018) T cell engaging bispecific antibody (T-BsAb): From technology to therapeutics. Pharmacol Ther 182:161–175

Wu L, Seung E, Xu L et al (2020) Trispecific antibodies enhance the therapeutic efficacy of tumor-directed T cells through T cell receptor co-stimulation. Nat Cancer 1:86–98

Wudhikarn K, King AC, Geyer MB et al (2022) Outcomes of relapsed B-cell acute lymphoblastic leukemia after sequential treatment with blinatumomab and inotuzumab. Blood Adv 6:1432–1443

Xu C, Haque F, Jasinski DL et al (2018) Favorable Biodistribution, Specific Targeting and Conditional Endosomal Escape of RNA Nanoparticles in Cancer Therapy. Cancer Lett 414:57–70

Xu Y, Pang L, Wang H et al (2019) Specific delivery of delta-5-desaturase siRNA via RNA nanoparticles supplemented with dihomo-gamma-linolenic acid for colon cancer suppression. Redox Biol 21:101085–101093

Yin H, Xiong G, Guo S et al (2019) Delivery of Anti-miRNA for triple-negative breast cancer therapy using RNA nanoparticles targeting stem cell marker CD133. Mol Ther 27:1252–1261

Zhang YN, Poon W, Tavares AJ et al (2016) Nanoparticle-liver interactions: Cellular uptake and hepatobiliary elimination. J Control Release 240:332–348

Zhang Y, Leonard M, Shu Y et al (2017) Overcoming tamoxifen resistance of human breast cancer by targeted gene silencing using multifunctional pRNA nanoparticles. ACS Nano 11:335–346

Zhang Y, Xie X, Yeganeh PN et al (2021) Immunotherapy for breast cancer using EpCAM aptamer tumor-targeted gene knockdown. Proc Natl Acad Sci USA 118:e2022830118

Zhao L, Li S, Wei X et al (2022) A novel CD19/CD22/CD3 trispecific antibody enhances therapeutic efficacy and overcomes immune escape against B-ALL. Blood 140:1790–1802

Zheng Z, Li Z, Xu C et al (2019) Folate-displaying exosome mediated cytosolic delivery of siRNA avoiding endosome trapping. J Control Release 311–312:43–49

Zhou J, Rossi J (2017) Aptamers as targeted therapeutics: current potential and challenges. Nat Rev Drug Discov 16:181–202

Zhu G, Chen X (2018) Aptamer-based targeted therapy. Adv Drug Deliv Rev 134:65–78

GPSR Compliance

The European Union's (EU) General Product Safety Regulation (GPSR) is a set of rules that requires consumer products to be safe and our obligations to ensure this.

If you have any concerns about our products, you can contact us on ProductSafety@springernature.com

In case Publisher is established outside the EU, the EU authorized representative is:

Springer Nature Customer Service Center GmbH
Europaplatz 3
69115 Heidelberg, Germany

Batch number: 10703851

Printed by Printforce, the Netherlands